Digital Communications

Second Edition

Ian A. Glover
UNIVERSITY OF BATH

Peter M. Grant
UNIVERSITY OF EDINBURGH

PEARSON
Prentice
Hall

Harlow, England • London • New York • Boston • San Francisco • Toronto
Sydney • Tokyo • Singapore • Hong Kong • Seoul • Taipei • New Delhi
Cape Town • Madrid • Mexico City • Amsterdam • Munich • Paris • Milan

Pearson Education Limited
Edinburgh Gate
Harlow
Essex CM20 2JE
England

and Associated Companies throughout the world

Visit us on the World Wide Web at:
www.pearsoned.co.uk

First published 1998 by Prentice Hall
Second edition 2004

ISBN 0 130 89399 4

British Library Cataloguing-in-Publication Data
A catalogue record for this book is available from the British Library

Library of Congress Cataloging-in-Publication Data
A catalog record for this book is available from the Library of Congress

10 9 8 7 6 5 4 3 2 1
08 07 06 05 04

Printed in Great Britain by Henry Ling Ltd., at the Dorset Press, Dorchester, Dorset.

The publisher's policy is to use paper manufactured from sustainable forests.

Contents

Part Two   Digital communications principles, 161

5 Sampling, multiplexing and PCM, 163

6 Baseband transmission and line coding, 204

Preface

Digital communications is a rapidly advancing applications area. Significant current activities are in the development of mobile communications equipment for personal use, in the expansion of the available bandwidth (and hence information carrying capacity) of the backbone transmission structure through developments in optical fibre, and in the ubiquitous use of networks for data communications.

The aim of this book is fourfold: (1) to present the mathematical theory of signals and systems as required to understand modern digital communications equipment and techniques, (2) to apply and extend these concepts to information transmission links which are robust in the presence of noise and other impairment mechanisms, (3) to show how such transmission links are used in fixed and mobile data communication systems for voice and video transmission, and (4) to introduce the operating principles of modern communications networks formed by the interconnection of many transmission links using a variety of topological structures.

The material is set in an appropriate historical context. Most of the chapters include substantive numerical examples to illustrate the material developed and conclude with problem questions which have been designed to help readers assess their comprehension of this material.

In Chapter 1, we summarise the history of communication systems and introduce some basic concepts such as accessing, modulation, multiplexing, coding and switching, for line and radio transmission. Chapter 1 also includes a review of the advantages of digital communications systems over the older analogue systems which they are now, largely, replacing.

The next 18 chapters are organised in four parts reflecting the four aims referred to above. Specifically Chapters 2 through 4 are devoted to a basic theory of periodic, transient and random signals and the concept of linear transmission systems. Chapters 5 through 13 cover the fundamentals of digital communications and include sampling and multiplexing, baseband line transmission, decision and information theory, cryptography and error control coding, including turbo coding. This second part also includes a description of the many bandpass modulation schemes used in modern systems, the calculation of received power and associated signal-to-noise ratio for a communications link, and an indication of how the performance of a system can be assessed by simulation, before any actual hardware construction is attempted.

Part Three, Chapters 14 through 16, describes how the principles of digital communications are applied in fixed point-to-point terrestrial, and satellite based, microwave systems, in mobile and cellular radio systems, and in video (TV) transmission and storage systems. The fourth part, Chapters 17 through 21, is devoted to communication networks. This starts with a discussion of network topologies, access techniques and their signalling and routing protocols and architectures before moving on to queueing theory. It then progresses naturally to public networks, SDH and ISDN, the internationally agreed standard for the worldwide digital telecommunications network before finally concluding with broadcast networks, both wired and wireless local area networks. This completely revised and extended networks section in the second edition introduces the reader to a range of rapidly evolving wireless networking techniques.

To assist the reader, the book includes a list of abbreviations and also a list of notations and conventions used for the mathematical material.

An extensive reference list including key WWW addresses, standards and a bibliography is provided at the end of the book, before the index. All publications referred to in the text are compiled in this list. Each reference is identified in the text by the name(s) of the author(s) and, where necessary, the year of publication in square brackets.

The book is aimed at readers who are completing a graduate level BEng/MEng degree, or starting a postgraduate level MSc degree in Communications, Electronics or Electrical Engineering. It is assumed that these readers will have competence in the mathematical concepts required to handle comfortably the material in Part One.

The book has been compiled from lecture notes associated with final year BEng/MEng/MSc core, and optional, courses in signal theory and digital communications as provided at the Universities of Bath, Bradford and Edinburgh from 1990 to date. We have deliberately extended our coverage, however, to include some practical aspects of the implementation of digital PCM, SDH, packet speech systems, and the capability of optical and microwave long haul communication systems. With this balance between theory, applications and systems implementation we hope that this text will be useful both in academia and in the rapidly growing communications industry.

To aid the instructor and the student we provide a current erratum plus outline solutions to the majority of the end of chapter problems on the World Wide Web at the Edinburgh server address: http://www.see.ed.ac.uk/~pmg/DIGICOMMS/index.html or via the Pearson Education website at www.booksites.net/glover

In addition we have some further software examples in the areas of filtering, transforms and adaptive processors which are available via the above server address.

Bath and Edinburgh   Ian Glover and Peter Grant
June 2003

Acknowledgements

Parts of this book have been developed from BEng, MEng and MSc courses provided at the Universities of Edinburgh and Bradford. Three of these courses were first taught by Dr James Dripps at Edinburgh, and Professor Peter Watson and Dr Neil McEwan at Bradford, and we acknowlegde their initial shaping of these courses which is reflected in the book's content and structure. We are grateful to Dr Dripps for having provided draft versions of Chapters 7 and 9 and also for giving us access to material which now forms parts of Chapters 6, 10, 17 and 18. We are grateful to Dr McEwan for providing the original versions of sections 2.5.1, 4.3.1, 4.3.2 and 4.3.3 in the form of his teaching notes. Some of the material in Chapters 2, 3, 4, 8 and 11 had its origins in notes taken during lectures delivered at Bradford by Professor Watson and Dr McEwan. We also acknowledge Dr Brian Flynn for assistance with parts of Chapter 19, Dr Angus McLachlan for providing initial thoughts on Chapter 12, Dr Tom Crawford (of Hewlett Packard, Telecomms Division, South Queensferry) for giving us access to further material for Chapter 19 and providing some initial insights into Chapter 6. We are grateful to Dr David Parish of Loughborough University of Technology, for providing an initial draft of Chapter 16, Professor Paddy Farrell (of Victoria University, Manchester) for helpful comments on Chapter 10 and Dr David Cruickshank at Edinburgh for assistance with the problem solutions which are provided on the WWW.

We would like to thank all those colleagues at the Universities of Bradford and Edinburgh who have provided detailed comments on sections of this text. Thanks must also go to the many students who have read and commented on earlier versions of this material, helped to refine the end of chapter problems and particularly Yoo-Sok Saw and Paul Antoszczyszn who generously provided figure material for Chapter 16.

Special thanks are due to Joan Burton, Liz Paterson, Diane Armstrong and Beverley Thomas for their perseverance over several years in typing the many versions of the individual chapters, as they have evolved from initial thoughts into their current form. We also acknowledge Bruce Hassall's generous assistance with the preparation of the final version of the text in the appropriate typefont and text format.

Finally we must thank our respective families, Nandini and Sonia, and Marjory, Lindsay and Jenny for the considerable time that we required to write this book.

Ian Glover and Peter Grant, 1998

Second edition

This second edition has been further developed from BEng, MEng and MSc courses provided at the Universities of Edinburgh, Bath and Bradford. We acknowledge Professor Keith Blow from the University of Birmingham for updates to Chapter 12, Professor Mike Woodward of Bradford University for preparing the revised Chapter 17 (now chapter 19), Professor Simon Shepherd also of Bradford University for reading and commenting on the new material on encryption in Chapter 9, Dr Robert Watson at Bath for preparing the new section in Chapter 10 on turbo coding and the Bluetooth section in Chapter 21, the generous assistance of John Martin, also from Bath, for providing access to all his material on networks for enhancing Part Four of this revised text and Dr David Cruickshank at Edinburgh for continued assistance with the problem solutions which are provided on the WWW.

We would like to thank all those colleagues at the Universities of Bath and Edinburgh who have again provided detailed comments on sections of this text. Thanks must also go to the many students who continue to read, comment and suggest improvements to the chapter contents and also the solutions to the problem questions. Thanks are also due to the many instructors worldwide who have emailed us with positive comments and suggestions.

Special thanks are due to Diane Armstrong, Caroline Saunders and Kim Orsi for their perseverance in typing the revised chapters and tables. We also acknowledge again Bruce Hassall, the IT Services Manager in the School of Engineering and Electronics at the University of Edinburgh and his staff, in particularly Michael Gordon, for their generous assistance with the typesetting, formatting, and figure editing to achieve the professional layout of the final text.

Finally we must thank our respective families, Nandini and Sonia, and Marjory for our time spent writing and revising this book.

Ian Glover and Peter Grant, 2003

Abbreviations

AAL	ATM adaptation layer
ABM	Asynchronous balanced mode
ABR	Available bit rate
AC	Alternating current (i.e. sinusoidal signal), access control, area code
ACF	Autocorrelation function
ACK	Acknowledgement
ACL	Asynchronous connectionless
ACSE	Association control service element
A/D or ADC	Analogue to digital converter
ADCCP	Advanced data communications control procedure
ADM	Add and drop multiplexer, adaptive delta modulation
ADPCM	Adaptive differential pulse code modulation
ADSL	Asymmetric digital subscriber line (transmissions)
AFI	Authority & format identifier
AGC	Automatic gain control
AI	Adaption interface
AIA	Active interference avoidance
AK-TPDU	Acknowledgement TPDU
ALOHA	(not an abbreviation but Hawaiian for 'hello')
AM	Amplitude modulation
AMI	Alternate mark inversion
AMPS	Advanced mobile phone system (USA)
AN	Access network
ANS	Abstract syntax notation
ANSI	American National Standards Institute
AP	Access point
APCO	(US) associated public safety comminications office
APD	Avalanche photodiode
APK	Amplitude/phase keying
ARM	Asynchronous response mode
ARPANET	Advanced Research Projects Agency Network
ARQ	Automatic repeat request
ASCII	American Standard Code for Information Interchange
ASIC	Application specific integrated circuit

ASK	Amplitude shift keying
ASN	Abstract syntax notation
ATM	Asynchronous transfer mode, automatic teller machine
AU	Administrative unit
AUG	AU group
AUI	Attachment unit interface
AWG	American wire gauge
BA	Basic (rate) access (in ISDN)
BASK	Binary amplitude shift keying
BCH	Bose–Chaudhuri–Hocquenghem
BCJR	Bahl, Cocke, Jelinek, Raviv (algorithm)
BER	Bit error ratio/rate
BFSK	Binary frequency shift keying
BFWA	Broadband fixed wireless access
BICI	Broadband (or B-ISDN) intercarrier interface
BIM	Broadcast interface module
BIS	Boundary/border IS
B-ISDN	Broadband ISDN
BL	Baseband layer
BMV	Branch metric value
BNA	Broadcast network adaptor
$BO_{i/o}$	Back-off (input/output)
BPI	Baseline privacy interface
B-PON	Broadband passive optical network
BPSK	Binary phase shift keying
BRL	Bluetooth radio layer
BRZ	Bipolar return to zero
BS	Base station
BSS	Broadcast satellite service, basic service set
BT	British Telecom
CAC	Connection access control
CAP	Carrierless amplitude and phase (modulation)
CASE	Common application service element
CATV	Community antenna TV
CC	Central controller
CCIR	Comité Consultatif International des Radiocommunications
CCITT	Comité Consultatif International Télégraphique et Téléphonique
CCK	Complementary code keying
CCRE	Commitment, concurrency and recovery element
CCS7	Common channel signalling system No. 7
CC-TPDU	Connection confirm TPDU
CD	Cumulative distribution, compact disc, collision detection
CDMA	Code division multiple access

CD-ROM	Compact disc read-only memory
CDT	Credit (flow control)
CDV	Cell delay variation
CELP	Codebook of excited linear prediction
CEPT	Confederation of European PTT Administrations
CFMSK	Continuous frequency minimum shift keying
CIR	Carrier to interference ratio
CLNP	Connectionless network layer (IP) protocol
CLNS	Connectionless network layer service
CLR	Cell loss ratio
CM	Cable modem
CMCI	Cable modem computer interface
CMI	Coded mark inversion
CMIP	Common management information protocol
CMIR	Carrier modulated IR
CMOS	Complementary metal oxide silicon (transistor)
CMRI	Cable modem return path interface
CMTRI	Cable modem telephone return path interface
CMTS	Cable modem termination system
CN	Core network
CNR	Carrier-to-noise ratio
CODEC	Coder/decoder
CONP	Connection-oriented network protocol
CONS	Connection-oriented network service
CPD	Centre point detection
CPN	Customer premises network
CP(S)M	Continuous phase (shift) modulation
CR	Call request
CRC	Cyclic redundancy check
CRT	Cathode-ray tube
CR-TPDU	Connection request TPDU
CS	Carrier sense, circuit switched, convergence sub-layer
CSDN	Circuit switched data network
CSMA/CD	Carrier sense multiple access/collision detection
CSPDN	Circuit switched packet data network
CTD	Cell transfer delay
CTS	Clear-to-send
CW	Continuous wave
D	Data
DA	Demand assigned
DAC	Digital to analogue converter
DASS	Digital access signalling system
DAT	Digital audio tape
DAVIC	Digital video broadcast-cable/digital audio video council

DBS	Direct broadcast satellite
DC	Direct current
D/C	Downconverter
DCCE	Digital cell centre exchange
DCE	Data communication equipment
DCF	Distributed coordination function
DCT	Discrete cosine transform
DDSSC	Digital delivered services switching centre
DECT	initially Digital European cordless telecommunications
	now Digital enhanced cordless telecommunications
DEPSK	Differentially encoded phase shift keying
DES	Data encryption standard
DFB	Distributed feedback (laser)
DFS	Discrete Fourier series, dynamic frequency selection
DFT	Discrete Fourier transform
DHCP	Dynamic host configuration protocol
DI	Distribution interface
DIUC	Downlink internal usage code
DLC	Data-link controller
DL-MAP	Downlink map
DM	Delta modulation
DMIR	Direct modulation IR
DMPSK	Differential M-symbol phase shift keying
DMSU	Digital main switching unit
DMT	Digital multitone
DNS	Domain name system
DOCIS	Data over cable service interface specification
DPCM	Differential pulse code modulation
DPNSS	Digital private network signalling system
DPRS	DECT packet radio service
DPSK	Differential phase shift keying
DQDB	Distributed queue dual bus
DRFSI	Downstream RF site interface
DSB	Double sideband
DSI	Digital speech interpolation
DSL	Digital subscriber line
DSMX	Digital system multiplexor
DSP	Digital signal processing, domain specific part
DSR	Data set ready
DSS1	Digital subscriber signalling No. 1
DSSS	Digital subscriber signalling system,
	direct sequence spread spectrum
DTE	Data terminal equipment
DTI	Department of Trade and Industry (UK)
DTP	Distributed transaction processing

DTR	Data terminal ready
DT-TPDU	Data TPDU
DUP	Data user part
DV	Data/voice (packet)
DVB	Digital video broadcast
DVD	Digital video disc
DVR	Digital video recorder
ECMA	European Computer Manufacturers Association
ED	End delimiter
EDFA	Erbium doped fibre amplifier
EDGE	Enhanced data rate for GPRS evolution
EFT	Electronic funds transfer
EFTPOS	Electronic funds transfer at point of sale
EIA	Electronic Industries Association
EIRP	Effective isotropic radiated power
EM	Encrypted message
EMI	Electromagnetic interference
ENQ	Enquiry
EOT	End of transmission
EOW	Engineering order wire
ER	Error reporting (flag)
ERD	End routing domain
ERMES	European Radio Message System
ES	End system
ESD	Energy spectral density
ES-IS	End system to intermediate system
ETSI	European Telecommunications Standards Institute (formerly CEPT)
ESS	Extended service set
EY-NPMA	Elimination yield non pre-emptive priority multiple access
FCC	Federal Communications Commission
FCS	Frame check sequence
FDDI	Fibre distributed data interface
FDM	Frequency division multiplex
FDMA	Frequency division multiple access
FECC	Forward error correction coding
FET	Field effect transistor
FEXT	Far end crosstalk
FFSK	Fast frequency shift keying
FFT	Fast Fourier transform
FH	Frequency hopped (transmission)
FHS	Frequency hop synchronisation
FH(SS)	Frequency hopped (spread spectrum)
FIFO	First in first out

FILO	First in last out
FIR	Finite impulse response
FIRO	First in random out
FM	Frequency modulation
FP	Final permutation
FPGA	Field programmable gate array
FPLMTS	Future public land mobile telecommunications system
FS	Fourier series
FSK	Frequency shift keying
FSPL	Free space path loss
FT	Fourier transform
FTAM	File transfer access and management
FTTB	Fibre to the building/business
FTTC	Fibre to the kerb
FTTCab	Fibre to the cabinet
FTTH	Fibre to the home
FZ	Fresnel zone
GAN	Global area network
GFI	General format identity
GMSK	Gaussian (filtered) minimum shift keying
GoS	Grade of service
GPRS	General packet radio system
GPS	Global positioning system
GSC	Group switching centre
GSM	originally Groupe Spéciale Mobile
	now Global System for Mobile communications
HACE	Higher order automatic cross-connect equipment
HALO	High altitude long operation
HAP	High altitude platform
HCI	Host controller interface
HDB	High density bipolar
HDLC	High level DLC
HDSL	High speed digital subscriber loop
HDTV	High definition television
HEO	High earth orbit
HF	High frequency
HFC	Hybrid fibre coax
HIHE	Highly inclined highly elliptical (orbit)
HIPERACCESS	ETSI HIPERLAN variant
HIPERLAN	High performance local area network
HomePNA	Home phone line network association
HPA	High power amplifier
HSCSD	High speed circuit switched data

HSLAN	High speed LAN
HUMAN	High rate unlicensed MAN
I	Inphase (signal component), information
ICSDS	Interactive channel satellite distribution system
I+D	Integrate and dump
ID	Identity
IDEA	International data encryption algorithm
IDI	Initial domain identifier
IDN	Integrated digital network
IDP	Initial domain port
IDRP	Interdomain routing protocol
IDSL	ISDN DSL
IEEE	Institute of Electrical and Electronics Engineers
IF	Intermediate frequency
IFA	Intermediate frequency amplifier
IFS	Inter frame space
IIM	Interactive network module
IIR	Infinite impluse response
ILD	Injection laser diode
INA	Interaction network adaptor
INMARSAT	International Maritime Satellite Consortium
INTELSAT	International Telecommunications Satellite Consortium
IP	Intermodulation product, internet protocol, initial permutation
IPSS	International packet switched service
IR	Infrared, interdomain routing
IS	Intermediate system
ISC	International switching centre
ISDN	Integrated services digital network
ISI	Intersymbol interference
IS-IS	Intermediate system to intermediate system
ISM	Industrial scientific and medical (frequency band)
ISO	International Organisation for Standardisation
ISO-PP	ISO presentation protocol
ISO-SP	ISO session (layer) protocol
ISP	Internet service provider, intermediate services port
ISUP	ISDN user part
ITU	International Telecommunication Union
IWU	Interworking unit
JANET	Joint Academic Network
JPEG	Joint Photographic Experts Group
JTAM	Job transfer access and management
L2CAP	Logical link control and adaption

LAN	Local area network
LAP-B/D	Link access protocol balanced/D-channel
LBT	Listen before talk
LC	Link controller
LCC	Logical link control
LCFS	Last come first served
LE	Local exchange
LED	Light emitting diode
LEO	Low earth orbit (satellite)
LI	Length indicator
LIFO	Last in first out
LLC	Logical link control
LLR	log-likelihood ratio
LM	Link manager
LMDS	Local multipoint distribution system
LMP	Link manager protocol
LNA	Low noise amplifier
LO	Local oscillator
LOH	Line overhead
LOS	Line of sight
LPC	Linear predictive coding
LPF	Low pass filter
LSA	Link state advertisment
LW	Long wave
LWT	Listen while talk
LZW	Lempel–Ziv (Welch) coding
MAC	Medium access control
MAN	Metropolitan area network
MAP	Manufacturers application protocol, maximum a posteriori (criterion)
MASK	M-symbol amplitude shift keying
MBC	Model based coding
MBITR	Multiband interteam radio
MCPC	Multiple channels per carrier
MCR	Minimum cell rate
MF	Multiple frequency
MFSK	Multiple frequency shift keying
MGF	Mask generation function
MHS	Message handling system
MMDS	Multichannel, multipoint distribution service
MMF	Multimode fibre
MMS	Manufacturing message service
MN	Mobile network
MODEM	Modulator/demodulator
MOS	mean opinion score (for speech quality assessment),

	metal oxide silicon (transistor)
MOTIS	Message oriented text interchange standard
MPDU	MAC layer PDU
MPE	Multipulse excitation
MPEG	Motion Picture Experts Group
MPSK	M-symbol phase shift keying
MQAM	M-symbol quadrature amplitude modulation
MS	More segments (flag)
MSC	Main switching centre
MSCIR	Multi-subcarrier modulated IR
MSK	Minimum shift keying
MSOH	Multiplexer section overhead
MSU	Message signal unit
MTA	Multimedia terminal adaptor
MTBF	Mean time between failures
MTP	Message transfer part
MVDS	Multipoint video distribution service
MW	Medium wave
NA	Not applicable
NAK	Negative acknowledgment
NASA	National Aeronautics and Space Administration
NATO	North Atlantic Treaty Organisation
NET	Network
NETS	Network entity titles
NEXT	Near end crosstalk
NICAM	Near instantaneous companded amplitude modulation
NIU	Network interface unit
NMC	Network management centres
NNI	Network–node interface
NPDU	Network protocol data unit
NPSD	Noise power spectral density
NRM	Normal response mode
NRZ	Non-return to zero
NSAP	Network service access point
NSC	Non-systematic convolutional (code)
NSDU	Network service data unit
NSF	National Science Foundation (USA)
NSI	Network side interface
NT	Network termination
NTSC	National Television Standards Committee (USA)
NUL	(packet with no user information, i.e. all zeros)
O&M	Operations and maintenance
OAEP	Optimal asymmetric encryption padding

OA&M	Operations, administration and maintenance
OAM&P	Operations, administration, maintenance and provisioning
ODN	Optical distribution network
O/E	Optical/electronic (conversion)
OFDM	Orthogonal frequency division multiplex
OLT	Optical line termination
OMC	Operation & maintenance centre
ONT	Optical network termination
ONU	Optical network unit
OOK	On–off keying
OPI	On-premises interface
OQPSK	Offset quadrature phase shift keying
OSI	Open systems interconnection
OSS	Operational support system
OSSI	Operational support system interface
PA	Preassigned, priority assertion
PABX	Private automatic branch exchange
PAD	Packet assembly and disassembly
PAL	Phase alternate line (TV)
PAM	Pulse amplitude modulation
PAN	Personal area network
PC	Personal computer, permuted choice, point code
PCF	Point coordination function
PCM	Pulse code modulation
PCN	Personal communications network
PCR	Peak cell rate
pdf	Probability density function
PDH	Plesiochronous digital hierarchy
PDN	Public data network
PDU	Protocol data unit
PEPL	Plane earth path loss
P/F	Poll/final
PGP	Pretty good privacy
PIN	Positive–intrinsic–negative (diode)
PLE	Principal local exchange
PLL	Phase locked loop
PLOAM	Physical layer OA&M
PLP	Packet level protocol
PM	Phase modulation
PMR	Private mobile radio
PMV	Path metric value
PN	Pseudo-noise
POCSAG	Post Office Code Standards Advisory Group
POF	Plastic optical fibre

POH	Path overhead
POLL	(polling request to identify if terminal is active or present)
PON	Passive optical network
POTS	Plain old telephone system
PPM	Pulse position modulation
PRBS	Pseudo-random bit sequence
PRK	Phase reversal keying
PS	Processor sharing, packet switch
PSD	Power spectral density
PSDN	Packet switched data network
PSE	Packet switch exchange
PSK	Phase shift keying
PSPDN	Packet switched public data network
PSRCS	Public safety radio communication service
PSS	Packet switched service
PSTN	Public switched telephone network
PTT	Post, telephone and telegraph
PVC	Permanent virtual circuit
PWM	Pulse width modulation
Q	Quadrature (signal component)
QA	Quasi-analytic
QAM	Quadrature amplitude modulation
QoS	Quality of service
QPR	Quadrative partial response
QPSK	Quadrature phase shift keying
RA	Random access
RACE	Research in Advanced Communications in Europe
RADSL	Rate adaptive ADSL
RCU	Remote concentrator unit
RDA	Remote database access
RDI	Routing domain identifier
REJ	Reject
RF(I)	Radio frequency (interference)
RFCOMM	Radio frequency communication
RGB	Red green blue
RJ-TDPU	Reject TPDU
RMS	Root mean square
RNR	Receiver not ready
ROSE	Remote operation service element
RR	Receiver ready
RS	Reed–Solomon (code)
RSA	Rivest, Shamir and Adelman
RSC	Recursive systematic convolutional (codes)

RSOH	Regenerator section overhead
RSSI	Received signal strength indication
RSU	Remote switching unit
RTS	Request to send, ready-to-send
RV	Random variable
RX(D)	Receive (data)
RZ	Return to zero
S	Supervisory, space (switch)
SAP	Service access point
SAR	Segmentation and reassembly
SBC	Sub-band coder
SCCP	Signalling connection control part
SCO	Synchronous connection oriented
SCP	Switching/service control point
SCPC	Single channel per carrier
SCSI	Small computer system interface
SD	Start delimiter
SDH	Synchronous digital hierarchy
SDLC	Synchronous data-link control
SDP	Service discovery protocol
SDR	Signal to distortion ratio
SDSL	Symmetric DSL
SE	Session entity
SECAM	Système en couleurs à mémoire
SEL	Selector (address)
SER	Symbol error rate
SERC	Science and Engineering Research Council
SHA	Secure hash algorithm
SHDSL	Single pair HDSL
SHF	Super high frequency
SI	Subnet identifier
SIG	Special interest group
SIL	Silence
SISO	Soft-in, soft-out (decision decoder)
SMF	Single mode fibre
SMI	System management interface
SNAP	Subnetwork point of attachment
SNDAP	Subnetwork dependent access protocol
SNDCP	Subnetwork dependent convergence protocol
SNI	Service network/node interface
SNICP	Subnetwork independent convergence protocol
SNPA	Subnetwork point address
SNR	Signal-to-noise ratio
SN_qR	Signal to quantisation noise ratio

SOH	Section overhead
SONET	Synchronous optical network
SOVA	Soft output Viterbi algorithm
SP	Segmentation parameter (flag)
SPC	Stored program control
SPDU	Session protocol data unit
SPE	Synchronous payload envelope
SPL	Sound pressure level
SREJ	Selective reject
SS	Subscriber station
SSAP	Service switched access point
SSB	Single sideband
SSCS	Specific-service convergence sub-layer
SSMA	Spread spectrum multiple access
SSP	Service switching point
STB	Set-top box
STM	Synchronous transfer mode, synchronous transport module
STP	Switching control point, signal transfer point, shielded twisted pair
STR	Symbol timing recovery
STS	Synchronous transfer structure, synchronous transport signal, space time switch
STU	Set-top unit
SWAP	Shared wireless access protocol
SYN	IA5 (synchronisation) control character
T	Terminal (interface), time (switch)
TA	Terminal adaptor
TACS	Total access communication system (AMPS derivative)
TASI	Time assigned speech interpolation
TB	Time–bandwidth (product)
TC	Transaction capability
TCAP	Transaction capability
TCM	Trellis coded modulation
TCP	Transport control protocol
TCU	Trunk coupling unit
TDM	Time division multiplex
TDMA	Time division multiple access
TE	Terminal equipment, transport entity
TEM	Transverse electromagnetic
TETRA	Trans European trunked radio (network)
TI	TPDU indicator
TILS	Time invariant linear system
TLK	Talkspurt
TOH	Transport overhead
TP	Transport (layer/level) protocol, twisted pair

TPC	Turbo product code, transmit power control
TPDU	Transport protocol data unit
TRD	Transmit routing domain
TSAP	Transport service access point
TSDU	Transport service data unit
TU	Terminal unit, tributary unit
TUG	Tributary unit group
TUP	Telephone user part
TV	Television
TWSLA	Travelling wave semiconductor laser amplifier
TWT	Travelling wave tube
TX(D)	Transmit (data)
U	Unnumbered, user (interface)
UBR	Unspecified bit rate
U/C	Upconverter
UHF	Ultra high frequency
UIUC	Uplink internal usage code
UL-MAP	Uplink map
UMTS	Universal mobile telecommunication service
UNI	User–network interface
URFSI	Upstream RF site interface
USB	Universal serial bus
UTP	Unshielded twisted pair
UTRA	UMTS terrestrial radio access
UWB	Ultra wide band
VAN	Value added network
VBR	Variable bit rate
VBR-NRT	Variable bit rate, non-real time
VBR-RT	Variable bit rate, real time
VC	Virtual circuit, virtual container (in SDH)
VCI	Virtual channel identifier
VCO	Voltage controlled oscillator
VDSL	Very high speed digital subscriber line
VDU	Video display unit
VHF	Very high frequency
VLSI	Very large scale integrated (circuit)
VO	Voice over, video object
VoIP	Voice over internet protocol
VOP	Video object plane
VPI	Virtual path identifier
VSAT	Very small aperture (satellite) terminal
VSB	Vestigial sideband
VT	Virtual tributary/terminal

WAN	Wide area network
WAP	Wireless applications protocol
WDM	Wavelength division multiplex
WLL	Wireless local loop
WPAN	Wireless PAN
WRAC	World Radio Administrative Conference
WWW	World Wide Web
XNI	X network interface
XTR	Crosstalk ratio
ZIP	File compression

Symbols

a	core radius in optical fibre; network transmission parameter
a_e	antenna effective area; effective earth radius
A	A-law PCM compander constant; total path attenuation; offered traffic (Erlangs)
A_e	effective aperture area of antenna
A_n	nth real Fourier coefficient
$\mathbf{B}$	magnetic flux density
B	signal bandwidth in Hz, blocking probability
B_n	equivalent noise bandwidth; nth imaginary (real valued) Fourier coefficient
$\mathbf{c}$	codeword vector
C	channel capacity; received carrier power level; constant; cipher text
$C_{node/link}$	network connectivity
$C(m\vert v)$	conditional cost in interpreting v as symbol m
$\bar{C}$	average cost of decisions
C_i	ith codeword; cost associated with decision error
$C_n; \tilde{C}_n$	nth complex Fourier coefficient
C_∞	channel capacity in infinite channel bandwidth
$^{N}C_J$	number of combinations of J digits from N-digit block
$\mathbf{d}$	message vector
$\mathbf{D}$	electric flux density
D	deterministic (impulsive) pdf; antenna diameter; average data rate
D_{ij}	Hamming distance between codewords i and j
$D_{\min}$	minimum Hamming distance; minimum degree
$D(\theta, \phi)$	antenna directivity
$\mathbf{e}$	error vector
e	number of detectable errors in codeword; water vapour partial pressure
$\mathbf{E}$	electric field strength
E	equivocation; energy; field strength
$\mathrm{E}[\ ]$	expectation operator
E_b	energy per bit

E_e	filter output error energy arising from DM bit error
$E_k(f)$	energy spectral density with $k = 1, 2$ for one-, two-sided spectra
E_{si}	normalised symbol i energy
$\mathrm{erf}(x)$	error function
$\mathrm{erfc}(x)$	complementary error function
f	noise factor (ratio)
$f(\ )$	cipher function
Δf	peak frequency deviation
f_b	bit rate
f_c	centre frequency
f_χ	filter cut-off frequency
f_H	highest frequency component
f_L	lowest frequency component
f_{LO}	local oscillator frequency
f_o	symbol clock frequency
f_s	sample frequency
f_{3dB}	half power bandwidth
$f(kT_o)$	decision circuit input voltage at sampling instant
$f(t)$	IF transmitted symbol; generalised function
F	noise figure (dB)
F_N	speech formant N frequency component
$F(v)$	cumulative distribution function for v
$F(x)$	non-linear PCM companding characteristic
$g(t)$	baseband information signal
$g(kT_s)$	sampled version of $g(t)$
$\tilde{g}(kT_s)$	estimate of $g(t)$
$\hat{g}(kT_s)$	prediction of $g(t)$
G	generator matrix
G	amplifier gain
$G_k(f)$	power spectral density with $k = 1, 2$ for 1, 2 sided spectra
G_l	(negative or fractional) gain of a lossy device
G_p	processing gain
$G_p(f)$	power spectral density of $p(t)$
G_R	receiver antenna gain
G_T	transmitter antenna gain
$G(kT_s)$	adaptive gain of ADM amplifier
$G(\theta, \phi)$	antenna directional gain
G/T	gain to system noise temperature ratio
G.n	G series (recommendation n) for telephony multiplex
h_B	height of earth's bulge above chord
$h(t)$	impulse response
H	parity check matrix; magnetic field strength

H	entropy
$H_{eff;max}$	effective; maximum entropy
$H(T\|C)$	entropy of plaintext T conditional on ciphertext C
$H(f)$	frequency response
i_m	length of codeword m
I_m	information content of message m
I_n	nth information bit in codeword
$I_o(z)$	modified Bessel function of first kind and order zero
$I(\theta, \phi)$	radiation intensity from an antenna
$\Im$	imaginary part of
$\mathbf{k}$	vector wave number
k	Boltzmann's constant; number of information digits in a codeword; earth profile k factor; scalar wave number
K	average number of (network) collisions
l_m	length of codeword m
L	loss in making incorrect decision; link path length; average codeword length; mean no. of packets in a system
m	mean value; number of hops; modulation index
m_i	ith bit of codeword
M	number of symbols in an alphabet or levels in PCM system
M	Markov (Poisson) pdf
n	number of bits in a codeword; constraint length in convolutional coder; refractive index; mode number; type of semiconductor material
$n(t)$	noise signal
N	noise power
N_e	noise power due to DM bit errors
N_o	unicity distance
N_0	noise power spectral density
$N(f)$	voltage spectrum of noise
$N(R)$	receive sequence number
$N(S)$	send sequence number
p	probability; type of semiconductor material
$p(v)$	probability density function of variable v
$p(v\|m)$	conditional probability density function of v given m
P	power; pressure
$P[\ ]$	permutation function
P_{ACK}	probability of ACK frame error
P_b	probability of bit error
P_B	blocking probability

P_D	probability of detection
P_e	probability of symbol error
P_{em}	probability of error for symbol m
P_f	probability of frame error
$P_{fc(s)}$	probability of correct frame (start)
P_{FA}	probability of false alarm
P_L	probability of packet loss
P_m	probability of a queue having length m
$P_{retrans}$	probability of retransmitting a data frame
P_T	transmitted power
P_1	first parity bit in codeword; probability of being in state 1
$P(j,i)$	joint probability of selecting (symbol) i followed by j
$P(k)$	probability of a queue of length k
$P(m)$	a priori transmission probability for symbol m
$P(m|v)$	a posteriori probability that symbol m was transmitted, given voltage v detected

q	quantisation step size
Q	quality factor; f_H/B; mean channel access delay

$\mathbf{r}$	received codeword data vector
r	radius
R	range; efficiency (rate) of a digital code; redundancy
R'	number of errors in codeword
R_b	bit rate
$R_{P;L}$	point; line rain rate
R_r	radiation resistance
R_s	symbol rate
$R_x(\tau)$	correlation function
RC	resistor–capacitor time constant τ_c
$\mathfrak{R}$	real part of

$\mathbf{s}$	syndrome codeword vector
s	standard deviation
S	normalised throughput
S_n	sum of geometric progression
$S_{(peak)}$	average (peak) signal power
S/N	signal-to-noise ratio
$\text{sgn}\,(x)$	signum function

t	time; number of correctable errors in codeword
t_g	guard time
t_i	frame duration
t_{out}	time out interval
t_p	one-way propagation delay

t_{pro}	processing delay
t_T	minimum time between data frames
$\mathbf{T}$	transition (coding) matrix
T	plaintext; temperature; frame length
T_A	antenna aperture temperature
T_B	brightness temperature
T_D	symbol duration
T_L	ring latency
T_e	effective noise temperature
T_o	symbol period
T_{ph}	physical temperature
T_q	noise temperature of q
T_s	sample period
T_0	reference temperature (290 K)
U	utilisation factor
$v_i(t)$	voltage waveform of symbol i
$v_n(t)$	noise voltage
$v_N(t)$	Nyquist symbol ISI-free pulse
v_{th}	receiver decision threshold voltage
$v(t)$	information signal
V	bipolar violation pulses
$V(f)$	voltage spectral density
$V_N(f)$	voltage spectrum of ISI-free pulse
V.n	V series recomendation n for data modems
W	power density; mean system delay
$\mathbf{x}$	vector x
x_{med}	median value of x
$X(f)$	voltage spectrum of $x(t)$
$|X(f)|$	amplitude spectrum
X.n	X series recommendation n for data networks
α	peak to mean power ratio; attenuation factor; normalised excess bandwidth; earth radius
δ	Dirac delta
$\delta(t)$	impulse function
Δ	stepsize
Δf	excess bandwidth
ΔV	voltage difference

ε	error voltage
ε_q	quantisation error
$\overline{\varepsilon_{de}^2}$	mean square (PCM) decoding error
$\overline{\varepsilon_q^2}$	mean square quantisation error
η	efficiency; normalised mean throughput
η_s	spectral efficiency
η_Ω	antenna ohmic efficiency
λ	free space wavelength; arrival rate
μ	μ-law PCM compander constant; service rate
ρ	normalised correlation coefficient; utilisation factor (λ/μ)
σ	Gaussian standard deviation
σ_g	standard deviation of $g(t)$
τ	time between successive arrivals; pulse width; overall network propagation delay
τ_c	RC time constant
χ^2	chi-square distribution

Special functions

[,]   scalar product
⟨ ⟩   time average
*   convolution operation
a^* complex conjugate of a

$$\prod(t) = \begin{cases} 1, & |t| < 0.5 \\ \tfrac{1}{2}, & t = 0.5 \\ 0, & |t| > 0.5 \end{cases}$$

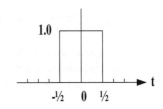

$$\prod\left(\frac{t-T}{\tau}\right) = \begin{cases} 1, & |t - T| < \tau/2 \\ \tfrac{1}{2}, & |t - T| = \tau/2 \\ 0, & |t - T| > \tau/2 \end{cases}$$

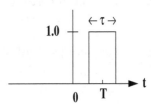

$$\Lambda(t) = \begin{cases} 1 - |t|, & |t| \le 1.0 \\ 0, & |t| > 1.0 \end{cases}$$

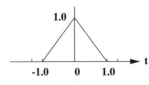

$$\Lambda\left(\frac{t}{\tau}\right) = \begin{cases} 1 - \dfrac{|t|}{\tau}, & |t| \le \tau \\ 0, & |t| > \tau \end{cases}$$

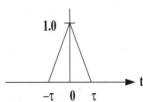

$$\text{sgn}(t) = \begin{cases} 1.0, & t > 0 \\ 0, & t = 0 \\ -1.0, & t < 0 \end{cases}$$

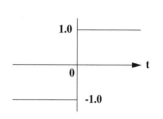

$$u(t) = \begin{cases} 1.0, & t > 0 \\ \frac{1}{2}, & t = 0 \\ 0, & t < 0 \end{cases}$$

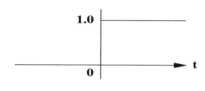

CHAPTER 1

Digital communications overview

1.1 Electronic communications

History, present requirements and future demands

Communication can be defined as the imparting or exchange of information [Hanks]. Telecommunication, which is more narrowly the topic of this book, refers to communication over a distance greater than would normally be possible without artificial aids. In the present day such aids are invariably electrical, electronic or optical and communication takes place by passing signals over wires, through optical fibres or by wireless transmission through space using electromagnetic waves.

Modern living demands that we have access to a reliable, economical and efficient means of communication. We use communication systems, particularly the public switched telephone network (PSTN), and its extension into cellular systems to contact people all around the world. Telephony is an example of point-to-point communication and normally involves a two-way flow of information. Another type of communication, which (traditionally) involves only one-way information flow, is broadcast radio and television. In these systems information is transmitted from one location but is received at many locations using many independent receivers. This is an example of point-to-multipoint communication.

Communication systems are now very widely applied. Navigation systems, for example, pass signals between a transmitter and a receiver in order to determine the location of a vehicle, or to guide and control its movement. Signalling systems for tracked vehicles, such as trains, are also simple communication systems.

Table 1.1 *Important events in the history of electronic communications.*

Year	Event	Originator	Information
1837	Line telegraphy perfected	Morse	Digital
1875	Telephone invented	Bell	Analogue
1887	Wireless telegraphy	Marconi	Digital
1897	Automatic exchange step by step switch	Strowger	
1905	Wireless telephony demonstrated	Fessenden	Analogue
1907	First regular radio broadcasts	USA	Analogue
1918	Superheterodyne radio receiver invented	Armstrong	Analogue
1921	First use of land based PMR	Detroit police	Analogue
1928	All electronic television demonstrated	Farnsworth	Analogue
1928	Telegraphy signal transmission theory	Nyquist	Digital
1928	Information transmission	Hartley	Digital
1931	Teletype		Digital
1933	FM demonstrated	Armstrong	Analogue
1934	Radar demonstrated	Kuhnold	
1937	PCM proposed	Reeves	Digital
1939	Voice coder	Dudley	Analogue
1939	Commercial TV broadcasting	BBC	Analogue
1940	Spread spectrum proposed		Digital
1943	Matched filtering proposed	North	Digital
1945	Geostationary satellite proposed	Clarke	
1946	ARQ systems developed	Duuren	Digital
1948	Mathematical theory of communications	Shannon	
1955	Terrestrial microwave relay	RCA	Analogue
1960	First laser demonstrated	Maiman	
1962	Satellite communications implemented	TELSTAR 1	Analogue
1963	Geostationary satellite communications	SYNCOM II	Analogue
1966	Optical fibres proposed	Kao & Hockman	
1966	Packet switching		Digital
1970	Medium scale data networks	ARPA/TYMNET	Digital
1970	LANs, WANs and MANs		Digital
1971	The term ISDN coined	CCITT	Digital
1974	Internet concept	Cerf & Kahn	Digital
1978	Cellular FDMA radio		Analogue
1978	Navstar GPS launched	Global	Digital
1980	OSI 7 layer reference model adopted	ISO	Digital
1981	HDTV demonstrated	NHK, Japan	Digital
1985	ISDN basic rate access in UK	BT	Digital
1986	SONET/SDH introduced	USA	Digital
1991	GSM TDMA cellular system	Europe	Digital
1991	MPEG video standards	International	Digital
1992	ETSI formed	Europe	
1993	PCN concept launched	Worldwide	Digital
1994	IS-95 CDMA specification	Qualcom	Digital
1995	ADSL transmission	International	Digital
1998	Wideband 3G CDMA	ITU Standards	Digital
2000	IMT 2000/UMTS	International	Digital

All early forms of communication system (e.g. smoke signals, semaphore, etc.) used digital traffic. The earliest form of electronic communications, telegraphy, was developed in the 1830s, Table 1.1. It was also digital in that the signals, transmitted over wires, were restricted to four types: dots and dashes, representing the Morse coded letters of the alphabet, letter spaces and word spaces. In the 1870s Alexander Graham Bell made analogue communications possible by inventing acoustic transducers to convert speech directly into (analogue) electrical signals.

This led quickly to the development of conventional telephony. Radio communications started around the turn of the century when Marconi patented the first wireless telegraphy system. This was quickly followed by the first demonstration of wireless (or radio) telephony and in 1918 Armstrong invented the superheterodyne radio receiver which is still an important component of much modern-day radio receiving equipment. In the 1930s Reeves proposed pulse code modulation (PCM) which laid the foundation for nearly all present-day digital communications systems.

Table 1.1 shows some of the principal events in the development of electronic communications over the last century and a half. The Second World War saw rapid, forced, developments in nearly all areas of engineering and technology. Electronics and communications benefited greatly and the new, but associated, discipline of radar became properly established.

In 1945 Arthur C. Clarke wrote his famous article proposing geostationary satellite communications and 1963 saw the launch of the first successful satellite of this type. In 1966 optical fibre communication was proposed by Kao and Hockman and, around the same time, public telegraph and telephone (PTT) operators introduced digital carrier systems.

The first, general purpose, large scale data networks (ARPANET and TYMNET) were developed around 1970, provoking serious commercial interest in packet switching (as an alternative to circuit switching).

The 1970s saw significant improvements in the performance of, and large increases in the volume of traffic carried by, telecommunications systems of all types. Optical fibre losses were dramatically reduced and the capacity of satellites dramatically increased. In the 1980s first analogue, and then digital, cellular radio became an important part of the PSTN. Micro-cellular and personal communications using both terrestrial and satellite based radio technology are now being developed. Wideband personal communications systems providing voice, data and video services have now become possible. Video delivery requires a broadband rather than a narrow (speech) bandwidth connection, Table 1.2.

Increasing demand for traditional services (principally analogue voice communications) has been an important factor in the development of telecommunications technologies. Such developments, combined with more general advances in electronics

Table 1.2 *Comparison of nominal bandwidths for several information signals.*

Information signal	Bandwidth
Speech telephony	4 kHz
High quality sound broadcast	15 kHz
TV broadcast (video)	6 MHz

and computing, have made possible the provision of entirely new (mainly digitally based) communications services. This in turn has stimulated demand still further. Figure 1.1 shows the past and predicted future growth of telecommunications traffic and Figure 1.2 shows the proliferation of services which have been, or are likely to be, offered over the same period.

In telecommunications there are various standards bodies which ensure interoperability of equipment. The International Telecommunication Union (ITU) is an important international communications standards body which only has the power to make recommendations for specifications. Within the ITU are the PTTs (post, telephone and telegraph organisations) from individual nations, e.g. British Telecom and Deutsche Bundespost. In Europe there was, until recently, the Confederation of European PTTs (CEPT), responsible for overseeing the actual implementation of technical standards. CEPT has now been replaced with the European Telecommunications Standards Institute (ETSI) [WWW, Temple].

The 1990s have seen tremendous advances in new digital transmission techniques. These include digital subscriber line (DSL) technology, which increases the maximum possible data rate over low bandwidth copper cables, MPEG standards for efficient video compression, and time division multiple access (TDMA) and code division multiple access (CDMA) mobile cellular communication systems. With the recent launch of third generation mobile cellular systems supporting both speech and data transmission these exciting advances are set to continue into the near future.

1.2 Sources and sinks of information

Sources of information can be either natural or man-made. An example of the former might be the air temperature at a given location. An example of the latter might be a set

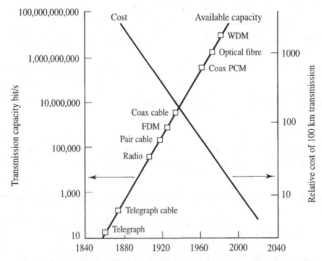

Figure 1.1 *Past and predicted growth of telecommunications traffic (source: Technical demographics, 1995, reproduced with permission of the IEE).*

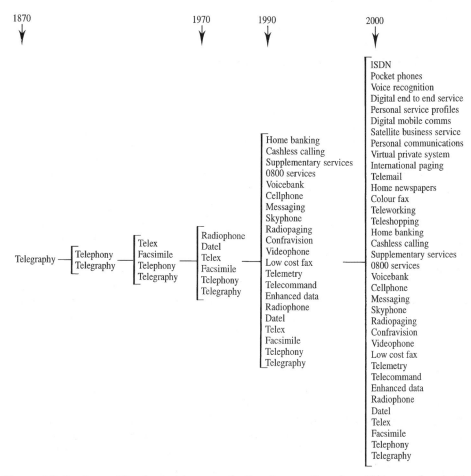

1870   1970   1990   2000

Figure 1.2 *Service proliferation in telecommunications (source: Earnshaw, 1991, reproduced with permission of Peter Peregrinus).*

of company accounts. (A third example, speech, falls, in some sense, into both categories.) Digital communications systems represent information, irrespective of its type or origin, by a discrete set of allowed symbols. It is this alphabet of symbols and the device or mechanism which selects them for transmission that is usually regarded here as the information source. The amount of information conveyed by each symbol, as it is selected and transmitted, is closely related to its selection probability. Symbols likely to be selected more often convey less information than those which are less likely to be selected. Information content (measured in bits) is thus related to symbol rarity.

Sinks of information are, ultimately, people although various types of information storage and display devices (computer disks, magnetic tapes, loudspeakers, VDUs, etc.) are usually involved as a penultimate destination.

Transmitters are the devices that impress source information onto an electrical wave (or carrier) appropriate to a particular transmission medium (e.g. optical fibre, cable, free space). Receivers are the devices which extract information from such carriers. They often also reproduce this information in the same form as it was originally generated (e.g. as speech).

1.3 Digital communications equipment

An important objective in the design of a communication system is often to minimise equipment cost, complexity and power consumption whilst also minimising the bandwidth occupied by the signal and/or transmission time. (Bandwidth is a measure of how rapidly the information bearing part of a signal can change and is therefore an important parameter for communication system design. Table 1.2 compares the nominal bandwidth of three common types of information signal.) Efficient use of bandwidth and transmission time ensures that as many subscribers as possible can be accommodated within the constraints of these limited, and therefore valuable, resources.

The component parts of a hypothetical digital communications transceiver (transmitter/receiver) are shown in Figure 1.3. Much of the rest of this book is concerned with the operating principles, performance and limitations of a communication system formed by a transmitter/receiver pair linked by a communications channel. Here, however, we give a qualitative overview of such a system, incorporating a brief account of what each block in Figure 1.3 does and why it might be required. (The transceiver in this figure has been chosen to include all the elements commonly encountered in digital communications systems. Not all transceivers will employ all of these elements of course.)

1.3.1 CODECs

At its simplest a transceiver CODEC (coder/decoder) consists of an analogue to digital converter (ADC) in the transmitter, which converts a continuous, analogue, signal into a sequence of codewords represented by binary voltage pulses, and a digital to analogue converter (DAC) in the receiver, which converts these voltage pulses back into a continuous, analogue, signal.

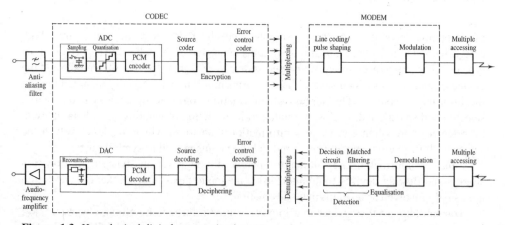

Figure 1.3 *Hypothetical digital communications transceiver.*

The ADC consists of a sampling circuit, a quantiser and a pulse code modulator (Figure 1.3). The sampling circuit provides discrete voltage samples taken, at regular intervals of time, from the analogue signal. The quantiser approximates these voltages by the nearest level from an allowed set of voltage levels. (It is the quantisation process which converts the analogue signal to a digital one.) The PCM encoder converts each quantised level to a binary codeword, digital ones and zeros each being represented by one of two voltages. An anti-aliasing filter is sometimes included prior to sampling in order to reduce distortion which can occur as a result of the sampling process.

In the receiver's DAC received binary voltage pulses are converted to quantised voltage levels by a PCM decoder which is then smoothed by a low pass filter to reconstruct (at least a good approximation to) the original, analogue, signal.

Digitisation of analogue signals usually increases the signal's transmission bandwidth but it permits reception at a lower signal-to-noise ratio than would otherwise be the case. This is an example of how one resource (bandwidth) can be traded off against another resource (transmitter power).

CODECs make widespread use of sophisticated digital signal processing techniques to encode efficiently the signal prior to transmission and also to decode the received signals when they are corrupted by noise, distortion and interference. This increases transceiver complexity, but allows higher fidelity, repeatable, almost error-free transmission to be achieved.

1.3.2 Source, security and error control coding

In addition to PCM encoding and decoding a CODEC may have up to three additional functions. Firstly (in the transmitter) it may reduce the number of binary digits (called bits, or sometimes binits) required to convey a given message. This is source coding and can be thought of as effectively removing redundant (i.e. unrequired or surplus) digits. Secondly it may encrypt the source coded digits using a cipher for security. This can yield both privacy (which assures the sender that only those entitled to the information being transmitted can receive it) and authentication (which assures the receiver that the sender is who he/she claims to be). Finally the CODEC may add extra digits to the (possibly source coded and/or encrypted) PCM signal which can be used at the receiver to detect, and possibly correct, errors made during symbol detection. This is error control coding and has the effect of incorporating binary digits at the transmitter which, from an information point of view, are redundant.

In some ways error control coding, which adds redundancy to the bit stream, is the opposite of source coding, which removes redundancy. Both processes may be employed in the same system, however, since the type of redundancy which occurs naturally in the information being transmitted is not necessarily the type best suited to detecting and correcting errors at the receiver.

The source, security and error control decoding operations in the receiver, Figure 1.3, are the inverse of those in the transmitter.

1.3.3 Multiplexers

In digital communications, multiplexing, to accommodate several simultaneous transmissions, usually means, more specifically, time division multiplexing (TDM). Time division multiplexers interleave either PCM codewords, or individual PCM binary digits, to allow more than one information link to share the same physical transmission medium. This can be cable, optical fibre or a radio frequency channel. If communication is to occur in real time this implies that the bit rate of the multiplexed signal is at least N times that of each of the N tributary PCM signals and this in turn implies an increased bandwidth requirement. The requirement for an increase in bandwidth comes from the fact that the transmitted signal now comprises shorter duration pulses which have a wider spectral response.

Demultiplexers split the received composite bit stream back into its component PCM signals.

1.3.4 MODEMs

MODEMs (modulators/demodulators) condition binary pulse streams so that the information they contain can be transmitted over a given physical medium, at a given rate, with an acceptable degree of distortion, in a specified or allocated frequency band. The modulator in the transmitter may change the voltage levels representing individual, or groups of, binary digits. Typically the modulator also shapes, or otherwise filters, the resulting pulses to restrict their bandwidth, and shifts the entire transmission to a convenient, allowed, frequency band. The input to a modulator is thus a baseband digital signal whilst the output is often a bandpass waveform.

The demodulator, in a receiver, reconverts the received waveform into a baseband signal. Equalisation corrects (as far as possible) signal distortion which may have occurred during transmission. Detection converts the demodulated baseband signal into a binary symbol stream. The matched filter, shown as one component of the detector in Figure 1.3, represents one type of signal processing which can be employed, prior to the final digital decision process, in order to improve error performance.

1.3.5 Multiple accessing

Multiple accessing refers to those techniques, and/or rules, which allow more than one transceiver pair to share a common transmission medium (e.g. one optical fibre, one satellite transponder or one piece of coaxial cable). Several different types of multiple accessing are currently in use, each type having its own advantages and disadvantages. The multiple accessing problem is essentially one of efficient and (in some sense) equitable sharing of the limited resource represented by the transmission medium.

1.4 Radio receivers

Many radio receivers (both digital and analogue) incorporate superheterodyning as part of their demodulation process. In these receivers (Figure 1.4) the incoming radio frequency (RF) signal, with carrier frequency f_{RF}, is mixed (i.e. multiplied) with the signal from a local oscillator (LO) of frequency f_{LO}. The sum $(f_{RF} + f_{LO})$ and difference $(f_{RF} - f_{LO})$ frequency products which appear at the mixer output are then filtered to select only the latter which is called the intermediate frequency or IF. The LO frequency is, therefore, always altered or tuned to ensure that the receiver operates with a fixed value of IF (i.e. $f_{RF} - f_{LO}$) irrespective of which RF channel is being received. This allows a considerable effort to be invested in the design of the receiver beyond this point,

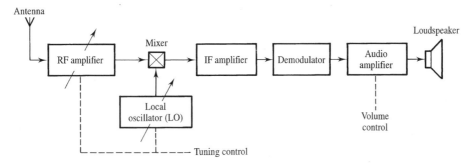

Figure 1.4 *Superheterodyne receiver.*

consisting typically of high gain (fixed frequency) IF amplifiers and high selectivity filters followed by an appropriate IF signal demodulator and/or detector.

The superheterodyne receiver can be made more sophisticated by using double frequency conversion in which there are two mixing stages. This enables higher gain and greater selectivity to be achieved in order to increase rejection of unwanted, interfering, signals.

The principal problem with the superheterodyne design is that the receiver is equally sensitive to RF bands centred on $f_{LO} + f_{IF}$ (which is the wanted band) and $f_{LO} - f_{IF}$ (which is an unwanted band). The unwanted 'image' band of frequencies, separated from the wanted RF band by twice the IF, represents a potentially serious source of RF interference and additional noise. A tunable image rejection filter (needing only modest selectivity) can be placed before the mixer in the RF amplifier of Figure 1.4 to attenuate or remove this unwanted band of frequencies.

1.5 Signal transmission

The communications path from transmitter to receiver may use lines or free space. Examples of the former are wire pairs, coaxial cables and optical fibres. The most important use of the latter is radio, although in some situations infrared and optical free space links are also possible (e.g. remote controls for TV, video and hi-fi equipment and also some security systems). Whatever the transmission medium, it is at this point that much of the attenuation, distortion, interference and noise is encountered.

Attenuation can be compensated for by introducing amplifiers or signal repeaters at intermediate points along the multiple hop link, Figure 1.5. Distortion may be compensated by equalisers and interference and noise can be minimised by using appropriate predetection signal processing (e.g. matched filters).

The nature and severity of transmission medium effects is one of the major influences on the design of transmitters, receivers and repeaters.

1.5.1 Line transmission

The essential advantages of line transmission are:

1. Path loss is usually modest.
2. Signal energy is essentially confined and interference between different systems is seldom severe and often negligible.
3. Path characteristics (e.g. attenuation and distortion) are usually stable and relatively easy to compensate for.
4. Capacity is unlimited in that bandwidth can always be reused by laying another line.

The disadvantages of line transmission are:

1. Laying cables in the ground or constructing overhead lines is generally expensive.
2. Extensive wayleaves and planning permission may be needed for underground cables and overhead wires.
3. Broadcasting requires a physical connection to a complex network for each subscriber.
4. Mobile communications services cannot be provided.
5. Networks cannot easily be added to, subtracted from, or otherwise reconfigured.

   The degree to which a signal is attenuated by a transmission line depends on the material from which the line is made, its physical construction, and the signal's frequency. Figures 1.6(a) to (e) show some typical attenuation/frequency characteristics for the most common types of line. Open wire has particularly low loss but it is expensive to maintain and susceptible to interference. Loaded cable, Figure 1.6(c) and Table 1.3, is only effective for speech bandwidth signals. Twisted pairs, as used underground, have higher installation cost but lower maintenance costs. (Low loss, circular, waveguides can also be used as a transmission medium but advances in optical fibre technology have, at least for the present, made this technology essentially redundant.) Optical fibre cables have an enormous information carrying capacity with

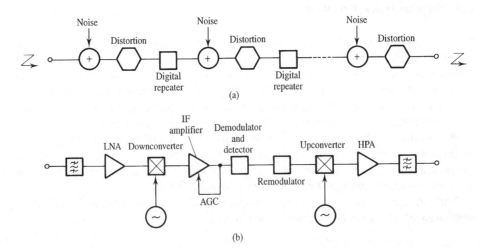

Figure 1.5 *(a) Digital communications (multi-hop) channel; (b) digital repeater (as typically used in terrestrial microwave relay applications).*

typical bandwidth-distance products of 0.5 GHzkm.

Table 1.3 *Nominal properties of selected transmission lines.*

	Frequency range	*Typical attenuation*	*Typical delay*	*Repeater spacing*
Open wire (overhead line)	0–160 kHz	0.03 dB/km @ 1 kHz	3.5 μs/km	40 km
Twisted pairs (multi-pair cables)	0–1 MHz	0.7 dB/km @ 1 kHz	5 μs/km	2 km
Twisted pairs (with *L* loading)	0–3.5 kHz	0.2 dB/km @ 1 kHz	50 μs/km	2 km (*L* spacing)
Coaxial cables	0–500 MHz	7 dB/km @ 10 MHz	4 μs/km	1–9 km
Optical fibres	λ = 1610–810 nm	0.2 to 0.5 dB/km	5 μs/km	100s km

Table 1.3 summarises the nominal frequency range of selected types of line, their typical attenuations and transmission delays, and typical repeater spacings. The useful bandwidths of the lines, which determine the maximum information transmission rate they can carry, are often, but not always, determined by their attenuation characteristics. Twisted wire pairs, for example, are normally limited to (line coded PCM) data rates of 2 Mbit/s. Coaxial cables, Figure 1.6(d), routinely carry 140 or 155 Mbit/s PCM signals but can handle symbol rates several times greater. Optical fibres have very large bandwidth potential but may be limited to a fraction of this by factors such as the spectral characteristics of optical sources and dispersion effects. Nevertheless, optical fibre PCM bit rates of Gbit/s are possible.

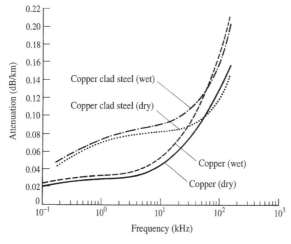

Figure 1.6(a) *Typical attenuation/frequency characteristics for aerial open wire pair lines.*

1.5.2 Radio transmission

The advantages of radio transmission are:

1. It is relatively cheap and quick to implement.
2. Wayleaves and planning permission are often only needed for the erection of towers to support repeaters and terminal stations.
3. It has an inherent broadcast potential.
4. It has an inherent mobile communications potential.
5. Communications networks can be quickly reconfigured and extra terminals or nodes easily introduced or removed.

The principal disadvantages of radio are:

1. Path loss is generally large due to the tendency of the transmitted signal energy to spread out, most of this energy effectively missing the receive antenna.

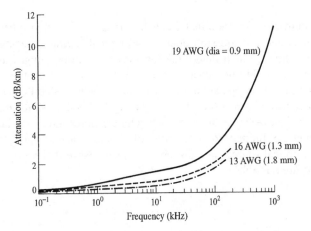

Figure 1.6(b) *Typical characteristics for twisted pair cable transmission lines.*

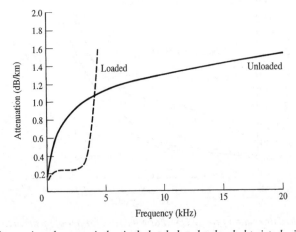

Figure 1.6(c) *Comparison between inductively loaded and unloaded twisted wire pairs.*

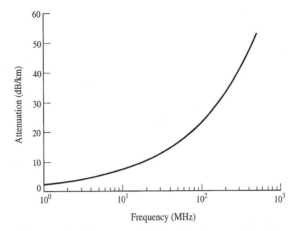

Figure 1.6(d) *Typical attenuation/frequency characteristic for coaxial cable.*

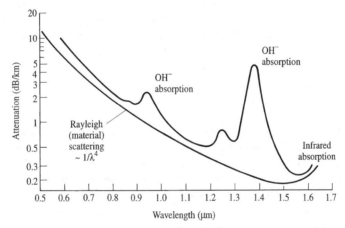

Figure 1.6(e) *Typical attenuation/wavelength characteristics for optical fibres (source: Young, 1994).*

2. The spreading of signal energy makes interference between different systems a potentially serious problem.
3. Capacity in a given locality is limited since bandwidth cannot be reused easily.
4. Path characteristics (i.e. attenuation and distortion) tend to vary with time, often in an unpredictable way, making equalisation more difficult and limiting reliability and availability.
5. The time varying nature of the channel can result in anomalous propagation of signals to locations well outside their normal range. This may cause unexpected interference between widely spaced systems.
6. Points 2 and 5 mean that frequency coordination is generally required when planning radio systems. Such coordination is difficult to achieve comprehensively and is expensive.

Table 1.4 *Frequency bands commonly used for radio communication.*

Band	Frequency	Wavelength	Propagation mechanism	Fading process	Noise process	Range	Applications
ELF	30–300 Hz	10^4–10^3 km	Waveguide modes	Diurnal variations due to D-layer	Man-made and atmospherics (lightning discharges)	Worldwide	Submarine
VF	300–3000 Hz	1000–100 km					
VLF	3–30 kHz	100–10 km	Surface waves	None			Standards/navigation
LF	30–300 kHz	10–1 km				1000s km	Maritime mobile, LW b'cast
MF	300–3000 kHz	1000–100 m	Sky waves	Surface/sky wave intf.		100s km	MW broadcast
HF	3–30 MHz	100–10 m		Complex ionospherics		4000 km/hop	Amateur
VHF	30–300 MHz	10–1 m	Line of sight	None	Galactic (synchrotron radiation)	Line of sight	FM broadcast
UHF	300–3000 MHz	100–10 cm		Ray bending and multipath	Cosmic background		TV broadcast, mobile, LOS
SHF	3–30 GHz	10–1 cm		Rain attenuation	Thermal noise from ground & atmosphere		Microwave LOS, satellite
EHF	30–300 GHz	10–1 mm				Short	

The appropriate radio propagation model for a communication system, the dominant fading and noise processes, and typical system range, all depend on frequency. Table 1.4 shows the electromagnetic spectrum used for radio transmissions and summarises these models and processes. At the lowest frequencies propagation is best modelled by oscillating electromagnetic modes which exist in the cavity between the concentric conducting spheres formed by the earth and its ionosphere. From a few kilohertz up to a few hundred kilohertz vertically polarised radio energy will propagate (by diffraction) around the curved surface of the earth for thousands of kilometres. This is called surface wave propagation and is the mechanism by which long wave radio broadcasts are received.

At slightly higher frequencies in the medium frequency (MF) band some radio energy propagates as a surface wave and some is reflected from the conducting ionosphere as a sky wave. The relative path lengths and phasing of these two signals may result in destructive interference causing fading of the received signal which will vary in severity as the relative strengths and phases of the sky wave and surface wave change. The exact condition of the ionosphere may be critical in this respect, making the quality of signal reception vary, for example, with the time of day or night.

In the high frequency (HF) band the sky wave is usually dominant and ranges of thousands of kilometres are possible, sometimes involving multiple reflections between the ionosphere and ground. At very high frequency (VHF), and above, signals propagate essentially along line-of-sight paths although reflection, refraction and, at the lower frequencies, diffraction can play an important role in the overall characteristics of the channel. At ultra high frequency (UHF) currently used for both TV transmissions and cellular radio communications, multipath (i.e. multiple path) propagation caused by reflections from, and diffraction around, buildings and other obstacles in urban areas is the principal cause of signal fading. In the super high frequency (SHF) band (usually called the microwave or centimetric wave band) applications tend to be point-to-point (or fixed-point) communications and the first order outdoor fading problem is often due to rain induced attenuation. These frequencies are also used for indoor radio local area networks (LANs). Extra high frequency (EHF) and higher frequencies are not yet widely used for communication systems, partly due to the significant gaseous background attenuation and large fades which occur in rain. However, millimetre wave wideband links are under consideration. As the electromagnetic spectrum becomes more congested, however, and as the demand for communications becomes yet greater (in terms of both traffic volume and service sophistication) the use of these higher frequency bands will almost certainly become both necessary and economic.

1.6 Switching and networks

Many modern communication systems are concerned exclusively with data traffic. One example is the Internet, over which users can transmit e-mail messages or browse distant information sources and transfer large data and text files. The data networks themselves, and the configuration of computer terminals on a user site, can be organised in many different ways using ring, star or bus connections. In order to ensure interoperability, standards have developed for these topologies and also for the signalling and switching protocols which control the assembly and routing of traffic.

The seven-layer ISO model is used throughout data communications networks as the standard hierarchical structure for organising data traffic. The data itself is usually sent as fixed length packets with associated overhead bits which provide addresses, timing or ordering information, and assist in error detection. The physical data interfaces follow evolved standards, e.g. X.25, IEEE 802, which develop progressively to higher data rates as the new high speed (wideband optical) transmission systems are introduced.

With packet data traffic there are inevitable delays while the packets are queued for access to the transmission system. These queues are not serious problems in simple mail networks but, if attempting to transmit speech or video traffic in real time, queue delays, and lost packets due to queue overflow in finite length buffers, can seriously degrade the operation of the communications link.

1.7 Advantages of digital communications

Digital communications systems usually represent an increase in complexity over the equivalent analogue systems. We therefore list here some of the reasons why digital communications have become the preferred option for most new systems and, in many instances, have replaced existing analogue systems.

1. Increased demand for data transmission.
2. Increased scale of integration, sophistication and reliability of digital electronics for signal processing, combined with decreased cost.
3. Facility to source code for data compression.
4. Possibility of channel coding (line, and error control, coding) to minimise the effects of noise and interference.
5. Ease with which bandwidth, power and time can be traded off in order to optimise the use of these limited resources.
6. Standardisation of signals, irrespective of their type, origin or the services they support, leading to an integrated services digital network (ISDN).

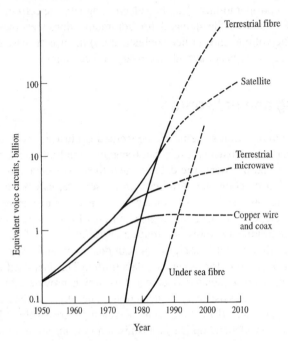

Figure 1.7 *Growth in world transmission capacity (source: Cochrane, 1990, reproduced with permission of British Telecommunications plc.).*

The increase in demand, for voice and data connections, is the principal driving force behind the growth in telecommunications. The traffic, in the backbone network, expressed as equivalent voice circuits, is shown in Figure 1.7. This figure not merely reflects the explosive growth in mobile communications for the final customer connections but shows world capacity for transmission.

1.8 Summary

The history of electronic communications over the last century and a half has demonstrated an essentially exponential growth in traffic and a continuously increasing demand for greater access to ever more sophisticated services. This trend shows no sign, at present, of changing.

Most modern telecommunications systems are digital and use some form of PCM irrespective of the origin of the information they convey. PCM signals are often coded themselves to improve system performance and/or provide security. Many PCM signals can be combined as a single (time division) multiplex to allow their simultaneous transmission over a single physical medium. Line coding and/or modulation can then be used to match the characteristics of the resulting multiplex to the transmission line or radio channel being used. Multiple accessing techniques allow many transceiver pairs to share a given transmission resource (e.g. cable, fibre, satellite transponder). Switching allows telecommunications networks to be designed which, at reasonable cost, can emulate a fully interconnected set of transceivers.

It is the purpose of this book to describe the operating principles and performance of modern digital communications systems. The description is presented at a systems, rather than a circuit, level and, in view of this, Part One of the book (Chapters 2 through 4) reviews some pertinent mathematical models and properties of signals, noise and systems. Part Two (Chapters 5 through 13) describes the analogue to digital conversion process, coding, and modulation techniques used to ensure adequate performance of a wide range of digital communications systems (Chapters 5 through 11); Chapter 12 is concerned with physical aspects of noise and the prediction of CNR at the end of a single or multi-hop transmission link; Chapter 13 discusses the computer simulation of communications systems. Part Three (Chapters 14 through 16) discusses modern digital telephony, terrestrial and satellite microwave systems, mobile cellular radio and video coding systems. Part Four (Chapters 17 through 21) describe switching and telecommunications networks including topologies, protocols, queuing theory and packet data transmission. It also includes a discussion of switched networks, wide area networks (WANs) and the public switched telephone network (PSTN) realised with the early plesiochronous digital heirarchy (PDH) and the subsequent synchronous digital hierarchy (SDH) which accommodates both telephony and data traffic. Finally broadcast networks and LANs are included.

Part One

Signals and systems theory

Signals and systems theory is the body of knowledge related to the definition and description of signals, and the behaviour of systems. In electrical engineering the study of signals is central to telecommunications, whilst the study of systems is probably more closely identified with control. It is obvious, however, that control engineers must be concerned with the signals which form the inputs and outputs of their systems, and conversely, communications engineers must be concerned with the systems which transmit, receive and otherwise process their signals. Nevertheless the closeness of the relationship between signal theory and communications means that the material presented in Part One is biased in favour of signals.

Chapter 2 presents the principal mathematical tools (Fourier series and Fourier transforms) normally used to describe, analyse and synthesise waveforms and transient signals. A unifying theme, here, is that of determinism, i.e. the waveforms and signals addressed all allow descriptions which permit their values to be determined, precisely, at any point in time. The choice of Fourier analysis as a technique for splitting complicated signals into their simpler (sinusoidal) component parts leads to the important concepts of spectrum and bandwidth.

Much of the communication engineer's effort is directed at conserving spectrum and utilising available signal bandwidth efficiently.

Chapter 2 also introduces ideas of signal orthogonality and correlation which relate to common sense notions of similarity. These concepts are important in communications since only signals which are in some way dissimilar can be assigned different meanings. In digital communications, especially, it is usually a requirement to generate signals which are easily distinguishable.

Chapter 3 deals with random signals (i.e. those which are not deterministic and are thus excluded from Chapter 2). Random signals are important, partly because information cannot be communicated by deterministic signals, and partly because unwanted random signals (constituting noise) always exist in a communications receiver. Such noise has the potential to modify, or obscure, wanted, information bearing signals. Due to their unpredictable nature random signals must be described in terms of their statistical properties. Chapter 3 therefore reviews probability theory and defines the mean, variance, covariance and other statistics, which are used to summarise the behaviour of random signals and noise. The similarity of a signal with a time shifted version of itself determines how rapidly the signal can change with time and provides information about the signal's spectrum. Chapter 3 makes the precise connection between self similarity (or autocorrelation) functions and the Fourier based spectral descriptions presented in Chapter 2.

Chapter 4 is concerned with systems, and in particular the effect that linear systems have on the spectral and autocorrelation properties of signals. The importance, and defining characteristics, of linear systems are discussed and use is made of the Fourier transform to link their equivalent time domain (impulse response) and frequency domain (frequency response) descriptions. The ways in which impulse and frequency responses are used to predict the effect of a system on both deterministic and random signals are thus developed.

The importance of systems to the communications engineer lies in the fact that signals conveying information must be processed many times by subsystems (filters, modulators, amplifiers, equalisers, etc.) before they reach their final destination. It is only through a thorough understanding of the modifying effect of these subsystems that one can ensure, in the presence of noise, that signals will remain adequately distinguished to achieve message reception without error.

CHAPTER 2

Periodic and transient signals

2.1 Introduction

Signals and waveforms are central to communications. A *signal* is defined [Hanks] as 'any sign, gesture, token, etc., that serves to communicate information'. It will be shown later that to communicate information such symbols must be in some sense unpredictable or random. The word signal, as applied to electronic communications, therefore implies an electrical quantity (e.g. voltage) possessing some characteristic (e.g. amplitude) which varies unpredictably. A *waveform* is defined as 'the shape of a wave or oscillation obtained by plotting the value of some changing quantity against time'. In electronic communications the term waveform implies an electrical quantity which varies *periodically*, and therefore predictably. Strictly this precludes a waveform from conveying information. However, a waveform can be adapted to convey information by varying one or more of its parameters in sympathy with a signal. Such waveforms are called carriers and typically consist of a sinusoid or pulse train modulated in amplitude, phase or frequency.

Fluctuating voltages and currents can be alternatively classified as either periodic or aperiodic. A periodic signal, if shifted by an appropriate time interval, is unchanged. An aperiodic signal does not possess this property. In this context the term periodic signal is clearly synonymous with waveform. In this chapter our principal concern is with periodic signals and one type of aperiodic signal, i.e. transients. A transient signal is one which has a well defined location in time. This does not necessarily mean it must be zero outside a certain time interval but it does imply that the signal at least tends to zero as time tends to $\pm\infty$. The one-sided decaying exponential function is an example of a transient signal which has a well defined start and tends to zero as $t \to \infty$.

If a signal's parameters (amplitude, shape and phase in the case of a periodic signal, amplitude, shape and location in the case of a transient signal) are known, then the signal is said to be deterministic. This means that, in the absence of noise, any future value of the signal can be determined precisely. Signals which are not deterministic must be described using probability theory, as discussed in Chapter 3.

2.2 Periodic signals

A periodic signal is defined as one which has the property:

$$f(t) = f(t \pm nT) \tag{2.1}$$

where n is any integer and T is the repetition period (or simply period) of the signal, Figure 2.1. A consequence of this definition is that periodic signals have no starting time or finishing time, i.e. they are eternal. The normalised power, P, averaged over any T second period, is:

$$P = \frac{1}{T} \int_t^{t+T} |f(t)|^2 \, dt \quad (\text{V}^2) \tag{2.2}$$

where the integral is the normalised energy per period. This is clearly a well defined finite quantity. The total energy, E, in a periodic signal, however, is infinite, i.e.:

$$E = \int_{-\infty}^{\infty} |f(t)|^2 \, dt = \infty \quad (\text{V}^2 \text{ s}) \tag{2.3}$$

For this reason periodic signals (along with some other types of signal) are sometimes called *power* signals. It also means that signals which are strictly periodic are unrealisable. The concept of a strictly periodic signal is, however, both simple and useful. Furthermore it is easy to generate signals which approximate very closely the conceptual ideal.

2.2.1 Sinusoids, cisoids and phasors

An especially simple and useful set of periodic signals is the set of sinusoids. These are generated naturally by projecting a point P, located on the circumference of a rotating disc (with unit radius), onto various planes, Figure 2.2.

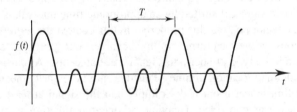

Figure 2.1 *Example of a periodic signal.*

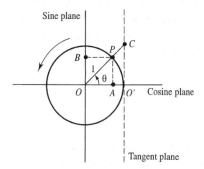

Figure 2.2 *Generation of sinusoids by projection of a radius onto perpendicular planes.*

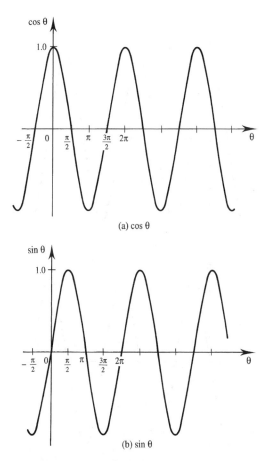

Figure 2.3 *Circular trigonometric functions plotted against phase: (a) cosine function of phase angle; (b) sine function of phase angle.*

If the length of *OA* in Figure 2.2 is plotted against angular position θ, then the result is the function cos θ, Figure 2.3(a). If the length of *OB* is plotted against θ, then the result is sin θ, Figure 2.3(b). (If the length of $O'C$ on the plane tangent to the disc is plotted against θ then the function tan θ results.) If the disc is not of unit radius then the normal (circular) trigonometric ratios are defined by:

$$\cos \theta = \frac{OA}{OO'} \qquad\qquad (2.4(a))$$

$$\sin \theta = \frac{OB}{OO'} \qquad\qquad (2.4(b))$$

The angle θ, expressed in degrees or radians, is called the phase of the function and can be related to the time period, T, taken for one revolution, i.e.:

$$\theta = 360 \, \frac{t}{T} \ \text{degrees} \qquad\qquad (2.5(a))$$

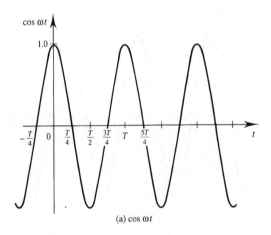

(a) cos ωt

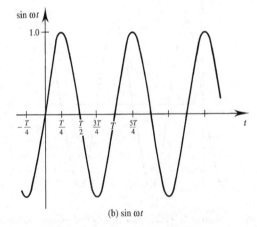

(b) sin ωt

Figure 2.4 *Circular trigonometric functions plotted against time: (a) cosine function of time; (b) sine function of time.*

$$\theta = 2\pi \; \frac{t}{T} \quad \text{radians} \tag{2.5(b)}$$

The angular velocity (or radian frequency), $\omega = d\theta/dt$, of the disc is therefore given by:

$$\omega = \frac{2\pi}{T} \quad \text{rad/s} \tag{2.6}$$

and angular position or phase by:

$$\theta = \omega t \quad \text{rad} \tag{2.7}$$

$1/T$ is the cyclical frequency of the disc in cycles/s or Hz. The sine and cosine functions plotted against time, t, are shown in Figure 2.4. The functions $\cos \theta$ and $\sin \theta$ are identical in shape but $\cos \theta$ reaches its peak value $T/4$ seconds (i.e. $\pi/2$ radians or 90°) *before* $\sin \theta$. $\cos \theta$ is therefore said to *lead* $\sin \theta$ by $\pi/2$ radians and, conversely, $\sin \theta$ is said to *lag* $\cos \theta$ by $\pi/2$ radians. The relationship between cosine and sine functions can be summarised by:

$$\cos \theta = \sin(\theta + \pi/2) \tag{2.8}$$

Notice that the cosine function and sine function have even and odd symmetry respectively about $t = 0$, i.e.:

$$\cos \theta = \cos(-\theta) \tag{2.9(a)}$$

$$\sin \theta = -\sin(-\theta) \tag{2.9(b)}$$

A cisoid is a general term which describes a rotating vector in the complex plane. Figure 2.5 shows a cisoid (which makes an angle ϕ with the plane's real axis at time $t = 0$) resolved onto real and imaginary axes. From the definition of the circular trigonometric functions it is clear that the component resolved onto the real axis is:

$$\Re[\text{cisoid}] = \cos(\omega t + \phi) \tag{2.10(a)}$$

and the component resolved onto the imaginary axis is:

$$\Im[\text{cisoid}] = \sin(\omega t + \phi) \tag{2.10(b)}$$

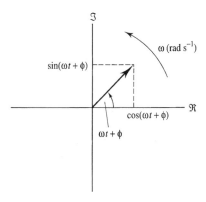

Figure 2.5 *Rotating vector or cisoid.*

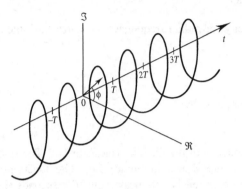

Figure 2.6 *Sketch of cisoid with time progressing perpendicular to the complex plane.*

Using Euler's formula (which relates geometrical and algebraic quantities) the real and imaginary components can be expressed together as:

$$\cos(\omega t + \phi) + j\,\sin(\omega t + \phi) = e^{j(\omega t + \phi)} \tag{2.11}$$

Equation (2.11) is the origin of the term cisoid which is a contraction of $(cos + i\,sin)$us*oid*, where $i = \sqrt{-1}$ replaces j. In three dimensions, with time (or phase) progressing along the axis perpendicular to the complex plane, the cisoid traces out a helical curve, Figure 2.6. For $\phi = 0$ the projection of this helix onto the imaginary/time plane is a sine wave and its projection onto the real/time plane is a cosine wave.

There is a satisfying symmetry relating real sinusoids and complex cisoids in that two, quadrature, sinusoids are required to generate a single cisoid and two, counter-rotating, cisoids are required to generate a single sinusoid. If the cisoids are a conjugate pair then the resulting sinusoid is purely real, Figure 2.7.

Phasors are cisoids which have had their time dependence suppressed. The phasor corresponding to $e^{j(\omega t+\phi)}$ is therefore $e^{j\phi}$ and corresponds to an instantaneous picture of the cisoid at the time $t = 0$, Figure 2.8. Another interpretation of phasors is that they represent a cisoid drawn in a plane which is itself rotating at the same angular frequency as the cisoid. The phasor is therefore stationary with respect to the complex plane. The

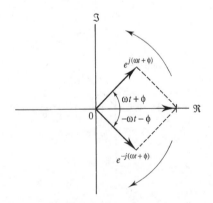

Figure 2.7 *Synthesis of real sinusoid wave from two counter-rotating, conjugate, cisoids.*

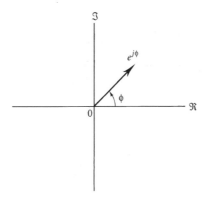

Figure 2.8 *Phasor corresponding to $e^{j(\omega t+\phi)}$.*

close relationship between cisoids and phasors is such that a distinction between them is rarely made in practice, the term phasor often being used to describe both.

2.2.2 Fourier series

Almost any periodic signal of practical interest can be approximated by adding together sinusoids with the correct frequencies, amplitudes and phases. An example of a sawtooth waveform approximated by a sum of sinusoids is shown in Figure 2.9. In general the error between the synthesised approximation and the actual waveform can be made as small as desired by including enough sinusoids in the sum. (This is not true at points of discontinuity, however: see section 2.2.4.) Only one sinusoid at each integer multiple of the fundamental frequency is required in the sum, providing that its amplitude and phase can be chosen freely. The fundamental frequency, f_1, is the reciprocal of the waveform's period, T, i.e.:

$$f_1 = 1/T \tag{2.12}$$

The sinusoid with frequency $f_n = nf_1$ is called the nth harmonic of the fundamental. If the waveform being approximated has a non-zero mean value then, in addition to the set of sinusoids, a 0 Hz, constant or DC term must be included in the sum. In general, then, the sinusoidal sum, which is called a Fourier series, is given by:

$$v(t) = C_0 + C_1 \cos(\omega_1 t + \phi_1) + C_2 \cos(\omega_2 t + \phi_2) + \cdots \tag{2.13}$$

where C_0 (V) is the DC term, $\omega_1 = 2\pi/T$ (rad/s) is the fundamental frequency and $\omega_2 = 2(2\pi/T)$ (rad/s) is the second harmonic frequency, etc. The series may be truncated after a finite number of terms or may extend indefinitely.

Trigonometric forms

The trigonometric form of the Fourier series, expressed by equation (2.13), can be written more compactly as:

$$v(t) = C_0 + \sum_{n=1}^{\infty} C_n \cos(\omega_n t + \phi_n) \tag{2.14(a)}$$

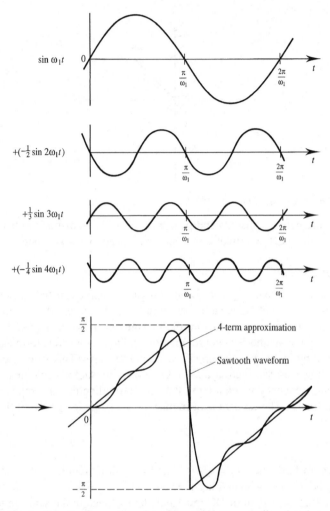

Figure 2.9 *Synthesis of sawtooth waveform by addition of harmonically related sinusoids.*

This is the *cosine* form of the series since each term is written as a cosine function (with an explicit phase angle, ϕ). Since each term in the periodic signal is a harmonic of the fundamental, equation (2.14(a)) can be rewritten as:

$$v(t) = C_0 + \sum_{n=1}^{\infty} C_n \cos(n\omega_1 t + \phi_n) \qquad (2.14(b))$$

A slightly different trigonometric form can be created by resolving each sinusoid into cosine and sine components (each with zero phase angle). This gives the *cosine–sine* form of the trigonometric Fourier series, i.e.:

$$v(t) = C_0 + \sum_{n=1}^{\infty} (A_n \cos \omega_n t - B_n \sin \omega_n t) \qquad (2.14(c))$$

Notice that the series is still specified by two real numbers per harmonic but in this case the numbers are cosine and sine amplitudes (or inphase and quadrature amplitudes) rather than amplitude and phase. (The use of a minus sign in equation (2.14(c)) may seem eccentric but its advantage will become clear later.)

If the amplitude, C_n, of the cosine Fourier series is plotted against frequency, $f_n = \omega_n/2\pi$ (Hz), the result is called a discrete, or line, amplitude spectrum, Figure 2.10(a). Similarly, if ϕ_n is plotted against f_n the result is a discrete phase spectrum, Figure 2.10(b). Notice that, for obvious reasons, the phase of the DC (0 Hz) component is not defined. Notice also that the height of the lines in the amplitude spectrum of Figure 2.10(a) represents the peak values of the sinusoidal components. It is possible, of course, to define an RMS amplitude spectrum which would be the same as the peak amplitude spectrum except that each line would be smaller by a factor of $1/\sqrt{2}$.

If the sinusoids of a cosine series are displayed in three dimensions, plotted against time and frequency, Figure 2.11, then the amplitude spectrum corresponds to a projection onto the amplitude–frequency plane. This gives a picture of the 'frequency content' of a signal.

Calculation of Fourier coefficients

Since C_0 is the DC, or average, value of the waveform being approximated it is clear that it can be calculated using:

$$C_0 = \frac{1}{T} \int_t^{t+T} v(t)\, dt \tag{2.15}$$

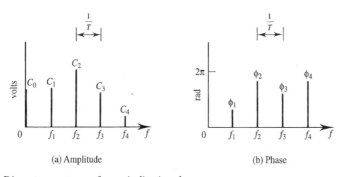

(a) Amplitude   (b) Phase

Figure 2.10 *Discrete spectrum of a periodic signal.*

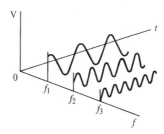

Figure 2.11 *Component sinusoids of a Fourier series plotted against time and frequency.*

In practice it is easier to calculate the A_n and B_n coefficients associated with the cosine–sine form of the Fourier series than to find the C_n and ϕ_n values of the cosine form. (C_n and ϕ_n can be easily calculated from A_n and B_n as will be shown later.) The essential task in calculating the value of A_1, for example, is to find out how much of the inphase fundamental component, $\cos \omega_1 t$, is contained in $v(t)$. In other words the similarity between $\cos \omega_1 t$ and $v(t)$ must be established. One way of quantifying this similarity is to find their mean product, i.e. $\langle v(t) \cos \omega_1 t \rangle$ where $\langle \ \rangle$ signifies a time average. If $v(t)$ tends to be positive when $\cos \omega_1 t$ is positive and negative when $\cos \omega_1 t$ is negative then $\langle v(t) \cos \omega_1 t \rangle$ will tend to be large and positive indicating a large degree of similarity, Figure 2.12. This would suggest that $v(t)$ contained a large $\cos \omega_1 t$ component. If, conversely, $v(t)$ tends to be negative when $\cos \omega_1 t$ is positive and vice versa then $\langle v(t) \cos \omega_1 t \rangle$ will tend to be large and negative. This would indicate extreme dissimilarity and the conclusion would be that $v(t)$ contained a large $-\cos \omega_1 t$ component. If there was little correlation between the polarity of $v(t)$ and $\cos \omega_1 t$ then $\langle v(t) \cos \omega_1 t \rangle$ would be close to zero and the conclusion would be that $v(t)$ contained almost no $\cos \omega_1 t$ component.

The normal way to find an average value is given by equation (2.15). Therefore:

$$\langle v(t) \cos \omega_1 t \rangle = \frac{1}{T} \int_t^{t+T} v(t) \cos \omega_1 t \ dt \tag{2.16}$$

To find the Fourier coefficient, A_1, however, the actual equation used is:

$$A_1 = \frac{2}{T} \int_t^{t+T} v(t) \cos \omega_1 t \ dt \tag{2.17}$$

This is because if $v(t)$ was *exactly* like $\cos \omega_1 t$ (i.e. $v(t) = \cos \omega_1 t$) then A_1 should be 1.0. Unfortunately:

$$\langle \cos^2 \omega_1 t \rangle = \langle \tfrac{1}{2}(1 + \cos 2\omega_1 t) \rangle = \tfrac{1}{2} \tag{2.18}$$

The factor of two in equation (2.17) is necessary to make $A_1 = 1$. The general formulae for calculating the cosine–sine Fourier coefficients are therefore:

$$A_n = \frac{2}{T} \int_t^{t+T} v(t) \cos \omega_n t \ dt \tag{2.19(a)}$$

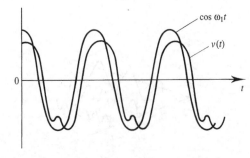

Figure 2.12 *Similar waveforms: $v(t)$ and $\cos \omega_1 t$.*

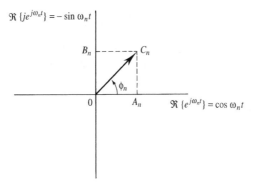

Figure 2.13 *Relationship between amplitude (C_n) of Fourier coefficients and inphase and quadrature components (A_n and B_n).*

$$B_n = -\frac{2}{T} \int_t^{t+T} v(t) \sin \omega_n t \; dt \qquad (2.19(b))$$

(B_n quantifies the similarity between $v(t)$ and $-\sin \omega_n t$.) If the cosine series is required the values of C_n and ϕ_n are found easily using simple trigonometry, Figure 2.13, i.e.:

$$C_n = \sqrt{(A_n^2 + B_n^2)} \qquad (2.20(a))$$

$$\phi_n = \tan^{-1}(B_n/A_n) \qquad (2.20(b))$$

A satisfying engineering interpretation of equations (2.19(a)) and (b) is that of 'filtering integrals'. If $v(t)$ is made up of many harmonically related sinusoids the average product of each of these sinusoids with $\cos \omega_n t$ or $\sin \omega_n t$ is only non-zero for that sinusoid which has the same frequency as $\cos \omega_n t$ or $\sin \omega_n t$. This *orthogonality* property is summarised mathematically by:

$$\frac{2}{T} \int_t^{t+T} \cos \omega_m t \cos \omega_n t \; dt = \begin{cases} 1, & m = n \\ 0, & m \neq n \end{cases} \qquad (2.21(a))$$

$$\frac{2}{T} \int_t^{t+T} \sin \omega_m t \sin \omega_n t \; dt = \begin{cases} 1, & m = n \\ 0, & m \neq n \end{cases} \qquad (2.21(b))$$

$$\frac{2}{T} \int_t^{t+T} \cos \omega_m t \sin \omega_n t \; dt = 0 \qquad (2.21(c))$$

These properties and their geometrical interpretation will be discussed further, in a more general context, in section 2.5.

EXAMPLE 2.1

Find the first two Fourier coefficients of a unipolar rectangular pulse train with amplitude 3 V, period 10 ms, duty cycle 20% and pulse leading edge at time $t = 0$. The pulse train $v(t)$ is shown in Figure 2.14.

The DC term is given by:

$$C_0 = \frac{1}{T} \int_t^{t+T} v(t)\, dt$$

$$= \frac{1}{0.01} \int_0^{0.002} 3\, dt$$

$$= 100\, [3\, t]_0^{0.002} = 0.6 \text{ V}$$

The inphase coefficients are given by:

$$A_1 = \frac{2}{T} \int_t^{t+T} v(t) \cos 2\pi f_1 t \, dt$$

$$= \frac{2}{0.01} \int_0^{0.002} 3 \cos\left(2\pi\, \frac{1}{0.01}\, t \right) dt$$

$$= 600 \left[\frac{\sin(2\pi\, 100\, t)}{2\pi\, 100} \right]_0^{0.002} = 0.9082 \text{ V}$$

$$A_2 = \frac{2}{T} \int_t^{t+T} v(t) \cos 2\pi f_2 t \, dt$$

$$= \frac{2}{0.01} \int_0^{0.002} 3 \cos\left(2\pi\, \frac{2}{0.01}\, t \right) dt$$

$$= 600 \left[\frac{\sin(2\pi\, 200\, t)}{2\pi\, 200} \right]_0^{0.002} = 0.2806 \text{ V}$$

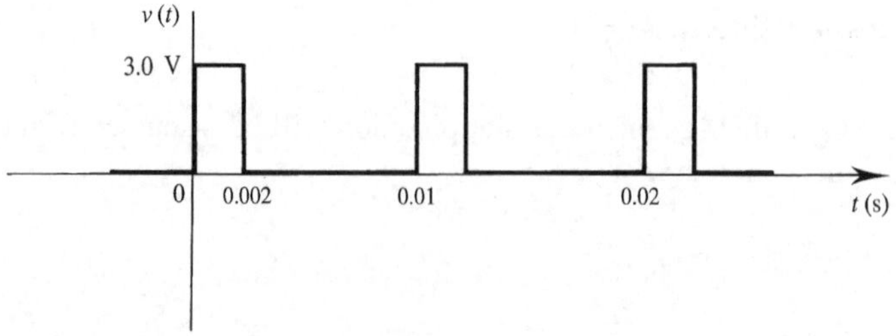

Figure 2.14 *Periodic rectangular pulse train for Example 2.1.*

The quadrature coefficients are given by:

$$B_1 = -\frac{2}{T} \int_{t}^{t+T} v(t) \sin 2\pi f_1 t \, dt$$

$$= -\frac{2}{0.01} \int_{0}^{0.002} 3 \sin\left(2\pi \frac{1}{0.01} t\right) dt$$

$$= -600 \left[\frac{-\cos(2\pi \, 100 \, t)}{2\pi \, 100}\right]_{0}^{0.002} = \frac{3}{\pi}[0.3090 - 1] = -0.6599 \quad V$$

$$B_2 = -\frac{2}{T} \int_{t}^{t+T} v(t) \sin 2\pi f_2 t \, dt$$

$$= -\frac{2}{0.01} \int_{0}^{0.002} 3 \sin\left(2\pi \frac{2}{0.01} t\right) dt$$

$$= -600 \left[\frac{-\cos(2\pi \, 200 \, t)}{2\pi \, 200}\right]_{0}^{0.002} = \frac{3}{2\pi}[-0.8090 - 1] = -0.8637 \quad V$$

The Fourier coefficient amplitudes are given in equation (2.20(a)) by:

$$C_n = \sqrt{(A_n^2 + B_n^2)}$$

i.e.:

$$C_0 = 0.6 \quad V$$

$$C_1 = \sqrt{(0.9082^2 + 0.6599^2)} = 1.1226 \quad V$$

$$C_2 = \sqrt{(0.2806^2 + 0.8637^2)} = 0.9081 \quad V$$

and the Fourier coefficient phases are given in equation (2.20(b)) by:

$$\phi_n = \tan^{-1}\left(\frac{B_n}{A_n}\right)$$

i.e.:

$$\phi_1 = \tan^{-1}\left(\frac{-0.6599}{0.9082}\right) = -0.6284 \text{ rad or} -36.0°$$

$$\phi_2 = \tan^{-1}\left(\frac{-0.8637}{0.2806}\right) = -1.257 \text{ rad or} -72.0°$$

Note that moving the pulse train to the right or left will change the phase spectrum but not the amplitude spectrum. For example, if the pulse train is moved 0.001 s to the left (such that it has even symmetry about $t = 0$) then the Fourier series will contain cosine waves only and the phase spectrum will be restricted to values of 0° and 180°.

Figure 2.9 shows the decomposition of a sawtooth wave into terms up to the fourth harmonic and also includes the wave reconstructed from these components.

Exponential form

As an alternative to calculating the A_n and B_n coefficients of the cosine–sine Fourier series separately they can be calculated together using:

$$A_n + jB_n = \frac{2}{T} \int_t^{t+T} v(t)(\cos \omega_n t - j \sin \omega_n t) \ dt \tag{2.22}$$

This corresponds to synthesising the function $v(t)$ from the real part of a set of harmonically related cisoids, i.e.:

$$v(t) = C_0 + \sum_{n=1}^{\infty} (A_n \cos \omega_n t - B_n \sin \omega_n t)$$

$$= C_0 + \Re \left\{ \sum_{n=1}^{\infty} \tilde{C}_n \ e^{j\omega_n t} \right\} \tag{2.23}$$

where $\tilde{C}_n = A_n + jB_n$. The tilde ($\tilde{\ }$) indicates that $\tilde{C}_n$ is generally complex.

Having a separate DC term, C_0, in equation (2.23) and being required to take the real part of the other terms is, at best, a little inelegant. This can be overcome, however, by using a pair of counter-rotating, conjugate cisoids to represent each real sinusoid in the series, i.e.:

$$v(t) = \sum_{n=-\infty}^{\infty} \tilde{C}'_n \ e^{j\omega_n t} \tag{2.24(a)}$$

where:

$$\tilde{C}'_n = \begin{cases} \tilde{C}_n/2 & \text{for } n > 0 \\ C_0 & \text{for } n = 0 \\ \tilde{C}^*_{-n}/2 & \text{for } n < 0 \end{cases} \tag{2.24(b)}$$

Thus, for example, the pair of cisoids corresponding to $n = \pm 3$ (i.e. the third harmonic cisoids) may look like those shown in Figure 2.15. Notice that the magnitude, $|\tilde{C}'_n|$, of each cisoid in the formulation of equations (2.24) is half that, $|\tilde{C}_n|$, of the corresponding cisoids in equation (2.23) or the corresponding sinusoids in equations (2.14(a)) and (b). Thus the formula for the calculation of Fourier (exponential) coefficients gives results only half as large as that for the trigonometric series, i.e.:

$$\tilde{C}'_n = \frac{1}{T} \int_t^{t+T} v(t) \ e^{-j\omega_n t} \ dt \tag{2.25}$$

$e^{-j\omega_n t}$, here, filters out that part of $v(t)$ which is identical to $e^{j\omega_n t}$ (since $(1/T)\int_0^T e^{j\omega_n t} e^{-j\omega_n t} \ dt = 1$). When n is positive, the positively rotating (i.e. anticlockwise) cisoids are obtained and when n is negative (remembering that $\omega_n = n\omega_1$) the negatively rotating (i.e. clockwise) cisoids are obtained. When $n = 0$ equation (2.25) gives the DC term.

Double sided spectra

If the amplitudes $|\tilde{C}'_n|$ of the cisoids found using equation (2.25) are plotted against frequency f_n the result is called a double (or two-) sided (voltage) spectrum. Such a spectrum is shown in Figure 2.16. If $v(t)$ is purely real, i.e. it represents a signal that can exist in a practical single channel system, then each positively rotating cisoid is matched by a conjugate, negatively rotating, cisoid which cancels the imaginary part to zero. The double sided amplitude spectrum thus has *even* symmetry about 0 Hz, Figure 2.16(a).

The single sided spectrum (positive frequencies only) representing the amplitudes of the real sinusoids in the trigonometric Fourier series can be found from the double sided spectrum by folding over the negative frequencies of the latter and adding them to the positive frequencies.

If the phase angles of the cisoids are plotted against f_n the result is the double sided phase spectrum which, due to the conjugate pairing of cisoids, will have *odd* symmetry for real waveforms, Figure 2.16(b).

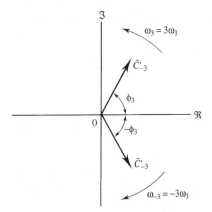

Figure 2.15 *Pair of counter-rotating, conjugate, cisoids (drawn for t = 0) corresponding to the (real) third harmonic of a periodic signal.*

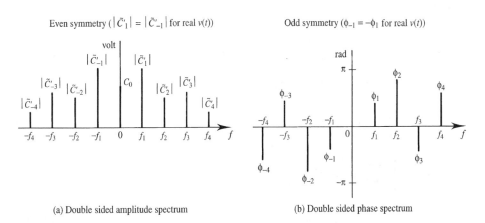

(a) Double sided amplitude spectrum   (b) Double sided phase spectrum

Figure 2.16 *Double sided voltage spectrum of a real periodic signal.*

The even and odd symmetries of the amplitude and phase spectra of real signals are summarised by:

$$|\tilde{C}'_n| = |\tilde{C}'_{-n}| \tag{2.26(a)}$$

and:

$$\arg(\tilde{C}'_n) = -\arg(\tilde{C}'_{-n}) \tag{2.26(b)}$$

Calculation of coefficients for waveforms with symmetry

For a waveform $v(t)$ with certain symmetry properties, the calculation of some, or all, of the Fourier coefficients is simplified. These symmetries and the corresponding simplifications for the calculation of C_0, A_n and B_n are shown in Table 2.1.

Discrete power spectra and Parseval's theorem

If the trigonometric coefficients are divided by $\sqrt{2}$ and plotted against frequency the result is a one-sided, RMS amplitude spectrum. If each RMS amplitude is then squared, the height of each spectral line will represent the normalised power (or power dissipated in $1\,\Omega$) associated with that frequency, i.e.:

$$\left\{ \frac{|\tilde{C}_n|}{\sqrt{2}} \right\}^2 = \frac{|\tilde{C}_n|^2}{2} = P_{n1} \ (\text{V}^2) \tag{2.27}$$

Such a one-sided power spectrum is illustrated in Figure 2.17(a). A double sided version of the power spectrum can be defined by associating half the power in each line, $P_{n2} = |\tilde{C}_n|^2/4$, with negative frequencies, Figure 2.17(b). The double sided power spectrum is therefore obtained from the double sided amplitude spectrum by squaring each cisoid amplitude, i.e. $P_{n2} = |\tilde{C}'_n|^2$. That the total power in an entire line spectrum is the sum of the powers in each individual line might seem an intuitively obvious statement. This is true, however, only because of the orthogonal nature of the individual sinusoids making up a periodic waveform. Obvious or not, this statement, which applies to any periodic signal, is known as Parseval's theorem. It can be stated in several forms, one of the most useful being:

$$\text{Total power, } P = \frac{1}{T} \int_t^{t+T} |v(t)|^2 \, dt = \sum_{n=-\infty}^{\infty} |\tilde{C}'_n|^2 \tag{2.28}$$

The proof of Parseval's theorem is straightforward and is given below:

$$P = \frac{1}{T} \int_t^{t+T} v(t)\, v^*(t)\, dt \tag{2.29(a)}$$

$$v^*(t) = \left[\sum_{n=-\infty}^{\infty} \tilde{C}'_n\, e^{j\omega_n t} \right]^*$$

$$= \sum_{n=-\infty}^{\infty} (\tilde{C}'^*_n\, e^{-j\omega_n t}) \tag{2.29(b)}$$

Table 2.1   *Fourier series formulae for waveforms with symmetry*

Type of Symmetry	Definition	Example $f(t)$	C_o	A_n	B_n	Non-zero terms
Zero mean	$\int_0^T f(t)\,dt = 0$		$C_o = 0$	$A_n = \dfrac{2}{T}\displaystyle\int_{-\frac{T}{2}}^{\frac{T}{2}} f(t)\cos\omega_n t\,dt$	$B_n = \dfrac{2}{T}\displaystyle\int_{-\frac{T}{2}}^{\frac{T}{2}} f(t)\sin\omega_n t\,dt$	A_n and B_n
Even	$f(t) = f(-t)$		$C_o = \dfrac{1}{T}\displaystyle\int_0^T f(t)\,dt$	$A_n = \dfrac{4}{T}\displaystyle\int_0^{\frac{T}{2}} f(t)\cos\omega_n t\,dt$	$B_n = 0$	A_n and C_o
Odd	$f(t) = -f(-t)$		$C_o = 0$	$A_n = 0$	$B_n = \dfrac{4}{T}\displaystyle\int_0^{\frac{T}{2}} f(t)\sin\omega_n t\,dt$	B_n
Half-wave	$f(t) = -f\!\left(t+\dfrac{T}{2}\right)$		$C_o = 0$	$A_n = \dfrac{4}{T}\displaystyle\int_0^{\frac{T}{2}} f(t)\cos\omega_n t\,dt$	$B_n = \dfrac{4}{T}\displaystyle\int_0^{\frac{T}{2}} f(t)\sin\omega_n t\,dt$	A_n and B_n, odd harmonics (n odd) only

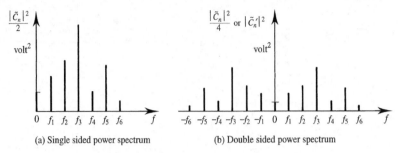

(a) Single sided power spectrum   (b) Double sided power spectrum

Figure 2.17 *Power spectra of a periodic signal.*

Therefore:

$$P = \frac{1}{T} \int\limits_{t}^{t+T} v(t) \sum_{n=-\infty}^{\infty} (\tilde{C}_n^{\prime *} e^{-j\omega_n t}) \, dt$$

$$= \sum_{n=-\infty}^{\infty} \left[\frac{1}{T} \int\limits_{t}^{t+T} v(t) \, e^{-j\omega_n t} \, dt \, \tilde{C}_n^{\prime *} \right]$$

$$= \sum_{n=-\infty}^{\infty} \tilde{C}_n^{\prime} \, \tilde{C}_n^{\prime *} \qquad\qquad (2.29(c))$$

EXAMPLE 2.2

Find the 0 Hz and first two harmonic terms in the double sided spectrum of the half-wave rectified sinusoid shown in Figure 2.18(a) and sketch the resulting amplitude and phase spectra. What is the total power in these components?

$$\tilde{C}_0^{\prime} = \frac{1}{T} \int\limits_{t}^{t+T} v(t) \, e^{-j2\pi \, 0 t} \, dt$$

$$= \frac{1}{0.02} \int\limits_{-0.0025}^{0.0075} \sin\left(2\pi \, \frac{1}{0.02} \, t + 2\pi \, \frac{0.0025}{0.02} \right) dt$$

$$= 50 \int\limits_{-0.0025}^{0.0075} \sin(100\pi t + 0.25\pi) \, dt$$

$$= 50 \int\limits_{-0.0025}^{0.0075} \sin(100\pi t) \cos(0.25\pi) + \cos(100\pi t) \sin(0.25\pi) \, dt$$

$$= 50 \, \frac{1}{\sqrt{2}} \left[\frac{-\cos 100\pi t}{100\pi} \right]_{-0.0025}^{0.0075} + 50 \, \frac{1}{\sqrt{2}} \left[\frac{\sin 100\pi t}{100\pi} \right]_{-0.0025}^{0.0075}$$

$$= \frac{50}{100\pi} \, \frac{1}{\sqrt{2}} \{ -[-0.7071 - 0.7071] + [0.7071 - (-0.7071)] \}$$

$$= \frac{50}{100\pi} \frac{1}{\sqrt{2}} 4 (0.7071) = \frac{1}{\pi} (= 0.3183 \text{ V})$$

$$\tilde{C}'_1 = \frac{1}{T} \int_t^{t+T} v(t) \, e^{-j2\pi f_1 t} \, dt$$

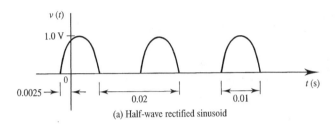

(a) Half-wave rectified sinusoid

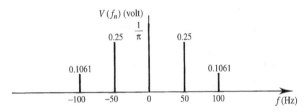

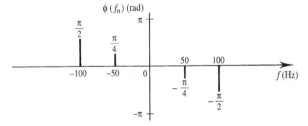

(b) Amplitude and phase spectra of DC and first two harmonics for waveform in Figure 2.18(a)

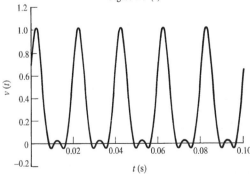

(c) Fourier series approximation to waveform in Example 2.2 (DC plus 2 harmonics)

Figure 2.18 *Waveform, spectra and Fourier series approximation for Example 2.2.*

$$
= \frac{1}{0.02} \int_{-0.0025}^{0.0075} \sin\left(2\pi \frac{1}{0.02} t + \frac{2\pi \, 0.0025}{0.02} \right) e^{-j2\pi \frac{1}{0.02} t} \, dt
$$

$$
= \frac{50}{\sqrt{2}} \int_{-0.0025}^{0.0075} \sin(100\pi t) \, (\cos 100\pi t - j \sin 100\pi t) \, dt
$$

$$
+ \frac{50}{\sqrt{2}} \int_{-0.0025}^{0.0075} \cos(100\pi t) \, (\cos 100\pi t - j \sin 100\pi t) \, dt
$$

$$
= \frac{25}{\sqrt{2}} \left[\int_{-0.0025}^{0.0075} \sin(200\pi t) \, dt + \int_{-0.0025}^{0.0075} 1 + \cos(200\pi t) \, dt \right]
$$

$$
- j \frac{25}{\sqrt{2}} \left[\int_{-0.0025}^{0.0075} (1 - \cos(200\pi t)) \, dt + \int_{-0.0025}^{0.0075} \sin(200\pi t) \, dt \right]
$$

$$
= \frac{25}{\sqrt{2}} \left\{ \left[\frac{-\cos(200\pi t)}{200\pi} \right]_{-0.0025}^{0.0075} + [\, t \,]_{-0.0025}^{0.0075} + \left[\frac{\sin(200\pi t)}{200\pi} \right]_{-0.0025}^{0.0075} \right.
$$

$$
\left. - j[\, t \,]_{-0.0025}^{0.0075} + j \left[\frac{\sin(200\pi t)}{200\pi} \right]_{-0.0025}^{0.0075} - j \left[\frac{-\cos(200\pi t)}{200\pi} \right]_{-0.0025}^{0.0075} \right\}
$$

$$
= \frac{25}{\sqrt{2}} \left\{ \left[\frac{0-0}{200\pi} \right] + [0.0075 + 0.0025] + \left[\frac{-1 - (-1)}{200\pi} \right] \right\}
$$

$$
- j \frac{25}{\sqrt{2}} \left\{ [0.0075 + 0.0025] - \frac{[-1 - (-1)]}{200\pi} + \left[\frac{0-0}{200\pi} \right] \right\}
$$

$$
= \frac{25}{\sqrt{2}} [0.01 - j0.01]
$$

$$
= 0.1768 - j0.1768
$$

$$
= 0.25 \text{ V} \text{ at} -45°
$$

Since $v(t)$ is real then:

$$
\tilde{C}'_{-1} = \tilde{C}'^{*}_{1}
$$

$$
= 0.1768 + j0.1768
$$

$$
= 0.25 \text{ V} \text{ at } 45°
$$

$$
\tilde{C}'_{2} = \frac{1}{T} \int_{t}^{t+T} v(t) \, e^{-j2\pi f_2 t} \, dt
$$

$$
= \frac{1}{0.02} \int_{-0.0025}^{0.0075} \sin\left(2\pi \frac{1}{0.02} t + \frac{2\pi \, 0.0025}{0.02} \right) e^{-j2\pi \frac{2}{0.02} t} \, dt
$$

$$= 0.1061 \ \text{V at} -90°$$

Since $v(t)$ is real then:

$$\tilde{C}'_{-2} = \tilde{C}'^{*}_2 = 0.1061 \ \text{V at } 90°$$

The amplitude and phase spectra are shown in Figure 2.18(b). The sum of the DC term, fundamental and second harmonic is shown in Figure 2.18(c). It is interesting to see that even with so few terms the Fourier series is a recognisable approximation to the half-wave rectified sinusoid. The total power in the DC, fundamental and second harmonic components is given by Parseval's theorem, equation (2.28), i.e.:

$$P = \sum_{n=-2}^{2} |\tilde{C}'_n|^2$$

$$= 0.1061^2 + 0.25^2 + (1/\pi)^2 + 0.25^2 + 0.1061^2$$

$$= 0.2488 \ \text{V}^2$$

Table 2.2 shows commonly encountered periodic waveforms and their corresponding Fourier series.

2.2.3 Conditions for existence, convergence and Gibb's phenomenon

The question might be asked: how do we know that it is possible to approximate $v(t)$ with a Fourier series and furthermore that adding further terms to the series continues to improve the approximation? Here we give the answer to this question without proof.

A function $v(t)$ has a Fourier series if the following conditions are met:

1. $v(t)$ contains a finite number of maxima and minima per period.
2. $v(t)$ contains a finite number of discontinuities per period.
3. $v(t)$ is absolutely integrable over one period, i.e.:

$$\int_{t}^{t+T} |v(t)| \, dt < \infty$$

The above conditions, called the Dirichlet conditions, are sufficient but not necessary. If a Fourier series does exist it converges (i.e. gets closer to $v(t)$ as more terms are added) at all points except points of discontinuity. Mathematically, this can be stated as follows:

$$\sum_{N} \text{series} \, |_{t_0} \rightarrow v(t_0) \text{ as } N \rightarrow \infty \text{ for all continuous points } t_0$$

At points of discontinuity the series converges to the arithmetic mean of the function value on either side of the discontinuity, Figure 2.19(a), i.e.:

$$\sum_{N} \text{series} \, |_{t_0} \rightarrow \frac{v(t_0^-) + v(t_0^+)}{2} \text{ as } N \rightarrow \infty \text{ for all discontinuous points } t_0$$

Table 2.2 *Fourier series of commonly occurring waveforms.*

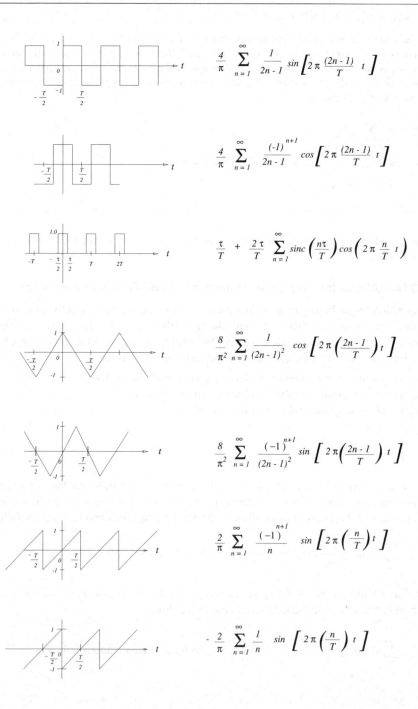

$$\frac{4}{\pi} \sum_{n=1}^{\infty} \frac{1}{2n-1} \sin\left[2\pi \frac{(2n-1)}{T} t\right]$$

$$\frac{4}{\pi} \sum_{n=1}^{\infty} \frac{(-1)^{n+1}}{2n-1} \cos\left[2\pi \frac{(2n-1)}{T} t\right]$$

$$\frac{\tau}{T} + \frac{2\tau}{T} \sum_{n=1}^{\infty} \text{sinc}\left(\frac{n\tau}{T}\right) \cos\left(2\pi \frac{n}{T} t\right)$$

$$\frac{8}{\pi^2} \sum_{n=1}^{\infty} \frac{1}{(2n-1)^2} \cos\left[2\pi\left(\frac{2n-1}{T}\right) t\right]$$

$$\frac{8}{\pi^2} \sum_{n=1}^{\infty} \frac{(-1)^{n+1}}{(2n-1)^2} \sin\left[2\pi\left(\frac{2n-1}{T}\right) t\right]$$

$$\frac{2}{\pi} \sum_{n=1}^{\infty} \frac{(-1)^{n+1}}{n} \sin\left[2\pi\left(\frac{n}{T}\right) t\right]$$

$$-\frac{2}{\pi} \sum_{n=1}^{\infty} \frac{1}{n} \sin\left[2\pi\left(\frac{n}{T}\right) t\right]$$

Table 2.2 ctd. *Fourier series of commonly occurring waveforms.*

$$\frac{2}{\pi} - \frac{4}{\pi} \sum_{n=1}^{\infty} \frac{1}{4n^2 - 1} \cos\left[2\pi\left(\frac{n}{T}\right) t \right]$$

$$\frac{2}{\pi} \left\{ \frac{1}{2} + \frac{\pi}{4} \cos\left[2\pi\left(\frac{1}{T}\right) t \right] - \sum_{n=1}^{\infty} \frac{(-1)^n}{4n^2 - 1} \cos\left[2\pi\left(\frac{2n}{T}\right) t \right] \right\}$$

$$\frac{1}{T} + \sum_{n=1}^{\infty} \frac{2}{T} \cos\left[2\pi\left(\frac{n}{T}\right) t \right]$$

At points on either side of a discontinuity the series oscillates with a period T_G given by:

$$T_G = 0.5 \, T/N \qquad\qquad (2.30(a))$$

where T is the period of $v(t)$ and N is the number of terms included in the series. The amplitude, Δ, of the overshoot on either side of the discontinuity is:

$$\Delta = 0.09A \qquad\qquad (2.30(b))$$

where A is the amplitude of the discontinuity, Figure 2.19(b). The overshoot, Δ, does not decrease as N increases, the resulting spikes sometimes being known as 'Gibb's ears'.

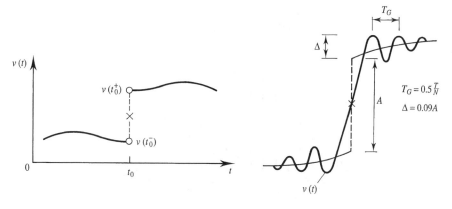

(a) Point of convergence (×) for Fourier series at a discontinuity   (b) Gibb's ears on either side of discontinuity

Figure 2.19 *Overshoot and undershoot of a truncated Fourier series at a point of discontinuity.*

2.2.4 Bandwidth, rates of change, sampling and aliasing

The bandwidth, B, of a signal is defined as the difference (usually in Hz) between two nominal frequencies f_{max} and f_{min}. Loosely speaking f_{max} and f_{min} are, respectively, the frequencies above and below which the spectral components are assumed to be small. It is important to realise that these frequencies are often chosen using some fairly arbitrary rule, e.g. the frequencies at which spectral components have fallen to $1/\sqrt{2}$ of the peak spectral component. It would therefore be wrong to assume always that the frequency components of a signal outside its quoted bandwidth are negligible for all purposes, especially if the precise definition being used for B is vague or unknown.

The $1/\sqrt{2}$ definition of B is a common one and is *usually* implied if no other definition is explicitly given. It is normally called the half power or 3 dB bandwidth since the factor $1/\sqrt{2}$ refers to the voltage spectrum and $20 \log_{10}(1/\sqrt{2}) \approx -3$ dB. The 3 dB bandwidth of a periodic signal is illustrated in Figure 2.20(a). For baseband signals (i.e. signals with significant spectral components all the way down to their fundamental frequency, f_1, or even DC) f_{min} is 0 Hz, *not* $-f_{max}$. This is important to remember when considering two-sided spectra. The physical bandwidth is measured using positive frequencies or negative frequencies only, not both, Figure 2.20(b).

In general, if a signal has no significant spectral components above f_H then it cannot change appreciably on a time scale much shorter than about $1/(8f_H)$. (This corresponds

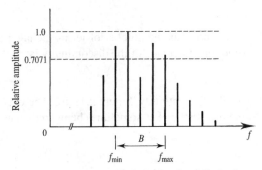

(a) 3 dB bandwidth of a (bandpass) periodic signal

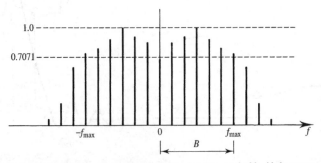

(b) 3 dB bandwidth of a (baseband) periodic signal shown on a double sided spectrum

Figure 2.20 *Definition of 3 dB signal bandwidth.*

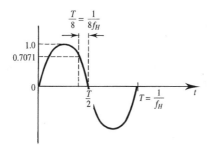

Figure 2.21 *Illustration of minimum time required for appreciable change of signal amplitude.*

to one-eighth of a period of the highest frequency sinusoid present in the signal, Figure 2.21.) A corollary of this is that signals with large rates of change must have high values of f_H. A rectangular pulse stream, for example, contains changes which occur (in principle) infinitely quickly. This implies that it must contain spectral components with infinite frequency. (In practice, of course, such pulse streams are, at best, only approximately rectangular and therefore their spectra can be essentially bandlimited.)

Sampling refers to the process of recording the values of a signal or waveform at (usually) regularly spaced instants of time. A schematic diagram of how this might be achieved is shown in Figure 2.22. It is a surprising fact that if a signal having no spectral components with frequencies above f_H is sampled rapidly enough then the original, continuous, signal can, in principle, be reconstructed from its samples *without error*. The minimum sampling rate or frequency, f_s, needed to achieve such ideal reconstruction is related to f_H by:

$$f_s \geq 2f_H \tag{2.31}$$

Equation (2.31) is called Nyquist's sampling theorem and is of central importance to digital communications. It will be discussed more rigorously in Chapter 5. Here, however, it is sufficient to demonstrate its reasonableness as follows.

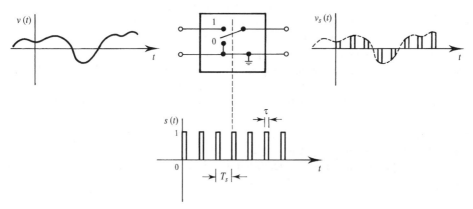

Figure 2.22 *Schematic illustration of sampling.*

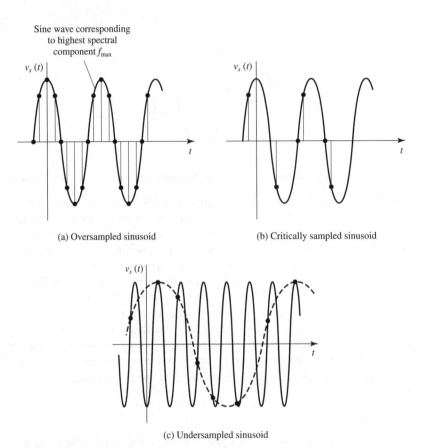

(a) Oversampled sinusoid

(b) Critically sampled sinusoid

(c) Undersampled sinusoid

Figure 2.23 *Demonstration of the sampling theorem and alias frequency.*

Figure 2.23(a) shows a sinusoid which represents the highest frequency spectral component in a certain waveform. The sinusoid is sampled in accordance with equation (2.31), i.e. at a rate higher than twice its frequency. (When $f_s > 2f_H$ the signal is said to be *oversampled*.) Nyquist's theorem essentially says that there is one, and only one, sinusoid which can be drawn through the given sample points. Figure 2.23(b) shows the same sinusoid sampled at a rate $f_s = 2f_H$. (This might be called critical, or Nyquist rate, sampling.) There is still only one frequency of sine wave which can be drawn through the samples. Figure 2.23(c) shows a sinusoid which is now *undersampled* (i.e. $f_s < 2f_H$). The samples could be (and usually are) interpreted as belonging to a sinusoid (shown dashed) of lower frequency than that to which they actually belong. The mistaken identity of the frequency of an undersampled sinusoid is called *aliasing* since the sinusoid inferred from the samples appears under the alias of a new and incorrect frequency. Aliasing is explained more fully later (section 5.3.3) with the aid of frequency domain concepts.

2.3 Transient signals

Signals are said to be transient if they are essentially localised in time. This obviously includes time limited signals which have a well defined start and stop time and which are zero outside the start–stop time interval. Signals with no start time, stop time, or either, are usually also considered to be transient, providing they tend to zero as time tends to ±∞ and contain finite total energy. Since the power of such signals averaged over all time is zero they are sometimes called *energy* signals. Since transient signals are not periodic they cannot be represented by an ordinary Fourier series. A related but more general technique, namely Fourier transformation, can, however, be used to find a frequency domain, or spectral, description of such signals.

2.3.1 Fourier transforms

The traditional way of approaching Fourier transforms is to treat them as a limiting case of a periodic signal Fourier series as the period, T, tends to infinity. Consider Figure 2.24. The waveform in this figure is periodic and pulsed with interpulse spacing, T_g. The amplitude and phase spectra of $v(t)$ are shown (schematically) in Figures 2.25(a) and (b) respectively. They are discrete (since $v(t)$ is periodic), have even and odd symmetry respectively (since $v(t)$ is real) and have line spacing $1/T$ Hz. If the interpulse spacing is now allowed to grow without limit (i.e. $T_g \to \infty$) then it follows that:

1. Period, $T \to \infty$.
2. Spacing of spectral lines, $1/T \to 0$.
3. The discrete spectrum becomes continuous (as $V(f)$ is defined at all points).
4. The signal becomes aperiodic (since only one pulse is left between $t = -\infty$ and $t = \infty$).

As the spectral lines become infinitesimally closely spaced the discrete quantities in the Fourier series:

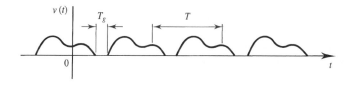

Figure 2.24 *Pulsed, periodic waveform.*

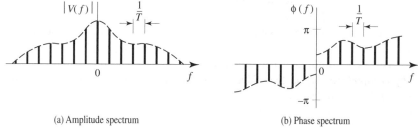

(a) Amplitude spectrum

(b) Phase spectrum

Figure 2.25 *Voltage spectrum of a pulsed, periodic, waveform.*

$$v(t) = \sum_{n=-\infty}^{\infty} \tilde{C}'_n \, e^{j2\pi f_n t} \tag{2.32}$$

become continuous, i.e.:

$$f_n \to f \quad \text{(Hz)}$$

$$\tilde{C}'_n(f_n) \to V(f) \, df \quad \text{(V)}$$

$$\sum_{n=-\infty}^{\infty} \to \int_{-\infty}^{\infty}$$

Since $\tilde{C}'_n(f_n)$, and therefore $V(f) \, df$, have units of V then $V(f)$ has units of V/Hz. $V(f)$ is called a *voltage spectral density*. The resulting 'continuous series', called an *inverse Fourier transform*, is:

$$v(t) = \int_{-\infty}^{\infty} V(f) \, e^{j2\pi ft} \, df \tag{2.33}$$

The converse formula, equation (2.25), which gives the (complex) Fourier coefficients for a Fourier series, is:

$$\tilde{C}'_n = \frac{1}{T} \int_{t-T/2}^{t+T/2} v(t) \, e^{-j2\pi f_n t} \, dt \tag{2.34}$$

If this is generalised in the same way as equation (2.32) by letting $T \to \infty$ then $\tilde{C}'_n \to 0$ for all n and $v(t)$. This problem can be avoided by calculating $T\tilde{C}'_n$ instead. In this case:

$$f_n \to f$$

$$T\tilde{C}'_n \to V(f)$$

$$t \pm \frac{T}{2} \to \pm\infty$$

(Note that $T\tilde{C}'_n$ and $V(f)$ have units of V s or equivalently V/Hz as required.) The *forward* Fourier transform therefore becomes:

$$V(f) = \int_{-\infty}^{\infty} v(t) \, e^{-j2\pi ft} \, dt \tag{2.35}$$

Equation (2.35) can be interpreted as finding that part of $v(t)$ which is identical to $e^{j2\pi ft}$. This is a cisoid (or rotating phasor) with frequency f and amplitude $V(f) \, df$ (V). For real signals there will be a conjugate cisoid rotating in the opposite sense located at $-f$. This pair of cisoids together constitute a sinusoid of frequency f and amplitude $2V(f) \, df$ (V). A one-sided amplitude spectrum can thus be formed by folding the negative frequency components of the two-sided spectrum, defined by equation (2.35), onto the positive frequencies and adding.

Sufficient conditions for the existence of a Fourier transform are similar to those for a Fourier series. They are:

1. $v(t)$ contains a finite number of maxima and minima in any finite time interval.
2. $v(t)$ contains a finite number of *finite* discontinuities in any finite time interval.
3. $v(t)$ must be absolutely integrable, i.e.:

$$\int_{-\infty}^{\infty} |v(t)| \, dt < \infty$$

2.3.2 Practical calculation of Fourier transforms

As with Fourier series simplification of practical calculations is possible if certain symmetries are present in the function being transformed. This is best explained by splitting the Fourier transform into cosine and sine transforms as follows:

$$V(f) = \int_{-\infty}^{\infty} v(t) \, e^{-j2\pi ft} \, dt$$

$$= \int_{-\infty}^{\infty} v(t) \cos 2\pi ft \, dt - j \int_{-\infty}^{\infty} v(t) \sin 2\pi ft \, dt \qquad (2.36)$$

The first term in the second line of equation (2.36) is made up of cosine components only. It therefore corresponds to a component of $v(t)$ which has *even* symmetry. Similarly the second term is made up of sine components only and therefore corresponds to an *odd* component of $v(t)$, i.e.:

$$V(f) = V(f)|_{even\ v(t)} + jV(f)|_{odd\ v(t)} \qquad (2.37(a))$$

where:

$$V(f)|_{even\ v(t)} = \int_{-\infty}^{\infty} v(t) \cos 2\pi ft \, dt \qquad (2.37(b))$$

and:

$$V(f)|_{odd\ v(t)} = - \int_{-\infty}^{\infty} v(t) \sin 2\pi ft \, dt \qquad (2.37(c))$$

It follows that if $v(t)$ is purely even (and real) then:

$$V(f) = 2 \int_{0}^{\infty} v(t) \cos 2\pi ft \, dt \qquad (2.38(a))$$

Conversely, if $v(t)$ is purely odd (and real) then:

$$V(f) = -2j \int_{0}^{\infty} v(t) \sin 2\pi ft \, dt \qquad (2.38(b))$$

That any function can be split into odd and even parts is easily demonstrated, as follows:

$$v(t) = \frac{v(t) + v(-t)}{2} + \frac{v(t) - v(-t)}{2} \qquad (2.39)$$

The first term on the right hand side of equation (2.39) is, by definition, even and the second term is odd. A summary of symmetry properties relevant to the calculation of Fourier transforms is given in Table 2.3 [after Bracewell].

Table 2.3 *Symmetry properties of Fourier transforms.*

Function	Transform
real and even	real and even
real and odd	imaginary and odd
imaginary and even	imaginary and even
imaginary and odd	real and odd
complex and even	complex and even
complex and odd	complex and odd
real and asymmetrical	complex and Hermitian
imaginary and asymmetrical	complex and anti-Hermitian
real even plus imaginary odd	real
real odd plus imaginary even	imaginary
even	even
odd	odd

EXAMPLE 2.3

Find and sketch the amplitude and phase spectrum of the transient signal $v(t) = 2e^{-|t|/\tau}$ (V) shown in Figure 2.26(a).

Since $v(t)$ is real and even:

$$V(f) = 2 \int_{0}^{\infty} v(t) \cos 2\pi ft \; dt$$

$$= 4 \int_{0}^{\infty} e^{-|t|/\tau} \cos 2\pi ft \; dt$$

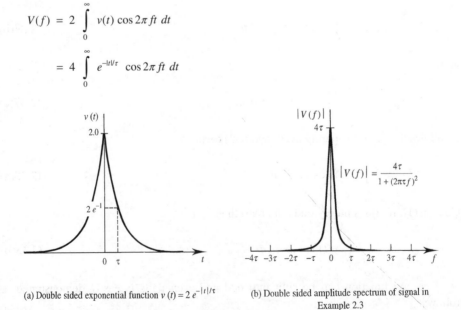

(a) Double sided exponential function $v(t) = 2 e^{-|t|/\tau}$

(b) Double sided amplitude spectrum of signal in Example 2.3

$$|V(f)| = \frac{4\tau}{1 + (2\pi\tau f)^2}$$

Figure 2.26 *Double sided exponential function and corresponding amplitude spectrum.*

Using the standard integral:

$$\int e^{ax} \cos bx \, dx = \frac{e^{ax}(a \cos bx + b \sin bx)}{a^2 + b^2}$$

$$V(f) = 4 \left[\frac{e^{-(1/\tau)t}[-1/\tau \cos 2\pi ft + 2\pi f \sin 2\pi ft]}{(1/\tau)^2 + (2\pi f)^2} \right]_0^\infty$$

$$= 4 \left[\frac{0 - [-1/\tau + 0]}{(1/\tau)^2 + (2\pi f)^2} \right] = \frac{4\tau}{1 + (2\pi \tau f)^2} \quad \text{(V/Hz)}$$

$|V(f)|$ is sketched in Figure 2.26(b) and since $V(f)$ is everywhere real and positive then $v(t)$ has a null phase spectrum.

2.3.3 Fourier transform pairs

The Fourier transform (for transient functions) and Fourier series (for periodic functions) provide a link between two quite different ways of describing signals. The more familiar description is the conventional time plot such as would be seen on an oscilloscope display. Applying the Fourier transform results in a frequency plot (amplitude and phase). These two descriptions are equivalent in the sense that there is one, and only one, amplitude and phase spectrum pair for each possible time plot. Given a complete time domain description, therefore, the frequency domain description can be obtained exactly and vice versa.

Comprehensive tables of Fourier transform pairs have been compiled by many authors. Table 2.4 lists some common Fourier transform pairs. The notation used here for several of the functions which occur frequently in communications engineering is included at the front of this text following the list of principal symbols. Owing to its central importance in digital communications, the Fourier transform of the rectangular function is derived from first principles below. Later the impulse function is defined as a limiting case of the rectangular pulse. The Fourier transform of the impulse is then shown to be a constant in amplitude and linear in phase.

Fourier transform of a rectangular pulse

The unit rectangular pulse, Figure 2.27(a), is represented here using the notation $\Pi(t)$ and is defined by:

$$\Pi(t) \triangleq \begin{cases} 1.0, & |t| < \tfrac{1}{2} \\ 0.5, & |t| = \tfrac{1}{2} \\ 0, & |t| > \tfrac{1}{2} \end{cases} \tag{2.40}$$

The voltage spectrum, $V_\Pi(f)$, of this pulse is given by its Fourier transform, i.e.:

$$V_\Pi(f) = \int_{-\infty}^{\infty} \Pi(t) \, e^{-j2\pi ft} \, dt$$

$$= \int_{-\frac{1}{2}}^{\frac{1}{2}} e^{-j2\pi ft} \, dt$$

Table 2.4 *Fourier transform pairs.*

Function	$x(t)$	$X(f)$
Rectangle of unit width	$\prod(t)$	$\text{sinc}(f)$
Delayed rectangle of width τ	$\prod\left(\dfrac{t-T}{\tau}\right)$	$\tau\,\text{sinc}(\tau f)e^{-j\omega T}$
Triangle of base width 2τ	$\Lambda\left(\dfrac{t}{\tau}\right)$	$\tau\,\text{sinc}^2(\tau\,f)$
Gaussian	$e^{-\pi(t/\tau)^2}$	$\tau e^{-\pi(\tau f)^2}$
One-sided exponential	$u(t)\,e^{-t/\tau}$	$\dfrac{\tau}{1+j2\pi\tau f}$
Two-sided exponential	$e^{-|t|/\tau}$	$\dfrac{2\tau}{1+(2\pi\tau f)^2}$
sinc	$\text{sinc}(2f_\chi t)$	$\dfrac{1}{2f_\chi}\prod\left(\dfrac{f}{2f_\chi}\right)$
Constant	1	$\delta(f)$
Phasor	$e^{j\,(\omega_c t+\phi)}$	$e^{j\phi}\delta(f-f_c)$
Sine wave	$\sin(\omega_c t+\phi)$	$\dfrac{1}{2j}\left[e^{j\phi}\delta(f-f_c)-e^{-j\phi}\delta(f+f_c)\right]$
Cosine wave	$\cos(\omega_c t+\phi)$	$\dfrac{1}{2}\left[e^{j\phi}\delta(f-f_c)+e^{-j\phi}\delta(f+f_c)\right]$
Impulse	$\delta(t-T)$	$e^{-j\omega T}$
Sampling	$\displaystyle\sum_{k=-\infty}^{\infty}\delta(t-kT_s)$	$\displaystyle f_s\sum_{n=-\infty}^{\infty}\delta(f-nf_s)$
Signum	$\text{sgn}(t)$	$\dfrac{1}{j\pi f}$
Heaviside step	$u(t)$	$\dfrac{1}{2}\delta(f)+\dfrac{1}{j2\pi f}$

$$
=\left[\frac{e^{-j2\pi ft}}{-j2\pi f}\right]_{-\frac{1}{2}}^{\frac{1}{2}}=\frac{1}{j2\pi f}\,[e^{j\pi f}-e^{-j\pi f}]
$$

$$
=\frac{j2\sin(\pi f)}{j2\pi f}=\frac{\sin(\pi f)}{\pi f} \tag{2.41}
$$

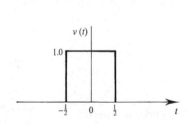

(a) Unit rectangular pulse, $\prod(t)$

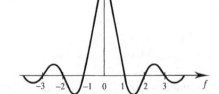

(b) Fourier transform of unit rectangular pulse centred on $t=0$

Figure 2.27 *Unit rectangular pulse and corresponding Fourier transform.*

The function sinc(x) is defined by:

$$\text{sinc}(x) \overset{\Delta}{=} \frac{\sin(\pi x)}{\pi x} \tag{2.42}$$

which means that the unit rectangular pulse and unit sinc function form a Fourier transform pair:

$$\Pi(t) \overset{\text{FT}}{\Leftrightarrow} \text{sinc}(f) \tag{2.43(a)}$$

The sinc(f) function is shown in Figure 2.27(b). Whilst in this case the voltage spectrum can be plotted as a single curve, in general the voltage spectrum of a transient signal is complex and must be plotted either as amplitude and phase spectra or as inphase and quadrature spectra. The amplitude and phase spectra corresponding to Figure 2.27(b) are shown in Figures 2.28(a) and (b).

It is left to the reader to show that the (complex) voltage spectrum of $\Pi[(t-T)/\tau]$ where T is the location of the centre of the pulse and τ is its width is given by:

$$\Pi\left(\frac{t-T}{\tau}\right) \overset{\text{FT}}{\Leftrightarrow} \tau\,\text{sinc}(\tau f)\,e^{-j2\pi fT} \tag{2.43(b)}$$

The impulse function and its Fourier transform

Consider a tall, narrow, rectangular voltage pulse of width τ seconds and amplitude $1/\tau$ volts occurring at time $t = T$, Figure 2.29. The area under the pulse (sometimes called its strength) is clearly 1.0 V s. The impulse function (also called the Dirac delta function) can be defined as the limit of this rectangular pulse as τ tends to zero, i.e.:

$$\delta(t-T) = \lim_{\tau \to 0} \left(\frac{1}{\tau}\right)\Pi\left(\frac{t-T}{\tau}\right) \tag{2.44}$$

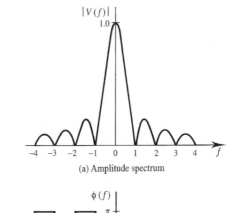

(a) Amplitude spectrum

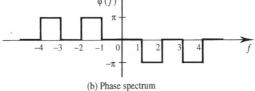

(b) Phase spectrum

Figure 2.28 *Voltage spectrum of unit rectangular pulse shown in Figure 2.27(a).*

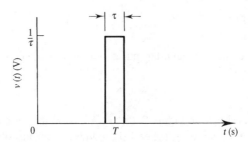

Figure 2.29 *Tall, narrow rectangular pulse of unit strength.*

This idea is illustrated in Figure 2.30. Whatever the value of τ the strength of the pulse remains unity. Mathematically the impulse might be described by:

$$\delta(t - T) = \begin{cases} \infty, & t = T \\ 0, & t \neq T \end{cases} \tag{2.45(a)}$$

$$\int_{-\infty}^{\infty} \delta(t - T)\, dt = \int_{T^-}^{T^+} \delta(t - T)\, dt = 1.0 \tag{2.45(b)}$$

More strictly the impulse is *defined* by its sampling, or *sifting*, property under integration, i.e.:

$$\int_{-\infty}^{\infty} \delta(t - T)\, f(t)\, dt = f(T) \tag{2.46(a)}$$

That equation (2.46(a)) is consistent with equations (2.45) is easily shown as follows:

$$\int_{-\infty}^{\infty} \delta(t - T)\, f(t)\, dt = \int_{T^-}^{T^+} \delta(t - T)\, f(t)\, dt$$

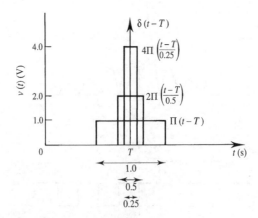

Figure 2.30 *Development of unit strength impulse, $\delta(t - T)$, as a limit of a sequence of unit strength rectangular pulses.*

$$= \int_{T^-}^{T^+} \delta(t - T) \, f(T) \, dt$$

$$= f(T) \int_{T^-}^{T^+} \delta(t - T) \, dt \; = \; f(T) \tag{2.46(b)}$$

Notice that if we insist that the strength of the impulse has units of V s, i.e. its amplitude has units of V, then the sampled quantity, $f(T)$, in equations (2.46) would have units of V^2 s (or joules in 1 Ω). In view of this the impulse is usually taken to have an amplitude measured in s^{-1} (i.e. to have dimensionless strength). This can be reconciled with an equivalent physical implementation of sampling using tall, narrow pulses, a multiplier and integrator by associating dimensions of V^{-1} with the multiplier (required for its output to have units of V) and dimensions of s^{-1} with the integrator (required for its output also to have units of V). Such an implementation, shown in Figure 2.31, is not, of course, used in practical sampling circuits.

As a rectangular pulse gets narrower its Fourier transform (which is a sinc function) gets wider, Figure 2.32. This reciprocal width relationship is a general property of all Fourier transform pairs. Using equation (2.43(b)) it can be seen that as $\tau \to 0$ then $\tau f \to 0$ and $\mathrm{sinc}(\tau f) \to 1.0$. It follows that:

$$\lim_{\tau \to 0} \mathrm{FT} \left\{ \frac{1}{\tau} \Pi \left(\frac{t - T}{\tau} \right) \right\} = e^{-j2\pi fT} \tag{2.47(a)}$$

i.e.:

$$\delta(t - T) \overset{\mathrm{FT}}{\Leftrightarrow} e^{-j2\pi fT} \tag{2.47(b)}$$

For an impulse occurring at the origin this reduces to:

$$\delta(t) \overset{\mathrm{FT}}{\Leftrightarrow} 1.0 \tag{2.47(c)}$$

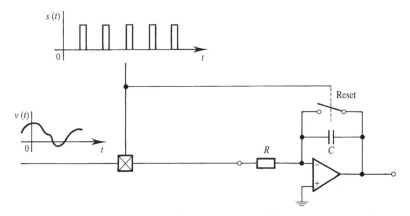

Figure 2.31 *Hypothetical sampling system reconciling physical units of impulse strength (V s) with units of sampled signal (V).*

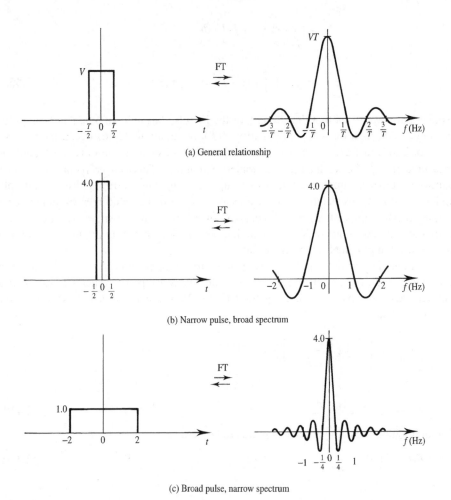

(a) General relationship

(b) Narrow pulse, broad spectrum

(c) Broad pulse, narrow spectrum

Figure 2.32 *Inverse width relationship between Fourier transform pairs.*

The amplitude spectrum of an impulse function is therefore a constant (measured in V/Hz if $\delta(t)$ has units of V). Such a spectrum is sometimes referred to as white, since all frequencies are present in equal quantities. This is analogous to white light. (From a strict mathematical point of view the impulse function, *as represented here*, does not have a Fourier transform owing to the infinite discontinuity which it contains. The impulse and constant in equation (2.47(c)) can be approximated so closely by tall, thin rectangular pulses and broad sinc pulses, however, that the limiting forms need not be challenged. In any event, if desired, the impulse can be derived as the limiting form of other pulse shapes which contain no discontinuity.)

2.3.4 Fourier transform theorems and convolution

Since signals can be fully described in either the time or frequency domain it follows that any operation on a signal in one domain has a precisely equivalent operation in the other domain. A list of equivalent operations on $v(t)$ and its transform, $V(f)$, is given in the form of a set of theorems in Table 2.5. Most of the operations in this list (addition, multiplication, differentiation, integration), whether applied to the functions themselves or their arguments, will be familiar. One operation, namely convolution, may be unfamiliar, however, and is therefore described below.

Table 2.5 *Fourier transform theorems.*

Linearity	$av(t) + bw(t)$	$aV(f) + bW(f)$
Time delay	$v(t - T)$	$V(f)e^{-j\omega T}$
Change of scale	$v(at)$	$\dfrac{1}{|a|} V\left(\dfrac{f}{a}\right)$
Time reversal	$v(-t)$	$V(-f)$
Time conjugation	$v^*(t)$	$V^*(-f)$
Frequency conjugation	$v^*(-t)$	$V^*(f)$
Duality	$V(t)$	$v(-f)$
Frequency translation	$v(t)e^{j\omega_c t}$	$V(f - f_c)$
Modulation	$v(t)\cos(\omega_c t + \phi)$	$\frac{1}{2}\left[e^{j\phi}V(f - f_c) + e^{-j\phi}V(f + f_c)\right]$
Time differentiation	$\dfrac{d^n}{dt^n} v(t)$	$(j2\pi f)^n V(f)$
Integration (1)	$\displaystyle\int_{-\infty}^{t} v(t')dt'$	$(j2\pi f)^{-1}V(f) + \frac{1}{2}V(0)\,\delta(f)$
Integration (2)	$\displaystyle\int_{0}^{t} v_e(t')\,dt' + \int_{-\infty}^{t} v_o(t')\,dt'$	$(j2\pi f)^{-1}V(f)$
Convolution	$v(t) * w(t)$	$V(f)W(f)$
Multiplication	$v(t)w(t)$	$V(f) * W(f)$
Frequency differentiation	$t^n v(t)$	$(-j2\pi)^{-n}\dfrac{d^n}{df^n} V(f)$

DC value	$V(0) = \displaystyle\int_{-\infty}^{\infty} v(t)\,dt$
Value at the origin	$v(0) = \displaystyle\int_{-\infty}^{\infty} V(f)\,df$
Integral of a product	$\displaystyle\int_{-\infty}^{\infty} v(t)w^*(t)\,dt = \int_{-\infty}^{\infty} V(f)W^*(f)\,df$

Convolution is normally denoted by * although $\otimes$ is also sometimes used. Applied to two time functions $z(t) = f(t) * g(t)$ is defined by:

$$z(t) = \int_{-\infty}^{\infty} f(\tau)\, g(t - \tau)\, d\tau \qquad\qquad (2.48(a))$$

Figure 2.33 illustrates time convolution graphically. It can be thought of as a five-step process:

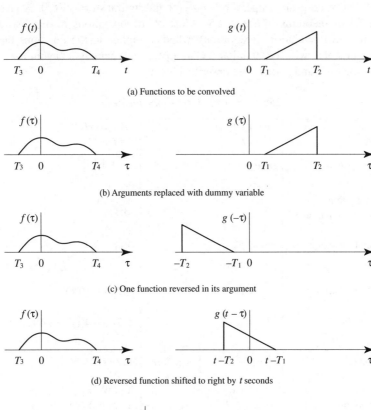

(a) Functions to be convolved

(b) Arguments replaced with dummy variable

(c) One function reversed in its argument

(d) Reversed function shifted to right by t seconds

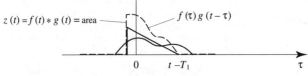

(e) Product formed for all possible values of t

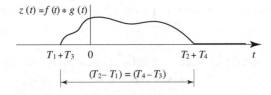

(f) Area of product plotted for all possible values of t

Figure 2.33 *Graphical illustration of time convolution.*

1. The arguments of the functions to be convolved are replaced with a dummy variable (in this case τ), Figure 2.33(b).
2. One of the functions (arbitrarily chosen here to be $g(\tau)$) is reversed in its argument (i.e. reflected about $\tau = 0$) giving $g(-\tau)$, Figure 2.33(c).
3. A variable time shift, t, is introduced into the argument of the reflected function giving $g(t - \tau)$. This is a version of $g(-\tau)$ shifted to the right by t s, Figure 2.33(d).
4. The product function $f(\tau)g(t - \tau)$ is formed for every possible value of t, Figure 2.33(e). (This function changes continuously as t varies and $g(t - \tau)$ slides across $f(\tau)$.)
5. The area under the product function is calculated (by integrating) for every value of t.

There are several important points to note about convolution:

1. The convolution integral, equation (2.48(a)), is sometimes called the superposition integral.
2. It is simply convention which dictates that dummy variables are used so that the result can be expressed as a function of the argument of f and g. There is no reason, in principle, why the alternative definition should not be used:

$$z(\tau) = \int_{-\infty}^{\infty} f(t)\, g(\tau - t)\, dt \qquad\qquad (2.48(b))$$

3. Convolution is not restricted to the time domain. It can be applied to functions of any variable, e.g. frequency, f, i.e.:

$$Z(f) = F(f) * G(f) = \int_{-\infty}^{\infty} F(\phi)G(f - \phi)\, d\phi \qquad\qquad (2.49(a))$$

or space, x, i.e.:

$$z(x) = f(x) * g(x) = \int_{-\infty}^{\infty} f(\lambda)\, g(x - \lambda)\, d\lambda \qquad\qquad (2.49(b))$$

4. The unitary operator for convolution is the impulse function $\delta(t)$ since convolution of $f(t)$ with $\delta(t)$ leaves $f(t)$ unchanged, i.e.:

$$f(t) * \delta(t) = \int_{-\infty}^{\infty} f(\tau)\, \delta(t - \tau)\, d\tau = f(t) \qquad\qquad (2.50(a))$$

(This follows directly from the sampling property of $\delta(t)$ under integration.)
5. Convolution in the time domain corresponds to multiplication in the frequency domain and vice versa.
6. Convolution is commutative, associative and distributive, i.e.:

$$f * g = g * f \qquad\qquad (2.50(b))$$

$$f * (g * h) = (f * g) * h \qquad\qquad (2.50(c))$$

$$f * (g + h) = f * g + f * h \qquad\qquad (2.50(d))$$

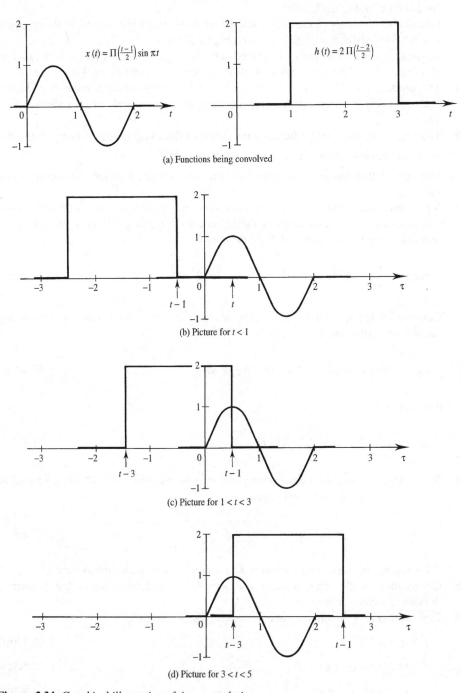

(a) Functions being convolved

(b) Picture for $t < 1$

(c) Picture for $1 < t < 3$

(d) Picture for $3 < t < 5$

Figure 2.34 *Graphical illustration of time convolution.*

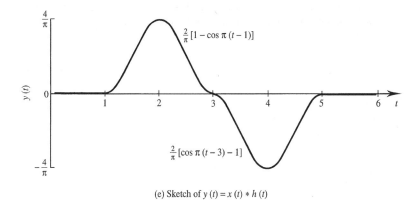

(e) Sketch of y (t) = x (t) * h (t)

Figure 2.34-ctd. *Graphical illustration of time convolution.*

7. The derivative of a convolution is the derivative of one function convolved with the other, i.e.:

$$\frac{d}{dt}[v(t) * w(t)] = \frac{dv(t)}{dt} * w(t) = v(t) * \frac{dw(t)}{dt} \qquad (2.50(e))$$

EXAMPLE 2.4

Convolve the two transient signals, $x(t) = \Pi[(t-1)/2] \sin \pi t$ and $h(t) = 2\ \Pi[(t-2)/2])$, shown in Figure 2.34(a).

For $t < 1$ the picture of the convolution process looks like Figure 2.34(b), i.e. $y(t) = x(t) * h(t) = 0$.

For $1 \le t \le 3$ the picture looks like Figure 2.34(c) and:

$$y(t) = \int_0^{t-1} 2 \sin(\pi \tau)\, d\tau$$

$$= 2\left[\frac{-\cos \pi \tau}{\pi} \right]_0^{t-1}$$

$$= \frac{2}{\pi}[1 - \cos \pi(t - 1)]$$

For $3 \le t \le 5$ the picture looks like Figure 2.34(d):

$$y(t) = \int_{t-3}^2 2 \sin(\pi \tau)\, d\tau$$

$$= 2\left[\frac{-\cos(\pi \tau)}{\pi} \right]_{t-3}^2$$

$$= \frac{2}{\pi}[\cos \pi(t - 3) - 1]$$

For $t > 5$ there is no overlapping of the two functions; therefore $y(t) = 0$. Figure 2.34(e) shows a sketch of $y(t)$. Note that the convolved output signal has a duration of 4 time units (i.e. the sum of the durations of the input signals) and, when the two input signals exactly overlap at $t = 3$, the output is 0 as expected.

EXAMPLE 2.5

Convolve the function $\Pi(t - \frac{1}{2})$ with itself and show that the Fourier transform of the result is the square of the Fourier transform of $\Pi(t - \frac{1}{2})$.

$$z(t) = \Pi(t - \tfrac{1}{2}) * \Pi(t - \tfrac{1}{2}) = \int_{-\infty}^{\infty} \Pi(\tau - \tfrac{1}{2}) \, \Pi(t - \tau - \tfrac{1}{2}) \, d\tau$$

For $t < 0$, $z(t) = 0$ (by inspection), Figure 2.35(a).

$$\text{For } 0 < t < 1, \ z(t) = \int_{0}^{t} (1 \times 1) \, d\tau \quad (\text{Figure 2.35(b)})$$

$$= [\, \tau \,]_{0}^{t} = t$$

$$\text{For } 1 < t < 2, \ z(t) = \int_{t-1}^{1} (1 \times 1) \, d\tau \quad (\text{Figure 2.35(c)})$$

$$= [\, \tau \,]_{t-1}^{1} = 2 - t$$

For $t > 2$, $z(t) = 0$ (by inspection), Figure 2.35(d).

Figure 2.35(e) shows a sketch of $z(t)$. This function, for obvious reasons, is called the triangular function, which, if centred on $t = 0$, is denoted by $\Lambda(t)$. (Note that the absolute width of $\Pi(t)$ is 1.0 whilst the width of $\Lambda(t)$ is 2.0.) Since the triangular function is centred here on $t = 1$ then we can use the time delay theorem and the tables of Fourier transform pairs to obtain:

$$\Lambda(t - 1) \overset{\text{FT}}{\Leftrightarrow} \text{sinc}^2(f) \, e^{-j\omega 1}$$

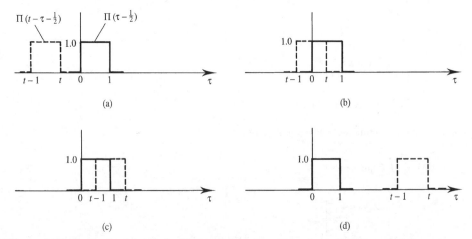

Figure 2.35 *Illustration (a)–(d) of self convolution of a rectangular pulse, Example 2.5.*

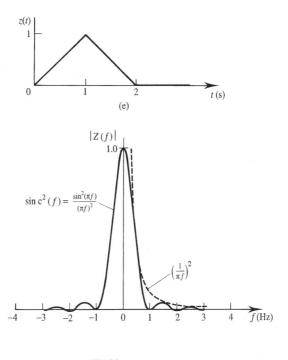

(e)

(f)

Figure 2.35-ctd. *Convolution result (e) and corresponding amplitude and phase spectra (f).*

The square of the Fourier transform of $\Pi(t - \frac{1}{2})$, using the time delay theorem and the table of pairs again, is given by:

$$\left[\mathrm{FT}\left\{ \Pi\left(t - \frac{1}{2}\right) \right\} \right]^2 = \left[\mathrm{sinc}(f)\, e^{-j\omega/2} \right]^2 = \mathrm{sinc}^2(f)\, e^{-j\omega}$$

The amplitude and phase spectra of $z(t)$ are given by:

$$|Z(f)| = \mathrm{sinc}^2(f) \quad \text{(V/Hz)}$$

$$\arg(Z(f)) = -2\pi f \quad \text{(rad)}$$

and are shown in Figure 2.35(f). (Notice that the phase spectrum really takes this form of a simple straight line with intercept zero, and gradient -2π (rad/Hz). It is conventional, however, to constrain its plot to the range $[-\pi, \pi]$ or sometimes $[0, 2\pi]$.)

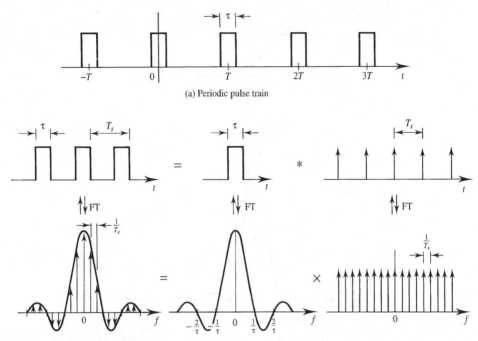

(a) Periodic pulse train

(b) Time and frequency domain representation of a periodic pulse train showing spectral lines arising from periodicity and spectral envelope arising from pulse shape

Figure 2.36 *Time and frequency domain representation of a periodic pulse train.*

Fourier transforms and Fourier series are, clearly, closely related. In fact by using the impulse function (section 2.3.3) a Fourier transform of a periodic function can be defined. Consider the periodic rectangular pulse stream $\sum_{n=-\infty}^{\infty} \Pi[(t - nT)/\tau]$, shown in Figure 2.36(a). This periodic waveform can be represented by the convolution of a transient signal (corresponding to the single period given by $n = 0$) with the periodic impulse train $\sum_{n=-\infty}^{\infty} \delta(t - nT)$, i.e.:

$$\sum_{n=-\infty}^{\infty} \Pi\left(\frac{t - nT}{\tau}\right) = \Pi\left(\frac{t}{\tau}\right) * \sum_{n=-\infty}^{\infty} \delta(t - nT) \tag{2.51(a)}$$

(Each impulse in the impulse train reproduces the rectangular pulse in the convolution process.) In the same way that the Fourier transform of a single impulse can be defined (as a limiting case) the Fourier transform of an impulse train is defined (in the limit) as another impulse train. There is the usual relationship between the width (or period) of the impulse train in time and frequency domains, i.e.:

$$\sum_{k=-\infty}^{\infty} \delta(t - kT_s) \overset{\text{FT}}{\Leftrightarrow} f_s \sum_{n=-\infty}^{\infty} \delta(f - nf_s) \tag{2.51(b)}$$

where the time domain period, T_s, is the reciprocal of the frequency domain period, f_s. (The subscript s is used because the impulse train in communications engineering is often employed as a 'sampling function', with T_s and f_s, as the sampling period and frequency respectively.) The voltage spectrum of a rectangular pulse train can therefore be obtained by taking the Fourier transform of equation (2.51(a)), Figure 2.36(b), i.e.:

$$FT\left\{\sum_{n=-\infty}^{\infty}\Pi\left(\frac{t-nT}{\tau}\right)\right\} = FT\left\{\Pi\left(\frac{t}{\tau}\right)\right\}FT\left\{\sum_{n=-\infty}^{\infty}\delta(t-nT)\right\}$$

$$= \tau\,\text{sinc}(\tau f)\frac{1}{T}\sum_{n=-\infty}^{\infty}\delta\left(f-\frac{n}{T}\right)\quad\text{(V/Hz)}\quad(2.51(c))$$

This shows that the spectrum is given by a periodic impulse train (i.e. a line spectrum) with impulse (or line) separation of $1/T$ and impulse (or line) strength of $(\tau/T)\text{sinc}(\tau f)$. $((\tau/T)\text{sinc}(\tau f)$ is usually called the spectrum envelope and, although real here, is potentially complex.)

The technique demonstrated here works for any periodic waveform, the separation of spectral lines being given by $1/T$ and the spectral envelope being given by $1/T$ times the Fourier transform of the single period contained in interval $[-T/2, T/2]$.

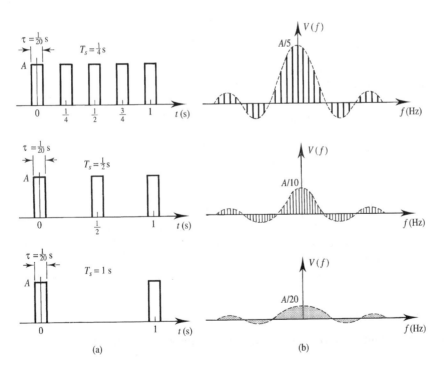

Figure 2.37 *Pulsed waveform and corresponding frequency spectra with specific values for pulse repetition period.*

EXAMPLE 2.6

Sketch the Fourier spectra for a rectangular pulse train comprising pulses of amplitude A V and width 0.05 s with the following pulse repetition periods: (a) ¼ s; (b) ½ s; (c) 1 s.

The spectral envelope is controlled by the rectangular pulses of width 0.05 s. The spectrum is sinc x shaped, Figure 2.32, with the first zeros at $\pm 1/0.05 = \pm 20$ Hz. In all cases the waveform is periodic so the frequency spectrum can be represented by a Fourier series in which the lines, Figure 2.37, are spaced by $1/T_s$ Hz where T_s is the period in seconds.

Thus for (a) the lines occur every 4 Hz and the 0 Hz component, C_0 in equation (2.15), has a magnitude of $A/20 \times 1/T_s = 4A/20 = A/5$. The other components $C_1, C_2, C_3, \cdots, C_n$ follow the sinc x envelope as shown in Figure 2.37(b).

For the case (b) where T_s is ½ s then the spectral lines are now $1/T_s = 2$ Hz apart which is half the spacing of the ¼ s period case in part (a). The envelope of the Fourier spectrum is unaltered but the C_0 term reduces in amplitude owing to the longer period. Thus $C_0 = A/20 \times 2 = A/10$ and the waveform and spectrum are shown in Figure 2.37(b).

For (c) the line spacing becomes 1 Hz and the amplitude at 0 Hz drops to $A/20$.

2.4 Power and energy spectra

As an alternative to plotting peak or RMS voltage against frequency the quantity:

$$G_1(f) = |V_{RMS}(f)|^2 \quad (V^2)$$
(2.52(a))

can be plotted for periodic signals. This is a line spectrum the ordinate of which has units of V^2 (or watts in a 1 Ω resistive load) representing *normalised* power. (The subscript 1 indicates that the spectrum is single sided.) If the impedance level, R, is not 1 Ω then the absolute (i.e. non-normalised) power spectrum is given by:

$$G_1(f) = \frac{|V_{RMS}(f)|^2}{R} \quad (W)$$
(2.52(b))

Figure 2.38(a) shows such a power spectrum for a periodic signal. Although each line in Figure 2.38(a) no longer represents a rotating phasor, two-sided power spectra are still often defined by associating half the power in each spectral line with a negative frequency, Figure 2.38(b). Notice that this means that the total power in a signal is the sum of the powers in all its spectral lines irrespective of whether a one- or two-sided spectral representation is being used.

For a transient signal the two-sided voltage spectrum $V(f)$ has units of V/Hz and the quantity:

$$E_2(f) = |V(f)|^2 \quad (V^2 \text{ s/Hz})$$
(2.53(a))

therefore has units of V^2/Hz^2 or V^2 s/Hz. The corresponding non-normalised spectrum is given by:

$$E_2(f) = \frac{|V(f)|^2}{R} \quad (J/Hz)$$
(2.53(b))

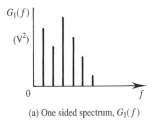
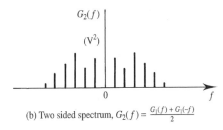

(a) One sided spectrum, $G_1(f)$

(b) Two sided spectrum, $G_2(f) = \dfrac{G_1(f) + G_1(-f)}{2}$

Figure 2.38 *Power spectra of a periodic signal.*

where R is load resistance (in Ω). The quantity $E_2(f)$ now has units of W s/Hz or J/Hz and is therefore called an *energy* spectral density. Like power spectra, energy spectra can be presented as two or one sided, Figures 2.39(a) and (b). Note that energy spectra are always continuous (never discrete) whilst power spectra can be either discrete (as is the case for periodic waveforms) or continuous (as is the case for random signals). Continuous power spectra, i.e. power spectral densities, are discussed in section 3.3.3 and, in the context of linear systems, in section 4.6.1.

2.5 Generalised orthogonal function expansions

Fourier series and transforms constitute a special case of a more general mathematical technique, namely the orthogonal function expansion. The concept of orthogonal functions is closely connected with that of orthogonal (i.e. perpendicular) vectors. For this reason the important characteristics and properties of vectors are now reviewed.

2.5.1 Review of vectors

Vectors possess magnitude, direction and sense. They can be added, Figure 2.40(a), i.e.:

$$\mathbf{a}, \mathbf{b} \rightarrow \mathbf{a} + \mathbf{b}$$

and multiplied by a scalar, λ, Figure 2.40(b), i.e.:

$$\mathbf{a} \rightarrow \lambda \mathbf{a}$$

Their properties [Spiegel] include commutation, distribution and association, i.e.:

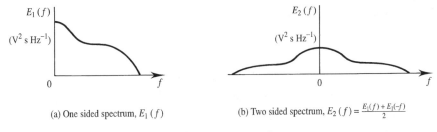

(a) One sided spectrum, $E_1(f)$

(b) Two sided spectrum, $E_2(f) = \dfrac{E_1(f) + E_1(-f)}{2}$

Figure 2.39 *Energy spectral densities of a transient signal.*

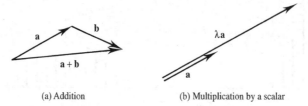

(a) Addition   (b) Multiplication by a scalar

Figure 2.40 *Fundamental vector operations.*

$$\mathbf{a} + \mathbf{b} = \mathbf{b} + \mathbf{a} \tag{2.54(a)}$$

$$\lambda(\mathbf{a} + \mathbf{b}) = \lambda\mathbf{a} + \lambda\mathbf{b} \tag{2.54(b)}$$

$$\lambda(\mu\mathbf{a}) = (\lambda\mu)\mathbf{a} \tag{2.54(c)}$$

A scalar product of two vectors, Figure 2.41, can be defined by:

$$\mathbf{a} \cdot \mathbf{b} = |\mathbf{a}|\,|\mathbf{b}|\cos\theta \tag{2.55}$$

where θ is the angle between the vectors and the modulus | | indicates their length or magnitude.

The distance between two vectors, i.e. their difference, is found by reversing the sense of the vector to be subtracted and adding, Figure 2.42:

$$\mathbf{a} - \mathbf{b} = \mathbf{a} + (-\mathbf{b}) \tag{2.56}$$

A vector in three dimensions can be specified as a weighted sum of any three non-coplanar *basis* vectors:

$$\mathbf{a} = a_1\mathbf{e}_1 + a_2\mathbf{e}_2 + a_3\mathbf{e}_3 \tag{2.57}$$

If the vectors are mutually perpendicular, i.e.:

$$\mathbf{e}_1 \cdot \mathbf{e}_2 = \mathbf{e}_2 \cdot \mathbf{e}_3 = \mathbf{e}_3 \cdot \mathbf{e}_1 = 0 \tag{2.58(a)}$$

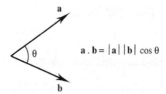

Figure 2.41 *Scalar product of vectors.*

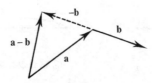

Figure 2.42 *Subtraction of vectors.*

then the three vectors are said to form an *orthogonal* set. If, in addition, the three vectors have unit length:

$$\mathbf{e}_1 . \mathbf{e}_1 = \mathbf{e}_2 . \mathbf{e}_2 = \mathbf{e}_3 . \mathbf{e}_3 = 1 \qquad (2.58(b))$$

then they are said to form an *orthonormal* set. The above concepts can be extended to vectors with any number of dimensions, N. Such a vector can be represented as the sum of N basis vectors, i.e.:

$$\mathbf{x} = \sum_{i=1}^{N} \lambda_i \mathbf{e}_i \qquad (2.59)$$

If $\mathbf{e}_i$ is an orthonormal basis it is easy to find the values of λ_i:

$$\mathbf{x} . \mathbf{e}_j = \sum_{i=1}^{N} \lambda_i \mathbf{e}_i . \mathbf{e}_j = \lambda_j \qquad (2.60)$$

(since $\mathbf{e}_i . \mathbf{e}_j = 0$ for $i \neq j$ and $\mathbf{e}_i . \mathbf{e}_j = 1$ for $i = j$). Scalar products of vectors expressed using orthonormal bases are also simple to calculate:

$$\mathbf{x} . \mathbf{y} = \sum_{i=1}^{N} \lambda_i \mathbf{e}_i . \sum_{j=1}^{N} \mu_j \mathbf{e}_j \qquad (2.61(a))$$

i.e.:

$$\mathbf{x} . \mathbf{y} = \sum_{i=1}^{N} \lambda_i \mu_i \qquad (2.61(b))$$

Equation (2.61(b)) is the most general form of Parseval's theorem, equation (2.28). A special case of this theorem is:

$$\mathbf{x} . \mathbf{x} = \sum_{i=1}^{N} \lambda_i^2 \qquad (2.61(c))$$

which has already been discussed in the context of the power contained in a periodic waveform (section 2.2.3).

If the number of basis vectors available to express an N-dimensional vector is limited to M ($M < N$) then, provided the basis is orthonormal, the best approximation (in a least square error sense) is given by:

$$\mathbf{x} \simeq \sum_{i=1}^{M} \lambda_i \mathbf{e}_i = \mathbf{x}_M \qquad (2.62)$$

where $\lambda_i = \mathbf{x} . \mathbf{e}_i$ as before. This is easily proved by calculating the squared error:

$$\begin{aligned}
|\mathbf{x} - \mathbf{x}_M|^2 &= \left| \mathbf{x} - \sum_{i=1}^{M} \lambda_i \mathbf{e}_i \right|^2 \\
&= |\mathbf{x}|^2 - 2 \sum_{i=1}^{M} \lambda_i (\mathbf{e}_i . \mathbf{x}) + \sum_{i=1}^{M} \lambda_i^2 \\
&= |\mathbf{x}|^2 + \sum_{i=1}^{M} (\lambda_i - \mathbf{e}_i . \mathbf{x})^2 - \sum_{i=1}^{M} (\mathbf{e}_i . \mathbf{x})^2 \qquad (2.63)
\end{aligned}$$

The right hand side of equation (2.63) is clearly minimised by putting $\lambda_i = \mathbf{x} \cdot \mathbf{e}_i$.
From the definition of the scalar product it is apparent that:

$$|\mathbf{x} \cdot \mathbf{y}| \leq |\mathbf{x}| \, |\mathbf{y}| \qquad (2.64)$$

This holds for any kind of vector providing that the scalar product is defined to satisfy:

$$\mathbf{x} \cdot \mathbf{x} \geq 0, \text{ for all } \mathbf{x} \qquad (2.65(a))$$

$$\mathbf{x} \cdot \mathbf{x} = 0, \text{ only if } \mathbf{x} = \mathbf{0} \qquad (2.65(b))$$

where $\mathbf{0}$ is a null vector. (Equation (2.64) is a particularly simple form of the Schwartz inequality, see equation 2.71(a).) In order to satisfy equation (2.65(a)) the scalar product of *complex* vectors must be defined by:

$$\mathbf{x} \cdot \mathbf{y} = \sum_{i=1}^{N} \lambda_i^* \mu_i \qquad (2.66)$$

2.5.2 Vector interpretation of waveforms

Nyquist's sampling theorem (section 2.2.5) asserts that a *periodic* signal having a highest frequency component located at f_H Hz is *fully* specified by N samples spaced $1/2f_H$ s apart, Figure 2.43. Since all the 'information' in a periodic waveform is contained in this set of independent samples (which repeat indefinitely), each sample value can be regarded as being the length of one vector belonging to an orthogonal basis set. A waveform requiring N samples per period for its specification can therefore be interpreted as an N-dimensional vector. For sampling at the Nyquist rate the intersample spacing is $T_s = 1/2f_H$ and:

$$N = T/T_s = 2f_H T \qquad (2.67(a))$$

where T is the period of the waveform. In a slightly more general form:

$$N = 2BT \qquad (2.67(b))$$

where B is waveform bandwidth. This is called the dimensionality theorem. For transient signals an infinite number of samples would be required to retain all the signal's information.

Functions can be added and scaled in the same way as vectors to produce new functions. A scalar product for certain periodic signals can be defined as a continuous version of equation (2.66) but with a factor $1/T'$ so that the scalar product has dimensions of V^2 and can be interpreted as a *cross-power*, i.e.:

Figure 2.43 *Nyquist sampling of a periodic function ($N = 8$).*

$$[f(t), g(t)] = \frac{1}{T'} \int_0^{T'} f^*(t) \, g(t) \, dt$$

$$= \langle f^*(t) \, g(t) \rangle \ (\text{V}^2) \tag{2.68(a)}$$

where T' is the period of the product $f^*(t)g(t)$. (The notation $[f(t), g(t)]$ is used here to denote the scalar product of functions rather than $\mathbf{f \cdot g}$ as used for vectors.) More generally, the definition adopted is:

$$[f(t), g(t)] = \lim_{T' \to \infty} \frac{1}{T'} \int_{-T'/2}^{T'/2} f^*(t) \, g(t) \, dt$$

$$= \langle f^*(t) \, g(t) \rangle \ (\text{V}^2) \tag{2.68(b)}$$

since this includes the possibility that $f^*(t)g(t)$ has infinite period. The corresponding definition for transient signals is a *cross-energy*, i.e.:

$$[f(t), g(t)] = \int_{-\infty}^{\infty} f^*(t) \, g(t) \, dt \quad (\text{V}^2 \text{ s}) \tag{2.68(c)}$$

The scalar products of periodic and transient signals with themselves, which represent average signal power and total signal energy respectively, are therefore:

$$[f(t), f(t)] \equiv \frac{1}{T} \int_0^{T} |f(t)|^2 \, dt \quad (\text{V}^2) \tag{2.69(a)}$$

$$[f(t), f(t)] \equiv \int_{-\infty}^{\infty} |f(t)|^2 \, dt \quad (\text{V}^2 \text{ s}) \tag{2.69(b)}$$

Interpreting the signals as vectors, equations (2.69) correspond to finding the vector's square magnitude, i.e.:

$$\mathbf{x \cdot x} = \sum_{i=1}^{N} \lambda_i^* \lambda_i$$

$$= \sum_{i=1}^{N} |\lambda_i|^2 = |\mathbf{x}|^2 \tag{2.70}$$

Using the definition of a scalar product for complex vectors, equation (2.66), the Schwartz inequality, equation (2.64), becomes:

$$\left| \int_{-\infty}^{\infty} f^*(t) \, g(t) \, dt \right| \leq \left[\int_{-\infty}^{\infty} |f(t)|^2 \, dt \right]^{1/2} \left[\int_{-\infty}^{\infty} |g(t)|^2 \, dt \right]^{1/2} \tag{2.71(a)}$$

where the equality holds if and only if:

$$g(t) = C \, f^*(t) \tag{2.71(b)}$$

Equations (2.71(a)) and (b) can be used to derive the optimum response of predetection filters in digital communications receivers.

2.5.3 Orthogonal and orthonormal signals

Consider a periodic signal, $f(t)$, with only three dimensions[1], i.e. with $N = 2BT = 3$, Figure 2.44. In sample space, Figure 2.45(a), the signal is represented by:

$$\mathbf{f} = \sum_{i=1}^{3} F_i \, \hat{\mathbf{a}}_i \tag{2.72(a)}$$

where $\hat{\mathbf{a}}_i$ represents an orthonormal sample set. The same function $\mathbf{f}$ could, however, be described in a second orthonormal coordinate system, Figure 2.45(b), rotated with respect to the first, i.e.:

$$\mathbf{f} = \sum_{i=1}^{3} C_i \, \hat{\mathbf{b}}_i \tag{2.72(b)}$$

Each unit vector $\hat{\mathbf{b}}_j$ itself represents a three-sample function:

$$\hat{\mathbf{b}}_j = \sum_{i=1}^{3} (\hat{\mathbf{b}}_j \cdot \hat{\mathbf{a}}_i) \hat{\mathbf{a}}_i \tag{2.73}$$

This demonstrates the important idea that a function or signal having N dimensions (in this case three) can be expressed in terms of a weighted sum of N other orthonormal, N-

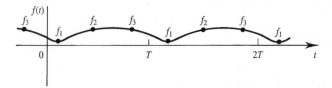

Figure 2.44 *Three-dimensional (i.e. three-sample) function.*

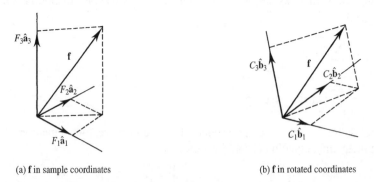

(a) **f** in sample coordinates   (b) **f** in rotated coordinates

Figure 2.45 *Vector interpretation of a 3-dimensional signal.*

---

[1] In reality the number of dimensions, N, must be even since for a periodic signal the maximum frequency, B Hz, must be an integer multiple, n, of the fundamental frequency $1/T$ Hz. The dimensionality theorem can therefore be written as $N = 2(n/T)T = 2n$. Choosing $N = 2$, however, trivialises this in that the orthogonal functions become phasors whilst choosing $N = 4$ precludes signal vector visualisation in 3-dimensional space.

dimensional, functions. (Actually these *basis* functions do not have to be orthonormal or even orthogonal providing none can be exactly expressed as a linear sum of the others.)

Generalising the vector notation to make it more appropriate for signals (which may include the case of transient signals where $N = \infty$) **f** is replaced by $f(t)$ and $\mathbf{b}_i$ (which represents an orthogonal but not necessarily orthonormal set) is replaced by $\phi_i(t)$. Equation (2.72(b)) then becomes:

$$f(t) = \sum_{i=1}^{N} C_i \, \phi_i(t) \tag{2.74}$$

If the basis function set $\phi_i(t)$ is orthonormal over an interval $[a, b]$ then:

$$\int_a^b \phi_i(t) \, \phi_j^*(t) \, dt = \begin{cases} 1, & i = j \\ 0, & i \neq j \end{cases} \tag{2.75}$$

which corresponds to the vector property:

$$\mathbf{b}_i \cdot \mathbf{b}_j = \begin{cases} 1, & i = j \\ 0, & i \neq j \end{cases} \tag{2.76}$$

If the functions are orthogonal but not orthonormal then the upper expression on the right hand side of equation (2.75) does not apply.

EXAMPLE 2.7

Consider the functions shown in Figure 2.46. Do these functions form an orthogonal set over the range $[-1, 1]$? Do these functions form an orthonormal set over the same range?

To determine whether the functions form an orthogonal set we use equation (2.75):

$$\int_a^b \phi_i(t) \, \phi_j^*(t) \, dt = 0 \quad (i \neq j)$$

(Since all the functions are real the conjugate symbol is immaterial here.)

First examine $x(t)$ and $y(t)$. Since the product is odd about zero the integral must be zero, i.e. $x(t)$ and $y(t)$ are orthogonal.

Now examine $x(t)$ and $z(t)$ noting that $x(t)$ and $z(t)$ are orthogonal.

$$\int_{-1}^{1} x(t) \, z(t) \, dt = \int_{-1}^{1} \left(\frac{3}{2} t^2 - \frac{1}{2} \right) dt = \frac{3}{2} \left[\frac{t^3}{3} \right]_{-1}^{1} - \frac{1}{2} [\, t \,]_{-1}^{1} = 0$$

Finally we examine $y(t)$ and $z(t)$. Here the product is again odd about zero and the integral is therefore zero by inspection, i.e. $y(t)$ and $z(t)$ are orthogonal.

To establish the normality or otherwise of the functions we test the square integral against 1.0:

$$\int_a^b |\phi_i(t)|^2 \, dt = 1 \quad \text{for normal } \phi_i(t)$$

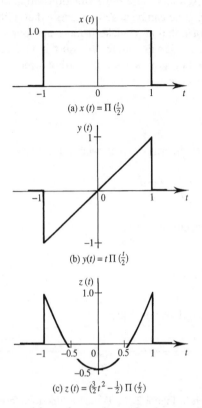

(a) $x(t) = \Pi\left(\frac{t}{2}\right)$

(b) $y(t) = t\,\Pi\left(\frac{t}{2}\right)$

(c) $z(t) = \left(\frac{3}{2}t^2 - \frac{1}{2}\right)\Pi\left(\frac{t}{2}\right)$

Figure 2.46 *Three functions tested for orthogonality and orthonormality in Example 2.7.*

The square integral of $x(t)$ is 2.0 by inspection. We need go no further, therefore, since if any function in the set fails this test then the set is not orthonormal.

2.5.4 Evaluation of basis function coefficients

Equation (2.74) can be multiplied by $\phi_j^*(t)$ to give:

$$f(t)\,\phi_j^*(t) = \sum_{i=1}^{N} C_i\,\phi_i(t)\,\phi_j^*(t) \tag{2.77}$$

Integrating and reversing the order of integration and summation on the right hand side:

$$\int_a^b f(t)\,\phi_j^*(t)\,dt = \sum_{i=1}^{N} C_i \int_a^b \phi_i(t)\,\phi_j^*(t)\,dt \tag{2.78}$$

Since the integral on the right hand side is zero for all $i \neq j$ equation (2.78) can be rewritten as:

$$\int_a^b f(t)\,\phi_j^*(t)\,dt = C_j \int_a^b |\phi_j(t)|^2\,dt \tag{2.79}$$

and rearranging equation (2.79) gives an explicit formula for C_j, i.e.:

$$C_j = \frac{\int_a^b f(t)\, \phi_j^*(t)\, dt}{\int_a^b |\phi_j(t)|^2\, dt} \tag{2.80}$$

If the basis functions $\phi_i(t)$ are orthonormal over the range $[a, b]$ then this reduces to:

$$C_j = \int_a^b f(t)\, \phi_j^*(t)\, dt \tag{2.81}$$

For the special case of $\phi_j(t) = e^{j2\pi f_j t}$, $a = t$ and $b = t + T$, equation (2.81) gives the coefficients for an orthogonal function expansion in terms of a set of cisoids. This, of course, is identical to $T\tilde{C}'_n$ in the Fourier series of equation (2.25).

2.5.5 Error energy and completeness

When a function is *approximated* by a superposition of N basis functions over some range, T, i.e.:

$$f(t) \approx f_N(t) \tag{2.82(a)}$$

where:

$$f_N(t) = \sum_{i=1}^{N} C_i\, \phi_i(t) \tag{2.82(b)}$$

then the 'error energy', E_e, is given by:

$$E_e = \int_t^{t+T} |f(t) - f_N(t)|^2\, dt \tag{2.83}$$

(Note that E_e/T is the mean square error.) The basis set $\phi_i(t)$ is said to be *complete* over the interval T, for a given class of signals, if $E_e \to 0$ as $N \to \infty$ for those signals. Calculation of the coefficients in an orthogonal function expansion using equation (2.80) or (2.81) results in a minimum error energy approximation.

EXAMPLE 2.8

The functions shown in Figure 2.47(a) are the first four elements of the orthonormal set of Walsh functions [Beauchamp, 1975, Harmuth]. Using these as basis functions find a minimum error energy approximation for the function, $f(t)$, shown in Figure 2.47(b). Sketch the approximation.

Each of the coefficients is found in turn using equation (2.81), i.e.:

$$C_0 = \int_0^1 f(t)\, w_0(t)\, dt = \int_0^1 t\, dt = \left[\frac{t^2}{2} \right]_0^1 = 0.5$$

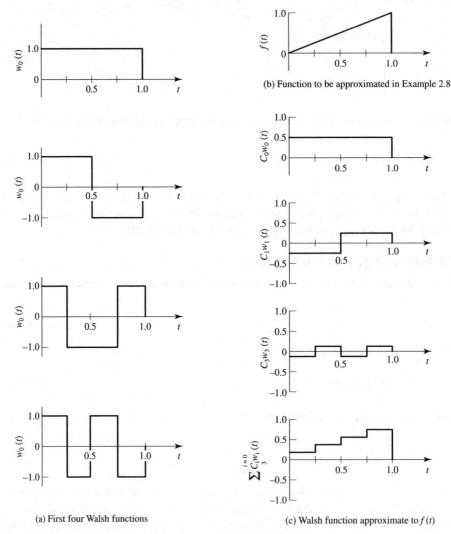

(a) First four Walsh functions

(b) Function to be approximated in Example 2.8

(c) Walsh function approximate to $f(t)$

Figure 2.47 *Orthonormal function expansion using Walsh functions.*

$$C_1 = \int_0^1 f(t)\, w_1(t)\, dt = \int_0^{0.5} t\, dt + \int_{0.5}^{1.0} -t\, dt$$

$$= \left[\frac{t^2}{2}\right]_0^{0.5} - \left[\frac{t^2}{2}\right]_{0.5}^{1.0} = -0.25$$

$$C_2 = \int_0^1 f(t)\, w_2(t)\, dt = \int_0^{0.25} t\, dt + \int_{0.25}^{0.75} -t\, dt + \int_{0.75}^{1.0} t\, dt$$

$$= \left[\frac{t^2}{2}\right]_0^{0.25} - \left[\frac{t^2}{2}\right]_{0.25}^{0.75} + \left[\frac{t^2}{2}\right]_{0.75}^{1.0} = 0$$

$$C_3 = \int_0^1 f(t)\, w_3(t)\, dt = \int_0^{0.25} t\, dt + \int_{0.25}^{0.5} -t\, dt + \int_{0.5}^{0.75} t\, dt + \int_{0.75}^{1.0} -t\, dt$$

$$= \left[\frac{t^2}{2}\right]_0^{0.25} - \left[\frac{t^2}{2}\right]_{0.25}^{0.5} + \left[\frac{t^2}{2}\right]_{0.5}^{0.75} - \left[\frac{t^2}{2}\right]_{0.75}^{1.0} = -0.125$$

The minimum error energy approximation is therefore given in equation (2.82(b)) by:

$$f_N(t) = \sum_{i=0}^{3} C_i\, w_i(t)$$

$$= 0.5w_0(t) - 0.25w_1(t) - 0.125w_3(t)$$

The approximation $f_N(t)$ is sketched in Figure 2.47(c).

2.6 Correlation functions

Attention is restricted here to *real* functions and signals. The scalar product of two transient signals, $v(t)$ and $w(t)$, defined by equation (2.68(c)), and repeated here for convenience, is therefore:

$$[v(t), w(t)] = \int_{-\infty}^{\infty} v(t)\, w(t)\, dt \qquad (2.84(a))$$

Since this quantity is a measure of similarity between the two signals it is usually called the *(cross) correlation* of $v(t)$ and $w(t)$, normally denoted by $R_{vw}(0)$, i.e.:

$$R_{vw}(0) = [v(t), w(t)] \qquad (2.84(b))$$

(Recall that $[v(t), w(t)]$ in section 2.5.2 was called a cross-energy.) More generally a cross correlation *function*, $R_{vw}(\tau)$, can be defined, i.e.:

$$R_{vw}(\tau) = [v(t), w(t - \tau)]$$

$$= \int_{-\infty}^{\infty} v(t)\, w(t - \tau)\, dt \qquad (2.85)$$

This is a measure of the similarity between $v(t)$ and a time shifted version of $w(t)$. The value of $R_{vw}(\tau)$ depends not only on the similarity of the signals, however, but also on their magnitude. This magnitude dependence can be removed by normalising both functions such that their associated normalised energies are unity, i.e.:

$$\rho_{vw}(\tau) = \frac{\displaystyle\int_{-\infty}^{\infty} v(t)\, w(t - \tau)\, dt}{\sqrt{\left(\displaystyle\int_{-\infty}^{\infty} |v(t)|^2\, dt\right)}\sqrt{\left(\displaystyle\int_{-\infty}^{\infty} |w(t)|^2\, dt\right)}} \qquad (2.86)$$

The normalised cross correlation function, $\rho_{vw}(\tau)$, has the following properties:

1. $-1 \le \rho_{vw}(\tau) \le 1$
2. $\rho_{vw}(\tau) = -1$ if, and only if, $v(t) = -kw(t - \tau)$
3. $\rho_{vw}(\tau) = 1$ if, and only if, $v(t) = kw(t - \tau)$

$\rho_{vw}(\tau) = 0$ indicates that $v(t)$ and $w(t - \tau)$ are orthogonal and hence have no similarity whatsoever. (Later, in Chapter 11, we will use two separate parallel channel carriers to construct a four-phase modulator. By using orthogonal carriers $\sin \omega t$ and $\cos \omega t$ we ensure that there is no interference between the parallel channels, equation (2.21(c)).)

For real *periodic* waveforms, $p(t)$ and $q(t)$, the correlation or scalar product, defined in equation (2.68(b)) and repeated here, is:

$$R_{pq}(0) = [p(t), q(t)]$$

$$= \lim_{T' \to \infty} \frac{1}{T'} \int_{-T'/2}^{T'/2} p(t)\, q(t)\, dt$$

$$= \langle p(t)\, q(t) \rangle \tag{2.87}$$

The generalisation to a cross correlation function is therefore:

$$R_{pq}(\tau) = [p(t), q(t - \tau)]$$

$$= \lim_{T' \to \infty} \frac{1}{T'} \int_{-T'/2}^{T'/2} p(t)\, q(t - \tau)\, dt \tag{2.88}$$

and the normalised cross correlation function, $\rho_{pq}(\tau)$, becomes:

$$\rho_{pq}(\tau) = \frac{\langle p(t)\, q(t - \tau) \rangle}{\sqrt{\langle |p(t)|^2 \rangle}\ \sqrt{\langle |q(t)|^2 \rangle}}$$

$$= \frac{\lim_{T' \to \infty} \left(\dfrac{1}{T'}\right) \int_{-T'/2}^{T'/2} p(t)\, q(t - \tau)\, dt}{\sqrt{\left(\left(\dfrac{1}{T}\right) \int_{t}^{t+T} |p(t)|^2\, dt\right)}\ \sqrt{\left(\left(\dfrac{1}{T}\right) \int_{t}^{t+T} |q(t)|^2\, dt\right)}} \tag{2.89}$$

(Notice that the denominator of equation (2.89) is the geometric mean of the normalised powers of $p(t)$ and $q(t)$. For periodic signals $\rho_{pq}(\tau)$ therefore represents $R_{pq}(\tau)$ after $p(t)$ and $q(t)$ have been normalised to an RMS value of 1.0.)

EXAMPLE 2.9

Find the normalised cross correlation function of the sinusoid and the square wave shown in Figure 2.48(a).

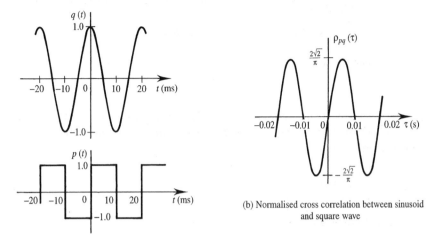

(a) Sinusoid and square wave referred to in Example 2.9

(b) Normalised cross correlation between sinusoid and square wave

Figure 2.48 *Normalised cross correlation of sinusoid and square wave.*

Since the periods of the two functions in this example are the same we can perform the averaging in equation (2.89) over the (finite) period of the product, T', i.e.:

$$
\rho_{pq}(\tau) = \frac{\left(\dfrac{1}{T'}\right)\displaystyle\int_{-T'/2}^{T'/2} p(t)q(t-\tau)\,dt}{\sqrt{\left(\dfrac{1}{T}\right)\displaystyle\int_{0}^{T} p^2(t)\,dt}\;\sqrt{\left(\dfrac{1}{T}\right)\displaystyle\int_{0}^{T} q^2(t)\,dt}}
$$

It is clear that here T' can be taken equal to the period, T, of the sinusoid and square wave. Furthermore the RMS values of the two functions are (by inspection) $1/\sqrt{2}$ and 1.0. The cross correlation function is therefore given by:

$$
\rho_{pq}(\tau) = \frac{\dfrac{1}{0.02}\displaystyle\int_{-0.01}^{0.01}\left[\Pi\left(\dfrac{t-0.005}{0.01}\right)-\Pi\left(\dfrac{t+0.005}{0.01}\right)\right]\cos[2\pi\,50(t-\tau)]\,dt}{(1/\sqrt{2})\times 1.0}
$$

$$
= \frac{\sqrt{2}}{0.02}\left\{\int_{0}^{0.01}\cos[2\pi\,50(t-\tau)]\,dt - \int_{-0.01}^{0}\cos[2\pi\,50(t-\tau)]\,dt\right\}
$$

Using change of variable $x = t - \tau$:

$$
\rho_{pq}(\tau) = \frac{\sqrt{2}}{0.02}\left\{\int_{-\tau}^{0.01-\tau}\cos(2\pi\,50x)\,dx - \int_{-\tau-0.01}^{-\tau}\cos(2\pi\,50x)\,dx\right\}
$$

$$
= \frac{\sqrt{2}}{0.02}\left\{\left[\frac{\sin(2\pi\,50x)}{2\pi\,50}\right]_{-\tau}^{0.01-\tau} - \left[\frac{\sin(2\pi\,50x)}{2\pi\,50}\right]_{-\tau-0.01}^{-\tau}\right\}
$$

$$= \frac{\sqrt{2}}{0.02 \ 2\pi 50} \{\sin[2\pi 50(0.01 - \tau)] + 2 \ \sin(2\pi 50\tau) - \sin[2\pi 50(\tau + 0.01)]\}$$

$$= \frac{1}{\sqrt{2} \ \pi} \ 4 \ \sin(2\pi \ 50\tau) = \frac{4}{\sqrt{2} \ \pi} \ \sin(2\pi \ 50\tau)$$

Figure 2.48(b) shows a sketch of $\rho_{pq}(\tau)$.

If the two signals being correlated are identical then the result is called the *auto*correlation function, $R_{vv}(\tau)$ or $R_v(\tau)$. For real transient signals the autocorrelation function is therefore defined by:

$$R_v(\tau) = \int_{-\infty}^{\infty} v(t) \ v(t - \tau) \ dt \qquad (2.90(a))$$

and for real periodic signals by:

$$R_p(\tau) = \frac{1}{T} \int_{t}^{t+T} p(t) \ p(t - \tau) \ dt \qquad (2.90(b))$$

Normalised autocorrelation functions can be defined by dividing equations (2.90(a)) and (b) by the energy in $v(t)$ and power in $p(t)$, respectively, dissipated in 1 Ω. There are several properties of (real signal) autocorrelation functions to note:

1. $R_x(\tau)$ is real.
2. $R_x(\tau)$ has even symmetry about $\tau = 0$, i.e.:

$$R_x(\tau) = R_x(-\tau) \qquad (2.91)$$

3. $R_x(\tau)$ has a maximum (positive) magnitude at $\tau = 0$, i.e.:

$$|R_x(\tau)| \leq R_x(0), \quad \text{for } any \ \tau \neq 0 \qquad (2.92)$$

4a.   If $x(t)$ is periodic and has units of V then $R_x(\tau)$ is also periodic (with the same period as $x(t)$) and has units of V^2 (i.e. normalised power).
4b.   If $x(t)$ is transient and has units of V then $R_x(\tau)$ is also transient with units of V^2 s (i.e. normalised energy).
5a.   The autocorrelation function of a transient signal and its (two-sided) energy spectral density are a Fourier transform pair, i.e.:

$$R_v(\tau) \stackrel{FT}{\Leftrightarrow} E_v(f) \qquad (2.93)$$

This theorem is proved as follows:

$$E_v(f) = |V(f)|^2 = V(f) \ V^*(f)$$

$$= FT \left\{ v(t) * v^*(-t) \right\} = FT \left\{ \int_{-\infty}^{\infty} v(t) \ v^*(-\tau + t) \ dt \right\}$$

$$= FT \{R_v(\tau)\} \quad (V^2 \text{ s/Hz}) \qquad (2.94(a))$$

i.e.:

$$E_v(f) = \int_{-\infty}^{\infty} R_v(\tau)\, e^{-j2\pi f \tau}\, d\tau \tag{2.94(b)}$$

(The last line of equation (2.94(a)) is obvious for real $v(t)$ but see also the more general definition of $R_v(\tau)$ given in equation (2.96(b)).)

5b.   The autocorrelation function of a periodic signal and its (two-sided) *power* spectral density (represented by a discrete set of impulse functions) are a Fourier transform pair, i.e.:

$$R_p(\tau) \overset{FT}{\Leftrightarrow} G_p(f) \tag{2.95}$$

(Since $R_p(\tau)$ is periodic and $G_p(f)$ consists of a set of discrete impulse functions, $G_p(f)$ could also be interpreted as a power spectrum derived as the Fourier series of $R_p(\tau)$.) Equation (2.95) also applies to stationary random signals which are discussed in Chapter 3.

EXAMPLE 2.10

What is the autocorrelation function, and decorrelation time, of the rectangular pulse shown in Figure 2.49(a)? From a knowledge of its autocorrelation function find the pulse's energy spectral density.

$$R_v(\tau) = \int_{-\infty}^{\infty} 2\, \Pi(t - 1.5)\ 2\, \Pi(t - 1.5 - \tau)\ dt$$

For $\tau = 0$ (see Figure 2.49(b)):

$$R_v(0) = \int_{1}^{2} 2^2\, dt = 4[\, t\,]_1^2 = 4\ (\text{V}^2\ \text{s})$$

For $0 < \tau < 1$ (see Figure 2.49(c)):

$$R_v(\tau) = \int_{1+\tau}^{2} 2^2\, dt = 4[\, t\,]_{1+\tau}^2 = 4(1 - \tau)\ (\text{V}^2\ \text{s})$$

For $-1 < \tau < 0$ (see Figure 2.49(d)):

$$R_v(\tau) = \int_{1}^{2+\tau} 2^2\, dt = 4[\, t\,]_1^{2+\tau} = 4(1 + \tau)\ (\text{V}^2\ \text{s})$$

For $|\tau| > 1$ the rectangular pulse and its replica do not overlap; therefore in these regions $R_v(\tau) = 0$. Figure 2.49(e) shows a sketch of $R_v(\tau)$. Notice that the location of this rectangular pulse in time (i.e. at $t = 1.5$) does not affect the location of $R_v(\tau)$ on the time delay axis. Notice also that the symmetry of $R_v(\tau)$ about $\tau = 0$ means that in practice the function need only be found for $\tau > 0$. The decorrelation time, τ_o, of $R_v(\tau)$ is given by:

$$\tau_o = 1\ \text{s}$$

(Other definitions for τ_o, e.g. ½ energy, $1/e$ energy, etc., could be adopted.)
   The energy spectral density, $E_v(f)$, is given by:

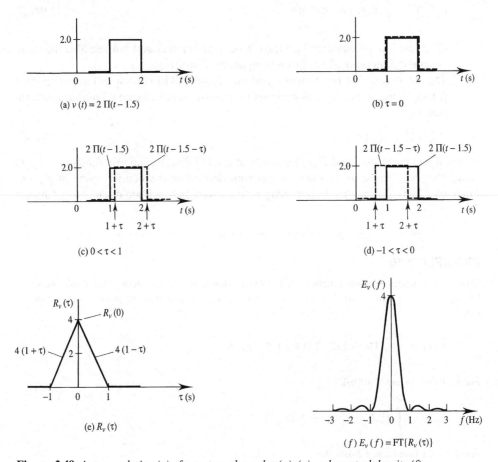

Figure 2.49 *Autocorrelation (e) of a rectangular pulse (a)–(e) and spectral density (f).*

$$E_v(f) = FT\{R_v(\tau)\} = FT\{4\,\Lambda(t)\}$$
$$= 4\,sinc^2(f) \ (V^2 \ s/Hz) \quad (\text{Using Table 2.4})$$

$E_v(f)$ is sketched in Figure 2.49(f). Notice that the area under $R_v(\tau)$ is equal to $E_v(0)$ and the area under $E_v(f)$ is equal to $R_v(0)$. This is a good credibility check on the answer to such problems. Also notice that the (first null) bandwidth of the rectangular pulse and the zero crossing definition of decorrelation time are consistent with the rule:

$$B \approx \frac{1}{\tau_o}$$

Note that auto and cross correlation functions are only necessarily real if the functions being correlated are real. If this is not the case then the more general definitions:

$$R_{pq}(\tau) = \lim_{T' \to \infty} \frac{1}{T'} \int_{-T'/2}^{T'/2} p(t)\,q^*(t - \tau)\,dt \qquad (2.96(a))$$

and:

$$R_{vw}(\tau) = \int_{-\infty}^{\infty} v(t)\, w^*(t - \tau)\,dt \qquad\qquad (2.96(b))$$

must be adopted.

2.7 Summary

Deterministic signals can be periodic or transient. Periodic waveforms are unchanged when shifted in time by nT seconds where n is any integer and T is the period of the waveform. They have discrete (line) spectra and, being periodic, exist for all time. Transient signals are aperiodic and have continuous spectra. They are essentially localised in time (whether or not they are strictly time limited).

All periodic signals of engineering interest can be expressed as a sum of harmonically related sinusoids. The amplitude spectrum of a periodic signal has units of volts, and the phase spectrum has units of radians or degrees.

Alternatively the amplitudes and phases of a set of harmonically related, counter-rotating, conjugate cisoids can be plotted against frequency. This leads naturally to two-sided amplitude and phase spectra. For purely real signals the two-sided amplitude spectrum has even symmetry about 0 Hz and the phase spectrum has odd symmetry about 0 Hz. If the power associated with each sinusoid in a Fourier series is plotted against frequency the result is a power spectrum with units of V^2 or W. Two-sided power spectra can be defined by associating half the total power in each line with a positive frequency and half with a negative frequency. The total power in a waveform is the sum of the powers in each spectral line. This is Parseval's theorem.

Bandwidth refers to the width of the frequency band in a signal's spectrum which contains significant power (or, in the case of transient signals, energy). Many definitions of bandwidth are possible, the most appropriate depending on the application or context. In the absence of a contrary definition, however, the half-power bandwidth is usually assumed. Signals with rapid rates of change have large bandwidth and those with slow rates of change small bandwidth.

The voltage spectrum of a transient signal is continuous and is given by the Fourier transform of the signal. Since the units of such a spectrum are V/Hz it is normally referred to as a voltage spectral density. A complex voltage spectral density can be expressed as an amplitude spectrum and a phase spectrum. The square of the amplitude spectrum has units of V^2 s/Hz and is called an energy spectral density. The total energy in a transient signal is the integral over all frequencies of the energy spectral density.

Fourier transform pairs are uniquely related (i.e. for each time domain signal there is only one, complex, spectrum) and have been extensively tabulated. Theorems allowing the manipulation of existing transform pairs and the calculation of new ones extend the usefulness of such tables. The convolution theorem is especially useful. It specifies the operation in one domain (convolution) which is precisely equivalent to multiplication in the other domain.

Basis functions other than sinusoids and cisoids can be used to expand signals and waveforms. Such generalised expansions are especially useful when the set of basis functions is orthogonal or orthonormal. Signals and waveforms can be interpreted as multidimensional vectors. In this context the concept of orthogonal functions is related to the concept of perpendicular vectors. The orthogonal property of a set of basis functions allows the optimum coefficients of the functions to be calculated independently. Optimum in this context means a minimum error energy approximation.

Correlation is the equivalent operation for signals to the scalar product for vectors and is a measure of signal similarity. The cross correlation function gives the correlation of two functions for all possible time shifts between them. It can be applied, with appropriate differences in its definition, to both transient and periodic functions. The energy and power spectral densities of transient and periodic signals respectively are the Fourier transforms of their autocorrelation functions. Chapter 3 extends correlation concepts to noise and other random signals. In Chapter 8 correlation is identified as an optimum signal processing technique, often employed in digital communications receivers.

2.8 Problems

2.1   Find the DC component and the first two non-zero harmonic terms in the Fourier series of the following periodic waveforms: (a) square wave with period 20 ms and magnitude +2 V from −5 ms to +5 ms and −2 V from +5 ms to +15 ms; (b) sawtooth waveform with a 2 s period and $y = t$ for $-1 \le t < 1$; and (c) triangular wave with 0.2 s period and $y = 1/3(1 - 10|t|)$ for $-0.1 \le t < 0.1$.

2.2   Find the proportion of the total power contained in the DC and first two harmonics of the waveforms shown in Table 2.2, assuming a 25% duty cycle for the third waveform.

2.3   Use Table 2.2 to find the Fourier series coefficients up to the third harmonic of the waveform with period 2 s, one period of which is formed by connecting the following points with straight lines: $(t, y) = (0,0), (0,0.5), (1,1), (1,0.5), (2,0)$. (Hint: decompose the waveform into a sum of waveforms which you recognise.)

2.4   The spectrum of a square wave, amplitude ±1.0 V and period 1.0 ms, is bandlimited by an ideal filter to 4.0 kHz such that frequencies below 4.0 kHz are passed (undistorted) and frequencies above 4.0 kHz are stopped. What is the normalised power (in V^2) and what is the maximum rate of change (in V/s) of the waveform at the output of the filter? [$0.90\ V^2$, 16×10^3 V/s]

2.5   How fast must the bandlimited waveform in Problem 2.4 be sampled if (in the absence of noise) it is to be reconstructed from the samples without error? [8.0 kHz]

2.6   Find (without using Table 2.4) the Fourier transforms of the following functions: (a) $\Pi((t - T)/\tau)$; (b) $\Lambda(t/2)$; (c) $3e^{-5|t|}$; (d) $[(e^{-at} - e^{-bt})/(b - a)]\ u(t)$.
(Hint: for (c) recall the standard integral –
$\int e^{ax} \cos bx\ dx = e^{ax}(a \cos bx + b \sin bx)/(a^2 + b^2))$.

2.7   Find (using Tables 2.4 and 2.5) the amplitude and phase spectra of the following transient signals: (a) a triangular pulse $3(1 - |t - 1|)\Pi((t - 1)/2)$; (b) a 'split phase' rectangular pulse having amplitude $y = 2$ V from $t = 0$ to $t = 1$ and $y = -2$ V from $t = 1$ to $t = 2$ and $y = 0$ elsewhere; (c) a truncated cosine wave $\cos(2\pi 20t)\Pi(t/0.2)$; (d) an exponentially decaying sinusoid $u(t)e^{-5t} \sin(2\pi 20t)$.

2.8   Sketch the following, purely real, frequency spectra and find the time domain signals to which they correspond: (a) $0.1\ \mathrm{sinc}(3f)$; (b) e^{-f^2}; (c) $\Lambda(f/2) + \Pi(f/4)$; and (d) $\Lambda(f - 10) + \Lambda(f + 10)$.

2.9 Convolve the following pairs of signals: (a) $\Pi(t/T_2)/T_2$ with $\Pi(t/T_1)/T_1$, $T_2 > T_1$; (b) $u(t)e^{-3t}$ with $u(t-1)$; (c) $\sin \pi t)\Pi((t-1)/2)$ with $2\Pi((t-2)/2)$; and (d) $\delta(t) - 2\delta(t-1) + \delta(t-2)$ with $\Pi(t-0.5) + 2\Pi(t-1.5)$.

2.10 Find and sketch the energy spectral densities of the following signals: (a) $10\Pi((t-0.05)/0.1)$; (b) $6\,e^{-6|t|}$; (c) $\mathrm{sinc}(100t)$; (d) $-\mathrm{sinc}(100t)$. What is the energy contained in signals (a) and (c), and how much energy is contained in signals (b) and (d) below a frequency of 6.0 Hz? [10 V^2 s, 0.01 V^2 s, 5.99 V^2 s, 1.2×10^{-3} V^2 s]

2.11 Demonstrate the orthogonality, or otherwise, of the function set: $(1/\sqrt{T})\,\Pi((t-T/2)/T)$; $(\sqrt{2/T})\cos((\pi/T)\,t)\,\Pi((t-T/2)/T)$; $2\Lambda((t-T/2)/(T/2)) - \Pi((t-T/2)T)$. Do these functions represent an orthonormal set?

2.12 Find the cross correlation function of the sine wave, $f(t) = \sin(2\pi 50t)$, with a half-wave rectified version of itself, $g(t)$.

2.13 Find, and sketch, the autocorrelation function of the 'split phase' rectangular pulse, where $x(t) = -V_0$ for $-T/2 < t < 0$ and $+V_0$ for $0 < t < T/2$.

2.14 What is the autocorrelation of $v(t) = u(t)e^{-t}$? Find the energy spectral density of this signal and the proportion of its energy contained in frequencies above 2.0 Hz. [5.1%]

CHAPTER 3

Random signals and noise

3.1 Introduction

The periodic and transient signals discussed in Chapter 2 are deterministic. This precludes them from conveying information since nothing new can be learned by receiving a signal which is entirely predictable. Unpredictability or randomness is a property which is essential for information bearing signals. (The definition and quantification of randomness is an interesting topic. Here, however, an intuitive and common sense notion of randomness is all that is required.)

Whilst one type of random signal creates information (i.e. increases knowledge) at a communication receiver another type of random signal destroys it (i.e. decreases knowledge). The latter type of signal is known as noise [Rice]. The distinction between signals and noise is therefore essentially one of their origin (i.e. an information source or elsewhere) and whether reception is intended or not. In this context interference (signals arising from information sources other than the one expected or intended) is a type of noise. From the point of view of describing information signals and noise mathematically, no distinction is necessary. Since signals and noise are random such descriptions must be, at least partly, in terms of probability theory.

3.2 Probability theory

Consider an experiment with three possible, random, outcomes A, B, C. If the experiment is repeated N times and the outcome A occurs L times then the probability of outcome A is defined by:

$$P(A) \overset{\Delta}{=} \lim_{N \to \infty} \left\{ \frac{L}{N} \right\} \tag{3.1}$$

Note that the error, ε, for N experimental trials does not tend to zero for large N but actually increases (on average) as $\sqrt{N}$, Figure 3.1. The ratio ε/N does tend to zero, however, for large N, Figure 3.2, i.e.:

$$\frac{\varepsilon}{N} \to 0 \quad \text{as} \quad N \to \infty$$

(Thus on tossing a coin, for example, the result of achieving close to 50% heads is much more likely to be achieved with a large number of samples, e.g. >50 individual tosses or trials.) Such an experiment could be performed N times with one set of (unchanging) apparatus or N times, simultaneously, with N sets of (identical) apparatus. The former is called a temporal experiment whilst the latter is called an ensemble experiment. If, after N trials, the outcome A occurs L times and the outcome B occurs M times, and if A and B are *mutually* *exclusive* (i.e. they cannot occur together), then the probability that A or B occurs is:

$$\begin{aligned} P(A \text{ or } B) &= \lim_{N \to \infty} \left\{ \frac{L + M}{N} \right\} \\ &= \lim_{N \to \infty} \left\{ \frac{L}{N} \right\} + \lim_{N \to \infty} \left\{ \frac{M}{N} \right\} \\ &= P(A) + P(B) \end{aligned} \tag{3.2}$$

This is the basic law of additive probabilities which can be used for any number of mutually exclusive events.

Since the outcome of the experimental trials described above is variable and random it is called (unsurprisingly) a random variable. Such random variables can be discrete or continuous. An example of the former would be the score achieved by the throw of a dice. An example of the latter would be the final position of a coin in a game of shove halfpenny. (In the context of digital communications relevant examples might be the voltage of a quantised signal source and the voltage of an unquantised noise source.)

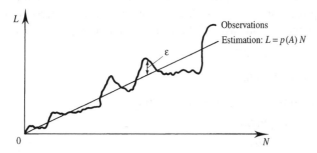

Figure 3.1 *Observations compared with estimation for L outcomes of A after N random trials.*

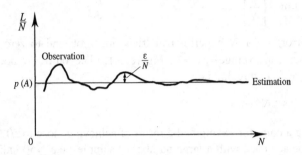

Figure 3.2 *Observation compared with estimation for the fraction L/N of outcomes A after N random trials.*

EXAMPLE 3.1

A dice is thrown once. What is the probability that: (i) the dice shows 3; (ii) the dice shows 6; (iii) the dice shows a number greater than 2; (iv) the dice does not show 5?

(i) Since all numbers between 1 and 6 inclusive are equiprobable, $P(3) = \dfrac{1}{6}$

(ii) As above, $P(6) = \dfrac{1}{6}$

(iii) $P(3) = P(4) = P(5) = P(6) = \dfrac{1}{6}$

$$P(3 \text{ or } 4 \text{ or } 5 \text{ or } 6) = P(3) + P(4) + P(5) + P(6) = \frac{4}{6} = \frac{2}{3}$$

(iv) $P(\text{any number but } 5) = 1 - P(5) = 1 - \dfrac{1}{6} = \dfrac{5}{6}$

because total probability, $\Sigma_{i=1}^{6} P(i)$, must sum to 1.

3.2.1 Conditional probabilities, joint probabilities and Bayes's rule

The probability of event A occurring given that event B is known to have occurred, $P(A|B)$, is called the conditional probability of A on B (or the probability of A conditional on B). The probability of A and B occurring together, $P(A, B)$, is called the joint probability of A and B. Joint and conditional probabilities are related by:

$$P(A, B) = P(B)P(A|B)$$

$$= P(A)P(B|A) \qquad (3.3)$$

Rearranging equation (3.3) gives Bayes's rule:

$$P(A|B) = \frac{P(A)P(B|A)}{P(B)} \qquad (3.4)$$

EXAMPLE 3.2

Four cards are dealt off the top of a shuffled pack of 52 playing cards. What is the probability that all the cards will be of the same suit?

Given that the first card is a spade the probability that the second card will be a spade is:

$$P(\text{2nd spade}) = \frac{12}{51} = 0.2353$$

and so on:

$$P(\text{3rd spade}) = \frac{11}{50} = 0.2200$$

$$P(\text{4th spade}) = \frac{10}{49} = 0.2041$$

Therefore $P(\text{4 spades}) = 0.2353 \times 0.2200 \times 0.2041$

$$= 0.01056$$

This is the probability of all the cards being from the same suit.

Notice that intuitively the suit of the first card has been ignored from a probability point of view. A more formal solution to this problem uses Bayes's rule explicitly:

$$P(\text{4 spades}) = P(\text{spade, 3 spades})$$

$$= P(\text{3 spades}) \, P(\text{spade} \mid 3 \text{ spades})$$

$$= P(\text{2 spades}) \, P(\text{spade} \mid 2 \text{ spades}) \, P(\text{spade} \mid 3 \text{ spades})$$

$$= P(\text{spade}) \, P(\text{spade} \mid \text{spade}) \, P(\text{spade} \mid 2 \text{ spades}) \, P(\text{spade} \mid 3 \text{ spades})$$

$$= \frac{13}{52} \times \frac{12}{51} \times \frac{11}{50} \times \frac{10}{49}$$

$$= 0.002641$$

$$P(\text{4 same suit}) = P(\text{4 spades or 4 clubs or 4 diamonds or 4 hearts})$$

$$= P(\text{4 spades}) + P(\text{4 clubs}) + P(\text{4 diamonds}) + P(\text{4 hearts})$$

$$= 0.002641 \times 4 = 0.01056$$

3.2.2 Statistical independence

Events A and B are statistically independent if the occurrence of one does not affect the probability of the other occurring, i.e.:

$$P(A|B) = P(A) \tag{3.5(a)}$$

and:

$$P(B|A) = P(B) \tag{3.5(b)}$$

It follows that for statistically independent events:

$$P(A, B) = P(A)P(B) \qquad (3.6)$$

EXAMPLE 3.3

Two cards are dealt one at a time, face up, from a shuffled pack of cards. Show that these two events are not statistically independent.

The unconditional probabilities for both events are:

$$P(A) = P(B) = \frac{1}{52}$$

The probability of event A is:

$$P(A) = \frac{1}{52}$$

The probability of event B is:

$$P(B|A) = \frac{1}{51}$$

The joint probability of the two events is therefore:

$$P(A, B) = P(A)P(B|A)$$

$$= \frac{1}{52} \times \frac{1}{51}$$

$$\neq P(A)\,P(B)$$

i.e. the events are not statistically independent.

3.2.3 Discrete probability of errors in a data block

When we consider the problem of performance prediction in a digital coding system we often ask what is the probability of having more than a given number of errors in a fixed length codeword? This is a discrete probability problem. Assume that the probability of single bit (binary digit) error is P_e, that the number of errors is R' and n is the block length, i.e. we require to determine the probability of having more than R' errors in a block of n digits. Now:

$$P(> R' \text{ errors}) = 1 - P(\leq R' \text{ errors}) \qquad (3.7(a))$$

because total probability must sum to 1. We also assume that errors are independent. The above equation may thus be expanded as:

$$P(> R' \text{ errors}) = 1 - [P(0 \text{ error}) + P(1 \text{ error}) + P(2 \text{ errors}) + \cdots + P(R' \text{ errors})] \quad (3.7(b))$$

These probabilities will be calculated individually starting with the probability of no errors. A block representing the codeword is divided into bins labelled 1 to n. Each bin corresponds to one digit in the n-digit codeword and it is labelled with the probability of the event in question.

Now considering the general case of j errors in n digits with a probability of error per digit of P_e we can generalise the above equations, i.e.:

$$P(j \text{ errors}) = (P_e)^j (1 - P_e)^{n-j} \, {}^nC_j \qquad (3.8)$$

where the binomial coefficient ${}^nC_j$ is given by:

$$ {}^nC_j = \frac{n!}{j!(n-j)!} = \binom{n}{j} \qquad (3.9(a)) $$

This is the probability of j errors in an n-digit codeword, but what we are interested in is the probability of having more than R' errors. We can write this using equation (3.7(b)) as:

$$ P(> R' \text{ errors}) = 1 - \sum_{j=0}^{R'} P(j) \qquad (3.9(b)) $$

For large n we have statistical stability in the sense that the number of errors in a given block will tend to the product $P_e \, n$. Furthermore, the fraction of blocks containing a number of errors that deviates significantly from this value will tend to zero, Figure 3.2. The above statistical stability gained from long blocks is very important in the design of effective error correction codewords and is discussed further in Chapter 10.

EXAMPLE 3.4

If the probability of single digit error is 0.01 and hence the probability of correct digit reception is 0.99 calculate the probability of 0 to 2 errors occurring in a 10-digit codeword.

If all events (receptions) are independent then the probability of having no errors in the block is $(0.99)^{10} = 0.904382$, Figure 3.3(a).

Now consider the probability of a single error. Initially assume that the error is in the first position. Its probability is 0.01. All other digits are received correctly, so their probabilities are 0.99 and there are 9 of them, Figure 3.3(b). However, there are 10 positions where the single error can occur in the 10-digit codeword and therefore the overall probability of a single error occurring is:

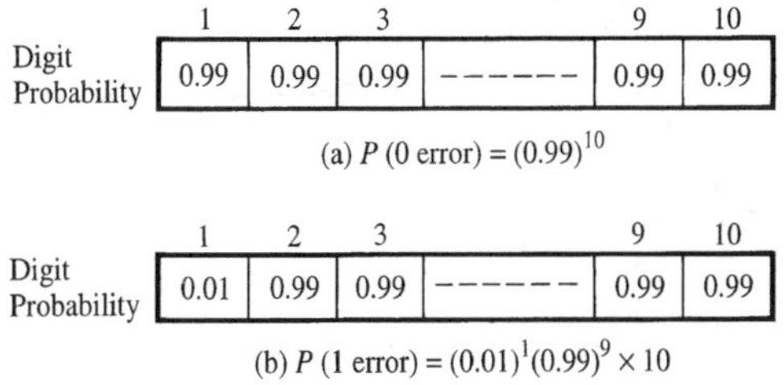

	1	2	3		9	10
Digit Probability	0.99	0.99	0.99	-------	0.99	0.99

(a) $P\,(0 \text{ error}) = (0.99)^{10}$

	1	2	3		9	10
Digit Probability	0.01	0.99	0.99	-------	0.99	0.99

(b) $P\,(1 \text{ error}) = (0.01)^1 (0.99)^9 \times 10$

Figure 3.3 *Discrete probability of codeword errors with a per digit error probability of 1%.*

$$P(1 \text{ error}) = (0.01)^1 (0.99)^9 \ ^{10}C_1 = 0.091352$$

where $^{10}C_1 = 10$ represents the number of combinations of 1 object from 10 objects. We can similarly calculate the probability of 2 errors as:

$$P(2 \text{ errors}) = (0.01)^2 (0.99)^8 \ ^{10}C_2 = 0.00415$$

Thus the probability of 3 or more errors, equation (3.7), is $1 - 0.904382 - 0.091352 - 0.00415 = 0.000116$. Notice how the probability of j errors in the data block falls off rapidly with j.

3.2.4 Cumulative distributions and probability density functions

A cumulative distribution (also called a probability distribution) is a curve showing the probability, $P_X(x)$, that the value of the random variable, X, will be less than or equal to some specific value, x, Figure 3.4(a), i.e.:

$$P_X(x) = P(X \le x) \tag{3.10}$$

(Conventionally, upper case letters are used for the name of a random variable and lower case letters for particular values of the random variable. Note, however, that the upper case subscript in the left hand side of equation (3.10) is often omitted.) Some properties of a cumulative distribution (CD) apparent from Figure 3.4(a) are:

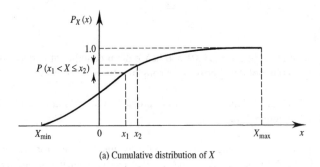

(a) Cumulative distribution of X

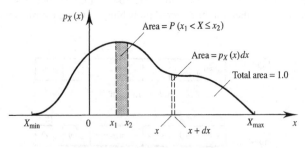

(b) Probability density function of X

Figure 3.4 *Descriptions of a continuous random variable (smooth CD and continuous pdf).*

1. $0 \leq P_X(x) \leq 1$ for $-\infty \leq x \leq \infty$
2. $P_X(-\infty) = 0$
3. $P_X(\infty) = 1$
4. $P_X(x_2) - P_X(x_1) = P(x_1 < X \leq x_2)$
5. $\dfrac{dP_X(x)}{dx} \geq 0$

Exceedance curves are also sometimes used to describe the probability behaviour of random variables. These curves are complementary to CDs in that they give the probability that a random variable exceeds a particular value, i.e.:

$$P(X > x) = 1 - P_X(x) \tag{3.11}$$

Probability density functions (pdfs) give the probability that the value of a random variable, X, lies between x and $x + dx$. This probability can be written in terms of a CD as:

$$P_X(x + dx) - P_X(x) = \frac{dP_X(x)}{dx}\, dx \tag{3.12}$$

It is the factor $[dP_X(x)]/dx$, normally denoted by $p_X(x)$, which is defined as the pdf of X. Figure 3.4(b) shows an example. Pdfs and CDs are therefore related by:

$$p_X(x) = \frac{dP_X(x)}{dx} \tag{3.13(a)}$$

$$P_X(x) = \int_{-\infty}^{x} p_X(x')\, dx' \tag{3.13(b)}$$

The important properties of pdfs are:

$$\int_{-\infty}^{\infty} p_X(x)\, dx = 1 \tag{3.14(a)}$$

and

$$\int_{x_1}^{x_2} p_X(x)\, dx = P(x_1 < X \leq x_2) \tag{3.14(b)}$$

Both CDs and pdfs can represent continuous, discrete or mixed random variables. Discrete random variables, as typified in Example 3.1, have stepped CDs and purely impulsive pdfs, Figure 3.5. Mixed random variables have CDs which contain discontinuities and pdfs which contain impulses, Figure 3.6.

Pdfs can be measured, in principle, by taking an ensemble of identical random variable generators, sampling them all at one instant and plotting a relative frequency histogram of samples. (The pdf is the limit of this histogram as the size interval shrinks to zero and the number of samples tends to infinity.) Alternatively, at least for ergodic processes (see section 3.3.1), the proportion of time the random variable spends in different value intervals of size Δx could be determined (see Figure 3.7). In this case:

$$p_X(x)\, \Delta x = \sum_{i=1}^{N} \frac{\Delta t_i}{\text{total observation time}} \tag{3.15}$$

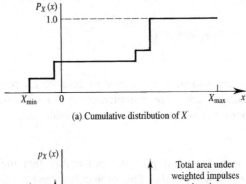

(a) Cumulative distribution of X

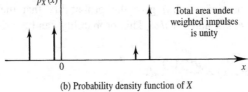

(b) Probability density function of X

Figure 3.5 *Probability descriptions of a discrete random variable (stepped CD and discrete pdf).*

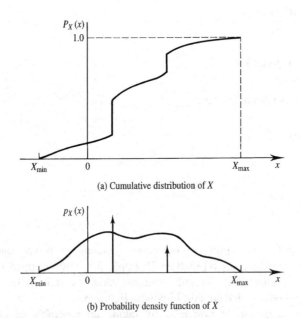

(a) Cumulative distribution of X

(b) Probability density function of X

Figure 3.6 *Probability descriptions of a mixed random variable.*

and $p_X(x)$ can be found as the limit of equation (3.15) as $\Delta x \to 0$ (and the observation time and $N \to \infty$). This allows one to calculate the time that the signal $x(t)$ lies between the voltages x and Δx.

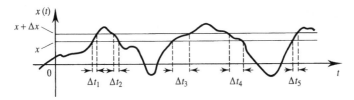

Figure 3.7 *Determination of pdf for an (ergodic) random process.*

EXAMPLE 3.5

A random voltage has a pdf given by:

$$p(V) = k\, u(V + 4)\, e^{-3(V + 4)} + 0.25\, \delta(V - 2)$$

where $u(\ )$ is the Heaviside step function. (i) Sketch the pdf; (ii) find the probability that $V = 2$ V; (iii) find the value of k; and (iv) find and sketch the CD of V.

(i) Figure 3.8(a) shows the pdf of V.
(ii) Area under impulse is 0.25 therefore $P(V = 2) = 0.25$.
(iii) If the area under impulse part of pdf is 0.25 then the area under exponential part of the pdf must be 0.75, i.e.:

$$\int_{-4}^{\infty} k\, e^{-3(V + 4)}\, dV = 0.75$$

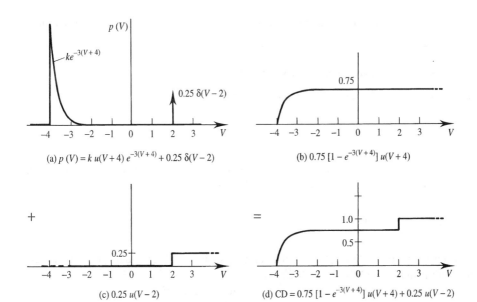

Figure 3.8 *Pdf of V in Example 3.5, (a), and components (b) and (c) of resulting CD, (d). (The step function is implied in the solution by the lower limit of integration.)*

Use change of variable:

$$V + 4 = x, \quad dx/dV = 1$$

when $V = -4$, $x = 0$, and when $V = \infty$, $x = \infty$

Therefore:

$$\int_0^\infty k \, e^{-3x} \, dx = 0.75$$

i.e.:

$$k = \frac{0.75}{[e^{-3x}/-3]_0^\infty}$$

$$= \frac{-3 \times 0.75}{[0-1]} = 2.25$$

(iv)

$$CD = \int_{-\infty}^V p(V') \, dV'$$

$$= \int_{-4}^V [2.25e^{-3(V'+4)} + 0.25\delta(V'-2)] \, dV'$$

Using the same substitution as in part (iii):

$$CD = 2.25 \left[\frac{e^{-3x}}{-3} \right]_0^{V+4} + 0.25u(V-2)$$

$$= -2.25/3 \times [e^{-3(V+4)} - 1] + 0.25u(V-2)$$

Figures 3.8(b) and (c) show these individual waveforms and (d) shows the combined result.

3.2.5 Moments, percentiles and modes

The first moment of a random variable X is defined by:

$$\bar{X} = \int_{-\infty}^\infty x \, p(x) \, dx \qquad (3.16)$$

where $\bar{X}$ denotes an ensemble *mean*. (Sometimes this quantity is called the expected value of X and is written $E[X]$.) The second moment is defined as:

$$\overline{X^2} = \int_{-\infty}^\infty x^2 \, p(x) \, dx \qquad (3.17(a))$$

and represents the mean square of the random variable. (The square root of the second moment is thus the RMS value of X.)

Higher order moments are defined by the general formula:

$$\overline{X^n} = \int_{-\infty}^\infty x^n \, p(x) \, dx \qquad (3.17(b))$$

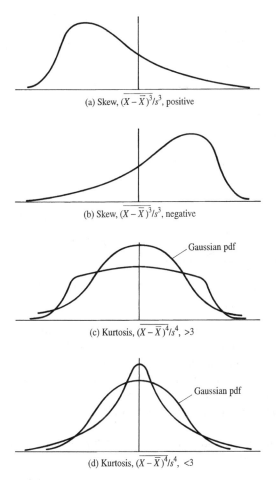

(a) Skew, $\overline{(X-\bar{X})^3}/s^3$, positive

(b) Skew, $\overline{(X-\bar{X})^3}/s^3$, negative

Gaussian pdf

(c) Kurtosis, $\overline{(X-\bar{X})^4}/s^4$, >3

Gaussian pdf

(d) Kurtosis, $\overline{(X-\bar{X})^4}/s^4$, <3

Figure 3.9 *Illustration of skew and kurtosis as descriptors of pdf shape.*

(The zeroth moment ($n = 0$) is always equal to 1.0 and is therefore not a useful quantity.)

Central moments are the net moments of a random variable taken about its mean. The second central moment is therefore given by:

$$\overline{(X-\bar{X})^2} = \int_{-\infty}^{\infty} (x-\bar{X})^2 \, p(x) \, dx \tag{3.18(a)}$$

and is usually called the *variance* of the random variable. The square root of the second central moment is the *standard deviation* of the random variable (general symbol s, although for a Gaussian random variable the symbol σ is commonly used). The variance or standard deviation provides a measure of random variable spread, or pdf width. Higher order central moments are defined by the general formula:

$$\overline{(X-\bar{X})^n} = \int_{-\infty}^{\infty} (x-\bar{X})^n \, p(x) \, dx \tag{3.18(b)}$$

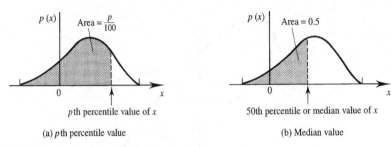

(a) p th percentile value　　　　　　　(b) Median value

Figure 3.10 *Illustration of percentiles and median.*

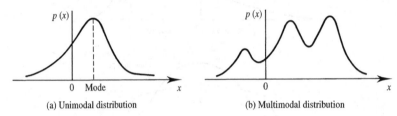

(a) Unimodal distribution　　　　　　　(b) Multimodal distribution

Figure 3.11 *Illustration of modal values.*

(The zeroth central moment is always 1.0 and the first central moment is always zero. Neither are of any practical use, therefore.) The 3rd and 4th central moments divided by s^3 and s^4 respectively are called *skew* and *kurtosis*[1]. These are a measure of pdf asymmetry and peakiness, the latter being in comparison to a Gaussian function, Figure 3.9. These higher order moments are of current interest for the analysis of non-stationary signals, such as speech, and they are also appropriate for the analysis of non-Gaussian signals.

The *p*th *percentile* is the value of X below which p % of the total area under the pdf lies, Figure 3.10(a), i.e.:

$$\int_{-\infty}^{x} p(x')\, dx' = \frac{p}{100} \tag{3.19}$$

where x is the *p*th percentile. In the special case of $p = 50$ (i.e. the 50th percentile) the corresponding value of x is called the *median* and the pdf is divided into two equal areas, Figure 3.10(b).

The *mode* of a pdf is the value of x for which $p(x)$ is a maximum, Figure 3.11(a). For a pdf with a single mode this can be interpreted as the most likely value of X. In general pdfs can be multimodal, Figure 3.11(b). Moments, percentiles and modes are all examples of *statistics*, i.e. numbers which in some way summarise the behaviour of a random variable. The difference between probabilistic and statistical models is that statistical models usually give an incomplete description of random variables.

There are some useful electrical interpretations of moments and central moments in the context of random voltages (and currents). These interpretations are summarised in

---

[1] If kurtosis is defined as $[\overline{(X - \bar{X})^4/s^4}] - 3$ then a Gaussian curve will have zero kurtosis.

the following table. The interpretations of 1st and 2nd moments are obvious. The interpretation of 2nd central moment, s^2, becomes clear when the left hand side of equation (3.18(a)) is expanded as shown below, i.e.:

$$s^2 = \langle [x(t) - \langle x(t) \rangle]^2 \rangle$$
$$= \langle x^2(t) - 2x(t) \langle x(t) \rangle + \langle x(t) \rangle^2 \rangle$$
$$= \langle x^2(t) \rangle - \langle x(t) \rangle^2 \qquad (3.20)$$

Moment	Familiar name	Interpretation
1st	Mean value	DC voltage (or current)
2nd	Mean square value	Total power[1]
2nd central	Variance	AC power[2]

Notes:
(1) This is the total normalised power, i.e. the power dissipated in a 1 Ω load.
(2) This is the power dissipated in a 1 Ω load by the fluctuating (i.e. AC) component of voltage.

(The angular brackets, $\langle \, \rangle$, here indicate time averages, equation (2.16), which for most random variables of engineering interest can be equated to ensemble averages, see section 3.3.1.) Equation (3.20) is an expression of the familiar statistical statement that variance is the mean square minus the square mean. This is clearly also equivalent to total power minus DC power which must be the AC (or fluctuating) power. Figure 3.12 illustrates these electrical interpretations.

EXAMPLE 3.6

A random voltage has a pdf given by:

$$p(V) = u(V)\, 3\, e^{-3V}$$

Find the DC voltage, the power dissipated in a 1 Ω load and the median value of voltage. What power would be dissipated at the output of an AC coupling capacitor?

Equation (3.16) defines the DC value:

$$\bar{V} = \int_{-\infty}^{\infty} V\, p(V)\, dV = 3 \int_{0}^{\infty} V\, e^{-3V}\, dV$$

Using the standard integral [Dwight, equation 567.9]:

$$\int x^n e^{ax} dx = e^{ax} \left[\frac{x^n}{a} - \frac{nx^{n-1}}{a^2} + \frac{n(n-1)x^{n-2}}{a^3} - \cdots (-1)^{n-1} \frac{n!x}{a^n} + (-1)^n \frac{n!}{a^{n+1}} \right], \quad n \geq 0$$

$$\bar{V} = 3 \left[e^{-3V} \left(\frac{V}{-3} - \frac{1 \times V^0}{(-3)^2} \right) \right]_{0}^{\infty}$$

$$= -3\, [0 - 1/9] = 1/3 \;\; V$$

Mean value: $\langle x\,(t)\rangle = \lim_{T\to\infty} \frac{1}{T}\int_{-\frac{T}{2}}^{\frac{T}{2}} x(t)\,dt$

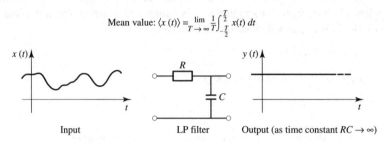

Input   LP filter   Output (as time constant $RC \to \infty$)

(a) Mean value represents DC component of signal

Mean square value $\langle x^2\,(t)\rangle = \lim_{T\to\infty} \frac{1}{T}\int_{-\frac{T}{2}}^{\frac{T}{2}} x^2(t)\,dt$

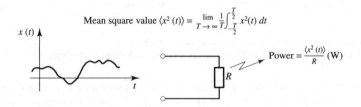

$\text{Power} = \dfrac{\langle x^2\,(t)\rangle}{R}\ (\text{W})$

(b) Mean square value represents power dissipated in 1 Ω

Variance: $\langle [x\,(t) - \langle x\,(t)\rangle]^2\rangle = \langle x^2\,(t)\rangle - \langle x\,(t)\rangle^2$

$\text{Power} = \dfrac{\langle x^2\,(t)\rangle - \langle x\,(t)\rangle^2}{R}$

DC block

(c) Variance represents AC, or fluctuation, power dissipated in 1 Ω

Figure 3.12 *Engineering interpretations of: (a) mean; (b) mean square; (c) variance.*

The power is given by equation (3.17):

$$\overline{V^2} = \int_{-\infty}^{\infty} V^2\, p(V)\, dV = \int_{0}^{\infty} V^2\, 3\, e^{-3V}\, dV$$

Again using [Dwight, equation 567.9]:

$$\overline{V^2} = 3\left[e^{-3V}\left(\frac{V^2}{-3} - \frac{2V^1}{(-3)^2} + \frac{2(1)\,V^0}{(-3)^3} \right) \right]_0^{\infty}$$

$$= 3\left[0 - \left(-\frac{2}{27} \right) \right] = \frac{6}{27}\ (\text{or } 0.2222)\ V^2$$

Median value = 50th percentile, i.e.:

$$\int\limits_{-\infty}^{X_{median}} p(V)\, dV = 0.5$$

Therefore:

$$\int\limits_{0}^{X_{median}} 3\, e^{-3V}\, dV = 0.5$$

i.e.:

$$3\left[\frac{e^{-3V}}{-3}\right]_{0}^{X_{median}} = 0.5$$

$$1 - e^{-3X_{median}} = 0.5$$

$$X_{median} = \frac{\ln(1 - 0.5)}{-3} = 0.2310 \text{ V}$$

The AC coupling capacitor acts as a DC block and the fluctuating or AC power is given by the variance of the random signal as defined in equation (3.20), i.e.:

$$P_{AC} = s^2$$

$$= \langle v^2(t)\rangle - \langle v(t)\rangle^2$$

$$= \overline{V^2} - \overline{V}^2 \quad \text{(signal assumed ergodic, see section 3.3.1)}$$

$$= \frac{6}{27} - \left(\frac{1}{3}\right)^2 = \frac{3}{27} \text{ (or 0.1111) } \text{V}^2$$

3.2.6 Joint and marginal pdfs, correlation and covariance

If X and Y are two random variables a joint probability density function, $p_{X,Y}(x, y)$, can be defined such that $p_{X,Y}(x, y)\, dx\, dy$ is the probability that X lies in the range x to $x + dx$ and Y lies in the range y to $y + dy$. The joint (or bivariate) pdf can be represented by a surface as shown in Figure 3.13(a). For quantitative work, however, it is often more convenient to display $p_{X,Y}(x, y)$ as a contour plot, Figure 3.13(b), and, when investigating bivariate random variables experimentally, sample values of (x, y) can be plotted as a scattergram, Figure 3.13(c). (Contours of constant point density in Figure 3.13(c) correspond, of course, to contours of constant probability density in Figure 3.13(b).)

Just as the total area under the pdf of a single random variable is 1.0, the volume under the surface representing a bivariate random variable is 1.0, i.e.:

$$\int\limits_{X}\int\limits_{Y} p_{X,Y}(x, y)\, dx\, dy = 1.0 \tag{3.21}$$

The probability of finding the bivariate random variable in any particular region, F, of the X,Y plane, Figure 3.13(b), is:

$$P([X, Y] \text{ lies within } F) = \int\limits_{F}\int p_{X,Y}(x, y)\, dx\, dy \tag{3.22}$$

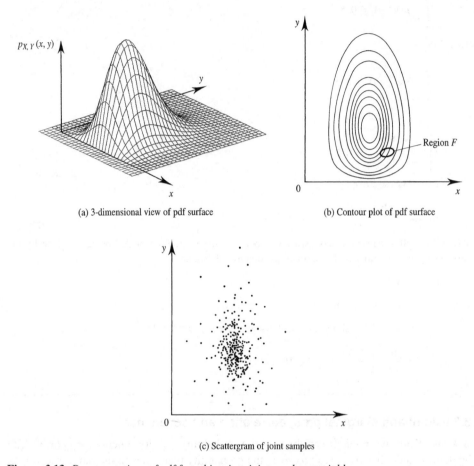

(a) 3-dimensional view of pdf surface   (b) Contour plot of pdf surface

(c) Scattergram of joint samples

Figure 3.13 *Representations of pdf for a bivariate joint random variable.*

If the joint pdf, $p_{X,Y}(x, y)$, of a bivariate variable is known then the probability that X lies in the range x_1 to x_2 (irrespective of the value of Y) is called a marginal probability of X and is found by integrating over all Y, Figure 3.14. The marginal pdf of X is therefore given by:

$$p_X(x) = \int_{-\infty}^{\infty} p_{X,Y}(x, y)\, dy \qquad\qquad (3.23(a))$$

Similarly, the marginal pdf of Y is given by:

$$p_Y(y) = \int_{-\infty}^{\infty} p_{X,Y}(x, y)\, dx \qquad\qquad (3.23(b))$$

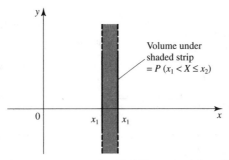

Figure 3.14 *Relationship between a marginal probability and a joint pdf.*

EXAMPLE 3.7

Two quantised signals have the following (discrete) joint pdfs:

		X			
		1.0	1.5	2.0	2.5
	−1.0	0.15	0.08	0.06	0.05
	−0.5	0.10	0.13	0.06	0.05
Y	0.0	0.04	0.07	0.05	0.05
	0.5	0.01	0.02	0.03	0.05

Find and sketch the marginal pdfs of X and Y.

$$p_X(x) = \int\limits_{-\infty}^{\infty} p_{X,Y}(x, y)\, dy$$

For a discrete joint random variable this becomes column summation:

$$P_X(x) = \sum_y P_{X,Y}(x, y)$$

$$P_X(1.0) = 0.15 + 0.10 + 0.04 + 0.01 = 0.30$$

$$P_X(1.5) = 0.08 + 0.13 + 0.07 + 0.02 = 0.30$$

$$P_X(2.0) = 0.06 + 0.06 + 0.05 + 0.03 = 0.20$$

$$P_X(2.5) = 0.05 + 0.05 + 0.05 + 0.05 = 0.20$$

See Figure 3.15(a) for the marginal pdf of X. Similarly:

$$P_Y(y) = \sum_x P_{X,Y}(x, y)$$

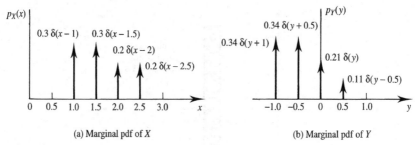

(a) Marginal pdf of X   (b) Marginal pdf of Y

Figure 3.15 *Marginal pdfs in Example 3.7.*

and by row summation:

$$P_Y(-1.0) = 0.15 + 0.08 + 0.06 + 0.05 = 0.34$$

$$P_Y(-0.5) = 0.10 + 0.13 + 0.06 + 0.05 = 0.34$$

$$P_Y(0.0) = 0.04 + 0.07 + 0.05 + 0.05 = 0.21$$

$$P_Y(0.5) = 0.01 + 0.02 + 0.03 + 0.05 = 0.11$$

Figure 3.15(b) shows the marginal pdf of Y.

3.2.7 Joint moments, correlation and covariance

The joint moments of $p_{X,Y}(x, y)$ are defined by:

$$\overline{X^n \, Y^m} = \int_{-\infty}^{\infty} \int_{-\infty}^{\infty} x^n \, y^m \, p_{X,Y}(x, y) \, dx \, dy \tag{3.24}$$

In the special case of $n = m = 1$ the joint moment is called the *correlation* of X and Y (see previous correlation definition in section 2.6 for transient and periodic signals). A large positive value of correlation means that when x is high then y, *on average*, will also be high. A large negative value of correlation means that when x is high y, on average, will be low. A small value of correlation means that x gives little information about the magnitude or sign of y. The effect of correlation between two random variables, on a scattergram, is shown in Figure 3.16.

The joint central moments of $p_{X,Y}(x, y)$ are defined by:

$$\overline{(X - \bar{X})^n \, (Y - \bar{Y})^m} = \int_{-\infty}^{\infty} \int_{-\infty}^{\infty} (x - \bar{X})^n \, (y - \bar{Y})^m \, p_{X,Y}(x, y) \, dx \, dy \tag{3.25}$$

In this case, when $n = m = 1$, the joint central moment is called the *covariance* of X and Y. This is because the mean values of X and Y have been subtracted before *correlating* the resulting zero mean variables. Covariance therefore refers to the correlation of the *varying* parts of X and Y. If X and Y are already zero mean variables then the correlation and covariance are identical.

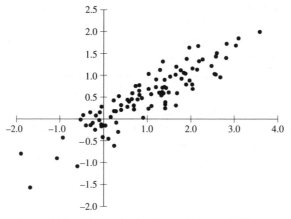

(a) Strongly correlated random variables ($\rho = 0.63$)

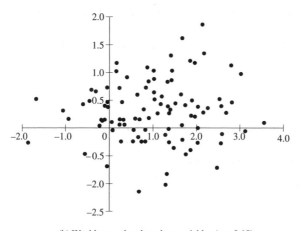

(b) Weakly correlated random variables ($\rho = 0.10$)

Figure 3.16 *Influence of correlation on a scattergram.*

The definitions of joint, and joint central, moments can be extended to multivariate (i.e. more than two) random variables in a straightforward way (e.g. $\overline{X^n Y^m Z^l}$ etc.).

The random variables X and Y are said to be uncorrelated if:

$$\overline{XY} = \bar{X}\,\bar{Y} \tag{3.26}$$

Notice that this implies that the *covariance* (not the correlation) is zero, i.e.:

$$\overline{(X - \bar{X})(Y - \bar{Y})} = 0 \tag{3.27}$$

It also follows that the correlation of X and Y can be zero, only if either $\bar{X}$ or $\bar{Y}$ is zero.

It is intuitively obvious that statistically independent random variables (i.e. random variables arising from physically separate processes) must be uncorrelated. It is not generally the case, however, that uncorrelated random variables must be statistically independent. Indeed the concept of correlation can be applied to deterministic signals such as $\cos \omega t$ and $\sin \omega t$. If concurrent samples are taken from these functions and the correlation is subsequently calculated the result will be zero (due to their orthogonality). It is clearly untrue to say that these are independent processes, however, since one can be derived from the other using a simple delay line. It follows that independence is a stronger statistical condition than uncorrelatedness, i.e.:

$$\text{Independence} \Rightarrow \text{uncorrelatedness}$$
$$\text{Uncorrelatedness} \nRightarrow \text{independence}$$

(An exception to the latter rule is when the random variables are Gaussian. In this special case uncorrelatedness does imply statistical independence: see section 3.2.8.)

The normalised correlation coefficient, ρ, between two random variables (as used in section 2.6, for transient or periodic signals) is the correlation between the corresponding standardised variables, standardisation in this context implying zero mean and unit standard deviation, i.e.:

$$\rho = \overline{\left(\frac{X - \bar{X}}{s_X} \right) \left(\frac{Y - \bar{Y}}{s_Y} \right)} \tag{3.28}$$

where s_X and s_Y are the standard deviations of X and Y respectively. (Since ρ can also be interpreted as the covariance of the random variables with normalised standard deviation then $\rho = 0$ can be viewed as the defining property for uncorrelatedness.)

EXAMPLE 3.8

Find the correlation and covariance of the discrete joint pdf described in Example 3.7.

Correlation is defined in equation (3.24) as:

$$\overline{XY} = \int_{-\infty}^{\infty} \int_{-\infty}^{\infty} xy \, p_{X,Y}(x, y) \, dx \, dy$$

For a discrete joint pdf this becomes:

$$\overline{XY} = \sum_x \sum_y xy \, P_{X,Y}(x, y)$$

$$
\begin{aligned}
&= (1.0)\,(-1.0)\,0.15 + (1.0)\,(-0.5)\,0.10 + (1.0)\,(0.0)\,0.04 + (1.0)\,(0.5)\,0.01 \\
&+ (1.5)\,(-1.0)\,0.08 + (1.5)\,(-0.5)\,0.13 + (1.5)\,(0.0)\,0.07 + (1.5)\,(0.5)\,0.02 \\
&+ (2.0)\,(-1.0)\,0.06 + (2.0)\,(-0.5)\,0.06 + (2.0)\,(0.0)\,0.05 + (2.0)\,(0.5)\,0.03 \\
&+ (2.5)\,(-1.0)\,0.05 + (2.5)\,(-0.5)\,0.05 + (2.5)\,(0.0)\,0.05 + (2.5)\,(0.5)\,0.05 \\
&= -0.6725
\end{aligned}
$$

Similarly covariance is obtained from equation (3.25) for the discrete pdf as:

$$\overline{(X - \bar{X})(Y - \bar{Y})} = \sum_x \sum_y (x - \bar{X})(y - \bar{Y}) \, P_{X,Y}(x, y)$$

Using the marginal pdfs found in Example 3.7:

$$\bar{X} = \frac{1}{N} \sum_x x \, P_X(x)$$

$$= \frac{1}{4} \,[(1.0)(0.3) + (1.5)(0.3) + (2.0)(0.2) + (2.5)(0.20)]$$

$$= 0.4125$$

$$\bar{Y} = \frac{1}{N} \sum_y y \, P_Y(y)$$

$$= \frac{1}{4} \left[(-1.0)(0.34) + (-0.5)(0.34) + (0.0)(0.21) + (0.5)(0.11) \right]$$

$$= -0.4550$$

and by using each of the 16 discrete probabilities in Example 3.7:

$$\overline{(X - \bar{X})(Y - \bar{Y})} = (1.0 - 0.4125)(-1.0 + 0.4550)\, 0.15$$

$$+ \ (1.0 - 0.4125)(-0.5 + 0.4550)\, 0.10$$

$$+ \ (1.0 - 0.4125)(0.0 + 0.4550)\, 0.04$$

$$+ \ (1.0 - 0.4125)(0.5 + 0.4550)\, 0.01$$

$$+ \ (1.5 - 0.4125)(-1.0 + 0.4550)\, 0.08$$

$$+ \ (1.5 - 0.4125)(-0.5 + 0.4550)\, 0.13$$

$$+ \ (1.5 - 0.4125)(0.0 + 0.4550)\, 0.07$$

$$+ \ (1.5 - 0.4125)(0.5 + 0.4550)\, 0.02$$

$$+ \ (2.0 - 0.4125)(-0.1 + 0.4550)\, 0.06$$

$$\cdots$$

$$+ \ (2.5 - 0.4125)(0.5 + 0.4550)\, 0.05$$

$$= 0.07825$$

3.2.8 Joint Gaussian random variables

A bivariate random variable is Gaussian if it can be reduced, by a suitable translation and rotation of axes, to the form:

$$p_{X,Y}(x, y) = \frac{1}{2\pi\sigma_X\sigma_Y} e^{-\left[\frac{x^2}{2\sigma_X^2} + \frac{y^2}{2\sigma_Y^2} \right]} \tag{3.29}$$

This idea is illustrated in Figure 3.17. The contours in the x, y plane, given by:

$$\frac{x''^2}{2\sigma_{X''}^2} + \frac{y''^2}{2\sigma_{Y''}^2} = \text{constant} \tag{3.30}$$

are ellipses and the double primed and unprimed coordinate systems are related by:

$$x'' = (x - x_0)\cos\theta + (y - y_0)\sin\theta \tag{3.31(a)}$$

$$y'' = -(x - x_0)\sin\theta + (y - y_0)\cos\theta \tag{3.31(b)}$$

where x_0, y_0 are the necessary translations and θ is the necessary rotation. If $\sigma_{X''} = \sigma_{Y''}$ then the ellipses become circles. The translation of axes removes any DC component in the random variables and the rotation has the effect of reducing the correlation between

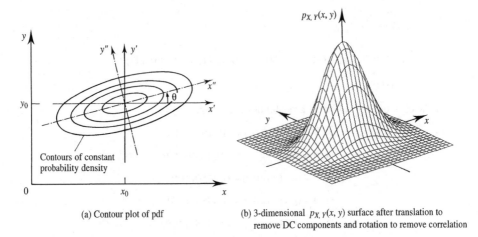

(a) Contour plot of pdf

(b) 3-dimensional $p_{X,Y}(x, y)$ surface after translation to remove DC components and rotation to remove correlation

Figure 3.17　*Joint Gaussian bivariate random variable.*

the random variables to zero. This can be seen by writing the probability density function in the original coordinate system, i.e.:

$$p_{X,Y}(x, y) = \frac{1}{2\pi\sigma_X\sigma_Y \sqrt{(1 - \rho^2)}}$$

$$\times e^{-\left[\frac{(x - \bar{X})^2}{2\sigma_X^2 (1 - \rho^2)} - 2\rho \frac{(x - \bar{X})(y - \bar{Y})}{2\sigma_X\sigma_Y(1 - \rho^2)} + \frac{(y - \bar{Y})^2}{2\sigma_Y^2 (1 - \rho^2)}\right]} \tag{3.32}$$

(for $\bar{X} = \bar{Y} = \rho = 0$ this reduces to equation (3.29)) which can then be written as a product of separate functions in x and y, i.e.:

$$p_{X,Y}(x, y) = \frac{1}{\sqrt{(2\pi)}\sigma_X} e^{-\frac{x^2}{2\sigma_X^2}} \frac{1}{\sqrt{(2\pi)}\sigma_Y} e^{-\frac{y^2}{2\sigma_Y^2}}$$

$$= p_X(x)\, p_Y(y) \tag{3.33}$$

Equation (3.33) is the necessary and sufficient condition for statistical independence of X and Y. For a multivariate Gaussian function, then, uncorrelatedness does imply independence.

3.2.9 Addition of random variables and the central limit theorem

If two random variables X and Y are added, Figure 3.18, and their joint pdf, $p_{X,Y}(x, y)$, is known, what is the pdf, $p_Z(z)$, of their sum, Z? To answer this question we note the following:

$$Z = X + Y \tag{3.34}$$

Therefore when $Z = z$ (i.e. Z takes on a particular value z) then:

$$Y = z - X \tag{3.35(a)}$$

and when $Z = z + dz$ then:

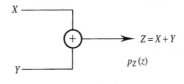

Figure 3.18 *Addition of random variables.*

$$Y = z + dz - X \qquad\qquad (3.35(b))$$

Equations (3.35), which both represent straight lines in the X, Y plane, are sketched in Figure 3.19. The probability that Z lies in the range z to $z + dz$ is given by the volume contained under $p_{X,Y}(x, y)$ in the strip between these two lines, i.e.:

$$P(z < Z \le z + dz) = \int_{strip} p_{X,Y}(x, y) \, ds \qquad\qquad (3.36(a))$$

or:

$$p_Z(z) \, dz = \int_{strip} p_{X,Y}(x, z - x) \, dx \, dz \qquad\qquad (3.36(b))$$

Therefore:

$$p_Z(z) = \int_{-\infty}^{\infty} p_{X,Y}(x, z - x) \, dx \qquad\qquad (3.37)$$

If X and Y are statistically independent then:

$$p_{X,Y}(x, z - x) = p_X(x) \, p_Y(z - x) \qquad\qquad (3.38)$$

and equation (3.37) becomes:

$$p_Z(z) = \int_{-\infty}^{\infty} p_X(x) p_Y(z - x) \, dx \qquad\qquad (3.39)$$

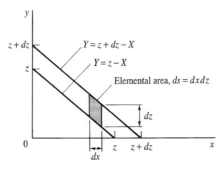

Figure 3.19 *Strip of integration in X, Y plane to find $P(z < Z \le z + dz)$.*

Equation (3.39) can be recognised as the convolution integral. *The pdf of the sum of independent random variables is therefore the convolution of their individual pdfs.*

The multiple convolution of pdfs which arises when many independent random variables are added has a surprising and important consequence. Since convolution is essentially an integral operation it almost always results in a function which is in some sense smoother (i.e. more gradually varying) than either of the functions being convolved. (This is true provided that the original functions are reasonably smooth which excludes, for instance, the case of impulse functions.) After surprisingly few convolutions this repeated smoothing results in a distribution which approximates a Gaussian function. The approximation gets better as the number of convolutions increases. The tendency for multiple convolutions to give rise to Gaussian functions is called the *central limit theorem* and accounts for the ubiquitous nature of Gaussian noise. It is illustrated for multiple self convolution of a rectangular pulse in Figure 3.20. In the context of statistics the central limit theorem can be stated as follows:

> *If N statistically independent random variables are added, the sum will have a probability density function which tends to a Gaussian function as N tends to infinity, irrespective of the original random variable pdfs.*

A second consequence of the central limit theorem is that the pdf of the *product* of N independent random variables will tend to a log-normal distribution as N tends to infinity, since multiplication of functions corresponds to addition of their logarithms.

If two *Gaussian* random variables are added their sum will also be a Gaussian random variable. In this case the result is exact and holds even if the random variables are correlated. The mean and variance of the sum are given by:

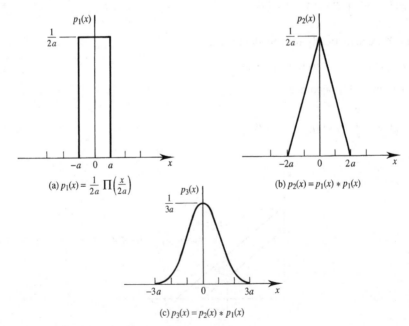

(a) $p_1(x) = \dfrac{1}{2a} \, \Pi\!\left(\dfrac{x}{2a}\right)$

(b) $p_2(x) = p_1(x) * p_1(x)$

(c) $p_3(x) = p_2(x) * p_1(x)$

Figure 3.20 *Multiple self convolution of a rectangular pulse.*

$$\bar{Z} = \bar{X} + \bar{Y} \tag{3.40}$$

and:

$$\sigma^2_{X\pm Y} = \sigma^2_X \pm 2\rho\sigma_X\sigma_Y + \sigma^2_Y \tag{3.41}$$

For uncorrelated (and therefore independent) Gaussian random variables the variances, like the means, are simply added. This is an especially easy case to prove since for independent variables the pdf of the sum is the convolution of two Gaussian functions. This is equivalent to multiplying the Fourier transforms of the original pdfs and then inverse Fourier transforming the result. (The Fourier transform of a pdf is called the *characteristic function* of the random variable.) When a Gaussian pdf is Fourier transformed the result is a Gaussian characteristic function. When Gaussian characteristic functions are multiplied the result remains Gaussian ($e^{-x^2}e^{-x^2} = e^{-2x^2}$). Finally when the Gaussian product is inverse Fourier transformed the result is a Gaussian pdf.

EXAMPLE 3.9

$x(t)$ and $y(t)$ are zero mean Gaussian random currents. When applied individually to 1 Ω resistive loads they dissipate 4.0 W and 1.0 W of power respectively. When both are applied to the load simultaneously the power dissipated is 3.0 W. What is the correlation between X and Y?

Since X and Y have zero mean their variance is equal to their normalised power, i.e.:

$$\sigma^2_X = 4.0 \quad \text{and} \quad \sigma^2_Y = 1.0$$

Their standard deviations are therefore:

$$\sigma_X = 2.0 \quad \text{and} \quad \sigma_Y = 1.0$$

Using equation (3.41):

$$\rho = \frac{\sigma^2_{X\pm Y} - \sigma^2_X - \sigma^2_Y}{\pm 2\, \sigma_X\, \sigma_Y}$$

We take the positive sign in the denominator since it is the sum (not difference) which dissipates 3.0 W:

$$\rho = \frac{3.0 - 4.0 - 1.0}{2\,(2.0)\,(1.0)} = -0.5$$

EXAMPLE 3.10

Two independent random voltages have a uniform pdf given by:

$$p(V) = \begin{cases} 0.5, & |V| \le 1.0 \\ 0, & |V| > 1.0 \end{cases}$$

Find the pdf of their sum.

$$p(V_1 + V_2) = p_1(V) * p_2(V)$$

Using χ to represent the characteristic function, i.e.:

$$\chi(W) = \text{FT}\{p(V)\}$$

then from the convolution theorem (Table 2.5) and the Fourier transform of a rectangular function (Table 2.4):

$$\chi(W) = \chi_1(W)\,\chi_2(W)$$

$$= 0.5\,[2\,\text{sinc}(2W)]\,0.5\,[2\,\text{sinc}(2W)]$$

$$= \text{sinc}^2(2W)$$

The pdf of the sum is then found by inverse Fourier transforming $\chi(W)$, i.e.:

$$p(V_1 + V_2) = 0.5\,\Lambda\left(\frac{V}{2}\right)$$

$$= \begin{cases} 0.5\left(1 - \dfrac{|V|}{2}\right), & |V| \le 2.0 \\[2mm] 0, & |V| > 2.0 \end{cases}$$

3.3 Random processes

The term random process usually refers to a random variable which is a function of time (or occasionally a function of position) and is strictly defined in terms of an ensemble (i.e. collection) of time functions, Figure 3.21. Such an ensemble of functions may, in

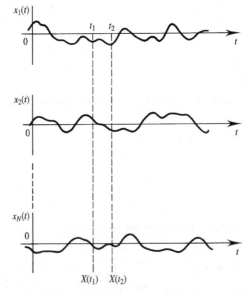

Figure 3.21 *Random process, $X(t)$, as ensemble of sample functions, $x_i(t)$.*

principle, be generated using many sets (perhaps an infinite number) of identical sources. The following notation for random processes is adopted here:

1. The random process (i.e. the entire ensemble of functions) is denoted by $X(t)$.
2. $X(t_1)$ or X_1 denotes an ensemble of samples taken at time t_1 and constitutes a random variable.
3. $x_i(t)$ is the ith sample function of the ensemble.

It is often the case, in practice, that only one sample function can be observed, the other sample functions representing what might have occurred (given the statistical properties of the process) but did not. It is also the case that each sample function $x_i(t)$ is *usually* a random function of time although this does not have to be so. (For example, $X(t)$ may be a set of sinusoids each sample function having random phase.)

Random processes, like other types of signal, can be classified in a number of different ways. For example, they may be:

1. Continuous or discrete.
2. Analogue or digital (or mixed).
3. Deterministic or non-deterministic.
4. Stationary or non-stationary.
5. Ergodic or non-ergodic.

The first category refers to continuity or discreteness in time or position. (Discrete time signals are also sometimes called a time series.) The second category could be (and sometimes is) referred to as continuous, discrete or mixed which in this context describes the pdf of the process, Figure 3.22. A deterministic random process seems, superficially, to be a contradiction in terms. It describes a process, however, in which each sample function is deterministic. An example of such a process, $x_i(t) = \sin(\omega t + \theta_i)$ where θ_i is a random variable with specified pdf, has already been given. Stationarity and ergodicity are concepts which are central to random processes and they are therefore discussed in some detail below.

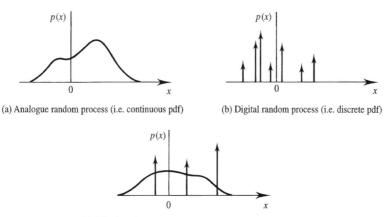

(a) Analogue random process (i.e. continuous pdf)

(b) Digital random process (i.e. discrete pdf)

(c) Mixed random process (i.e. continuous plus discrete pdf)

Figure 3.22 *Pdfs of analogue, digital and mixed random processes.*

3.3.1 Stationarity and ergodicity

Stationarity relates to the time independence of a random process's statistics. There are two definitions:

(a) A random process is said to be stationary in the *strict* (sometimes called narrow) sense if all its pdfs (joint, conditional and marginal) are the same for any value of t, i.e. if none of its statistics change with time.

(b) A random process is said to be stationary in the *loose* (sometimes called wide) sense if its mean value, $\overline{X(t)}$, is independent of time, t, and the correlation, $\overline{X(t_1)X(t_2)}$, depends only on time difference $\tau = t_2 - t_1$.

Ergodicity relates to the equivalence of ensemble and time averages. It implies that each sample function, $x_i(t)$, of the ensemble has the same statistical behaviour as any set of ensemble values, $X(t_j)$, Figure 3.23. Thus for an ergodic process:

$$\langle x_i^n(t) \rangle = \lim_{T \to \infty} \frac{1}{T} \int_{-T/2}^{T/2} x_i^n(t)\, dt$$

$$= \int_{-\infty}^{\infty} X_j^n\, p(X_j)\, dx$$

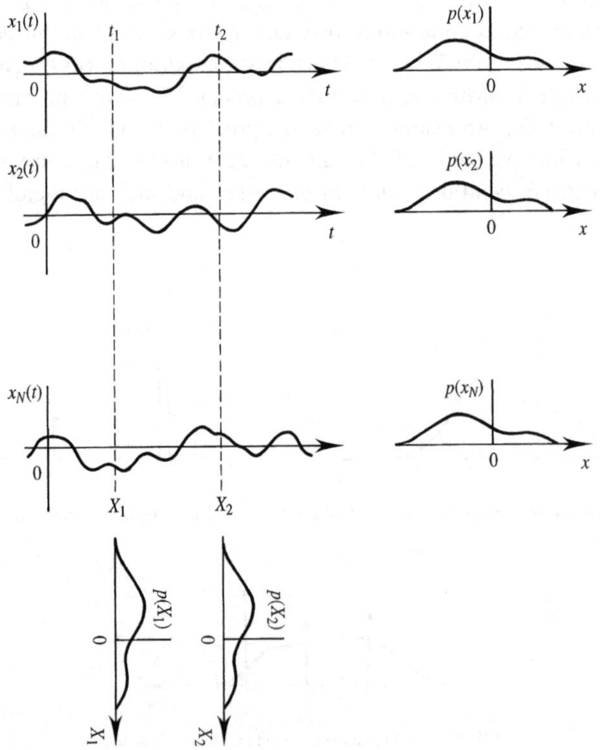

Figure 3.23 *Identity of sample function pdfs, $p(x_i)$, and ensemble random variable pdfs, $p(X_i)$, for an ergodic process.*

$$= \overline{X^n(t_j)}, \quad \text{for } any \text{ } i \text{ and } j \tag{3.42}$$

It is obvious that an ergodic process must be statistically stationary. The converse is not true, however, i.e. stationary processes need not be ergodic. Ergodicity is therefore a stronger (more restrictive) condition on a random process than stationarity, i.e.:

$$\text{Ergodicity} \Rightarrow \text{stationarity}$$
$$\text{Stationarity} \nRightarrow \text{ergodicity}$$

3.3.2 Strict and loose sense Gaussian processes

A sample function, $x_i(t)$, is said to belong to a Gaussian random process, $X(t)$, in the *strict* sense if the random variables $X_1 = X(t_1)$, $X_2 = X(t_2)$, $\cdots$, $X_N = X(t_N)$ have an N-dimensional joint Gaussian pdf, Figure 3.24. For an ergodic process the strict sense Gaussian condition can be defined in terms of a single sample function. In this case if the joint pdf of multiple sets of N-tuple samples, taken with fixed time intervals between the samples of each N-tuple, is N-variate Gaussian then the process is Gaussian in the strict sense. This definition is illustrated in Figures 3.25(a) and (b) for multiple sets of sample pairs (i.e. $N = 2$).

A sample function, $x_i(t)$, is said to belong to a Gaussian random process in the *loose* sense if isolated samples taken from $x_i(t)$ come from a Gaussian pdf, Figure 3.26. The following points can be made about strict and loose sense Gaussian processes:

1. Being a strict sense Gaussian process is a very strong statistical condition, much stronger than being a loose sense Gaussian process. All strict sense Gaussian processes are, therefore, also loose sense Gaussian processes.

2. Examples do exist of processes which are Gaussian in the loose sense but not the strict sense. They are rare in practice, however.

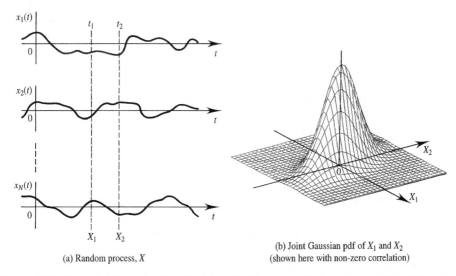

(a) Random process, X

(b) Joint Gaussian pdf of X_1 and X_2 (shown here with non-zero correlation)

Figure 3.24 *Example (drawn for $N = 2$) of the joint Gaussian pdf of random variables (X_1 and X_2) taken from a strict sense Gaussian process.*

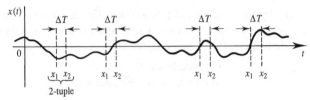

(a) *N*-tuple (*N* = 2) samples with constant sample separation (Δ*T*) taken at
random times from a sample function , *x*(*t*), of the random process *X*(*t*)

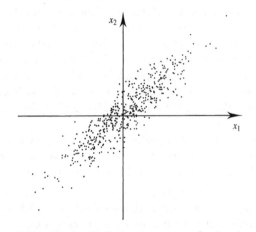

(b) Joint, *N*-variate (*N* = 2), Gaussian scattergram for $p(x_1, x_2)$

Figure 3.25 *Single sample function definition of ergodic, strict sense Gaussian, random process.*

3. A strict sense Gaussian process is the most structureless, random or unpredictable statistical process possible. It is also one of the most important processes in the context of communications since it describes thermal noise which is present to some degree in all practical systems.

4. A strict sense *N*-dimensional Gaussian pdf is specified completely by its first and second order moments, i.e. its means, variances and covariances, as all higher moments (Figure 3.9) of the Gaussian pdf are zero.

3.3.3 Autocorrelation and power spectral density

A simple pdf is obviously insufficient to fully describe a random signal (i.e. a sample function from a random process) because it contains no information about the signal's rate of change.

Such information would be available, however, in the joint pdfs, $p(X_1, X_2)$, of random variables, $X(t_1)$ and $X(t_2)$, separated by $\tau = t_2 - t_1$. These joint pdfs are not usually known in full but partial information about them is often available in the form of the correlation, $\overline{X(t_2)X(t_2 - \tau)}$. For ergodic signals the ensemble average taken at any time is equal to the temporal average of any sample function, i.e.:

$$\overline{X(t)\,X(t - \tau)} = \langle x(t)\,x(t - \tau)\rangle$$

$$= \lim_{T \to \infty} \frac{1}{T} \int_{-T/2}^{T/2} x(t) \, x(t - \tau) \, dt$$

$$= R_x(\tau) \ (\text{V}^2) \tag{3.43}$$

The autocorrelation function, $R_x(\tau)$, of a sample function, $x(t)$, taken from a real, ergodic, random process has the following properties:

1. $R_x(\tau)$ is real.
2. $R_x(\tau)$ has even symmetry (see Figure 3.27), i.e.:

$$R_x(-\tau) = R_x(\tau) \tag{3.44}$$

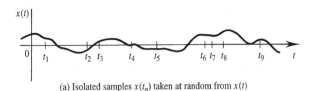

(a) Isolated samples $x(t_n)$ taken at random from $x(t)$

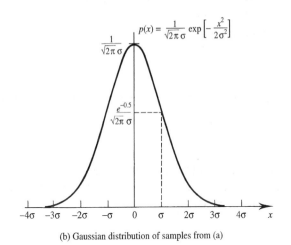

$$p(x) = \frac{1}{\sqrt{2\pi}\,\sigma} \exp\left[-\frac{x^2}{2\sigma^2}\right]$$

(b) Gaussian distribution of samples from (a)

Figure 3.26 *Definition of a loose sense Gaussian process.*

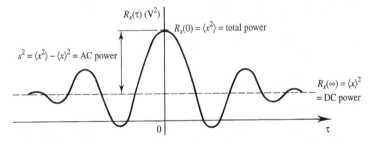

Figure 3.27 *General behaviour of $R_x(\tau)$ for a random process.*

3. $R_x(\tau)$ has a maximum (positive) magnitude at $\tau = 0$ which corresponds to the mean square value of (or normalised power in) $x(t)$, i.e.:

$$\langle x^2(t) \rangle = R_x(0) > |R(\tau)|, \qquad \text{for } all \ \tau \neq 0 \tag{3.45}$$

4. If $x(t)$ has units of V then $R_x(\tau)$ has units of V^2 (i.e. normalised power).
5. $R_x(\infty)$ is the square mean value of (or normalised DC power in) $x(t)$, i.e.:

$$R_x(\infty) = \langle x(t) \rangle^2 \tag{3.46}$$

6. $R_x(0) - R_x(\infty)$ is the variance, s^2, of $x(t)$, i.e.:

$$R_x(0) - R_x(\infty) = \langle x^2(t) \rangle - \langle x(t) \rangle^2 = s^2 \tag{3.47}$$

7. The autocorrelation function and *two-sided* power spectral density of $x(t)$ form a Fourier transform pair, i.e.:

$$R_x(\tau) \overset{\text{FT}}{\Leftrightarrow} G_x(f) \tag{3.48}$$

(This is the Wiener–Kintchine theorem [Papoulis] which, although not proved here, can be readily accepted since a similar theorem has been proved in Chapter 2 for transient signals.)   Properties of the corresponding power spectral density (most of which are corollaries of the above) include the following:

1. $G_x(f)$ has even symmetry about $f = 0$, i.e.:

$$G_x(-f) = G_x(f) \tag{3.49}$$

2. $G_x(f)$ is real.
3. The area under $G_x(f)$ is the mean square value of (or normalised power in) $x(t)$, i.e.:

$$\int_{-\infty}^{\infty} G_x(f) \, df = \langle x^2(t) \rangle \tag{3.50}$$

4. If $x(t)$ has units of V then $G_x(f)$ has units of V^2/Hz.
5. The area under any impulse in $G_x(f)$ occurring at $f = 0$ is the square mean value of (or normalised DC power in) $x(t)$, i.e.:

$$\int_{0-}^{0+} G_x(f) \, df = \langle x(t) \rangle^2 \tag{3.51}$$

6. The area under $G_x(f)$, excluding any impulse function at $f = 0$, is the variance of $x(t)$ or the normalised power in the fluctuating component of $x(t)$, i.e.:

$$\int_{-\infty}^{0-} G_x(f) \, df + \int_{0+}^{\infty} G_x(f) \, df = \langle x^2(t) \rangle - \langle x(t) \rangle^2 \tag{3.52}$$

7. $G_x(f)$ is positive for all f, i.e.:

$$G_x(f) \geq 0, \qquad \text{for } all \ f \tag{3.53}$$

(White noise is a random signal with particularly extreme spectral and autocorrelation properties. It has no self similarity with any time shifted version of itself so its autocorrelation function consists of a single impulse at zero delay, and its power spectral density is flat.)

A normalised autocorrelation function, $\rho_x(\tau)$, can be defined by subtracting any DC value present in $x(t)$, dividing by the resulting RMS value and autocorrelating the result. This is equivalent to:

$$\rho_x(\tau) = \frac{\langle x(t)x(t-\tau) \rangle - \langle x(t) \rangle^2}{\langle x^2(t) \rangle - \langle x(t) \rangle^2} \tag{3.54}$$

The normalised function, Figure 3.28, is clearly an extension of the normalised correlation coefficient (equation (3.28)) and has the properties:

$$\rho_x(0) = 1 \tag{3.55(a)}$$

$$\rho_x(\pm\infty) = 0 \tag{3.55(b)}$$

It can be interpreted as the fraction of $x(t-\tau)$ which is contained in $x(t)$ neglecting DC components. This is easily demonstrated as follows.

Let $f(t)$ be a zero mean stationary random process, i.e.:

$$f(t) = x(t) - \langle x(t) \rangle \tag{3.56}$$

If the new function:

$$g(t) = f(t) - \rho f(t-\tau) \tag{3.57}$$

is formed then the value of ρ which minimises $\langle g^2(t) \rangle$ will be the fraction of $f(t-\tau)$ contained in $f(t)$. Expanding $\langle g^2(t) \rangle$:

$$\begin{aligned} \langle g^2(t) \rangle &= \langle [f(t) - \rho f(t-\tau)]^2 \rangle \\ &= \langle f^2(t) - 2\rho f(t)f(t-\tau) + \rho^2 f^2(t-\tau) \rangle \\ &= \langle f^2(t) \rangle - 2\rho \langle f(t)f(t-\tau) \rangle + \rho^2 \langle f^2(t-\tau) \rangle \end{aligned} \tag{3.58}$$

The value of ρ which minimises $\langle g^2(t) \rangle$ is found by solving $d\langle g^2(t) \rangle/d\rho = 0$, i.e.:

$$0 - 2R_f(\tau) + 2\rho \langle f^2(t-\tau) \rangle = 0 \tag{3.59}$$

giving:

$$\rho = \frac{R_f(\tau)}{\langle f^2(t) \rangle} = \frac{\langle [x(t) - \langle x(t) \rangle][x(t-\tau) - \langle x(t) \rangle] \rangle}{\langle [x(t) - \langle x(t) \rangle]^2 \rangle}$$

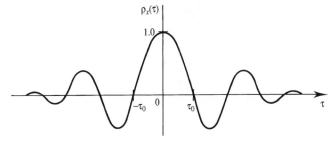

Figure 3.28 *General behaviour of $\rho_x(\tau)$ for a random process. (Shows first null definition of decorrelation time, τ_0.)*

$$= \frac{\langle x(t)x(t-\tau)\rangle - \langle x(t)\rangle^2}{\langle x^2(t)\rangle - \langle x(t)\rangle^2} \qquad (3.60)$$

EXAMPLE 3.11

Find and sketch the autocorrelation function of the stationary random signal whose power spectral density is shown in Figure 3.29(a).

Using the triangular function, Λ, with f measured in Hz:

$$G_x(f) = 3.0\left[\Lambda\left(\frac{f-5000}{1000}\right) + \Lambda\left(\frac{f+5000}{1000}\right)\right]$$

$$= 3.0\,\Lambda\left(\frac{f}{1000}\right) * [\delta(f-5000) + \delta(f+5000)]$$

$$R_x(\tau) = \text{FT}^{-1}\{G_x(f)\}$$

$$= 3.0\,\text{FT}^{-1}\left\{\Lambda\left(\frac{f}{1000}\right)\right\}\text{FT}^{-1}\{\delta(f-5000) + \delta(f+5000)\}$$

Using Tables 2.4 and 2.5:

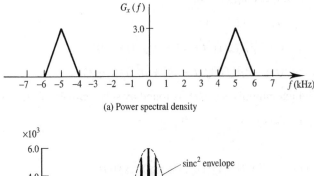

(a) Power spectral density

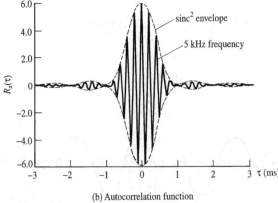

(b) Autocorrelation function

Figure 3.29 *Spectral and temporal characteristics of stationary random signal, Example 3.11.*

$$R_x(\tau) = 3.0 \times 1000 \, \text{sinc}^2(1000\tau) \, 2\cos(2\pi \, 5000\tau)$$

Figure 3.29(b) shows a sketch for the solution of Example 3.11.

3.3.4 Signal memory, decorrelation time and white noise

It is physically obvious that practical signals must have a finite memory, i.e. samples taken close enough together must be highly correlated. The decorrelation time, τ_0, of a signal provides a quantitative measure of this memory and is defined as the minimum time shift, τ, required to reduce $\rho_x(\tau)$ to some predetermined, or reference, value, Figure 3.28. The reference value depends on the application and/or preference and can be somewhat arbitrary in the same way as the definition of bandwidth, B (section 2.2.5). Popular choices, however, are $\rho_x(\tau_0) = 1/\sqrt{2}$, 0.5, $1/e$ and 0. Due to the Wiener–Kintchine theorem there is clearly a relationship between B and τ_0, i.e.:

$$B \propto \frac{1}{\tau_0} \, \text{Hz} \qquad (3.61)$$

(The constant of proportionality depends on the exact definitions adopted but for reasonably consistent choices is of the order of unity.) Equation (3.61) requires a careful interpretation if the random signal has a passband spectrum (see Example 3.12).

For a random signal or noise with a white power spectral density, equation (3.61) implies that $\tau_0 = 0$, i.e. that the signal has zero memory. In particular the autocorrelation function of white noise will be impulsive, i.e.:

$$R_x(\tau) = C\delta(\tau) \qquad (3.62)$$

This means that adjacent samples taken from a white noise process are uncorrelated no matter how closely the samples are spaced. As this is physically impossible it means that white noise, whilst important and useful conceptually, is not practically realisable. (The same conclusion is obvious when considering the total power in a white noise process.)

The common assumption of white, Gaussian, noise processes sometimes gives the impression that Gaussianness and whiteness are connected. This is not true. Noise may be Gaussian or white, or both, or neither. If noise is Gaussian *and* white (and thermal noise, for example, is often modelled in this way) then the fact that adjacent samples from the process are uncorrelated (irrespective of separation) means that they are also independent.

EXAMPLE 3.12

Stating the definition you use, find the decorrelation time of the random signal described in Example 3.11.

Referring to Figure 3.29(b) and using $\rho_x(\tau_0) = 0$, the decorrelation time of $x(t)$ in Example 3.11 is 0.25 cycles of the 5 kHz signal, i.e.:

$$\rho_x(\tau_0) = 0.25 \, T = 0.25 \, \frac{1}{f_c} = \frac{0.25}{5000}$$

$$= 5 \times 10^{-5} \text{ s} \quad (\text{or } 50 \text{ } \mu\text{s})$$

(Note that the decorrelation time of the *envelope* of $R_x(\tau)$ in Figure 3.29(b) is 1.0 ms and it is this quantity which is of the order of the reciprocal of the bandwidth.)

3.3.5 Cross correlation of random processes

The cross correlation of functions, taken from two *real* ergodic random processes, is:

$$R_{xy}(\tau) = \langle x(t)y(t - \tau) \rangle$$

$$= \lim_{T \to \infty} \frac{1}{T} \int_{-T/2}^{T/2} x(t)y(t - \tau) \, dt \tag{3.63}$$

Some of the properties of this cross correlation function are:

1. $R_{xy}(\tau)$ is real. $\hfill$ (3.64)

2. $R_{xy}(-\tau) = R_{yx}(\tau)$ $\hfill$ (3.65)

   (Note that, in general, $R_{xy}(-\tau) \neq R_{xy}(\tau)$.)

3. $[R_x(0)R_y(0)]^{1/2} > |R_{xy}(\tau)| \quad$ for *all* τ $\hfill$ (3.66)

   (Note that the maximum value of $R_{xy}(\tau)$ can occur anywhere.)
4. If $x(t)$ and $y(t)$ have units of V then $R_{xy}(\tau)$ has units of V^2 (i.e normalised power) and, for this reason, it is sometimes called a *cross-power*.

5. $[R_x(0) + R_y(0)]/2 > |R_{xy}(\tau)|, \quad$ for *all* τ $\hfill$ (3.67)

   (This follows from property 3 since the geometric mean of two real numbers cannot exceed their arithmetic mean.)
6. For *statistically independent* random processes:

   $$R_{xy}(\tau) = R_{yx}(\tau) \tag{3.68(a)}$$

   and if either process has zero mean then:

   $$R_{xy}(\tau) = R_{yx}(\tau) = 0 \quad \text{for *all* } \tau \tag{3.68(b)}$$

7. The Fourier transform of $R_{xy}(\tau)$ is often called a cross-power spectral density, $G_{xy}(f)$, since its units are V^2/Hz:

   $$R_{xy}(\tau) \overset{\text{FT}}{\Leftrightarrow} G_{xy}(f) \tag{3.69}$$

If the functions $x(t)$ and $y(t)$ are complex then the cross correlation is defined by:

$$R_{xy}(\tau) = \langle x(t)y^*(t - \tau) \rangle = \langle x^*(t)y(t + \tau) \rangle \tag{3.70}$$

and many of the properties listed above do not apply.

3.4 Summary

Variables are said to be random if their particular value at specified future times cannot be predicted. Information about their probable future values is often available, however, from a probability model. The (unconditional) probability that any of the events, belonging to a subset of mutually exclusive possible events, occurs as the outcome of a random experiment or trial is the sum of the individual probabilities of the events in the subset. The joint probability of a set of statistically independent events is the product of their individual probabilities. A conditional probability is the probability of an event given that some other, specified, event is known to have occurred. Bayes's rule relates joint, conditional and unconditional probabilities.

Cumulative distributions give the probability that a random variable will be less than, or equal to, any particular value. Pdfs are the derivative of the cumulative distribution. The definite integral of a pdf is the probability that the random variable will lie between the integral's limits. Exceedances are the complement of cumulative distributions.

Moments, central moments and modes are statistics of random variables. In general they give partial information about the shape and location of pdfs. Joint pdfs (on definite integration) give the probability that two or more random variables will concurrently take particular values between the specified limits. A marginal pdf is the pdf of one random variable irrespective of the value of any other random variable. The correlation of two random variables is their mean product. The covariance is the mean product of their fluctuating (zero mean) components only, being zero for uncorrelated signals. Statistically independent random variables are always uncorrelated but the converse is not true. The (normalised) correlation coefficient of two random variables is the correlation of their fluctuating components (i.e. covariance) after the standard deviations of both variables have been normalised to 1.0. The pdf of the sum of independent random variables is the convolution of their individual pdfs and, for the sum of many independent random variables, this results in a Gaussian pdf. This is called the central limit theorem. If the random variables are independent then the mean of their sum is the sum of their means and the variance of their sum is the sum of their variances.

Random processes are random variables which change with time (or spatial position). They are defined strictly by an ensemble of functions. Both ensemble and temporal (or spatial) statistics can therefore be defined. A random process is said to be (statistically) stationary in the strict, or narrow, sense if all its statistics are invariant with time (or space). It is said to be stationary in the loose, or wide, sense if its ensemble mean is invariant with time and the correlation between its random variables at different times depends only on time difference. A random process is said to be ergodic if its ensemble and time averages are equal. Random processes which are ergodic are statistically stationary but the converse is not necessarily true.

Gaussian processes are extremely common and important due to the action of the central limit theorem. A process is said to be Gaussian in the strict sense if any pair of (ensemble) random variables has a joint Gaussian pdf. Any sample function of a random process is said to be Gaussian in the loose sense if samples from it are Gaussianly distributed. Not all loose sense Gaussian sample functions belong to strict sense Gaussian processes. Gaussian processes are completely specified by their first and second order moments.

Signal memory is characterised by the signal's autocorrelation function. This function gives the correlation between the signal and a time shifted version of the signal for all possible time shifts. The decorrelation time of a signal is that time shift for which the autocorrelation function has fallen to some prescribed fraction of its peak value. The Wiener–Kintchine theorem identifies the power and energy spectral densities of power and energy signals with the Fourier transform of these signals' autocorrelation functions. The normalised autocorrelation can be interpreted as the fraction of a signal contained within a time shifted version of itself. Signal memory (i.e. decorrelation time) and signal bandwidth are inversely proportional. White noise, with an impulsive autocorrelation function, is physically unrealisable and is memoryless.

Cross correlation relates to the similarity between a pair of different functions, one offset from the other by a time shift. The Fourier transform of a cross correlation function is a cross-energy, or power, spectral density depending on whether the function pair represent energy or power signals.

3.5 Problems

3.1   A box contains 30 resistors: 15 of the resistors have nominal values of 1.0 kΩ, 10 have nominal values of 4.7 kΩ and 5 have nominal values of 10 kΩ; 3 resistors are taken at random and connected in series. What is the probability that the 3-resistor combination will have a nominal resistance of: (i) 3 kΩ; (ii) 15.7 kΩ; and (iii) 19.4 kΩ? [0.1121, 0.1847, 0.0554]

3.2   A transceiver manufacturer buys power amplifiers from three different companies (A, B, C). Assembly line workers pick power amplifiers from a rack at random without noticing the supplier. Customer claims, under a one-year warranty scheme, show that 8% of all power amplifiers (irrespective of supplier) fail within one year and that 25%, 35% and 40% of all failed power amplifiers were supplied by companies A, B and C respectively. The purchasing department records that power amplifiers have been supplied by companies A, B and C in the proportions 50:40:10 respectively. What is the probability of failure within one year of amplifiers supplied by each company? [0.04, 0.07, 0.32]

3.3   The cumulative distribution function for a continuous random variable, X, has the form:

$$P_X(x) = \begin{cases} 0, & -\infty < x \le -2 \\ a\,(1 + \sin(bx)), & -2 < x \le 2 \\ c, & x > 2 \end{cases}$$

Find: (a) the values of a, b and c that make this a valid CD; (b) the probability that x is negative; and (c) the corresponding probability density function.

3.4   A particular random variable has a cumulative distribution function given by:

$$P_X(x) = \begin{cases} 0, & -\infty < x \le 0 \\ 1 - e^{-x}, & 0 \le x < \infty \end{cases}$$

Find: (a) the probability that $x > 0.5$; (b) the probability that $x \le 0.25$; and (c) the probability that $0.3 < x \le 0.7$. [0.6065, 0.2212, 0.2442]

3.5   The power reflected from an aircraft of complicated shape that is received by a radar can be described by an exponential random variable, w. The pdf of w is:

$$p(w) = \begin{cases} (1/w_o)e^{-w/w_o}, & \text{for } w > 0 \\ 0, & \text{for negative } w \end{cases}$$

where w_o is the average amount of received power. What is the probability that the power received by the radar will be greater than the average received power? [0.368]

3.6   An integrated circuit manufacturer tests the propagation delays of all chips of one particular batch. It discovers that the pdf of the delays is well approximated by a triangular distribution with mean value 8 ns, maximum value 12 ns and minimum value 4 ns. Find: (a) the variance of this distribution; (b) the standard deviation of the distribution; and (c) the percentage of chips which will be rejected if the specification for the device is 10 ns. [2.66, 1.63, 12.5%]

3.7   A bivariate random variable has the joint pdf:

$$p(x, y) = A(x^2 + 2xy) \prod(x) \prod(y)$$

Find: (a) the value of A which makes this a valid pdf; (b) the correlation of X and Y; (c) the marginal pdfs of X and Y; (d) the mean values of X and Y; and (e) the variances of X and Y. [12, 0.05, 1, 0, 0.15, 0.0833]

3.8   (a) For the zero mean Gaussian pdf $p_X = (1/\sqrt{2\pi}\sigma)e^{-x^2/(2\sigma^2)}$, prove explicitly that the RMS value $\sqrt{\overline{X^2}}$ is σ.

(Hint: $\displaystyle\int_a^b x^2 e^{-x^2} \, dx = -\frac{d}{d\lambda}\left[\int_a^b e^{-\lambda x^2} \, dx\right]_{\lambda = 1}$

and remember that a complete integral (with limits $\pm\infty$) may be found using the Fourier transform DC value theorem, Table 2.5), and (b) show that, for the Gaussian pdf as defined in part (a), $\overline{X^4} = 3\sigma^4$.

(Hint: $\displaystyle\int_0^\infty x^{n-1}e^{-x} \, dx = \Gamma(n),$ any $n > 0$

and the gamma function $\Gamma(n)$ has the properties:

$$\Gamma(n + 1) = n\Gamma(n) \quad \text{and} \quad \Gamma(\tfrac{1}{2}) = \sqrt{\pi})$$

3.9   A random signal with uniform pdf, $p_X(x) = \prod(x)$, is added to a second, independent, random signal with one-sided exponential pdf, $p_Y(y) = 3 u(y)e^{-3y}$. Find the pdf of the sum.

3.10 $v(t)$, $w(t)$, $x(t)$ and $y(t)$ are independent random signals which have the following pdfs:

$$P_V(v) = \frac{2}{1 + (2\pi v)^2} \qquad P_W(w) = \frac{1}{1 + (\pi w)^2}$$

$$P_X(x) = \frac{2/3}{1 + [(2/3)\pi x]^2} \qquad P_Y(y) = \frac{1/2}{1 + (\tfrac{1}{2}\pi y)^2}$$

Use characteristic functions to find the pdf of their sum.

3.11   For a tossed dice:

(a) Use convolution to deduce the probabilities of the sum of two thrown dice being 2, 3, etc.

(b) The largest sum possible on throwing 4 dice is 24. What is the probability of this event from the joint probability of independent events? Check your answer by convolution.

(c) What is the most probable sum for 4 dice? Use convolution to find the probability of this event.

(d) A box containing 100 dice is spilled on the floor. Make as many statements as you can about the sum of the uppermost faces by extending the patterns you see developing in the convolution in parts (a), (b) and (c). [(a) 1/36, 2/36, $\cdots$, 6/36, 5/36, 4/36, $\cdots$, 1/36]; [(b) 7.7 × 10^{-4}]; [(c) 14, 1.13 × 10^{-1}]; [(d) $\Sigma_{min} = 100$, $\Sigma_{max} = 600$, 501 possible sums, most likely $\Sigma = 350$, pdf is truncated Gaussian approximation]

3.12 Two, independent, zero mean, Gaussian noise sources (X and Y) each have an RMS output of 1.0 V. A cross-coupling network is to be used to generate two noise signals (U and V), where $U = (1 - \alpha)X + \alpha Y$ and $V = (1 - \alpha)Y + \alpha X$, with a correlation coefficient of 0.2 between U and V. What must the (voltage) cross-coupling ratio, α, be? [0.8536 or 0.1465]

3.13 A periodic time function, $x(t)$, of period T is defined as a sawtooth waveform with a random 'phase' (i.e. positive gradient zero crossing point), τ, over the period nearest the origin, i.e.:

$$x(t) = \frac{2V}{T}(t - \tau), \qquad -T/2 + \tau \leq t < T/2 + \tau$$

The pdf of the random variable is:

$$p_T(\tau) = \begin{cases} 1/T, & |\tau| \leq T/2 \\ 0, & |\tau| > T/2 \end{cases}$$

Show that the function is ergodic.

3.14 Consider the following time function:

$$X(t) = A \cos(\omega t - \Theta)$$

The phase angle, Θ, is a random variable whose pdf is given as:

$$p_\Theta(\theta) = \frac{1}{2\pi} \qquad \text{for } 0 \leq \theta < 2\pi \quad \text{and zero elsewhere}$$

Find the mean value and variance of Θ and of $X(t)$.

3.15 Given that the autocorrelation function of a certain stationary process is:

$$R_{xx}(\tau) = 25 + \frac{4}{1 + 6\tau^2}$$

Find: (a) the mean value, and (b) the variance of the process. [± 5, 4]

3.16 $X(t)$ is a deterministic random process defined by:

$$X(t) = \cos(2\pi ft + \Theta) + 0.5$$

where Θ is a uniformly distributed random variable in the range $[-\pi, \pi]$, but remains fixed for a given sample waveform of the random process. Calculate $R_{xx}(\tau)$, and identify the source of each term in your answer.

3.17 A stationary random process has an autocorrelation function given by:

$$R(\tau) = \begin{cases} 10(1 - |\tau|/0.05), & |\tau| \leq 0.05 \\ 0, & \text{elsewhere} \end{cases}$$

Find: (a) the variance; and (b) the power spectral density of this process. State the relation between bandwidth and decorrelation time of this random process, both being defined by the first zero crossing (or touching) point in their respective domains.

3.18 A stationary random process has a power spectral density given by:

$$G(f) = \begin{cases} 5, & 10/2\pi \leq |f| \leq 20/2\pi \\ 0, & \text{elsewhere} \end{cases}$$

Find: (a) the mean square value; and (b) the autocorrelation function of the process. (If you have access to simple plotting software plot the autocorrelation function.)

3.19 A stationary random process has a bilateral (i.e. double sided) power spectral density given by:

$$G_{xx}(\omega) = \frac{32}{\omega^2 + 16}$$

Find: (a) the average power (on a per-ohm basis) of this random process; and (b) the average power (on a per-ohm basis) of this random process in the range −4 rad/s to 4 rad/s. [4, 2]

3.20 A random variable, $Z(t)$, is defined to be:

$$Z(t) = X(t) + X(t + \tau)$$

$X(t)$ is a stationary process whose autocorrelation function is:

$$R_{xx}(\tau) = e^{-\tau^2}$$

Derive an expression for the autocorrelation of the random process $Z(t)$.

3.21 Show that the autocorrelation function of a non zero mean random process, $X(t)$, may be written as:

$$R_{xx}(\tau) = R_{x'x'}(\tau) + E[(X(t))^2]$$

where $R_{x'x'}(\tau)$ is the autocorrelation function of a zero mean random process and E[.] is the expectation operator as defined in equation (3.16).

3.22 The stationary random process $X(t)$ has a power spectral density $G_{xx}(f)$. What is the power spectral density of $Y(t) = X(t − T)$?

3.23 Two jointly stationary random processes are defined by:

$$X(t) = 5 \cos(10t + \Theta) \quad \text{and} \quad Y(t) = 20 \sin(10t + \Theta)$$

where Θ is a random variable that is uniformly distributed from 0 to 2π. Find the cross correlation function $R_{xy}(\tau)$ of these two processes.

CHAPTER 4

Linear systems

4.1 Introduction

The word system is defined [Hanks] as 'a group or combination of inter-related, inter-dependent, or interacting elements forming a collective entity'. In the context of a digital communications system the interacting elements, such as electronic amplifiers, mixers, detectors, etc., are themselves subsystems made up of components such as resistors, capacitors and transistors. An understanding of how systems behave, and are described, is therefore important to the analysis of electronic communications equipment.

This chapter reviews the properties of the most analytically tractable, but also most important, class of system (i.e. linear systems) and applies concepts developed in Chapters 2 and 3 to them. In particular convolution is used to provide a time domain description of the effect of a system on a signal and the convolution theorem is used to link this to the equivalent description in the frequency domain. Towards the end of the chapter the effect of *memoryless non*-linear systems on the pdf of signals and noise is briefly discussed.

4.2 Linear systems

Linear systems constitute one, restricted, class of system. Electronic communications equipment is predominantly composed of interconnected linear subsystems.

4.2.1 Properties of linear systems

Before becoming involved in the mathematical description of linear systems there are two important questions which should be answered:

1. What is a linear system?
2. Why are linear systems so important?

In answering the first question it is almost as important to say what a linear system is not, as to say what it is. Figure 4.1 shows the input/output characteristic of a system which is specified mathematically by the straight line equation:

$$y(t) = mx(t) + C \tag{4.1}$$

This system, perhaps surprisingly, is non-linear (providing $C \neq 0$). A definition of a linear system can be given as follows:

A system is linear if its response to the sum of any two inputs is the sum of its responses to each of the inputs alone.

This property is usually called the principle of *superposition* since responses to component inputs are superposed at the output. In this context linearity and superposition are synonymous. If $x_i(t)$ are inputs to a system and $y_i(t)$ are the corresponding outputs then superposition can be expressed mathematically as:

$$y(t) = \sum_i y_i(t) \tag{4.2(a)}$$

when:

$$x(t) = \sum_i x_i(t) \tag{4.2(b)}$$

Proportionality (also called homogeneity) is a property which follows directly from linearity. It is defined by:

$$y(t) = my_1(t) \tag{4.3(a)}$$

when:

$$x(t) = mx_1(t) \tag{4.3(b)}$$

The system described by Figure 4.1 and equation (4.1) would have this property if $C = 0$ and in this special case is, therefore, linear. For $C \neq 0$, however, the system does not obey

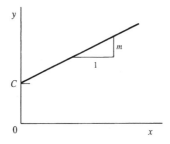

Figure 4.1 *Input/output characteristic of a non-linear system.*

proportionality and therefore cannot be linear. (Equations (4.3) represent a necessary and sufficient condition for linearity providing the system is *memoryless*, i.e. its instantaneous output depends only on its instantaneous input.) A further property which systems often have is *time invariance*. This means that the output of a system does not depend on when the input is applied (except in so far as its location in time). More precisely time invariance can be defined by:

$$y(t) = y_1(t - T) \qquad\qquad (4.4(\text{a}))$$

when:

$$x(t) = x_1(t - T) \qquad\qquad (4.4(\text{b}))$$

The majority of communications subsystems obey both equations (4.2) and (4.4) and are therefore called *time invariant linear systems* (TILS).

Time invariant linear systems can be defined using a single formula which also explicitly recognises proportionality, i.e.:

$$\text{If } y_1(t) = S\{x_1(t)\} \text{ and } y_2(t) = S\{x_2(t)\} \qquad\qquad (4.5(\text{a}))$$

$$\text{then } S\{ax_1(t - T) + bx_2(t - T)\} = ay_1(t - T) + by_2(t - T) \qquad\qquad (4.5(\text{b}))$$

where $S\{\ \}$ represents the functional operation of the system.

4.2.2 Importance of linear systems

The importance of linear systems in engineering cannot be overstated. It is interesting to note, however, that (like periodic signals) linear systems constitute a conceptual ideal that cannot be strictly realised in practice. This is because any device behaves non-linearly if excited by signals of large enough amplitude. An obvious example of this in electronics is the transistor amplifier which saturates when the amplitude of the output approaches the power supply rail voltages (Figure 4.2). Such an amplifier is at least approximately linear, however, over its normal operating range. It is ironic, therefore, that whilst no systems are linear if driven by large enough signals, many non-linear systems are at least approximately linear when driven by small enough signals. This is because the transfer characteristic of a non-linear (memoryless) system can normally be represented by a polynomial of the form:

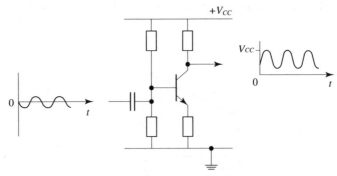

Figure 4.2 *Non-linear behaviour of a simple transistor amplifier.*

$$y(t) = ax(t) + bx^2(t) + cx^3(t) + \cdots \tag{4.5(c)}$$

For small enough input signals (and providing $a \neq 0$) only the first term in equation (4.5(c)) is significant and the system therefore behaves linearly.

An important property of linear systems is that they respond to sinusoidal inputs with sinusoidal outputs of the same frequency (i.e. they conserve the shape of sinusoidal signals).

Other compelling reasons for studying and using linear systems are that:

1. The electric and magnetic properties of free space are linear, i.e.:

$$\mathbf{D} = \varepsilon_o \mathbf{E} \quad (\text{C/m}^2) \tag{4.6(a)}$$

$$\mathbf{B} = \mu_o \mathbf{H} \quad (\text{Wb/m}^2) \tag{4.6(b)}$$

(Since free space is memoryless, proportionality is sufficient to imply linearity.)

2. The electric and magnetic properties of many materials are linear over a large range of field strengths, i.e.:

$$\mathbf{D} = \varepsilon_r \varepsilon_o \mathbf{E} \quad (\text{C/m}^2) \tag{4.7(a)}$$

$$\mathbf{B} = \mu_r \mu_o \mathbf{H} \quad (\text{Wb/m}^2) \tag{4.7(b)}$$

$$\mathbf{J} = \sigma \mathbf{E} \quad (\text{A/m}^2) \tag{4.7(c)}$$

where ε_r, μ_r and σ are constants. (There are notable exceptions to this, of course, e.g. ferromagnetic materials.)

3. Many general mathematical techniques are available for describing, analysing and synthesising linear systems. This is in contrast to non-linear systems for which few, if any, general techniques exist.

EXAMPLE 4.1

Demonstrate the linearity or otherwise of the systems represented by the diagrams in Figure 4.3.

(a) For input $x_1(t)$ output is $y_1(t) = x_1(t) + f(t)$
For input $x_2(t)$ output is $y_2(t) = x_2(t) + f(t)$
For input $x_1(t) + x_2(t)$ output is $x_1(t) + x_2(t) + f(t) \neq y_1(t) + y_2(t)$
i.e. superposition does not hold and system (a) is, therefore, not linear.

(b) For input $x(t) = x_1(t) + x_2(t)$ the output $y(t)$ is:

$$y(t) = f(t)[x_1(t) + x_2(t)]$$

$$= f(t)x_1(t) + f(t)x_2(t) = y_1(t) + y_2(t)$$

which is the superposition of the outputs due to $x_1(t)$ and $x_2(t)$ alone. System (b) is, therefore, linear.

(c) For input $x(t) = x_1(t) + x_2(t)$ output $y(t)$ is:

$$y(t) = \frac{d}{dt}\left[x_1(t) + x_2(t)\right]$$

$$= \frac{d}{dt} x_1(t) + \frac{d}{dt} x_2(t)$$

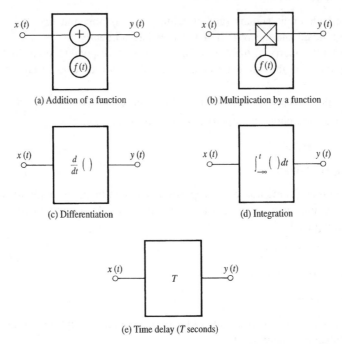

(a) Addition of a function

(b) Multiplication by a function

(c) Differentiation

(d) Integration

(e) Time delay (*T* seconds)

Figure 4.3 *Systems referred to in Example 4.1.*

$$= y_1(t) + y_2(t)$$

i.e. system (c) is linear.

(d) For input $x(t) = x_1(t) + x_2(t)$ output $y(t)$ is:

$$y(t) = \int_{-\infty}^{t} [x_1(t') + x_2(t')]\, dt'$$

$$= \int_{-\infty}^{t} x_1(t')\, dt' + \int_{-\infty}^{t} x_2(t')\, dt' = y_1(t) + y_2(t)$$

i.e. system (d) is linear.

(e) For input $x(t) = x_1(t) + x_2(t)$ output is:

$$y(t) = x_1(t - T) + x_2(t - T)$$

$$= y_1(t) + y_2(t)$$

i.e. system (e) is linear.

There is an apparent paradox involved in the consideration of the linearity of additive and multiplicative systems. This is that the operation of addition is, by definition, linear, i.e. if two inputs are added the output is the sum of each input by itself. The reason system (a) is not linear is that $f(t)$ is considered to be part of the system, not an input. Conversely multiplication is a non-linear operation in the sense that if two inputs are multiplied then the output is not the sum of the outputs due to each input alone (which would both be zero since each input alone would be multiplied by zero). When $f(t)$ in Figure 4.3(b) is considered to be part of the system, however, superposition holds and the system is therefore linear.

Another point to note is that systems must be either non-linear or time varying (or both) in order to generate frequency components at the output which do not appear at the input. Multiplying by $f(t)$ in Figure 4.3(b) will result in new frequencies at the output providing $f(t)$ is not a constant. This is because in this case we have a time varying linear system. If $f(t) = $ constant then system (b) is a time invariant linear system (in fact a linear amplifier) and no new frequencies are generated.

4.3 Time domain description of linear systems

Just as signals can be described in either the time or frequency domain, so too can systems. In this section time domain descriptions are addressed and the close relationship between linear systems and linear equations is demonstrated.

4.3.1 Linear differential equations

Any system which can be described by a linear differential equation, of the form:

$$a_0 y + a_1 \frac{dy}{dt} + a_2 \frac{d^2 y}{dt^2} + \cdots + a_{N-1} \frac{d^{N-1} y}{dt^{N-1}}$$
$$= b_0 x + b_1 \frac{dx}{dt} + b_2 \frac{d^2 x}{dt^2} + \cdots + b_{M-1} \frac{d^{M-1} x}{dt^{M-1}} \tag{4.8}$$

always obeys the principle of superposition and is therefore linear. If the coefficients a_i and b_i are constants then the system is also time invariant. The response of such a system to an input can be defined in terms of two components. One component, the *free response*, is the output, $y_{free}(t)$, when the input (or forcing function) $x(t) = 0$. (Since $x(t)$ is zero for all t then all the derivatives $d^n x(t)/dt^n$ are also zero.) The free response is therefore the solution of the homogeneous equation:

$$a_0 y + a_1 \frac{dy}{dt} + a_2 \frac{d^2 y}{dt^2} + \cdots + a_{N-1} \frac{d^{N-1} y}{dt^{N-1}} = 0 \tag{4.9}$$

subject to the value of the output, and its derivatives, at $t = 0$, i.e.:

$$y(0), \frac{dy}{dt} \Big|_{t=0}, \frac{d^2 y}{dt^2} \Big|_{t=0}, \cdots, \frac{d^{N-1} y}{dt^{N-1}} \Big|_{t=0}$$

These values are called the *initial conditions*. The second component, the *forced response*, is the output, $y_{forced}(t)$, when the input, $x(t)$, is applied but the initial conditions are set to zero, i.e. it is the solution of equation (4.8) when:

$$y(0) = \frac{dy}{dt} \Big|_{t=0} = \frac{d^2 y}{dt^2} \Big|_{t=0} = \cdots = \frac{d^{N-1} y}{dt^{N-1}} \Big|_{t=0} = 0 \tag{4.10}$$

The total response of the system (unsurprisingly, since superposition holds) is the sum of the free and forced responses, i.e.:

$$y(t) = y_{free}(t) + y_{forced}(t) \tag{4.11}$$

An alternative decomposition of the response of a linear system is in terms of its steady state and transient responses. The steady state response is that component of $y(t)$ which does not decay (i.e. tend to zero) as $t \rightarrow \infty$. The transient response is that component of $y(t)$ which does decay as $t \rightarrow \infty$, i.e.:

$$y(t) = y_{steady}(t) + y_{transient}(t) \tag{4.12}$$

4.3.2 Discrete signals and matrix algebra

Consider a linear system with discrete (or sampled) input $x_1, x_2, x_3, \cdots, x_N$ and discrete output $y_1, y_2, y_3, \cdots, y_M$ as shown in Figure 4.4. Each output is then given by a weighted sum of all the inputs [Spiegel]:

$$\begin{bmatrix} y_1 \\ y_2 \\ \cdot\cdot \\ \cdot\cdot \\ y_M \end{bmatrix} = \begin{bmatrix} G_{11} \ G_{12} \ .... \ G_{1N} \\ G_{21} \ G_{22} \ .... \ G_{2N} \\ \cdot\cdot \ \ \cdot\cdot \ .... \ \cdot\cdot \\ \cdot\cdot \ \ \cdot\cdot \ .... \ \cdot\cdot \\ G_{M1} \ \ .. \ .... \ G_{MN} \end{bmatrix} \begin{bmatrix} x_1 \\ x_2 \\ \cdot\cdot \\ \cdot\cdot \\ x_N \end{bmatrix} \tag{4.13}$$

i.e.:

$$y_i = \sum_{j=1}^{N} G_{ij} x_j \tag{4.14}$$

(If the system is a physical system operating in real time then $G_{ij} = 0$ for all values of x_j occurring after y_i.)

4.3.3 Continuous signals, convolution and impulse response

If the discrete input and output of equation (4.14) are replaced with continuous equivalents, i.e.:

$$y_i \rightarrow y(t)$$

$$x_j \rightarrow x(\tau)$$

(the reason for keeping the input and output variables, τ and t, separate will become clear later) then the discrete summation becomes continuous integration giving:

$$y(t) = \int_{0}^{N\Delta\tau} G(t, \tau)x(\tau)\, d\tau \tag{4.15}$$

The limits of integration in equation (4.15) assume that x_1 occurs at $\tau = 0$, and the N

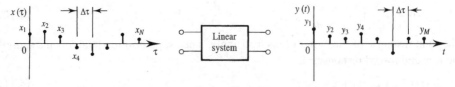

Figure 4.4 *Linear systems with discrete input and output.*

input samples are spaced by $\Delta\tau$ seconds. Once again, for physical systems operating in real time, it is obvious that future values of input do not contribute to current, or past, values of output. The upper limit in the integral of equation (4.15) can therefore be replaced by t without altering its value, i.e.:

$$y(t) = \int_0^t G(t, \tau)x(\tau)\, d\tau \tag{4.16}$$

Furthermore, if input signals are allowed which start at a time arbitrarily distant in the past then:

$$y(t) = \int_{-\infty}^t G(t, \tau)x(\tau)\, d\tau \tag{4.17}$$

Systems described by equations (4.16) and (4.17) are called *causal* since only past and current input values affect (or cause) outputs. Equations (4.15) to (4.17) are all examples of integral transforms (of $x(\tau)$) in which $G(t, \tau)$ is the transform kernel. Replacing the input to the system described by equation (4.17) with a (unit strength) impulse, $\delta(\tau)$, results in:

$$h(t) = \int_{-\infty}^t G(t, \tau)\delta(\tau)\, d\tau \tag{4.18}$$

(The symbol $h(t)$ is traditionally used to represent a system's impulse response.) If the impulse is applied at time $\tau = T$ then, assuming the system is time invariant, the output will be:

$$h(t - T) = \int_{-\infty}^t G(t, \tau)\delta(\tau - T)\, d\tau \tag{4.19}$$

The sampling property of $\delta(\tau - T)$ under integration means that $G(t, T)$ can be interpreted as the response to an impulse applied at time $\tau = T$, i.e.:

$$h(t - T) = G(t, T) \tag{4.20}$$

the surface $G(t, \tau)$ therefore representing the responses for impulses applied at all possible times (Figure 4.5). Replacing T with τ in equation (4.20) (which is a change of notation only) and substituting into equation (4.17) gives:

$$y(t) = \int_{-\infty}^t h(t - \tau)x(\tau)\, d\tau \tag{4.21}$$

If non-causal systems are allowed then equation (4.21) is rewritten as:

$$y(t) = \int_{-\infty}^\infty h(t - \tau)x(\tau)\, d\tau \tag{4.22}$$

Equations (4.21) and (4.22) can be recognised as convolution, or superposition, integrals. The output of a time invariant linear system is therefore given by the convolution of the

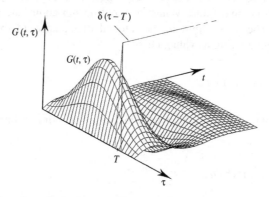

Figure 4.5 *G(t, τ) for a hypothetical system. Response for an impulse applied at time τ = T is the curve formed by the intersection of G(t, τ) with the plane containing δ(τ − T).*

system's input with its impulse response, i.e.:

$$y(t) = h(t) * x(t) \tag{4.23}$$

Note that the commutative property of convolution means that equations (4.22) and (4.23) can also be written as:

$$y(t) = x(t) * h(t)$$

$$= \int_{-\infty}^{\infty} h(\tau)x(t - \tau) \, d\tau \tag{4.24}$$

Note also that equations (4.21) to (4.24) are consistent with the definition of an impulse response since in this case:

$$y(t) = h(t) * \delta(t) = h(t) \tag{4.25}$$

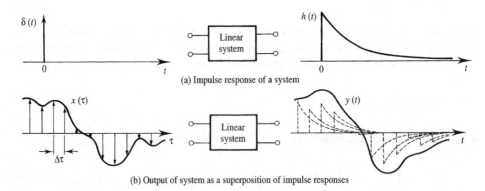

(a) Impulse response of a system

(b) Output of system as a superposition of impulse responses

Figure 4.6 *Decomposition of input into (orthogonal) impulse functions and output formed as a sum of weighted impulse responses.*

4.3.4 Physical interpretation of $y(t) = h(t) * x(t)$

The input signal $x(t)$ can be considered to consist of many closely spaced impulses, each impulse having a strength, or weight, equal to the value of $x(t)$ at the time the impulse occurs times the impulse spacing. The output is then simply the sum (i.e. superposition) of the responses to all the weighted impulses. This idea is illustrated schematically in Figure 4.6. It essentially represents a decomposition of $x(t)$ into a set of (orthogonal) impulse functions. Each impulse function is operated on by the system to give a (weighted, time shifted) impulse response and the entire set of impulse responses is then summed to give the (reconstituted) response of the system to the entire input signal. In this context equation (4.22) can be reinterpreted as:

$$y(t) = \int_{-\infty}^{\infty} h(t - \tau)[x(\tau) \, d\tau] \tag{4.26}$$

where $[x(\tau)d\tau]$ is the weight of the impulse occurring at the input at time τ and $h(t - \tau)$ is the 'fractional' value to which $[x(\tau)d\tau]$ has decayed at the system output by time t (i.e. $t - \tau$ seconds after the impulse occurred at the input). As always, for causal systems, the upper limit in equation (4.26) could be replaced by t corresponding to the condition (see Figure 4.7):

$$h(t - \tau) = 0, \qquad \text{for } t < \tau \tag{4.27}$$

EXAMPLE 4.2

Find the output of a system having a rectangular impulse response (amplitude A volts, width τ seconds) when driven by an identical rectangular input signal.

Figure 4.8 shows the evolution of the output as $h(t - \tau)$ moves through several different values of t. The result is a triangular function. For a discretely sampled input signal using the standard z-transform notation [Mulgrew *et al.*], where $h(n) = x(n) = A + Az^{-1} + Az^{-2} + Az^{-3}$, the output signal is discretely sampled with values $A^2z^{-1} + 2A^2z^{-2} + 3A^2z^{-3} + 4A^2z^{-4} + 3A^2z^{-5}$, etc. Note in this example that, as the impulse response is symmetrical, time reversal of $h(\tau)$ to form $h(-\tau)$ produces a simple shift along the τ-axis.

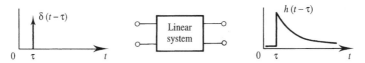

Figure 4.7 *Causal impulse response of a baseband system.*

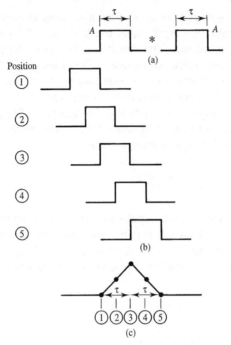

Figure 4.8 *Convolution of two rectangular pulses: (a) pulses; (b) movement of second pulse with respect to first; and (c) values of the convolved output.*

EXAMPLE 4.3

Sketch the convolution of the binary coded waveforms $f_1(t)$ and $f_2(t)$ shown in Figure 4.9(a).

This is obtained by time reversing one waveform, e.g. $f_2(t)$, and then sliding it past $f_1(t)$. As each time unit overlaps then, as we are using rectangular pulses, the convolution result is piecewise linear. The waveforms have been deliberately chosen so that when $f_1(t)$ is time aligned with $f_2(-t)$ then they are identical. The convolution result is shown in Figure 4.9(b). It takes 7 time units to reach the maximum value and another 7 time units to decay again to the final zero value. If $f_2(t)$ is the impulse response of a filter, this represents an example of a matched filter receiver. This type of optimum receiver is discussed in detail later (section 8.3.1).

4.3.5 Step response

Consider the system impulse response shown in Figure 4.10. If the system is driven with a step signal, $u(t)$ (sometimes called the Heaviside step), defined by:

$$u(t) = \begin{cases} 1.0, & t > 0 \\ 0.5, & t = 0 \\ 0, & t < 0 \end{cases} \qquad (4.28)$$

then the output of the system (i.e. its step response) is given by:

$$q(t) = \int_{-\infty}^{\infty} h(\tau)u(t - \tau)\, d\tau \qquad (4.29)$$

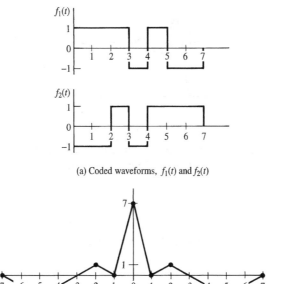

(a) Coded waveforms, $f_1(t)$ and $f_2(t)$

(b) Convolution result, $f_1(t) * f_2(t)$

Figure 4.9 *Convolution of two coded waveforms (Example 4.3).*

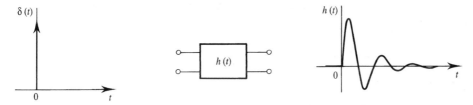

Figure 4.10 *Causal impulse response of a bandpass system.*

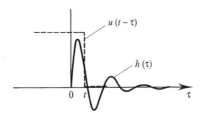

Figure 4.11 *Elements of integrand in equation (4.29).*

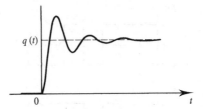

Figure 4.12 *Step response corresponding to impulse response in Figure 4.10.*

A graphical interpretation of the integrand in equation (4.29) is shown in Figure 4.11. Since $u(t - \tau) = 0$ for $\tau > t$ and $h(\tau) = 0$ for $\tau < 0$, equation (4.29) can be rewritten as:

$$q(t) = \int_0^t h(\tau)u(t - \tau)\, d\tau \tag{4.30}$$

Furthermore, in the region $0 < \tau < t$, $u(t - \tau) = 1.0$, i.e.:

$$q(t) = \int_0^t h(\tau)\, d\tau \tag{4.31}$$

The step response is therefore the integral of the impulse response, Figure 4.12. Conversely, of course, the impulse response is the derivative of the step response, i.e.:

$$h(t) = \frac{d}{dt}q(t) \tag{4.32}$$

Equation (4.32) is particularly useful if the step response of a system is more easily measured than its impulse response.

EXAMPLE 4.4

Find and sketch the impulse response of the system which has the step response $\Lambda(t - 1)$.

The step response is:

$$u(t) = \Lambda(t - 1) = \begin{cases} t, i & 0 \le t \le 1 \\ 2 - t, & 1 \le t \le 2 \\ 0, & \text{elsewhere} \end{cases}$$

Therefore the impulse response is:

$$h(t) = \frac{d}{dt}[\Lambda(t - 1)] = \begin{cases} 1, & 0 < t < 1 \\ -1, & 1 < t < 2 \\ 0, & \text{elsewhere} \end{cases}$$

A sketch of $h(t)$ is shown in Figure 4.13.

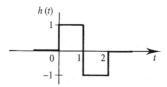

Figure 4.13 *Impulse response of a system with triangular step response (Example 4.4).*

4.4 Frequency domain description

In the time domain the output of a time invariant linear system is the convolution of its input and its impulse response, i.e.:

$$y(t) = h(t) * x(t) \tag{4.33}$$

The equivalent frequency domain expression is found by taking the Fourier transform of both sides of equation (4.33) and using the convolution theorem (see Table 2.5):

$$FT\{y(t)\} = FT\{h(t) * x(t)\}$$

$$= FT\{h(t)\} \, FT\{x(t)\} \tag{4.34}$$

$$\text{i.e. } Y(f) = H(f) \, X(f) \tag{4.35}$$

In equation (4.35), $Y(f)$ is the output voltage spectrum, $X(f)$ is the input voltage spectrum and $H(f)$ is the frequency response of the system. All three quantities are generally complex and can be plotted as either amplitude and phase or real and imaginary components. At a particular frequency, f_o, the frequency response is a single complex number giving the voltage gain (or attenuation) and phase shift of a sinusoid of frequency f_o as it passes from system input to output, i.e.:

$$H(f_o) = A(f_o)e^{j\phi(f_o)} \tag{4.36}$$

For a sinusoidal input, $x(t) = \cos 2\pi f_o t$, the output is therefore given by:

$$y(t) = A(f_o) \cos[2\pi f_o t + \phi(f_o)] \tag{4.37}$$

It follows directly from the Fourier transform relationship between $H(f)$ and $h(t)$ that the frequency responses of systems with real impulse responses have *Hermitian* symmetry, i.e.:

$$\Re\{H(f)\} = \Re\{H(-f)\} \tag{4.38(a)}$$

$$\Im\{H(f)\} = -\Im\{H(-f)\} \tag{4.38(b)}$$

where $\Re/\Im$ indicate real/imaginary parts. Equivalently:

$$|H(f)| = |H(-f)| \tag{4.38(c)}$$

$$\phi(f) = -\phi(-f) \tag{4.38(d)}$$

EXAMPLE 4.5

A linear system with the impulse response shown in Figure 4.14(a) is driven by the input signal shown in Figure 4.14(b). Find (i) the voltage spectral density of the input signal, (ii) the frequency response of the system, (iii) the voltage spectral density of the output signal, (iv) the (time domain) output signal.

(i) The input signal is given by the difference between two rectangular functions:

$$v_{in}(t) = 3.0 \, \Pi\left(\frac{t-1}{2}\right) - 3.0 \, \Pi\left(\frac{t-3}{2}\right)$$

The voltage spectral density of the input is given by the Fourier transform of this:

$$V_{in}(f) = FT\{v_{in}(t)\}$$

$$= 3.0 \, FT\left\{\Pi\left(\frac{t-1}{2}\right)\right\} - 3.0 \, FT\left\{\Pi\left(\frac{t-3}{2}\right)\right\}$$

$$= 3.0 \left[2 \, \text{sinc}(2f) \, e^{-j\omega 1} \right] - 3.0 \left[2 \, \text{sinc}(2f) \, e^{-j\omega 3} \right]$$

$$= 6 \, \text{sinc}(2f) \left[e^{j\omega} - e^{-j\omega} \right] e^{-j\omega 2}$$

$$= 6 \, \text{sinc}(2f) \, 2j \sin \omega \, e^{-j2\omega} = 12j \, \text{sinc}(2f) \sin (2\pi f) \, e^{-j4\pi f}$$

$$= 12 \, \frac{\sin^2 (2\pi f)}{2\pi f} \, e^{-j\left(4\pi f - \frac{\pi}{2}\right)}$$

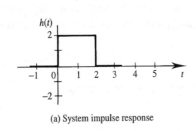

(a) System impulse response

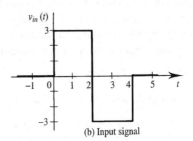

(b) Input signal

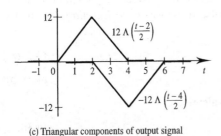

(c) Triangular components of output signal

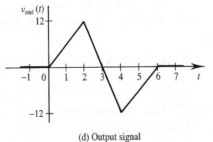

(d) Output signal

Figure 4.14 *Functions for Example 4.5.*

(ii) The frequency response of the system is:

$$H(f) = \text{FT}\{h(t)\}$$

$$= \text{FT}\left\{2.0\,\Pi\left(\frac{t-1}{2}\right)\right\}$$

$$= 2.0\left[2\,\text{sinc}(2f)\,e^{-j\omega 1}\right] = 4\,\text{sinc}(2f)\,e^{-j2\pi f}$$

(iii) The voltage spectral density of the output signal is given by:

$$V_{out}(f) = V_{in}(f)\,H(f)$$

$$= j12\,\frac{\sin^2(2\pi f)}{2\pi f}\,e^{-j4\pi f}\,4\,\frac{\sin(2\pi f)}{2\pi f}\,e^{-j2\pi f}$$

$$= 12\,\frac{\sin^3(2\pi f)}{(\pi f)^2}\,e^{-j\left(6\pi f - \frac{\pi}{2}\right)} = 12\,\frac{\sin^3(2\pi f)}{(\pi f)^2}\,e^{-j\pi(6f - 0.5)}$$

(iv) The time domain output signal could be found as the inverse Fourier transform of $V_{out}(f)$. It is easier in this example, however, to find the output by convolving the input and impulse response, i.e.:

$$v_{out}(t) = v_{in}(t) * h(t) = \int_{-\infty}^{\infty} v_{in}(\tau)h(t-\tau)\,d\tau$$

Furthermore the problem can be simplified if $v_{in}(t)$ is split into its component parts:

$$v_{out}(t) = \left[3\,\Pi\left(\frac{t-1}{2}\right) - 3\,\Pi\left(\frac{t-3}{2}\right)\right] * 2\Pi\left(\frac{t-1}{2}\right)$$

$$= 6\left[\Pi\left(\frac{t-1}{2}\right) * \Pi\left(\frac{t-1}{2}\right)\right] - 6\left[\Pi\left(\frac{t-3}{2}\right) * \Pi\left(\frac{t-1}{2}\right)\right]$$

We know that the result of convolving two rectangular functions of equal width gives a triangular function. Furthermore the peak value of the triangular function is numerically equal to the area under the product of the aligned rectangular functions and occurs at the time shift of the reversed function which gives this alignment. The half width of the triangular function is the same as the width of the rectangular function. Thus:

$$v_{out}(t) = 6\left[2\Lambda\left(\frac{t-2}{2}\right)\right] - 6\left[2\Lambda\left(\frac{t-4}{2}\right)\right]$$

$$= 12\Lambda\left(\frac{t-2}{2}\right) - 12\Lambda\left(\frac{t-4}{2}\right)$$

The two triangular functions making up $v_{out}(t)$ are shown in Figure 4.14(c) and their sum, $v_{out}(t)$, is shown in Figure 4.14(d).

4.5 Causality and the Hilbert transform

All physically realisable systems must be causal, i.e.:

$$h(t) = 0, \qquad \text{for } t < 0 \tag{4.39(a)}$$

This is intuitively obvious since physical systems should not respond to inputs before the inputs have been applied. An equivalent way of expressing equation (4.39(a)) is:

$$h(t) = u(t)h(t) \tag{4.39(b)}$$

where $u(t)$ is the Heaviside step function. The frequency response of a causal system with real impulse response must therefore satisfy:

$$H(f) = \text{FT}\{u(t)\} * \text{FT}\{h(t)\}$$

$$= \left[\frac{1}{2}\,\delta(f) + \frac{1}{j2\pi f} \right] * H(f)$$

$$= \frac{1}{2}H(f) + \left[\frac{1}{j2\pi f} * H(f) \right] \tag{4.40(a)}$$

$$\text{i.e. } H(f) = \frac{1}{j\pi f} * H(f) \tag{4.40(b)}$$

Equation (4.40(b)) is precisely equivalent to equation (4.39(b)).

A necessary and sufficient condition for an amplitude response, $A(f) = |H(f)|$, to be *potentially* causal is:

$$\int_{-\infty}^{\infty} \frac{|\ln A(f)|}{1 + f^2}\, df < \infty \tag{4.41}$$

The expression potentially causal, in this context, means that a system satisfying this criterion will be causal *given that it has a suitable phase response*. Equation (4.41) is called the Paley–Wiener criterion. It has the important implication that a causal system can only have isolated zeros in its amplitude response, i.e. $A(f)$ cannot be zero over a finite band of frequencies.

Returning to the causality condition of equation (4.40(b)), if $H(f)$ is expressed as real and imaginary parts:

$$H_\Re(f) + j\,H_\Im(f) = \frac{1}{j\pi f} * [H_\Re(f) + j\,H_\Im(f)]$$

$$= \left[\frac{1}{j\pi f} * H_\Re(f) \right] + \left[\frac{1}{\pi f} * H_\Im(f) \right] \tag{4.42}$$

and real and imaginary parts are equated, then:

$$H_\Re(f) = \frac{1}{\pi f} * H_\Im(f) \tag{4.43(a)}$$

$$H_\Im(f) = -\frac{1}{\pi f} * H_\Re(f) \tag{4.43(b)}$$

The relationship between real and imaginary parts of $H(f)$ in equation (4.43(a)) is called the *inverse* (frequency domain) *Hilbert* transform which can be written explicitly as:

$$H_\Re(f) = \frac{1}{\pi} \int\limits_{-\infty}^{\infty} \frac{H_\Im(f')}{f - f'} \, df' \qquad (4.44)$$

Equation (4.43(b)) is the *forward* Hilbert transform often denoted by:

$$H_\Im(f) = \hat{H}_\Re(f) \qquad (4.45)$$

In the time domain (since the real part of $H(f)$ transforms to the even part of $h(t)$ and the imaginary part of $H(f)$ transforms to the odd part of $h(t)$) the equivalent operations to equations (4.43) are:

$$h_{even}(t) = j \, \text{sgn}(t) h_{odd}(t) \qquad (4.46)$$

$$h_{odd}(t) = -j \, \text{sgn}(t) h_{even}(t) \qquad (4.47)$$

Notice that, unlike the Fourier transform, the Hilbert transform does *not* change the domain of the function being transformed. It can therefore be applied either in the frequency domain (as in equation (4.43(b))) or in the time domain. Table 4.1 summarises the frequency and time domain Hilbert transform relationships.

Table 4.1 *Summary of frequency and time domain Hilbert transform relationships.*

$-j \, \text{sgn}(t)x(t)$	$\overset{FT}{\rightleftarrows}$	$\hat{X}(f) = \dfrac{-1}{\pi f} * X(f)$
$\downarrow\uparrow$ HT$_f$		HT$_f$ $\uparrow\downarrow$
$x(t)$	$\overset{FT}{\rightleftarrows}$	$X(f)$
$\uparrow\downarrow$ HT$_t$		HT$_t$ $\downarrow\uparrow$
$x(t) = \dfrac{-1}{\pi t} * x(t)$	$\overset{FT}{\rightleftarrows}$	$+j \, \text{sgn}(f)X(f)$

(HT$_t$ *is the time domain Hilbert transform,* HT$_f$ *is the frequency domain Hilbert transform.*)

The time domain Hilbert transform is sometimes called the quadrature filter since it represents an all-pass filter which shifts the phase of positive frequency components by +90° and negative frequency components by −90°. This operation is useful in the representation of bandpass signals and systems as equivalent baseband processes (see section 13.2). It also makes obvious the property that a function and its Hilbert transform are orthogonal.

EXAMPLE 4.6

Establish which of the following systems are causal and which are not: (i) $h(t) = \Lambda(t-3)$; (ii) $h(t) = e^{-(t-10)^2}$; (iii) $h(t) = u(t)e^{-t}$; (iv) $H(f) = e^{-f^2}$; (v) $H(f) = \Pi(f)$; (vi) $H(f) = \Lambda(f-3) + \Lambda(f+3)$; (vii) $H(f) = (1 - jf)/(1 + f^2)$.

(i) $\Lambda(t-3)$ represents a triangular function which is centred on $t = 3$ and which is zero for $t < 2$ and $t > 4$. It is therefore a causal impulse response.

(ii) $e^{-(t-10)^2}$ represents a Gaussian function centred on $t = 10$. Since it only tends to zero as $t \to \pm\infty$ it represents an acausal impulse response.

(iii) $u(t)e^{-t}$ is causal by definition since the Heaviside factor ensures it is zero for $t < 0$.

(iv) e^{-f^2} represents a Gaussian frequency response. Since Gaussian functions in one domain transform to Gaussian functions in the other domain the impulse response of this system is Gaussian. The system is therefore acausal as in (ii).

(v) $\Pi(f)$ is a strictly bandlimited frequency response. The impulse response cannot therefore be time limited and is thus acausal. (The impulse response is, of course, sinc(t).)

(vi) $\Lambda(f-3) + \Lambda(f+3)$ represents a bandpass triangular amplitude response. It is strictly bandlimited and therefore an acausal system as in (v).

(vii) To test whether $H(f)$ is causal we can find out if $H_3(f)$ is the Hilbert transform of $H_\Re(f)$.

$$H_3(f) = \frac{-f}{1+f^2}, \quad H_\Re(f) = \frac{1}{1+f^2}$$

In the absence of Hilbert transform tables:

$$\hat{H}_\Re(f) = -\frac{1}{\pi f} * \frac{1}{1+f^2} = -\int_{-\infty}^{\infty} \frac{1}{\pi\phi} \frac{1}{1+(f-\phi)^2} \, d\phi$$

$$= -\frac{1}{\pi} \int_{-\infty}^{\infty} \frac{1}{\phi\,(\phi^2 - 2f\phi + f^2 + 1)} \, d\phi$$

Using a table of standard integrals (e.g. Dwight, 4th edition, Equation 161.11):

$$\hat{H}_\Re(f) = -\frac{1}{\pi}\left[\frac{1}{2(f^2+1)} \ln\left(\frac{\phi^2}{\phi^2 - 2f\phi + f^2 + 1} \right) + \frac{2f}{2(f^2+1)} \int \frac{1}{\phi^2 - 2f\phi + f^2 + 1} \, d\phi \right]_{-\infty}^{\infty}$$

The logarithmic factor in the first term in square brackets above tends to zero as $\phi \to \pm\infty$. The integral in the second term is also standard (e.g. Dwight, 4th edition, equation (160.01)) giving:

$$\hat{H}_\Re(f) = -\frac{1}{\pi}\left[\frac{f}{(f^2+1)} \frac{2}{\sqrt{4(f^2+1)-4f^2}} \tan^{-1}\left(\frac{(2\phi - 2f)}{\sqrt{4(f^2+1)-4f^2}} \right) \right]_{-\infty}^{\infty}$$

$$= \frac{-1}{\pi} \frac{f}{f^2+1}\left[\tan^{-1}(\phi - f) \right]_{-\infty}^{\infty} = -\frac{f}{f^2+1} = H_3(f)$$

Thus $H_3(f)$ is the Hilbert transform of $H_\Re(f)$ and $H(f)$ therefore represents a causal system.

4.6 Random signals and linear systems

The effect of a linear system on a deterministic signal is specified completely by:

$$y(t) = h(t) * x(t) \tag{4.48}$$

or, alternatively, by:

$$Y(f) = H(f)X(f) \tag{4.49}$$

Random signals cannot, by definition, be specified as deterministic functions either in the time domain or in the frequency domain. It follows that neither equation (4.48) nor equation (4.49) is particularly useful for information bearing signals or noise. In practice, however, the two properties of such signals which must most commonly be specified are their power spectra and their probability density functions. The effects of linear systems on these signal characteristics are now described.

4.6.1 Power spectral densities and linear systems

The most direct way of deriving the relationship between the power spectral density at the input and output of a linear system is to take the square magnitude of equation (4.49), i.e.:

$$|Y(f)|^2 = |H(f)X(f)|^2$$

$$= |H(f)|^2|X(f)|^2 \tag{4.50}$$

Since $|Y(f)|^2$ and $|X(f)|^2$ are power spectral densities equation (4.50) can be rewritten as:

$$G_y(f) = |H(f)|^2 G_x(f) \quad (V^2/Hz) \tag{4.51}$$

If the system input is an energy signal then the power spectral densities, equations (2.52) and (3.49), are replaced by energy spectral densities, equation (2.53):

$$E_y(f) = |H(f)|^2 E_x(f) \quad (V^2 \, s/Hz) \tag{4.52}$$

The equivalent time domain description is obtained by taking the inverse Fourier transform of equation (4.50):

$$\text{FT}^{-1}\{Y(f)\,Y^*(f)\} = \text{FT}^{-1}\{H(f)H^*(f)\} * \text{FT}^{-1}\{X(f)X^*(f)\} \tag{4.53}$$

Using the Wiener–Kintchine theorem (or equivalently the conjugation and time reversal Fourier transform theorems, Table 2.5):

$$R_{yy}(\tau) = R_{hh}(\tau) * R_{xx}(\tau) \tag{4.54}$$

where R is the correlation function and the double subscript emphasises auto or self correlation. It is almost always the frequency domain description which is the most convenient in practice. As an example of the application of equations (4.51) and (4.52) the noise power spectral density and total noise power at the output of an RC filter are now calculated for the case of white input noise.

EXAMPLE 4.7

Find the output power spectral density for a simple *RC* filter when it is driven by white noise. What is the total noise power at the filter's output?

Figure 4.15 shows the problem schematically. The power spectral density at the filter output is:

$$G_o(f) = |H(f)|^2 \, G_i(f)$$

The frequency response of the filter is given by:

$$H(f) = \frac{1}{1 + j(f/f_{3dB})}$$

where f_{3dB} is the filter −3 dB, or cut-off, frequency. Substituting:

$$G_o(f) = \left| \frac{1}{1 + j(f/f_{3dB})} \right|^2 G_i(f)$$

$$= \frac{1}{1 + (f/f_{3dB})^2} G_i(f)$$

Interpreting $G_i(f)$ and $G_o(f)$ as one sided, the total noise power, N, at the filter output is:

$$N = \int_0^\infty G_o(f) \, df$$

$$= \int_0^\infty \frac{1}{1 + (f/f_{3dB})^2} G_i(f) \, df$$

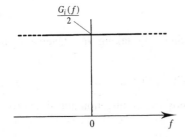
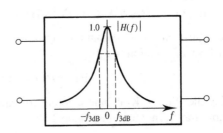

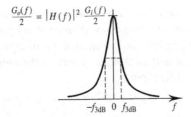

Figure 4.15 *NPSD at output of single pole RC filter driven by white noise. ($G(f)$ is the one-sided power spectral density.)*

Using the change of variable $u = f/f_{3dB}$ and remembering that the input noise is white (i.e. $G_i(f)$ is a constant, Figure 4.15) then:

$$N = G_i \int_0^\infty \frac{1}{1 + u^2} f_{3dB} \, du$$

$$= G_i f_{3dB} \int_0^\infty \frac{1}{1 + u^2} \, du = G_i \, f_{3dB} \, [\tan^{-1} u]_0^\infty$$

$$= G_i \, f_{3dB} \, \pi/2 \quad (\text{V}^2)$$

4.6.2 Noise bandwidth

The noise bandwidth, B_N, of a filter is defined as that width which a rectangular frequency response would need to have to pass the same noise power as the filter, given identical white noise at the input to both. This definition is illustrated in Figure 4.16. It can be expressed mathematically as:

$$B_N = \int_0^\infty \frac{|H(f)|^2}{|H(f_p)|^2} \, df \qquad (4.55)$$

where f_p is the frequency of peak amplitude response. Notice that noise bandwidth is not equal, in general, to the −3 dB bandwidth. (For the single pole low pass filter noise bandwidth is larger than the −3 dB bandwidth by a factor $\pi/2$ (see Example 4.7). In this case, for white noise calculations, the use of the 3 dB bandwidth in place of noise bandwidth would therefore lead to a noise power error of 2 dB.)

4.6.3 Pdf of filtered noise

Not only the power spectral density of a random signal is changed when it is filtered but, in general, so is its pdf. The effect of memoryless systems on the pdf of a signal is discussed in section 4.7. Most communications subsystems, however, have non-zero memory. Unfortunately in this case there is no general, analytical, method of deriving the pdf of the output from the pdf of the input. There is, however, an important exception to this, for which a general result can be derived. This is the pdf of filtered Gaussian noise. Consider Figure 4.17. The output noise, $n_o(t)$, is given by the convolution of the input noise, $n_i(t)$, with the impulse response of the filter or system, i.e.:

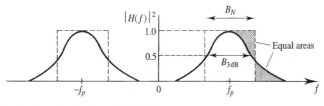

Figure 4.16 *Illustration of noise bandwidth, B_N.*

$$n_o(t) = \int_{-\infty}^{t} h(t - \tau)\, n_i(\tau)\, d\tau \tag{4.56}$$

Notice that the output can be interpreted as a sum of weighted input impulses of strength $n_i(\tau)d\tau$, the weighting factor, $h(t - \tau)$, depending on the time at which the individual input impulses occurred. If $n_i(t)$ is white Gaussian noise then the adjacent impulses are independent Gaussian random variables (since the autocorrelation of $n_i(t)$ is impulsive). The output noise at any instant, e.g. $n_o(t_1)$, is therefore a linear sum of Gaussian random variables and is consequently, itself, a Gaussian random variable. Thus:

Filtered white Gaussian noise is Gaussian.

The above result is easily generalised in the following way.

Consider the frequency response in Figure 4.17 to be split into two parts as shown in Figure 4.18. White Gaussian noise at the input of $H_1(f)$ has been shown to result in (non-white) Gaussian noise at the output of $H_2(f)$. Applying the same reasoning,

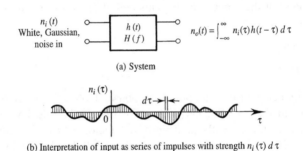

(a) System

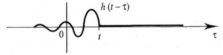

(b) Interpretation of input as series of impulses with strength $n_i(\tau)\,d\tau$

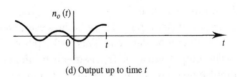

(c) Weighting factor, at time t, for input impulses

(d) Output up to time t

Figure 4.17 *Output as linear sum of many independent, Gaussian, random impulses. ($n_i(\tau)$ is white, Gaussian, noise but it can only be drawn as a bandlimited process.)*

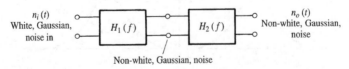

Non-white, Gaussian, noise

Figure 4.18 *Reinterpretation of $H(f)$ in Figure 4.17 as two cascaded sections.*

however, the input to $H_2(f)$ is (non-white) Gaussian noise. It follows, by considering the input and output of $H_2(f)$ that:

Filtered Gaussian noise is Gaussian.

The above result is exact and, to some extent, obvious in that if it were not true then the Gaussian nature of thermal noise, for example, would be obscured by the many filtering processes it is normally subjected to before it is measured.

4.6.4 Spectrum analysers

Spectrum analysers are instruments which are used to characterise signals in the frequency domain. If the signal is periodic the characterisation is partial in the sense that phase information is not usually displayed. If the signal is random, as is the case for noise, the spectral characterisation is essentially complete. Figure 4.19 shows a simplified block diagram of a spectrum analyser and Figure 4.20 shows an alternative conceptual implementation of the same instrument. At a given frequency the display shows either the RMS voltage or mean square voltage which is passed by the filter when it is centred on that frequency. In practice the display y-axis is usually calibrated in dBμ given by $20 \log_{10}(V_{RMS}/10^{-6})$ or dBm given by $10 \log_{10}[(V_{RMS}^2/R_{in})/10^{-3}]$ where dBμ indicates dB with respect to 1 μV and dBm indicates dB with respect to 1 mW. (R_{in}, often 50 Ω, is the input impedance which converts V^2 to W.)

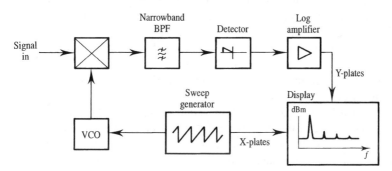

Figure 4.19 *Simplified block diagram of real time, analogue, spectrum analyser.*

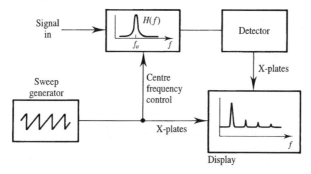

Figure 4.20 *Conceptual model of spectrum analyser.*

If the signal being measured is periodic, and the filter has a bandwidth which adequately resolves the resulting spectral lines, then the (dBm) display is a faithful representation of the signal's discrete power spectrum. If the signal being measured is a stationary random process, however, then its power spectral density is continuous and the resulting display is the actual power spectral density *correlated* with the filter's squared amplitude response, $|H(f)|^2$ (see section 2.6 and equation (2.85)). If the bandwidth of the filter is narrow compared with the frequency scale over which the signal's power spectral density changes significantly, then the smearing of the spectrum in the correlation process is small and the shape of the spectrum is essentially unchanged. In this case the signal's power spectral density in W/Hz can be found by dividing the displayed spectrum (in W) by the noise bandwidth, B_N, of the filter (in Hz). On a dB scale this corresponds to:

$$G(f) \text{ (dB mW/Hz)} = \text{Display (dBm)} - 10 \log_{10} B_N \text{ (dB Hz)} \tag{4.57}$$

4.7 Non-linear systems and transformation of random variables

Non-linear systems are, in general, difficult to analyse. This is principally because superposition no longer applies. As a consequence complicated input signals cannot be decomposed into simple signals (on which the effect of the system is known) and the resulting modified components recombined at the output.

There is one signal characteristic, however, which can often be found at the output of *memoryless* non-linear systems without too much difficulty. This is the signal's pdf. Mathematically this problem is called a transformation of random variables. An outline of this technique is given below.

Consider a pair of bivariate random variables X, Y and S, T which are related in some deterministic way. Every point in the x, y plane can be mapped into the s, t plane as shown in Figure 4.21. Now consider all the points $(x_1, y_1), (x_2, y_2), \cdots$ in the x, y plane which map into the rectangle centred on s_1, t_1. (There may be none, one or more than one such point.) Each one of these points (x_n, y_n) has its own small area dA_n in the x, y plane which maps into the rectangle in (s, t). The probability that X, Y lies in any of the areas (x_n, y_n) is equal to the probability that S, T lies in the rectangle at (s_1, t_1), i.e.:

$$\sum_n p_{X,Y}(x_n, y_n) \, dA_n = p_{S,T}(s_1, t_1) \, ds \, dt \tag{4.58}$$

Equation (4.58) can be interpreted as a conservation of probability law.

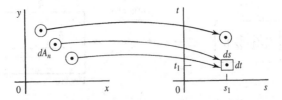

Figure 4.21　*Mapping of random variables using a transformation $(x, y) \rightarrow (s, t)$.*

4.7.1 Rayleigh pdf

When Gaussian noise (Figure 4.22) is present at the input of an envelope (i.e. amplitude) detector the pdf of the noise at the output is Rayleigh distributed (Figure 4.23). The derivation of this distribution, given below, is a transformation of random variables and uses the conservation of probability law given in equation (4.58).

Let X, Y (quadrature noise components) be independent Gaussian random variables with equal standard deviations, σ, and zero means. Equation (3.33) then simplifies to:

$$p_{X,Y}(x, y) = \frac{1}{\sigma\sqrt{2\pi}} e^{\left(\frac{-x^2}{2\sigma^2}\right)} \frac{1}{\sigma\sqrt{2\pi}} e^{\left(\frac{-y^2}{2\sigma^2}\right)} \tag{4.59}$$

Let R, Θ (noise amplitude and phase) be a new pair of random variables related to X, Y:

$$r = \sqrt{x^2 + y^2} \tag{4.60(a)}$$

$$\theta = \tan^{-1}(y/x) \tag{4.60(b)}$$

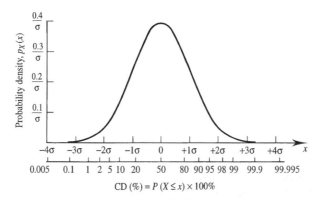

Figure 4.22 *Gaussian pdf, $p_X(x) = [1/(\sigma\sqrt{2\pi})]e^{-(x^2/2\sigma^2)}$ and CD in percent.*

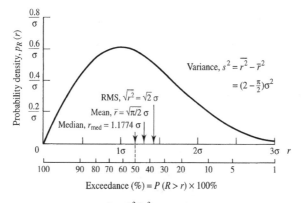

Figure 4.23 *Rayleigh pdf, $p_R(r) = [r/\sigma^2]e^{-(r^2/2\sigma^2)}$, where σ is the standard deviation of either component in the parent bivariate Gaussian pdf.*

$(r, \theta$ can be interpreted as the polar coordinates of the point x, y as shown in Figure 4.24.) The area $d\theta dr$ in the R, Θ plane corresponds to an area $dA = r \, d\theta \, dr$ in the X, Y plane, Figure 4.25. Conservation of probability requires that:

$$p_{R,\Theta}(r, \theta) \, dr \, d\theta \; = \; p_{X,Y}(x, y) \, r \, d\theta dr \qquad (4.61)$$

Therefore:

$$p_{R,\Theta}(r, \theta) \; = \; p_{X,Y}(x, y) \, r$$

$$= \; \frac{r}{\sigma^2 2\pi} \, e^{\left(-\frac{x^2 + y^2}{2\sigma^2}\right)} \; = \; \frac{r}{2\pi\sigma^2} \, e^{\left(\frac{-r^2}{2\sigma^2}\right)} \qquad (4.62)$$

Equation (4.62) gives the joint pdf of R and Θ. The (marginal) pdf of R is now given by:

$$p_R(r) \; = \; \int_0^{2\pi} p_{R,\Theta}(r, \theta) \, d\theta$$

$$= \; \frac{r}{2\pi\sigma^2} \, e^{\left(\frac{-r^2}{2\sigma^2}\right)} \int_0^{2\pi} d\theta \qquad (4.63)$$

i.e. $\; p_R(r) \; = \; \frac{r}{\sigma^2} \, e^{\left(\frac{-r^2}{2\sigma^2}\right)} \qquad (4.64)$

Equation (4.64) is the Rayleigh pdf shown in Figure 4.23. Since $p_{R,\Theta}(r, \theta)$ has no θ dependence the marginal pdf of Θ is uniform, Figure 4.26, i.e.:

$$p_\Theta(\theta) \; = \; \frac{1}{2\pi} \qquad (4.65)$$

(Strictly the right hand side of equations (4.64) and (4.65) should be multiplied by the Heaviside step function, $u(r)$, and the rectangular function, $\Pi(\theta/2\pi)$, respectively since the probability densities are zero outside these ranges.)

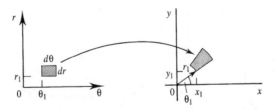

Figure 4.24 *Relationship between (r, θ) and (x, y).*

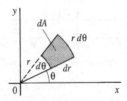

Figure 4.25 *Area in x, y corresponding to rectangle $drd\theta$ in r, θ.*

4.7.2 Chi-square distributions

Another transformation of random variables common in electronic communications systems occurs when Gaussian noise is present at the input to a square law device. Let the random variable X representing noise at the input of a square law detector be Gaussianly distributed, i.e.:

$$p_X(x) = \frac{1}{\sigma\sqrt{2\pi}} e^{\left(\frac{-x^2}{2\sigma^2}\right)} \qquad (4.66)$$

The detector (Figure 4.27) is characterised by:

$$Y = X^2 \qquad (4.67)$$

which means that two probability areas in X (i.e. $p_X(x)dx$ and $p_X(-x)dx$) both transform to the same probability area $p_Y(y)dy$. This is illustrated in Figure 4.28. Conservation of probability therefore requires that:

$$p_X(x)\,dx + p_X(-x)\,dx = p_Y(y)\,dy \qquad (4.68)$$

and by symmetry this means that:

$$2\,p_X(x)\,dx = p_Y(y)\,dy \qquad (4.69)$$

Thus:

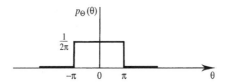

Figure 4.26 *Uniform distribution of* Θ.

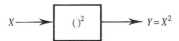

Figure 4.27 *Square law detector.*

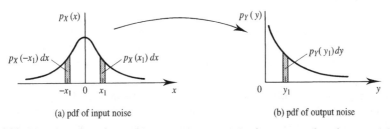

(a) pdf of input noise   (b) pdf of output noise

Figure 4.28 *Mapping of two input points to one output point for a square law detector: (a) pdf of input noise; (b) pdf of output noise.*

$$p_Y(y) \ = \ 2 \, p_X(x) \, \frac{dx}{dy} \tag{4.70}$$

and since $y = x^2$ then:

$$\frac{dy}{dx} \ = \ 2x \tag{4.71(a)}$$

and:

$$\frac{dx}{dy} \ = \ \frac{1}{2x} \tag{4.71(b)}$$

Therefore:

$$p_Y(y) \ = \ 2 \, p_X(x) \, \frac{1}{2x}$$

$$= \ \frac{1}{\sigma\sqrt{2\pi}} \, \frac{1}{x} \, e^{\left(-\frac{x^2}{2\sigma^2}\right)} \tag{4.72}$$

Using $x = \sqrt{y}$ gives:

$$p_Y(y) \ = \ \frac{1}{\sigma\sqrt{2\pi}} \, \frac{1}{\sqrt{y}} \, e^{\left(\frac{-y}{2\sigma^2}\right)}, \quad \text{for } y \geq 0 \tag{4.73}$$

(For $y < 0$, $P_Y(y) = 0$.) Equation (4.73) is, in fact, the special case for $N = 1$ of a more general distribution which results from the transformation $Y = \sum_{i=1}^{N} X_i^2$ where X_i are independent Gaussian random variables with equal variance, σ^2, and zero mean. N, here, is the number of degrees of freedom of the distribution. The mean and variance of this generalised chi-square, χ^2, distribution are, respectively:

$$\bar{Y} \ = \ N\sigma^2 \tag{4.74(a)}$$

and

$$\sigma_Y^2 \ = \ 2N\sigma^4 \tag{4.74(b)}$$

The pdf for a χ^2 distribution with various degrees of freedom and $\sigma^2 = 1$ is shown in Figure 4.29.

4.8 Summary

Linear systems obey the principle of superposition. Many of the subsystems used in the design of digital communications systems are linear over their normal operating ranges. Linear systems are useful and important because they can be described by linear differential equations.

It is a property of a linear system that its time domain output is given by its time domain input convolved with its impulse response. The impulse response of a linear system is the time derivative of its step response. The output (complex) voltage spectrum of a linear system is the voltage spectrum of its input multiplied by its (complex) frequency response. A system's impulse response and frequency response form a Fourier

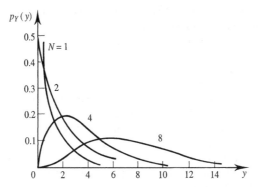

Figure 4.29 *Pdf of a chi-square distribution for several different degress of freedom ($\sigma^2 = 1$).*

transform pair.

All physically realisable systems are causal, i.e. their outputs do not anticipate their inputs. The real and imaginary parts of the frequency response of a causal system form a (frequency domain) Hilbert transform pair. The PSD of a random signal at the output of a linear system is given by the PSD at its input multiplied by its squared amplitude response. The autocorrelation of a random signal at the output of a linear system is the convolution of the input signal's autocorrelation with the autocorrelation of the system's impulse response. The noise bandwidth of a system is equal to the width of an ideal rectangular amplitude response which passes the same noise power as the system, for identical white noise at the inputs.

The pdf at the output of memoryless systems, both linear and non-linear, can be found by the method of transformation of random variables. (Linear memoryless systems have a trivial effect since they represent simple scaling factors.) No general, analytical methods are currently known for predicting the pdf at the output of systems with memory. A useful result for the special case of Gaussian noise, however, is that filtered Gaussian noise is Gaussian.

Spectrum analysers are widely used to display the power spectrum of signals. For periodic signals a properly adjusted spectrum analyser will display the signal's discrete line spectrum, usually on a decibel scale of power. For random signals the spectrum analyser displays a good approximation to the signal's PSD. In this case care is needed in interpreting the absolute magnitude of the spectrum (in W/Hz) if this information is important. (Often only the shape of the spectrum is required, as the total power in the signal is already known.)

4.9 Problems

4.1 Classify the following systems (input $x(t)$, output $y(t)$, impulse response $h(t)$) as: linear or non-linear, time varying or time invariant, causal or non-causal, memoryless or non-zero memory. (N.B. $u(t)$ is the Heaviside step function.)

(a) $y(t) = 3.7x(t)$, (b) $y(t) = 3.7x(t - 6.2)$, (c) $y(t) = 3.7x(t + 10^{-20})$,

(d) $y(t) = 3.7[x(t - 6.2) + 0.01]$, (e) $y(t) = x(t)\cos(2\pi 50t)$, (f) $y(t) = x^{1.1}(t)e^{-t}$,

(g) $y(t) = \cos(2\pi 50t)[x(t) + x(t - 1)]$, (h) $h(t) = u(t)\cos[2\pi 100(t + 4)]e^{-t}$,

(i) $y(t) = x(t)x(t - 2)$, (j) $y(t) = d/dt\ [x(t + 1)]$, (k) $y(t) = x(t) * u(t)e^{-t}$, (l) $y(t) = x(t/3)$,
(m) $y(t) = \int_0^t t'x(t')\ dt'$, (n) $y(t) = 1/(1 + x(t))$, (o) $y(t) = x(t) + y(t - 1)$,
(p) $h(t) = [1 - u(-t)]e^{-t^2}$, (q) $y(t) = \text{sgn}[x(t)]$

4.2   A circuit is described by the linear differential equation:

$$Ry(t) + 2L\frac{dy(t)}{dt} + RLC\frac{d^2y(t)}{dt^2} + L^2C\frac{d^3y(t)}{dt^3} = R\ x(t)$$

where R, L and C are constants, $x(t)$ is the input and $y(t)$ is the output. Find, by taking the Fourier transform of the differential equation, term by term, an expression for the frequency response of the system. What is the amplitude multiplication factor, and phase shift, of a sinusoidal input at the frequency $f = 1/(2\pi\sqrt{LC})$?

4.3   A linear system has the impulse response $h(t) = u(t) - u(t - 2)$. Sketch $h(t)$ and find the system output when its input is $x(t) = \frac{1}{2}\Pi((t - 1)/2) - \frac{1}{2}\Pi((t - 3)/2)$. What is the system's step response and what is its frequency response?

4.4   How might a system with the impulse response given in Problem 4.3 be implemented using integrators, delay lines, invertors (i.e. amplifiers with voltage gain $G_V = -1.0$) and adders?

4.5   The impulse response of a system is given by $h(t) = \Pi((t - 2)/2)$ and the system's input signal is given by $x(t) = (2/3)\ t\ \Pi((t - 1.5)/3)$. Find, and sketch, the system's output.

4.6   The impulse response of a time invariant linear system is $h(t) = u(t)/(1 + t^2)$ where $u(t)$ is the Heaviside step function. Find, and sketch, the response of this system to a rectangular pulse of unit height and width.

4.7   The amplitude response of a rectangular low pass filter is given by $|H(f)| = \Pi(f/(2f_\chi))$. Find, and sketch, the impulse response of this filter if its phase response is: (a) $\phi(f) = 0$; and (b) $\phi(f) = \text{sgn}(f)\pi/2$.
       (N.B. Problems 4.8 and 4.9 presuppose some knowledge of elementary circuit theory.)

4.8   Find the frequency response and impulse response of: (a) an LR; (b) an RL; and (c) a CR filter. (The input is across both components in series and the output is across the second, earthed, component alone. L, R and C denote inductors, resistors and capacitors respectively.)

4.9   An electrical system consists of an RC potential divider (input across series combination of RC, output across C alone) followed by an ideal differentiator described by $y(t) = d/dt\ [x(t)]$. For an impulse applied at the input to the potential divider find and sketch: (i) the response at the potential divider output; and (ii) the response at the differentiator output. Find the frequency response of the entire system and use convolution to calculate the system output when the system input is $x(t) = \Pi((t - T/2)/T)$. Sketch the output if the input pulse width equals the time constant of the potential divider, i.e. $T = RC$.
       (N.B. Problem 4.10 presupposes some knowledge of elementary circuit theory and electronics.)

4.10   An ideal operational amplifier is driven at its non-inverting input by a signal via an R_1C potential divider. It is driven at its inverting input by the same signal via a series resistor R_2. Negative feedback is applied using a resistor R_3 connected across the operational amplifier's output and inverting input. Find the impulse response and frequency response of this electronic circuit which is commonly used in signal processing.

4.11   A system has an impulse response $h(t) = u(t)\ e^{-t/\tau}$ and an applied input signal $x(t) = \Pi((t - \tau/2)/\tau)t$. Find the system's output signal.

4.12   A raised cosine filter has the amplitude response:

$$|H(f)| = \begin{cases} \frac{1}{2}[1 + \cos(\pi f/2f_\chi)], & |f| \leq 2f_\chi \\ 0, & |f| > 2f_\chi \end{cases}$$

Explain (in a few words) why (strictly) this filter is not physically realisable.

4.13 An electrical system consists of 10 cascaded *RC* filters. (Each filter is a potential divider with input across the series combination of *R* and *C* and output across *C* alone.) If operational amplifier impedance buffers are inserted between all *RC* filters deduce (without elaborate calculations) the approximate *shape* of the system's amplitude response. (The impedance buffers merely reduce the loading effect of each *RC* stage on the preceding stage to a negligible level.)

4.14 What is the −3 dB bandwidth of the system with a one-sided exponential impulse response $h(t) = u(t)e^{-5t}$? If white Gaussian noise with one-sided NPSD of 2.0×10^{-9} V^2/Hz is applied to the input of this system what is the PSD of the noise at the system output? What is the total noise power at the system output? What is the output noise power within the system's −3 dB bandwidth? What is the pdf of the noise at the system output?

4.15 If the noise at the output of the system described in Problem 4.14 is applied (after impedance buffering to avoid loading effects) to a second, identical, system, what will be the total noise power at the (second) system output? What proportion of this total noise power resides in the frequency band below 1.0 Hz? [6.4×10^{-12} V^2, 20%]

4.16 A mobile communications system, consisting of a transmitting mobile and receiving fixed base station, experiences noise at the receiving antenna. Assuming that the noise is spectrally white, and has variance σ^2, calculate the coherence (i.e. autocorrelation) function for the output of a low pass *RC* filter attached to the antenna output when the spectral density of the transmitted signal is given by:

$$S_{xx}(\omega) = \begin{cases} \frac{1}{2}[1 + \cos(\pi\omega/10)], & |\omega| < 10 \\ 0, & \text{otherwise} \end{cases}$$

and the square magnitude of the channel frequency response is given by:

$$|H(\omega)|^2 = \frac{A^2}{B^2 + \omega^4}$$

What information does this give you about the system? How might it be measured in practice?

4.17 Find the equivalent noise bandwidth of the finite-time integrator whose impulse response is given by:

$$h(t) = \frac{1}{T}[u(t) - u(t - T)] \qquad \left(\text{Hint:} \int_0^\infty \frac{\sin^2(ax)}{x^2} \, dx = |a| \frac{\pi}{2} \right)$$

4.18 A linear system has the following impulse response: $h(t) = e^{-5t}$, when $t \geq 0$ and $h(t) = 0$ at other times. The input signal to the above system is a sample function from a random process which has the form:

$$X(t) = M, \qquad -\infty < t < \infty$$

in which *M* is a random variable that is uniformly distributed from −6 to +18. Find: (a) an expression for the output sample function; (b) the mean value of the output; and (c) the variance of the output. [$M/5$, 0.2, 1.92]

4.19 Find the cross correlation function $R_{xy}(\tau)$ for a single stage low pass *RC* filter when the input $x(t)$ has the following autocorrelation function:

$$R_{xx}(\tau) = \frac{\beta N_0}{2} e^{-\beta|\tau|}, \qquad -\infty < \tau < \infty$$

[0.80 Hz, 1.0×10^{-10} V^2, 5.0×10^{-11} V^2]

4.20 A random variable x has a pdf $p_X(x) = u(x)5e^{-5x}$ and a statistically independent random variable y has a pdf $p_Y(y) = 2u(y)e^{-2y}$. For the random variable $Z = X + Y$ find: (a) $p_Z(0)$; (b) the modal value of z (i.e. that for which $p_Z(z)$ is a maximum); and (c) the probability that $z > 1.0$. [$3.33\ e^{-2z})\ (1 - e^{-3z})$, 0, 0.305, 0.22]

4.21 Find the pdf of noise at the output of a full wave rectifier if Gaussian noise with a variance of $1\ V^2$ is present at its input. (N.B. the output $y(t)$ of a full wave rectifier is related to its input $x(t)$ by $y(t) = |x(t)|$.)

4.22 A signal with uniform pdf $p_X(x) = 0.5\ \Pi((x - 1)/2)$ is processed by a non-linear memoryless system with input/output characteristic $y(t) = 5x(t) + 2$. What is the pdf of the output signal?

4.23 A signal with pdf $p_X(x) = 1/[\pi(1 + x^2)]$ is processed by a square law detector (characteristic $Y = X^2$). What is the pdf of the processed signal?

Part Two

Digital communications principles

Part Two, by far the largest part of the book, uses the theoretical concepts of Part One to describe and analyse communications links which are robust in the presence of noise and other impairment mechanisms.

Chapter 5 starts with a discussion of sampling and aliasing and demonstrates the practical problems associated with representing an analogue signal in digital (pulse code modulated) form. This highlights the care that must be taken to achieve accurate reconstruction, without distortion, of the original analogue signal in a receiver. Chapter 5 also describes a variety of techniques by which the bandwidth of a PCM signal may be reduced in order to allow the effective use of bandlimited channels. Chapter 6 addresses the fundamentals of binary baseband transmission, covering important aspects of practical decision theory, and describes how the spectral properties of baseband digital signals can be altered by the use of different line coding schemes. Receiver equalisation, employed to overcome transmission channel distortion, is also discussed in the context of examples including digital subscriber line local loop transmission. Probability theory is applied to receiver detection and decision processes in Chapter 7 along with a discussion of the Bayes and Neyman–Pearson decision criteria. Chapter 8 investigates optimum pulse shaping at the transmitter, and optimum filtering at the receiver, designed to minimise transmitted signal bandwidth whilst maximising the probability of correct symbol decisions.

Chapter 9 presents the fundamentals of information theory and source coding, introducing the important concepts of entropy and coding efficiency. It is shown how redundancy present in source data may be minimised, using variable length coding schemes to achieve efficient, low bit rate, digital speech, and other signals. This also includes a basic discussion on cryptographic security techniques. Chapter 10 describes the converse technique, in which transmitted data redundancy is increased to achieve error correction, or detection, in the presence of noise and includes turbo coding.

Chapter 11 analyses the bandpass binary modulation schemes which employ amplitude, frequency or phase shift keying. Variants, and hybrid combinations, of these schemes are then examined which are especially spectrally efficient (e.g. QAM), are especially power efficient (e.g. MFSK), or have some other desirable property. Following a detailed discussion of the sources of noise in electronic circuits, Chapter 12 outlines the calculation of received signal power, noise power and associated signal-to-noise ratio, for simple communications links. Finally Chapter 13 indicates how the performance of a complex communication system can be predicted by simulation before any hardware prototyping is attempted.

Sampling, multiplexing and PCM

5.1 Introduction

Chapter 1 provided an overview of a digital communications system and Chapters 2, 3 and 4 were reviews of some important concepts in the theories of signals, systems and noise. Chapter 5 is the starting point for a more detailed examination of the major functional blocks which make up a complete digital communications system. The discussion of these blocks will primarily be at systems level (i.e. it will concentrate on the inputs, outputs and functional relationships between them) rather than at the implementation level (the design of the electronic circuits which realise these relationships). Referring to Figure 1.3, the principal transmitter subsystems with which this chapter is concerned are the anti-aliasing filter, the sampling circuit, the quantiser, the PCM encoder and the baseband channel multiplexer. In the receiver we are concerned with the demultiplexer, PCM decoder and reconstruction filter. First, however, we give a brief review of pulse modulation techniques. This is mainly because pulse amplitude modulation (in particular) can be identified with the sampling operation preceding quantisation in Figure 1.3 and, as such, constitutes an important part of a digital communications transmitter. In addition, however, pulse modulations (generally) can be used as modulation schemes in their own right for analogue communications systems.

5.2 Pulse modulation

Pulse modulation describes the process whereby the amplitude, width or position of individual pulses in a periodic pulse train are varied (i.e. modulated) in sympathy with the amplitude of a baseband information signal, $g(t)$, Figures 5.1(a) to (d) (adapted from

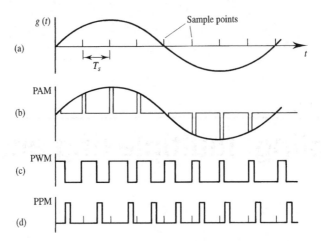

Figure 5.1 *Illustration of pulse amplitude, width and position modulation: (a) input signal.*

[Stremler]). Pulse modulation may be an end in itself allowing, for example, many separate information carrying signals to share a single physical channel by interleaving the individual signal pulses as illustrated in Figure 5.2 (adapted from [Stremler]). Such pulse interleaving is called time division multiplexing (TDM) and is discussed in detail in section 5.4. Pulse modulation also, however, represents an intermediate stage in the generation of digitally modulated signals. (It is important to realise that pulse modulation is not, in itself, a digital but an analogue technique.) The minimum pulse rate representing each information signal must be twice the highest frequency present in the signal's spectrum. This condition, called the Nyquist sampling criterion, must be satisfied if proper reconstruction of the original continuous signal from the pulses is to be possible. Sampling criteria, and the distortion (aliasing) which is introduced when they are not satisfied, are discussed further in section 5.3.

Since pulse amplitude modulation (PAM) relies on changes in pulse amplitude it requires larger signal-to-noise ratio (SNR) than pulse position modulation (PPM) or pulse width modulation (PWM) [Lathi]. This is essentially because a given amount of additive noise can change the amplitude of a pulse (with rapid rise and fall times) by a greater fraction than the position of its edges (Figure 5.3). PWM is particularly attractive in analogue remote control applications because the reconstructed control signal can easily be obtained by integrating (or averaging) the transmitted PWM signal.

All pulse modulated signals have wider bandwidth than the original information signal since their spectrum is determined solely by the pulse shape and duration. The bandwidth and filtering requirements for pulsed signals are discussed in Chapter 9. If the pulses are short compared with the reciprocal of the information signal bandwidth (or equivalently short with respect to the decorrelation time of the information signal) then the original continuous signal can be reconstructed by low pass filtering. If the pulse duration is not sufficiently short then equalisation may be necessary after low pass filtering. The need for, and effect of, equalisation are discussed in the context of sampling in section 5.3.1. Equalisation implementations are described in section 8.5.

5.3 Sampling

The process of selecting or recording the ordinate values of a continuous (usually analogue) function at specific (usually equally spaced) values of its abscissa is called sampling. If the function is a signal which varies with time then the samples are sometimes called a time series. This is the most common type of sampling process encountered in electronic communications although spatial sampling of images is also important.

There are obvious similarities between sampling and PAM. In fact, in many cases, the two processes are indistinguishable, for instance if the pulse duration of the PAM signal is very short. There are, however, two processes both commonly referred to as

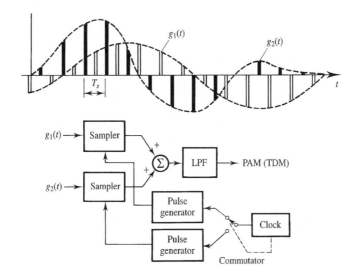

Figure 5.2 *Time division multiplexing of two pulse amplitude modulated signals.*

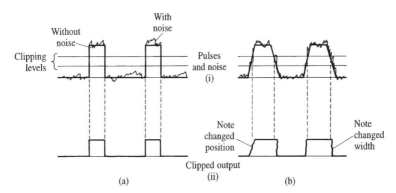

Figure 5.3 *Effects of noise on pulses: (a) noise induced position and width errors completely absent for ideal pulse; (b) small, noise induced, position and width errors for realistic pulse.*

sampling, which should be distinguished. These are flat topped sampling (which is identical to PAM) and natural sampling.

The fundamental property of sampling is that, for a sampling frequency f_s, a constant voltage DC input signal and a periodic input signal with a fundamental frequency at integer multiples of f_s both give (to within a multiplicative constant) the same sampled output values. A consequence of this, in the frequency domain, is that the sampled baseband spectrum repeats at f_s and multiples of this sampling frequency, Figure 5.4.

5.3.1 Natural and flat topped sampling

A naturally sampled signal is produced by multiplying the baseband information signal, $g(t)$, by the periodic pulse train, shown previously in Figure 2.36. This is illustrated in Figures 5.4(a), (c) and (e). The important point to note in this case is that the pulse tops follow the variations of the signal being sampled. The spectrum of the information signal is shown in Figure 5.4(b). (The spectrum is, of course, the Fourier transform of $g(t)$ and would normally be complex but is represented here only by its amplitude.) The spectrum of the periodic pulse train (Figure 5.4(d)) is discrete and consists of a series of equally spaced, weighted, Dirac delta or impulse functions. The spacing between impulses is $1/T_s$ where T_s is the pulse train period and the envelope of weighted impulses is the Fourier transform of a single time domain pulse. (In this case the spectrum has a $\mathrm{sinc}(\tau f)$ shape since the time domain pulses are rectangles of width τ seconds.) Since multiplication in the time domain corresponds to convolution in the frequency domain the spectrum of the sampled signal is found by convolving Figure 5.4(b) with Figure 5.4(d).

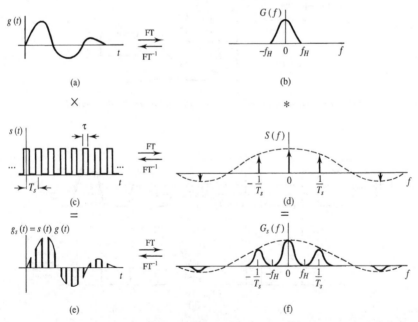

Figure 5.4 *Time and frequency domain illustrations of natural sampling: (a) signal $g(t)$; (b) signal spectrum; (c) sampling function; (d) spectrum of sampling function; (e) sampled signal; (f) spectrum of sampled signal.*

It is a property of the impulse that, under convolution (section 2.3.4) with a second function, it replicates the second function about the position of the impulse. Each impulse in Figure 5.4 therefore replicates the spectrum of $g(t)$ at a frequency corresponding to its own position. The replicas have the same amplitude weightings as the impulses producing them. The sampled signal spectrum is shown in Figure 5.4(f). It is clear that appropriate low pass filtering will pass only the baseband spectral version of $g(t)$. It follows that $g(t)$ will appear undistorted (and, in the absence of noise, without error) at the output of the low pass filter. For obvious reasons such a filter is sometimes called a *reconstruction* filter.

If the pulses produced by the process described above are artificially flattened we have a true PAM signal or flat topped sampling. This can be modelled by assuming that natural sampling proceeds using an impulse train, Figures 5.5(a), (c), (e) (this is sometimes called impulse, or ideal, sampling), and the resulting time series of weighted impulses is convolved with a rectangular pulse, Figures 5.5(e), (g), (i). The resulting spectrum is that of the ideally sampled information signal (Figure 5.5(f)) multiplied with the sinc(τf) spectrum of a single rectangular pulse, Figures 5.5(h), (j). A baseband spectral version of $g(t)$ can be recovered by low pass filtering but this must then be multiplied by a function which is the inverse of the pulse spectrum ($1/\text{sinc}(\tau f)$, Figure 5.5(k)) if $g(t)$ is to be restored exactly. This process, which is not required for reconstruction of naturally sampled signals, is called equalisation. For proper equalisation the inverse frequency response in Figure 5.5(k) need only exist over the signal bandwidth, $0 \rightarrow f_H$.

5.3.2 Baseband sampling and Nyquist's criterion

The spectral replicas of the information signal in Figures 5.4(f) and 5.5(j) are spaced by $f_s = 1/T_s$ Hz. The baseband spectrum can therefore be recovered by simple low pass filtering provided that the width of the spectral replicas is less than the spacing, as defined in equation (2.31), i.e.:

$$f_s \geq 2f_H \quad \text{(Hz)} \tag{5.1}$$

where f_H is the highest frequency component in the information signal. If, however, the spacing between spectral replicas is less than their width then they will overlap and reconstruction of $g(t)$ using low pass filtering will no longer be possible. Equation (5.1) is a succinct statement of Nyquist's sampling criterion. In words this criterion could be stated as follows:

A signal having no significant spectral components above a frequency f_H Hz is specified completely by its values at uniform spacings, no more than $1/(2f_H)$ s apart.

Whilst this sampling criterion is valid for any signal it is usually only used in the context of baseband signals. Another less stringent (but more complicated) criterion which can be used for bandpass signals is discussed in section 5.3.5. In the context of sampling a strict distinction between baseband and bandpass signals can be made as follows:

For baseband signals, $B \geq f_L$ (Hz) $\tag{5.2(a)}$

For bandpass signals, $B < f_L$ (Hz) $\tag{5.2(b)}$

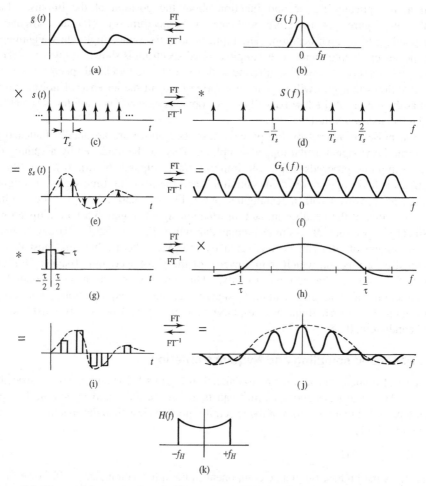

Figure 5.5 *Time and frequency domain illustrations of PAM or flat topped sampling: (a) signal; (b) signal spectrum; (c) sampling function; (d) spectrum of (c); (e) sampled signal; (f) spectrum of (e); (g) finite width sample; (h) spectrum of (g); (i) sampled signal; (j) spectrum of (i); (k) receiver equalising filter to recover g(t).*

B in equations (5.2) is the signal's (absolute) bandwidth and f_L is the signal's lowest frequency component. These definitions are illustrated in Figure 5.6.

5.3.3 Aliasing

Figure 5.7 shows the spectrum of an undersampled baseband signal ($f_s < 2f_H$). The baseband spectrum of $g(t)$ clearly cannot be recovered exactly, even with an ideal rectangular low pass filter. The best achievable, in terms of separating the baseband spectrum from the adjacent replicas, would be to use a rectangular low pass filter with a cut-off frequency of $f_s/2$. The filtered signal will then be, approximately, that of the original signal $g(t)$ but with the frequencies above $f_s/2$ folded back so that they actually

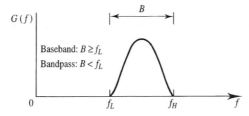

Figure 5.6 *Definitions of baseband and bandpass signals.*

appear below $f_s/2$. (The approximation becomes better as the width of the sampling pulses gets smaller. In the limit of ideal (impulse) sampling the approximation becomes exact.) The spectral components originally representing high frequencies now appear under the alias of lower frequencies. Thus, as in Figure 2.23(c), the sampling represents a high frequency component by a lower frequency sinusoid.

To avoid aliasing a low pass anti-aliasing filter with a cut-off frequency of $f_s/2$ is often placed immediately before the sampling circuit. Whilst this filter may remove high frequency energy from the information signal the resulting distortion is generally less than that introduced if the same energy is aliased to incorrect frequencies by the sampling process.

5.3.4 Practical sampling, reconstruction and signal to distortion ratio

Two practical points need to be appreciated when choosing the sampling rate in real systems. The first is that f_H must usually be interpreted as the highest frequency component *with significant spectral amplitude*. This is because practical signals start and stop in time and therefore, in principle, have spectra which are not bandlimited in an absolute sense. (For the case of voice signals, Figure 5.8, f_H, in Europe, is usually assumed to be 3.4 kHz.) The second is that whilst it is theoretically possible to reconstruct the original continuous signal from its samples if the sampling rate is exactly twice the highest frequency in its spectrum, in practice it is necessary to sample at a slightly faster rate. This is because ideal, rectangular, anti-aliasing and reconstruction filters (with infinitely steep skirts) are not physically realisable. A practical version of the baseband sampling criterion might therefore be expressed as:

$$f_s \geq 2.2f_H \quad \text{(Hz)} \tag{5.3}$$

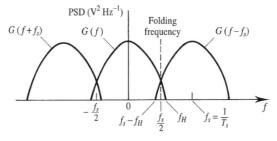

Figure 5.7 *Spectrum of undersampled information signal showing interference between spectral replicas and the folding frequency.*

to allow for the transition, or roll-off, into the filter stopband.

A quantitative measure of the distortion introduced by aliasing can be defined as the ratio of unaliased to aliased power in the reconstructed signal. If the reconstruction filter is ideal with rectangular amplitude response then, in the absence of an anti-aliasing filter, the signal to distortion ratio (SDR) is given by:

$$\text{SDR} = \frac{\displaystyle\int_{0}^{f_s/2} G(f)\,df}{\displaystyle\int_{f_s/2}^{\infty} G(f)\,df} \tag{5.4}$$

where $G(f)$ is the (two-sided) power spectral density of the (real) baseband information signal $g(t)$. More generally, if a filter with a frequency response $H(f)$ is used for reconstruction then the integral limits are extended (see Figure 5.7) to give:

$$\text{SDR} \simeq \frac{\displaystyle\int_{0}^{\infty} G(f)\,|H(f)|^2\,df}{\displaystyle\int_{0}^{\infty} G(f-f_s)\,|H(f)|^2\,df} \tag{5.5}$$

An approximation sign is used in equation (5.5) for two reasons. Firstly, and most importantly, spectral replicas centred on $2f_s$ Hz and above are assumed to be totally suppressed by $|H(f)|$ in equation (5.5). Secondly, ideal (impulsive) sampling is assumed such that there is no $\text{sinc}(\tau f)$ roll-off in the spectrum of the sampled signal as discussed in section 5.3.1. (In the absence of $1/\text{sinc}(\tau f)$ equalisation then this additional spectral roll-off should be cascaded with $H(f)$ in equation (5.5) if flat topped sampling is

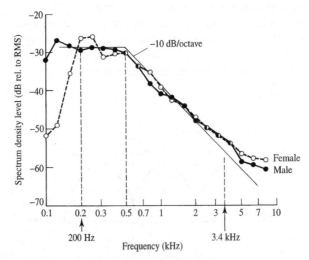

Figure 5.8 *Long term averaged speech spectra for male and female speakers (source: reprinted from Furui, 1989, by courtesy of Marcel Dekker, Inc.).*

employed.) It should also be appreciated that SDR in equations (5.4) and (5.5) is specifically defined to represent distortion due to aliasing of frequency components. Thus, for example, phase distortion due to the action of a reconstruction filter (with non-linear phase) on the baseband signal's voltage spectrum is not included.

EXAMPLE 5.1

Consider a signal with the power spectral density shown in Figure 5.9(a). Find the alias induced SDR if the signal is sampled at 90% of its Nyquist rate and the reconstruction filter has: (i) an ideal rectangular amplitude response; and (ii) an *RC* low pass filter response with a 3 dB bandwidth of $f_s/2$.

The sampling rate is given by:

$$f_s = 0.9 \times 2 \, f_H$$

$$= 0.9 \times 2 \times 100 = 180 \ \text{kHz}$$

The folding frequency (Figure 5.9(b)) is thus:

$$\frac{f_s}{2} = 90 \ \text{kHz}$$

(i) For an ideal rectangular reconstruction filter, using equation (5.4):

$$\text{SDR} = \frac{\displaystyle\int_0^{f_s/2} G(f) \, df}{\displaystyle\int_{f_s/2}^{\infty} G(f) \, df} = \frac{\displaystyle\int_0^{f_s/2} (1 - 10^{-5}f) \, df}{\displaystyle\int_{f_s/2}^{10^5} (1 - 10^{-5}f) \, df}$$

For such a simple $G(f)$, however, we can evaluate the integrals from the area under the triangles in Figure (5.9(b)), i.e.:

$$\text{SDR} = \frac{\frac{1}{2} \, (100 \times 1) - \frac{1}{2} \, (10 \times 0.1)}{\frac{1}{2} \, (10 \times 0.1)}$$

$$= 99 = 20.0 \ \text{dB}$$

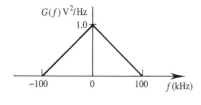

(a) Input signal spectrum

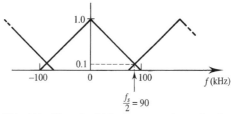

(b) As (a) but illustrating folding frequency and spectral replicas

Figure 5.9 *Signal power spectral density for Example 5.1.*

(ii) For the *RC* filter:

$$H(f) = \frac{1}{1 + j\,2\pi\,RC\,f}$$

$$|H(f)|^2 = \frac{1}{1 + (2\pi\,RC)^2\,f^2}$$

$$f_{3dB} = \frac{f_s}{2} = 90\text{ kHz}$$

$$RC = \frac{1}{2\pi\,f_{3dB}} = \frac{1}{2\pi\,9\times 10^4} = 1.768\times 10^{-6}\text{ s}$$

Reducing the upper limit in the numerator of equation (5.5) to f_H Hz, as this is the maximum signal frequency in the baseband spectrum, and replacing the limits in the denominator with $f_s - f_H$ and $f_s + f_H$ since these are lower and upper frequency limits in the first spectral replica:

$$SDR = \frac{\displaystyle\int_0^{f_H} (1 - 10^{-5}f)\,[1 + (2\pi\,1.768\times 10^{-6})^2\,f^2]^{-1}\,df}{\displaystyle\int_{f_s - f_H}^{f_s + f_H} [1 - 10^{-5}\,|f - f_s|]\,[1 + (2\pi\,1.768\times 10^{-6})^2\,f^2]^{-1}\,df}$$

Evaluating the numerator:

$$Num = \int_0^{10^5} \frac{1 - 10^{-5}f}{1 + a^2 f^2}\,df$$

(where $a = 2\pi\,1.768\times 10^{-6} = 1.111\times 10^{-5}$)

$$= \int_0^{10^5} \frac{1}{1 + a^2 f^2}\,df - 10^{-5}\int_0^{10^5} \frac{f}{1 + a^2 f^2}\,df$$

Put $x = af$, $\dfrac{dx}{df} = a$; when $f = 0$, $x = 0$ and when $f = 10^5$, $x = 1.111$:

$$Num = \int_0^{1.111} \frac{1}{1 + x^2}\frac{dx}{a} - 10^{-5}\int_0^{1.111} \frac{1}{a}\frac{x}{1 + x^2}\frac{dx}{a}$$

$$= \frac{1}{a}\Big[\tan^{-1} x\Big]_0^{1.111} - \frac{10^{-5}}{a^2}\Big[\frac{1}{2}\ln(1 + x^2)\Big]_0^{1.111}$$

$$= 7.542\times 10^4 - 3.256\times 10^4 = 4.286\times 10^4$$

Evaluating the denominator:

$$Denom = \int_{f_s - f_H}^{f_s} \frac{1 - 10^{-5}f_s + 10^{-5}f}{1 + a^2 f^2}\,df + \int_{f_s}^{f_s + f_H} \frac{1 + 10^{-5}f_s - 10^{-5}f}{1 + a^2 f^2}\,df$$

$$= (1 - 10^{-5}\times 180\times 10^3)\int_{80\times 10^3}^{180\times 10^3} \frac{1}{1 + a^2 f^2}\,df + 10^{-5}\int_{80\times 10^3}^{180\times 10^3} \frac{f}{1 + a^2 f^2}\,df$$

$$+ (1 + 10^{-5} \times 180 \times 10^3) \int_{180 \times 10^3}^{280 \times 10^3} \frac{1}{1 + a^2 f^2} \, df - 10^{-5} \int_{180 \times 10^3}^{280 \times 10^3} \frac{f}{1 + a^2 f^2} \, df$$

Using $x = af$, when $f = 80 \times 10^3$, $x = 0.8888$, when $f = 180 \times 10^3$, $x = 2.000$, and when $f = 280 \times 10^3$, $x = 3.111$:

$$\text{Denom} = -0.8 \frac{1}{a} \left[\tan^{-1}(x) \right]_{0.8888}^{2.0} + 10^{-5} \frac{1}{a^2} \left[\frac{1}{2} \ln(1 + x^2) \right]_{0.8888}^{2.0}$$

$$+ 2.8 \frac{1}{a} \left[\tan^{-1}(x) \right]_{2.0}^{3.111} - 10^{-5} \frac{1}{a^2} \left[\frac{1}{2} \ln(1 + x^2) \right]_{2.0}^{3.111}$$

$$= -7.201 \times 10^4 \, [1.1071 - 0.7266] + \frac{8.102 \times 10^4}{2} \, [1.6094 - 0.5822]$$

$$+ 2.520 \times 10^5 [1.2598 - 1.1071] - \frac{8.102 \times 10^4}{2} \, [2.3682 - 1.6094]$$

$$= 2.195 \times 10^4$$

$$\text{SDR} = \frac{\text{Num}}{\text{Denom}} = \frac{4.286 \times 10^4}{2.195 \times 10^4} = 1.95 = 2.9 \text{ dB}$$

Comparing the results for (i) and (ii) shows that if significant aliased energy is present a good multipole filter with a steep skirt is essential for signal reconstruction if the SDR is to be kept to tolerable levels. Furthermore the SDR calculated in part (ii) is extremely optimistic since spectral replicas centred on $2f_s$ Hz and above have been ignored. For reconstruction filters with only gentle roll-off equation (5.5) should really be used only as an upper bound on SDR, see Problem 5.3.

___

5.3.5 Bandpass sampling

In some applications it is desirable to sample a bandpass signal in which the centre frequency is many times the signal bandwidth. Whilst, in principle, it would be possible to sample this signal at twice the highest frequency component in its spectrum and reconstruct the signal from its samples by low pass filtering it is usually possible to retain all the information needed to reconstruct the original signal whilst sampling at a much lower rate. If advantage is taken of this then there exists one or more frequency *bands* in which the sampling frequency should lie. Thus when sampling a bandpass signal there is generally an upper limit for proper sampling as well as a lower limit. The bandpass sampling criterion can be expressed as follows:

> *A bandpass signal having no spectral components below f_L Hz or above f_H Hz is specified uniquely by its values at uniform intervals spaced $T_s = 1/f_s$ s apart provided that:*

$$2B \left\{ \frac{Q}{n} \right\} \leq f_s \leq 2B \left\{ \frac{Q-1}{n-1} \right\} \tag{5.6}$$

> *where $B = f_H - f_L$, $Q = f_H/B$, n is a positive integer and $n \leq Q$.*

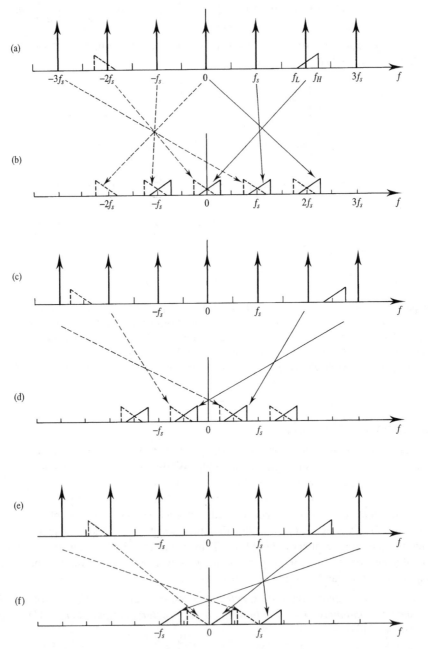

Figure 5.10 *Criteria for correct sampling of bandpass signals: (a) spectrum of signal where $G(f)$ straddles $2 \times f_s$; (c) spectrum of signal where $G(f)$ straddles $2.5 \times f_s$; (b) & (d) overlapped spectra after bandpass sampling; (e) bandpass spectrum avoiding the straddling in (a) & (c); (f) baseband or bandpass spectra, recoverable by filtering.*

The following comments are made to clarify the use of equation (5.6) and its relationship to the Nyquist baseband sampling criterion.

1. If $Q = f_H/B$ is an integer then $n \leq Q$ allows us to choose $n = Q$. In this case $f_s = 2B$ and the correct sampling frequency is exactly twice the signal bandwidth.

2. If $Q = f_H/B$ is not an integer then the lowest allowed sampling rate is given by choosing $n = int(Q)$ (i.e. the next lowest integer from Q). Lower values of n will still allow reconstruction of the original signal but the sampling rate will be unnecessarily high. (Lower values of n may, however, give a wider band of allowed f_s.)

3. If $Q < 2$ (i.e. $f_H < 2B$ or, equivalently, $f_L < B$) then $n \leq Q$ means that $n = 1$. In this case:

$$2BQ \leq f_s \leq \infty \quad \text{(Hz)}$$

and since $BQ = f_H$ we have:

$$2f_H \leq f_s \leq \infty \quad \text{(Hz)}$$

This is a statement of the Nyquist (baseband) sampling criterion.

The validity of the bandpass sampling criterion is most easily demonstrated using convolution (section 2.3.4) for the following special cases. (Convolution results in the sampled signal being replicated at DC and multiples of the sample frequency.)

1. When the spectrum of the bandpass signal $g(t)$ straddles nf_s, i.e. $G(f)$ straddles any of the lines in the spectrum of the sampling signal (Figure 5.10(a)), then convolution results in interference between the positive and negative frequency spectral replicas (Figure 5.10(b)).

2. When the spectrum of $g(t)$ straddles $(n + \frac{1}{2})f_s$, i.e. $G(f)$ straddles any odd integer multiple of $f_s/2$ (Figure 5.10(c)), then similar interference occurs (Figure 10(d)).

3. When the spectrum of $g(t)$ straddles neither nf_s nor $(n + \frac{1}{2})f_s$ (Figure 5.10(e)), then no interference between positive and negative frequency spectral replicas occurs (Figure 5.10(f)) and the baseband (or bandpass) spectrum can be obtained by filtering.

Summarising we have the following conditions for proper sampling:

$$f_H \leq n\frac{f_s}{2} \quad \text{(Hz)} \tag{5.7}$$

$$f_L \geq (n-1)\frac{f_s}{2} \quad \text{(Hz)} \tag{5.8}$$

Using $f_L = f_H - B$ we have:

$$\frac{2}{n} f_H \leq f_s \leq \frac{2}{n-1}(f_H - B) \quad \text{(Hz)} \tag{5.9}$$

Defining $Q = f_H/B$ gives the bandpass sampling criterion of equation (5.6).

EXAMPLE 5.2

The following examples illustrate the use and significance of equation (5.6). First consider a signal with centre frequency 9.5 kHz and bandwidth 1.0 kHz.

The highest and lowest frequency components in this signal are:

$$f_L = 9.0 \text{ kHz} \qquad f_H = 10.0 \text{ kHz}$$

Quotient Q is thus:

$$Q = f_H/B = 10.0/1.0 = 10.0$$

Applying the bandpass sampling criterion of equation (5.6):

$$2 \times 10^3 \left\{ \frac{10}{n} \right\} \le f_s \le 2 \times 10^3 \left\{ \frac{10 - 1}{n - 1} \right\} \quad \text{(Hz)}$$

Since Q is an integer the lowest allowed sampling rate is given by choosing $n = Q = 10$, i.e.:

$$2.0 \le f_s \le 2.0 \text{ (kHz)}$$

The significant point here is that there is zero tolerance in the sampling rate if distortion is to be completely avoided. If n is chosen to be less than its maximum value, e.g. $n = 9$, then:

$$2.222 \le f_s \le 2.250 \text{ (kHz)}$$

The sampling rate in this case would be chosen to be 2.236 ± 0.014 kHz. The accuracy required of the sampling clock is therefore $\pm 0.63\%$. Now consider a signal with centre frequency 10.0 kHz and bandwidth 1.0 kHz. The quotient $Q = 10.5$ is now no longer an integer. The lowest allowed sampling rate is therefore given by $n = int(Q) = 10.0$. The sampling rate is now bound by:

$$2.100 \le f_s \le 2.111 \text{ (kHz)}$$

This gives a required sampling rate of 2.106 ± 0.006 kHz or 2.106 kHz $\pm 0.26\%$.

5.4 Analogue pulse multiplexing

In many communications applications different information signals must be transmitted over the same physical channel. The channel might, for example, be a single coaxial cable, an optical fibre or, in the case of radio, the free space between two antennas. In order for the signals to be received independently (i.e. without crosstalk) they must be sufficiently separated in some sense.

This quality of separateness is usually called orthogonality, section 2.5.3. Orthogonal signals can be received independently of each other whilst non-orthogonal signals cannot. There are many ways in which orthogonality between signals can be provided. An intuitively obvious way, sometimes used in microwave radio communications, is to use two perpendicular polarisations for two independent information channels. Vertical and horizontal, linear, antenna polarisations are usually used but right and left hand circular polarisations may also be used. Such signals are orthogonal in polarisation.

The traditional way of providing orthogonality in analogue telephony and broadcast applications is to transmit different information signals using different carrier frequencies. Such signals (provided their spectra do not overlap) are disjoint in frequency and can be received separately using filters. Tuning the local oscillator in a superheterodyne receiver, Figure 1.4, also allows one signal to be separated from others with disjoint (and therefore orthogonal) frequency spectra. Using different carriers, or frequency bands, to isolate signals from each other in this way is called frequency division multiplexing (FDM).

In FDM telephony, 3.2 kHz bandwidth telephone signals, Figure 5.8, are stacked in frequency at 4 kHz spacings with small frequency guard bands between them to allow separation using practical filters. Figure 5.11 shows how an FDM signal can be generated and Figure 5.12 shows a schematic representation of an FDM signal spectrum generated using single (lower) sideband filters. FDM was the original multiplexing technique for analogue communications and is now experiencing a resurgence in fibre optic systems in which different wavelengths are used for simultaneous transmission of many information signals. In this particular context the term wavelength division multiplexing (WDM) is usually used in preference to FDM.

Orthogonality can be provided in a quite different way for pulse modulated signals. Instead of occupying separate frequency bands (as in FDM) the signals occupy separate time slots. This technique, illustrated in Figure 5.2, is called (analogue) time division multiplexing (TDM). In telephony parlance the separate inputs in Figure 5.2 are called tributaries (see Chapter 20).

TDM obviously increases the overall sample rate and therefore the required bandwidth for transmission. It is, however, possible to reduce the bandwidth of the TDM signal dramatically by appropriate filtering since, strictly, it is only necessary that the TDM signal provides the correct amplitude information at the sampling instants. Nyquist's sampling theorem says that a waveform with a bandwidth of $2f_H$ Hz exists which passes through these required points. This is illustrated in Figure 5.13. (At points between the sampling instants the filtered TDM signal is made up of a complicated sum of contributions arising from many pulses. At the sampling points themselves, however, the waveform amplitude is due to a single TDM pulse only.) If a minimum bandwidth

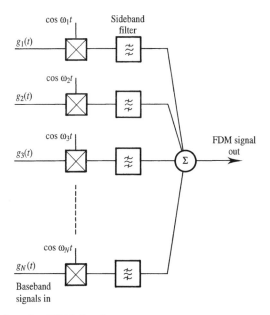

Figure 5.11 *Generation of an FDM signal.*

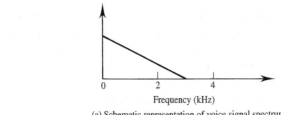

(a) Schematic representation of voice signal spectrum

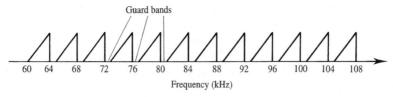

(b) Schematic representation of FDM signal spectrum

Figure 5.12 *FDM example for multiplexed speech channels.*

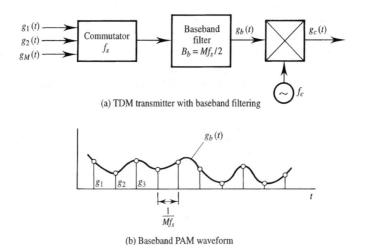

(a) TDM transmitter with baseband filtering

(b) Baseband PAM waveform

Figure 5.13 *Filtered TDM waveform.*

TDM signal is formed by filtering then sampling accuracy becomes critical in that samples taken at times other than the correct instant will result in crosstalk between channels.

Crosstalk can also occur between the channels of a TDM signal even if filtering is not explicitly applied. This is because the transmission medium itself may bandlimit the signal. Such bandlimiting effects may often be at least approximated by RC low pass filtering. In this case the response of the medium results in pulses with exponential rising and falling edges as shown in Figure 5.14. If the guard time between rectangular pulses (i.e. the time between the trailing edge of one pulse and the rising edge of the next) is t_g, and the time constant of the transmission channel is RC, then the amplitude of each pulse will decay to a fraction $e^{-t_g/RC}$ of its peak value by the time the next pulse starts. For the RC characteristic the channel bandwidth is $f_{3dB} = 1/(2\pi RC)$, therefore the crosstalk ratio (XTR) in dB at the pulse trailing edge (i.e. the optimum XTR sampling instant) is:

$$\text{XTR} = 20 \log_{10} e^{2\pi f_{3dB}(t_g + \tau)}$$

$$= 54.6 \, f_{3dB} \, (t_g + \tau) \quad \text{(dB)} \tag{5.10}$$

Equation (5.10) therefore gives an estimate of the required guard time to maintain a desired crosstalk ratio in a bandlimited channel, for a given rectangular pulse width, τ, at the input. (If sampling occurs at the centre of the τ s nominal pulse slot, rather than at its end, then XTR is reduced by approximately 6 dB when τ is much less than RC.)

The discussion of TDM in this chapter has been mainly in the context of analogue pulse multiplexing. In Chapter 20 it is shown how (digital) TDM forms the basis of the telephone hierarchy for transmitting multiple simultaneous telephone calls over high speed 2, or 140, Mbit/s data links.

5.5 Quantised PAM

An information signal which is pulse amplitude modulated becomes discrete (in time) rather than continuous but nevertheless remains analogue in nature since all pulse amplitudes within a specified range are allowed. An alternative way of expressing the

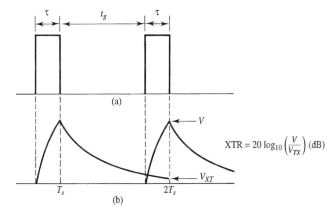

Figure 5.14 *Crosstalk between tributary channels of a TDM signal: (a) signal at channel input; (b) signal at channel output.*

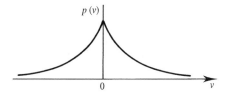

Figure 5.15 *Continuous pdf of typical analogue PAM signal.*

analogue property of a PAM signal is to say that the probability density function (pdf) of pulse amplitudes is continuous (Figure 5.15). If a PAM signal is quantised, i.e. each pulse is adjusted in amplitude to coincide with the nearest of a finite set of allowed amplitudes (Figure 5.16), then the resulting signal is no longer analogue but digital and as a consequence has a discrete pdf as illustrated in Figure 5.17 (and previously in Figure 3.4).

This digital signal can be represented by a finite set of symbols – the obvious set consisting of one symbol for each quantisation level. Other symbol sets (or alphabets) can be conceived, however, and unique mappings from one set to another established. Probably the simplest and most important alphabet is the binary set consisting of two symbols only, usually denoted by 0 and 1. The rest of this chapter is primarily concerned with the quantisation process and the subsequent efficient coding of quantised levels into binary symbols. In terms of Figure 1.3 the subsystems of principal importance here are the quantiser, pulse code modulation (PCM) encoder and source coder, although digital pulse multiplexing will also be discussed briefly. To some extent the separation of quantiser, PCM encoder and source coder might be misleading since they do not necessarily exist as identifiably separate pieces of hardware in all digital communications systems. For example, quantisation and PCM encoding are implemented together as a binary A/D converter (which may then be followed by a parallel to series converter) in many systems. Similarly some source coders (e.g. delta modulators) effectively *replace* the PCM encoder whilst others take a PCM signal and recode the binary symbols. In some systems (e.g. delta PCM) the source coder precedes the PCM encoder.

Quantising PAM signals is usually a precursor to generating PCM which has some significant advantages over other baseband modulation types. The quantisation process in itself, however, actually degrades the quality of the information signal. This is easy to see

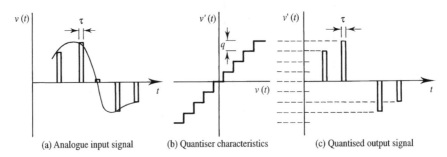

(a) Analogue input signal   (b) Quantiser characteristics   (c) Quantised output signal

Figure 5.16 *Quantisation of a PAM signal.*

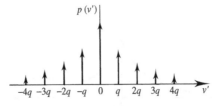

Figure 5.17 *Discrete pdf of quantised PAM signal.*

since the quantised PAM signal no longer exactly represents the original, continuous, analogue signal but a distorted version of it. Figure 5.18 (which is drawn with PAM pulse width, τ, equal to the sampling period, T_s) shows that the quantised signal can be decomposed into the sum of the analogue signal and the difference between the quantised and the analogue signals. The difference signal is essentially random and can therefore be thought of as a special type of noise process. Like any other signal the power or RMS value of this *quantisation noise* can be calculated or measured. This leads to the concept of a *signal to quantisation noise ratio* (SN$_q$R).

5.6 Signal to quantisation noise ratio (SN$_q$R)

To calculate the SN$_q$R of a quantised signal it is convenient to make the following assumptions:

1. Linear quantisation (i.e. equal increments between quantisation levels).
2. Zero mean signal (i.e. symmetrical pdf about 0 V).
3. Uniform signal pdf (i.e. all signal levels equally likely).

The probability density function $p(v)$ of allowed levels is illustrated in the left hand side of Figure 5.18(a). (Narrow rectangles are used to represent the delta functions in the pdf to make interpretation easy.) Each rectangle has an area of $1/M$ where M is the (even) number of quantisation levels. (This is a result of assumption 3.) If the distance between adjacent quantisation levels is q V then the pdf of allowed levels is given by:

$$p(v) = \sum_{k=-M}^{M} (1/M)\, \delta(v - qk/2) \tag{5.11}$$

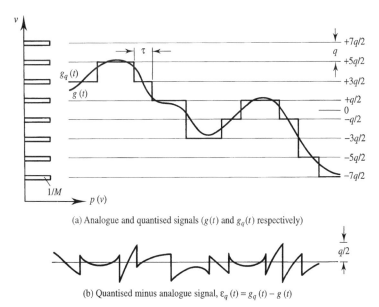

(a) Analogue and quantised signals ($g(t)$ and $g_q(t)$ respectively)

(b) Quantised minus analogue signal, $\varepsilon_q(t) = g_q(t) - g(t)$

Figure 5.18 *Quantisation error interpreted as noise, i.e.* $g_q(t) = g(t) + \varepsilon_q(t)$.

where k takes on odd values only. The mean square signal after quantisation is:

$$\overline{v^2} = \int_{-\infty}^{+\infty} v^2 p(v) \, dv$$

$$= \frac{2}{M} \left[\int_0^\infty v^2 \, \delta(v - q/2) \, dv + \int_0^\infty v^2 \, \delta(v - 3q/2) \, dv + \cdots \right]$$

$$= \frac{2}{M} \left(\frac{q}{2} \right)^2 [1^2 + 3^2 + 5^2 + \cdots + (M-1)^2]$$

$$= \frac{2}{M} \left(\frac{q}{2} \right)^2 \left[\frac{M(M-1)(M+1)}{6} \right] \tag{5.12}$$

i.e.:

$$\overline{v^2} = \frac{M^2 - 1}{12} q^2 \quad (\mathrm{V}^2) \tag{5.13}$$

Denoting the quantisation error (i.e. the difference between the unquantised and quantised signals) as ε_q, Figure 5.18(b), then it follows from assumption 3 that the pdf of ε_q is uniform:

$$p(\varepsilon_q) = \begin{cases} 1/q, & -q/2 \le \varepsilon_q < q/2 \\ 0, & \text{elsewhere} \end{cases} \tag{5.14}$$

The mean square quantisation error (or noise) is:

$$\overline{\varepsilon_q^2} = \int_{-q/2}^{+q/2} \varepsilon_q^2 p(\varepsilon_q) \, d\varepsilon_q \tag{5.15}$$

i.e.:

$$\overline{\varepsilon_q^2} = \frac{q^2}{12} \quad (\mathrm{V}^2) \tag{5.16}$$

The $\mathrm{SN_q R}$ is therefore given by:

$$\mathrm{SN_q R} = \overline{v^2}/\overline{\varepsilon_q^2} = M^2 - 1 \tag{5.17}$$

For large $\mathrm{SN_q R}$ the approximation $\mathrm{SN_q R} = M^2$ is often used.

Equation (5.17) represents the average signal to quantisation noise (power) ratio. Since the *peak* signal level is $Mq/2$ V then the peak signal to quantisation noise (power) ratio is:

$$(\mathrm{SN_q R})_{peak} = \frac{(Mq/2)^2}{\overline{\varepsilon_q^2}} \tag{5.18}$$

$$= 3M^2$$

Expressed in dB the $\mathrm{SN_q Rs}$ are:

$$SN_qR = 20 \log_{10} M \quad \text{(dB)} \tag{5.19}$$

$$(SN_qR)_{peak} = 4.8 + SN_qR \quad \text{(dB)} \tag{5.20}$$

5.7 Pulse code modulation

After a PAM signal has been quantised the possibility exists of transmitting not the pulse itself but a number indicating the height of the pulse. Usually (but not necessarily always) the pulse height is transmitted as a binary number. As an example, if the number of allowed quantisation levels were eight then the pulse amplitudes could be represented by the binary numbers from zero (000) to seven (111). The binary digits are normally represented by two voltage levels (e.g. 0 V and 5 V). Each binary number is called a codeword and, since each quantised pulse is represented by a codeword, the resulting modulation is called pulse code modulation (PCM). Figures 5.19(a) to (e) show the relationship between an information, PAM, quantised PAM and PCM signal. Figures 5.19(b) and (c) also illustrate the difference between a pulsed signal with duty cycle (τ/T_s) less than 1.0 and a pulsed signal with duty cycle equal to 1.0. The former are often referred to as return to zero (RZ) signals and the latter as non-return to zero (NRZ) signals.

There is clearly a bandwidth penalty to pay for PCM, if information is to be transmitted in real time, since, in the example given above, three binary pulses are transmitted instead of one quantised PAM pulse. (The penalty here is a factor of three since, for the same pulse duty cycle, each PCM pulse must be one-third the duration of the PAM pulse.) The advantage of PCM is that for a given transmitted power the difference between adjacent voltage levels is much greater than for quantised PAM. This means that for a given RMS noise voltage the total voltage (signal plus noise) at the receiver is less likely to be interpreted as representing a level other than that which was transmitted. PCM signals are therefore said to have greater noise immunity than PAM signals.

5.7.1 SN$_q$R for linear PCM

Whilst it is true that PCM signals are more tolerant of noise than the equivalent quantised PAM signals it is also true that both suffer the same degradation due to quantisation noise. For a given number of quantisation levels, M, the number of binary digits required for each PCM codeword is $n = \log_2 M$. The PCM peak signal to quantisation noise ratio, $(SN_qR)_{peak}$, is therefore:

$$(SN_qR)_{peak} = 3M^2 = 3(2^n)^2 \tag{5.21}$$

If the ratio of peak to mean signal power, $v_{peak}^2/\overline{v^2}$, is denoted by α then the average SN$_q$R is:

$$SN_qR = 3(2^{2n})(1/\alpha) \tag{5.22}$$

Expressed in dB this becomes:

$$SN_qR = 4.8 + 6n - \alpha_{dB} \tag{5.23}$$

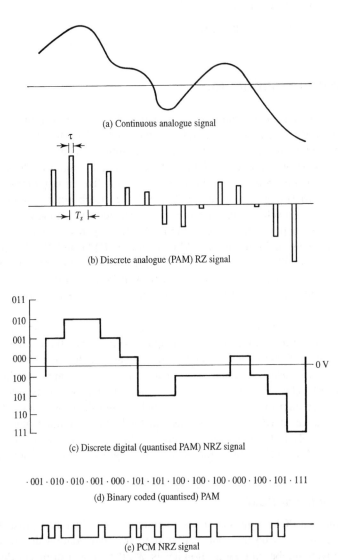

(a) Continuous analogue signal

(b) Discrete analogue (PAM) RZ signal

(c) Discrete digital (quantised PAM) NRZ signal

· 001 · 010 · 010 · 001 · 000 · 101 · 101 · 100 · 100 · 100 · 000 · 100 · 101 · 111

(d) Binary coded (quantised) PAM

(e) PCM NRZ signal

Figure 5.19 *Relationship between PAM, quantised PAM and PCM signal.*

For a sinusoidal signal $\alpha = 2$ (or 3 dB). For a (clipped) Gaussianly distributed random signal (with $v_{peak}/\sigma_g = 4$ where σ_g is the signal's standard deviation, or RMS value, as in section 3.2.5) $\alpha = 16$ (or 12 dB), and for speech $\alpha = 10$ dB. The SN_qR for an n-bit PCM voice system can therefore be estimated using the rule of thumb $6(n - 1)$ dB.

EXAMPLE 5.3

A digital communications system is to carry a single voice signal using linearly quantised PCM. What PCM bit rate will be required if an ideal anti-aliasing filter with a cut-off frequency of 3.4 kHz is used at the transmitter and the SN_qR is to be kept above 50 dB?

From equation (5.23):

$$SN_qR = 4.8 + 6n - \alpha_{dB}$$

For voice signals $\alpha = 10$ dB, i.e.:

$$n = \frac{50 + 10 - 4.8}{6} = 9.2$$

10 bit/sample are therefore required. The sampling rate required is given by Nyquist's rule, $f_s \geq 2f_H$. Taking a practical version of the sampling theorem, equation (5.3), gives:

$$f_s = 2.2 \times 3.4 \text{ kHz} = 7.48 \text{ kHz} \text{ (or k samples/s)}$$

The PCM bit rate (or more strictly binary baud rate) is therefore:

$$R_b = f_s n$$
$$= 7.48 \times 10^3 \times 10 \text{ bit/s}$$
$$= 74.8 \text{ kbit/s}$$

5.7.2 SNR for decoded PCM

If all PCM codewords are received and decoded without error then the SNR of the decoded signal is essentially equal to the SN_qR, as given in equations (5.21) to (5.23). In the presence of channel and/or receiver noise, however, it is possible that one or more symbols in a given codeword will be changed sufficiently in amplitude to be interpreted in error. For binary PCM this involves a digital 1 being interpreted as a 0 or a digital 0 being interpreted as a 1. The effect that such an error has on the SNR of the decoded signal depends on which symbol is detected in error. The least significant bit in a binary PCM word will introduce an error in the decoded signal equal to one quantisation level. The most significant bit would introduce an error of many quantisation levels.

The following reasonably simple analysis gives a useful expression for the SNR performance of a PCM system in the presence of noise.

We first assume that the probability of more than one error occurring in a single n-bit PCM codeword is negligible. We also assume that all bits in the codeword have the same probability (P_e) of being detected in error. Using subscripts $1, 2, \cdots, n$ to denote the significance of PCM codeword bits (1 corresponding to the least significant, n corresponding to the most significant) then the possible errors in the decoded signal are:

$$\varepsilon_1 = q$$
$$\varepsilon_2 = 2q$$
$$\varepsilon_3 = 4q \tag{5.24}$$
$$\cdots$$
$$\varepsilon_n = 2^{n-1}q$$

The mean square decoding error, $\overline{\varepsilon_{de}^2}$, is the mean square of the possible errors multiplied by the probability of an error occurring in a codeword, i.e.:

$$\overline{\varepsilon_{de}^2} = nP_e(1/n)[(q)^2 + (2q)^2 + \cdots + (2^{n-1}q)^2]$$
$$= P_e(q)^2[4^0 + 4 + 4^2 + 4^3 + \cdots + 4^{(n-1)}] \tag{5.25}$$

The square bracket is the sum of a geometric progression with the form:

$$S_n = a + ar + ar^2 + \cdots + ar^{n-1} = \frac{a(r^n - 1)}{r - 1} \tag{5.26}$$

where $a = 1$ and $r = 4$. Thus:

$$\overline{\varepsilon_{de}^2} = P_e q^2 (4^n - 1)/3 \quad (\mathrm{V}^2) \tag{5.27}$$

Since the error or noise which results from incorrectly detected bits is statistically independent of the noise which results from the quantisation process we can add them together on a power basis, i.e.:

$$\mathrm{SNR} = \frac{\overline{v^2}}{\overline{\varepsilon_q^2} + \overline{\varepsilon_{de}^2}} \tag{5.28}$$

where $\overline{v^2}$ is the received signal power. Using equation (5.13) for $\overline{v^2}$ and equation (5.16) for $\overline{\varepsilon_q^2}$, and remembering that the number of quantisation levels $M = 2^n$, we have:

$$\mathrm{SNR} = \frac{M^2 - 1}{1 + 4(M^2 - 1)P_e} \tag{5.29}$$

Equation (5.29) allows us to calculate the average SNR of the decoded PCM signal including both quantisation noise and the decoding noise which occurs due to corruption of individual PCM bits by channel or receiver noise. In Chapter 6 expressions are developed which relate P_e to channel SNR. If we denote the channel SNR using the subscript *in* and the decoded PCM SNR using the subscript *out* then for binary, polar, NRZ signalling, using simple centre point decisions (see section 6.2), we have:

$$\mathrm{SNR}_{out} = \frac{M^2 - 1}{1 + 4(M^2 - 1) \tfrac{1}{2} \, \mathrm{erfc} \, (\tfrac{1}{2}\mathrm{SNR}_{in})^{\frac{1}{2}}} \tag{5.30}$$

The SNRs in equation (5.30) are linear ratios (not dB values) and the function erfc(x) is the complementary error function (defined later in equation (6.3)). Equation (5.30) is sketched for various values of $n = \log_2 M$ in Figure 5.20. The noise immunity advantage of PCM illustrated by this figure is clear. The x-axis is the SNR of the received PCM signal. The y-axis is the SNR of the reconstructed (decoded) information signal. If the SNR of the received PCM signal is very large then the total noise is dominated by the quantisation process and the output SNR is limited to $\mathrm{SN_qR}$. In practice, however, PCM

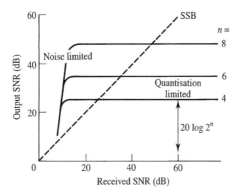

Figure 5.20 *Input/output SNR for PCM.*

systems are operated at lower input SNR values near the knee or threshold of the curves in Figure 5.20. The output SNR is then significantly greater than the input SNR. At very low input SNR, when the noise is of comparable amplitude to the PCM pulses, then the interpretation of codewords starts to become unreliable. Since even a single error in a PCM codeword can change its numerical value by a large amount then the output SNR in this region (i.e. below threshold) decreases very rapidly.

EXAMPLE 5.4

Find the overall SNR for the reconstructed analogue voice signal in Example 5.3 if receiver noise induces an error rate, on average, of one in every 10^6 PCM bits.

From equations (5.29) and (5.17):

$$SNR_{out} = \frac{SN_qR}{1 + 4\,SN_qR\,P_e}$$

$$SN_qR = 4.8 + 6n - \alpha_{dB} = 4.8 + (6 \times 10) - 10$$

$$= 54.8 \text{ dB} \quad (\text{or } 3.020 \times 10^5)$$

$$SNR_{out} = \frac{3.020 \times 10^5}{1 + 4\,(3.020 \times 10^5)(1 \times 10^{-6})}$$

$$= 1.368 \times 10^5 = 51.4 \text{ dB}$$

The SNR (as a ratio) available with PCM systems increases with the square of the number of quantisation levels while the baud rate, and equivalently the bandwidth, increases with the logarithm of the number of quantisation levels. Thus bandwidth can be exchanged for SNR (as in analogue frequency modulation). Close to threshold PCM is superior to all analogue forms of pulse modulation (and it is also marginally superior to FM) at low SNR. However, all practical PCM systems have a performance which is an order of magnitude below their theoretical optimum.

As PCM signals contain no information in their pulse amplitude they can be regenerated using non-linear processing at each repeater in a long haul system. Such digital, regenerative, repeaters allow accumulated noise to be removed and essentially noiseless signals to be retransmitted to the next repeater in each section of the link. The probability of error does accumulate from hop to hop, however. This is discussed in Chapter 6 (section 6.3).

5.7.3 Companded PCM

The expressions for SN_qR derived in section 5.6 (e.g. equations (5.19) and (5.20)) assume that the information signal has a uniform pdf, i.e. that all quantisation levels are used equally. For most signals this is not a valid assumption. If the pdf of the information signal is not uniform but is nevertheless known, and is constant with time, then it is intuitively obvious that to optimise the average SN_qR those quantisation levels used most should introduce least quantisation noise. One way to arrange for this to occur is to adopt non-linear quantisation or, equivalently, companding. Non-linear quantisation is illustrated in Figure 5.21(a). If the information signal pdf has small amplitude for a large fraction of time and large amplitude for a small fraction of time (as is usually the case) then the step size between adjacent quantisation levels is made small for low levels and larger for higher levels. Companding (*comp*ressing–exp*anding*) achieves the same result by compressing the information signal using a non-linear amplitude characteristic (Figure 5.22(a)) prior to linear quantisation and then expanding the reconstructed information signal with the inverse characteristic (Figure 5.22(b)). Ideally the companding characteristic would result in a signal which has a precisely uniform pdf. Whilst this ideal is unlikely to be achieved the compressed signal will have a *more nearly* uniform pdf and therefore better SN_qR than the uncompressed signal.

A rather different problem arises if the information signal has an unknown pdf or if its pdf (measured on some relevant time scale) changes with time. In the case of voice signals, for example, the pdf arising from an individual speaker is usually fairly constant and the *shape* of pdfs arising from different users is usually similar. (A typical voice exceedance curve is shown in Figure 5.23.) The gross signal level, however, can vary widely between speakers, with perhaps a man who habitually shouts whilst using the

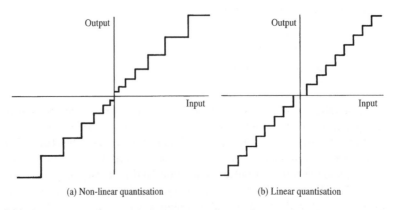

(a) Non-linear quantisation   (b) Linear quantisation

Figure 5.21 *Quantisation characteristics.*

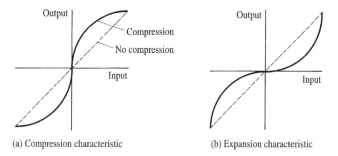

(a) Compression characteristic   (b) Expansion characteristic

Figure 5.22 *Typical compression and expansion (compander) characteristics.*

telephone at one extreme, and a woman who is especially softly spoken at the other extreme. In these cases the companding strategy is normally to maintain, as nearly as possible, a constant SN_qR for all signal levels. (This is quite a different strategy from maximising the average SN_qR.) Since quantisation noise power is proportional to q^2 then RMS quantisation noise voltage is proportional to q. If SN_qR is to be constant for all signal levels then q must clearly be proportional to signal level, i.e. v/q must be a constant. If uniform quantisation is used then the signal should be compressed such that increasing the input signal by a given *factor* increases the output signal by a corresponding *additional constant*. Thus equal quantisation increments in the output signal correspond to quantisation increments in the input signal which are equal *fractions* of the input signal. The function which converts multiplicative factors into additional constants is the logarithm and the constant SN_qR compression characteristic is therefore of the form $y = \log x$ (Figure 5.24(a)).

Since information signals can usually take on negative as well as positive values the logarithmic compression characteristic must be reflected to form an odd symmetric function (Figure 5.24(b)). Furthermore the characteristic must obviously be continuous across zero volts and so the two logarithmic functions are modified and joined by a linear section as shown in Figure 5.24(c). The actual compression characteristic used in Europe is the *A*-law defined by:

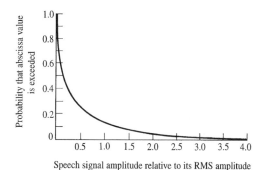

Figure 5.23 *Statistical distribution of single talker speech signal amplitude.*

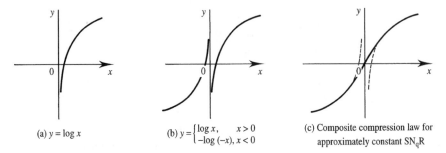

(a) $y = \log x$

(b) $y = \begin{cases} \log x, & x > 0 \\ -\log(-x), & x < 0 \end{cases}$

(c) Composite compression law for approximately constant SN_qR

Figure 5.24 *Development of constant SN_qR compression characteristic.*

$$F(x) = \begin{cases} \mathrm{sgn}(x)\, \dfrac{1 + \ln(A|x|)}{1 + \ln A}, & 1/A < |x| < 1 \\[2ex] \mathrm{sgn}(x)\, \dfrac{A|x|}{1 + \ln A}, & 0 < |x| < 1/A \end{cases} \qquad (5.31)$$

where $|x| = |v/v_{peak}|$ is the normalised input signal to the compressor, $F(x)$ is the normalised output signal from the compressor and $\mathrm{sgn}(x)$ is the signum function which is $+1$ for $x > 0$ and -1 for $x < 0$.

The parameter A in equation (5.31) defines the curvature of the compression characteristic with $A = 1$ giving a linear law (Figure 5.25). The commonly adopted value is $A = 87.6$ which gives a 24 dB improvement in SN_qR over linear PCM for small signals ($|x| < 1/A$) and an (essentially) constant SN_qR of 38 dB for large signals ($|x| > 1/A$) [Dunlop and Smith]. The dynamic range of the logarithmic (constant SN_qR) region of this characteristic is $20 \log_{10}[1/(1/A)] \approx 39$ dB. The overall effect is to allow 11 bit (2048 level) linear PCM, which would be required for adequate voice signal quality, to be reduced to 8 bit (256 level) companded PCM. A 4 kHz voice channel sampled at its Nyquist rate (i.e. 8 kHz) therefore yields a companded PCM bit rate of 64 kbit/s.

The A-law characteristic is normally implemented as a 13-segment piecewise linear approximation to equation (5.31) (in practice 16 segments but with 4 segments near the origin co-linear as illustrated in Figure 5.26). For 8 bit PCM 1 bit gives polarity, 3 bits

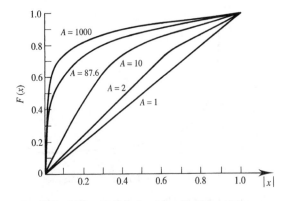

Figure 5.25 *A-law compression characteristic for several values of A.*

indicate which segment the sample lies on and 4 bits provide the location on the segment.

In the USA and Japan a similar logarithmic compression law is used. This is the μ-law given by:

$$F(x) = \text{sgn}(x)\ \frac{\ln\ (1 + \mu|x|)}{\ln\ (1 + \mu)}\ ,\qquad 0 \le |x| \le 1 \tag{5.32}$$

The μ-law (with $\mu = 255$) tends to give slightly improved SN_qR for voice signals when compared with the A-law but it has a slightly smaller dynamic range. In practice, like the A-law, the μ-law is usually implemented as a piecewise linear approximation.

A- and μ-law 64 kbit/s companded PCM has been adopted by ITU-T as the international toll quality standards (recommendation G.711) for digital coding of voice frequency signals. The sampling rate is 8 kHz and the encoding law uses 8 binary digits per sample. The A- and μ-laws are implemented as 13- and 15-segment piecewise linear curves respectively. For communications between countries using A and μ companding laws conversion from one to the other is the responsibility of the country using the μ-law.

G.711 PCM communication has a *mean opinion score* (MOS) speech quality, measured by subjective testing, of 4.3 on a scale of 0 to 5. An MOS of 4 allows audible but not annoying degradation, 3 implies that the degradation is slightly annoying, 2 is annoying and 1 is very annoying. If narrow bandwidth is important then other coding techniques are employed (see section 5.8).

5.7.4 PCM multiplexing

Multiplexing of analogue PAM pulses has already been described in section 5.4. If such a multiplexed pulse stream is fed to a PCM encoder the resulting signal is a codeword (or *character*) interleaved digital TDM signal. This is illustrated in Figure 5.27(a). In this figure the TDM signal is generated by *first* multiplexing and *then* PCM coding. It is of course possible to generate a character interleaved TDM signal by *first* PCM coding the information signals and *then* interleaving the codewords, Figure 5.27(b). There are two reasons for drawing attention to this alternative method of TDM generation:

1. As digital technology has become cheaper and more reliable there has been a tendency for analogue signals to be digitally coded as near to their source as possible.

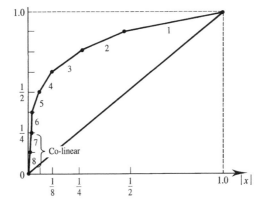

Figure 5.26 *The 13-segment compression A-law realised by piecewise linear approximation.*

2. Time division multiplexing of PCM signals (rather than PCM encoding of TDM PAM signals) suggests the possibility of interleaving the individual bits of the PCM codewords rather than the codeword (bytes) themselves.

The TDM concept can be extended to higher levels of multiplexing by time division multiplexing two or more TDM signals, see later PCM multiplex hierarchies, Chapter 20.

5.8 Bandwidth reduction techniques

Bandwidth is a limited, and therefore valuable, resource. This is because all physical transmission lines have characteristics which make them suitable for signalling only over a finite band of frequencies. (The problem of constrained bandwidth is especially severe for the local loop which connects individual subscribers to their national telephone network.) Bandwidth (or more strictly spectrum) is even limited in the case of radio communications since the transmission properties of the earth's atmosphere are highly variable as a function of frequency (see Chapter 14). Furthermore, co-channel (same frequency) radio transmissions are difficult to confine spatially and tend to interfere with each other, even when such systems are widely spaced geographically.

A given installation of transmission lines (wire-pairs, coaxial cables, optical fibres, microwave links and others) therefore represents a finite spectral resource and since adding to this installation (by laying new cables, for instance) is expensive, there is great advantage to be gained in using the existing installation efficiently. This is the real

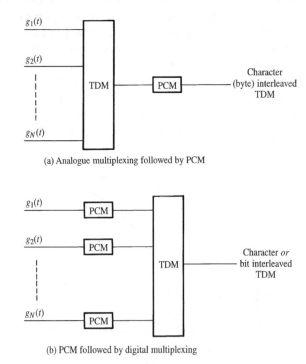

(a) Analogue multiplexing followed by PCM

(b) PCM followed by digital multiplexing

Figure 5.27 *Generation of bit and byte interleaved PCM–TDM signals.*

incentive to develop spectrally efficient (i.e. reduced bandwidth) signalling techniques for encoding speech signals.

5.8.1 Delta PCM

One technique to reduce the bandwidth of a PCM signal is to transmit information about the changes between samples instead of sending the sample values themselves. The simplest such system is delta PCM which transmits the difference between adjacent samples as conventional PCM codewords. The difference between adjacent samples is, generally, significantly less than the actual sample values, which allows the differences to be coded using fewer binary symbols per word than conventional PCM would require. (This reflects the fact that the adjacent samples derived from most, naturally generated, information signals are not usually independent but correlated, see section 2.6.) Block diagrams of a delta PCM transmitter and receiver are shown in Figure 5.28. It can be seen from this figure that the delta PCM transmitter simply represents an adjacent sample differencing operation followed by a conventional (usually reduced word length) pulse code modulator. Similarly the receiver is a pulse code demodulator followed by an adjacent 'sample' summing operation. The reduced number of bits per PCM codeword translates directly into a saving of bandwidth. (This saving could, of course, then be traded against signal power and/or transmission time.) The delta PCM system cannot, however, accommodate rapidly varying transient signals as well as a conventional PCM system.

5.8.2 Differential PCM

The correlation between closely spaced samples of information signals originating from natural sources has already been referred to. An alternative way of expressing this phenomenon is to say that the signal contains redundant information, i.e. the same or similar information resides in two or more samples. An example of this is the redundancy present in a picture or image. To transmit an image over a digital communications link it is normally reduced to a 2-dimensional array of picture cells

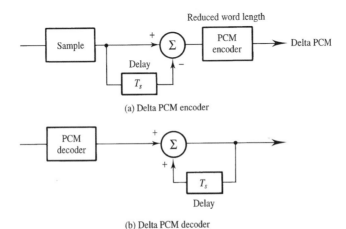

(a) Delta PCM encoder

(b) Delta PCM decoder

Figure 5.28 *Delta PCM transmitter and receiver.*

(pixels). These pixels are quantised in terms of colour and/or brightness. Nearly all naturally occurring images have very high average correlations, section 2.6, between the characteristics of adjacent pixels. Put crudely, if a given pixel is black then there is a high probability that the adjacent pixels will be at least nearly black. In this context the bandwidth of a random signal, the rate at which it is sampled, the correlation between closely spaced samples and its information redundancy are all closely related. One consequence of redundancy in a signal is that its future values can be predicted (within certain confidence limits) from its current and past values (see Chapter 16).

Differential PCM (DPCM) uses an algorithm to predict an information signal's future value. It then waits until the actual value is available for examination and subsequently transmits a signal which represents a correction to the predicted value. The correction signal is therefore a distillation of the information signal, i.e. it represents the information signal's surprising or unpredictable part. DPCM thus reduces the redundancy in a signal and allows the information contained in it to be transmitted using fewer symbols, less spectrum, shorter time and/or lower signal power. Figure 5.29(a) shows a block diagram of a DPCM transmitter. $g(t)$ is a continuous analogue information signal and $g(kT_s)$ is a sampled version of $g(t)$. k represents the (integer) sample number. $\varepsilon(kT_s)$ is the error between the actual value of $g(kT_s)$ and the value, $\hat{g}(kT_s)$, predicted from previous samples. It is this error which is quantised, to form $\varepsilon_q(kT_s)$, and encoded to give the DPCM signal which is transmitted. $\tilde{g}(kT_s) = \hat{g}(kT_s) + \varepsilon_q(kT_s)$ is an estimate of $g(kT_s)$ and is the predicted value $\hat{g}(kT_s)$, *corrected* by the addition of the *quantised* error $\varepsilon_q(kT_s)$. From the output of the sampling circuit to the input of the encoder the system constitutes a type of source coder. The DPCM receiver is shown in Figure 5.29(b) and is identical to the predictor loop in the transmitter. The reconstructed signal at the receiver is therefore the estimate, $\tilde{g}(t)$, of the original signal $g(t)$.

The predictor in DPCM systems is often a linear weighted sum of previous samples (i.e. a transversal digital filter) implemented using shift registers [Mulgrew *et al.*]. A schematic diagram of such a predictor is shown in Figure 5.30.

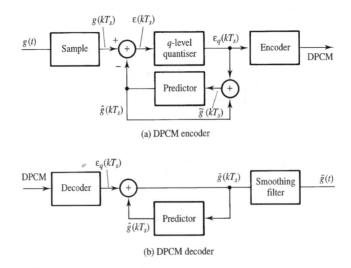

(a) DPCM encoder

(b) DPCM decoder

Figure 5.29 *DPCM transmitter and receiver.*

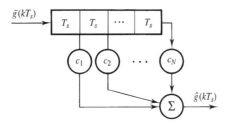

Figure 5.30 *DPCM predictor implemented as a transversal digital filter.*

5.8.3 Adaptive DPCM

Adaptive DPCM (ADPCM) is a more sophisticated version of DPCM. In this scheme the predictor coefficients (i.e. the weighting factors applied to the shift register elements) are continuously modified (i.e. adapted) to suit the changing signal statistics. (The values of these coefficients must, of course, also be transmitted over the communications link.) ITU has adopted ADPCM as a reduced bit rate standard. The ITU-T ADPCM encoder takes a 64 kbit/s companded PCM signal (G.711) and converts it to a 32 kbit/s ADPCM signal (G.721). The G.721 encoder and decoder are shown in Figure 5.31. The encoder uses 15 level, 4-bit, codewords to transmit the quantised difference between its input and estimated signal. The subjective quality of error-free 32 kbit/s ADPCM voice signals is only slightly inferior to 64 kbit/s companded PCM. For probabilities of error greater than

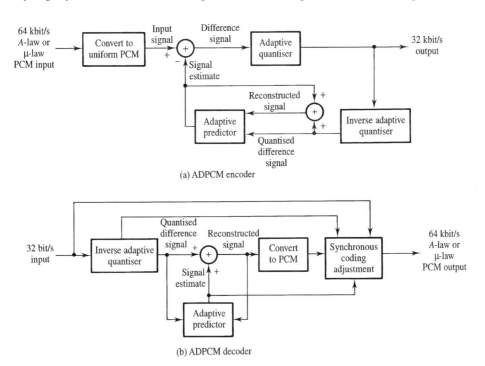

Figure 5.31 *ITU G.721 adaptive differential PCM (ADPCM) transmitter and receiver.*

10^{-4} its subjective quality is actually better than 64 kbit/s PCM, and 32 kbit/s ADPCM can achieve network quality speech with an MOS of 4.1 at error ratios of 10^{-3} to 10^{-2} for a complexity (measured by a logic gate count) of 10 times simple PCM. It allows an ITU-T 30 + 2 TDM channel (see Chapter 20) to carry twice the number of voice signals which are possible using 64 kbit/s companded PCM. Other specifications are defined by ITU-T, G.726 and G.727, for ADPCM with transmission rates of 16 to 40 kbit/s.

5.8.4 Delta modulation

If the quantiser of a DPCM system is restricted to 1 bit (i.e. two levels only, $\pm\Delta$) then the resulting scheme is called delta modulation (DM). This can be implemented by replacing the DPCM differencing block and quantiser with a comparator as shown in Figure 5.32(a). An especially simple prediction algorithm assumes that the next sample value will be the same as the last sample value, i.e. $\hat{g}(kT_s) = \tilde{g}[(k-1)T_s]$. This is called a previous sample predictor and is implemented as a one-sample delay. The DM decoder is shown in Figure 5.32(b). Specimen signal waveforms at different points in the DM system are shown in Figure 5.33. Slope overload noise, which occurs when $g(t)$ changes too rapidly for $\tilde{g}(kT_s)$ to follow faithfully, and quantisation noise (also called granular noise), are illustrated in Figure 5.33. There is a potential conflict between the requirements for acceptable quantisation noise and acceptable slope overload noise. To reduce the former the step size Δ should be small and to reduce the latter Δ should be large. One way of keeping both types of noise within acceptable limits is to make Δ small, but sample much faster than the normal minimum (i.e. Nyquist) rate. Typically the DM sampling rate will be many times the Nyquist rate. (This does mean that the bandwidth saving which DM systems potentially offer may be partially or completely eroded.)

If the information signal $g(t)$ remains constant for a significant period of time then $\tilde{g}(kT_s)$ displays a hunting behaviour and the resulting quantisation noise becomes a square wave with a period twice that of the sampling period. When this occurs the quantisation noise is called idling noise. Since the fundamental frequency of this noise is half the sampling frequency, which is usually itself many times the highest frequency in the information signal, then much of the idling noise will be removed by the smoothing filter in the DM receiver.

A realistic analysis of the SNR performance of DM systems is rather complicated. Here we quote results adapted from [Schwartz, 1980] for a random information signal with Gaussian pdf and white spectrum bandlimited to f_H Hz. In this case the signal to

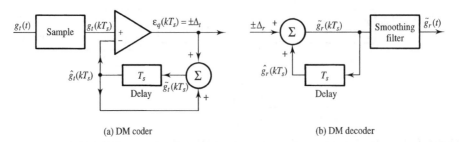

(a) DM coder   (b) DM decoder

Figure 5.32 *DM transmitter (a) and receiver (b) using previous sample (or one-tap) prediction.*

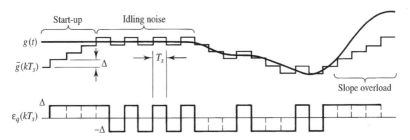

Figure 5.33 *DM signal waveforms illustrating slope overload and quantisation noise.*

slope overload noise ratio is:

$$\frac{S}{N_{ov}} = \frac{\sigma_g^2}{\sigma_{ov}^2} = 1.2 \left(\frac{1}{2\pi} \frac{\Delta}{4\sigma_g} \frac{f_s}{f_H} \right)^5 e^{1.5 \left(\frac{1}{2\pi} \frac{\Delta}{4\sigma_g} \frac{f_s}{f_H} \right)^2} \tag{5.33}$$

where σ_g is the standard deviation of the (Gaussian) information signal (equal to its RMS value if the DC component is zero), $N_{ov} = \sigma_{ov}^2$ is the variance (i.e. mean square) of the slope overload error, Δ is the DM step size, f_s is the DM sampling rate and f_H is the highest frequency in the baseband information signal (usually the information signal bandwidth). The SN_qR (neglecting the effect of the smoothing filter) is:

$$SN_qR = \frac{\sigma_g^2}{N_q} = 1.5 \left(\frac{4\sigma_g}{\Delta} \right)^2 \frac{f_s}{f_H} \tag{5.34}$$

For a given peak signal level (assumed here to be $4\sigma_g$) the DM step size Δ can be reduced by a factor of 2 for each factor of 2 increase in f_s without introducing any more slope overload noise. DM SN_qR therefore potentially increases with f_s^3 leading to a $30 \log_{10} 2$ or 9 dB improvement for each octave increase in sampling frequency.

Assuming that the slope overload noise, N_{ov}, and quantisation noise, N_q, are statistically independent then they can be added power-wise to give the total SNR, i.e.:

$$SNR = \frac{S}{N_{ov} + N_q} = \frac{1}{(S/N_{ov})^{-1} + (S/N_q)^{-1}} \tag{5.35}$$

Equation (5.35) neglects the effects of channel and/or receiver noise which, if severe, might cause $+\Delta$ and $-\Delta$ symbols to be received in error. Figure 5.34 shows SNR plotted against the normalised step size–sampling frequency product, $(\Delta/4\sigma_g)(f_s/f_H)$, for various values of normalised sampling frequency, f_s/f_H ($4\sigma_g$ is taken, for practical purposes, to be the maximum amplitude of the Gaussian information signal). The peaky shape of the curves in Figure 5.34 reflects the fact that slope overload noise dominates for small DM step size and quantisation noise dominates for large step size.

If the SNR is not sufficiently high then the DM receiver will occasionally interpret a received symbol in error (i.e. $+\Delta$ instead of $-\Delta$ or the converse). This is equivalent to the addition of an error of 2Δ (i.e. the difference in the analogue signal represented by $+\Delta$ and $-\Delta$) to the accumulated signal at the DM receiver. The estimated signal $\tilde{g}_r(kT_s)$ at the receiver thereafter follows the variation of the estimated signal at the transmitter $\tilde{g}_t(kT_s)$ but with a constant offset of 2Δ. This situation continues until another error occurs which

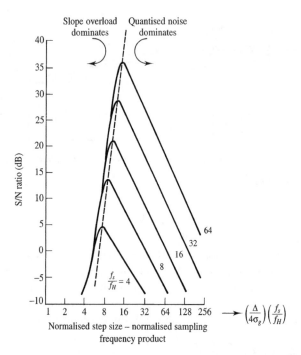

Figure 5.34 *SNR versus normalised step size–sampling frequency product for DM systems (parameter is normalised sampling frequency, f_s/f_H) (source: O'Neal, 1966, reproduced with permission of ATT Technical Journal).*

either cancels the first error or doubles it. The offset is therefore a stepped signal, step transitions occurring at random sample times with an average occurrence rate equal to the BER or P_e times the symbol rate. This is illustrated in Figure 5.35 which shows the result of errors being received in the DM signal and the consequent deviation from the ideal output. On average the error (or noise) represented by such a signal increases without limit as errors accumulate. In practice, however, most of the power in this 'pseudo DC signal' is contained at low frequencies (assuming P_e is small). Provided the frequency response of the post accumulator (or smoothing) filter goes to zero at 0 Hz then most of this noise can be removed. A simple way of showing that this is the case is to consider a post accumulator filter which, close to 0 Hz, has a high pass RC characteristic. The response of this filter to a step change of 2Δ volts at its input is $2\Delta e^{-t/\tau_c}$ where the time constant $\tau_c = RC$. The energy E_e (dissipated in 1 Ω) at the filter output due to a single step transition at its input is:

$$E_e = \int_0^\infty [2\Delta\, e^{-t/\tau_c}]^2\, dt$$

$$= 2\Delta^2 \tau_c \text{ (joules/error)} \tag{5.36}$$

The average 'noise' power, N_e, in the post filtered signal due to errors is therefore:

$$N_e = \text{BER} \times E_e \quad \text{(W)} \tag{5.37}$$

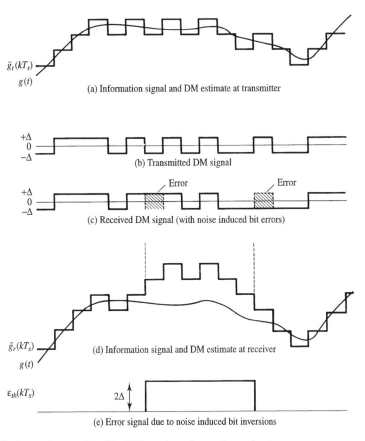

$\tilde{g}_t(kT_s)$
$g(t)$

(a) Information signal and DM estimate at transmitter

$+\Delta$
0
$-\Delta$

(b) Transmitted DM signal

Error   Error

$+\Delta$
0
$-\Delta$

(c) Received DM signal (with noise induced bit errors)

$\tilde{g}_r(kT_s)$
$g(t)$

(d) Information signal and DM estimate at receiver

$\varepsilon_{th}(kT_s)$ 2Δ

(e) Error signal due to noise induced bit inversions

Figure 5.35 *Stepped error signal in DM receiver due to thermal noise.*

where BER is the bit (or more strictly symbol) error rate. Using BER $= P_e f_s$, where P_e is the probability of error and f_s is the DM sampling rate, and putting $\tau_c = 1/2\pi f_L$, where f_L is the lowest frequency component with significant amplitude present in the information signal, we have:

$$N_e = \frac{\Delta^2}{\pi} \frac{f_s}{f_L} P_e \tag{5.38}$$

Thus as f_L is increased more of the noise due to bit errors is removed. If an ideal rectangular high pass characteristic is assumed in place of the *RC* characteristic the result is smaller by a factor of $2/\pi$ (i.e. approximately 2 dB) [Taub and Schilling].

5.8.5 Adaptive delta modulation

In conventional DM the problem of keeping both quantisation noise and slope overload noise acceptably low is solved by oversampling, i.e. keeping the DM step size small and sampling at many times the Nyquist rate. The penalty incurred is the loss of some, or all, of the saving in bandwidth which might be expected with DM. An alternative strategy is to make the DM step size *variable*, making it larger during periods when slope overload noise would otherwise dominate and smaller when quantisation noise might dominate. Such systems are called adaptive DM (ADM) systems.

A block diagram of an ADM transmitter is shown in Figure 5.36. The gain block, $G(kT_s)$, controls the variable step size represented by the *constant* amplitude pulses $\pm\Delta$. The step size is varied or adapted according to the history of ε_q. For example, if $\varepsilon_q = +\Delta$ for several adjacent samples it can be inferred that $g(t)$ is rising more rapidly than $\tilde{g}(kT_s)$ is capable of tracking it. Under this condition an ADM system increases the step size to reduce slope overload noise. Conversely if ε_q alternates between $+\Delta$ and $-\Delta$ then the inference is that $g(t)$ is changing slowly and slope overload is not occurring. The step size would therefore be decreased in order to reduce quantisation error. Figure 5.37 illustrates both these conditions. A simple ADM step size adjustment algorithm would be:

$$G(kT_s) = \begin{cases} G[(k-1)T_s]C, & \text{if } \varepsilon_q(kT_s) = \varepsilon_q[(k-1)T_s] \\ G[(k-1)T_s]/C, & \text{if } \varepsilon_q(kT_s) = -\varepsilon_q[(k-1)T_s] \end{cases} \tag{5.39}$$

where C is a constant > 1.

ADM typically performs with between 8 and 14 dB greater SNR than standard DM. Furthermore, since ADM uses large step sizes for wide variations in signal level and small step sizes for small variations in signal level it has wider dynamic range than standard DM. ADM can therefore operate at much lower bit rates than standard DM,

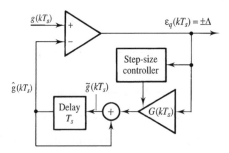

Figure 5.36 *Adaptive delta modulation (ADM) transmitter.*

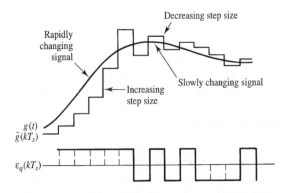

Figure 5.37 *ADM signal waveforms illustrating variable step size.*

typically 32 kbit/s and exceptionally 16 kbit/s (16 kbit/s ADM allows reduced quality digital speech to be transmitted directly over radio channels which are allocated 25 kHz channel spacings).

Other speech and audio coders, based on the use of source (entropy) coding techniques, are described in section 9.7. Here the source statistics enable a further reduction in the transmitted bit rate with improved decoded speech quality.

5.9 Summary

Continuous, analogue, information signals are converted to discrete, analogue, signals by the process of sampling. Pre-sampling anti-aliasing filters are used to limit any resulting distortion to acceptable levels. A naturally sampled signal can be converted to a PAM signal by flattening the pulse tops to give rectangular pulses. An ideal, impulse, sampled signal is similarly converted to a PAM signal by replacing the sample values with rectangular pulses of equivalent amplitude. PWM and PPM have a significant SNR advantage over PAM.

The minimum sampling rate required for an information signal with specified bandwidth is given by Nyquist's sampling theorem, if the signal is baseband, and by the bandpass sampling theorem if the signal is bandpass. In the bandpass case there is one or more allowed sampling rate bands rather than a simple sampling rate minimum. Reconstruction of a continuous information signal from a sampled, or PAM, signal can be achieved using a low pass filter. In the absence of noise such reconstruction can be essentially error free. Practical filter designs usually mean that modest oversampling is required if reconstruction is to be ideal.

Pulse modulated signals can be multiplexed together by interleaving the samples of the tributary information signals. The resulting TDM signal has narrower pulses, and therefore greater bandwidth, than is necessary for each of the tributary signals alone. It has the advantage, however, of allowing a single physical channel to carry many real time tributary signals, essentially simultaneously. Guard times are normally required between the adjacent pulses in a TDM signal to keep crosstalk, generated by the bandlimited channel, to acceptable levels.

Quantisation, the process which converts an analogue signal to a digital signal, results in the addition of quantisation noise, which decreases as the number of quantisation levels increases. Pulse code modulation replaces M quantisation levels with M codewords each comprising $n = \log_2 M$ binary digits. A PCM signal then represents each binary digit by a voltage pulse which can have either of two possible amplitudes. The SN_qR of an n-bit, linear, PCM voice signal is approximately $6(n-1)$ dB.

Companding of voice signals prior to PCM encoding increases the SN_qR of the decoded signal by increasing the resolution of the quantiser for small signals at the expense of the resolution for large signals. The ITU-T standard for digitally modulated companded voice signals is 8000 sample/s, 8 bit/sample giving a PCM bit rate of 64 kbit/s. Redundancy in transmitted PCM signals can be reduced by using DPCM techniques and its variants. ADPCM is an ITU standard for reduced bandwidth digital (32 kbit/s) voice transmission which gives received signal quality comparable to standard PCM. ADM, which is a variation on ADPCM, uses 1 bit quantisation.

5.10 Problems

5.1   (a) Sketch the design of TDM and FDM systems each catering for 12 voice channels of 4 kHz bandwidth.
(b) In the case of TDM, indicate how the information would be transmitted using: (i) PAM; (ii) PPM; (iii) PCM.
(c) Calculate the bandwidths required for the above FDM and TDM systems. You may assume the TDM system uses PPM with 2% resolution. [FDM 48 kHz; TDM 2.4 MHz]

5.2   Two low pass signals, each bandlimited to 4 kHz, are to be multiplexed into a single channel using pulse amplitude modulation. Each signal is impulse sampled at a rate of 10 kHz. If the time-multiplexed signal waveform is filtered by an ideal low pass filter (LPF) before transmission:
(a) What is the minimum clock frequency (or baud rate) of the system? [20 kHz]
(b) What is the minimum cut-off frequency of the LPF? [10 kHz]

5.3   Rewrite equation (5.5) such that it takes all spectral replicas into account. Hence find the aliasing induced SDR for Example 5.1 (part (ii)) accounting for the spectral replicas centred on f_s and $2f_s$ Hz.

5.4   An analogue bandpass signal has a bandwidth of 40 kHz and a centre frequency of 10 MHz. What is the minimum theoretical sampling rate which will avoid aliasing? What would be the best practical choice for the nominal sampling rate? Would an oscillator with a frequency stability of 1 part in 10^6 be adequate for use as the sampling clock? [80.16 kHz, 80.1603 kHz]

5.5   A rectangular PAM signal, with pulse widths of 0.1 μs, is transmitted over a channel which can be modelled by an RC low pass filter with a half-power bandwidth of 1.0 MHz. If an average XTR of 25 dB or better is to be maintained, estimate the required guard time between pulses. (Assume that pulse sampling at the channel output occurs at the optimum instants.) [0.36 μs]

5.6   Explain why the XTR in Problem 5.5 is degraded if the pulses are sampled (prematurely) at the midpoint of the nominal, rectangular pulse, time slots. Quantify this degradation for this particular system. [7.5 dB]

5.7   Twenty-five input signals, each bandlimited to 3.3 kHz, are each sampled at an 8 kHz rate and then time multiplexed. Calculate the minimum bandwidth required to transmit this multiplexed signal in the presence of noise if the pulse modulation used is: (a) PAM; (b) quantised PPM with a required level resolution of 5%; and (c) binary PCM with a required level resolution of > 0.5%. (This higher resolution requirement on PCM is normal for speech-type signals because the quantisation noise is quite objectionable.) [(a) 100 kHz, (b) 2 MHz, (c) 800 kHz]

5.8   A hi-fi music signal has a bandwidth of 20 kHz. Calculate the bit rate required to transmit this as a linearly quantised PCM signal maintaining an SN_qR of 55 dB. What is the minimum (baseband) bandwidth required for this transmission? (Assume that the signal's peak to mean ratio is 20 dB.) [480 kbit/s, 240 kHz]

5.9   Information is to be transmitted as a linearly quantised, 8 bit, signal over a noisy channel. What probability of error can be tolerated in the detected PCM bit stream if the reconstructed information signal is to have an SNR of 45 dB? [4.1 × 10^{-6}]

5.10   Show that the peak signal to quantisation noise ratio, $(SN_qR)_{peak}$, for an n-bit, μ-law companded, communications system is given by:

$$(SN_qR)_{peak} = \frac{3 \times 2^{2n}}{[\ln(1 + \mu)]^2}$$

Hence calculate the degraded output SNR in dB for large signal amplitudes in an $n = 8$ bit companded PCM system with typical μ value when compared with a linear PCM system operating at the same channel transmission rate. [15 dB]

5.11   An 8 bit *A*-law companded PCM system is to be designed with piecewise linear approximation as shown in Figure 5.26; 16 segments are employed (with 4 co-linear near the origin) and the segments join at 1/2, 1/4, etc., of the full scale value, as shown in Figure 5.26. Calculate the approximate SNR for full scale and small signal values.

[Full scale 36.3 dB, small signal 71.9 dB]

5.12   A DM communication system must achieve an SNR of 25 dB. The DM step size is 1.0 V. The (zero mean) baseband information signal has a bandwidth of 3.0 kHz and a Gaussian pdf with an RMS value of 0.2 V. Use Figure 5.34 to estimate both the optimum sampling frequency and the gain of the amplifier which is required immediately prior to the modulator input. [96 kHz, 10.5 dB]

CHAPTER 6

Baseband transmission and line coding

6.1 Introduction

This chapter addresses two issues central to baseband signal transmission, digital signal detection and line coding. The detection process described here is restricted to a simple implementation of the decision circuit in Figure 1.3. The analysis of detection processes is important since this leads to the principal objective measure of digital communications systems quality, i.e. the probability of symbols being detected in error, as quantified by probability theory (section 3.2). Appropriate coding of the transmitted symbol pulse shape can minimise the probability of error by ensuring that the spectral characteristics of the digital signal are well matched to the transmission channel. The line code must also permit the receiver to extract accurate PCM bit and word timing signals directly from the received data, for proper detection and interpretation of the digital pulses.

6.2 Baseband centre point detection

The detection of digital signals involves two processes:

1. Reduction of each received voltage pulse (i.e. symbol) to a single numerical value.
2. Comparison of this value with a reference voltage (or, for multisymbol signalling, a set of reference voltages) to determine which symbol was transmitted.

In the case of symbols represented by different voltage levels the simplest way of achieving 1 is to sample the received signal plus noise; 2 is then implemented using one or more comparators. In the case of equiprobable, binary, symbols (zero and one) represented by two voltage levels (e.g. 0 V and 3 V) intuition tells us that a sensible

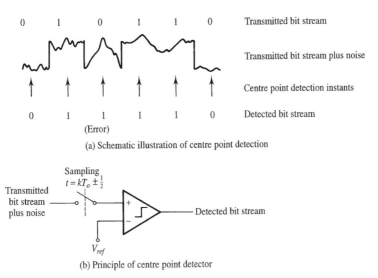

(a) Schematic illustration of centre point detection

(b) Principle of centre point detector

Figure 6.1 *Centre point detection.*

strategy would be to set the reference, V_{ref} (Figure 6.1(b)), mid-way between the two voltage levels (i.e. at 1.5 V). Decisions would then be made at the centre of each symbol period on the basis of whether the instantaneous voltage (signal plus noise) is above or below this reference. Sampling the instantaneous signal plus noise voltage somewhere near the middle of the symbol period is called *centre point detection*, Figure 6.1(a).

The noise present during detection is often either Gaussian or approximately Gaussian. (This is always the case if thermal noise is dominant, but may also be the case when other sources of noise dominate owing to operation of the central limit theorem.) Since noise with a Gaussian probability density function (pdf) is common and analytically tractable, the bit error rate (BER) of a communications system is often modelled assuming this type of noise alone.

6.2.1 Baseband binary error rates in Gaussian noise

Figure 6.2(a) shows the pdf of a binary information signal which can take on voltage levels V_0 and V_1 only. Figure 6.2(b) shows the pdf of a zero mean Gaussian noise process, $v_n(t)$, with RMS value σ V. (Since the process has zero mean the RMS value and standard deviation are identical.) Figure 6.2(c) shows the pdf of the sum of the signal and noise voltages. (Whilst Figure 6.2(c) is perhaps no surprise – we can think of the 'quasi-DC' symbol voltages biasing the mean value of the noise to V_0 and V_1 – recall, from Chapter 3, that when independent random variables are added their pdfs are convolved. Figure 6.2(c) is thus the convolution of Figures 6.2(a) and 6.2(b).)

For equiprobable symbols the optimum decision level is set at $(V_0 + V_1)/2$. (This is not the optimum threshold if the symbols are not equiprobable: see section 7.4.4.) Given that the symbol 0 is transmitted (i.e. a voltage level V_0) then the probability, P_{e1}, that the signal plus noise will be above the threshold at the decision instant is given by twice[1] the

---

[1] $p_0(v_n)$ and $p_1(v_n)$, as defined here, each represent a total probability of 0.5. Strictly speaking they are not, therefore, pdfs although their sum is the total signal plus noise pdf irrespective of whether a one or zero is transmitted. The pdf of the signal plus noise conditional on a zero being transmitted is $2p_0(v_n)$.

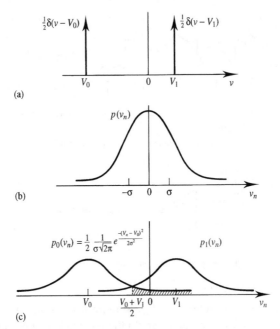

Figure 6.2 *Probability density function of: (a) binary symbol; (b) noise; (c) signal plus noise.*

shaded area under the curve $p_0(v_n)$ in Figure 6.2(c), i.e.:

$$P_{e1} = \int_{(V_0+V_1)/2}^{\infty} \frac{1}{\sigma\sqrt{(2\pi)}} e^{\frac{-(v_n-V_0)^2}{2\sigma^2}} \, dv_n \tag{6.1}$$

Using the change of variable $x = (v_n - V_0)/\sqrt{2}\sigma$ this becomes:

$$P_{e1} = \frac{1}{\sqrt{\pi}} \int_{(V_1-V_0)/2\sqrt{2}\sigma}^{\infty} e^{-x^2} \, dx \tag{6.2}$$

The incomplete integral in equation (6.2) cannot be evaluated analytically but can be recast as a complementary error function, erfc(x), defined by:

$$\text{erfc}(z) \overset{\Delta}{=} \frac{2}{\sqrt{\pi}} \int_{z}^{\infty} e^{-x^2} \, dx \tag{6.3}$$

Thus equation (6.2) becomes:

$$P_{e1} = \frac{1}{2} \text{erfc} \left(\frac{V_1 - V_0}{2\sigma\sqrt{2}} \right) \tag{6.4}$$

Alternatively, since erfc(z) and the error function, erf(z), are related by:

$$\text{erfc}(z) \equiv 1 - \text{erf}(z) \tag{6.5}$$

then P_{e1} can also be written as:

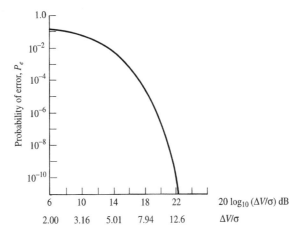

Figure 6.3 *Probability of error versus $\Delta V/\sigma$ (polar NRZ SNR $= 20 \log_{10}(\Delta V/\sigma) - 6$ dB, while unipolar SNR $= 20 \log_{10}(\Delta V/\sigma) - 3$ dB).*

$$P_{e1} = \frac{1}{2}\left[1 - \text{erf}\left(\frac{V_1 - V_0}{2\sigma\sqrt{2}}\right)\right] \tag{6.6}$$

The advantage of using the error (or complementary error) function in the expression for P_{e1} is that this function has been extensively tabulated[2] (see Appendix A).

If the digital symbol one is transmitted (i.e. a voltage level V_1) then the probability, P_{e0}, that the signal plus noise will be below the threshold at the decision instant is:

$$P_{e0} = \int_{-\infty}^{(V_0+V_1)/2} \frac{1}{\sigma\sqrt{(2\pi)}} e^{\frac{-(v_n - V_1)^2}{2\sigma^2}} \, dv_n \tag{6.7}$$

It is clear from the symmetry of this problem that P_{e0} is identical to both P_{e1} and the probability of error, P_e, irrespective of whether a one or zero was transmitted. Noting that the probability of error depends on only symbol voltage difference, and not absolute voltage levels, P_e can be rewritten in terms of $\Delta V = V_1 - V_0$, i.e.:

$$P_e = \frac{1}{2}\left[1 - \text{erf}\left(\frac{\Delta V}{2\sigma\sqrt{2}}\right)\right] \tag{6.8}$$

Equation (6.8) is valid for both *unipolar* signalling (i.e. symbols represented by voltages of 0 and ΔV) and *polar signalling* (i.e. symbols represented by voltages of $\pm\Delta V/2$). In fact, it is valid for all pulse levels and shapes providing that ΔV represents the voltage *difference at the sampling instant*. Figure 6.3 shows P_e versus the voltage ratio $\Delta V/\sigma$ expressed both in dB and as a ratio.

---

[2] The Q-function, which represents the area under the tail of a (zero mean, unit variance) Gaussian pdf, is defined by $Q(z) = \int_z^\infty (1/\sqrt{2\pi})e^{-(x^2/2)} \, dx$. This function is often used as an alternative to erfc(z) in the formulation of P_e problems and is related to it by $Q(z) = \frac{1}{2}\text{erfc}(z/\sqrt{2})$ or erfc $(z) = 2Q(z\sqrt{2})$.

Whilst the x-axis of Figure 6.3 is clearly related to signal-to-noise ratio it is not identical to it for all pulse levels and shapes. This is partly because it involves the signal only at the sampling instant (whereas a conventional SNR uses a time averaged signal power) and partly because the use of ΔV neglects any transmitted DC component. For NRZ (non-return to zero, see section 6.4), unipolar, rectangular pulse signalling, Figure 6.4, the normalised peak signal power is $S_{peak} = \Delta V^2$ and the average signal power is $S = \Delta V^2/2$. The normalised Gaussian noise power is $N = \sigma^2$. We therefore have:

$$\frac{\Delta V}{\sigma} = \left(\frac{S}{N}\right)_{peak}^{1/2} = \sqrt{2}\left(\frac{S}{N}\right)^{1/2} \tag{6.9}$$

where $(S/N)_{peak}$ indicates peak signal power divided by average (or expected) noise power. (The peak noise power would in principle be infinite for Gaussian noise.) Substituting for $\Delta V/\sigma$ in equation (6.8) we therefore have:

$$P_e = \frac{1}{2}\left[1 - \operatorname{erf}\frac{1}{2\sqrt{2}}\left(\frac{S}{N}\right)_{peak}^{1/2}\right] \tag{6.10(a)}$$

or

$$P_e = \frac{1}{2}\left[1 - \operatorname{erf}\frac{1}{2}\left(\frac{S}{N}\right)^{1/2}\right] \tag{6.10(b)}$$

For NRZ, polar, rectangular pulse signalling with the same voltage spacing as in the unipolar case, Figure 6.5, the peak and average signal powers are identical, i.e. $S_{peak} = S = (\Delta V/2)^2$ and we can therefore write:

$$\frac{\Delta V}{\sigma} = 2\left(\frac{S}{N}\right)_{peak}^{1/2} = 2\left(\frac{S}{N}\right)^{1/2} \tag{6.11}$$

Substituting into equation (6.8) we now have:

$$P_e = \frac{1}{2}\left[1 - \operatorname{erf}\frac{1}{\sqrt{2}}\left(\frac{S}{N}\right)^{1/2}\right] \tag{6.12}$$

where a distinction between peak and average SNR no longer exists. Equations (6.10) and (6.12) are specific to a particular signalling format. In contrast equation (6.8) is general and, therefore, more fundamental. For statistically independent, equiprobable

Figure 6.4 *Unipolar rectangular pulse signal.*

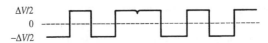

Figure 6.5 *Polar rectangular pulse signal.*

binary symbols, such as are being discussed here, each symbol carries 1 bit of information (see Chapter 9). The probability of symbol error, P_e, in this special case is therefore identical to the probability of bit error, P_b.

Whilst probabilities of symbol and bit error are the quantities usually *calculated*, it is the symbol error rate (SER) or bit error rate[3] (BER) which is usually *measured*. These are simply the number of symbol or bit errors occurring per unit time (usually 1 second) measured over a convenient (and sometimes specified) period. The symbol error rate is clearly related to the probability of symbol error by:

$$\text{SER} = P_e R_s \qquad (6.13)$$

where R_s is the symbol rate in symbol/s or baud. Bit error rate is related to the probability of bit error by:

$$\text{BER} = P_b R_s n = P_b R_b \qquad (6.14)$$

where n is the number of binary digits mapped to each symbol and R_b is the bit rate.

EXAMPLE 6.1

Find the BER of a 100 kbaud, equiprobable, binary, polar, rectangular pulse signalling system assuming ideal centre point decisions, if the measured SNR at the detector input is 12.0 dB.

$$\frac{S}{N} = 10^{\frac{12}{10}} = 15.85$$

Using equation (6.12):

$$P_e = P_b = \frac{1}{2}\left[1 - \text{erf}\,\frac{1}{\sqrt{2}}(15.85)^{\frac{1}{2}}\right]$$

$$= \frac{1}{2}\,[1 - \text{erf}(2.815)] = 3.45 \times 10^{-5}$$

Using equation (6.14):

$$\text{BER} = P_b R_s n$$

$$= 3.45 \times 10^{-5} \times 100 \times 10^3 \times 1 = 3.45 \text{ bit errors/s}$$

6.2.2 Multilevel baseband signalling

Figure 6.6 shows a schematic diagram of a multilevel (or multisymbol) signal. If the number of equally spaced, allowed levels is M then the symbol plus noise pdfs look like those in Figure 6.7 (drawn for $M = 4$). The probability of symbol error for the $M - 2$ inner symbols (i.e. any but the symbols represented by the lowest or highest voltage level) is now twice that in the binary case. This is because the symbol can be in error if the signal plus noise voltage is too high *or* too low, i.e.:

---

[3] SER and BER are also used to indicate symbol error *ratio* and bit error *ratio*. These are then synonymous with P_e and P_b respectively.

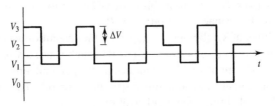

Figure 6.6 *Illustration of waveform for four-level baseband signalling.*

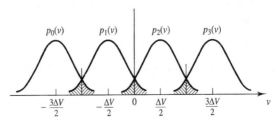

Figure 6.7 *Conditional pdfs for the four-level baseband signalling system in Figure 6.6 assuming Gaussian noise ($p_i(v) = p(v|V_i)$).*

$$P_{eM}\big|_{inner\ symbols} = 2P_e \tag{6.15}$$

The symbol error for the two outer levels (i.e. the symbols represented by the lowest and highest voltage levels) is identical to that for the binary case, i.e.:

$$P_{eM}\big|_{outer\ symbols} = P_e \tag{6.16}$$

Once again assuming equiprobable symbols, the average probability of symbol error is:

$$P_{eM} = \frac{M-2}{M}\, 2P_e + \frac{2}{M}\, P_e$$

$$= \frac{2(M-1)}{M}\, P_e \tag{6.17}$$

Substituting for P_e from equation (6.8) we have:

$$P_{eM} = \frac{M-1}{M}\left[1 - \mathrm{erf}\left(\frac{\Delta V}{2\sigma\sqrt{2}}\right)\right] \tag{6.18}$$

where ΔV is the difference between the equally spaced, adjacent, voltage levels, Figure 6.6. (Multilevel signalling is extended from baseband to bandpass systems in Chapter 11 where the different symbols are represented by differences in signal frequency, amplitude or phase.)

EXAMPLE 6.2

A four-level, equiprobable, baseband signalling system uses NRZ rectangular pulses. The attenuation between transmitter and receiver is 15 dB and the noise power at the 50 Ω input of an ideal centre point decision detector is 10 μW. Find the average signal power which must be transmitted to maintain a symbol error probability of 10^{-4}.

The standard deviation of the noise voltage is equal to the RMS noise voltage (since the noise has zero mean).

$$\sigma = \sqrt{P\,R}$$

$$= \sqrt{1 \times 10^{-5} \times 50} = 2.236 \times 10^{-2} \ \ (V)$$

Rearranging equation (6.18):

$$\Delta V = 2\sigma\,\sqrt{2}\ \text{erf}^{-1}\left[1 - \frac{M\,P_{eM}}{M-1}\right]$$

$$= 2 \times 2.236 \times 10^{-2} \times \sqrt{2}\ \text{erf}^{-1}\left[1 - \frac{4 \times 10^{-4}}{4-1}\right]$$

$$= 6.324 \times 10^{-2}\ \text{erf}^{-1}(0.999867) = 0.171 \ \ (V)$$

Thus symbol levels are ± 85.5 mV, ± 256 mV in Figures 6.6 and 6.7.

Two of the symbols are represented by received power levels of $0.0855^2/50 = 1.46 \times 10^{-4}$ W and two by received power levels of $0.256^2/50 = 1.31 \times 10^{-3}$ W. Since the symbols are equiprobable the average received power, S_R, must be:

$$S_R = \tfrac{1}{2}(0.146 + 1.31)10^{-3}\ \text{W} = 0.728\ \text{mW}$$

The transmitted power, P_T, is therefore:

$$P_T = S_R \times 10^{\frac{15}{10}}$$

$$= 0.728 \times 31.62 = 23\ \text{mW}$$

6.3 Error accumulation over multiple hops

All signal transmission media (e.g. cables, waveguides, optical fibres) attenuate signals to a greater or lesser extent. This is even the case for the space through which radio waves travel (although for free space, the use of the word attenuate, to describe this effect, might be disputed – see Chapters 12 and 14). For long communication paths attenuation might be so severe that the sensitivity of normal receiving equipment would be inadequate to detect the signal. In such cases the signal is boosted in amplitude at regular intervals along the transmitter–receiver path. The equipment which boosts the signal is called a *repeater* and the path between adjacent repeaters is called a *hop*. (A repeater along with its associated, preceding, hop is called a *section*.) Thus long distance communication is usually achieved using multiple hops (see also later, section 20.5.2).

Repeaters used on multi-hop links can be divided into essentially two types. These are *amplifying* repeaters and *regenerative* repeaters. For analogue communications linear amplifiers are required. For digital communications either type of repeater could be used but normally regenerative repeaters are employed.

Figure 6.8 shows a schematic diagram of an *m*-hop link. (*m* hops imply a transmitter, a receiver and *m* −1 repeaters.) If a binary signal with voltage levels $\pm\Delta V/2$ is transmitted then the voltage received at the input of the first amplifying repeater is $\pm\alpha\Delta V/2 + n_1(t)$ where α is the linear voltage attenuation factor (or voltage gain factor < 1) and $n_1(t)$ is the random noise (with standard deviation, or RMS value, σ) added during the first hop. In a well designed system the voltage gain, G_V, of the repeater will be just adequate to compensate for the attenuation over the first hop, i.e. $G_V = 1/\alpha$. At the output of the first repeater the signal is restored to its original level (i.e. $\pm\Delta V/2$) but the noise signal is also amplified to a level $G_V n_1(t)$ and will now have an RMS value of $G_V\sigma$. (Chapter 12 contains a discussion on the origin of such noise.)

Assuming that each hop incurs the same attenuation the signal voltage at the input to the second repeater will again have fallen to $\pm\alpha\Delta V/2$ and the noise voltage from the first hop will have fallen to $n_1(t)$. A similar noise voltage, $n_2(t)$, arising from the second hop will also, however, be added and providing it is statistically independent from $n_1(t)$ will add to it on a power basis. The total noise power (in 1 Ω) will therefore be the sum of the noise variances, i.e. $2\sigma^2$ assuming equal noise is added on each hop. It can therefore be seen that with amplifying repeaters the noise power after *m* hops will be *m* times the noise power after one hop whilst the signal voltage at the receiver will be essentially the same as at the input to the first repeater. The probability of error for an *m*-hop link is therefore given by equation (6.8) with the RMS noise voltage, σ, replaced by $\sigma\sqrt{m}$, i.e.:

$$P_e\big|_{m\ hops} = \frac{1}{2}\left[1 - \mathrm{erf}\left(\frac{\Delta V}{2\sigma\sqrt{2m}}\right)\right] \tag{6.19}$$

If the linear amplifiers are replaced with regenerative repeaters, Figure 6.9, the situation

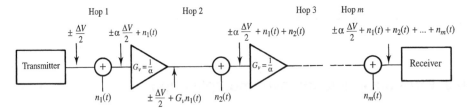

Figure 6.8 *Multi-hop link utilising linear amplifiers as signal boosters.*

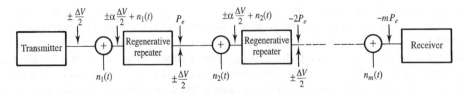

Figure 6.9 *Multi-hop link utilising regenerative repeaters as signal boosters.*

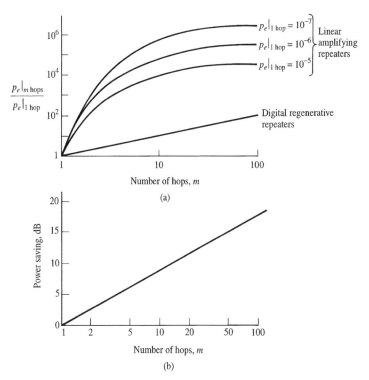

Figure 6.10 *(a) P_e degradation due to multiple hops for different repeater types (source: Stremler, 1990, reproduced with the permission of Addison-Wesley) and (b) saving in transmission power using digital regenerative, instead of linear, repeaters for $P_e = 10^{-5}$ (source: Carlson, 1986, reproduced with the permission of McGraw-Hill).*

changes dramatically. The repeater now uses a decision process to establish whether a digital 0 or 1 is present at its input and a new, and noiseless, pulse is generated for transmission to the next repeater. Noise does not therefore accumulate from repeater to repeater as it does in the case of linear amplifiers. There is, however, an equivalent process in that symbols will be detected in error at each repeater with a probability P_e given by equation (6.8). Providing this probability is small (specifically $mP_e \ll 1$) the probability of any given symbol being detected in error (and therefore inverted) more than once over the m hops of the link can be neglected. In this case the probability of error (rather than the noise power) accumulates linearly over the hops and after m hops we have:

$$P_e\big|_{m\ hops} = mP_e = \frac{m}{2}\left[1 - \mathrm{erf}\left(\frac{\Delta V}{2\sigma\sqrt{2}}\right)\right] \tag{6.20}$$

where P_e is the one-hop probability given in equation (6.8).

Figure 6.10(a) illustrates the increase in $P_e\big|_{m\ hops}$ as the number of hops increases for both amplifying and regenerative repeaters, clearly showing the benefit of digital regeneration. Figure 6.10(b) shows the typical saving in transmitter power (per repeater) realised using digital regeneration. (The power saving shown is the square of the factor by which ΔV in equation (6.19) must be larger than ΔV in equation (6.20) for $P_e = 10^{-5}$ after m hops.)

EXAMPLE 6.3

If 15 link sections, each identical to that described in Example 6.1, are cascaded to form a 15-hop link find the probability of bit error when the repeaters are implemented as (i) linear amplifiers and (ii) digital regenerators.

(i) Using equation (6.11):

$$\frac{\Delta V}{\sigma} = 2\left(\frac{S}{N}\right)^{\frac{1}{2}} = 2(15.85)^{\frac{1}{2}} = 7.962$$

Now using equation (6.19):

$$P_e = \frac{1}{2}\left[1 - \text{erf}\left(\frac{7.962}{2\sqrt{2}\sqrt{15}}\right)\right]$$

$$= \frac{1}{2}\left[1 - \text{erf}(0.727)\right] = 0.152$$

(ii) Since $mP_e = 15 \times 3.45 \times 10^{-5} \ll 1.0$ we can use equation (6.20), i.e.:

$$P_e|_{15\ hops} \approx 15 \times P_e|_{1\ hop} = 5.175 \times 10^{-4}$$

The advantage of regenerative repeaters in this example is clear.

6.4 Line coding

The discussion of baseband transmission up to now has centred on the BER performance of unipolar, and polar, rectangular pulse representations of the binary symbols 0 and 1. Binary data can, however, be transmitted using many other pulse types. The choice of a particular pair of pulses to represent the symbols 1 and 0 is called line coding and selection is usually made on the grounds of one or more of the following considerations:

1. Presence or absence of a DC level.
2. Power spectral density – particularly its value at 0 Hz.
3. Spectral occupancy (i.e. bandwidth).
4. BER performance (i.e. relative immunity from noise).
5. Transparency (i.e. the property that any arbitrary symbol, or bit, pattern can be transmitted and received).
6. Ease of clock signal recovery for symbol synchronisation.
7. Presence or absence of inherent error detection properties.

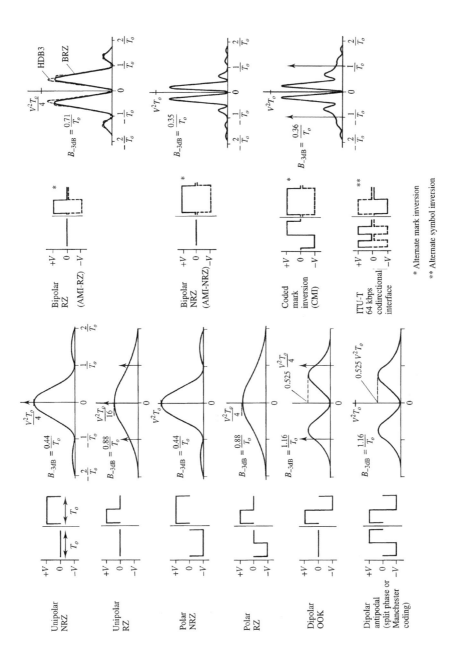

Figure 6.11 *Selection of commonly used line code symbols (0, 1) and associated spectra.*

Line coding is usually thought of as the selection, or design, of pulse pairs which retain sharp transitions between voltage levels. Figure 6.11, in which T_o represents the symbol period, shows a variety of pulse types and spectra corresponding to several commonly used line codes. Figure 6.12 shows the transmitted waveforms of these (and a

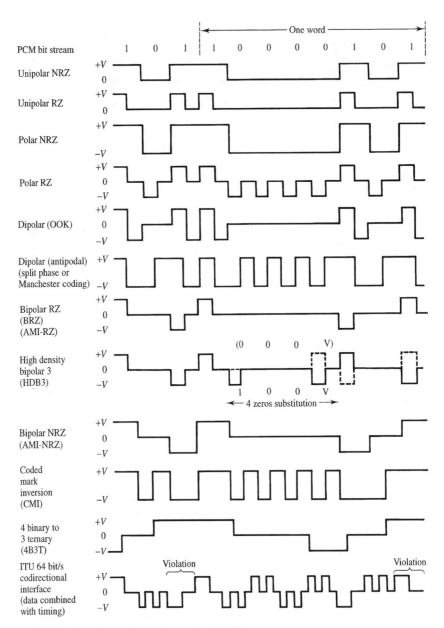

Figure 6.12 *Various line code waveforms for PCM bit sequence.*

few other) line codes for an example sequence of binary data and Table 6.1 compares some of their important properties.

After line coding the pulses may be filtered or otherwise shaped to further improve their properties: for example, their spectral efficiency and/or immunity to intersymbol interference (see Chapter 8). The distinction between line coding and pulse shaping is not always easy to make, and in some cases, it might be argued, artificial. Here, however,

Table 6.1 *Comparison of line (baseband) code performance.*

| | *Timing extraction* | *Error detection* | *Relative transmitter power (single point decisions)* | | *First null bandwidth* | *AC coupled* | *Trans- parent* |
			Average	*Peak*			
Unipolar (NRZ)	Difficult	No	2	4	f_o	No	No
Unipolar (RZ)	Simple	No	1	4	$2f_o$	No	No
Polar (NRZ)	Difficult	No	1	1	f_o	No	No
Polar (RZ)	Rectify	No	½	1	$2f_o$	No	No
Dipolar – OOK	Simple	No	2	4	$2f_o$	Yes	No
Dipolar – split ϕ	Difficult	No	1	1	$2f_o$	Yes	Yes
Bipolar (RZ)	Rectify	Yes	1	4	f_o	Yes	No
Bipolar (NRZ)	Difficult	Yes	2	4	$f_o/2$	Yes	No
HDB3	Rectify	Yes	1	4	f_o	Yes	Yes
CMI	Simple	Yes	See note	See note	$2f_o$	Yes	Yes

Note for CMI transmission at least two samples per symbol are required (e.g. taken 0.25 and 0.75 of the way through the symbol pulse). The difference between these samples results in a detected voltage which is twice the transmitted level for a digital zero, and a voltage of zero for a digital 1. The RMS noise, σ, is increased by a factor of $\sqrt{2}$ due to the addition of the independent noise samples. On the (same) equal BER basis as the re- quired transmitter powers given in the table this results in a required CMI relative transmitter power (both av- erage and peak) of 2. To compare this fairly with the other line codes, the relative powers required in the table must all be reduced by a factor of 0.5 to reflect the factor of 2 improvement in SNR due to double sampling.

we make the distinction if for no other reason than to subdivide our discussion of baseband transmission into manageable parts. The line codes included in Figures 6.11 and 6.12, and Table 6.1, do not form a comprehensive list, there being many other baseband signalling formats. One notable line code not included here, for example, is the Miller code, which has a particularly narrow spectrum centred on $0.4/T_o$ Hz [Stremler, Sklar].

For the simple binary signalling formats (with statistically independent symbols), shown on the left hand side of Figure 6.11, the power spectral densities can be found using [Stremler]:

$$G(f) = p(1-p)\frac{1}{T_o}|F_1(f) - F_2(f)|^2$$

$$+ \frac{1}{T_o^2}\sum_{n=-\infty}^{\infty}|pF_1(nf_o) + (1-p)F_2(nf_o)|^2\,\delta(f - nf_o) \tag{6.21}$$

where p is the probability of symbol 1, $f_o = 1/T_o$ represents the symbol rate and $F(f)$ is the symbol voltage spectrum.

The codes which appear in Figures 6.11, 6.12 and Table 6.1 are now described.

6.4.1 Unipolar signalling

Unipolar signalling (also called on–off keying, OOK) refers to a line code in which one binary symbol (denoting a digital 0, for example) is represented by the absence of a pulse (i.e. a space) and the other binary symbol (denoting a digital 1) is represented by the presence of a pulse (i.e. a mark). There are two common variations on unipolar signalling, namely non-return to zero (NRZ) and return to zero (RZ). In the former case the duration (τ) of the mark pulse is equal to the duration (T_o) of the symbol slot. In the latter case τ is less than T_o.

Typically RZ pulses fill only the first half of the time slot, returning to zero for the second half. (The mark duty cycle, τ/T_o, would be 50% in this case although other duty cycles can be, and are, used.) The power spectral densities of both NRZ and RZ signals have a $[(\sin x)/x]^2$ shape where $x = \pi\tau f$. RZ signals (assuming a 50% mark duty cycle) have the disadvantage of occupying twice the bandwidth of NRZ signals (see Figure 6.11). They have the advantage, however, of possessing a spectral line at the symbol rate, $f_o = 1/T_o$ Hz (and its odd integer multiples), which can be recovered for use as a symbol timing clock signal, Table 6.1. Non-linear processing must be used to recover a clock waveform from an NRZ signal (see section 6.7).

Both NRZ and RZ unipolar signals have a non-zero average (i.e. DC) level represented in their spectra by a line at 0 Hz, Figure 6.11. Transmission of these signals over links with either transformer or capacitor coupled (AC) repeaters results in the removal of this line and the consequent conversion of the signals to a polar format. Furthermore, since the continuous part of both the RZ and NRZ signal spectrum is non-zero at 0 Hz then AC coupling results in distortion of the transmitted pulse shapes. If the AC coupled lines behave as high pass RC filters (which is typically the case) then the distortion takes the form of an exponential decay of the signal amplitude after each transition. This effect, referred to as signal 'droop', is illustrated in Figure 6.13 for an NRZ signal. Although the long term DC component is zero, after AC coupling 'short term' DC levels accumulate with long strings of ones or zeros. The accumulated DC level is most apparent for the first few symbols after a string represented by a constant voltage. Neither variety of unipolar signal is therefore suitable for transmission over AC coupled lines.

Since unipolar voltage levels of 0 or V volts are equivalent (in terms of BER) to polar levels of $\pm V/2$ volts (section 6.2.1) then unipolar signalling requires twice the average, and four times the peak, transmitter power when compared with polar signalling, Table 6.1.

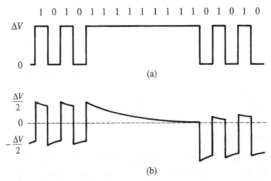

(a)

(b)

Figure 6.13 *Distortion due to AC coupling of unipolar NRZ signal: (a) input; (b) output.*

6.4.2 Polar signalling

In polar signalling systems a binary one is represented by a pulse $g_1(t)$ and a binary zero by the opposite (or *antipodal*) pulse $g_0(t) = -g_1(t)$, Figure 6.11. Figure 6.12 compares polar and unipolar signals for a typical data stream. The NRZ and RZ forms of polar signals have identically shaped spectra to the NRZ and RZ forms of unipolar signals except that, due to the opposite polarity of the one and zero symbols, neither contain any spectral lines. Polar signals have the same bandwidth requirements as their equivalent unipolar signals and suffer the same distortion effects (in particular signal droop) if transmitted over AC coupled lines, Table 6.1.

As pointed out in section 6.4.1 polar signalling has a significant power (or alternatively BER) advantage over unipolar signalling. Fundamentally this is because the pulses in a unipolar scheme are only *orthogonal* whilst the pulses in a polar scheme are *antipodal*. Another way of explaining the difference in performance is to observe that the average or DC level transmitted with unipolar signals contains no information and is therefore wasted power.

Polar binary signalling also has the advantage that, providing the symbols are equiprobable, the decision threshold is 0 V. This means that no automatic gain control (AGC) is required in the receiver.

6.4.3 Dipolar signalling

Dipolar signalling is designed to produce a spectral null at 0 Hz. This makes it especially well suited to AC coupled transmission lines. The symbol interval, T_o, is split into positive and negative pulses each of width $T_o/2$ s, Figure 6.11. This makes the total area under either pulse type equal to zero which results in the desirable DC null in the signal's spectrum. Both OOK and antipodal forms of dipolar signalling are possible, the latter being called split phase or Manchester coding (Figure 6.11). A spectral line at the clock frequency ($1/T_o$ Hz) is present in the OOK form but absent in the antipodal form.

Manchester coding is widely used for the distribution of clock signals within VLSI circuits, for magnetic recording and for Ethernet LANs (see Chapter 21).

6.4.4 Bipolar alternate mark inversion signalling

Bipolar signalling (also called alternate mark inversion, AMI) uses three voltage levels $(+V, 0, -V)$ to represent two binary symbols (0 and 1) and is therefore a pseudo-ternary line code. Zeros, as in unipolar signalling, are represented by the absence of a pulse (i.e. 0 V) and ones (or *marks*) are represented alternately by voltage levels of $+V$ and $-V$. Both RZ and NRZ forms of bipolar signalling are possible, Figure 6.11, although the RZ form is more common. Alternating the mark voltage level ensures that the bipolar spectrum has a null at DC and that signal droop on AC coupled lines is avoided. The alternating mark voltage also gives bipolar signalling a single error detection capability and reduces its bandwidth over that required for the equivalent unipolar or polar format (see Figure 6.11).

6.4.5 Pulse synchronisation and HDB*n* coding

Pulse synchronisation is usually required at a repeater or receiver to ensure that the samples, on the basis of which symbol decisions are made, are taken at the correct instants in time. In principle those line codes (such as unipolar RZ and dipolar OOK) which possess a spectral line at $1/T_o$ Hz have an inherent pulse synchronisation capability since all that is required to regenerate a clock signal is for this spectral line to be extracted using a filter or phase locked loop. Other line codes which do not possess a convenient spectral line can often be processed, e.g. by rectification, in order to generate one (see later Figures 6.26 to 6.29). This is the case for the bipolar RZ (BRZ) line code which is often used in practical PCM systems.

Although rectification of a BRZ signal results in a unipolar RZ signal and therefore a spectral line at $1/T_o$ Hz, in practice there is a problem if long strings of zeros are transmitted. In this case pulse synchronisation might be lost due, for instance, to loss of lock of the pulse timing phase locked loop, Figure 6.30. To prevent this, many BRZ systems use high density bipolar (HDB*n*) substitution. Here, when the number of continuous zeros exceeds *n* then they are replaced by a special code. $n = 3$ (HDB3) is the code recommended (G.703) by ITU-T for PCM systems at multiplexed bit rates of 2, 8 and 34 Mbit/s (see section 19.2). In HDB3 a string of four zeros is replaced by either $000V$ or $100V$. Here V is a binary 1 with sign chosen to *violate* the alternating mark rule so that it can be detected as the special sequence representing the all-zero code, Figure 6.12. Furthermore, consecutive violation (V) pulses alternate in polarity to avoid introducing a DC component. (This is achieved by having the two possibilities $000V$ or $100V$. The selection depends on the number of digital ones since the last code insertion.) The HDB spectrum has minor variations compared with the BRZ signal from which it is derived (Figure 6.11). HDB3 is sometimes referred to as B4ZS denoting bipolar signalling with four-zeros substitution.

6.4.6 Coded mark inversion (CMI)

CMI is a polar NRZ code which uses both amplitude levels (each for half the symbol period) to represent a digital 0 and either amplitude level (for the full symbol period) to represent a digital 1. The level used alternates for successive digital ones. CMI is therefore a combination of dipolar signalling (used for digital zeros) and NRZ AMI (used for digital ones). CMI is the code recommended (G.703) by ITU-T for 140 Mbit/s multiplexed PCM (see Chapter 20). The ITU codirectional interface at 64 kbit/s uses a refinement of CMI, Figure 6.11, in which the polarity of consecutive symbols (irrespective of whether they are ones or zeros) is alternated. Violations of the alternation rule are then used every eighth symbol to denote the last bit of each (8 bit) PCM codeword, Figure 6.12.

6.4.7 *n*B*m*T coding

*n*B*m*T is a line code in which *n* binary symbols are mapped into *m* ternary symbols. A coding table for $n = 4$ and $m = 3$ (i.e. 4B3T) is shown as Table 6.2. This code lengthens the transmitted symbols to reduce the signal bandwidth. In Table 6.2 the top six outputs are balanced and hence are fixed. The lower 10 have a polarity imbalance and need occasionally to be inverted to avoid the running digital sum causing DC wander. In the table the left hand column of coded signals or symbols (which sum to a zero or positive value) is used if the preceding running sum is negative. The right hand column, in which the symbols sum to zero or a negative value, is used if the preceding running sum is positive. With three ternary symbols we have $3^3 = 27$ possible states and by not transmitting (0, 0, 0) we have 26, comprising the 6 + 10 + 10 unique states in Table 6.2. As $2^4 = 16$ this conveniently matches the 4 bit binary code requirements.

Table 6.2 *4B3T coding example showing uncoded binary and coded ternary signals.*

Binary input signal	Ternary output signal Running digital sum at end of preceding word equal to	
	−2, −1 or 0	1, 2 or 3
0000	+ 0 −	+ 0 −
0001	− + 0	− + 0
0010	0 − +	0 − +
0011	+ − 0	+ − 0
0100	0 + −	0 + −
0101	− 0 +	− 0 +
0110	0 0 +	0 0 −
0111	0 + 0	0 − 0
1000	+ 0 0	− 0 0
1001	+ + −	− − +
1010	+ − +	− + −
1011	− + +	+ − −
1100	0 + +	0 − −
1101	+ 0 +	− 0 −
1110	+ + 0	− − 0
1111	+ + +	− − −

The 4B3T spectrum is further modified from that of HDB3 skewing the energy towards low frequencies and 0 Hz [Flood and Cochrane]. A further development of this concept is 2B1Q where two binary bits are converted into one four-level (quaternary) symbol. This is an example of *M*-ary (compare with *bin*ary, *tern*ary, etc.) signalling. *M*-ary coding is examined further in sections 11.4 and 11.5 in the context of carrier based coded signals.

6.5 Multiplex telephony

PCM is used in conjunction with TDM to realise multichannel digital telephony. The internationally agreed *European* ITU standard provides for the combining of 30 speech channels, together with two subsidiary channels for signal and system monitoring. Each speech channel signal is sampled at 8 kHz and non-linearly quantised (or companded) into 8 bit words (see Chapter 5). The binary symbol rate per speech channel is therefore 64 kbit/s, and for the composite 30 + 2 channel signal multiplex is 32×64 kbit/s = 2.048 Mbit/s. For convenience this is often referred to as a 2 Mbit/s signal. A key advantage of the 2 Mbit/s TDM multiplex is that it is readily transmitted over 2 km sections of twisted pair (copper) cables which originally carried only one analogue voice signal. Now, 2 Mbit/s signals are also transmitted over optical fibre circuits. In the USA and Japan the multiplex combines fewer speech channels into a 1.5 Mbit/s signal.

The 2 Mbit/s data transmission system is now examined to highlight the problems associated with transmitting and receiving such signals over a typical twisted pair, wire cable. The principal requirement is that signal fidelity should be sufficiently high to allow the receiver electronics to synchronise to the noisy and distorted incoming data signal and make correct data decisions.

6.6 Digital signal regeneration

The key facet of digital transmission systems is that after one, physically short, section in a link there is usually sufficiently high SNR to detect reliably the received binary data, and (possibly after error correction) regenerate an almost error-free data stream, Figure 6.14, for retransmission over the next stage or section. Regeneration allows an increase in overall communications path length with negligible decrease in message quality provided each regenerator operates at an acceptable point on the error rate curve, Figure 6.3. If the path loss on each section is approximately 40 dB then a 3 V transmitted polar signal is received as 30 mV and, allowing for near end crosstalk noise (see later, Figure 6.23) rather than thermal noise, the SNR is typically 18 dB. At this SNR the P_e, found using equation (6.12) and the approximation:

$$\text{erf}(x) \simeq 1 - \left[\frac{e^{-x^2}}{\sqrt{\pi}x}\right], \quad \text{for } x \geq 4 \tag{6.22}$$

is 10^{-15}. (For NRZ polar signalling $(\Delta V/\sigma)$ dB $= (S/N) + 6$ dB, see equation (6.11).) For identical calculations on each section then the P_e on each section will be 10^{-15} and the total error will be the sum of the errors on each section. Thus, on the two-section link shown in Figure 6.14, the total error probability is 2×10^{-15}. This was discussed previously in section 6.3 where for an m-section link $(P_e|_{m \text{ hops}})$ was shown to be m times the error on a single hop or regenerative section. (In practice the signal attenuation, and noise introduced, is rarely precisely equal for all sections. Since the slope of the P_e curve in Figure 6.3, in the normal operating region, is very steep it is often the case that

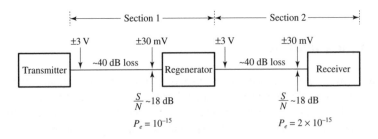

Figure 6.14 *Principle of regenerative repeater in long communications system.*

the P_e performance of a multi-hop link is dominated by the performance of its worst section.)

Provided the error rate on each section is acceptable then the cumulative or summed rate for the link is low, compared with the error rate when there is no regeneration and the single section loss is very high. The input SNR must typically be 18 dB in a section design, to ensure that there is an extremely low BER on each individual section. Regeneration thus permits long distance transmission with high message quality provided the link is properly sectioned with the appropriate loss on each section. The principal component of such a digital line system is the digital regenerative repeater or regenerator, Figure 6.15.

6.6.1 PCM line codes

Line coding is used, primarily, to match the spectral characteristics of the digital signal to its channel and to provide guaranteed recovery of timing signals. ITU-T recommends HDB3 for (multiplexed) PCM bit rates of 2, 8 and 34 Mbit/s and CMI for 140 Mbit/s. For the unmultiplexed bit rate (64 kbit/s) ITU-T G.703 has three different line code recommendations, one for each of three PCM equipment interface standards. The code used depends on the type of equipment connected to either side of the PCM interface. Three types of equipment interface are defined, namely codirectional, centralised clock and contradirectional. The interface type depends on the origin of 64 kHz and 8 kHz timing signals and their transmission direction with respect to that of the information or data (Figure 6.16).

For a codirectional interface (Figure 6.16(a)) there is one transmission line for each transmission direction. In order to incorporate the timing signals into the data the 64 kbit/s pulse or bit period is subdivided into four 'quarter-bit' intervals. Binary ones are unipolar NRZ coded as 1100 and binary zeros are coded as 1010. The quarter-bits are then converted to a three-level signal by alternating the polarity of consecutive four quarter-bit blocks. (Each quarter-bit has a nominal duration of 3.9 μs.) The alternation in polarity of the blocks is violated every eighth block. These violations represent the last PCM bit of each 8 bit PCM word (Figure 6.12).

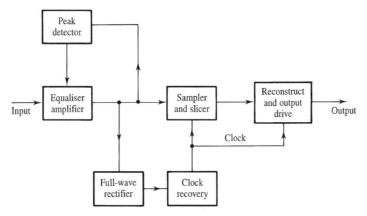

Figure 6.15 *Simplified block diagram of pulse code modulation (PCM) regenerator.*

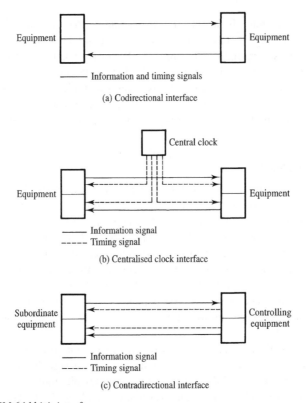

Figure 6.16 *PCM 64 kbit/s interfaces.*

For a centralised clock interface (Figure 6.16(b)) there is one transmission line for each transmission direction to carry PCM data and one transmission line from the central clock to the equipment on each side of the interface to carry 64 kHz and 8 kHz timing signals. The data line uses a bipolar code with a 100% duty cycle (i.e. AMI NRZ). The timing signal line code is bipolar with a duty cycle between 50% and 80% (i.e. AMI RZ) with polarity violations aligned with the last (eighth) bit of each PCM codeword.

For a contradirectional interface (Figure 6.16(c)) there are two transmission lines for each transmission direction, one for PCM data and one for the timing signal. The data line code (AMI NRZ) and timing signal (all ones) line code (AMI RZ) are identical to those for a centralised clock interface except that the timing signal has a 50% duty cycle.

6.6.2 Equalisation

A significant problem in PCM cable, and many other, communication systems is that considerable amplitude and phase distortion may be introduced by the transmission medium. For a 2 Mbit/s PCM system the RZ bipolar pulse has a width of $\frac{1}{4}$ μs and hence a bandwidth of approximately 2 MHz. When this is compared with typical metallic cable characteristics, Figure 6.17, it can be seen that the received pulse will be heavily distorted and attenuated. A potentially serious consequence of this distortion is that the pulse will be stretched in time as shown in Figure 6.18.

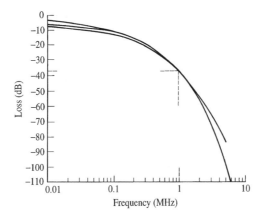

Figure 6.17 *Typical frequency responses for a 2 km length of cable.*

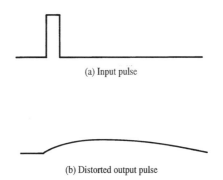

(a) Input pulse

(b) Distorted output pulse

Figure 6.18 *(a) Input and (b) output 2 Mbit/s pulse for a 2 km length of cable.*

Distortion also results when wideband signals are transmitted over the local loop telephone connection from the exchange to the subscriber. Here the cable bandwidth is only several kHz. With the high speed digital subscriber line (HDSL) [Baker, Young *et al.*] we can transmit at Mbit/s provided there is still adequate SNR at the receiver. HDSL relies on efficient data coding, such as 2B1Q, i.e. four-level signalling (i.e. an extension of the coding in Table 6.2) to reduce transmitted signal bandwidth, combined with techniques to equalise (i.e. compensate for) distortion introduced by the restricted network bandwidth. HDSL thus achieves high rate transmission on existing networks without requiring replacement of the copper cables by optical transmission media.

When we move from considering individual pulses to a data stream, Figure 6.19, the long time domain tails from the individual received symbols cause intersymbol interference (ISI). This is overcome by applying an equalising filter [Mulgrew *et al.*] in the receiver which has the inverse frequency response, Figure 6.20, to the raw line or channel characteristic of Figure 6.17. Cascading the effect of the line with the equaliser provides a flat overall response which reduces the distortion (time stretching) of the pulses as shown in Figure 6.19(c). A detailed discussion of ISI and one aspect of equalisation (Nyquist filtering), which is used to minimise it, is given in Chapter 8.

6.6.3 Eye diagrams

A useful way of quickly assessing the adequacy of a digital communications system is to display its *eye diagram* on an oscilloscope. This is achieved by triggering the oscilloscope with the recovered symbol timing clock signal and displaying the symbol stream using a sweep time sufficient to show two or three symbol periods. Figure 6.21(a) shows a noiseless, but distorted, binary bit stream and Figure 6.21(b) shows the resulting two-level eye pattern when the pulses are overlaid on top of one another. Distortion of the digital signal causes ISI which results in variations of the pulse amplitudes at the sampling instants. The eye opening indicates how much tolerance the system possesses to noise before incorrect decisions will be made.

It can be seen that there is no ISI-free instant in this case which could be chosen for ideal sampling. In addition to partial closing of the eye, distortion also results in a narrowing of the eye in the horizontal (time) dimension. This is because the symbol timing signal is derived from the zero crossings of the distorted bit stream resulting in symbol timing *jitter*. This effectively means that some symbols will be sampled at non-optimum instants. Figure 6.22 shows examples of eye patterns for NRZ binary and

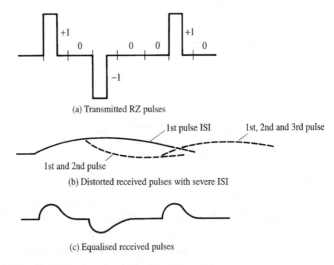

(a) Transmitted RZ pulses

(b) Distorted received pulses with severe ISI

(c) Equalised received pulses

Figure 6.19 *Intersymbol interference in a pulse train.*

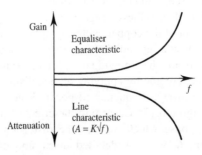

Figure 6.20 *Line equaliser frequency responses.*

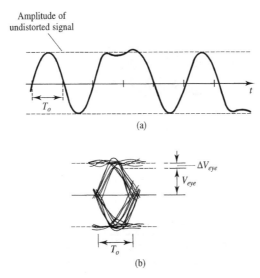

Figure 6.21 *(a) Noiseless but distorted signal; and (b) corresponding eye diagram.*

multilevel systems. The optimum sampling time clearly occurs at the position of maximum eye opening. Both distortion of the digital signal and additive noise contribute to closing of the eye. The slope of the eye pattern between the maximum eye opening and the eye corner is a measure of sensitivity to timing error.

The trade-off possible between ISI and noise can be estimated from the noise-free eye diagram. For example, in Figure 6.21(b) the degradation in eye opening, $\Delta V_{eye}/V_{eye}$, is about 20%. (Notice that ΔV_{eye} is about half the thickness of the eye pattern at the point of maximum opening because distortion results in traces above as well as below the undistorted signal level.) If the signal voltage is increased by a factor $1/(1 - 0.2) = 1.25$ then the eye opening will be restored to the value it would have in the absence of ISI. An increase in SNR of $20 \log_{10} 1.25 = 1.9$ dB would therefore, at least approximately, compensate the BER degradation due to ISI.

In a practical regenerator, the major contribution to ISI generally arises from the residual amplitude of the pulse at the preceding and succeeding sampling instants and ISI from other sampling instants can be ignored.

Before regenerator design is examined in more detail we need to discuss one further impairment, i.e. that caused by signal crosstalk.

6.6.4 Crosstalk

Two types of crosstalk arise when bidirectional signals are transmitted across a bundled cable comprising many individual twisted pairs. With approximately 40 dB of insertion loss across the cable a 3 V transmitted pulse is received as a 30 mV signal at the far end. Near end crosstalk (NEXT) results from the capacitive coupling of the 3 V transmitted pulse on an outgoing pair interfering with the 30 mV received pulse on an incoming pair, Figure 6.23.

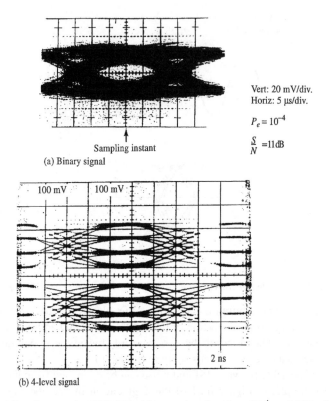

Vert: 20 mV/div.
Horiz: 5 μs/div.

$P_e = 10^{-4}$

$\frac{S}{N} = 11 dB$

↑
Sampling instant
(a) Binary signal

100 mV   100 mV

2 ns

(b) 4-level signal

Figure 6.22 *(a) Eye diagram for NRZ digital binary signal at $P_e = 10^{-4}$ and SNR of 11 dB (source: Feher, 1983, reproduced with his permission); and (b) 4-level signal at 200 Mbit/s (source: Feher, 1981, reproduced with his permission).*

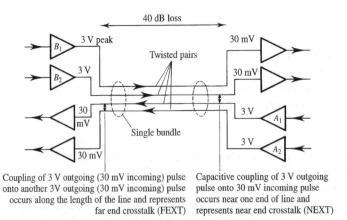

40 dB loss

B_1 3 V peak   30 mV

Twisted pairs

B_2 3 V   30 mV

30 mV

Single bundle   3 V A_1

3 V A_2

30 mV

Coupling of 3 V outgoing (30 mV incoming) pulse onto another 3V outgoing (30 mV incoming) pulse occurs along the length of the line and represents far end crosstalk (FEXT)

Capacitive coupling of 3 V outgoing pulse onto 30 mV incoming pulse occurs near one end of line and represents near end crosstalk (NEXT)

Figure 6.23 *Near and far end crosstalk (NEXT and FEXT) due to capacitive cable coupling.*

Far end crosstalk (FEXT), on the other hand, occurs due to coupling of a transmitted pulse on one outgoing pair with a pulse on another outgoing pair. NEXT thus refers to crosstalk between signals travelling in opposite directions and *effectively* takes place near the cable ends whilst FEXT refers to crosstalk between signals travelling in the same direction and takes place over the cable's entire length.

Assuming the transmitted pulse spectrum has a $|(\sin x)/x|$ shape and the coupling can be modelled as a high pass RC filter, which introduces a coupling gain of 6 dB/octave, then NEXT has a distorted spectrum as shown in Figure 6.24, in which the spectral magnitude, relative to the normal spectrum, increases with increasing frequency.

Crosstalk and ISI reduction therefore demands a composite equaliser response which differs from Figure 6.20 as shown in Figure 6.25. This has a low frequency portion

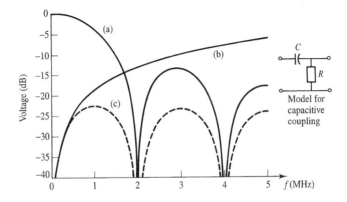

(a) Transmitted spectrum, $V_{dB} = 20 \log_{10} \left| \dfrac{\sin (2\pi T_o f)}{2\pi T_o f} \right|$, $T_o = 0.25$ μs

(b) Coupling loss (reducing at 6 dB/octave for high frequency), $L = 20 \log_{10} \left(\dfrac{2\pi f RC}{\sqrt{1 + (2\pi f RC)^2}} \right)$, $RC = 0.02$ μs

(c) Resultant NEXT spectrum, $V_{dB} + L$

Figure 6.24 *Spectrum of NEXT signal.*

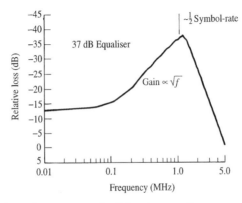

Figure 6.25 *Combined equaliser response for ISI and crosstalk.*

where loss decreases with the square root of frequency, as in Figure 6.20. The high frequency portion of the spectrum, however, is shaped to reduce crosstalk effects. This is a compromise, as it does increase ISI and sensitivity to jitter. The equaliser normally has lowest loss at half the symbol rate (i.e. 1 MHz on a 2 Mbit/s link).

6.7 Symbol timing recovery

After effective equalisation has been implemented it is necessary to derive a receiver clock, from the received signal, in order to time, accurately, the sampling and data recovery process. Symbol timing recovery (STR) is required since most digital systems are self timed from the received signal to avoid the need for a separate timing channel. It can be achieved by first filtering and then rectifying, or squaring, a bipolar RZ line coded signal, Figure 6.26. Rectifying or squaring removes the alternating pulse format which approximately doubles the received signal's bandwidth. It removes the notch at the symbol rate, $f_o = 1/T_o$, and introduces an f_o clock signal component which can be extracted with a resonant circuit.

As the resonant circuit oscillates at its natural frequency it fills in the gaps left by the zero data bits for which no symbol is transmitted, Figure 6.27. The frequency and phase of this recovered clock must be immune from transmission distortions in, and noise on, the received signal. A problem arises in the plesiochronous multiplex in that extra (justification) bits are added or removed at the individual multiplexers to obtain the correct bit rates on the transmission links (see section 20.5.1 and Figure 20.20). This means that symbol timing can become irregular introducing timing jitter. Timing jitter manifests itself as frequency modulation on the recovered clock signal, Figure 6.28.

The passive, tuned *LC* circuit, Figure 6.29, has a low Q-factor (30 to 100) which does not give good noise suppression. Its wide bandwidth, however, means it is relatively tolerant to small changes in precise timing of the received signals (i.e. jitter). A high Q (1000 to 10000) phase locked loop gives good noise reduction but is no longer so tolerant to jitter effects. In Figure 6.28 we see suppression of the high frequency jitter

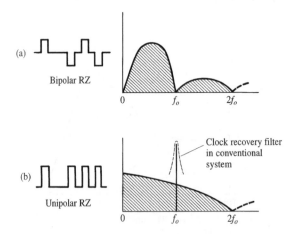

Figure 6.26 *Clock recovery by full wave rectification.*

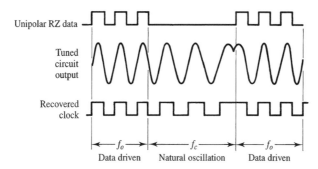

Figure 6.27 *Clock recovery oscillator operation when a string of no symbols (spaces) is received.*

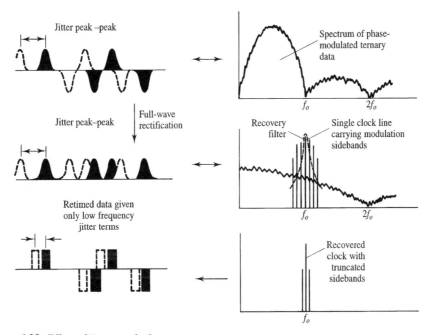

Figure 6.28 *Effect of jitter on clock recovery.*

terms. In both systems an output comparator clips the sinusoid to obtain the recovered clock waveform.

In some systems timing extraction is obtained by detecting the zero crossings of the waveform. The 'filter and square' operation is also used for STR in QPSK receivers, see later Figure 11.24. (Another QPSK STR technique is to delay the received symbol stream by half a symbol period and then form a product with the undelayed signal. This gives a periodic component at the symbol rate which can be extracted using a PLL.)

We may define peak-to-peak jitter (in s, or bits) as the maximum peak-to-peak displacement of a bit or symbol with respect to its position in a hypothetical unjittered reference stream. One effect of the jitter is that the regenerated clock edge varies with

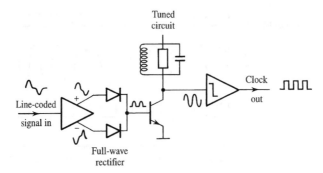

Figure 6.29 *Passive oscillator for clock recovery.*

respect to the correct timing point in the eye diagram, increasing the P_e. Clipping removes the noise but the remaining phase modulation still contributes to the jitter.

A significant problem with jitter is that it accumulates over a multi-hop system. Timing jitter in the incoming data introduces eye jitter, while the clock recovery process in each repeater introduces more clock edge jitter. To minimise this problem data scramblers can be used to stop the accumulation of data dependent jitter. Limits on jitter are specified in the ITU-T G.823/4 recommendations.

6.8 Repeater design

In the receiver a matched filter (see section 8.3) for data bit detection is usually by far the most computationally demanding task. It is typically implemented using a finite impulse response filter [Mulgrew *et al.*] to obtain the linear phase requirement. In comparison with this, timing extraction, phase/frequency error determination and correction usually account for only 10 to 20% of the computational load in the receiver.

A more detailed block diagram than that shown in Figure 6.15, for a complete PCM regenerative repeater, is shown in Figure 6.30. Power is fed across the entire PCM link, and each repeater AC couples the HDB3 signal, with the power extracted, from the primary winding of the coupling transformer. For an *m* repeater link total supply voltage is *m* times the single repeater requirement, e.g. $m \times 5$ V. (The duplicated sampling and reconstruction stages in Figure 6.30 accommodate separately the positive and negative bipolar pulses.) Complete repeaters are now available as single integrated circuits which only require transformer connection to the PCM system.

6.9 Digital transmission in local loop

The equalisation techniques of section 6.6.2 have recently been applied to greatly extend the transmission rate for data over the conventional twisted pair cables used to provide the local loop telephone network connection in the so-called plain old telephone system (POTS).

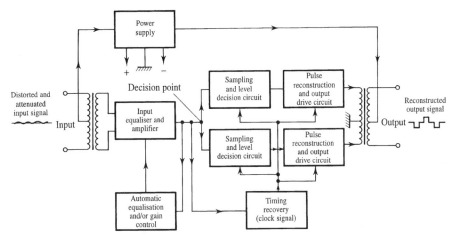

Figure 6.30 *Complete PCM regenerator.*

Internet use has provided the incentive to achieve high speed data connections for domestic personal computers (PCs). One way to transmit data is by using data modem techniques (described in section 11.6) where bit to symbol mapping (see section 9.2.2) permits data at 56 kbit/s to be transmitted over a 3.2 kHz bandwidth, analogue POTS, local loop connection. This is achieved by keeping the signalling rate in the low ksymbol/s range for compatiblity with the POTS line bandwidth. For Internet use a 10 Mbyte file, representing a graphics rich slide set or 5 to 10 photographs, takes 23 minutes to download over a 56 kbit/s connection, ignoring any protocol overheads. For faster downloading digital subscriber line (DSL) techniques [Star *et al.*] utilise the existing installed copper pair cables (of which there are 790 million worldwide) for broadband data delivery at rates up to 10 Mbit/s and beyond. The current European local loop connection from exchange to subscribers is shown in Figure 6.31.

The overall distance from the subscriber to the exchange is typically less than 5 km. The cables are arranged in bundles of approximately 50 twisted pairs with occasional junction boxes, cross-connections or flexibility points, typically located in street mounted cabinets.

DSL, see also section 20.11.1, is an overlay technology that adds a broadband high speed network connection on top of the existing copper cables using advanced digital signal processing, modulation, coding and equalisation techniques. DSL and analogue telephony can co-exist, Figure 6.32, since DSL signals are transmitted above the analogue baseband spectral allocation [Czajkowski].

DSL has evolved from the high speed digital subscriber line (HDSL) introduced in 1993, through asymmetric DSL (ADSL) [Bingham, WWW] to the very high speed digital subscriber line (VDSL). xDSL [WWW] is the generic term now used to describe these services. Table 6.3 summarises these and the integrated services digital network (ISDN), see Chapter 20, which provided the first (1987) data transmission standard for individual subscriber use at 144 kbit/s. The asymmetric xDSL services provide the high rate downstream channel which is required for users to access Internet data pages rapidly.

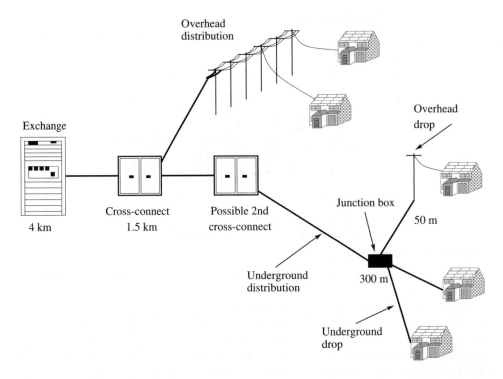

Figure 6.31 *Local loop connections with typical customer distances (source: Czajkowski, 1999, reproduced with permission of the IEE).*

In xDSL systems there is a trade-off between the length of the copper connection and the achievable bit rate. High rate VDSL transmissions, see VDSL standards, require short connections or alternatively wideband optical fibre delivery from the exchange to the cross-connectors in Figure 6.31.

Table 6.3 *Overview of xDSL, compared to ISDN.*

Standard	ISDN	HDSL	ADSL	VDSL
Introduction	1987	1993	1995-9	2000?
Spectral allocation (MHz)	DC–0.16	DC–0.78	0.025–1.1	0.3–10/30
Transmissions	symmetric	symmetric	asymmetric	asymmetric
Upstream rate (Mbit/s)	0.144	< 2	< 1	< 2
Downstream rate (Mbit/s)	0.144	< 2	< 8	< 52
Line codes	4B3T, 2B1Q	2B1Q, CAP	DMT, CAP/QAM	DMT, QAM

The key problem with xDSL is that the transmitted spectrum can extend to 10 MHz which overlaps the amateur radio and medium wave broadcast bands, Table 1.4. It thus suffers from crosstalk, section 6.6.4, and impulsive noise. In the USA, the bridged tap connections further degrade the frequency response of Figure 6.17 introducing nulls.

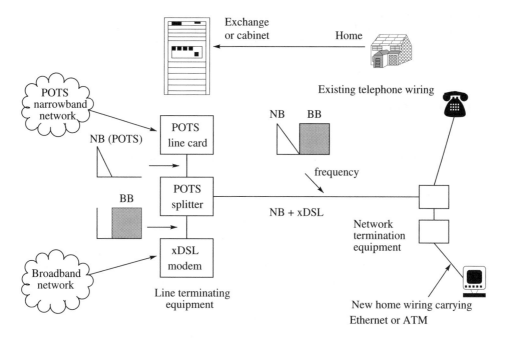

Figure 6.32 *xDSL as a broadband (BB) overlay technology on the narrowband (NB) POTS local loop (source: Czajkowsji, 1999, reproduced with permission of the IEE).*

In common with the data modems described in section 11.6, xDSL relies on multi-symbol modulation techniques to reduce signalling rate and, hence, occupied bandwidth. One favoured approach is to use the quadrature amplitude modulation (QAM) signal constellations and waveforms of Figures 11.19 and 11.20 and the closely related technique of carrierless amplitude and phase (CAP) modulation. CAP modulation replaces the quadrature modulators of Figures 11.20(g) and 11.23, for example, by complex digital filters to provide an identical transmitted waveform. Another xDSL modulation technique is digital multitone (DMT), Table 6.3, which is a form of OFDM transmission, see section 11.5.2. This is an extension of frequency shift keying, Chapter 11, using many simultaneous orthogonal carriers. The precise modulation used, however, varies from carrier to carrier depending on receiver SNR and interfering signal levels. Low frequency carriers will have a high SNR, Figure 6.17, and hence they can support a high level QAM constellation, Figure 11.21, with a correspondingly high number of bit/s per carrier of data throughput. DMT has the added advantage that carriers can be selectively disabled to avoid narrowband interference from broadcast and other sources.

The major interference mechanism in all xDSL systems is impulsive noise and characterisation of this is an ongoing activity. The key attraction of xDSL transmission is that sophisticated data modulation techniques, CAP, QAM and DMT, combined with equalisation and the forward error correction coding (FECC) techniques described in Chapter 10 do permit broadband carrier based data to be transmitted over the POTS analogue network and received in sophisticated digital signal processing based CODECs located in, or alongside, the street mounted cross-connect cabinets of Figure 6.31. This gives an extremely cost effective solution to providing the broadband downstream data rates which are required for rapidly accessing the pages of data needed for Internet searching and retrieval without the prohibitive cost that would be incurred in the replacement of the entire local loop network with optical fibre transmission.

6.10 Summary

The simplest form of baseband digital detection uses *centre point sampling* to reduce the received symbol plus noise to a single voltage, and comparators to test this voltage against appropriate references. A formula giving the single hop probability of error for symbols represented by uniformly spaced voltages in the presence of Gaussian noise has been derived. The probability of error after *m identical* hops is *m* times greater than that after a single hop providing that *regenerative* repeaters are used between hops. The main polar and bipolar signalling techniques have been outlined and those which are preferred for various transmission systems, used in the PCM multiplex hierarchy, discussed. This has involved consideration of the spectral properties of these line codes to ensure that they adequately match the transmission channel frequency response.

This chapter has also addressed some specific problems of data transmission in a metallic wire, twisted pair, cable. Distortion due to the cable frequency response and its compensation via equalisation have been examined. The effects of crosstalk interference and imperfect timing recovery have also been discussed to show their importance in practical data transmission systems. Although we have principally addressed the 2 Mbit/s metallic wire system, these signal degradation mechanisms are present in practically all digital communications systems. The chapter has shown how use of the analogue local loop can be extended to carry high data rate digital traffic.

6.11 Problems

6.1   A baseband, NRZ, rectangular pulse signal is used to transmit data over a single section link. The binary voltage levels adopted for digital 0s and 1s are -2.5 V and $+3.5$ V respectively. The receiver uses a centre point decision process. Find the probability of symbol error in the presence of Gaussian noise with an RMS amplitude of 705 mV. By how much could the transmitter power be reduced if the DC component were reduced to its optimum value? [0.12 dB]

6.2   Find the mean and peak transmitter power required for a single section, 2 km, link which employs unipolar RZ (50% mark space ratio), rectangular pulse signalling and centre point detection if the probability of bit error is not to exceed 10^{-6}. The specific attenuation of the (perfectly equalised) link transmission line is 1.8 dB/km and the (Gaussian) noise power at the receiver decision circuit input is 1.8 mW. (Assume that the (real) impedance level at transmitter output and decision circuit input are both equal to the characteristic impedance of the transmission line.) [93 mW, 0.37 W]

6.3   A 1.0 Mbaud, baseband, digital signal has eight allowed voltage levels which are equally spaced between $+3.5$ V and -3.5 V. If this signal is transmitted over a 26-section, regenerative repeater link, and assuming that all the sections of the link are identical, find the RMS noise voltage which could be tolerated at the end of each hop whilst maintaining an overall link symbol error rate of 1 error/s. [91.4 mV]

6.4 What are the two factors which control the power spectral density (PSD) of a line coded communications signal? Using the fact that the Fourier transform of a rectangular pulse has a characteristic $(\sin x)/x$ shape, approximate the PSD and consequent channel bandwidth requirements for a return to zero, on–off keyed, line coded signal incorporating a positive pulse whose width equals one-third the symbol interval. Compare this with a non-return to zero signal, sketching the line coded waveforms and PSDs for both cases.

How do these figures compare with the minimum theoretical transmission bandwidth? What are the practical disadvantages of the on–off keyed waveform?

6.5 A binary transmission scheme with equiprobable symbols uses the absence of a voltage pulse to represent a digital 0 and the presence of the pulse $p(t)$ to represent a digital 1 where $p(t)$ is given by:

$$p(t) = \Pi\{t/(T_o/3)\} - 0.5\,\Pi\{[t - (T_o/3)]/(T_o/3)\} - 0.5\,\Pi\{[t + (T_o/3)]/(T_o/3)\}$$

Sketch the pulse, $p(t)$, and comment on the following aspects of this line code: (a) its DC level; (b) its suitability for transmission using AC coupled repeaters; (c) its (first null) bandwidth; (d) its P_e performance, using ideal centre point decision, compared with unipolar NRZ rectangular pulse signalling; (e) its self clocking properties (referring to equation (6.21) if you wish); and (f) its error detection properties.

6.6 Justify the entries in the last (transparency) column of Table 6.1.

6.7 The power spectral density of a bipolar NRZ signal (Figure 6.11) is given by:

$$G(f) = V^2 T_o\,\text{sinc}^2(T_o f)\,\sin^2(2\pi T_o f)$$

Use this to help verify the power spectral density of the coded mark inversion (CMI) signal given in Figure 6.11. (A plot of your result is probably necessary to verify the PSD.) (Hint: since the correlation of the mark and space parts of the CMI signal is zero for all values of time shift, then the power spectra of these parts can be found separately and summed to get the power spectral density of the CMI signal.)

6.8 Draw a perfectly equalised prototype positive pulse for an AMI signal. Use it to construct the eye diagram for the signal $+1, -1, +1, -1, + \cdots$. Add and label the remaining trajectories for a random signal.

6.9 Explain why a non-linear process is required to recover clock timing from an HDB3 signal.

6.10 For Gaussian variables use the tables of erf(x) supplied in Appendix A to find:

(a) The probability that $x \le 5.5$ if x is a Gaussian random variable with mean $\mu = 3$ and standard deviation $\sigma = 2$.

(b) The probability that $x > 5.5$.

(c) Assuming the height of clouds above the ground at some location is a Gaussian random variable x with $\mu = 1830$ metres and $\sigma = 460$ metres, find the probability that clouds will be higher than 2750 metres.

(d) Find the probability that a Gaussian random variable with standard deviation of σ will exceed: (i) σ, (ii) 2σ, and (iii) 3σ. [(a) 0.89, (b) 0.11, (c) 0.023, (d) 0.159, 0.023, 0.0014]

6.11 Explain near end crosstalk. Provide a diagram to show the shape of the crosstalk spectrum at the regenerator output. How does this spectrum differ at the equaliser output?

CHAPTER 7

Decision theory

7.1 Introduction

In section 6.2 the detection of baseband binary symbols was described and analysed for the special case of Gaussian noise and equiprobable symbols. The decision reference voltage, against which detected symbols were tested, was set intuitively (and correctly) mid-way between those voltages which would have been detected, if the binary symbols were received without noise. In this chapter a more systematic analysis of symbol decision theory is given, applying the descriptions of random signals and noise developed in Chapter 3. This allows the optimum levels for receiver decision voltages to be found in the more general case of arbitrary noise distributions and unequal symbol probabilities.

For a binary data stream there are two principal types of decision:

1. Soft (multilevel) decisions.
2. Hard (two-level) decisions.

Soft decision receivers, Figure 7.1(a), quantise the decision instant signal plus noise voltage using several allowed levels, each represented by a decision word of a few binary bits. Figure 7.1(c) illustrates an eight-level (3 bit) soft decision process which is typical. Each soft decision contains not only information about the most likely transmitted symbol (000 to 011 indicating a likely 0 and 100 to 111 indicating a likely 1) but also information about the confidence or likelihood which can be placed on this decision. The soft decisions are converted to final or hard decisions using an algorithm which inspects a sequence of several PCM words and makes decisions accounting for the confidence levels they represent, in conjunction with the error control decoding rules. The algorithm tends to be heuristic and requires tailoring to the particular application. Although such algorithms can now be implemented using high speed VLSI decoders, these techniques

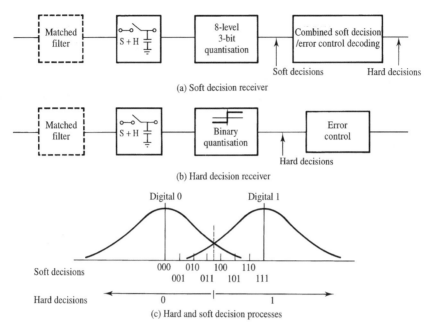

(a) Soft decision receiver

(b) Hard decision receiver

(c) Hard and soft decision processes

Figure 7.1 *Comparison of hard and soft decision receivers.*

will not be discussed further here.

Hard decisions (Figure 7.1(b)) are more common than soft decisions. The two major decision criteria used in this case are the Bayes and Neyman–Pearson. The Bayes decision criterion is used extensively in binary communications while the Neyman–Pearson criterion is used more in radar applications. The principal difference between the two is that the Bayes decision rule assumes known a priori source statistics for the occurrence of digital ones and zeros while the Neyman–Pearson criterion makes no such assumption. This is appropriate to radar applications because the a priori probability of the appearance of a target is not normally known.

7.2 A priori, conditional and a posteriori probabilities

There are four probabilities and two probability density functions, see Chapter 3, associated with symbol transmission and reception. These are:

1. $P(0)$ – a priori probability of transmitting the digit 0.
2. $P(1)$ – a priori probability of transmitting the digit 1.
3. $p(v|0)$ – conditional probability density function of detecting voltage v given that a digital 0 was transmitted.
4. $p(v|1)$ – conditional probability density function of detecting voltage v given that a digital 1 was transmitted.
5. $P(0|v)$ – a posteriori probability that a digital 0 was transmitted given that voltage v was detected.

6. $P(1|v)$ – a posteriori probability that a digital 1 was transmitted given that voltage v was detected.

The terms a priori and a posteriori imply reasoning from cause to effect and reasoning from effect to cause respectively. The cause of a communication event, in this context, is the symbol transmitted and the effect is the voltage detected. The a priori probabilities (in communications applications) are usually known in advance. The conditional probabilities are dependent on the data (1 or 0) transmitted. The a posteriori probabilities can only be established after many events (i.e. symbol transmissions and receptions) have been completed. (Note that the lower case notation is used for conditional probability density functions and upper case for a posteriori probabilities. Also, note that the position of the transmitted and detected quantities is interchanged between these cases.)

7.3 Symbol transition matrix

When communications systems operate in the presence of noise (which is always the case) there is the possibility of transmitted symbols being sufficiently corrupted to be interpreted, at the receiver, in error. If the characteristics of the noise are precisely enough known, or many observations of symbol transmissions and receptions are made, then the probability of each transmitted symbol being interpreted at the receiver as any of the other symbols can be found. These *transition* probabilities are usually denoted using the conditional probability notation, $P(i|j)$, where i represents the symbol received and j represents the symbol transmitted. $P(i|j)$ is therefore the probability that a symbol will be interpreted as i given that j was transmitted. The symbol transition probabilities can be arranged as the elements of a matrix to describe the end-to-end properties of a communications channel. Examples of such matrices are given below.

7.3.1 Binary symmetric channel

In a binary channel four types of communication events can occur. These events are:

> 0 transmitted and 0 received
> 0 transmitted and 1 received
> 1 transmitted and 1 received
> 1 transmitted and 0 received

Denoting transmitted symbols with subscript *TX* and received symbols with subscript *RX* (for extra clarity) the binary channel can be represented schematically as shown in Figure 7.2(a). The corresponding symbol transition matrix is:

$$\begin{bmatrix} P(0_{RX}|0_{TX}) & P(0_{RX}|1_{TX}) \\ P(1_{RX}|0_{TX}) & P(1_{RX}|1_{TX}) \end{bmatrix}$$

If the probability, p, of a transmitted 0 being received in error as a 1 is equal to the probability of a transmitted 1 being received as a 0, then the binary channel is said to be symmetric and the transition matrix can be written as:

$$\begin{bmatrix} 1-p & p \\ p & 1-p \end{bmatrix}$$

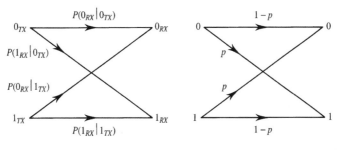

(a) Schematic of binary channel   (b) Schematic of binary symmetric channel

Figure 7.2 *Schematic representations of binary channels.*

Figure 7.2(b) shows the equivalent schematic diagram. The conditional probabilities in the matrix must sum vertically to unity, i.e.:

$$P(0_{RX}|0_{TX}) + P(1_{RX}|0_{TX}) = 1 \tag{7.1}$$

For the binary symmetric case:

$$P(0_{RX}|0_{TX}) = P(1_{RX}|1_{TX}) = 1 - p \tag{7.2}$$

The unconditional probability of receiving a 0 is therefore:

$$P_{0_{RX}} = P(0_{TX})P(0_{RX}|0_{TX}) + P(1_{TX})P(0_{RX}|1_{TX}) \tag{7.3}$$

which can be rewritten for the binary symmetric channel as:

$$P_{0_{RX}} = P(0)(1 - p) + P(1)p \tag{7.4}$$

Similarly:

$$P_{1_{RX}} = P(1)(1 - p) + P(0)p \tag{7.5}$$

Equations (7.4) and (7.5) relate the observed probabilities of received symbols to the a priori probabilities of transmitted symbols and the error probabilities of the channel.

EXAMPLE 7.1 – Multisymbol transmission

For an M-symbol communication alphabet the 2×2 transition matrix of the binary channel must be extended to an $M \times M$ matrix. Consider a source with the $M = 6$ symbols, $A\ B\ C\ D\ E\ F$, and a 6-ary receiver which can distinguish these symbols. The transition matrix, **T**, for this system contains 6×6 transition probabilities which are:

$$T = \begin{bmatrix} \dfrac{1}{2} & 0 & \dfrac{1}{24} & 0 & 0 & \dfrac{1}{8} \\[6pt] \dfrac{1}{4} & \dfrac{1}{2} & \dfrac{1}{6} & \dfrac{1}{3} & \dfrac{1}{4} & \dfrac{1}{12} \\[6pt] \dfrac{1}{8} & \dfrac{1}{8} & \dfrac{1}{4} & \dfrac{1}{6} & \dfrac{0}{1} & \dfrac{1}{12} \\[6pt] \dfrac{0}{1} & \dfrac{1}{3} & \dfrac{1}{4} & \dfrac{1}{3} & \dfrac{1}{4} & \dfrac{1}{4} \\[6pt] \dfrac{1}{8} & \dfrac{0}{1} & \dfrac{1}{6} & \dfrac{1}{12} & \dfrac{1}{2} & \dfrac{1}{8} \\[6pt] \dfrac{0}{1} & \dfrac{1}{24} & \dfrac{1}{8} & \dfrac{1}{12} & \dfrac{0}{1} & \dfrac{1}{3} \end{bmatrix}$$

(For a transmitted symbol B the probabilities of receiving symbols $A, \cdots, F$ are given by the $M = 6$ entries in the matrix's second column.)

If the a priori transmission probabilities of the six symbols are:

$P(A) = 0.1$

$P(B) = 0.15$

$P(C) = 0.4$

$P(D) = 0.05$

$P(E) = 0.2$

$P(F) = 0.1$

then find the probability of error when (i) only the symbol D is transmitted and (ii) when a random string of symbols is transmitted. Also find, for this latter case, the probability of receiving the symbol C.

The probability of error, P_e, when the symbol D is transmitted is given by the sum of all probabilities in the fourth (i.e. D transmitted) column, excluding the fourth (i.e. D received) element:

$$P_e = \sum_{m=A}^{F} P(m_{RX}|D_{TX}) \quad \text{(excluding the } m = D \text{ term)}$$

$$= 0 + 1/3 + 1/6 + 1/12 + 1/12 = 2/3$$

This is identical to $1 - P(D_{RX}|D_{TX})$, i.e.:

$$P_e = 1 - 1/3 = 2/3$$

(Note that the leading diagonal elements, $P(m|m)$, in the transition matrix are the probabilities of correct symbol reception and $1 - P(m|m)$ are therefore the probabilities of symbol error.)

The probability of error when a (long) random string of symbols $(A, \cdots, F)$ is transmitted is a weighted sum of the probabilities of error for each transmitted symbol where the weighting factors are the a priori probabilities of transmission, i.e.:

$$P_e = \sum_{m=A}^{F} P(m)\,[1 - P(m_{RX}|m_{TX})]$$

$$= 0.1 \times \frac{1}{2} + 0.15 \times \frac{1}{2} + 0.4 \times \frac{3}{4} + 0.05 \times \frac{2}{3} + 0.2 \times \frac{1}{2} + 0.1 \times \frac{2}{3}$$

$$= 0.625$$

In addition to probabilities of error, the various probabilities of symbol reception can also be found from the transition matrix. The probability of receiving symbol C when symbol D is transmitted is given by inspection of the appropriate matrix element, i.e.:

$$P(C_{RX}|D_{TX}) = 1/6$$

The (unconditional) probability of receiving symbol C when a (long) random string of symbols $(A, \cdots, F)$ is transmitted is the a priori weighted sum of all elements in the third (i.e. C received) row of the transition matrix:

$$P(C_{RX}) = \sum_{m=A}^{F} P(C_{RX}|m_{TX}) \, P(m)$$

$$= 0.1 \times \frac{1}{8} + 0.15 \times \frac{1}{8} + 0.4 \times \frac{1}{4} + 0.05 \times \frac{1}{6} + 0.2 \times 0 + 0.1 \times \frac{1}{12}$$

$$= 0.148$$

7.4 Bayes's decision criterion

This is the most widely applied decision rule in communications systems and, as a consequence, it will be discussed in detail. In essence it operates so as to minimise the average *cost* (in terms of errors or lost information) of making decisions.

7.4.1 Decision costs

In binary transmission there are two ways to lose information:

1. Information is lost when a transmitted digital 1 is received in error as a digital 0.
2. Information is lost when a transmitted digital 0 is received in error as a digital 1.

The cost in the sense of lost information due to mechanisms 1 and 2 is denoted here by C_0 and C_1 respectively. There is no cost (i.e. no information is lost) when correct decisions are made.

7.4.2 Expected conditional decision costs

The expected conditional cost, $C(0|v)$, incurred when a detected voltage v is interpreted by a decision circuit as a digital 0 is given by:

$$C(0|v) = C_0 \, P(1|v) \tag{7.6(a)}$$

where C_0 is the cost if the decision is in error and $P(1|v)$ is the (a posteriori) probability that the decision is in error. By symmetry the corresponding equation can be written:

$$C(1|v) = C_1 \, P(0|v) \tag{7.6(b)}$$

This is the expected conditional cost incurred when v is interpreted as a digital 1.

7.4.3 Optimum decision rule

A rational decision rule to adopt is to interpret each detected voltage, v, as either a 0 or a 1, so as to minimise the expected conditional cost, i.e.:

$$C(1|v) \underset{\substack{> \\ 0}}{\overset{1}{<}} C(0|v)$$ (7.7)

The interpretation of inequality (7.7) is 'if the upper inequality holds then decide 1, if the lower inequality holds then decide 0'. Substituting equations (7.6) into the inequality (7.7) and cross dividing gives:

$$\frac{P(0|v)}{P(1|v)} \underset{\substack{> \\ 0}}{\overset{1}{<}} \frac{C_0}{C_1}$$ (7.8)

If the costs of both types of error are the same, or unknown (in which case the rational assumption must be $C_0 = C_1$), then (7.8) represents a *maximum a posteriori* (MAP) probability decision criterion. Inequality (7.8) is one form of Bayes's decision criterion. It uses a posteriori probabilities, however, which are not usually known. The criterion can be transformed to a more useful form by using Bayes's theorem, equation (3.4), i.e.:

$$P(0|v) = \frac{p(v|0)\, P(0)}{p(v)}$$ (7.9(a))

and:

$$P(1|v) = \frac{p(v|1)\, P(1)}{p(v)}$$ (7.9(b))

Figure 7.3 illustrates the conditional probability density functions for the case of zero mean Gaussian noise. Dividing equation (7.9(a)) by (7.9(b)):

$$\frac{P(0|v)}{P(1|v)} = \frac{p(v|0)\, P(0)}{p(v|1)\, P(1)}$$ (7.10)

and substituting equation (7.10) into (7.8):

$$\frac{p(v|0)\, P(0)}{p(v|1)\, P(1)} \underset{\substack{> \\ 0}}{\overset{1}{<}} \frac{C_0}{C_1}$$ (7.11)

Rearranging the inequality (7.11) gives Bayes's criterion in a form using conditional probability density functions and the usually known a priori source probabilities:

$$\frac{p(v|0)}{p(v|1)} \underset{\substack{> \\ 0}}{\overset{1}{<}} \frac{C_0\, P(1)}{C_1\, P(0)}$$ (7.12)

The left hand side of the above inequality is called the *likelihood ratio* (which is a function of v) and the right hand side is a likelihood threshold which will be denoted by

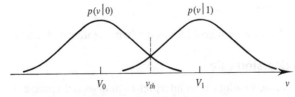

Figure 7.3 *Probability distributions for binary transmissions V_0 or V_1.*

L_{th}. If $C_0 = C_1$ and $P(0) = P(1)$ (or, more generally, $C_0 P(1) = C_1 P(0)$), or if neither costs nor a priori probabilities are known (in which case $C_0 = C_1$ and $P(0) = P(1)$ are the most rational assumptions), then $L_{th} = 1$ and equation (7.12) is called a *maximum likelihood* decision criterion. This type of hypothesis testing is used by the receivers described in Chapter 11, e.g. Figure 11.7. Table 7.1 illustrates the relationship between maximum likelihood, MAP and Bayes's decision criteria.

Table 7.1 *Comparison of receiver types.*

Receiver	A priori prob-abilities known	Decision costs known	Assumptions	Decision criterion
Bayes	Yes	Yes	None	$\dfrac{p(v\vert0)}{p(v\vert1)} \underset{0}{\overset{1}{\underset{>}{<}}} \dfrac{C_0\, P(1)}{C_1\, P(0)}$
MAP	Yes	No	$C_0 = C_1$	$\dfrac{p(v\vert0)}{p(v\vert1)} \underset{0}{\overset{1}{\underset{>}{<}}} \dfrac{P(1)}{P(0)}$
Max. likelihood	No	No	$C_0 P(1) = C_1 P(0)$	$\dfrac{p(v\vert0)}{p(v\vert1)} \underset{0}{\overset{1}{\underset{>}{<}}} 1$

7.4.4 Optimum decision threshold voltage

Bayes's decision criterion represents a general solution to setting the optimum reference or threshold voltage, v_{th}, in a receiver decision circuit. The threshold voltage which minimises the expected conditional cost of each decision is the value of v which satisfies:

$$\frac{p(v\vert0)}{p(v\vert1)} = \frac{C_0\, P(1)}{C_1\, P(0)} = L_{th} \tag{7.13}$$

This is illustrated for two conditional pdfs in Figure 7.4. (These pdfs actually correspond to those obtained for envelope detection of OOK rectangular pulses, see Figure 6.11, $p(v\vert0)$ being a Rayleigh distribution (see section 4.7.1) and $p(v\vert1)$ being Rician.) If $L_{th} = 1.0$, such as would be the case for statistically independent, equiprobable symbols with equal error costs, then the voltage threshold would occur at the intersection of the two conditional pdfs. (This is exactly the location of the decision threshold selected intuitively for centre point detection of equiprobable, rectangular binary symbols in section 6.2.1.)

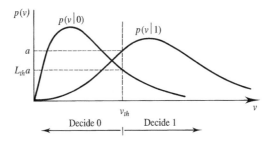

Figure 7.4 *Optimum decision threshold voltage for envelope detected OOK signal.*

In practice, a single threshold, such as that shown in Figure 7.4, is normally adequate against which to test detected voltages. It is possible in principle, however, for the conditional pdfs to intersect at more than one point. This situation is illustrated in Figure 7.5. Assuming the symbols are equiprobable the decision thresholds are then set at the intersection of the conditional pdfs. Here, however, this implies more than one threshold, with several decision regions delineated by these thresholds. It is important to appreciate that despite their unconventional appearance the conditional pdfs in equation (7.13) behave in the same way as any other function. Equation (7.13), which can be rewritten in the form:

$$p(v|0) = L_{th}\, p(v|1) \tag{7.14}$$

can therefore be solved using any of the normal techniques including, where necessary, numerical methods.

7.4.5 Average unconditional decision cost

It is interesting to note (and perhaps self evident) that as well as minimising the cost of each decision, Bayes's criterion also minimises the average cost of decisions (i.e. the costs averaged over all decisions irrespective of type). This leads to the following alternative derivation of Bayes's criterion.

The average cost, $\bar{C}$, of making decisions is given by:

$$\bar{C} = C_1\, P(0)P(1_{RX}|0_{TX}) + C_0\, P(1)P(0_{RX}|1_{TX}) \tag{7.15}$$

where $P(0)P(1_{RX}|0_{TX})$ is the probability that any given received symbol is a digital 1 which has been detected in error and $P(1)P(0_{RX}|1_{TX})$ is the probability that any given received symbol is a digital 0 also detected in error. From Figure 7.3 it can be seen that:

$$P(1_{RX}|0_{TX}) = \int_{v_{th}}^{\infty} p(v|0)\, dv \tag{7.16(a)}$$

$$P(0_{RX}|1_{TX}) = \int_{-\infty}^{v_{th}} p(v|1)\, dv \tag{7.16(b)}$$

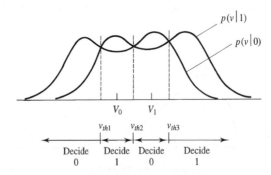

Figure 7.5 *Optimum decision thresholds for equiprobable binary voltages in noise with multimodal pdf.*

where v_{th} is the receiver decision threshold voltage. Substituting equations (7.16) into equation (7.15) the average cost of making decisions can be written as:

$$\bar{C} = C_1\, P(0) \int_{v_{th}}^{\infty} p(v|0)\, dv + C_0\, P(1) \int_{-\infty}^{v_{th}} p(v|1)\, dv \tag{7.17}$$

Bayes's decision criterion sets the decision threshold v_{th} so as to minimise $\bar{C}$. To find this optimum threshold $\bar{C}$ must be differentiated with respect to v_{th} and equated to zero. This involves differentiating the integrals in equation (7.17) which is most easily done using [Dwight, equations 69.1 and 69.2], i.e.:

$$\frac{d}{dx} \int_{c}^{x} f(t)\, dt = f(x) \tag{7.18(a)}$$

$$\frac{d}{dx} \int_{x}^{c} f(t)\, dt = -f(x) \tag{7.18(b)}$$

We therefore have:

$$\frac{d\bar{C}}{dv_{th}} = C_1\, P(0)\, [-p(v_{th}|0)] + C_0\, P(1)\, [p(v_{th}|1)] \tag{7.19}$$

Setting equation (7.19) equal to zero for minimum (or maximum) $\bar{C}$:

$$C_0\, P(1)\, p(v_{th}|1) = C_1\, P(0)\, p(v_{th}|0) \tag{7.20(a)}$$

and rearranging, gives:

$$\frac{P(v_{th}|1)}{P(v_{th}|0)} = \frac{C_1\, P(0)}{C_0\, P(1)} \tag{7.20(b)}$$

Solution of equations (7.20) for v_{th} gives Bayes's criterion (i.e. optimum) decision voltage.

EXAMPLE 7.2 – Binary transmission

Consider a binary transmission system subject to additive Gaussian noise which has a mean value of 1.0 V and a standard deviation of 2.5 V. (Note here that the standard deviation of the noise is not the same as its RMS value.) A digital 1 is represented by a rectangular pulse with amplitude 4.0 V and a digital 0 by a rectangular pulse with amplitude −4.0 V. The costs (i.e. information lost) due to each type of error are identical (i.e. $C_0 = C_1$) but the a priori probabilities of symbol transmission are different, $P(1)$ being twice $P(0)$, as the symbols are not statistically independent. Find the optimum decision threshold voltage and the resulting probability of symbol error.

Using Bayes's decision criterion the optimum decision threshold voltage is given by solving equation (7.14), i.e.:

$$p(v_{th}|0) = L_{th}\, p(v_{th}|1)$$

where:

$$L_{th} = \frac{C_0}{C_1}\frac{P(1)}{P(0)}$$

Since $C_1 = C_0$ and $P(1) = 2P(0)$ then:

$$L_{th} = 2.0$$

The noise is Gaussian and its mean value adds to the symbol voltages. An equivalent signalling system therefore has symbol voltages of -3 V and 5 V, and noise with zero mean.

The conditional probability density functions are:

$$p(v|0) = \frac{1}{\sigma_0\sqrt{(2\pi)}} e^{\frac{-(v-V_0)^2}{2\sigma_0^2}}$$

$$p(v|1) = \frac{1}{\sigma_1\sqrt{(2\pi)}} e^{\frac{-(v-V_1)^2}{2\sigma_1^2}}$$

where V_0 and V_1 are the signal voltages representing digital 0s and 1s respectively. Substituting these equations into the above version of equation (7.14) with $\sigma_0 = \sigma_1 = 2.5$, $V_0 = -3$ and $V_1 = 5$:

$$\frac{1}{2.5\sqrt{(2\pi)}} e^{\frac{-(v+3)^2}{2(2.5)^2}} = 2.0\,\frac{1}{2.5\sqrt{(2\pi)}} e^{\frac{-(v-5)^2}{2(2.5)^2}}$$

Cancelling and taking logs:

$$\frac{-(v+3)^2}{2(2.5)^2} = \ln(2.0) - \frac{(v-5)^2}{2(2.5)^2}$$

$$\frac{-(v^2+6v+9)+(v^2-10v+25)}{2(2.5)^2} = \ln(2.0)$$

$$16 - 16v = 2(2.5)^2\ln(2.0)$$

or:

$$v = \frac{16 - [2(2.5)^2\ln(2.0)]}{16} = 0.46 \text{ V}$$

Figure 7.6 illustrates the *un*conditional probability density functions of symbols plus noise and the location of the optimum threshold voltage at $v_{th} = 0.46$ V for this example. (The conditional probability density functions would both enclose unit area.)

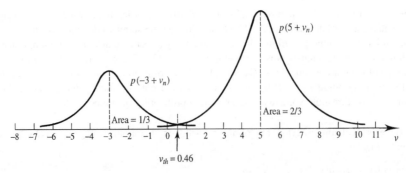

Figure 7.6 *Unconditional symbol plus noise voltage pdf for Example 7.2; $p(V + v_n)$ denotes voltage pdf of symbol voltage V plus noise.*

The probability of error for each symbol can be found by integrating the error tails of the normalised ($\bar{v} = 0$, $\sigma = 1$) Gaussian probability density function. The probability of a transmitted 0 being received in error is therefore:

$$P_{e1} = \int\limits_{\frac{v_{th}-V_0}{\sigma_0}}^{\infty} \frac{1}{\sqrt{(2\pi)}} e^{-\frac{v^2}{2}} dv = \int\limits_{1.384}^{\infty} \frac{1}{\sqrt{(2\pi)}} e^{-\frac{v^2}{2}} dv$$

Using the substitution $x = v/\sqrt{2}$:

$$P_{e1} = \frac{1}{\sqrt{\pi}} \int\limits_{1.384/\sqrt{2}}^{\infty} e^{-x^2} dx = 0.5 \operatorname{erfc}(1.384/\sqrt{2}) = 0.0831$$

Similarly the probability of a transmitted 1 being received in error is:

$$P_{e0} = \int\limits_{-\infty}^{\frac{v_{th}-V_1}{\sigma_1}} \frac{1}{\sqrt{(2\pi)}} e^{-\frac{v^2}{2}} dv$$

From symmetry P_{e0} can also be written:

$$P_{e0} = \int\limits_{-\frac{(v_{th}-V_1)}{\sigma_1}}^{\infty} \frac{1}{\sqrt{(2\pi)}} e^{-\frac{v^2}{2}} dv = \int\limits_{1.816}^{\infty} \frac{1}{\sqrt{(2\pi)}} e^{-\frac{v^2}{2}} dv$$

Using the same substitution as before this becomes:

$$P_{e0} = 0.5 \operatorname{erfc}(1.816/\sqrt{2}) = 0.0347$$

The overall probability of symbol error is therefore:

$$P_e = P(0)P_{e1} + P(1)P_{e0}$$

$$= \frac{1}{3}(0.0831) + \frac{2}{3}(0.0347) = 0.0508$$

Optimum thresholding, as discussed above, can be applied to centre point detection processes, such as described in Chapter 6, or after predetection signal processing (e.g. matched filtering), as discussed in Chapter 8. The important point to appreciate is that the sampling instant signal plus noise pdfs, used to establish the optimum threshold voltage(s), are those at the decision circuit input irrespective of whether predetection processing has been applied or not.

7.5 Neyman–Pearson decision criterion

The Neyman–Pearson decision criterion requires only a posteriori probabilities. Unlike Bayes's decision rule it does not require a priori source probability information. This criterion is particularly appropriate to pulse detection in Gaussian noise, as occurs in radar applications, where the source statistics (i.e. probabilities of presence, and absence, of a target) are unknown. It also works well when $C_0 \gg C_1$ or, in radar terms, when the

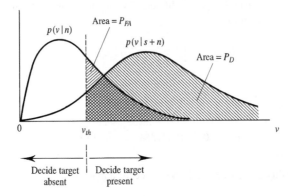

Figure 7.7 *Conditional pdfs and threshold voltage for Neyman–Pearson radar signal detector.*

information cost of erroneously deciding a target is present is much less than the information cost of erroneously deciding a target is absent. In this context the important probabilities are those for target detection, P_D, and target false alarm, P_{FA}. The conditional pdfs for these two decisions and a selected decision threshold voltage are shown in Figure 7.7. The two probabilities are:

$$P_D = \int_{v_{th}}^{\infty} p(v|s + n)\, dv \tag{7.21}$$

$$P_{FA} = \int_{v_{th}}^{\infty} p(v|n)\, dv \tag{7.22}$$

where s denotes signal (arising due to reflections from a target) and n denotes noise. The optimum detector consists of a linear filter followed by a threshold detector [Blahut, 1987] which tests the hypothesis as to whether there is noise only present or a signal pulse plus noise. The filter is matched to the pulse (see section 8.3). In the Neyman–Pearson detector the threshold voltage, v_{th}, is chosen to give an acceptable value of P_{FA} and the detection probability then follows the characteristic shown in Figure 7.8. Note here that detection performance is dependent on both the choice of P_{FA} and the ratio of received pulse energy to noise power spectral density, E/N_0.

7.6 Summary

Soft and hard decision processes are both used in digital communications receivers but hard decision processes are simpler to implement and therefore more common. Transition probabilities $P(i|j)$ describe the probability with which transmitted symbol j will be corrupted by noise sufficiently to be interpreted at the receiver as symbol i. These probabilities can be assembled into a matrix to describe the end to end error properties of a communications channel.

Bayes's decision criterion is the criterion most often used in digital communications receivers. It is optimum in the sense that it minimises the cost of (i.e. reduces the

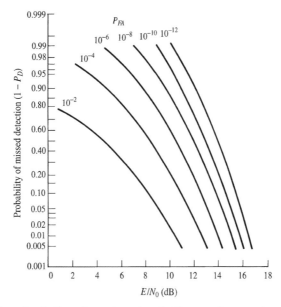

Figure 7.8 *Pulse detection in Gaussian noise using the Neyman–Pearson detection criterion (source: Blahut, 1987).*

information lost when) making decisions. Bayes's criterion allows the optimum threshold voltage(s), which delineate(s) decision regions, to be found. Maximum a posteriori probability (MAP) and maximum likelihood detectors are special cases of a Bayes's criterion detector, appropriate when decision costs, or decision costs and a priori probabilities, are unknown.

The Neyman–Pearson decision criterion is normally used in radar applications. It has the advantage over Bayes's criterion that a priori symbol probabilities, whilst known to be very different, need not be quantified. The threshold level of the Neyman–Pearson criterion is set to give acceptable probabilities of false alarm and the probability of detection, given a particular received E/N_0 ratio, follows from this.

7.7 Problems

7.1 Derive the Bayes decision rule from first principles.

7.2 Under what assumptions does the Bayes receiver become a maximum likelihood receiver? Illustrate your answer with the appropriate equations.

7.3 In a baseband binary transmission system a 1 is represented by +5 V and a 0 by −5 V. The channel is subject to additive, zero mean, Gaussian noise with a standard deviation of 2 V. If the a priori probabilities of 1 and 0 are 0.5 and the costs C_0 and C_1 are equal, calculate the optimum position for the decision boundary, and the probability of error for this optimum position. [0 V, 6.23×10^{-3}]

7.4 What degenerate case of the Bayes receiver design is being implemented in Problem 7.3? Assume that the decision threshold in Problem 7.3 is incorrectly placed by being moved up from its optimum position by 0.5 V. Calculate the new probability of error associated with this suboptimum threshold. [7.4125×10^{-3}]

7.5   Design a Bayes detector for the a priori probabilities $P(1) = 2/3, P(0) = 1/3$ with costs $C_0 = 1$ and $C_1 = 3$. The conditional pdfs for the received variable v are given by:

$$p(v|0) = \frac{1}{\sqrt{(2\pi)}} e^{-v^2/2}$$

$$p(v|1) = \frac{1}{\sqrt{(2\pi)}} e^{-(v-1)^2/2} \qquad\qquad [v = 0.905]$$

7.6   A radar system operates in zero mean, Gaussian noise with variance 5 V. If the probability of false alarm is to be 10^{-2} calculate the detection threshold. If the expected return from a target at extreme range is 4 V, what is the probability that this target will be detected? [5.21, 0.2945]

CHAPTER 8

Optimum filtering for transmission and reception

8.1 Introduction

There are two signal filtering techniques which are of basic importance in digital communications. The first is concerned with filtering for transmission in order to minimise signal bandwidth. The second is concerned with filtering at the receiver in order to maximise the SNR at the decision instant (and consequently minimise the probability of symbol error). This chapter examines each of these problems and establishes criteria to be met by those filters providing optimum solutions.

8.2 Pulse shaping for optimum transmissions

Spectral efficiency, η_s, is defined as the rate of information transmission per unit of occupied bandwidth[1], i.e.:

$$\eta_s = R_s H / B \quad \text{(bits/s/Hz)} \tag{8.1}$$

where R_s is the symbol rate, H is entropy, i.e. the average amount of information (measured in bits) conveyed per symbol, and B is occupied bandwidth. (For an alphabet containing M, statistically independent, equiprobable symbols, $H = \log_2 M$ bit/symbol, see Chapter 9.) The same term is also sometimes used for the quantity R_s / B which has

---

[1] In cellular radio applications the term spectral efficiency is also used in a more general sense, incorporating the spatial spectrum 'efficiency'. This quantity is variously ascribed units of voice channels/MHz/km^2, Erlangs/MHz/km^2 or voice channels/cell. (Typically these spectral 'efficiencies' are much greater than unity!) The quantity called spectral efficiency here is then referred to as bandwidth efficiency.

units of symbol/s/Hz or baud/Hz. Since spectrum is a limited resource it is often desirable to minimise the bandwidth occupied by a signal of given baud rate. Nyquist's sampling theorem (section 5.3.2) limits the transmission rate of independent samples (or symbols) in a baseband bandwidth B to:

$$R_s \le 2B \quad \text{(symbol/s)} \tag{8.2}$$

The essential pulse shaping problem is therefore one of how to shape transmitted pulses to allow signalling at, or as close as possible to, the maximum (Nyquist) rate of $2B$ symbol/s.

8.2.1 Intersymbol interference (ISI)

Rectangular pulse signalling, in principle, has a spectral efficiency of 0 bit/s/Hz since each rectangular pulse, strictly speaking, has infinite bandwidth. In practice, of course, rectangular pulses can be transmitted over channels with finite bandwidth if a degree of distortion can be tolerated.

In digital communications it might appear that distortion is unimportant since a receiver must only distinguish between pulses which have been distorted in the same way. This might be thought especially true for OOK signalling in which only the presence or absence of pulses is important. This is not, in fact, the case since, if distortion is severe enough, then pulses may overlap in time. The decision instant voltage might then arise not only from the current symbol pulse but also from one or more preceding pulses. The smearing of one pulse into another is called *intersymbol interference* and is illustrated in Figure 8.1 for the case of rectangular baseband pulses, distorted by an *RC* low pass channel (as discussed previously in section 5.4).

8.2.2 Bandlimiting of rectangular pulses

The nominal bandwidth of a baseband, unipolar, NRZ signal with baud rate $R_s = 1/T_o$ symbol/s was taken in section 6.4.1 to be $B = 1/T_o$ Hz. (This corresponds to the positive frequency width of the signal's main spectral lobe.) It is instructive to see the effect of limiting a rectangular pulse to this bandwidth before transmission (Figure 8.2). The filtered pulse spectrum, Figure 8.2(f), is then restricted to the main lobe of the rectangular pulse spectrum, Figure 8.2(b). The filtered pulse shape, Figure 8.2(e), is the rectangular pulse convolved with the filter's $\text{sinc}(2Bt)$ impulse response. Figure 8.2(c) shows the rectangular pulse superimposed on the filter's impulse response for two values of time offset, 0 and T_o seconds. At zero offset the convolution integral gives the received pulse

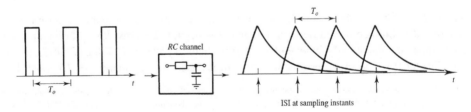

Figure 8.1 *Pulse smearing due to distortion in an RC channel.*

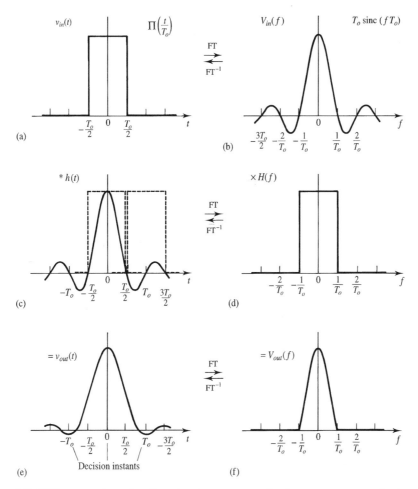

Figure 8.2 *NRZ rectangular pulse distortion due to rectangular frequency response filtering.*

peak at $t = 0$. At an offset of $T_o/2$ (not shown) the convolution integral clearly gives a reduced, but still large, positive result. At an offset of T_o the convolution gives a negative value (since the negative first sidelobe of the sinc function is larger than its positive second sidelobe). The zero crossing point of the filtered pulse therefore occurs a little before T_o seconds, Figure 8.2(e). At the centre point sampling instants $(\cdots, -T_o, 0, T_o, 2T_o, \cdots)$ receiver decisions would therefore be based not only on that pulse which should be considered but also, erroneously, on contributions from adjacent pulses. These unwanted contributions have the potential to degrade BER performance.

8.2.3 ISI-free signals

The decision instants marked on Figure 8.2(e) illustrate an important point, i.e.:

> *Only decision instant ISI is relevant to the performance of digital communications systems. ISI occurring at times other than the decision instants does not matter.*

If the signal pulses could be persuaded to pass through zero at every decision instant (except, of course, one) then ISI would no longer be a problem. This suggests a definition for an ISI-free signal, i.e.:

An ISI-free signal is any signal which passes through zero at all but one of the sampling instants.

Denoting the ISI-free signal by $v_N(t)$ and the sampling instants by nT_o (where n is an integer and T_o is the symbol period) this definition can be expressed mathematically as:

$$v_N(t) \sum_{n=-\infty}^{\infty} \delta(t - nT_o) = v_N(0)\, \delta(t) \tag{8.3}$$

The important property of ISI-free signals, summarised by equation (8.3), is illustrated in Figure 8.3. Such signals suppress all the impulses in a sampling function except one (in this case the one occurring at $t = 0$). A good example of an ISI-free signal is the sinc pulse, Figure 8.4. These pulses have a peak at one decision instant and are zero at all other decision instants as required. (They also have the minimum (Nyquist) bandwidth for a given baud rate.) An OOK, multilevel, or analogue PAM, system could, in principle, be implemented using sinc pulse signalling. In practice, however, there are two problems:

1. sinc pulses are not physically realisable.
2. sinc pulse sidelobes (and their rates of change at the decision instants) are large and decay only with $1/t$.

The obvious way to generate sinc pulses is by shaping impulses with low pass rectangular filters. The first problem could be equally well stated, therefore, as 'linear phase low pass rectangular filters are not realisable'. The second problem means that extremely accurate decision timing would be required at the receiver to keep decision instant ISI to tolerable levels.

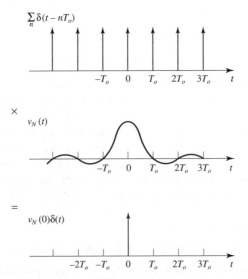

Figure 8.3 *Impulse suppression property of ISI-free pulse.*

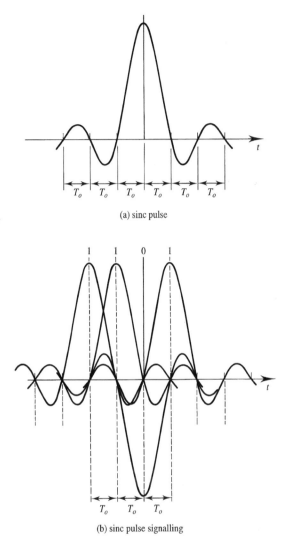

(a) sinc pulse

(b) sinc pulse signalling

Figure 8.4 *ISI-free transmission using sinc pulses.*

A more practical signal pulse shape would retain the desirable feature of sinc pulses (i.e. regularly spaced zero crossing points) but have an envelope with much more rapid roll-off. This can be achieved by multiplying the sinc pulse with a rapidly decaying monotonic function, Figures 8.5(a) to (c). In the frequency domain this corresponds to convolving the sinc pulse rectangular spectrum, Figure 8.5(d), with the spectrum of the decaying function, Figure 8.5(e), to obtain the final spectrum, Figure 8.5(f). As long as the decaying function is real and even its spectrum will be real and even, which implies that the modified pulse spectrum will have *odd* symmetry about the sinc pulse's cut-off frequency, f_χ, Figure 8.5(f). This suggests an alternative definition for ISI-free (baseband) signals, i.e.:

An ISI-free baseband signal has a voltage spectrum which displays odd symmetry about $1/(2T_o)$ Hz.

A more general statement, which includes ISI-free bandpass signals, can be made by considering frequency translation of the baseband spectrum using the modulation theorem (Figure 8.6), i.e.:

An ISI-free signal has a voltage spectrum which displays odd symmetry between its centre frequency, f_c, and $f_c \pm 1/T_o$ Hz.

This property can also be demonstrated by Fourier transforming equation (8.3), replacing the product by a convolution (*), section 2.3.4, to give:

$$V_N(f) * \frac{1}{T_o} \sum_{n=-\infty}^{\infty} \delta\left(f - \frac{n}{T_o}\right) = v_N(0) \tag{8.4}$$

which, using the replicating action of delta functions under convolution, gives:

$$\frac{1}{T_o} \sum_{n=-\infty}^{\infty} V_N\left(f - \frac{n}{T_o}\right) = v_N(0) \tag{8.5}$$

Figure 8.7 gives a pictorial interpretation. The spectrum of the ISI-free signal is such that if replicated along the frequency axis with periodic spacing $1/T_o$ Hz the sum of all replicas is a constant. This demands a mathematical precise spectral symmetry.

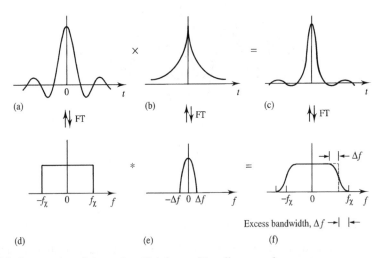

(a)   (b)   (c)

(d)   (e)   (f)

Excess bandwidth, Δf

Figure 8.5 *Suppression of sinc pulse sidelobes and its effect on pulse spectrum.*

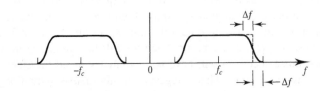

Figure 8.6 *Amplitude spectrum of a bandpass ISI-free signal.*

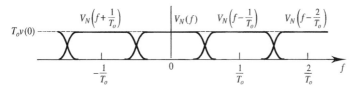

Figure 8.7 *Constant sum property of replicated ISI-free signal spectra.*

EXAMPLE 8.1

Specify a baseband Nyquist channel which has a piecewise linear amplitude response, an absolute bandwidth of 10 kHz, and is appropriate for a baud rate of 16 kbaud. What is the channel's excess bandwidth?

The cut-off frequency of the parent rectangular frequency response is given by:

$$f_\chi = R_s/2 = 16 \times 10^3/2 = 8 \times 10^3 \text{ Hz}$$

The simplest piecewise linear roll-off therefore starts at $8 - 2 = 6$ kHz, is 6 dB down at $f_\chi = 8$ kHz and is zero ($-\infty$ dB down) at $8 + 2 = 10$ kHz. (The amplitude response, below the start of roll-off, is flat and the phase is linear.) Thus:

$$|H_N(f)| = \begin{cases} 1.0, & |f| < 6000 \text{ (Hz)} \\ 2.5 - 0.25 \times 10^{-3}f, & 6000 \leq |f| \leq 10000 \text{ (Hz)} \\ 0, & |f| > 10000 \text{ (Hz)} \end{cases}$$

The channel's excess bandwidth is 10 kHz -8 kHz = 2 kHz.

8.2.4 Nyquist's vestigial symmetry theorem

Nyquist's vestigial symmetry theorem defines a symmetry condition on $H(f)$ which must be satisfied to realise an ISI-free baseband impulse response. It can be stated as follows:

> *If the amplitude response of a low pass rectangular filter with linear phase and bandwidth f_χ is modified by the addition of a real valued function having odd symmetry about the filter's cut-off frequency, then the resulting impulse response will retain at least those zero crossings present in the original sinc($2f_\chi t$) response, i.e. it will be an ISI-free signal.*

This 'recipe' for deriving the whole family of Nyquist filters from a low pass rectangular prototype is illustrated in Figure 8.8. The theorem requires no further justification since it follows directly from the spectral properties of ISI-free signals (section 8.2.3). The theorem can be generalised to include filters with ISI-free bandpass impulse responses in an obvious way.

8.2.5 Raised cosine filtering

The family of raised cosine filters is an important and popular subset of the family of Nyquist filters. The odd symmetry of their amplitude response is provided using a cosinusoidal, half cycle, roll-off (Figure 8.9). Their (low pass) amplitude response therefore has the following piecewise form:

$$|H(f)| = \begin{cases} 1.0, & |f| \le (f_\chi - \Delta f) \\ \frac{1}{2}\left\{1 + \sin\left[\frac{\pi}{2}\left(1 - \frac{|f|}{f_\chi}\right)\frac{f_\chi}{\Delta f}\right]\right\}, & (f_\chi - \Delta f) < |f| < (f_\chi + \Delta f) \\ 0, & |f| \ge (f_\chi + \Delta f) \end{cases}$$ (8.6)

and their phase response is linear (implying the need for finite impulse response filters [Mulgrew *et al.*]). f_χ in equation (8.6) is the cut-off frequency of the prototype rectangular low pass filter (and the −6 dB frequency of the raised cosine filter). f_χ is related to the symbol period, T_o, by $f_\chi = R_s/2 = 1/(2T_o)$. Δf is the excess (absolute) bandwidth of the filter over the rectangular low pass prototype. The normalised excess bandwidth, α, given by:

$$\alpha = \frac{\Delta f}{f_\chi}$$ (8.7)

is called the roll-off factor and can take any value between 0 and 1. Figure 8.10(a) shows

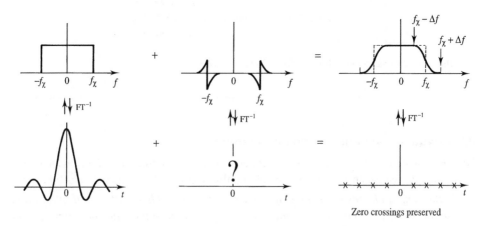

Figure 8.8 *Illustration of Nyquist's vestigial symmetry theorem.*

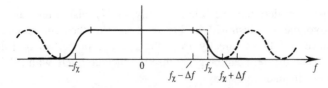

Figure 8.9 *Amplitude response of (linear phase) raised cosine filter ($f_\chi = R_s/2$).*

the raised cosine amplitude response for several values of α. When $\alpha = 1$ the characteristic is said to be a *full* raised cosine and in this case the amplitude response simplifies to:

$$|H(f)| = \begin{cases} \dfrac{1}{2}\left[1 + \cos\left(\dfrac{\pi f}{2f_\chi}\right)\right], & |f| \le 2f_\chi \\[3mm] 0, & |f| > 2f_\chi \end{cases}$$

$$= \begin{cases} \cos^2\left(\dfrac{\pi f}{4f_\chi}\right), & |f| \le 2f_\chi \\[3mm] 0, & |f| > 2f_\chi \end{cases} \tag{8.8}$$

(The power spectral density (PSD) of an ISI-free signal generated using a full raised cosine filter therefore has a $\cos^4[\pi f/(4f_\chi)]$ shape, Figure 8.11.) The impulse response of a full raised cosine filter (found from the inverse Fourier transform of equation (8.8)) is:

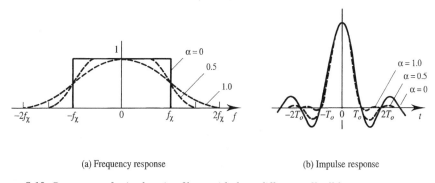

(a) Frequency response   (b) Impulse response

Figure 8.10 *Responses of raised cosine filters with three different roll-off factors.*

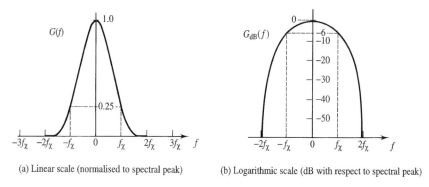

(a) Linear scale (normalised to spectral peak)   (b) Logarithmic scale (dB with respect to spectral peak)

Figure 8.11 *PSD of an ISI-free signal generated using a full raised cosine filter.*

$$h(t) = 2f_\chi \frac{\sin 2\pi f_\chi t}{2\pi f_\chi t} \frac{\cos 2\pi f_\chi t}{1 - (4f_\chi t)^2} \tag{8.9}$$

This is shown in Figure 8.10(b) along with the impulse responses of raised cosine filters with other values of α. The first part of equation (8.9) represents the sinc impulse response of the protype rectangular filter. The second part modifies this with extra zeros (due to the numerator) and faster decaying envelope ($1/t^3$ in total due to the denominator). The absolute bandwidth of a baseband filter (or channel) with a raised cosine frequency response is:

$$B = \frac{1}{2T_o} (1 + \alpha)$$

$$= \frac{R_s}{2} (1 + \alpha) \quad \text{(Hz)} \tag{8.10}$$

where R_s is the symbol (or baud) rate. For a bandpass raised cosine filter the bandwidth is twice this, i.e.:

$$B = R_s (1 + \alpha) \quad \text{(Hz)} \tag{8.11}$$

(This simply reflects the fact that when baseband signals are converted to bandpass (double sideband) signals by amplitude modulation, their bandwidth doubles.) Impulse signalling over a raised cosine *baseband* channel has a spectral efficiency of 2 symbol/s/Hz when $\alpha = 0$ and 1 symbol/s/Hz when $\alpha = 1$. For binary signalling systems (assuming equiprobable, independent, symbols) this translates to 2 bit/s/Hz and 1 bit/s/Hz respectively (see Chapter 9). For bandpass filters and channels these efficiencies are halved, due to the double sideband spectrum.

EXAMPLE 8.2

What absolute bandwidth is required to transmit an information rate of 8.0 kbit/s using 64 level baseband signalling over a raised cosine channel with a roll-off factor of 40%?

The bit rate, R_b, is given by the number of bits per symbol (H) times the number of symbols per second (R_s), i.e.:

$$R_b = H \times R_s$$

Therefore:

$$R_s = R_b/H$$

$$= \frac{8 \times 10^3}{\log_2 64} = 1.333 \times 10^3 \quad \text{(symbol/s)}$$

$$B = \frac{R_s}{2} (1 + \alpha)$$

$$= \frac{1.333 \times 10^3}{2} (1 + 0.4) = 933.1 \quad \text{(Hz)}$$

This illustrates the bandwidth efficiency of a multilevel signal.

8.2.6 Nyquist filtering for rectangular pulses

It is possible to generate ISI-free signals by shaping impulses with Nyquist filters. A more usual requirement, however, is to generate such signals by shaping rectangular pulses. The appropriate pulse shaping filter must then have a frequency response:

$$H(f) = \frac{V_N(f)}{\text{sinc}(\tau f)} \tag{8.12}$$

where $V_N(f)$ is the voltage spectrum of an ISI-free pulse and $\text{sinc}(\tau f)$ is the frequency response of a hypothetical filter which converts impulses into rectangular pulses with width τ. If $V_N(f)$ is chosen to have a full raised cosine shape ($\alpha = 1$) then:

$$H(f) = \frac{\pi f \tau}{\sin(\pi f \tau)} \cos^2\left(\frac{\pi f}{4f_\chi}\right) \tag{8.13}$$

Figure 8.12 ($\alpha = 1$) shows the frequency response corresponding to equation (8.13). Responses corresponding to other values of α are also shown.

8.2.7 Duobinary signalling

One of the problems associated with the use of sinc pulses, for ISI-free signalling at the Nyquist rate, is the construction of a linear phase, low pass rectangular filter. Such a filter (or rather an adequate approximation) is required to shape impulses. Duobinary signalling uses not a rectangular filter for pulse shaping but a cosine filter (not to be confused with the raised cosine filter). The amplitude response of a baseband cosine

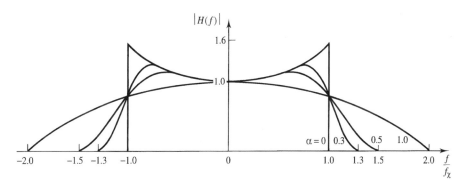

Figure 8.12 *Amplitude response of Nyquist filter for rectangular pulse shaping.*

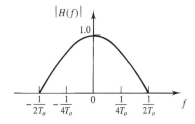

Figure 8.13 *Amplitude response of cosine filter.*

filter, Figure 8.13, is given by:

$$|H(f)| = \begin{cases} \cos \pi f T_o, & |f| \le 1/(2T_o) \\ 0, & |f| \le 1/(2T_o) \end{cases}$$

(8.14(a))

and its (linear) phase response is usually taken to be:

$$\phi(f) = -\frac{\omega T_o}{2} \quad (\text{rad})$$

(8.14(b))

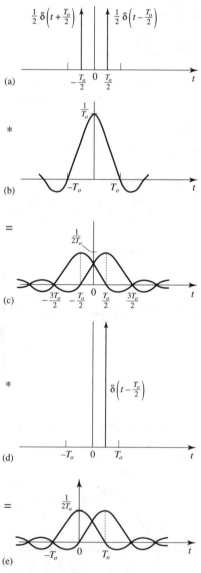

Figure 8.14 *Derivation of cosine filter impulse response.*

It has the same absolute bandwidth, $1/(2T_o)$ Hz, as the rectangular filter used for sinc signalling, and duobinary signalling therefore proceeds at the same maximum baud rate. Since its amplitude response has fallen to a low level at the filter band edge the linearity of the phase response in this region is not critical. This makes the cosine filter relatively easy to approximate. The impulse response of the cosine filter is most easily found by expressing its frequency response as a product of cosine, rectangular low pass and phase factors, i.e.:

$$H(f) = \cos(\pi f T_o) \, \Pi\left(\frac{f}{2f_\chi}\right) e^{-j\pi fT_o} \tag{8.15}$$

where $f_\chi = 1/(2T_o) = f_o/2$, and Π is the rectangular gate function. The product of the first two factors corresponds, in the time domain, to the convolution of a pair of delta functions with a sinc function, Figures 8.14(a) to (c). The third factor simply shifts the resulting pair of sinc functions to the right by $T_o/2$ seconds, Figures 8.14(d) and (e). The impulse response, shown in Figure 8.15, is therefore:

$$h(t) = \frac{1}{2T_o} [\text{sinc } f_o t + \text{sinc } f_o(t - T_o)] \tag{8.16}$$

It can be seen that the more gradual roll-off of the cosine filter (when compared to the

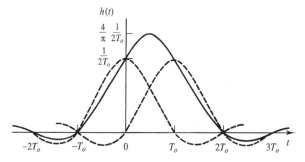

Figure 8.15 *Impulse response (solid curve) for the cosine filter.*

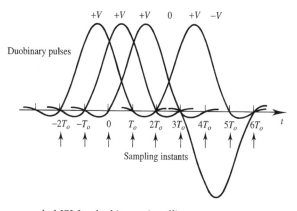

Figure 8.16 *Adjacent symbol ISI for duobinary signalling.*

rectangular filter) has been obtained at the expense of a significantly lengthened impulse response, which results in severe ISI. Duobinary signalling is important, however, because sampling instant interference occurs between *adjacent* symbols only, Figure 8.16, and is of predictable magnitude.

A useful model of duobinary signalling (which does *not* correspond to its normal implementation) is suggested by equation (8.16). It is clear that this is a superposition of two low pass rectangular filter impulse responses, one delayed with respect to the other by T_o seconds. The cosine filter could therefore be implemented using a one-symbol delay device, an adder and a rectangular filter, Figure 8.17. This implementation is obvious if equation (8.15) is rewritten as:

$$H(f) = \frac{1}{2}(1 + e^{-j2\pi fT_o}) \prod\left(\frac{f}{2f_\chi}\right) \tag{8.17}$$

Duobinary signalling can therefore be interpreted as adjacent pulse summation followed by rectangular low pass filtering. Figure 8.18 shows the composite pulses which arise from like and unlike combinations of binary impulse pairs. At the receiver, sampling can take place at the centre of the composite pulses (i.e. at the point of maximum ISI midway between the response peaks of the original binary impulses). This results in *three* possible levels at each decision instant, i.e. $+V$, 0 and $-V$, the level observed depending on whether the binary pulse pair are both positive, both negative or of opposite sign,

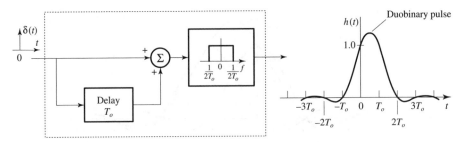

Figure 8.17 *Equivalent model of cosine filter for duobinary signalling.*

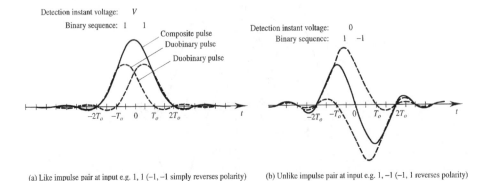

(a) Like impulse pair at input e.g. 1, 1 (−1, −1 simply reverses polarity)   (b) Unlike impulse pair at input e.g. 1, −1 (−1, 1 reverses polarity)

Figure 8.18 *Composite pulses arising from like and unlike combinations of input impulse pairs.*

Figure 8.19. Like bipolar line coding (Chapter 6), duobinary signalling is therefore a form of pseudo-ternary signalling [Lender]. The summing of adjacent pulse pairs at the transmitter can be described explicitly using the notation:

$$z_k = y_k + y_{k-1} \tag{8.18}$$

where z_k represents the kth (ternary) symbol after duobinary coding and y_k represents the kth (binary) symbol before coding. The decoding process after detection at the receiver is therefore the inverse of equation (8.18), i.e.:

$$\hat{y}_k = \hat{z}_k - \hat{y}_{k-1} \tag{8.19}$$

(where the hats (^) distinguish between detected and transmitted symbols to allow for the possibility of errors). The block diagram of the receiver is shown in Figure 8.20(a).

The essential advantage of duobinary signalling is that it permits signalling at the Nyquist rate without the need for linear phase, rectangular, low pass filters (or their equivalent). The disadvantages are:

1. The ternary nature of the signal requires approximately 2.5 dB greater SNR when compared with ideal binary signalling for a given probability of error.
2. There is no one-to-one mapping between detected ternary symbols and the original binary digits.
3. The decoding process, $\hat{y}_k = \hat{z}_k - \hat{y}_{k-1}$, results in the propagation of errors.
4. The duobinary power spectral density (which is $\cos^2(\pi T_o f) \, \Pi[f/(2f_\chi)]$, i.e. the square of the cosine filter amplitude response) has a maximum at 0 Hz making it unsuitable for use with AC coupled transmission lines.

Problems 2 and 3 can be solved by adding the following precoding algorithm to the bit stream prior to duobinary pulse shaping:

$$y_k = x_k \oplus y_{k-1} \tag{8.20}$$

where y_k is the kth precoded bit, x_k is the uncoded bit and $\oplus$ represents modulo-2 addition. Figure 8.20(b) shows the block diagram corresponding to equation (8.20). The

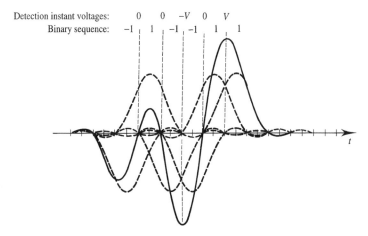

Figure 8.19 *Duobinary waveform arising from an example binary sequence.*

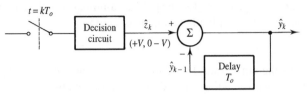

(a) Duobinary receiver (order of decision circuit and decoder could be reversed)

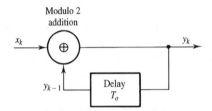

(b) Precoding for duobinary signalling

Figure 8.20 *Duobinary receiver and precoder.*

effect of precoding plus duobinary coding can be simplified, i.e.:

$$z_k = (x_k \oplus y_{k-1}) + (x_{k-1} \oplus y_{k-2})$$
$$= (x_k \oplus y_{k-1}) + y_{k-1} \tag{8.21}$$

The truth table for equation (8.21) is shown in Table 8.1. The important property of this table is that $z_k = 1$ when, and only when, $x_k = 1$. The precoded duobinary signal can therefore be decoded on a bit-by-bit basis (i.e. without the use of feedback loops). One-to-one mapping, of received ternary symbols to original binary symbols, is thus re-established and error propagation is eliminated.

Table 8.1 *Precoded duobinary truth table.*

x_k	y_{k-1}	$x_k \oplus y_{k-1}$	z_k
0	0	0	0
1	0	1	1
0	1	1	2
1	1	0	1

EXAMPLE 8.3

Find the output data sequence of a duobinary signalling system (a) without precoding and (b) with precoding if the input data sequence is: 1 1 0 0 0 1 0 1 0 0 1 1 1.

(a) Using equation (8.18):

input data, y_k 1100010100111

y_{k-1} ?1100010100111

duobinary data, z_k ?210011110122?

(b) Using equations (8.20) and (8.21) or Table 8.1, and assuming that the initial precoded bit is a digital 1:

x_k 1100010100111

y_{k-1} (1)011110011101

z_k 1122210122111

Assuming that, after precoding, the (y_k) binary digits (1,0) are represented by positive and negative impulses, then the duobinary detection algorithm is:

$$\hat{x}_k = \begin{cases} 1, & f(kT_s) = 0 \\ 0, & f(kT_s) = \pm V \end{cases} \qquad (8.22)$$

where $f(kT_s)$ is the received voltage at the appropriate decision instant.

Problem 4 (i.e. the large DC value of the duobinary PSD) can be addressed by replacing the transmitter 1 bit delay and adder in Figure 8.17 by a 2 bit $(2T_o)$ delay and subtractor, Figure 8.21(a). This results in *modified duobinary* signalling. The PSD has a null at 0 Hz but is still strictly bandlimited to $1/(2T_o)$ Hz. Figure 8.21(b) is a block diagram of the modified duobinary receiver. The frequency response of the pulse shaping filter, Figure 8.22(a), is:

$$H(f) = \tfrac{1}{2}(1 - e^{-j4\pi fT_o}) \Pi\!\left(\frac{f}{2f_\chi}\right) \qquad (8.23)$$

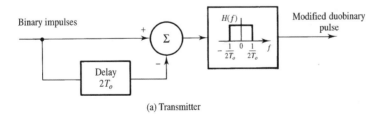

(a) Transmitter

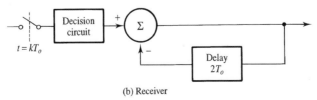

(b) Receiver

Figure 8.21 *Equivalent model for modified duobinary signalling.*

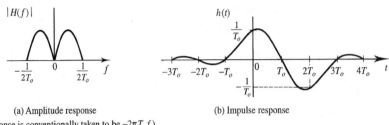

(a) Amplitude response
(Phase response is conventionally taken to be $-2\pi T_o f$)

(b) Impulse response

Figure 8.22 *Characteristics of pulse shaping filter for modified duobinary signalling.*

where, once again, the filter cut-off frequency is $f_\chi = 1/(2T_o)$ Hz. The corresponding impulse response is shown in Figure 8.22(b). The amplitude and phase response are:

$$|H(f)| = \begin{cases} |\sin(2\pi fT_o)|, & |f| \le 1/(2T_o) \\ 0, & |f| > 1/(2T_o) \end{cases} \qquad (8.24)$$

$$\phi(f) = -\omega T_o, \qquad |f| \le 1/(2T_o) \qquad (8.25)$$

Since $(1 - e^{-j4\pi fT_o})$ can be factorised into $(1 - e^{-j2\pi fT_o})(1 + e^{-j2\pi fT_o})$, modified duobinary pulse shaping can be realised by cascading a 1 bit delay and subtractor (to implement the first factor) with a conventional duobinary pulse shaping filter (to implement the second factor). The block diagram corresponding to this implementation is shown in Figure 8.23. Precoding can be added to avoid modified duobinary error propagation. The appropriate precoding and post-decoding algorithms are illustrated as block diagrams in Figure 8.24.

8.2.8 Partial response signalling

Partial response signalling is a generalisation of duobinary signalling in which the single element transversal filter of Figure 8.17 is replaced with an N-element (tap weighted) filter. This produces a multilevel signal with non-zero correlation between symbols over an $N + 1$ symbol window. Since the ISI introduced as a result of this correlation is of a prescribed form it can be, as in duobinary signalling, effectively cancelled at the receiver. Partial response signalling is also known as correlative coding.

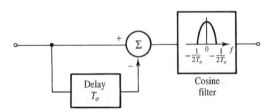

Figure 8.23 *Practical implementation of modified duobinary pulse shaping filter.*

8.3 Pulse filtering for optimum reception

Formulae were derived in sections 6.2.1 and 6.2.2 for the probability of bit error expected when equiprobable, rectangular, baseband symbols are detected, using a centre point decision process in the presence of Gaussian noise. Since this process compares a single sample value of signal plus noise with an appropriate threshold the following question might be asked. *If several samples of the signal plus noise voltage are examined at different time instants within the duration of a single symbol (as illustrated in Figure 8.25) is it not possible to obtain a more reliable (i.e. lower P_e) decision?* The answer to this question is normally yes since, at the very least, majority voting of multiple decisions associated with a given symbol could be employed to reduce the probability of error. Better still, if n samples were examined, an obvious strategy would be to add the samples together and compare the result with n times the appropriate threshold for a single sample. If this idea is extended to its limit (i.e. $n \to \infty$) then the discrete summation of symbol plus noise samples becomes continuous integration of the symbol plus noise voltage. The post-integration decision threshold then becomes $\frac{1}{2} (\int_0^{T_o} v_0 \, dt + \int_0^{T_o} v_1 \, dt)$ where v_0 and v_1 are the voltage levels representing binary zeros and ones respectively.

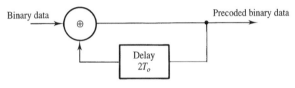

Binary data   Precoded binary data

Delay $2T_o$

(a) Precoding ($\oplus$ signifies modulo 2 addition)

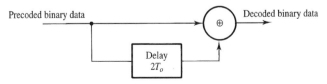

Precoded binary data   Decoded binary data

Delay $2T_o$

(b) Decoding ($\oplus$ signifies modulo 2 addition)

Figure 8.24 *Precoding and de-precoding for modified duobinary signalling.*

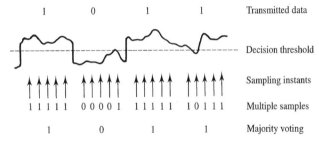

| 1 | 0 | 1 | 1 | Transmitted data |

Decision threshold

Sampling instants

| 1 1 1 1 1 | 0 0 0 0 1 | 1 1 1 1 1 | 1 0 1 1 1 | Multiple samples |

| 1 | 0 | 1 | 1 | Majority voting |

Figure 8.25 *Multiple sampling of single symbols.*

After each symbol the integrator output would be reset to zero ready for the next symbol. This signal processing technique, Figure 8.26, is a significant improvement on centre point sampling and is called integrate and dump (I+D) detection. It is the optimum detection process for baseband rectangular pulses in that the resulting probability of error is a minimum. It is also easy to implement as shown in Figure 8.27. I+D is a special case of a general and optimum type of detection process, which can be applied to any pulse shape, called *matched filtering*.

8.3.1 Matched filtering

A matched filter can be defined as follows:

> *A filter which immediately precedes the decision circuit in a digital communications receiver is said to be matched to a particular symbol pulse, if it maximises the output SNR at the sampling instant when that pulse is present at the filter input.*

The criteria which relate the characteristics (amplitude and phase response) of a filter to those of the pulse to which it is matched can be derived as follows.

Consider a digital communications system which transmits pulses with shape $v(t)$, Figure 8.28(a). The pulses have a (complex) voltage spectrum $V(f)$, Figure 8.28(b), and a normalised energy spectral density (ESD) $|V(f)|^2$ V^2s/Hz, Figure 8.28(d). If the noise power spectral density (NPSD) is white, it can be represented as a constant ESD per pulse period as shown in Figure 8.28(d). If the spectrum is divided into narrow frequency bands it can be seen from this figure that some bands (such as j) have large SNR and some (such as r) have much smaller SNR. Any band which includes signal energy should clearly make a contribution to the decision process (otherwise signal is being discarded). It is intuitively obvious, however, that those bands with high SNR should be correspondingly more influential in the decision process than those with low SNR. This

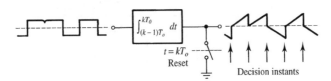

Figure 8.26 *Integrate and dump detection for rectangular pulses.*

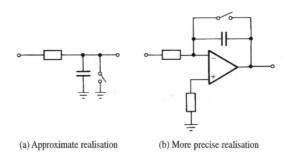

(a) Approximate realisation   (b) More precise realisation

Figure 8.27 *Simple circuit realisations for integrate and dump (I+D) detection.*

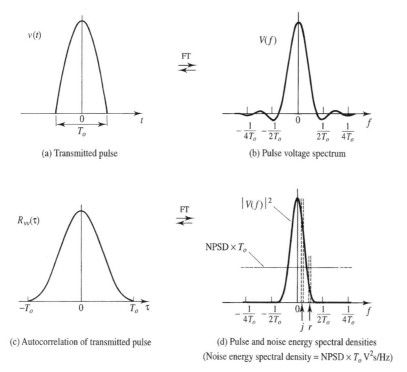

(a) Transmitted pulse

(b) Pulse voltage spectrum

(c) Autocorrelation of transmitted pulse

(d) Pulse and noise energy spectral densities
(Noise energy spectral density = NPSD $\times\, T_o$ V²s/Hz)

Figure 8.28 *Relationship between energy spectral densities of signal pulse and white noise to illustrate matched filtering amplitude criterion.*

suggests forming a weighted sum of the individual sub-band signal and noise energies where the weighting is in direct proportion to each band's SNR. Since the NPSD is constant with frequency the SNR is proportional to $|V(f)|^2$. Remembering that the power or energy density passed by a filter is proportional to $|H(f)|^2$, this argument leads to the following statement of the amplitude response required for a matched filter assuming white noise:

> *The square of the amplitude response of a matched filter has the same shape as the energy spectral density of the pulse to which it is matched.*

Now consider the pulse spectrum in Figure 8.28 to be composed of many closely spaced and harmonically related spectral lines (Figures 8.29(c), (d)). The amplitude and phase spectra give the amplitude and phase of each of the cosine waves into which a periodic version of the pulse stream has been decomposed (Figures 8.29(a), (b)). If it can be arranged for all the cosine waves to reach a peak simultaneously in time then the signal voltage (and therefore the signal power) will be a maximum at that instant.

The filter which achieves this has a phase response which is the opposite (i.e. the negative) of the pulse phase spectrum, Figure 8.29(e). The post-filtered pulse would then have a null phase spectrum, Figure 8.29(g), and the component cosine waves would peak together at time $t = 0, T_o, 2T_o, \cdots$ (Figure 8.29(h)). In practice a linear phase shift, $e^{-j\omega T_o}$, corresponding to a time delay of T_o seconds, must be added (or rather included) in

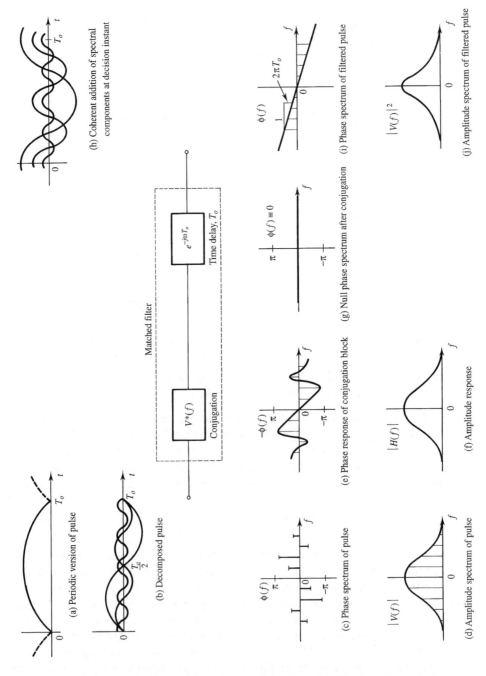

Figure 8.29 *Schematic illustration demonstrating origin of matched filtering phase criterion.*

the matched filter's frequency response to make it realisable[1]. This gives us a statement of the phase response required for a matched filter, i.e.:

The phase response of a matched filter is the negative of the phase spectrum of the pulse to which it is matched plus an additional linear phase of $-2\pi fT_o$ rad.

The matched filtering amplitude and phase criteria can be expressed mathematically as:

$$|H(f)|^2 = k^2|V(f)|^2 \tag{8.26(a)}$$

$$\phi(f) = -\phi_v(f) - 2\pi T_o f \quad \text{(rad)} \tag{8.26(b)}$$

where $\phi_v(f)$ is the phase spectrum of the expected pulse and k is a constant. Equations (8.26) can be combined into a single matched filtering criterion, i.e.:

$$H(f) = kV^*(f)\, e^{-j\omega T_o} \tag{8.27}$$

where the superscript * indicates complex conjugation.

Matched filtering essentially takes advantage of the fact that the pulse or signal frequency components are coherent in nature whilst the corresponding noise components are incoherent. It is therefore possible, using appropriate processing, to add spectral components of the signal voltage-wise whilst the same processing adds noise components only power-wise. The extension of the above arguments to pulses buried in non-white noise is straightforward in which case the matched filtering amplitude response generalises to:

$$|H(f)| = \frac{k|V(f)|}{\sqrt{G_n(f)}} \tag{8.28}$$

where $G_n(f)$ is the NPSD (see Chapter 3). The phase response is identical to that for white noise.

EXAMPLE 8.4

Find the frequency response of the filter which is matched to the triangular pulse $\Lambda(t-1)$.

The voltage spectrum of the pulse is given by:

$$V(f) = \text{FT}\{\Lambda(t-1)\}$$

$$= \text{sinc}^2(f)\, e^{-j2\pi f 1}$$

The frequency response of the matched filter is therefore:

$$H(f) = V^*(f)\, e^{-j\omega T_o}$$

$$= \text{sinc}^2(f)\, e^{+j2\pi f}\, e^{-j2\pi f 2}$$

$$= \text{sinc}^2(f)\, e^{-j2\pi f}$$

---

[1] This is to shift the *(single)* instant of constructive interference, between the (elemental) component sinusoids of the *(single)* aperiodic symbol, from $t = 0$ to $t = T_o$.

8.3.2 Correlation detection

We now apply correlation (described in section 2.6) to receiver design. The impulse response of a filter is related to its frequency response by the inverse Fourier transform, i.e.:

$$h(t) = \int_{-\infty}^{\infty} H(f) \, e^{j2\pi ft} \, df \tag{8.29}$$

This equation can therefore be used to transform the matched filtering criterion described by equation (8.27) into the time domain:

$$h(t) = \int_{-\infty}^{\infty} k \, V^*(f) \, e^{j2\pi f(t-T_o)} \, df$$

$$= k \left[\int_{-\infty}^{\infty} V(f) \, e^{j2\pi f(T_o-t)} \, df \right]^* \tag{8.30}$$

i.e.:

$$h(t) = k \, v^*(T_o - t) \tag{8.31}$$

Equation (8.31) is a statement of the matched filtering criterion in the time domain. For a filter matched to a purely real pulse it can be expressed in words as follows:

The impulse response of a matched filter is a time reversed version of the pulse to which it is matched, delayed by a time equal to the duration of the pulse.

Figure 8.30 illustrates equation (8.31) pictorially. The time delay, T_o, is needed to ensure causality (section 4.5) and corresponds to the need for the linear phase factor in equation (8.27).

Equation (8.31) allows the output pulse of a matched filter to be found directly from its input. The output of any time invariant linear filter is its input convolved with its impulse response. The convolution process involves reversing one of the functions (either input or impulse response), sliding the reversed over the non-reversed function and integrating the product. Since the impulse response of the matched filter is a time reversed copy of the expected input, and since convolution requires a further reversal, then the output is given by the integrated sliding product of *either* the input *or* the impulse response with an *un*reversed version of *itself*. This is illustrated in Figure 8.31. The output is thus the *autocorrelation* (section 2.6) of either the input pulse *or* the impulse response. An algebraic proof for a real signal pulse is given below.

Let $v_{in}(t)$, $v_{out}(t)$ and $h(t)$ be the input, output and impulse response of a filter. Then by convolution (section 4.3.4):

$$v_{out}(t) = v_{in}(t) * h(t) \tag{8.32}$$

If the filter is matched to $v_{in}(t)$ then:

$$v_{out}(t) = v_{in}(t) * k \, v_{in}(T_o - t)$$

$$= k \int_{-\infty}^{\infty} v_{in}(t') \, v_{in}(T_o + t' - t) \, dt' \tag{8.33}$$

Putting $T_o - t = \tau$:

$$v_{out}(t) = v_{out}(T_o - \tau) = k \int_{-\infty}^{\infty} v_{in}(t') \, v_{in}(t' + \tau) \, dt'$$

$$= k \, R_{v_{in}v_{in}}(\tau)$$

$$(= k \, R_{hh}(\tau)) \qquad\qquad (8.34)$$

Equation (8.34) can be expressed in words as follows:

The output of a filter driven by, and matched to, a real input pulse is, to within a multiplicative constant, k, and a time shift, T_o, the autocorrelation of the input pulse.

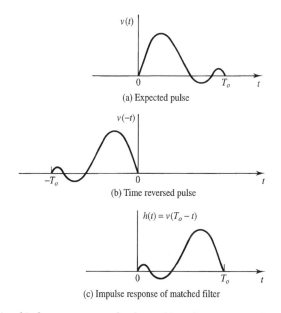

(a) Expected pulse

(b) Time reversed pulse

(c) Impulse response of matched filter

Figure 8.30 *Relationship between expected pulse and impulse response of matched filter.*

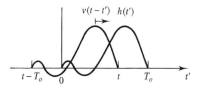

Figure 8.31 *Equivalence of $v(t) * h(t)$ and $R_{hh}(t')$.*

EXAMPLE 8.5

What will be the output of a filter matched to rectangular input pulses with width 1.0 ms?

The output pulse is the autocorrelation of the input pulse (equation (8.34)). Thus the output pulse will be triangular with width 2.0 ms.

The correlation property of a matched filter can be realised directly in the time domain. A block diagram of a classical correlator is shown in Figure 8.32. The correlator input pulse, $v_{in}(t)$, is distinguished from the reference pulse by a subscript since the input is strictly the sum of the signal pulse plus noise, i.e.:

$$v_{in}(t) = v(t) + n(t) \tag{8.35}$$

In digital communications the variable delay, τ, is usually unnecessary (Figure 8.33) since the pulse arrival times are normally known. Furthermore, it is only the peak value of the correlation function, $R_{v_{in}v}(0)$, which is of importance. The correlator output reaches this maximum value at the end of the input pulse, i.e. after T_o seconds. This, therefore, represents the correct sampling instant which leads to an optimum (i.e. maximum) decision SNR. (A matched filter response for a baseband coded waveform using the receiver correlation operation was demonstrated previously in Figure 4.9.)

It is interesting to note that an analogous argument to that used in the frequency domain to derive the amplitude response of a matched filter (equation (8.26(a))) can be used to demonstrate the optimum nature of correlation detection. Specifically, if the noise is stationary then its expected amplitude throughout the duration of the signal pulse

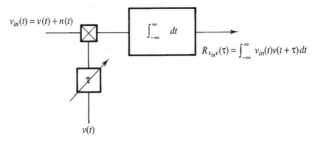

Figure 8.32 *Block diagram of signal correlator.*

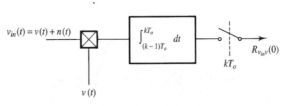

Figure 8.33 *Signal correlator for digital communications receiver.*

will be constant. The expected signal to RMS noise *voltage* ratio during pulse reception is therefore proportional to $v(t)$. (The expected signal to noise *power* ratio is proportional to $|v(t)|^2$.) It seems entirely reasonable, then, to weight each instantaneous value of signal plus noise voltage, $v_{in}(t)$, by the corresponding value of $v(t)$ and then to add (i.e. integrate) the result. (This corresponds to weighting each value of signal plus noise *power* by $|v(t)|^2$.)

If the reference signal, $v(t)$, is approximated by n sample values at regularly spaced time instants, $v(\Delta t), v(2\Delta t), \cdots, v((n-1)\Delta t), v(n\Delta t)$, the correlator can be implemented using a shift register, a set of weighting coefficients (the sample values) and an adder (Figure 8.34). This particular implementation has the form of a finite impulse response digital filter [Mulgrew *et al.*] and illustrates, clearly, the equivalence of matched filtering and correlation detection.

A single matched filter, or correlation channel, is obviously adequate as a detector in the case of on–off keyed (OOK) systems since the output will be a maximum when a pulse is present and essentially zero (ignoring noise) when no pulse is present. For binary systems employing two, non-zero, pulses a possible implementation would include two filters or correlators, one matched to each pulse type as shown in Figure 8.35. (This configuration will be used later for FSK detection, Chapter 11.) If the filter output is denoted by $f(t)$ then the possible sampling instant output voltages are:

$$f(kT_o) = \begin{cases} \displaystyle\int_0^{T_o} v_0^2(t)\, dt - \int_0^{T_o} v_0(t)v_1(t)\, dt, & \text{if symbol 0 is present} \\[4mm] \displaystyle -\int_0^{T_o} v_1^2(t)\, dt + \int_0^{T_o} v_1(t)\, v_0(t)\, dt, & \text{if symbol 1 is present} \end{cases} \tag{8.36}$$

If the signal pulses $v_0(t)$ and $v_1(t)$ are orthogonal but contain equal normalised energy, E_s V^2 s (i.e. joules of energy dissipated in a 1 Ω load) then the sampling instant voltages will be $\pm E_s$. If the pulses are antipodal (i.e. $v_1(t) = -v_0(t)$) then the sampling instant voltages will be $\pm 2E_s$. The same output voltages can be generated, however, for all (orthogonal, antipodal or other) binary pulse systems using only one filter or correlator by matching to the pulse difference signal, $v_1(t) - v_0(t)$, as shown in Figure 8.36.

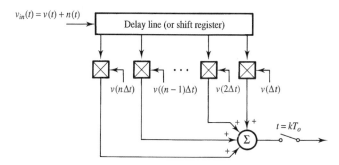

Figure 8.34 *Shift register implementation of matched filter illustrating relationship to correlation.*

For multisymbol signalling the number of channels in the matched filter or correlation receiver can be extended in an obvious way, Figure 8.37. (If antipodal signal pairs are used in an M-ary system only $M/2$ detection channels are needed.)

There is a disquieting aspect to equation (8.36) in that it seems dimensionally unsound. $f(kT_o)$ is a voltage yet the right hand side of the equation has dimensions of normalised energy (V^2 s). This apparant paradox is resolved by considering the implementation of a correlator in more detail. The upper channel of the receiver in Figure 8.35, for example, contains a multiplier and an integrator. The multiplier must have an associated constant k_m with dimensions of V^{-1} if the output is to be a voltage (which it certainly is). Strictly, then, the output of the multiplier is $k_m v_0^2(t)$ volts when a

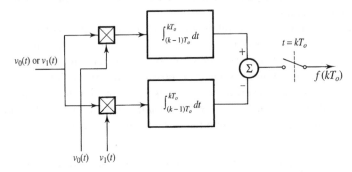

Figure 8.35 *Two-channel, binary symbol correlator.*

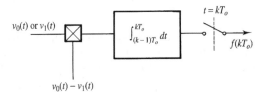

Figure 8.36 *One-channel, binary symbol correlator.*

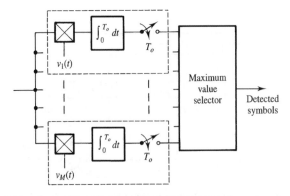

Figure 8.37 *Multichannel correlator for reception of M-ary signals.*

digital 0 is present at its input. Similarly the integrator has a constant k_I which has dimensions of s^{-1}. (If this integrator is implemented as an operational amplifier, for example, with a resistor, R, in series with its inverting input and a capacitor, C, as its negative feedback element then $k_I = 1/(RC)$ (s^{-1}).) Since these constants affect signal voltages and noise voltages in identical ways they are usually ignored. This is equivalent to arbitrarily assigning to them a numerical value of 1.0 (resulting in an overall 'conversion' constant of 1.0 V/V^2 s) and then (for orthogonal symbols) equating the numerical value of voltage at the correlator output with normalised symbol energy at the correlator input.

8.3.3 Decision instant SNR

A clue to the SNR performance of ideal matched filters and correlation detectors comes from equation (8.36). For orthogonal signal pulses the second (cross) terms in these equations are (by definition) zero. This leaves the first terms which represent the normalised energy, E_s, contained in the signal pulses. (The minus signs in equations (8.36) arise due to the subtractor placed after the integrators.) It is important to remember that it is the correlator output *voltage* which is numerically equal (assuming a 'conversion' constant of 1.0 V/V^2 s) to the normalised symbol energy E_s, i.e.:

$$f(kT_o) = E_s \text{ (V)} \tag{8.37(a)}$$

The sampling instant normalised signal power at the correlator output is therefore:

$$|f(kT_o)|^2 = E_s^2 \text{ (V}^2) \tag{8.37(b)}$$

Since the noise at the correlator input is a random signal it must properly be described by its autocorrelation function (ACF) or its PSD. At the input to the multiplier the ACF of $n(t)$ is:

$$R_{nn}(\tau) = \langle n(t) \, n(t + \tau) \rangle \text{ (V}^2) \tag{8.38}$$

Assuming that $n(t)$ is white with double sided PSD $N_0/2$ (V^2/Hz) then its ACF can be calculated by taking the inverse Fourier transform (Table 2.4) to obtain:

$$R_{nn}(\tau) = \frac{N_0}{2} \, \delta(\tau) \text{ (V}^2) \tag{8.39}$$

The ACF of the noise after multiplication with $v(t)$ is:

$$R_{xx}(\tau) = \langle x(t) \, x(t + \tau) \rangle$$
$$= \langle n(t)v(t) \, n(t + \tau)v(t + \tau) \rangle \text{ (V}^2) \tag{8.40}$$

where $x(t) = n(t)v(t)$ and a 'multiplier constant' of 1.0 V/V^2 has been adopted. Since $n(t)$ and $v(t)$ are independent processes equation (8.40) can be rewritten as:

$$R_{xx}(\tau) = \langle n(t) \, n(t + \tau) \rangle \langle v(t) \, v(t + \tau) \rangle$$
$$= \frac{N_0}{2} \, \delta(\tau) \, R_{vv}(\tau) \text{ (V}^2) \tag{8.41}$$

$\delta(\tau)$ is zero everywhere except at $\tau = 0$, therefore:

$$R_{xx}(\tau) = \frac{N_0}{2} \, \delta(\tau) \, R_{vv}(0) \text{ (V}^2) \tag{8.42}$$

$R_{vv}(0)$ is the mean square value of $v(t)$. Thus:

$$R_{xx}(\tau) = \frac{N_0}{2} \delta(\tau) \frac{1}{T_o} \int_0^{T_o} v^2(t) \, dt$$

$$= \frac{N_0}{2} \delta(\tau) \frac{E_s}{T_o} \quad (\text{V}^2) \tag{8.43}$$

Using the Wiener–Kintchine theorem (equation (3.48)) the two sided PSD of $x(t) = n(t)v(t)$ is the Fourier transform of equation (8.43), i.e.:

$$G_x(f) = \frac{N_0}{2} \frac{E_s}{T_o} \quad (\text{V}^2/\text{Hz}) \tag{8.44}$$

The impulse response of a device which integrates from 0 to T_o seconds is a rectangle of unit height and T_o seconds duration (i. e. $\Pi[(t - T_o/2)/T_o]$). A good conceptual model of such a device is shown in Figure 8.38. The frequency response of this time windowed integrator (sometimes called a moving average filter) is the Fourier transform of its impulse response, i.e.:

$$H(f) = T_o \, \text{sinc}(T_o f) \, e^{-j\omega T_o/2} \tag{8.45}$$

and its amplitude response is therefore:

$$|H(f)| = T_o \, |\text{sinc}(T_o f)| \tag{8.46}$$

The NPSD at the integrator output is:

$$G_y(f) = G_x(f) \, |H(f)|^2$$

$$= \frac{N_0}{2} E_s T_o \, \text{sinc}^2(T_o f) \tag{8.47}$$

and the total noise power at the correlator output is:

$$N = \frac{N_0}{2} E_s T_o \int_{-\infty}^{\infty} \text{sinc}^2(T_o f) \, df$$

$$= \frac{N_0}{2} E_s \quad (\text{V}^2) \tag{8.48}$$

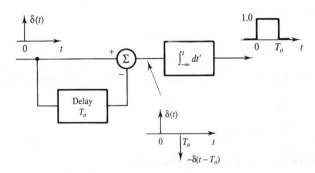

Figure 8.38 *Time windowed integrator (or moving average filter).*

(The integral of the sinc^2 function can be seen by inspection to be $1/T_o$ using the Fourier transform 'value at the origin' theorem, Table 2.5.) The standard deviation of the noise at the correlator output (or, equivalently, its RMS value since its mean value is zero) is:

$$\sigma = \sqrt{N} = \sqrt{\left(\frac{N_0}{2} E_s\right)} \text{ (V)} \tag{8.49}$$

Equations (8.37(a)) and (8.49) give a decision instant signal to RMS noise voltage ratio of:

$$\frac{f(T_o)}{\sigma} = \sqrt{\left(\frac{2E_s}{N_0}\right)} \tag{8.50}$$

or, alternatively, a decision instant signal to noise power ratio of:

$$\frac{S}{N} = \frac{|f(T_o)|^2}{\sigma^2} = \frac{2E_s}{N_0} \tag{8.51}$$

The important point here is that *the decision instant SNR at the output of a correlation receiver (or matched filter) depends only on pulse energy and input NPSD. It is independent of pulse shape.*

EXAMPLE 8.6

What is the sampling instant signal-to-noise ratio at the output of a filter matched to a triangular pulse of height 10 mV and width 1.0 ms if the noise at the input to the filter is white with a power spectral density of 10 nV^2/Hz?

Energy in the input pulse is given by:

$$E_s = \int_0^{T_o} v^2(t)\, dt = \int_0^{\frac{T_o}{2}} v^2(t)\, dt + \int_{\frac{T_o}{2}}^{T_o} v^2(t)\, dt$$

$$= \int_0^{0.5 \times 10^{-3}} [20\, t]^2\, dt + \int_{0.5 \times 10^{-3}}^{1 \times 10^{-3}} [2 \times 10^{-2} - 20\, t]^2\, dt$$

$$= 400 \left[\frac{t^3}{3}\right]_0^{0.5 \times 10^{-3}} + 4 \times 10^{-4} \left[\frac{t}{1}\right]_{0.5 \times 10^{-3}}^{1 \times 10^{-3}}$$

$$- 80 \times 10^{-2} \left[\frac{t^2}{2}\right]_{0.5 \times 10^{-3}}^{1 \times 10^{-3}} + 400 \left[\frac{t^3}{3}\right]_{0.5 \times 10^{-3}}^{1 \times 10^{-3}}$$

$$= 0.33 \times 10^{-7} \text{ (V}^2 \text{ s)}$$

Using equation (8.51) the sampling instant SNR when a pulse is present at the filter input is:

$$\frac{S}{N} = \frac{2E_s}{N_0} = \frac{2 \times 0.33 \times 10^{-7}}{10 \times 10^{-9}} = 6.67 = 8.2 \text{ dB}$$

8.3.4 BER performance of optimum receivers

A general formula giving the probability of symbol error for an optimum binary receiver (matched filter or correlator) is most easily derived by considering a single channel correlator matched to the binary symbol difference, $v_1(t) - v_0(t)$. When a binary 1 is present at the receiver input the decision instant voltage at the output is given by equation (8.36) as:

$$f(kT_o) = \int_{(k-1)T_o}^{kT_o} v_1(t) [v_1(t) - v_0(t)] \, dt$$

$$= E_{s1} - \int_{(k-1)T_o}^{kT_o} v_1(t)v_0(t) \, dt \tag{8.52}$$

where E_{s1} is the binary 1 symbol energy. When a binary 0 is present at the receiver input the decision instant voltage is:

$$f(kT_o) = -E_{s0} + \int_{(k-1)T_o}^{kT_o} v_1(t)v_0(t) \, dt \tag{8.53}$$

where E_{s0} is the binary 0 symbol energy. The second term of both equations (8.52) and (8.53) represents the correlation between symbols. Defining the normalised correlation coefficient to be:

$$\rho = \frac{1}{\sqrt{(E_{s0} E_{s1})}} \int_{(k-1)T_o}^{kT_o} v_1(t) \, v_0(t) \, dt \tag{8.54}$$

then equations (8.52) and (8.53) can be written as:

$$f(kT_o) = \begin{cases} E_{s1} - \rho\sqrt{(E_{s0}E_{s1})}, & \text{for binary 1} \\ -E_{s0} + \rho\sqrt{(E_{s0}E_{s1})}, & \text{for binary 0} \end{cases} \tag{8.55}$$

(The proper interpretation of equation (8.54) when $v_0(t) = 0$ and therefore $E_{s0} = 0$ (i.e. OOK signalling, section 6.4.1) is $\rho = 0$.) The difference in decision instant voltages representing binary 1 and 0 is:

$$\Delta V = E_{s1} + E_{s0} - 2\rho\sqrt{(E_{s1} E_{s0})} \tag{8.56}$$

Equation (8.56) also represents the energy, E_s', in the reference pulse of the single channel correlator, i.e.:

$$E_s' = \int_0^{T_o} [v_1(t) - v_0(t)]^2 \, dt = \Delta V \tag{8.57}$$

The RMS noise voltage at the output of the receiver is given by equation (8.49), i.e.:

$$\sigma = \sqrt{\left(\frac{N_0}{2} E_s'\right)}$$

$$= \left[\frac{N_0}{2} \left(E_{s1} + E_{s0} - 2\rho\sqrt{(E_{s1} \, E_{s0})} \right) \right]^{1/2} \qquad (8.58)$$

The quantity $\Delta V/\sigma$ is therefore:

$$\frac{\Delta V}{\sigma} = \left[\frac{2}{N_0} \left(E_{s1} + E_{s0} - 2\rho\sqrt{(E_{s1} \, E_{s0})} \right) \right]^{1/2} \qquad (8.59)$$

For binary symbols of equal energy, E_s, this simplifies to:

$$\frac{\Delta V}{\sigma} = 2\sqrt{\left[\frac{E_s}{N_0} (1 - \rho) \right]} \qquad (8.60)$$

Equation (8.59) or (8.60) can be substituted into the centre point sampling formula (equation (6.8)) to give the ideal correlator (or matched filter) probability of symbol error. In the (usual) equal symbol energy case this gives:

$$P_e = \frac{1}{2} \left\{ 1 - \mathrm{erf} \sqrt{\left[\frac{E_s}{2N_0} (1 - \rho) \right]} \right\} \qquad (8.61)$$

Although strictly speaking equations (8.60) and (8.61) are valid only for equal energy binary symbols they also give the correct probability of error for OOK signalling *providing* that E_s is interpreted as the average energy per symbol (i.e. half the energy of the non-null symbol). For all orthogonal signalling schemes (including OOK) $\rho = 0$. For all antipodal schemes (in which $v_1(t) = - v_0(t)$) $\rho = -1$.

EXAMPLE 8.7

A baseband binary communications system transmits a positive rectangular pulse for digital ones and a negative triangular pulse for digital zeros. If the (absolute) widths, peak pulse voltages, and noise power spectral density at the input of an ideal correlation receiver are all identical to those in Example 8.6 find the probability of bit error.

The energy in the triangular pulse has already been calculated in Example 8.6:

$$E_{s0} = 0.33 \times 10^{-7} \ \mathrm{V}^2 \ \mathrm{s}$$

The energy in the rectangular pulse is:

$$E_{s1} = v^2 T_o = (10 \times 10^{-3})^2 \times 1 \times 10^{-3} = 1 \times 10^{-7} \ \mathrm{V}^2 \ \mathrm{s}$$

Using equation (8.54):

$$\rho = \frac{1}{\sqrt{E_{s0} \, E_{s1}}} \int_0^{T_o} v_1(t) \, v_0(t) \, dt$$

$$= \frac{1}{\sqrt{0.33 \times 10^{-7} \times 1 \times 10^{-7}}} \int_0^{10^{-3}} \left[-10 \times 10^{-3} \, \prod \left(\frac{t - 0.5 \times 10^{-3}}{10^{-3}} \right) \right.$$

$$\times 10 \times 10^{-3} \Lambda \left(\frac{t - 0.5 \times 10^{-3}}{5 \times 10^{-4}} \right) \right] dt$$

$$= -1.74 \times 10^3 \times 2 \left[2 \times 10^3 \ t^2/2 \right]_0^{0.5 \times 10^{-3}} = -0.87$$

Using equation (8.59):

$$\frac{\Delta V}{\sigma} = \left[\frac{2}{N_0} \left(E_{s1} + E_{s0} - 2 \ \rho \ \sqrt{(E_{s1} \ E_{s0})} \right) \right]^{1/2}$$

$$= \left[\frac{2}{10 \times 10^{-9}} (0.33 \times 10^{-7} + 1.0 \times 10^{-7} - 2 \ (-0.87) \ \sqrt{0.33 \times 10^{-14}}) \right]^{1/2}$$

$$= 6.83$$

Using equation (6.8):

$$P_e = \frac{1}{2} \left[1 - \text{erf} \ \frac{\Delta V}{2 \ \sqrt{2} \ \sigma} \right]$$

$$= \frac{1}{2} \left[1 - \text{erf} \ \frac{6.83}{2\sqrt{2}} \right] = 3.2 \times 10^{-4}$$

8.3.5 Comparison of baseband matched filtering and centre point detection

Equation (8.59) can be used to compare the performance of a baseband matched filter receiver with simple centre point detection of rectangular pulses as discussed in Chapter 6. For unipolar NRZ transmission equation (8.59) shows that the detection instant $\Delta V/\sigma$ after matched filtering is related to that for simple centre point detection (CPD) by:

$$\left(\frac{\Delta V}{\sigma} \right)_{MF} = \left(\frac{2 \ E_{s1}}{N_0} \right)^{1/2}$$

$$= \left(\frac{2 \ \Delta V^2 T_o}{\sigma^2/B} \right)^{1/2}$$

$$= \sqrt{2}(T_o B)^{1/2} \left(\frac{\Delta V}{\sigma} \right)_{CPD} \tag{8.62}$$

where T_o is the rectangular pulse duration and B is the CPD predetection (rectangular) bandwidth. It may be disturbing to recognise that if rectangular pulse CPD transmission is interpreted literally then B must be infinite to accommodate infinitely fast rise and fall times. However, in this (literal) case $(\Delta V/\sigma)_{CPD}$ is zero due to the infinite noise power implied by a white noise spectrum. In practice, the CPD predetection bandwidth B is limited to a finite value (say two or three times $1/T_o$) and T_o is interpreted as the symbol period to allow the resulting spreading of the symbol in time. The saving of transmitter power (or allowable increase in NPSD) that matched filtering provides for compared with CPD is therefore:

$$\frac{(\Delta V/\sigma)_{MF}}{(\Delta V/\sigma)_{CPD}} = \sqrt{2}(T_o B)^{\frac{1}{2}}$$

$$= 3.0 + (T_o B)_{dB} \quad \text{(dB)} \tag{8.63}$$

A CPD predetection bandwidth of three times the baud rate $(B = 3/T_o)$, for example, therefore gives a power saving of 7.8 dB. (Although equation (8.63) has been derived for unipolar transmission it is also correct for the polar transmission case.)

For $B = 0.5/T_o$ (the minimum bandwidth consistent with ISI-free transmission), then the performance of matched filtering and CPD is the same. This apparent paradox is resolved by appreciating that this minimum bandwidth implies sinc pulse transmission in which case the rectangular predetection CPD filter is itself precisely matched to the expected symbol pulse shape.

8.3.6 Differences between matched filtering and correlation

Although matched filters and correlation detectors give identical detection instant signal and noise voltages at their outputs for identical inputs (and therefore have identical P_e performance) they do not *necessarily* give the same pulse shapes at their outputs. This is because in the case of the correlator (Figure 8.33) the received pulse and reference pulse are aligned in time throughout the pulse duration, whereas in the case of the filter (Figure 8.34) the received pulse slides across the reference pulse giving the true ACF (neglecting noise) of the input pulse. Specifically the pulse at the output of the correlator is given by:

$$f(t) = \int_0^t v^2(t') \, dt' \tag{8.64}$$

whilst the pulse at the output of the matched filter (see equation (8.33)) is given by:

$$f(t) = \int_0^t v(t') \, v(t' + T_o - t) \, dt' \tag{8.65}$$

(The lower limits in the integrals of equations (8.64) and (8.65) assume that the pulses start at $t = 0$.) This difference has no influence on P_e providing there are no errors in symbol timing. If, however, decision instants are not perfectly timed then there is the possibility of a discrepancy in matched filter and correlator performance. This is well illustrated by the case of rectangular RF pulse signalling. Figure 8.39 shows the detector output pulses for a matched filter and a correlator. It is clear that, providing the timing instant never occurs after $t = T_o$, the matched filter would suffer greater performance degradation due to symbol timing errors, for this type of pulse, than the correlator.

8.4 Root raised cosine filtering

Nyquist filtering and matched filtering have both been identified as optimum filtering techniques, the former because it results in ISI-free signalling in a bandlimited channel and the latter because it results in maximum SNR at the receiver decision instants. Whilst Nyquist filtering was discussed in the context of transmitter filters it is important to realise that the Nyquist frequency response which gives ISI-free detection includes the

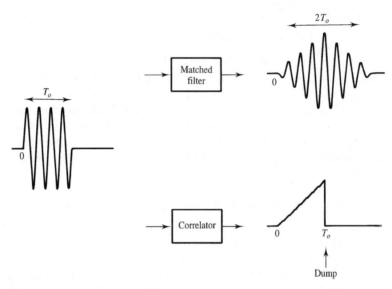

Figure 8.39 *Output pulses of matched filter and correlator for a rectangular RF input pulse.*

pulse shaping filter at the transmitter, the frequency response of the transmission medium and any filtering in the receiver prior to the decision circuit, i.e.:

$$H_N(f) = H_T(f) \, H_{ch}(f) \, H_R(f) \tag{8.66}$$

where the subscripts denote the Nyquist, transmitter, channel and receiver frequency responses respectively. Assuming that the channel introduces negligible distortion ($H_{ch}(f) = 1$) then it is clear that the Nyquist frequency response can be split in any convenient way between the transmitter and receiver. It is also clear that if the transmitter and receiver filter are related by:

$$H_T^*(f) = H_R(f) \tag{8.67}$$

then the spectrum of the transmitted pulses (assuming impulses prior to filtering) will be the conjugate of the frequency response of the receiver. Apart from a linear phase factor this is precisely the requirement for matched filtering. It is therefore possible, by judicious splitting of the overall system frequency response, to satisfy both the Nyquist and matched filtering criteria simultaneously. A popular choice for $H_T(f)$ and $H_R(f)$ is the root raised cosine filter (Figure 8.40(a)) derived from equation (8.8), i.e.:

$$H_T(f) = H_R(f)$$

$$= \begin{cases} \sqrt{\cos^2(\pi f / 4 f_\chi)}, & f \le 2f_\chi \\ 0, & f > 2f_\chi \end{cases} \tag{8.68}$$

where $f_\chi = 1/(2T_o)$. The overall frequency response then has a full raised cosine characteristic giving ISI-free detection. The impulse response of the root raised cosine filter, given by the inverse Fourier transform of equation (8.68), is:

$$h(t) = \frac{8 f_\chi}{\pi} \frac{\cos(4\pi f_\chi t)}{1 - 64 f_\chi^2 t^2} \tag{8.69}$$

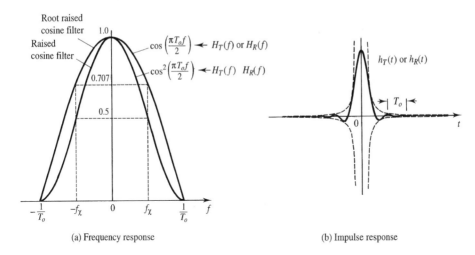

(a) Frequency response   (b) Impulse response

Figure 8.40 *Root raised cosine filter responses.*

This impulse response, Figure 8.40(b), is, of course, the transmitted pulse shape.

The similarity between the root raised cosine filter and the cosine filter used in duobinary signalling is obvious. The difference is in their bandwidth. The bandwidth, B, of the root raised cosine filter is $B = 1/T_o$ Hz. The bandwidth of the (duobinary) cosine filter is $B = 1/(2T_o)$ Hz.

8.5 Equalisation

When a digital signal is transmitted over a realistic channel it can be severely distorted. The communications channel including transmitter filters, multipath effects and receiver filters can be modelled by a finite impulse response (FIR) filter, with the same structure as that shown in Figure 8.34 [Mulgrew *et al.*]. The transmitted data can often be effectively modelled as a discrete random binary sequence, $x(kT_o)$, which can take on values of, say, ± 1 V. Gaussian noise samples, $n(kT_o)$, are added to the FIR filter (i.e. channel) output resulting in the received samples, $f(kT_o)$. In the simplest case all the coefficients of the FIR filter would be zero except for one tap which would have weight h_0. The received signal samples would then be:

$$f(kT_o) = \pm h_0 + n(kT_o) \tag{8.70}$$

and we could tell what data was being transmitted by simply testing whether $f(kT_o)$ is greater than or less than zero, i.e.:

$$\text{if } f(kT_o) \geq 0 \quad \text{then } x(kT_o) = +1 \quad \text{else } x(kT_o) = -1 \tag{8.71}$$

The received sample $f(kT_o)$ is more often, however, a function of several transmitted bits or symbols as defined by the number of taps with significant weights. This merging of samples represents ISI, as discussed in section 8.2.

Each channel has a particular frequency response. If we knew this frequency response we could include a filter in the receiver with the 'opposite' or inverse frequency response, as discussed in section 6.6.2. Everywhere the channel had a peak in its frequency response the inverse filter would have a trough and vice versa. The frequency response of the channel in cascade with the inverse filter would ideally have a wideband, flat amplitude response and a linear phase response, i.e. as far as the transmitted signal was concerned the cascade of the channel and the inverse filter together would look like a simple delay. Effectively we would have equalised the frequency response of the channel.

Note that the equaliser operation is fundamentally wideband, compared to the matched filter of Figure 8.34. When a signal is corrupted by white noise the matched filter detector possesses a frequency response which is *matched accurately* to the expected signal characteristic. Consequently the matched filter bandwidth equals the signal bandwidth. The equalising filter bandwidth is typically much greater than the signal bandwidth, however, to achieve a commensurate narrower duration output pulse response, than with the matched filter operation.

In practice when we switch on our digital mobile radio, or telephone modem, we have no idea what the frequency response of the channel between the transmitter and the receiver will be. In this type of application the receiver equaliser must, therefore, be adaptive. If such equalisers are to approximate the optimal filter or estimator they require explicit knowledge of the signal environment in the form of correlation functions, power delay profiles, etc. In most situations such functions are unknown and/or time varying. The equaliser must therefore employ a closed loop (feedback) arrangement in which its frequency response is adapted, or controlled, by a feedback algorithm. This permits it to compensate for time varying distortions and still achieve performance close to the optimal estimator function.

Adaptive filters [Mulgrew *et al.*] use an adjustable or programmable filter whose impulse response is controlled to pass the desired components of each signal sample and to attenuate the undesired components in order to compensate distortion present in the input signal. This may be achieved by employing a known data sequence or training signal, Figure 8.41. An input data plus noise, sample sequence, $f(kT_o)$, is convolved with a time varying FIR sequence, $h_i(kT_o)$. The output of the N-tap filter $\hat{x}(kT_o)$, is given by the discrete convolution operation (see section 13.6):

$$\hat{x}(kT_o) = \sum_{i=0}^{N-1} h_i(kT_o)f((k-i)T_o) \tag{8.72}$$

The filter output, $\hat{x}(kT_o)$, is used as the estimate of the training signal, $x(kT_o)$, and is subtracted from this signal to yield an error signal:

$$e(kT_o) = x(kT_o) - \hat{x}(kT_o) \tag{8.73}$$

The error is then used in conjunction with the input signal, $f(kT_o)$, to determine the next set of filter weight values, $h_i((k+1)T_o)$.

The impulse response of the adaptive filter, $h_i(kT_o)$, is thus progressively altered as more of the observed, and training, sequences become available, such that the output $\hat{x}(kT_o)$ converges to the training sequence $x(kT_o)$ and hence the output of the optimal filter. Adaptive filters again employ an FIR structure, as this is more stable than other filter forms, such as recursive infinite impulse response designs.

Figure 8.42 illustrates how this technique can be applied for practical data communications. Here the input, $f(kT_o)$, is genuine data, $x(kT_o)$, convolved with the communication channel impulse response, plus additive noise. When the transmitter is switched on, however, it sends a training sequence prior to the data. The objective here is to make the output, $\hat{x}(kT_o)$, approximate the training sequence and in so doing 'teach' the adaptive algorithm the required impulse response of the inverse filter.

The adaptive algorithm can then be switched off and genuine data transmitted. On conventional telephone lines the channel response does not change with time once the circuit has been established. Having trained the equaliser the adaptive algorithm can, therefore, be disconnected until a new line is dialled up.

In digital mobile radio applications, for example, this is not the case since the channel between the transmitter and receiver will change with time, and also with the position and velocity of the mobile transceiver. Thus the optimum filter is required to track changes in the channel.

One way of tackling this problem is to use what is called 'decision directed' operation, Figure 8.43. If the equaliser is working well then $\hat{x}(kT_o)$ will just be the binary data plus noise. We can remove the noise in a decision circuit which simply tests whether $\hat{x}(kT_o)$ is greater or less than zero. The output of the decision circuit is then identical to the transmitted data. We can use this data to continue training the adaptive

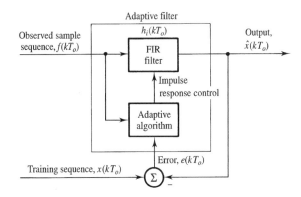

Figure 8.41 *Adaptive filter operation.*

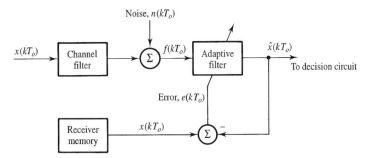

Figure 8.42 *Use of adaptive filter for adaptive equalisation.*

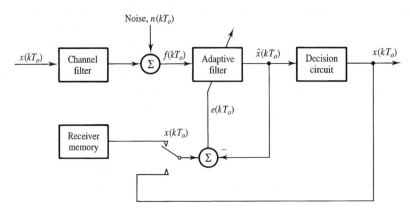

Figure 8.43 *Decision directed adaptive equaliser operation.*

filter by changing the switch in Figure 8.43 to its lower position. Here the error is the difference between the binary data and the filter output. This 'decision directed' system will work well provided the receiver continues to make correct decisions about the transmitted data. The decision directed equaliser thus operates effectively in slowly changing channels where the adaptive filter feedback loop can track these changes. If the filter cannot track the signal then the switch in Figure 8.43 must be moved to the upper position and the training sequence retransmitted to initialise the filter, as previously in Figure 8.42. The adaptive filter concept is fundamental to the operation of digital cellular systems which must overcome the channel fading effects described in section 15.2.

A useful way of quickly assessing the performance of a digital communications equaliser is to display its output as an eye diagram on an oscilloscope as described in section 6.6.3.

8.6 Summary

Two types of optimum filtering are important to digital communications. Nyquist filtering constrains the bandwidth of a signal whilst avoiding sampling instant ISI at the decision circuit input. Matched filtering maximises the sampling instant SNR at the decision circuit input. Both types of filtering are optimum, therefore, in the sense that they minimise the probability of bit error. Nyquist filtering can be implemented entirely at the transmitter if no distortion occurs in the transmission channel or receiver. Implementation is often, however, distributed between transmitter and receiver, distortion in the channel being cancelled by a separate equaliser. The amplitude response of a Nyquist filter has odd symmetry about its −6 dB frequency points and its phase response is linear.

Duobinary signalling uses pulses, transmitted at a baud rate of $1/(2B)$ symbol/s, which suffer from severe, but predictable, ISI. The predictability of the ISI means that it can be cancelled by appropriate signal, or symbol, processing at the receiver, thus effectively allowing ISI-free transmission at the maximum theoretical baud rate. Partial response signalling represents a generalisation of the duobinary technique to multilevel signalling. In this case the (predictable) ISI extends across a window of several adjacent symbols.

Matched filtering is implemented at the receiver. The amplitude response of this filter is proportional to the amplitude spectrum of the symbol to which it is matched and its phase response is, to within a linear phase factor, opposite to the phase spectrum of the symbol. Correlation detection is matched filtering implemented in the time domain.

Root raised cosine filters (applied to impulse signalling at the transmitter and the transmitted symbols at the receiver) satisfy both Nyquist and matched filtering criteria, assuming a distortionless, or perfectly equalised, channel. Channels which are time varying may require adaptive equalisation.

8.7 Problems

8.1   A binary information source consists of statistically independent, equiprobable, symbols. If the bandwidth of the baseband channel over which the symbols are to be transmitted is 3.0 kHz what baud rate will be necessary to achieve a spectral efficiency of 2.5 bit/s/Hz? Is ISI-free reception at this baud rate possible? What must be the minimum size of the source symbol alphabet to achieve ISI-free reception and a spectral efficiency of 16 bit/s/Hz? [7.5 kbaud, no, 256]

8.2   What is the Nyquist filtering criterion expressed in: (a) the time domain; and (b) the frequency domain?

8.3   State Nyquist's vestigial symmetry theorem. Why is this theorem useful in the context of digital communications?

8.4   Which is the more general, the family of Nyquist filters or the family of raised cosine filters? Sketch the amplitude response of: (a) a baseband raised cosine filter with a normalised excess bandwidth of 0.3; and (b) a bandpass full raised cosine filter.

8.5   Given that a Nyquist filter has odd symmetry about its parent rectangular filter's cut-off frequency, demonstrate that the impulse response of the Nyquist filter retains those zeros present in the impulse response of the top hat filter. (Hint: consider how the odd symmetry of the Nyquist filter's frequency response could be obtained by convolving the rectangular function with an even function.)

8.6   Justify Nyquist's vestigial symmetry theorem, as encapsulated in Figure 8.8, directly (i.e. without recourse to the argument used in Problem 8.5).

8.7   A baseband binary (two-level) PCM system is used to transmit a single 3.4 kHz voice signal. If the sampling rate for the voice signal is 8 kHz and 256 level quantisation is used, calculate the bandwidth required. Assume that the total system frequency response has a full raised cosine characteristic. [64 kHz]

8.8   An engineer proposes to use sinc pulse signalling for a baseband digital communications system on the grounds that the sinc pulse is a special case of a Nyquist signal. Briefly state whether you support or oppose the proposal and on what grounds your case is based.

8.9   A baseband transmission channel has a raised cosine frequency response with a roll-off factor $\alpha = 0.4$. The channel has an (absolute) bandwidth of 1200 kHz. An analogue signal is converted to binary PCM with 64 level quantisation before being transmitted over the channel. What is the maximum limit on the bandwidth of the analogue signal? What is the maximum possible spectral efficiency of this system? [143 kHz, 1.43 bit/s/Hz]

8.10 A voice signal is restricted to a bandwidth of 3.0 kHz by an ideal anti-aliasing filter. If the bandlimited signal is then oversampled by 33% find the ISI-free bandwidth required for transmission over a channel having a Nyquist response with a roll-off factor of 50%, for the following schemes: (a) PAM; (b) eight-level quantisation, binary PCM; and (c) 64 level quantisation, binary PCM. [6 kHz, 18 kHz, 36 kHz]

8.11 A four-level PCM communications system has a bit rate of 4.8 kbit/s and a raised cosine total system frequency response with a roll-off factor of 0.3. What is the minimum transmission bandwidth required? (N.B. A four-level PCM system is one in which information signal samples are coded into a four-level symbol stream rather than the usual binary symbol stream.) [1.56 kHz]

8.12 Demonstrate that duobinary signalling suffers from error propagation whilst precoded duobinary signalling does not.

8.13 (a) What is the principal objective of matched filtering?
(b) What is the (white noise) matched filtering criterion expressed in: (i) the time domain; and (ii) the frequency domain?
(c) How is the sampling instant SNR at the output of a matched filter related to the energy of the expected symbol and NPSD at the filter's input?

8.14 Sketch the impulse response of the filter which is matched to the pulse:

$$f(t) = \prod(t - 0.5) + (2/3) \prod(t - 1.5)$$

What is the output of this filter when the pulse $f(t)$ is present at its input?

8.15 The transmitted pulse shape of an OOK, baseband, communication system is $1000t\prod([t - 0.5 \times 10^{-3}]/10^{-3})$. What is the impulse response of the predetection filter which maximises the sampling instant SNR (i.e. the matched filter)? If the noise at the input to the predetection filter is white and Gaussian with a one-sided power spectral density of 2.0×10^{-5} V^2/Hz, what probability of symbol error would you expect in the absence of intersymbol interference? [2×10^{-4}]

8.16 A polar binary signal consists of +1 or −1 V pulses during the interval $(0, T)$. Additive white Gaussian noise having a two-sided NPSD of 10^{-6} W/Hz is added to the signal. If the received signal is detected with a matched filter, determine the maximum bit rate that can be sent with an error probability, P_e, of less than or equal to 10^{-3}. Assume that the impedance level is 50 Ω. What is the sampling instant SNR in dB at the output of the filter? Can you say anything about the SNR at the filter input? [2.1 kbit/s, 9.8 dB]

8.17 The time domain implementation of a matched filter is called a *correlation* detector. In view of the fact that the output of a filter is the *convolution* of its input with its impulse response explain why this terminology is appropriate.

8.18 A binary baseband communications system employs the transmitted pulses $v_0(t) = -\Lambda(t/(0.5 \times 10^{-3}))$ V and $v_1(t) = \prod(t/10^{-3})$ V to represent digital zeros and ones respectively. What is the ideal, single channel, correlator reference signal for this system? If the loss from transmitter to receiver is 40.0 dB what value of noise power spectral density at the correlator input can be tolerated whilst maintaining a probability of symbol error of 10^{-6}? [5.15×10^{-9} V^2/Hz]

8.19 A digital communications receiver uses root raised cosine filtering at both its transmitter and receiver. Show that the transmitted pulse has the form:

$$v(t) = \frac{4}{\pi} R_s \frac{\cos(2\pi R_s t)}{1 - 16R_s^2 t^2}$$

where R_s is the baud rate.

CHAPTER 9

Information theory, source coding and encryption

9.1 Introduction

There may be a number of reasons for wishing to change the form of a digital signal as supplied by an information source prior to transmission. In the case of English language text, for example, we start with a data source consisting of about 40 distinct symbols (the letters of the alphabet, integers and punctuation). In principle we could transmit such text using a signal alphabet consisting of 40 distinct voltage waveforms. This would constitute an M-ary system where $M = 40$ unique signals. It may be, however, that for one or more of the following reasons this approach is inconvenient, difficult or impossible:

1. The transmission channel may be physically unsuited to carrying such a large number of distinct signals.
2. The relative frequencies with which different source symbols occur will vary widely. This will have the effect of making the transmission inefficient in terms of the time it takes and/or bandwidth it requires.
3. The data may need to be stored and/or processed in some way before transmission. This is most easily achieved using binary electronic devices as the storage and processing elements.

For all these reasons (and perhaps others) sources of digital information are almost always converted as soon as possible into binary form, i.e. each symbol is encoded as a binary word. After appropriate processing the binary words may then be transmitted directly, as either baseband or bandpass signals, or may be recoded into another multi-symbol alphabet. (In the latter case it is unlikely that the transmitted symbols map directly onto the original source symbols.) The body of knowledge, information theory, concerned with the representation of information by symbols gives theoretical bounds on the performance of communication systems and permits assessment of practical system efficiency. The landmarks in information theory were developed by Hartley and Nyquist in the 1920s and are summarised in [Shannon, 1948].

The simplified communication system shown in Figure 9.1 may include several encoders and decoders to implement one or more of the following processes:

- Formatting (which transforms information from its original, or natural, form to a well defined, and standard, digital form, e.g. PCM, see Chapter 6).
- Source coding (which reduces the average number of symbols required to transmit a given message).
- Encryption (which codes messages using a cipher to prevent unauthorised reception or transmission).
- Error control coding (which allows a receiver to detect, and sometimes correct, symbols which are received in error, see Chapter 10).
- Line coding/pulse shaping (which ensures the transmitted symbol waveforms are well suited to the characteristics of the channel, see Chapters 6 and 8).

It is the second of these processes, source coding, which is the principal concern of this chapter.

9.2 Information and entropy

9.2.1 The information measure

The concept of information content is related to predictability or scarcity value. That is, the more predictable or probable a particular message, the less information is conveyed by transmitting that message. For example, the football score Manchester United 7, Bradford Academicals 0, contains little information. The information content of Bradford Academicals 7, Manchester United 0, on the other hand is enormous![1]

In essence highly probable messages contain little information and we can write:

$P(message) = 1$ carries zero information

$P(message) = 0$ carries infinite information

The definition of information content for a symbol m should be such that it

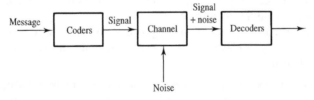

Figure 9.1 *The communications channel model.*

[1]Readers must be aware that Manchester United is a premier division team while Bradford Academicals is a non-league team and thus the chances are very remote for the second score to occur.

monotonically decreases with increasing message probability, $P(m)$, and it goes to zero for a probability of unity. Another desirable property is that of additivity. If one were to communicate two (independent) messages in sequence, the total information content should be equal to the sum of the individual information contents of the two messages. We know that the total probability of the composite message is the product of the two individual, independent, probabilities. Therefore, the definition of information must be such that when probabilities are multiplied information is added.

The required properties of an information measure are summarised in Table 9.1.

Table 9.1 *Information measures.*

Message	P (message)	Information content
m_1	$P(m_1)$	I_1
m_2	$P(m_2)$	I_2
$(m_1 + m_2)$	$P(m_1)P(m_2)$	$I_1 + I_2$

The logarithm operation clearly satisfies these requirements. We thus *define* ($\overset{\Delta}{=}$) the information content, I_m, of a message, m, as:

$$I_m \overset{\Delta}{=} \log \frac{1}{P(m)} \equiv -\log P(m) \tag{9.1}$$

This definition satisfies the additivity requirement, the monotonicity requirement, and for $P(m) = 1$, $I_m = 0$. Note that this is true regardless of the base chosen for the logarithm. Base 2 is usually chosen, however, the resulting quantity of information being measured in bits:

$$I_1 + I_2 = -\log_2 P(m_1) - \log_2 P(m_2) = -\log_2 [P(m_1) P(m_2)] \quad \text{(bits)} \tag{9.2}$$

Table 9.2 *Word length and symbol probabilities.*

Vocabulary size	No. binary digits	Symbol probability
2	1	1/2
4	2	1/4
8	3	1/8
•	•	•
•	•	•
•	•	•
128	7	1/128

9.2.2 Multisymbol alphabets

Consider, initially, vocabularies of equiprobable message symbols represented by fixed length binary codewords. Thus a vocabulary size of four symbols is represented by the binary digit pairs 00, 01, 10, 11. The binary word length and symbol probabilities for other vocabulary sizes are shown in Table 9.2. The ASCII vocabulary used in teleprinters contains 128 symbols and therefore uses a $\log_2 128 = 7$ digit (fixed length) binary codeword to represent each symbol. ASCII symbols are not equiprobable, however, and for this particular code each symbol does not, therefore, have a selection probability of 1/128.

9.2.3 Commonly confused entities

Several information-related quantities are sometimes confused. These are:

- **Symbol**   A member of a source alphabet. May or may not be binary, e.g. 2 symbol binary, 4 symbol PSK (see Chapter 11), 128 symbol ASCII.
- **Baud**   Rate of symbol transmission, i.e. 100 baud = 100 symbol/s.
- **Bit**   Quantity of information carried by a symbol with selection probability $P = 0.5$.
- **Bit rate**   Rate of information transmission (bit/s). (In the special, but common, case of signalling using independent, equiprobable, binary symbols, the bit rate equals the baud rate.)
- **Message**   A meaningful sequence of symbols. (Also often used to mean a source symbol.)

9.2.4 Entropy of a binary source

Entropy (H) is defined as the average amount of information conveyed per symbol. For an alphabet of size 2 and assuming that symbols are statistically independent:

$$H \triangleq \sum_{m=1}^{2} P(m) \log_2 \frac{1}{P(m)} \quad \text{(bit/symbol)} \tag{9.3}$$

For the two-symbol alphabet (0, 1) if we let $P(1) = p$ then $P(0) = 1 - p$ and:

$$H = p \log_2 \frac{1}{p} + (1-p) \log_2 \frac{1}{1-p} \quad \text{(bit/symbol)} \tag{9.4}$$

The entropy is maximised when the symbols are equiprobable as shown in Figure 9.2. The entropy definition of equation (9.3) holds for all alphabet sizes. Note that, in the binary case, as either of the two messages becomes more likely, the entropy decreases. When either message has probability 1, the entropy goes to zero. This is reasonable since, at these points, the outcome of the transmission is certain. Thus, if $P(0) = 1$, we know the symbol 0 will be sent repeatedly. If $P(0) = 0$, we know the symbol 1 will be sent repeatedly. In these two cases, no information is conveyed by transmitting the

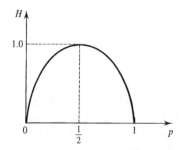

Figure 9.2 *Entropy for binary data transmission versus the selection probability, p, of a digital 1.*

symbols.

For the ASCII alphabet, source entropy would be given by $H = -\log_2 (1/128) = 7$ bit/symbol if all the symbols were both *equiprobable* and *statistically independent*. In practice H is less than this, i.e.:

$$H \overset{\Delta}{=} \sum_{m=1}^{128} P(m) \log_2 \frac{1}{P(m)} \; < 7 \text{ bit/symbol} \tag{9.5}$$

since the symbols are neither, making the code less than 100% efficient. Entropy thus indicates the *minimum* number of binary digits required per symbol (averaged over a long sequence of symbols).

9.3 Conditional entropy and redundancy

For sources in which each symbol selected is not statistically independent from all previous symbols (i.e. sources with memory) equation (9.3) is insufficiently general to give the entropy correctly. In this case the joint and conditional statistics (section 3.2.1) of symbol sequences must be considered. A source with a memory of one symbol, for example, has an entropy given by:

$$H = \sum_i \sum_j P(j,i) \log_2 \frac{1}{P(j|i)} \quad \text{(bit/symbol)} \tag{9.6}$$

where $P(j,i)$ is the probability of the source selecting i and j and $P(j|i)$ is the probability that the source will select j given that it has previously selected i. Bayes's theorem, equation (3.3), can be used to re-express equation (9.6) as:

$$H = \sum_i P(i) \sum_j P(j|i) \log_2 \frac{1}{P(j|i)} \quad \text{(bit/symbol)} \tag{9.7}$$

(For independent symbols $P(j|i) = P(j)$ and equation (9.7) reduces to equation (9.3).) The effect of having dependency between symbols is to increase the probability of selecting some symbols at the expense of others *given a particular symbol history*. This reduces the average information conveyed by the symbols, which is reflected in a reduced entropy. The difference between the actual entropy of a source and the (maximum) entropy, $H_{\max}$, the source could have if its symbols were independent and equiprobable is called the *redundancy* of the source. For an M-symbol alphabet redundancy, R, is therefore given by:

$$R = H_{\max} - H$$

$$= \log_2 (M) - H \quad \text{(bit/symbol)} \tag{9.8}$$

(It is easily shown that the quantity $R/(H^2 + RH)$ relates to the number of symbols per bit of information which are transmitted unnecessarily from an information theory point of view. There may well be good reasons, however, to transmit such redundant symbols, e.g. for error control purposes as discussed in Chapter 10.)

EXAMPLE 9.1

Find the entropy, redundancy and information rate of a four-symbol source (A, B, C, D) with a baud rate of 1024 symbol/s and symbol selection probabilities of 0.5, 0.2, 0.2 and 0.1 under the following conditions:

(i) The source is memoryless (i.e. the symbols are statistically independent).
(ii) The source has a one-symbol memory such that no two consecutively selected symbols can be the same. (The long term relative frequencies of the symbols remain unchanged, however.)

(i)

$$H = \sum_{m=1}^{M} p(m) \log_2 \frac{1}{p(m)}$$

$$= 0.5 \log_2 \left(\frac{1}{0.5} \right) + 2 \times 0.2 \log_2 \left(\frac{1}{0.2} \right) + 0.1 \log_2 \left(\frac{1}{0.1} \right)$$

$$= 0.5 \frac{\log_{10} 2}{\log_{10} 2} + 0.4 \frac{\log_{10} 5}{\log_{10} 2} + 0.1 \frac{\log_{10} 10}{\log_{10} 2}$$

$$= 1.761 \quad \text{bit/symbol}$$

$$R = H_{\max} - H$$

$$= \log_2 M - H$$

$$= \log_2 4 - 1.761 = 2.0 - 1.761 = 0.239 \quad \text{bit/symbol}$$

$$R_i = R_s H = 1024 \times 1.761 = 1.803 \times 10^3 \quad \text{bit/s}$$

where R_i is the information rate and R_s is the symbol rate.

(ii) The appropriate formula to apply to find the entropy of a source with one-symbol memory is equation (9.7). First, however, we must find the conditional probabilities which the formula contains. If no two consecutive symbols can be the same then:

$$P(A|A) = P(B|B) = P(C|C) = P(D|D) = 0$$

Since the (unconditional) probability of A is unchanged, $P(A) = 0.5$, then every, and only every, alternate symbol must be A, $P(\bar{A}|A) = P(A|\bar{A}) = 1.0$, where $\bar{A}$ represents not A. Furthermore if every alternate symbol is A then no two non-A symbols can occur consecutively, $P(\bar{A}|\bar{A}) = 0$. Writing the above three probabilities explicitly:

$$P(B|A) + P(C|A) + P(D|A) = 1.0$$

$$P(A|B) = P(A|C) = P(A|D) = 1.0$$

$$P(B|C) = P(B|D) = P(C|B) = P(C|D) = P(D|B) = P(D|C) = 0$$

Since the (unconditional) probabilities of B, C and D are to remain unchanged, i.e.:

$$P(B) = P(C) = 0.2 \quad \text{and} \quad P(D) = 0.1$$

the conditional probability $P(B|A)$ must satisfy:

$$P(B) = P(B|A)P(A) + P(B|B)P(B) + P(B|C)P(C) + P(B|D)P(D)$$

$$= P(B|A)0.5 + 0 + 0 + 0$$

i.e. $P(B|A) = \dfrac{0.2}{0.5} = 0.4$

Similarly:

$$P(C) = P(C|A)P(A) + P(C|B)P(B) + P(C|C)P(C) + P(C|D)P(D)$$

$$= P(C|A)\, 0.5 + 0 + 0 + 0$$

i.e. $P(C|A) = \dfrac{0.2}{0.5} = 0.4$

$$P(D) = P(D|A)P(A) + P(D|B)P(B) + P(D|C)P(C) + P(D|D)P(D)$$

$$= P(D|A)\, 0.5 + 0 + 0 + 0$$

i.e. $P(D|A) = \dfrac{0.1}{0.5} = 0.2$

We now have numerical values for all the conditional probabilities which can be substituted into equation (9.7):

$$H = \sum_i P(i) \sum_j P(j|i) \log_2 \frac{1}{P(j|i)} \quad \text{(bit/symbol)}$$

$$= P(A)\left[P(A|A) \log_2 \frac{1}{P(A|A)} + P(B|A) \log_2 \frac{1}{P(B|A)} + P(C|A) \log_2 \frac{1}{P(C|A)} + P(D|A) \log_2 \frac{1}{P(D|A)} \right]$$

$$+ P(B)\left[P(A|B) \log_2 \frac{1}{P(A|B)} + P(B|B) \log_2 \frac{1}{P(B|B)} + P(C|B) \log_2 \frac{1}{P(C|B)} + P(D|B) \log_2 \frac{1}{P(D|B)} \right]$$

$$+ P(C)\left[P(A|C) \log_2 \frac{1}{P(A|C)} + P(B|C) \log_2 \frac{1}{P(B|C)} + P(C|C) \log_2 \frac{1}{P(C|C)} + P(D|C) \log_2 \frac{1}{P(D|C)} \right]$$

$$+ P(D)\left[P(A|D) \log_2 \frac{1}{P(A|D)} + P(B|D) \log_2 \frac{1}{P(B|D)} + P(C|D) \log_2 \frac{1}{P(C|D)} + P(D|D) \log_2 \frac{1}{P(D|D)} \right]$$

$$= 0.5\left[0.0 \log_2 (1/0) + 0.4 \log_2 2.5 + 0.4 \log_2 2.5 + 0.2 \log_2 5.0 \right]$$

$$+ 0.2\left[1.0 \log_2 1.0 + 0 + 0 + 0 \right] + 0.2\left[1.0 \log_2 1.0 + 0 + 0 + 0 \right]$$

$$+ 0.1\left[1.0 \log_2 1.0 + 0 + 0 + 0 \right]$$

$$= 0.5\left[0.4 \log_{10} 2.5 + 0.4 \log_{10} 2.5 + 0.2 \log_{10} 5.0 \right] / \log_{10} 2 = 0.761 \text{ bit/symbol}$$

$$R = H_{\text{max}} - H$$

$$= \log_2 4 - 0.761 = 1.239 \text{ bit/symbol}$$

(Note here the symbol *A* carries no information at all since its occurrences are entirely predictable.)

$$R_i = R_s H = 1024 \times 0.761 = 779 \text{ bit/s}$$

9.4 Information loss due to noise

The information transmitted by a memoryless source (i.e. a source in which the symbols selected are statistically independent) is related only to the source symbol probabilities by equation (9.1), i.e.:

$$I_{TX}(i_{TX}) = \log_2 \frac{1}{P(i_{TX})} \quad \text{(bits)} \tag{9.9}$$

where $I_{TX}(i_{TX})$ is the information transmitted when the ith source symbol is selected for transmission and $P(i_{TX})$ is the (a priori) probability that the ith symbol will be selected (see section 7.2 for the definition of a priori probability). The use of a subscript TX with i, although not really necessary, will help with what can be a confusing notation later in the chapter.

For a noiseless channel there is no doubt on detecting a given received symbol, i_{RX}, which source symbol was selected for transmission. This is because the transmitted symbol (assuming adequate equalisation) will always be the same as the received symbol. The information gained at the receiver is, therefore, identical to the information transmitted, i.e.:

$$I_{RX}(i_{RX}) = I_{TX}(i_{TX})$$

$$= \log_2 \frac{1}{P(i_{TX})} \quad \text{(bits)} \tag{9.10}$$

For a noisy channel, however, there is some uncertainty about which symbol was actually selected at the source. This uncertainty is related to the conditional probability, $P(i_{TX}|j_{RX})$, that symbol i was transmitted given that symbol j was received. (For the noiseless channel $P(i_{TX}|j_{RX})$ is unity for $i = j$ and zero otherwise.) Intuition tells us that for a noisy channel the information received (i.e. the effective entropy), H_{eff}, should be less than that transmitted by an amount related to the uncertainty (or equivocation, E) in the receiver's knowledge of the transmitted source symbol, i.e.:

$$H_{eff} = H - E \quad \text{(bit/symbol)} \tag{9.11}$$

The equivocation, which can be thought of as representing spurious (or negative) information contributed by noise, is given by:

$$E = \sum_i \sum_j P(i_{TX}, j_{RX}) \log_2 \frac{1}{P(i_{TX}|j_{RX})} \quad \text{(bit/symbol)} \tag{9.12}$$

The similarity between equations (9.12) and (9.3) is notable, the essential difference being that $P(i_{TX}|j_{RX})$ relates, for $i \neq j$, to probabilities of symbol error and, for $i = j$, to probabilities of correct symbol reception. The spurious entropies associated with these two cases can, therefore, be written separately as:

$$E' = \sum_{i,j \ (i \neq j)} P(i_{TX}, j_{RX}) \log_2 \frac{1}{P(i_{TX}|j_{RX})} \quad \text{(bit/symbol)} \tag{9.13}$$

and:

$$E'' = \sum_i P(i_{TX}, i_{RX}) \log_2 \frac{1}{P(i_{TX}|i_{RX})} \quad \text{(bit/symbol)} \tag{9.14}$$

where:

$$E = E' + E''$$ (9.15)

Using $P(i_{TX}, j_{RX}) = P(j_{RX})P(i_{TX}|j_{RX})$ equation (9.12) can also be written as:

$$E = \sum_j P(j_{RX}) \sum_i P(i_{TX}|j_{RX}) \log_2 \frac{1}{P(i_{TX}|j_{RX})} \quad \text{(bit/symbol)}$$ (9.16)

where $P(j_{RX})$ is the (unconditional) symbol reception probability.

EXAMPLE 9.2

Consider a three-symbol source A, B, C with the following transition matrix:

	A_{TX}	B_{TX}	C_{TX}
A_{RX}	0.6	0.5	0
B_{RX}	0.2	0.5	0.333
C_{RX}	0.2	0	0.667

(Note that the elements of the transition matrix are the conditional probabilities $P(i_{RX}|j_{TX})$ and that columns and rows are labelled i_{TX} and i_{RX} only for extra clarity.) For the specific a priori (i.e. transmission) probabilities $P(A) = 0.5$, $P(B) = 0.2$, $P(C) = 0.3$, find: (i) the (unconditional) symbol reception probabilities $P(i_{RX})$ from the conditional and a priori probabilities; (ii) the equivocation; (iii) the source entropy; and (iv) the effective entropy.

(i) The symbol reception probabilities are found as follows:

$$\begin{aligned}
P(A_{RX}) &= P(A_{TX}) P(A_{RX}|A_{TX}) + P(B_{TX}) P(A_{RX}|B_{TX}) + P(C_{TX}) P(A_{RX}|C_{TX}) \\
&= 0.5 \times 0.6 + 0.2 \times 0.5 + 0.3 \times 0 = 0.4
\end{aligned}$$

$$P(B_{RX}) = 0.5 \times 0.2 + 0.2 \times 0.5 + 0.3 \times 0.333 = 0.3$$

$$P(C_{RX}) = 0.5 \times 0.2 + 0.2 \times 0 + 0.3 \times 0.667 = 0.3$$

(ii) To find the equivocation from equation (9.16) we require the probabilities $P(i_{TX}|j_{RX})$ which can be found in turn from $P(i_{RX}|j_{TX})$ and $P(i_{RX})$ using Bayes's rule, equation (3.3):

$$P(A_{TX}|A_{RX}) = \frac{P(A_{TX}, A_{RX})}{P(A_{RX})} = \frac{P(A_{RX}|A_{TX}) P(A_{TX})}{P(A_{RX})} = \frac{0.6 \times 0.5}{0.4} = 0.75$$

$$P(B_{TX}|A_{RX}) = \frac{P(B_{TX}, A_{RX})}{P(A_{RX})} = \frac{P(A_{RX}|B_{TX})P(B_{TX})}{P(A_{RX})} = \frac{0.5 \times 0.2}{0.4} = 0.25$$

$$P(C_{TX}|A_{RX}) = \frac{P(A_{RX}|C_{TX})P(C_{TX})}{P(A_{RX})} = \frac{0 \times 0.3}{0.4} = 0$$

$$P(A_{TX}|B_{RX}) = \frac{P(B_{RX}|A_{TX})P(A_{TX})}{P(B_{RX})} = \frac{0.2 \times 0.5}{0.3} = 0.333$$

$$P(B_{TX}|B_{RX}) = \frac{P(B_{RX}|B_{TX})P(B_{TX})}{P(B_{RX})} = \frac{0.5 \times 0.2}{0.3} = 0.333$$

$$P(C_{TX}|B_{RX}) = \frac{P(B_{RX}|C_{TX}) P(C_{TX})}{P(B_{RX})} = \frac{0.333 \times 0.3}{0.3} = 0.333$$

$$P(A_{TX}|C_{RX}) = \frac{P(C_{RX}|A_{TX})P(A_{TX})}{P(C_{RX})} = \frac{0.2 \times 0.5}{0.3} = 0.333$$

$$P(B_{TX}|C_{RX}) = \frac{P(C_{RX}|B_{TX})P(B_{TX})}{P(C_{RX})} = \frac{0 \times 0.2}{0.3} = 0$$

$$P(C_{TX}|C_{RX}) = \frac{P(C_{RX}|C_{TX}) \, P(C_{TX})}{P(C_{RX})} = \frac{0.667 \times 0.3}{0.3} = 0.667$$

Equation (9.16) now gives the equivocation:

$$E = P(A_{RX}) \left[P(A_{TX}|A_{RX}) \log_2 \frac{1}{P(A_{TX}|A_{RX})} + P(B_{TX}|A_{RX}) \log_2 \frac{1}{P(B_{TX}|A_{RX})} \right]$$

$$+ P(A_{RX}) \left[P(C_{TX}|A_{RX}) \log_2 \frac{1}{P(C_{TX}|A_{RX})} \right] + P(B_{RX}) \left[P(A_{TX}|B_{RX}) \log_2 \frac{1}{P(A_{TX}|B_{RX})} \right]$$

$$+ P(B_{RX}) \left[P(B_{TX}|B_{RX}) \log_2 \frac{1}{P(B_{TX}|B_{RX})} + P(C_{TX}|B_{RX}) \log_2 \frac{1}{P(C_{TX}|B_{RX})} \right]$$

$$+ P(C_{RX}) \left[P(A_{TX}|C_{RX}) \log_2 \frac{1}{P(A_{TX}|C_{RX})} + P(B_{TX}|C_{RX}) \log_2 \frac{1}{P(B_{TX}|C_{RX})} \right]$$

$$+ P(C_{RX}) \left[P(C_{TX}|C_{RX}) \log_2 \frac{1}{P(C_{TX}|C_{RX})} \right]$$

$$= 0.4 \left[0.75 \log_2 \left(\frac{1}{0.75} \right) + 0.25 \log_2 \left(\frac{1}{0.25} \right) + 0 \right]$$

$$+ 0.3 \left[0.333 \log_2 \left(\frac{1}{0.333} \right) + 0.333 \log_2 \left(\frac{1}{0.333} \right) + 0.333 \log_2 \left(\frac{1}{0.333} \right) \right]$$

$$+ 0.3 \left[0.333 \log_2 \left(\frac{1}{0.333} \right) + 0 + 0.667 \log_2 \left(\frac{1}{0.667} \right) \right]$$

$$= 0.4 \, [0.311 + 0.5] + 0.3 \, [0.528 + 0.528 + 0.528] + 0.3 \, [0.528 + 0.390]$$

$$= 1.08 \text{ bit/symbol}$$

(iii) The source entropy H is:

$$H = \sum_i P(i_{TX}) \log_2 \frac{1}{P(i_{TX})}$$

$$= 0.5 \log_2 \left(\frac{1}{0.5} \right) + 0.2 \log_2 \left(\frac{1}{0.2} \right) + 0.3 \log_2 \left(\frac{1}{0.3} \right) = 1.48 \text{ bit/symbol}$$

(iv) The effective entropy, equation (9.11), is therefore:

$$H_{eff} = H - E$$

$$= 1.48 - 1.08 = 0.4 \text{ bit/symbol}$$

9.5 Source coding

Source coding does not change or alter the source entropy, i.e. the average number of information bits per source symbol. In this sense source entropy is a fundamental property of the source. Source coding does, however, alter (usually increase) the entropy of the source coded symbols. It may also reduce fluctuations in the information rate from the source and avoid symbol 'surges' which could overload the channel when the message sequence contains many high probability (i.e. frequently occurring, low entropy) symbols.

9.5.1 Code efficiency

Recall the definition of entropy (equation (9.3)) for a source with statistically independent symbols:

$$H = \sum_{m=1}^{M} P(m) \log_2 \frac{1}{P(m)} \quad \text{(bit/symbol)} \tag{9.17}$$

The maximum possible entropy, H_{max}, of this source would be realised if all symbols were equiprobable, $P(m) = 1/M$, i.e.:

$$H_{max} = \log_2 M \quad \text{(bit/symbol)} \tag{9.18}$$

A code efficiency can therefore be defined as:

$$\eta_{code} = \frac{H}{H_{max}} \times 100\% \tag{9.19}$$

If source symbols are coded into another symbol set, Figure 9.3, then the new code efficiency is given by equation (9.19) where H and H_{max} are the entropy and maximum possible entropy of this new symbol set.

   If source symbols are coded into binary words then there is a useful alternative interpretation of η_{code}. For a set of symbols represented by binary codewords with lengths l_m (binary) digits, an overall code length, L, can be defined as the average codeword length, i.e.:

$$L = \sum_{m=1}^{M} P(m) l_m \quad \text{(binary digits/symbol)} \tag{9.20}$$

The code efficiency can then be found from:

$$\eta_{code} = \frac{H}{L} \quad \text{(bit/binary digit)} \times 100\% \tag{9.21}$$

Equations (9.21) and (9.19) are seen to be entirely consistent when it is remembered that the maximum information conveyable per L digit binary codeword is given by:

Message — Transformation — Message
(Symbol set 1) — coder — (Symbol set 2)

Figure 9.3 *Translating a message between different alphabets.*

$$H_{\text{max}} \;\; (\text{bit/symbol}) \;=\; L \;\; (\text{bit/codeword}) \qquad\qquad (9.22)$$

EXAMPLE 9.3

A scanner converts a black and white document, line by line, into binary data for transmission. The scanner produces source data comprising symbols representing runs of up to six similar image pixel elements with the probabilities as shown below:

No. of consecutive pixels	1	2	3	4	5	6
Probability of occurrence	0.2	0.4	0.15	0.1	0.06	0.09

Determine the average length of a run (in pixels) and the corresponding effective information rate for this source when the scanner is traversing 1000 pixel/s.

$$H = \sum_{m=1}^{6} P(m) \log_2 \frac{1}{P(m)}$$

$$= 0.2 \times 2.32 + 0.4 \times 1.32 + 0.15 \times 2.74 + 0.1 \times 3.32 + 0.06 \times 4.06 + 0.09 \times 3.47$$

$$= 2.29 \;\; \text{bit/symbol}$$

Average length:

$$L = \sum_{m=1}^{6} P(m) \, l_m \; = 0.2 + 0.8 + 0.45 + 0.4 + 0.3 + 0.54 \; = 2.69 \;\; \text{pixels}$$

At 1000 pixel/s scan rate we generate $1000/2.69 = 372$ symbol/s. Thus the source information rate is $2.29 \times 372 = 852$ bit/s.

We are generally interested in finding a more efficient code which represents the same information using fewer digits on average. This results in different lengths of codeword being used for different symbols. The problem with such variable length codes is in recognising the start and end of the symbols.

9.5.2 Decoding variable length codewords

The following properties need to be considered when attempting to decode variable length codewords:

(1) Unique decoding.

This is essential if the received message is to have only a single possible meaning. Consider an $M = 4$ symbol alphabet with symbols represented by binary digits as follows:

$$A = 0$$
$$B = 01$$
$$C = 11$$
$$D = 00$$

If we receive the codeword 0011 it is not known whether the transmission was D, C or A, A, C. This example is not, therefore, uniquely decodable.

(2) Instantaneous decoding.

Consider now an $M = 4$ symbol alphabet, with the following binary representation:

$$A = 0$$
$$B = 10$$
$$C = 110$$
$$D = 111$$

This code can be instantaneously decoded using the decision tree shown in Figure 9.4 since no complete codeword is a prefix of a larger codeword. This is in contrast to the previous example where A is a prefix of both B and D. The latter example is also a 'comma code' as the symbol zero indicates the end of a codeword except for the all-ones word whose length is known. Note that we are restricted in the number of available codewords with small numbers of bits to ensure we achieve the desired decoding properties.

Using the representation:

$$A = 0$$
$$B = 01$$
$$C = 011$$
$$D = 111$$

the code is identical to the example just given but the bits are time reversed. It is thus still uniquely decodable but no longer instantaneous, since early codewords are now prefixes of later ones.

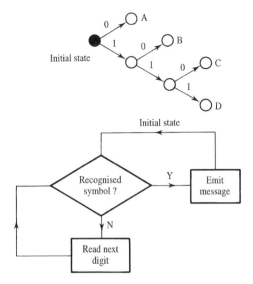

Figure 9.4 *Algorithm for decision tree decoding and example of practical code tree.*

9.6 Variable length coding

Assume an $M = 8$ symbol source $A, \cdots, H$ having probabilities of symbol occurrence:

m	A	B	C	D	E	F	G	H
$P(m)$	0.1	0.18	0.4	0.05	0.06	0.1	0.07	0.04

The source entropy is given by:

$$H = \sum_m P(m) \log_2 \frac{1}{P(m)} = 2.55 \ \text{bit/symbol} \tag{9.23}$$

and, at a symbol rate of 1 symbol/s, the information rate is 2.55 bit/s. The maximum entropy of an eight-symbol source is $\log_2 8 = 3$ bit/symbol and the source efficiency is therefore given by:

$$\eta_{source} = \frac{2.55}{3} \times 100\% = 85\% \tag{9.24}$$

If the symbols are each allocated 3 bits, comprising all the binary patterns between 000 and 111, the coding efficiency will remain unchanged at 85%.

Shannon–Fano coding [Blahut, 1987], in which we allocate the regularly used or highly probable messages fewer bits, as these are transmitted more often, is more efficient. The less probable messages can then be given the longer, less efficient bit patterns. This yields an improvement in efficiency compared with that before source coding was applied. The improvement is not as great, however, as that obtainable with another variable length coding scheme, namely Huffman coding, which is now described.

9.6.1 Huffman coding

The Huffman coding algorithm comprises two steps – reduction and splitting. These steps can be summarised by the following instructions:

(1) Reduction: List the symbols in descending order of probability. Reduce the two least probable symbols to one symbol with probability equal to their combined probability. Reorder in descending order of probability at each stage, Figure 9.5. Repeat the reduction step until only two symbols remain.

(2) Splitting: Assign 0 and 1 to the two final symbols and work backwards, Figure 9.6. Expand or lengthen the code to cope with each successive split and, at each stage, distinguish between the two split symbols by adding another 0 and 1 respectively to the codeword.

The result of Huffman encoding the symbols $A, \cdots, H$ in the previous example (Figures 9.5 and 9.6) is to allocate the symbols codewords as follows:

Symbol	C	B	A	F	G	E	D	H
Probability	0.40	0.18	0.10	0.10	0.07	0.06	0.05	0.04
Codeword	1	001	011	0000	0100	0101	00010	00011

The code length is now given by equation (9.20) as:

$$L = 1(0.4) + 3(0.18 + 0.10) + 4(0.10 + 0.07 + 0.06) + 5(0.05 + 0.04) = 2.61 \tag{9.25}$$

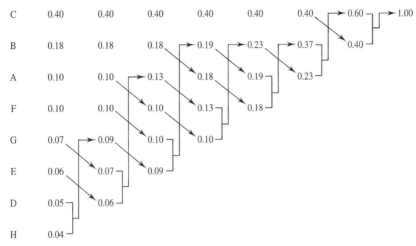

Figure 9.5 *Huffman coding of an eight-symbol alphabet – reduction step.*

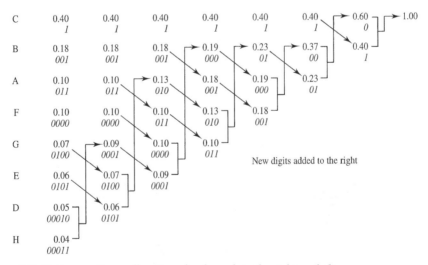

Figure 9.6 *Huffman coding – allocation of codewords to the eight symbols.*

and the code efficiency, given by equation (9.21), is:

$$\eta_{code} \; = \; \frac{H}{L} \times 100\% \; = \; \frac{2.55}{2.61} \times 100\% \; = \; 97.7\% \tag{9.26}$$

The 85% efficiency without coding would have been improved to 96.6% using Shannon–Fano coding but Huffman coding at 97.7% is even better. (The maximum efficiency is obtained when symbol probabilities are all negative, integer, powers of two, i.e. $1/2^n$.) Note that the Huffman codes are formulated to minimise the average codeword length. They do not necessarily possess error detection properties but are uniquely, and instantaneously, decodable, as defined in section 9.5.2.

9.7 Source coding examples

An early example of source coding occurs in Morse where the most commonly occurring letter 'e' (dot) is allocated the shortest code (1 bit) and less used consonants, such as 'y' (dash dot dash dash), are allocated 4 bits. A more recent example is facsimile (fax) transmission where an A4 page is scanned at 3.85 scan lines/mm in the vertical dimension with 1728 pixels across each scan line. If each pixel is then binary quantised into black or white this produces 2 Mbit of data per A4 page. If transmitted at 4.8 kbit/s over a telephone modem (see Chapter 11), this takes approximately 7 min/page for transmission.

In Group 3 fax runs of pixels of the same polarity are examined. (Certain run lengths are more common than others, Figure 9.7.) The black letters, or drawn lines, are generally not more than 10 pixels wide while large white areas are much more common. Huffman coding is employed in the ITU-T standard for Group 3 fax transmission, Figure 9.8, to allocate the shortest codes to the most common run lengths. The basic scheme is

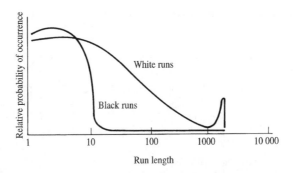

Figure 9.7 *Relative probability of occurrence of run lengths in scanned monochrome text.*

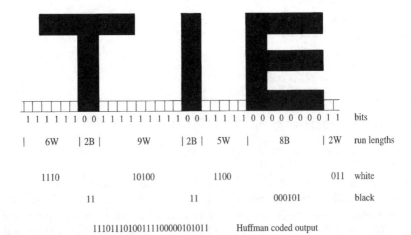

Figure 9.8 *Redundancy removal by source coding of printed character data (source: Pugh, 1991, reproduced with the permission of the IEE).*

modified to include a unique end of line code for resynchronisation which determines whether lines are received in error, as these should be always 1728 pixels apart.

In this application, Huffman coding improves the efficiency and reduces the time to transmit the page to less than 1 minute. Prior to email, facsimile was very significant with 25% of international telephone traffic representing fax transmissions between the 15 million terminals in use worldwide. Continued improvements in the Group 3 fax have made the transition to (the more advanced) Group 4 digital system unattractive. There is now an ITU-T V.17 data modem standard (see section 11.6) which transmits fax signals at 14.4 kbit/s and which is very similar to the V.32 9.6 kbit/s standard used to achieve the Figure 11.55(c) result.

There is current interest in run length codes aimed at altering their properties to control the spectral shape of the coded signal. In particular it is desirable to introduce a null at DC, as discussed in section 6.4, and/or place nulls or minima at those locations where there is expected to be a null in the channel's frequency response. Such techniques are widely used to optimise the performance of optical and magnetic recording systems [Schouhammer-Immink]. Run length codes can also be modified to achieve some error control properties, as will be described in the next chapter.

9.7.1 Source coding for speech signals

Vocoders (voice coders) are simplified coding devices which extract, in an efficient way, the significant components in a speech waveform, exploiting speech redundancy, to achieve low bit rate (< 2.4 kbit/s) transmission. Speech basically comprises four *formants* (Figure 9.9), and the vocoder analyses the input signal to find how the positions, F_1, F_2, F_3 and F_4, and magnitudes, A_1, A_2, A_3 and A_4, of the formants vary with time. In this instance we fit a model to the input spectrum and transmit the model parameters rather than, for example, the ADPCM quantised error samples, $\varepsilon_q(kT_s)$, of section 5.8.3.

Figure 9.10 shows a 1.2 s segment of voiced speech and the corresponding formants as they vary with time. This shows how the second formant frequency changes from 1 kHz to 2 kHz during the speech utterance. It is this movement which the linear predictive

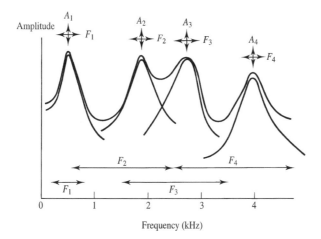

Figure 9.9 *Instantaneous spectral representation of speech as a set of four formant frequencies.*

coding (LPC) tracks by measuring the LPC coefficients every 20 ms.

Two major vocoder designs exist at present, the channel vocoder and the LPC. The channel vocoder is basically a non-linearly spaced filterbank spectrum analyser with 19 distinct filters. The LPC vocoder is usually implemented as a cascade of linear prediction error filters which remove from the speech signal the components that can be predicted [Jayant and Noll, Gray and Markel] from its previous history by modelling the vocal tract as an all-pole filter. (See standard signals and systems texts such as [Jackson, Mulgrew *et al.*] for a discussion of the properties of poles and zeros and their effect on system, and filter, frequency responses.) The LPC vocoder extends the DPCM system described in section 5.8.2 to perform a *full modelling* of the speech production mechanism and

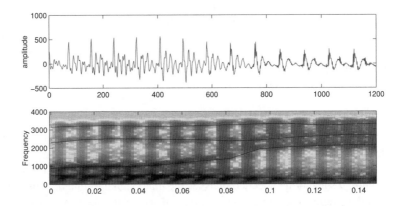

Figure 9.10 *Upper trace comprises a time domain plot from a voiced speech waveform, time scale in ms, which shows the regular pitch excitation. Lower trace shows the movement of the four formants in a spectrogram display, time scale in s, frequency scale in Hz.*

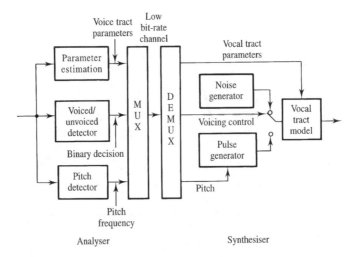

Figure 9.11 *Block diagram of the linear predictive vocoder.*

removes the necessity for the quantised error sample transmissions.

In the LPC vocoder, Figure 9.11, the analyser and encoder normally process the signal in 20 ms frames and subsequently transmit the coarse spectral information via the filter coefficients. The residual error output from the parameter estimation operation is not transmitted. The error signal is used to provide an estimate of the input power level which is sent, along with the pitch information and a binary decision as to whether the input is voiced or unvoiced (Figure 9.11). The pitch information can be ascertained by using autocorrelation (see sections 2.6 and 3.3.3) or zero crossing techniques on either the input signal or the residual error signal, depending on the sophistication and cost of the vocoder implementation. The decoder and synthesiser apply the received filter coefficients to a synthesising filter which is excited with impulses at the pitch frequency if voiced, or by white noise if unvoiced. The excitation amplitude is controlled by the input power estimate information. This excitation with a synthetic signal reduces the transmission bit rate requirement but also reduces the speech quality. (Speech quality is assessed using the mean opinion score (MOS) scale (see section 5.7.3).)

With delays in the vocal tract of about 1 ms and typical speech sample rates of 8 to 10 kHz the number of predictor stages is normally in the range 8 to 12; 10 is the number adopted in the integrated NATO LPC vocoder standard (LPC-10, also US Federal Standard 1015) which transmits at a rate of 2.4 kbit/s and achieves an MOS of approximately 2, Figure 9.12 and Table 9.3, for a complexity which is 40 times that of PCM. The MOS score is low due to the synthesiser's excitation with noise or regenerated pitch information. This causes LPC to lose all background sounds and the speech to have a characteristic metallic sound. The basic spoken words, however, are still quite intelligible. These vocoders can be implemented as single chip DSP designs. In addition to vocoder applications, LPC synthesisers are used in several commercial speech systems, including the Texas Instruments speak and spell children's toy.

Much current research is in progress to improve the quality of vocoders. This includes studies of hybrid coder systems such as LPC excited from a codebook of possible signal vectors (CELP) or excited with multiple pulses (MPE) rather than a simple pitch generator, as in Figure 9.11. US Federal Standard 1016 defines a 4.8 kbit/s

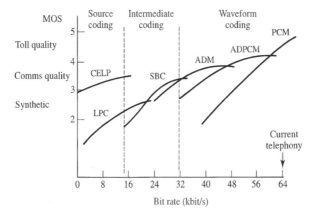

Figure 9.12 *Comparison of quality and transmission rate for various speech coding systems, using the mean opinion score (MOS) rating.*

CELP implementation for secure systems. In developing these systems low coder delay (not greater than 2 ms) is an important design goal.

CELP systems achieve an MOS of 3.2 at 4 kbit/s but they are 50 to 100 times as complex as the basic PCM coder. For higher quality transmission ITU-T G.728 defines a 16 kbit/s low delay CELP implementation. CELP and MPE, which also involve feedback to compare the synthesised speech with the original speech and minimise the difference, are being applied in mobile systems such as GSM and CDMA 2000, see Chapter 15.

In order to achieve spectrally efficient coding with toll quality speech, coders designed at the launch of GSM employed a variant of CELP, with residual pulse excitation and a long term prediction [Gray and Markel]. Here the speech is processed in 20 ms blocks to generate 260 encoded bits, resulting in an overall bit rate of 13 kbit/s. There is an inherent long delay in the coder, however, of 95 ms which necessitates echo cancellation. GSM was subsequently enhanced to accommodate half-rate speech coders which double the system capacity while maintaining similar speech quality.

Table 9.3 *Rate, performance and complexity comparison for various speech coder designs.*

Class	Technique	Bit rate (kbit/s)	MOS quality	Relative complexity
Waveform coders	PCM G.711	64	4.3	1
	ADPCM G.721	32	4.1	10
	DM	16	3	0.3
Intermediate coder	SBC (2) G.722 (7 kHz)	64–48	4.3	30
Enhanced source coder	CELP/MPE for ½ rate GSM	4.8	3.2	50–100
Source coder	LPC-10 vocoder	2.4	2	40

9.7.2 High quality speech coders

Other audio coding techniques are based on frequency domain approaches where the DC to 3.4 kHz speech bandwidth is split into 2 to 16 individual sub-bands by a filterbank or a discrete Fourier transform (DFT) (see section 13.5), Figure 9.13, to form a sub-band coder (SBC) [Vetterli and Kovacevic]. Band splitting is used to exploit the fact that the individual bands do not all contain signals with the same energy. This permits the accuracy or number of bits in the individual quantisers in the encoder to be reduced, in bands with low energy signals, saving on the overall coder transmission rate.

While this was initially applied to achieve low bit rate intermediate quality speech at 16 kbit/s, the same technique is now used in the ITU-T G.722, 0 to 7 kHz, high quality audio coder which employs only two sub-bands to code this wideband signal at 64/56/48 kbit/s with an MOS between 3.7 and 4.3. This is a low complexity coder, implemented in one DSP microprocessor, which can operate at a P_e of 10^{-4} for ISDN teleconferencing and telephone loudspeaker applications. The MOS drops to 3.0 at a P_e of 10^{-3}. Compact disc player manufacturers are also investigating an extension of this technique with 32 channels within the 20 kHz hi-fi bandwidth. Simple 16 bit PCM at 44.1 ksample/s requires 768 kbit/s but the DFT based coder offers indistinguishable music quality at 88 kbit/s or only 2 bit/sample. This development of the ISO/MPEG standard is very

Upper sub-band

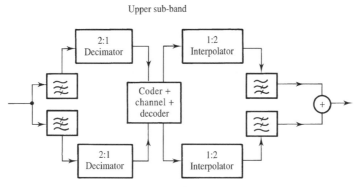

Figure 9.13 *Sub-band (speech) coder with two, equal width, sub-bands and separate encoding of sub-band signals after sample rate reduction.*

important for the digital storage of broadcast quality signals, e.g. HDTV and digital audio (see Chapter 16).

Table 9.4 *PCM coding rates for speech and audio signals.*

	Frequency band (kHz)	*Sample rate (kHz)*	*Bits/sample*	*Rate (kbit/s)*
Telephony	DC–3.4	8	8	64
Wideband speech	0.05–7	16	8	128
Audio	0.01–11	24	16	384
Hi-fi audio	0.01–20	44.1/48	16	706/768

Finally it has been shown that variants of these techniques can achieve very high quality transmission at 64 kbit/s, which is also very significant for reducing the storage requirements of digital memory systems.

Similar techniques are applied to encode video signals but due to the large information content the bit rates are much higher. Video is usually encoded at 34 or 140 Mbit/s with the lower rate involving some of the prediction techniques discussed here. In video there is also redundancy in the vertical and horizontal dimensions, and hence spatial Fourier, and other, transformation techniques [Clarke, 1998] are attractive for reducing the bit rate requirements in, for example, video telephony applications (see Chapter 16).

9.7.3 Audio coders

In addition to the coding of speech signals, stereo music and audio signals can also be encoded more efficiently than by simply employing the linear PCM methods represented by Figure 5.19. The near instantaneous block companded multiplex (NICAM) European TV audio coder (ITU-R, recommendation 660) uses 32 kHz sampling with 14 bit actual quantisation followed by compression of 1 ms blocks into a 10 bit code resulting eventually in a stereo rate of 728 kbit/s. More efficient audio coding relies on exploiting the frequency characteristics or the perceptual response of the human ear where the short encoder delay constraint of speech coders can also be relaxed. When a high level (sinusoidal or narrowband) signal is present then this masks adjacent lower level sinusoids, as shown in Figure 9.14. This shows how the minimum detectable sound pressure level in dB at the ear, with respect to a reference level, falls with increasing frequency due to the hearing threshold before encountering a masking signal which renders lower level tones inaudible. At high frequencies, approaching 10 to 20 kHz, the sound pressure level threshold for the human ear increases again.

Perceptual coding exploits these characteristics and adjusts the quantiser accuracy at the outputs of an analysis filterbank so that the quantisation noise in each band is increased to just below the level where it would become apparent to achieve an overall minimised bit rate. Thus all the quantisation errors are effectively hidden within the hearing/masking thresholds but the penalty is the need to split the signal into sub-bands and continuously update the allocated number of quantisation levels in each sub-band.

The ideal filterbank would be logarithmically spaced with increasing bandwidth at higher frequencies; 16 filters are required to model the DC to 3.4 kHz speech bandwidth and a further 9 to 10 are required to extend the speech to 20 kHz for hi-fi audio applications. This can be achieved by extending Figure 9.13 into a tree structure but often a discrete cosine transform (DCT), wavelet transform [Vetterli and Kovacevic] or other overlapped filterbank processor is used, typically with 32 individual channels. The transmission encoder takes the form shown in Figure 9.15 where a parallel DFT processor is used to evaluate the masking thresholds and then perform the adaptive bit allocation strategy on each frame of transformed data.

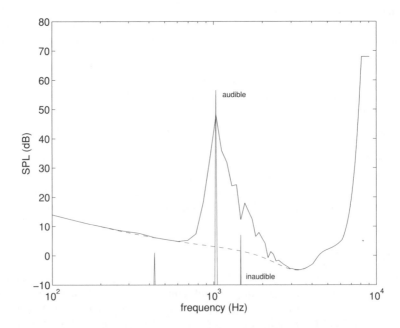

Figure 9.14 *Inaudible tones masked by sound pressure level (SPL) threshold (solid plot) resulting from the audible tone at 1 kHz.*

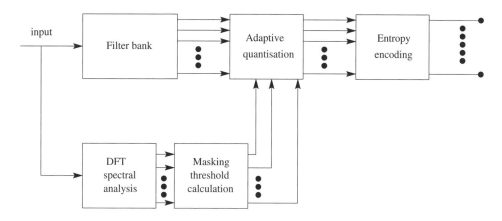

Figure 9.15 *Transmission encoder schematic for a perceptual speech coder.*

Encoding delay is typically 20 to 60 ms. Some of the lower sub-bands will usually be allocated a 6 or 7 bit quantisation while the majority of the bands need only be allocated 3 or 2 bit quantisation.

The audio standard No. 1, defined by the Moving Pictures Experts Group (MPEG), which is formally defined as ISO/IEC 11172, has three distinct layers or parts which increase in processor complexity and give a progressively larger compression, Table 9.5. In comparison PCM audio coders sample at compact disc (CD) rates of 44.1 kHz with 16 bits allocated per mono channel, i.e. 1.41 Mbit/s overall stereo transmission rate, Table 9.4.

Table 9.5 *MPEG-1 stereo coder bit rates.*

	Bit rate (kbit/s)	*Compression factor cf. CD*
Layer I	384	4
Layer II	192	8
Layer III	128	12

All the MPEG-1 layer I to III audio encoders, Table 9.5, accommodate sample rates of 32, 44.1 and 48 kHz and the careful choice of perceptual coding technique ensures that the stereo output is indistinguishable from CD quality sound. CD stereo signals at 1.41 Mbit/s are further expanded with error correction, synchronisation and other overhead data to 4.32 Mbit/s while the corresponding digital audio tape (DAT) rate is 3.08 Mbit/s. (See Chapter 10 for discussion on error correction coding.) The MPEG-2 standard is aimed at multichannel audio encoding which is backwards compatible with MPEG-1 with additional advanced audio coding that is no longer compatible with MPEG-1 signals. The MP3 downloadable digital audio player uses layer III of the MPEG-2 standard and gives a quality slightly below CD players.

9.7.4 String coding

This family of coding methods [Sayood] is fundamentally different from the previous Huffman arithmetic coding techniques of section 9.6. It uses the fact that small strings within messages often repeat locally, i.e. the source has memory. One such Lempel–Ziv method uses a history buffer and looks for matches. If a match is found then the position and length of the match within the buffer is transmitted instead of the character string. Commonly encountered word strings, e.g. 000, 0000 or 'High Street', experience compression when they are allocated their 12 bit codes, especially for long (> 10 character) strings. Thus the scheme relies on redundancy in the character occurrence, i.e. individual character string repetitions, but it does not exploit, in any way, positional redundancy.

EXAMPLE 9.4

Encode the message below using a code with history buffer of length 16 and maximum match length of eight characters:

tam_eht   _no_tas_tac_eht_

to be transmitted   history buffer (already sent)

If we have this string and we wish to encode 'tam_eht', then we would transmit '_eht' as a match to '_eht' at the end of our history buffer using 8 bits: 1 bit as a match flag, 4 bits to give the position of the match and 3 bits for the length of the match. (Note that we only need to use 3 bits to encode a maximum match length of 8 as a match length of 0 is not a match.)

So encoding '_eht' gives:

1 *to indicate that a match has been found*

1011 *to indicate that the match starts at position* 11, as the first position is zero

011 *to indicate the length of the match is 4 (000 indicates length of* 1, 001 2, 010 3, 011 4, *etc.)*

Thus '_eht' encodes as 11011011 to give:

tam   _eht_no_tas_tac

to be transmitted   history buffer (already sent)

There is no match for 'm'. We would thus transmit 'm' as 9 bits, 1 bit (0) to indicate no match found followed by an 8 bit ASCII code for 'm'. Thus encoding 'm' gives:

0 *to indicate no match*

01011101 *8 bit ASCII code for the m*

and 'm' encodes as 001011101 to give:

ta   m_eht_no_tas_tac

to be transmitted   history buffer (already sent)

We would transmit 'ta' as a match to the 'ta' at history buffer position 9 (or 13) using the same type of 8 bit pattern as we used for '_eht'. Thus encoding 'ta' gives:

1 *to indicate a match has been found*

1001 *to indicate that the match starts at position* 9

001 *to indicate that the length of the match is 2 characters*

and 'ta' encodes as 11001001.

Overall, this means we have encoded seven 8-bit characters into 25 bits, giving a compression factor of 25/56, which is typical for this type of coder. Normally, the history buffer is much longer, around 256 characters, to enhance the compression factor. The above implementation of Lempel–Ziv is not very fast because of the requirement to search history buffers.

This technique has many advantages over arithmetic coding methods such as Huffman. It is more efficient (often exceeding 100% by our previous memoryless definition) and can adapt to changes in data statistics. However, errors can be disastrous, the coding time is longer (as searching is needed to identify the matches) and it does not work effectively for short bursts of data. This technique is used for compression of computer files for transport (ZIP), for saving space on hard disks (compress, stacker, etc.) and in some modem standards. Lempel–Ziv coding's most common practical implementation, the LZW algorithm [Welch, 1984], builds up a table of these strings to identify commonly occurring repeat patterns. Each string in the table is then allocated a unique 12 bit code.

LZW compression operates typically on blocks of 10000 to 30000 symbols and achieves little compression during the adaptive phase when the table is being constructed. It uses simplified logic, which operates at three clock cycles per symbol. After 2000 to 5000 words have been processed typical compression is 1.8:1 on text, 1.5:1 on object files and > 2:1 on data or program source code. There is no compression, however, on floating point arrays which present noise-like inputs.

9.8 Data encryption

Encryption is concerned with:

1. Preventing the unauthorised reception of a message (i.e. secrecy or privacy).
2. Verifying the message's origin (i.e. authentication).
3. Establishing that a received message has not been altered (i.e. integrity).

The following 'locked box' analogy is a useful device for illustrating some basic encryption ideas.

9.8.1 The locked box analogy

A message is kept secret by locking it in a secure box and couriering the box to its intended recipient. If the box and its lock are sufficiently strong, and the key is delivered only to the intended recipient, then the message is secure, even if it falls into the hands of a third (unauthorised) party. Secrecy, in this scheme, relies on none but the intended recipient (and the message's originator) having access to the key.

The classical problem of encryption, referred to as the key distribution problem, is that of securely delivering the key to the intended recipient. (Once the key is correctly delivered messages could continue to be transported using identical locks that share the same key. To guard against the key being covertly copied, however, the key, and lock, might be periodically changed. This makes key distribution a continuing, rather than a one time only, problem.)

In an encryption scheme the locked box is replaced by the transformation of an un-coded message or plaintext into a coded message or ciphertext. Deciphering is only possible if the recipient has knowledge of the deciphering algorithm and key. (The tacit assumption in cryptography is that all unauthorised recipients know the deciphering algorithm but none know the key.) The simplest example of such an encryption scheme is the Caesar cipher in which each letter of the 26-character alphabet is replaced with the corresponding letter of a cyclically shifted version of the alphabet (e.g. $a = d, b = e, c = f, \cdots, x = a, y = b, z = c$). The key to this cipher is the number of the cyclical shift (3 in the example).

Until relatively recently, encryption and deciphering were always effected using forward and reverse forms of a single transformation (in the locked box analogy the key simply being turned in different directions): +3 and −3 in the case of the Caesar cipher given above. (Sometimes the forward and reverse transforms are identical, e.g. a Caesar cipher with a cyclical shift of 13.) Since security depends on knowledge of a single key being restricted to the sender and authorised recipient this type of encryption scheme is called a symmetric key or private key system. Private key encryption suffers from the classical key distribution problem referred to above.

The key distribution problem can be circumvented if the intended recipient of a message (rather than the message sender) takes responsibility for secrecy. In the locked box analogy the recipient can do this by sending a 'snap-shut' pad-lock (to which he/she retains the only key) to the message sender. The pad-lock corresponds to a publicly available encryption process (distributed freely by the message recipient) that produces ciphertext. The key (always in the possession of the message recipient) corresponds to a secret de-encryption process, different in form (not only direction) from the encryption process. This is called an asymmetric key or public key system.

9.8.2 Secrecy

A basic measure of cryptographic security or strength is the mutual information, I_M, that exists between the plaintext, T, and the ciphertext, C. This is defined by:

$$I_M(T, C) = H(T) - H(T|C) \tag{9.27}$$

where $H(T)$ is the uncertainty (or entropy) of the plaintext and $H(T|C)$ is the uncertainty (entropy) of the plaintext conditional on the ciphertext. Perfect secrecy is achieved when the ciphertext and plaintext are statistically independent and $I_M(T, C) = 0$. This condition is equivalent to:

$$H(T|C) = H(T) \tag{9.28}$$

which expresses the intuitively reasonable proposition that for perfect secrecy the uncertainty in T should be unaffected by knowledge of C (for all but the intended recipient of the message).

The uncertainty in T given C clearly becomes zero when the key needed to decipher the message is known. It follows that for perfect secrecy the uncertainty in the key, K, must be at least as large as the uncertainty in T, i.e.:

$$H(K) = H(T) \tag{9.29}$$

Equation (9.29) is Shannon's bound for perfect secrecy. In order to minimally satisfy this bound the key for a cipher must contain the same number of symbols as the plaintext (assuming key and plaintext share a common alphabet size and redundancy).

One-time pad

The one-time pad (or Vernam cipher) satisfies the condition for perfect secrecy. Its name arises from the fact that the key is applied only once, i.e. it is used to encrypt only a single character and is then discarded. For a 100-character message the keys might, for example, be 100 random numbers each of which specifies the cyclical shift number (key) of a Caesar cipher. A practical implementation for characters coded as binary digits operates by modulo-2 summing a plaintext bit sequence with a randomly generated binary key of the same length. (The entire random sequence is usually referred to as the key but each bit in the sequence can be thought of as an independent key, changing or not changing the binary value of the plaintext bit to which it is applied. Since the key for each character is new, i.e. independent of the key for all other characters, this algorithm qualifies as a one-time pad.) The random bit stream is clearly statistically independent of the plaintext in this case and, being of the same length as the plaintext, has (at least) equal entropy to the plaintext. Deciphering is achieved by modulo-2 summing with the same key as that used for encryption. (This is another example of the forward and reverse transformations being identical.)

9.8.3 Substitution and permutation

Cryptanalytic attacks on ciphertext typically rely on a statistical analysis of the ciphertext characters. It is therefore desirable that a cipher should remove or obscure any character patterns that occur in the plaintext. This can be achieved by applying source coding to remove redundancy prior to encryption and/or using substitution and permutation in the encryption process.

Substitution of characters making up the plaintext obscures the relationship between the statistics of ciphertext characters and the statistics of plaintext characters. One-to-one mapping of alphabet characters (e.g. $a = w, b = f, c = k, \cdots, z = b$) is the simplest possible example of substitution although the level of security afforded is clearly too limited to be of modern-day value, despite substitutions with random mapping yielding 26! different possible keys. (The Caesar cipher is a special case of one-to-one character substitution with only 26 keys, including the trivial 0 shift or 26 shift.) More complex substitutions can be realised by mapping plaintext character strings to ciphertext character strings. Taking n-character strings, from an M-character plaintext alphabet, gives a substitution alphabet size of M^n. The substitution pattern (i.e. the ways the connections are made between all possible pairs of n-character strings) is the key to the cipher, the number of possible keys being $M^n!$. This is a huge number even for binary strings ($M = 2$) of modest length. Figure 9.16 is a schematic representation of a substitution algorithm (often called an S-box) that takes eight binary symbols (i.e. 1 byte) at a time. This requires $2^8 = 256$ substitution connections and gives 256! possible keys.

Permutation (i.e. transposition or rearrangement) of characters also complicates the relationship between the statistics of plaintext and ciphertext characters. (In a pure permutation process the statistics of single characters are unaltered but the joint and conditional statistics of character strings are changed.) Figure 9.17 shows a permutation (or P-) box for a binary alphabet that illustrates the process.

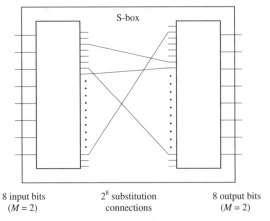

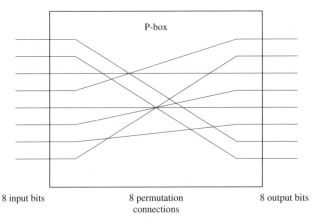

Figure 9.16 *Example of binary substitution algorithm (S-box).*

Figure 9.17 *Example of binary permutation algorithm (P-box).*

Practical encryption algorithms typically use repeated rounds of alternating substitution and permutation in order to create an overall plaintext to ciphertext algorithm of sufficient complexity to defy cryptanalysis.

9.8.4 Confusion, diffusion and the unicity distance

Substitution and permutation relate, respectively, to Shannon's general concepts of confusion and diffusion [Shannon, 1949]. Substitution of characters (or character strings) confuses the cryptanalyst by making the relationship between the statistics of the ciphertext characters and the description of the key complex. This is useful because cryptanalysis typically uses a statistical analysis of the ciphertext to restrict the (usually large) multidimensional space containing all possible keys (i.e. the key-space) to a smaller sub-space making an exhaustive search for the key feasible. If one statistic of the ciphertext limits the key-space to region X and another statistic limits it to region Y, for example, then the key must lie in the overlapping space. If the relationship between the ciphertext statistics and the key is very complicated, however, the key-space will itself be complicated and the overlapping region will be hard to define and difficult to use. Figure 9.18 illustrates this idea schematically.

Permutation (when combined in alternating rounds with substitution) spreads out (diffuses) the influence of each bit in the plaintext over many bits of the ciphertext. This is illustrated in Figure 9.19 which shows a set of alternating permutation and substitutions for a 16-character block.

The first round of substitutions (S-boxes) in this example takes a set of four characters

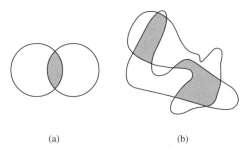

(a)   (b)

Figure 9.18 *Schematic illustration of overlapping key sub-spaces where relationship between ciphertext statistics and key is (a) simple, (b) complex.*

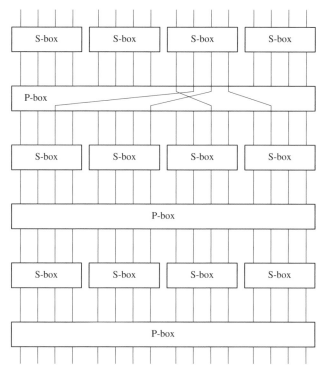

Figure 9.19 *Use of permutation in iterative rounds alternating with substitution for diffusion of plaintext character influence on ciphertext characters. (A product cipher.)*

from the plaintext and makes one substitution from 27^4 possible substitutions. (The possible substitutions include the trivial 'no change' substitution and a space character is included in this example making the alphabet size 27.) The individual output characters from the substitution blocks are then permuted (in a P-box) which has the effect of distributing the influence of a single character in the plaintext (M say) over more than one (all four in the case shown) substitution box inputs in the next round. As the number of rounds increases the influence of a given plaintext character becomes evermore uniformly distributed over the ciphertext characters.

The effect of diffusion is to dissipate the statistical structure of the plaintext over long strings of ciphertext characters. This results in the relative frequencies of single ciphertext characters being more equal than those of the plaintext characters but does not alter overall redundancy. (The redundancy is simply distributed more evenly over longer character strings.)

A quite different (but classic) method of diffusion [Shannon, 1949] ensures that the influence of plaintext characters is distributed over multiple ciphertext characters by explicitly adding plaintext character strings to form each ciphertext character using a transformation of the type:

$$C_n = \sum_{i=1}^{N} T_{n+i} \ (\text{modulo} - M) \tag{9.30}$$

where C_n is the nth ciphertext character, T_n is the nth plaintext character, M is the size of the plaintext alphabet (e.g. 26 for minimal English language plaintext excluding spaces and punctuation) and N is the number of plaintext characters summed (modulo-M) to form the ciphertext.

Incorporating diffusion in a cipher makes cryptanalysis difficult because:

1. A large amount (quantified as the *unicity distance*) of ciphertext must be intercepted (in order to capture its long range statistical structure).
2. A large amount of ciphertext analysis is necessary to make use of this structure since it is distributed over a large number of individual statistics.

The unicity distance, N_o, which represents the minimum amount of ciphertext required by a cryptanalyst to find the cipher key is defined such that $H(K|N_o) = 0$. Under a reasonable set of cryptographic assumptions the unicity distance is well estimated by the ratio of key entropy and plaintext redundancies (i.e. $N_o = H(K)/R_{plaintext}$).

9.8.5 Block ciphers and stream ciphers

Block ciphers are those which split plaintext into fixed length fragments (or blocks) and then operate on each fragment separately to produce a corresponding ciphertext fragment. The S-box and P-box transformations described above are examples of block ciphers. All such ciphers result in ciphertext blocks that are a (deterministic) function of the corresponding plaintext blocks and the cipher key only. This means that the transformation of a block of plaintext is fixed in the sense that if a message contained two identical plaintext blocks then the corresponding ciphertext blocks would also be identical.

Stream ciphers operate on plaintext bit by bit (or, more generally, character by character) rather than on plaintext blocks. A common type of stream cipher is a version of the binary one-time pad. Instead of a random bit pattern, however, a pseudo-random key-stream is generated using a shift register with feedback. The key in this case is the pattern of feedback connections on the key-stream shift register. Figure 9.20 is a schematic illustration of a stream cipher coder and decoder. (Details of pseudo-random bit stream generation using shift registers with feedback are given in section 13.4.2.)

If the key-stream is generated independently of the ciphertext then the stream cipher is said to be synchronous. Some method of synchronising the key-stream at the message source and destination is needed in this case in order for deciphering to proceed correctly. An advantage of such synchronous stream ciphers is that each ciphertext character depends only on the key and a single plaintext character, which means that errors in the received ciphertext do not propagate. A stream cipher employing a linear feedback shift register (as illustrated in Figure 9.20) has only weak encryption properties, however, and is especially vulnerable to attack if a plaintext and ciphertext message pair are known. In this case the corresponding plaintext and ciphertext bit sequences need only be twice as long as the number of stages employed in the key-sequence generator shift register.

A stream cipher in which the key for the current plaintext character is derived from past ciphertext characters is said to have cipher feedback. It is self synchronising and therefore does not need the explicit synchronisation mechanism required by a synchronous stream cipher. If the previous k bits of ciphertext are used to obtain the key for the current bit then errors will generally propagate for k bits after a single error in the received ciphertext. If multiple errors occur the deciphered plaintext will be correct only after k correct ciphertext bits have been received. Generating the key-stream directly from the ciphertext would make cipher feedback systems too vulnerable to cryptanalytic attack. Vulnerability is reduced, therefore, by interposing a block encryption algorithm between the key-stream shift register and the key/plaintext character combiner. (The combiner is a modulo-2 adder for binary data implementations.) Figure 9.21 shows the encryption and decoding processes for a stream cipher with ciphertext feedback.

9.8.6 Product ciphers

Block ciphers that rely on concatenating a number of (individually simple) permutation and substitution operations to produce an overall level of complexity that defies cryptanalytic attack are called product ciphers. The iterative cipher used to describe the diffusion properties of permutation when combined with alternating rounds of

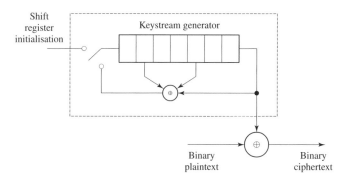

Figure 9.20 *Stream cipher encoder. (Decoder is identical.)*

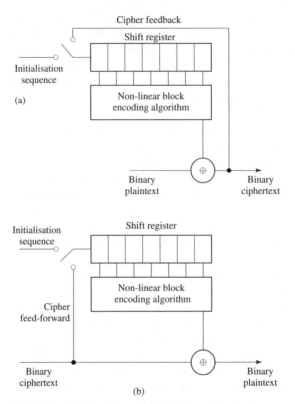

Figure 9.21 *Cipher feedback encryption (a) and deciphering (b) processes.*

substitution (Figure 9.19) is an example of a product cipher. (In practice, the substitutions and permutation in modern ciphers are not of the letters of a conventional written language but of the binary digits that represent these letters.)

9.8.7 Data encryption standard

Data encryption standard (DES) is a practical private key (product) block cipher that works as follows:

1. A 56 bit random number key, K, is generated.
2. K is divided into eight 7 bit words. Each word is supplemented by an odd parity check bit (placed in the last, i.e. 8th bit position) to form an 8 bit byte. These 8 bytes (64 bits) constitute an initial (augmented) key, K_0.
3. Sixteen 'keylets' ($K_1, K_2, K_3, \cdots, K_{16}$) are calculated by first permuting K_0 using the *permuted choice 1* function shown in Table 9.6.
4. The resulting 56 bit block is split into two 28 bit blocks referred to as C_0 (the first 28 bits corresponding to rows 1–4 of Table 9.6) and D_0 (the last 28 bits corresponding to rows 5–8 of Table 9.6).
5. First iterations of C and D (C_1 and D_1) are found from the initial values (C_0 and D_0) using the algorithm:

$$C_i = \text{Leftshift}_i(C_{i-1}) \tag{9.31(a)}$$

$$D_i = \text{Leftshift}_i(D_{i-1}) \tag{9.31(b)}$$

Leftshift$_i$( ), here, denotes the left shift operation, i.e. a movement of all bits to the left with bits wrapping around from position 1 to position 28. The number of positions shifted is either one or two depending on the value of i as specified by Table 9.7.

Table 9.6 *Permuted choice 1 (PC-1) for the DES key schedule calculation.*

57	49	41	33	25	17	9
1	58	50	42	34	26	18
10	2	59	51	43	35	27
19	11	3	60	52	44	36
63	55	47	39	31	23	15
7	62	54	46	38	30	22
14	6	61	53	45	37	29
21	13	5	28	20	12	4

(Table 9.6 is read left to right and top to bottom such that the 57th bit of K_0 becomes the 1st bit of permuted initial key (K_0'), the 49th bit of K_0 becomes the 2nd bit of K_0', $\cdots$, the 1st bit of K_0 becomes the 8th bit of K_0', $\cdots$, and the 4th bit of K_0 becomes the last bit of K_0'. Note that the length of K_0' is 56 bits since the function PC-1 ignores the parity bits in bit locations 8, 16, 24. 32, 40, 48, 56 and 64 of K_0.)

Table 9.7 *Number of left shifts in the operation leftshift$_i$().*

Iteration no. (*i*)	1	2	3	4	5	6	7	8	9	10	11	12	13	14	15	16
No. of left shifts	1	1	2	2	2	2	2	2	1	2	2	2	2	2	2	1

6. K_i is found from C_i and D_i by transforming the concatenation $C_i \| D_i$ (i.e. C_i bits followed by D_i bits) using the *permuted choice 2 (PC-2)* function as defined in Table 9.8.

Table 9.8 *Permuted choice 2 (PC-2) for DES.*

14	17	11	24	1	5
3	28	15	6	21	10
23	19	12	4	26	8
16	7	27	20	13	2
41	52	31	37	47	55
30	40	51	45	33	48
44	49	39	56	34	53
46	42	50	36	29	32

Note that that the output of PC-2 is a 48 bit sequence. Figure 9.22 shows a flow diagram of the algorithm to obtain the 16 keylets. This algorithm is called the key schedule calculation.

7. The plaintext is split into 64 bit blocks $(T_1 T_2 T_3 \cdots T_i \cdots T_{64})$.
8. The *initial permutation (IP)* specified in Table 9.9 is applied to the plaintext block. (The 58th bit in the plaintext data block becomes bit 1 in the permuted data block, bit 50 in the plaintext becomes bit 2, $\cdots$, bit 60 becomes bit 9, $\cdots$, and bit 7 becomes bit 64.)

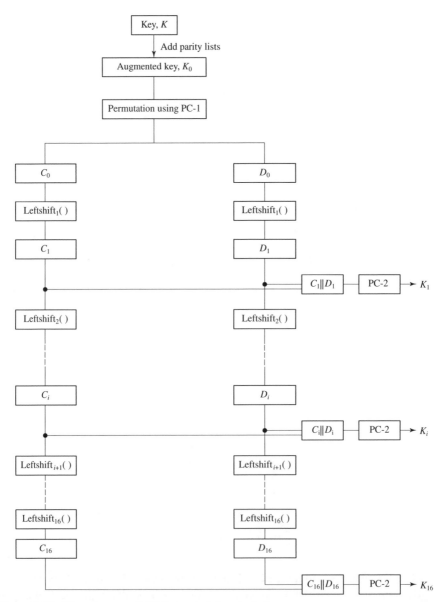

Figure 9.22 *Flow diagram of DES key schedule calculation.*

Table 9.9 *Initial permutation (IP) of plaintext for DES.*

58	50	42	34	26	18	10	2
60	52	44	36	28	20	12	4
62	54	46	38	30	22	14	6
64	56	48	40	32	24	16	8
57	49	41	33	25	17	9	1
59	51	43	35	27	19	11	3
61	53	45	37	29	21	13	5
63	55	47	39	31	23	15	7

9. The 64 bit permuted data block is then divided into a 32 bit *left half* (L_0) and a 32 bit *right half* (R_0).
10. The initial left and right halves (L_0 and R_0) are transformed to *first round* left and right halves (L_1 and R_1) using the algorithm:

$$L_i = R_{i-1} \qquad\qquad (9.32(\text{a}))$$

and:

$$R_i = L_{i-1} \oplus f(R_{i-1}, K_i) \qquad\qquad (9.32(\text{b}))$$

($\oplus$ in equation (9.32(b)) denotes modulo-2 addition, K_i is a 48 bit keylet obtained from the key schedule calculation, and the *cipher function f* (described below) is a function of R_{i-1} and K_i that results in a 32 bit string.)
11. The transformations defined by equations (9.32(a) and (b)) are repeated 15 times (i.e. applied 16 times in total), ultimately producing L_{16} and R_{16}.
12. A 64 bit data block is then reformed but with round-16 left and right halves interchanged (i.e. R_{16} bits followed by L_{16} bits)
13. A final permutation (FP), Table 9.10 (the inverse of the initial permutation), is applied to form the ciphertext.

Figure 9.23 shows a flow chart of the encryption process.

Table 9.10 *Final permutation (FP = IP^{-1}) producing ciphertext for DES.*

40	8	48	16	56	24	64	32
39	7	47	15	55	23	63	31
38	6	46	14	54	22	62	30
37	5	45	13	53	21	61	29
36	4	44	12	52	20	60	28
35	3	43	11	51	19	59	27
34	2	42	10	50	18	58	26
33	1	41	9	49	17	57	25

The cipher function, f, operates on K_i (a 48 bit number) and R_{i-1} (a 32 bit number). In order to make the two numbers compatible it turns R_{i-1} into a 48 bit number using the bit selection process (*E*-function) defined by Table 9.11. (Notice that some input bits appear more than once in the output of E.)

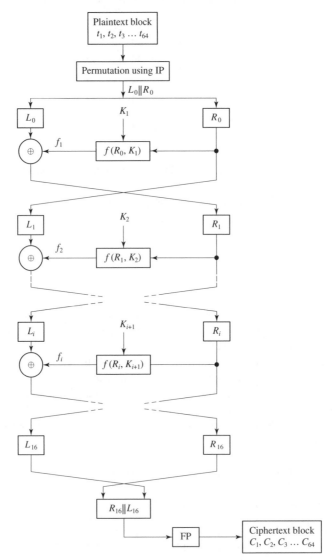

Figure 9.23 *DES encryption process.*

The 48 bit sequence, $E(R_{i-1})$, is then modulo-2 added to K_i and the result is divided into eight 6 bit words (B_1, B_2, $\cdots$, B_8). Each word (B_n) is operated on by a bit selection function, S_n, which interprets the first and last bits of B_n as a 2 bit row number from 0 to 3 and the middle 4 bits of B_n as a 4 bit column number from 0 to 15. B_n is then replaced with the 4 bit binary number corresponding to the entry in the S_n function table (Tables 9.12(a)–(h)) and the eight 4 bit S_ns concatenated to form l_i.

Table 9.11 *E-function for DES. (Part of cipher function, f.)*

32	1	2	3	4	5
4	5	6	7	8	9
8	9	10	11	12	13
12	13	14	15	16	17
16	17	18	19	20	21
20	21	22	23	24	25
24	25	26	27	28	29
28	29	30	31	32	1

Table 9.12(a) *Selection function $S_1(B_1)$. (Part of the cipher function, f.)*

Row no.	Column number															
	0	**1**	**2**	**3**	**4**	**5**	**6**	**7**	**8**	**9**	**10**	**11**	**12**	**13**	**14**	**15**
0	14	4	13	1	2	15	11	8	3	10	6	12	5	9	0	7
1	0	15	7	4	14	2	13	1	10	6	12	11	9	5	3	8
2	4	1	14	8	13	6	2	11	15	12	9	7	3	10	5	0
3	15	12	8	2	4	9	1	7	5	11	3	14	10	0	6	13

Table 9.12(b) *Selection function $S_2(B_2)$. (Part of the cipher function, f.)*

Row no.	Column number															
	0	**1**	**2**	**3**	**4**	**5**	**6**	**7**	**8**	**9**	**10**	**11**	**12**	**13**	**14**	**15**
0	15	1	8	14	6	11	3	4	9	7	2	13	12	0	5	10
1	3	13	4	7	15	2	8	14	12	0	1	10	6	9	11	5
2	0	14	7	11	10	4	13	1	5	8	12	6	9	3	2	15
3	13	8	10	1	3	15	4	2	11	6	7	12	0	5	14	9

Table 9.12(c) *Selection function $S_3(B_3)$. (Part of the cipher function, f.)*

Row no.	Column number															
	0	**1**	**2**	**3**	**4**	**5**	**6**	**7**	**8**	**9**	**10**	**11**	**12**	**13**	**14**	**15**
0	10	0	9	14	6	3	15	5	1	13	12	7	11	4	2	8
1	13	7	0	9	3	4	6	10	2	8	5	14	12	11	15	1
2	13	6	4	9	8	15	3	0	11	1	2	12	5	10	14	7
3	1	10	13	0	6	9	8	7	4	15	14	3	11	5	2	12

Table 9.12(d) *Selection function $S_4(B_4)$. (Part of the cipher function, f.)*

Row no.	Column number															
	0	**1**	**2**	**3**	**4**	**5**	**6**	**7**	**8**	**9**	**10**	**11**	**12**	**13**	**14**	**15**
0	7	13	14	3	0	6	9	10	1	2	8	5	11	12	4	15
1	13	8	11	5	6	15	0	3	4	7	2	12	1	10	14	9
2	10	6	9	0	12	11	7	13	15	1	3	14	5	2	8	4
3	3	15	0	6	10	1	13	8	9	4	5	11	12	7	2	14

Table 9.12(e) *Selection function $S_5(B_5)$. (Part of the cipher function, f.)*

Row no.	Column number															
	0	**1**	**2**	**3**	**4**	**5**	**6**	**7**	**8**	**9**	**10**	**11**	**12**	**13**	**14**	**15**
0	2	12	4	1	7	10	11	6	8	5	3	15	13	0	14	9
1	14	11	2	12	4	7	13	1	5	0	15	10	3	9	8	6
2	4	2	1	11	10	13	7	8	15	9	12	5	6	3	0	14
3	11	8	12	7	1	14	2	13	6	15	0	9	10	4	5	3

Table 9.12(f) *Selection function $S_6(B_6)$. (Part of the cipher function, f.)*

Row no.	Column number															
	0	**1**	**2**	**3**	**4**	**5**	**6**	**7**	**8**	**9**	**10**	**11**	**12**	**13**	**14**	**15**
0	12	1	10	15	9	2	6	8	0	13	3	4	14	7	5	11
1	10	15	4	2	7	12	9	5	6	1	13	14	0	11	3	8
2	9	14	15	5	2	8	12	3	7	0	4	10	1	13	11	6
3	4	3	2	12	9	5	15	0	11	14	1	7	6	0	8	13

Table 9.12(g) *Selection function $S_7(B_7)$. (Part of the cipher function, f.)*

Row no.	Column number															
	0	**1**	**2**	**3**	**4**	**5**	**6**	**7**	**8**	**9**	**10**	**11**	**12**	**13**	**14**	**15**
0	13	11	2	14	15	0	8	13	3	12	9	7	5	10	6	1
1	1	0	11	7	4	9	1	10	14	3	5	12	2	15	8	6
2	7	4	11	13	12	3	7	14	10	15	6	8	0	5	9	2
3	2	11	13	8	1	4	10	7	9	5	0	15	14	2	3	12

Table 9.12(h) *Selection function $S_8(B_8)$. (Part of the cipher function, f.)*

Row no.	Column number															
	0	**1**	**2**	**3**	**4**	**5**	**6**	**7**	**8**	**9**	**10**	**11**	**12**	**13**	**14**	**15**
0	13	2	8	4	6	15	11	1	10	9	3	14	5	0	12	7
1	1	15	13	8	10	3	7	4	12	5	6	11	0	14	9	2
2	7	11	4	1	9	12	14	2	0	6	10	13	15	3	5	8
3	2	1	14	7	4	10	8	13	15	12	9	0	3	5	6	11

The final operation in f transforms the 32 bit data block $l_i = S_1 S_2 S_3 \cdots S_8$ into a 32 bit data block, f_i, using the *permutation function, P* (Table 9.13).

Table 9.13 *Permutation function, P, for DES. (Part of the cipher function, f.)*

16	7	20	21
29	12	28	17
1	15	23	26
5	18	31	10
2	8	24	14
32	27	3	9
19	13	30	6
22	11	4	25

Thus the 16th bit of l becomes the 1st bit of P, the 7th bit becomes the 2nd bit, $\cdots$, and the 25th bit becomes the 32nd (last) bit.

The cipher function can be expressed as:

$$f(R_{i-1}, K_i) = P[S_1(B_1)S_2(B_2)S_3(B_3) \cdots S_8(B_8)] \qquad (9.33(\text{a}))$$

where:

$$B_1 B_2 B_3 \cdots B_8 = E(R_{i-1}) \oplus K_i \qquad (9.33(\text{b}))$$

and i takes values from 1 to 16. Figure 9.24 shows a flow chart for the cipher function, $f(\ )$.

Deciphering of DES ciphertext is achieved by applying the encryption algorithm a second time to the ciphertext data blocks. The same keylet is used in each deciphering round as was used in the corresponding encryption round. Thus $R_{16} \| L_{16}$ (i.e. R_{16} bits followed by L_{16} bits) is the permuted input data block and $L_0 \| R_0$ is the input data block to the final permutation (Figure 9.25).

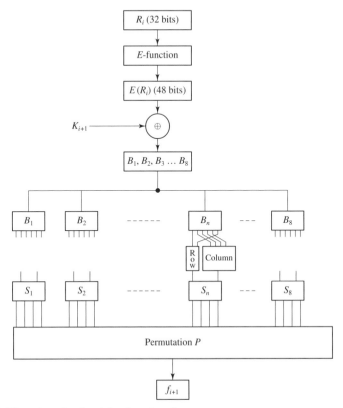

Figure 9.24 *Flow chart for the cipher function, f.*

9.8.8 Public key encryption

Good private key encryption processes such as DES represent extremely strong ciphers making the probability of successful cryptanalytic attack very small. They rely for their security, however, on keeping the key secret and their real vulnerability is therefore to an unauthorised recipient gaining access to the key. This particular weakness is addressed by public (sometimes called public–private) key systems.

The principle of public key encryption was first proposed in 1970 [Ellis] and the first, practical, public key encryption algorithm was developed in 1973 [Cocks]. Neither piece of work was made public, however, until 1997 by which time the possibility of public key

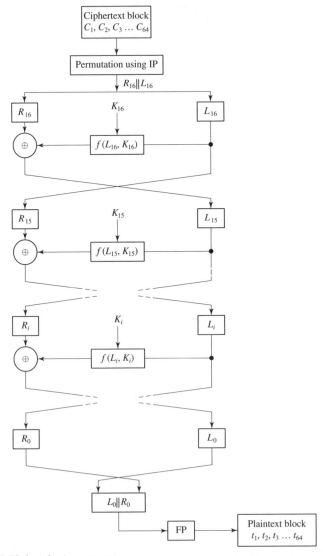

Figure 9.25 *DES deciphering process.*

encryption had been independently proposed [Diffie and Hellman] and a practical algorithm published [Rivest, Shamir and Adelman]. The 1973 algorithm illustrates the public key principle particularly well, however, and so is described below.

1. The intended message recipient chooses two (large) prime numbers, P and Q, which are also prime to $P - 1$ and $Q - 1$ (i.e. P is not a factor of $Q - 1$ and Q is not a factor of $P - 1$).
2. The recipient then transmits the product $N = PQ$ (but not P or Q) to the message sender.
3. The sender's plaintext is represented by a numerical value, T $(0 < T < N)$, and the plaintext is encrypted by transforming T to its ciphertext value, C, using:

$$C = T^N (\text{modulo-}N) \tag{9.34}$$

4. C is transmitted to the recipient.
5. The recipient finds two numbers P' and Q' which satisfy $PP' = 1$ (modulo-$(Q - 1)$) and $QQ' = 1$ (modulo-$(P - 1)$).
6. The ciphertext, C, is then transformed back to the plaintext, T, using either:

$$T = C^{P'} (\text{modulo-}Q) \tag{9.35(a)}$$

or:

$$T = C^{Q'} (\text{modulo-}P) \tag{9.35(b)}$$

If the encrypted message is intercepted by an unauthorised third party then a (potentially) exhaustive search for all factors of N (i.e. all possible P and Q) would need to be undertaken to find the cipher key. If P and Q are chosen to be large numbers (e.g. several hundred digits long) then the exhaustive search for factors will take a prohibitively long time even using the fastest processing resources available.

EXAMPLE 9.5

Show that the (unrealistically small) numbers $P = 3$ and $Q = 5$ satisfy the criteria required by Cocks's encryption algorithm (described above) and use them to encrypt the plaintext $T = 2$. Find the ciphertext and show that both of equations (9.35(a)) and (9.35(b)) successfully decipher the encrypted message.

P and Q are both prime numbers and both $(Q - 1)/P$ and $(P - 1)/Q$ leave non-zero remainders. $N = PQ = 3 \times 5 = 15$ and, therefore, $0 < T < N$. $P = 3$ and $Q = 5$ thus satisfy Cocks's encryption requirements.

Using equation (9.34) the ciphertext is given by:

$$C = 2^{15} (\text{modulo-}15) = 8$$

To decipher the encrypted message requires P' or Q' which are found as follows:

$$3P' = 1 (\text{modulo-}4)$$

i.e.:

$$P' = 3, 7, 11, \cdots$$

and:

$$5Q' = 1(\text{modulo-2})$$

i.e.:

$$Q' = 1, 3, 5, \cdots$$

The ciphertext is decoded with equation (9.35(a)) using any allowed value of P', e.g.:

$$T = 8^3(\text{modulo-5}) = 2$$

or with equation (9.35(b)) using any allowed value of Q', e.g.:

$$T = 8^1(\text{modulo-3}) = 2$$

The RSA (Rivest, Shamir and Adleman) public key cipher is a generalisation of the algorithm described above. The RSA public and private keys are each represented by a number pair ($[N, A]$ and $[N, B]$ respectively). They are calculated, by the prospective message recipient, from two large positive prime numbers (P and Q) as follows:

$$N = PQ$$
$$\phi(N) = (P - 1)(Q - 1)$$

Integers A and B are selected that satisfy:

A not a factor of $\phi(N)$
A not a multiple of $\phi(N)$
$AB = 1 \quad (\text{modulo-}\phi)$

Note that P and Q must be chosen such that N equals or exceeds the number of possible plaintext messages (or characters) to be encrypted.

Plaintext is encrypted using the public key $[N, A]$ by transforming the numerical value, T, of each word or character into cipher text using:

$$C = T^A(\text{modulo-}N) \tag{9.36}$$

Ciphertext is decrypted using the private key $[N, B]$ by transforming the numerical value, C, of each word or character into plaintext using:

$$T = C^B(\text{modulo-}N) \tag{9.37}$$

Secrecy is assured providing the recipient of the message does not reveal P, Q or B and chooses these numbers to be sufficiently large such that a search for the keys by systematic factorisation of N is not practically possible.

Evaluating equations (9.36) and (9.37) can be a significant processing task. The following formula (cast in the form for equation (9.36)), iterated A times, from $i = 1$ to $i = A$, and in which $C_0 = 1$, represents an efficient alternative to direct evaluation:

$$C_i = remainder\left\{\frac{C_{i-1}T}{N}\right\} \tag{9.38}$$

EXAMPLE 9.6

Create public and private keys using the RSA algorithm outlined above and use these to encrypt, character by character, the plaintext message: account. Use the algorithm to de-encrypt the resulting cipher text. Use the International Alphabet No. 5 (IA5) given in Appendix C to convert plaintext characters to plaintext numbers. (Octets, i.e. 8 bit words, will be needed to represent the encoded characters.)

Since the 7 bit words of IA5 provide for 128 characters (0000000 to 1111111) then $N = PQ$ must be greater than 128.

In order to make the example easy to follow the unrealistically small prime numbers $P = 11$, $Q = 13$ are chosen.

$$N = 11 \times 13 = 143$$
$$\phi = 10 \times 12 = 120$$

The factors of $\phi = 120$ are: 1, 2, 3, 4, 5, 6, 8, 10, 12, 15, 20, 24, 30, 40, 60.
The allowed values of A are therefore: 7, 9, 11, 13, 14, 16, 17, 18, 19, 21, $\cdots$.
Trying $A = 7$ an integer value of B must be found that satisfies:

$$remainder\left\{\frac{B7}{120}\right\} = 1, \ \text{i.e. } 7B = 121, 241, 361, 481, 601, 721, \cdots$$

Testing each of these in turn the first allowed (integer) value is $B = 103$ (i.e. $B = 721/7$).
From Appendix C the plaintext characters are represented by the following plaintext numbers:

Character	Binary representation	Decimal representation
a	1100001	97
c	1100011	99
o	1101111	111
u	1110101	117
n	1101110	110
t	1110100	116

Encrypting the first character a:

$$C(97) = remainder\left\{\frac{97^7}{143}\right\}$$

Using the iterative formula to find $remainder\left\{\dfrac{97^7}{143}\right\}$:

$$C_1(97) = remainder\left\{\frac{1 \times 97}{143}\right\} = 97$$

$$C_2(97) = remainder\left\{\frac{97 \times 97}{143}\right\} = 114$$

$$C_3(97) = remainder\left\{\frac{114 \times 97}{143}\right\} = 47$$

$$C_4(97) = remainder\left\{\frac{47 \times 97}{143}\right\} = 126$$

$$C_5(97) = remainder\left\{\frac{126 \times 97}{143}\right\} = 67$$

$$C_6(97) = remainder\left\{\frac{67 \times 97}{143}\right\} = 64$$

$$C_7(97) = remainder\left\{\frac{64 \times 97}{143}\right\} = 59$$

i.e.:

$$C(97) = remainder\left\{\frac{97^7}{143}\right\} = 59$$

Encrypting the other characters of the plaintext allows the ciphertext to be added to the table as shown below.

Plaintext charater	IA5 plaintext binary value	IA5 plaintext decimal value	Ciphertext decimal value	Ciphertext binary value
a	1100001	97	59	00111011
c	1100011	99	44	00101100
o	1101111	111	45	00101101
u	1110101	117	39	00100111
n	1101110	110	33	00100001
t	1110100	116	129	10000001

The complete binary plaintext of the message (inserting spaces between characters for clarity) is therefore:

1100001 1100011 1100011 1101111 1110101 1101110 1110100

and the complete binary ciphertext of the message is:

00111011 00101100 00101100 00101101 00100111 00100001 10000001

To decipher the first number of the ciphertext (59) equation (9.37) is used, i.e.:

$$T = remainder\left\{\frac{59^{103}}{143}\right\}$$

which gives $T = 97$.

Clearly the remainder in this case cannot easily be calculated manually and so computer code must be used. Example code is given below in the form of a MATLAB program extract that provides the required value of T in the array element $T(104)$.

```
N=143;
C=59;
B=103;
T(1)=1.0;
For n=2: B+1
T(n)=rem(T(n-1)*C, N);
end
```

The other characters are deciphered in the same way to give the original plaintext numbers.

Example 9.6, which illustrates the RSA algorithm, is unrealistic not only in the size of P and Q but also in that the encryption algorithm is implemented in isolation. A given character, therefore, is always encoded as the same octet. The patterns (single and joint letter statistics) found in 'natural' types of plaintext (such as conventional written languages) can thus be exploited to 'break' the code. Character patterns can be weakened, and cryptanalysis made correspondingly more difficult, by compressing (i.e. source coding) the plaintext prior to encryption and/or by preceding the RSA with another type of encoding algorithm (e.g. OAEP, see below) that disrupts the one-to-one mapping between plaintext and ciphertext character strings.

9.8.9 Hash functions and OAEP

A hash function takes a message or data block and produces a fixed length hash (or 'digest') of the message that is usually much shorter than the message itself. A one-way hash function is constructed such that it is virtually impossible to find the original message from the hash. The secure hash algorithm-1 (SHA-1), for example, which is popular in encryption applications, produces a 160 bit (20 byte) hash from an original message of up to 2^{64} (i.e. a little less than 18.5 quintillion) bits long. Finding two messages with the same SHA-1 value or obtaining a message from its SHA-1 value are both computationally infeasible. (SHA-1 is a more secure version of the original SHA algorithm.)

The optimal asymmetric encryption padding (OAEP) algorithm [Bellare and Rogaway, Johnson and Matyas] transforms a plaintext data block (T) to an encrypted message data block (EM) using a random string (r), and hash functions G and H. The term 'encrypted message' (which is used in published descriptions of OAEP) is adopted here to distinguish it from the *ciphertext* that is produced when OAEP is combined with RSA in RSA-OAEP.

Basic OAEP generates the encrypted message, EM, using the algorithm:

$$EM = \{T \oplus G(r)\} \| \{r \oplus H(T \oplus G(r))\} \tag{9.39}$$

where $\|$ denotes concatenation (i.e. the bits generated by the first curly bracket followed by the bits generated by the second curly bracket). The ciphertext produced by the OAEP encoding algorithm is determined, uniquely, by the plaintext, T, and the random string, r. Figure 9.26 shows a flow diagram for basic OAEP. An OAEP decoding algorithm implements the inverse transformation mapping EM back to T (without requiring knowledge of r).

The most important quality of OAEP is that it introduces randomness to the encryption process so that different instances of identical plaintext are transformed to different encrypted messages.

Figures 9.27 and 9.28 show the OAEP encoding and decoding processes, respectively, as used in the RSA Laboratories' implementation of their RSAES-OAEP encryption scheme. The mask generation function (MGF) in these figures is an algorithm that includes a hash operation as one of its constituent parts. The padding (P) also incorporates the hash value of some encoding parameters that can be used to verify the success of the decryption process.

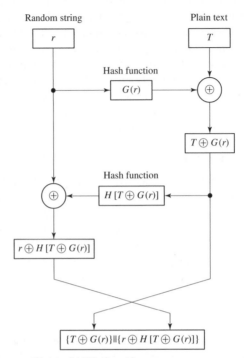

Figure 9.26 *Flow diagram of basic OAEP algorithm.*

9.8.10 Hybrid public key/private key encryption and PGP

The advantage of public key, over private key, cryptography with respect to the key distribution problem is obvious. Public key systems have the disadvantage, however, that the encryption process is computationally more demanding and therefore takes longer (for a given processing speed). The speed advantage of private key encryption and the key distribution advantage of public key encryption can be combined by using the former to produce ciphertext from a randomly generated private key (called a one-time session key), and using the latter to encrypt the private key. Both the privately encrypted ciphertext and publicly encrypted private key can be transmitted to the recipient who first decodes the private key (using the public key deciphering process) and then uses the private key to decode the ciphertext. Pretty good privacy (PGP) is an example of a practical encryption algorithm that uses this approach.

PGP creates a random session key that is used to encrypt the plaintext using the international data encryption algorithm (IDEA). The session key is then encrypted using RSA. (IDEA is similar to DES and uses a 128 bit key to encrypt 64 bit blocks of plaintext. The plaintext is divided into four 16 bit sub-blocks that are then processed iteratively in eight rounds. During each round the sub-blocks are combined, using modulo-2 summation, addition and multiplication, with each other and with six 16 bit keylets derived from the key. The middle two sub-blocks are interchanged between each round.)

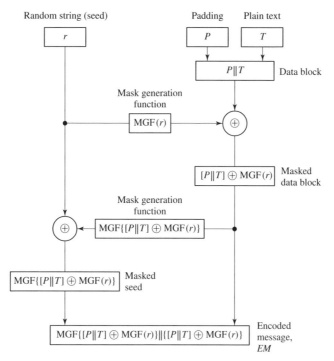

Figure 9.27 *Flow diagram of basic OAEP encoding process as used in RSAES-OAEP.*

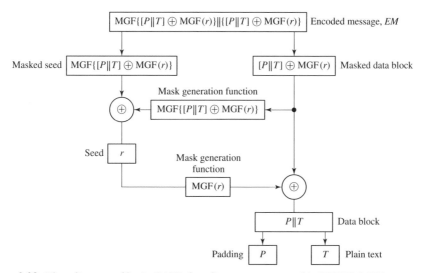

Figure 9.28 *Flow diagram of basic OAEP decoding process as used in RSAES-OAEP.*

9.9 Authentication

Authentication is the process of verifying the identity of a message's originator. It might be thought that possession of a secret key (known only to the originator and intended recipient) would be sufficient to authenticate a message. This is not the case, however, since an old encrypted message might have been intercepted and recorded by a third party, for example, and then simply retransmitted to the intended recipient at some later (and inappropriate) time. An authentication scheme that incorporates some type of challenge and response prevents this type of personation and is generally much more secure.

Public key encryption can be used to provide a challenge/response authentication scheme as follows.

The potential recipient of a message generates a 'challenge' consisting of a random cryptogram (i.e. a random bit stream using the characters of the ciphertext). The potential message originator deciphers the random cryptogram using his/her private key. The originator then responds to the challenge by transmitting back to the intended message recipient the 'deciphered' cyptogram (which is meaningless but could only have been generated by the possessor of the originator's private key) along with his/her (private key encrypted) message. The message recipient then 'encrypts' the deciphered cryptogram with the message originator's public key. Providing the resulting bit stream is identical to the original random cryptogram then the message originator must be authentic and 'current' (on the grounds that he/she possesses the genuine originator's private key and has responded correctly to the recent random cryptogram challenge).

9.10 Integrity

Hash functions can be used to establish message integrity. The hash function is designed such that the probability of two different messages producing the same hash value is extremely small. If a message hash is generated at source and sent (in encrypted form) with the message itself then the intended recipient can recalculate the hash from the received message, decipher the hash that arrived with the message, and compare the two. If they are the same it is very improbable that the message has been changed, i.e. it gives good confidence in the message's integrity. This method of integrity confirmation is especially advantageous if the integrity of the message is important but secrecy is not, since in this case the message itself can remain unencrypted reducing processing requirements considerably.

9.11 Digital signatures

Digital signatures fulfil the same role as convention signatures, i.e. they allow a document (and any assurances or undertakings it contains) to be attributed to an individual. It provides both authentication and integrity. Public key encryption can be used as a signing device as follows.

The originator of a document 'deciphers' the document's hash using his/her private key that then becomes the signature. The document plus the signature are then encrypted with the intended recipient's public key and transmitted to the recipient. The recipient deciphers the message and hash with his/her private key, verifying the signature by applying the originator's public key and comparing the result to the (locally generated) hash of the deciphered message. Providing there is an exact match the document can be safely attributed to its originator.

9.12 Summary

The information content of a symbol is defined as the negative logarithm of the symbol's selection probability. If 2 is chosen as the base of the logarithm then the unit of information is the bit. The entropy of an information source is the average information per selected, or transmitted, symbol. The entropy of a binary source with statistically independent, equiprobable, symbols is 1 bit/binary digit. Statistical dependence of symbols, and unequal selection probabilities, both reduce the entropy of a source below its maximum value. Redundancy is the difference between a source's maximum possible entropy and its actual entropy. Noise at the receiver makes the decision process uncertain and thus reduces the information content of the received symbol stream. The information content of the symbols after detection is called the effective entropy and is equal to the entropy on transmission minus the equivocation introduced by the noise. In this sense noise may be interpreted as negative information added to the transmitted symbol stream.

Source coding removes redundancy, making transmission and/or storage of information more efficient. Source codes must, in general, be uniquely decodable. Source coding of discrete information sources is, normally, also loss-less, i.e. the coding algorithm, in the absence of symbol errors, is precisely reversible. Codes may, in addition, be instantaneously decodable, i.e. contain no codewords which form a prefix of any other codewords. The Huffman code, used for Group 3 fax transmission, is an example of a unique, loss-less, instantaneous code.

Vocoders use models of the physical mechanism of speech production to transmit intelligible, but low bit rate, speech signals. In essence it is the model parameters, in the form of adaptive filter coefficients, which are transmitted instead of the voice signals themselves. The adaptive filters are excited at the receiver by a pulse generator, of the correct frequency (pitch), for voiced sounds or white noise for unvoiced sounds. These vocoders can also be extended to higher complexity and rate, e.g. CELP, to achieve an even higher quality speech coder design with low transmission rates.

Sub-band coding splits the transmitted signal into two or more frequency sub-bands using hardware or (DFT) software. This allows the quantisation accuracy for each sub-band to reflect the importance of that band as measured by the signal power it contains. Other transform coding techniques can be used to encode speech and images. Transform coding is particularly significant in 2-dimensional image transmission as it minimises redundancy which takes the form of correlation in intensity or luminance values between closely spaced pixels. Vocoder and transform coding techniques are lossy in that there is an inevitable (but often small) loss in the information content of the encoded data. Perceptual coding techniques are widely applied in high quality MPEG speech coders.

String coding algorithms are particularly efficient at compressing character-based data but coding times are long since commonly occurring strings must be initially searched for in the data. String coding is used for compression of computer files in order to speed up transport between machines and save disk space.

Encryption is concerned with message secrecy, authentication and integrity. Perfect secrecy is achieved when the ciphertext, in the absence of the cipher's key, does not alter the entropy (uncertainty) of the plaintext message and requires that the entropy of the key is at least as large as that of the plaintext. A one-time pad is an example of a cipher with perfect secrecy.

Symmetric, or private, key ciphers employ a single key for encryption and deciphering and therefore rely on key secrecy for their security properties. The classical problem of private key systems is key distribution. Private key block ciphers operate on fixed size fragments of plaintext and produce corresponding fragments of ciphertext. They often use alternating rounds of substitution and permutation to confuse the relationship between the statistics of plaintext character strings and the statistics of ciphertext characters. They also complicate the relationship between ciphertext and cipher key structure making the use of ciphertext analysis to restrict the likely key-space a difficult task. DES is an example of a practical private key block cipher

Stream ciphers operate on plaintext character by character. One-time pads using a pseudo-random key stream generator are stream ciphers that have the advantage of not suffering from error propagation. They are not strong ciphers, however, and are vulnerable to known plaintext attack. They also require key-stream synchronisation. Stream ciphers using ciphertext feedback in the key-stream generator suffer from error propagation but are self synchronising.

Asymmetric, or public, key ciphers allow the recipient of a message to take responsibility for secrecy, thus circumventing the key distribution problem. They typically rely on the impracticality of finding the decryption key by systematically searching for all prime factors of a large number. RSA is an example of a practical public key cipher.

RSA is a deterministic algorithm, identical strings of plaintext being encrypted as identical strings of ciphertext. Patterns in plaintext character strings are therefore a source of weakness in a pure RSA scheme. This can be addressed by aggressively source coding the plaintext prior to encryption (thus removing the plaintext character patterns) and/or by explicitly introducing randomness into the encryption process. OAEP is an example of a coding process that uses a random number stream or seed to introduces randomness into an encrypted message but which can be decoded without knowledge of this random number. Hash functions, which produce an effectively unique fixed length digest of a message, form an important component of OAEP, and similar, algorithms. An encryption scheme that applies OAEP followed by RSA has the theoretical encryption strength of RSA combined with the randomness properties of OAEP. Public key encryption generally requires greater processing resources than private key systems.

PGP combines the processing efficient secrecy properties of private key systems with the key distribution advantage of public key systems. It uses private key encryption for the message but public key encryption for the private encryption key. In addition to achieving secrecy public key encryption can be used to authenticate the source of a message and (with the assistance of hash operations) check a message's integrity.

9.13 Problems

9.1 (a) Consider a source having an $M = 3$ symbol alphabet where $P(x_1) = \frac{1}{2}$, $P(x_2) = P(x_3) = \frac{1}{4}$ and symbols are statistically independent. Calculate the information conveyed by the receipt of the symbol x_1. Repeat for x_2 and x_3. [$I_{x_1} = 1$ bit, $I_{x_2} = I_{x_3} = 2$ bits]

(b) Consider a source whose statistically independent symbols consist of all possible binary sequences of length k. Assume all symbols are equiprobable. How much information is conveyed on receipt of any symbol? [k bits]

(c) Determine the information conveyed by the specific message $x_1 x_3 x_2 x_1$ when it emanates from each of the following, statistically independent, symbol sources: (i) $M = 4$; $P(x_1) = \frac{1}{2}$, $P(x_2) = \frac{1}{4}$, $P(x_3) = P(x_4) = 1/8$ [7 bits]; (ii) $M = 4$; $P(x_1) = P(x_2) = P(x_3) = P(x_4) = \frac{1}{4}$. [8 bits]

9.2 (a) Calculate the entropy of the source in Problem 9.1(a). [$1\frac{1}{2}$ bit/symbol]

(b) Calculate the entropy of the sources in Problem 9.1(c). [$1\frac{3}{4}$ bit/symbol, 2 bit/symbol]

(c) What is the maximum entropy of an eight-symbol source and under what conditions is this situation achieved? What are the entropy and redundancy if $P(x_1) = \frac{1}{2}$, $P(x_i) = 1/8$ for $i = 2, 3, 4$ and $P(x_i) = 1/32$ for $i = 5, 6, 7, 8$? [3 bit/symbol, 2.25 bit/symbol]

9.3 Find the entropy, redundancy and code efficiency of a three-symbol source A, B, C, if the following statistical dependence exists between symbols. There is a 20% chance of each symbol being succeeded by the next symbol in the *cyclical* sequence $A\ B\ C$ and a 30% chance of each symbol being succeeded by the previous symbol in this sequence. [1.485 bit/symbol; 0.1 bit/symbol; 93.7%]

9.4 Show that the number of redundant symbols per bit of information transmitted by an M-symbol source with code efficiency η_{code} is given by $(1 - \eta_{code})/(\eta_{code} \log_2 M)$ symbol/bit.

9.5 Estimate the maximum information content of a black and white television picture with 625 lines and an aspect ratio of 4/3. Assume that 10 brightness values can be distinguished and that the picture resolution is the same along a horizontal line as along a vertical line. What maximum data rate does a picture rate of 25 picture/s correspond to and what, approximately, must be the bandwidth of the (uncoded and unmodulated) video signal if it is transmitted using binary symbols? (If necessary you should consult Chapter 16 to obtain TV scanning format information.) [4 bit/symbol and 2.112 Mbit/picture; 52.8 Mbit/s; 26.4 MHz]

9.6 Calculate the loss in information due to noise, per transmitted digit, if a random binary signal is transmitted through a channel, which adds zero mean Gaussian noise, with an average signal-to-noise ratio of: (a) 0 dB; (b) 5 dB; (c) 10 dB. [0.6311; 0.2307; 0.0094 bit/binit]

9.7 An information source contains 100 different, statistically independent, equiprobable symbols. Find the maximum code efficiency, if, for transmission, all the symbols are represented by binary codewords of equal length. [7 bit words and 94.9%]

9.8 (a) Apply Huffman's algorithm to deduce an optimal code for transmitting the source defined in Problem 9.1(c)(i) over a binary channel. Is your code unique?

(b) Define the efficiency of a code and determine the efficiency of the code devised in part (a).

(c) Construct another code for the source of part (a) and assign equal length binary words irrespective of the occurrence probability of the symbols. Calculate the efficiency of this source. [(a) 0, 10, 110, 111, Yes; (b) 100%; (c) 87.5%]

9.9 Design a Lempel–Ziv coding scheme with a history buffer of 16 8 bit characters and a maximum match length of four characters. Use your coding scheme to encode the character sequence '1415927' where the next character to be transmitted is to the right. You should assume the history buffer contains '1414213617920408' with the last character to have been transmitted to the left. Calculate the compression factor obtained by your code on this character sequence. [0.536]

9.10 Decrypt the ciphertext string: *uggc://jjj.ongu.np.hx/ryrp-rat/*. What does this imply about good cipher design?

9.11 A product cipher takes 16 bit blocks of (binary) plaintext and employs four (4-input, 4-output) S-boxes followed by one (16-input, 16-output) P-box in each round of its encryption algorithm. How many possible keys are there in eight rounds for a binary plaintext/ciphertext alphabet?

9.12 Substitution and permutation operations are applied alternately in typical multi-round private block key ciphers. Explain why it would not be wise to implement all the substitutions first followed by all the permutations.

9.13 Use RSA to encrypt the plaintext message number 13. Choose the smallest possible values of P and Q. Decipher the resulting ciphertext to check your answer.

9.14 The brute force method of attacking a public key encryption scheme is to systematically try all possible decryption keys after finding all possible factors of the chosen prime product N. If order $\exp[\ln(N)\ln(\ln N)]^{0.5}$ processing operations are required by even the most efficient algorithms to find (all) the required factors, estimate how long it would take to break the RSA algorithm using the currently recommended minimum length for N of 1024 bits.

9.15 Explain the principles of public key encryption. How can such encryption schemes be used to authenticate a message and check its integrity?

CHAPTER 10

Error control coding

10.1 Introduction

The fundamental resources at the disposal of a communications engineer are signal power, time and bandwidth. For a given communications environment (summarised in this context by an effective noise power spectral density) these three resources can be traded against each other. The basis on which the trade-offs are made will depend on the premium attached to each resource in a given situation. A general objective, however, is often to achieve maximum data transfer, in a minimum bandwidth, *while maintaining an acceptable quality of transmission*. The quality of transmission, in the context of digital communications, is essentially concerned with the probability of bit error, P_b, at the receiver. (There are other factors which determine transmission quality, in its widest sense, of course, but focusing on P_b makes the discussion, and more especially the analysis, tractable.)

The Shannon–Hartley law (see section 11.4.1) for the capacity of a communications channel demonstrates two things. Firstly it shows (quantitatively) how bandwidth and signal power may be traded in an ideal system, and secondly it gives a theoretical limit for the transmission rate of (reliable, i.e. error-free) data from a transmitter of given power, over a channel with a given bandwidth, operating in a given noise environment. In order to realise this theoretical limit, however, an appropriate coding scheme (which the Shannon–Hartley law assures us exists) must be found. (It should, perhaps, be noted at this point that there is one more quantity which must be traded in return for the advantage which such a coding scheme confers, i.e. time delay which results from the coding process.)

In practice, the objective of the design engineer is to realise the required data rate (often determined by the service being provided) within the bandwidth constraint of the available channel and the power constraint of the particular application. (In a mobile radio application, for example, bandwidth may be determined by channel allocations or frequency coordination considerations, and maximum radiated power may be determined by safety considerations or transceiver battery technology.) Furthermore this data rate must be achieved with an *acceptable* BER and time delay. If an essentially uncoded PCM transmission cannot achieve the required BER within these constraints then the application of error control coding may be able to help, providing the constraints are not such as to violate the Shannon–Hartley law.

Error control coding (also referred to as channel coding) is used to detect, and often correct, symbols which are received in error. A taxonomy of coding for error control, spectral shaping, source coding, etc., is provided in [Burr]. Error *detection* can be used as the initial step of an error *correction* technique by, for example, triggering a receiving terminal to generate an automatic repeat request (ARQ) signal which is carried, by the return path of a duplex link, to the originating terminal. A successful retransmission of the affected data results in the error being corrected. The operation of ARQ techniques is described more fully later in section 18.3.2. If ARQ techniques are inconvenient, as is the case, for example, when the propagation delay of the transmission medium is large, then forward error correction coding (FECC) may be appropriate [Blahut, 1983, Clark and Cain, MacWilliams and Sloane]. FECC incorporates extra information (i.e. redundancy) into the transmitted data which can then be used not only to detect errors but also to correct them without the need for any retransmissions.

Table 10.1 *A taxonomy of error control codes.*

ARQ				FECC				
Stop & wait	Continuous ARQ (pipelining)			Block codes				Convolutional codes
	Go-back-N	Selective repeat	Others (non-linear)		Group (linear)			
				Others (non-cyclic)		Polynomially generated (cyclic)		
					Golay		BCH	
						Reed-Solomon	Binary BCH	
							Hamming ($e = 1$)	$e > 1$

This chapter begins with a general discussion of error rate control, in its widest sense, as may be applied in digital communications systems. Five particular error control methods are identified and briefly described. It is one of these methods, namely FECC, which is then treated in detail during the remainder of the chapter. Some typical applications of FECC are outlined and the threshold phenomenon is highlighted. The Hamming distance between a pair of codewords and the weight of a codeword are defined. The discussion of FEC codes, that follows, is structured loosely around the taxonomy of codes shown in Table 10.1 starting with a description of block codes and including special mention of linear group codes, cyclic codes, the Golay code, BCH codes, Reed–Solomon codes and Hamming codes. The Hamming bound on the performance of a block code is derived and strategies for nearest neighbour, or maximum likelihood, decoding are discussed. Convolution coding is treated towards the end of the chapter. Tree, trellis and state transition diagrams are used to illustrate the encoding process. The significance of constraint length and decoding window length are discussed and Viterbi decoding is illustrated using a trellis diagram. Decoding is accomplished by finding the most likely path through the trellis and relating this back to the transmitted data sequence.

Coding concepts are presented, here, with an essentially non-mathematical treatment, in the context of specific examples of group, cyclic and convolutional codes. The chapter concludes with a brief discussion of practical coders.

10.1.1 Error rate control concepts

The normal measure of error performance is bit error rate (BER) or the probability of bit error (P_b). P_b is simply the probability of any given transmitted binary digit being in error. The bit error rate is, strictly, the average rate at which errors occur and is given by the product $P_b R_b$, where R_b is the bit transmission rate in the channel. Typical long term P_b for linear PCM systems is 10^{-7} while, for companded PCM, it is 10^{-5} and for ADPCM (Chapter 5) it is 10^{-4}. If the error rate of a particular system is too large then what can be done to make it smaller? The first and most obvious solution is to increase transmitter power, but this may not always be desirable; for example, in portable systems where the required extra battery weight may be unacceptable.

A second possible solution, which is especially effective against burst errors caused by signal fading, is to use diversity. There are three main types of diversity: space diversity, frequency diversity and time diversity. All these schemes incorporate redundancy in that data is, effectively, transmitted twice, i.e. via two paths, at two frequencies or at two different times. In space diversity two or more antennas are used which are sited sufficiently far apart for fading at their outputs to be decorrelated. Frequency diversity employs two different frequencies to transmit the same information. (Frequency diversity can be in-band or out-band depending upon the frequency spacing between the carriers.) In time diversity systems the same message is transmitted more than once at different times.

A third possible solution to the problem of unacceptable BER is to introduce full duplex transmission, implying simultaneous two-way transmission. Here when a transmitter sends information to a receiver, the information is 'echoed' back to the transmitter on a separate feedback channel. Information echoed back which contains errors can then be retransmitted. This technique requires twice the bandwidth of single direction (simplex) transmission, however, which may be unacceptable in terms of spectrum utilisation.

A fourth method for coping with poor BER is ARQ. Here a simple error *detecting* code is used and, if an error is detected in a given data block, then a request is sent via a feedback channel to retransmit that block. There are two major ARQ techniques. These are *stop and wait*, in which each block of data is positively, or negatively, acknowledged by the receiving terminal as being error free before the next data block is transmitted, and *continuous* ARQ, in which blocks of data continue to be transmitted without waiting for each previous block to be acknowledged. (In stop and wait ARQ data blocks are *timed out* if neither a positive nor negative acknowledgement is received within a predetermined time window. After timing out the appropriate data block is retransmitted in the same way as if it had been negatively acknowledged.) Continuous ARQ can, in turn, be divided into two variants. In the go-back-N version, data blocks carry a sequence or reference number, N. Each acknowledgment signal contains the reference number of a data block and effectively acknowledges all data blocks up to $N - 1$. When a negative acknowledgement is received (or a data block timed out) all data blocks starting from the reference number in the last acknowledgement signal are retransmitted. In the *selective repeat* version only those data blocks explicitly negatively acknowledged (or timed out) are retransmitted (necessitating data block reordering buffers at the receiver). Go-back-N ARQ has a well defined storage requirement whilst selective repeat ARQ, although very efficient, has a less well defined, and potentially much lager, storage requirement, especially when deployed on high speed links. ARQ is very effective, e.g. in facsimile transmission, Chapter 9. ARQ go-back-N techniques are described more comprehensively later in Chapter 18. On long links with fast transmission rates, however, such as is typical in satellite communications, ARQ can be very difficult to implement.

The fifth technique for coping with high BER is to employ FECC. In common with three of the other four techniques FECC introduces redundancy, this time with data check bits interleaved with the information traffic bits. It relies on the number of errors in a long block of data being close to the statistical average and, being a forward technique, requires no return channel. The widespread adoption of FECC was delayed, historically, because of its complexity and high cost of implementation relative to the other possible solutions. Complexity is now less of a problem following the proliferation of VLSI custom coder/decoder chips.

FECC exploits the difference between the transmission rate or information bit rate R_b and the channel capacity R_{max} as given by the Shannon–Hartley law (see equation (11.38)). P_b can be reduced, at the expense of increasing the transmission delay [Schwartz, 1987], by using FECC with a sufficiently long block or constraint length. The increased transmission delay arises due to the need to assemble the data blocks to be transmitted and the time spent in examining received data blocks to correct errors. The

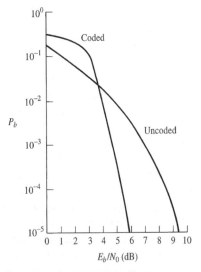

Figure 10.1 *The threshold phenomenon in FECC systems.*

benefits of error control, however, usually outweigh the inherent FECC processor delay disadvantages.

10.1.2 Threshold phenomenon

Figure 10.1 illustrates the error rate for an uncoded system in which P_b increases gradually as SNR decreases, as shown previously in Figure 6.3. Figure 10.1 is plotted as the ratio of bit energy to noise power spectral density (E_b/N_0), which is defined later in Chapter 11. With FECC the P_b versus E_b/N_0 curve is steeper. If the SNR is above a certain value, which here corresponds to an E_b/N_0 of around 6 dB, the error rate will be virtually zero. Below this value system performance degrades rapidly until the coded system is actually poorer than the corresponding uncoded system. (The reason for this is that there is a region of low E_b/N_0 where, in attempting to correct errors, the decoder approximately doubles the number of errors in a decoded codeword.) This behaviour is analogous to the threshold phenomenon in wideband frequency modulation. A coding gain can be defined, for a given P_b, by moving horizontally in Figure 10.1 from the uncoded to the coded curve. The value of the coding gain in dB is relatively constant for $P_b \leq 10^{-5}$, and is dependent on the precise details of the FECC system deployed.

10.1.3 Applications for error control

Compact disc players provide a growing application area for FECC. In CD applications the powerful Reed–Solomon code is used since it works at a symbol level, rather than at a bit level, and is very effective against burst errors, particularly when combined with interleaving to randomise the bursts. The Reed–Solomon code is also used in computers for data storage and retrieval. Cosmic particles create, on average, one error every two to three days in a 4 Mbyte memory, although small geometry devices are helping to reduce this probability. Digital audio and video systems are also areas in which FECC is applied. Error control coding, generally, is applied widely in control and communications systems for aerospace applications, in mobile (GSM) cellular telephony and for enhancing security in banking and barcode readers.

10.2 Hamming distance and codeword weight

Before embarking on a detailed discussion of code performance the following definitions are required. The Hamming distance between two codewords is defined as the number of places, bits or digits in which they differ. This distance is important since it determines how easy it is to change or alter one valid codeword into another. The weight of a binary codeword is defined as the number of ones which it contains.

EXAMPLE 10.1

Calculate the Hamming distance between the two codewords 11100 and 11011 and find the minimum codeword weight.

The two codewords 11100 and 11011 have a Hamming distance of 3 corresponding to the differences in the third-, fourth- and fifth-digit positions. Thus with three appropriately positioned errors in these locations the codeword 11100 could be altered to 11011.

In this example, 11011 has a weight of 4 due to the four ones and 11100 has a weight of 3. The minimum weight is thus 3. Hamming distance and weight will be used later to bound the error correcting performance of codewords.

10.3 *(n, k)* **block codes**

Figure 10.2 illustrates a block coder with k information digits going into the coder and n digits coming out after the encoding operation. The n-digit codeword is thus made up of k information digits and $(n-k)$ redundant *parity check* digits. The rate, or efficiency, for this code (R) is k/n, representing the ratio of information digits to the total number of digits in the codeword. Rate is normally in the range ½ to unity. (Unlike source coding in which data is compressed, here redundancy is deliberately added, to achieve error detection.) This is an example of a *systematic* code in that the information digits are explicitly transmitted together with the parity check digits, Figure 10.2. In a non-systematic code the n-digit codeword may not contain any of the information digits explicitly. There are two definitions of systematic codes in the literature. The stricter of the two definitions assumes that, for the code to be systematic, the k information digits must be transmitted contiguously as a block, with the parity check digits making up the codeword as another contiguous block. The less strict of the two definitions merely stipulates that the information digits must be included in the codeword but not necessarily in a contiguous block. The latter definition is the one which is adopted here.

10.3.1 Single parity check code

The example in Figure 10.3 will be familiar to many as an option in ASCII coded data transmission (see Chapter 8). Consider the data sequence 1101000, to which a single parity check digit (P) is added. For even parity in this example sequence, P will be 1. For odd parity, P will be 0. Since the seven information digits contain three ones, another 1 must be added giving an even number of ones to achieve even parity. Alternatively if a zero is added, ensuring an odd number of ones, i.e. odd parity, the transmission of the all-zero codeword is avoided. Even parity is, however, much more common.

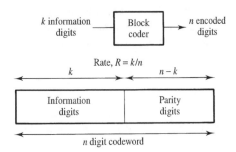

Figure 10.2 *(n, k) systematic block code.*

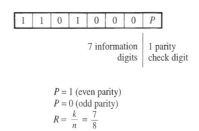

| 1 | 1 | 0 | 1 | 0 | 0 | 0 | 0 | P |

7 information   1 parity
digits   check digit

$P = 1$ (even parity)
$P = 0$ (odd parity)
$R = \dfrac{k}{n} = \dfrac{7}{8}$

Figure 10.3 *Example of a single parity check digit codeword.*

Here the rate, $R = k/n$, is seven-eighths which represents a very low level of redundancy. This scheme can only identify an odd number of errors because an even number of errors will not violate the chosen parity rule. Single error detection, as illustrated in Figure 10.4, is often used to extend a 7 bit word, with a checksum bit, into an 8 bit codeword. Another example, used in libraries, is the 10-digit ISBN codeword, Figure 10.5. This uses a modulo 11, weighted, checksum in which the weightings are 10 for the first digit, 9 for the second digit, etc., down to 2 for the ninth digit (and 1 for the checksum). The weighting can be applied either left to right or vice versa and a checksum digit of 10 is represented by the symbol C.

Single parity checks are also used on rows and columns of simple 2-dimensional data arrays, Figure 10.6. Single errors in the array will be detected *and located* via the corresponding row and column parity bits. Such errors can therefore be corrected. Double errors can be detected but not necessarily corrected as several error patterns can produce the same parity violations. When the data is an array of ASCII characters the row and column check words can also be sent as ASCII characters.

In the English language there is a high level of redundancy. This is why spelling mistakes can be corrected and abbreviations expanded. There is, in fact, an approximate correspondence between the words of a language and codewords as being discussed here,

Example: Even parity

Tx | 1 | 1 | 0 | 1 | 0 | 0 | 0 | 1 | ✓

Rx | 1 | 1 | 0 | 1 | 1 | 0 | 0 | 1 | ✗
• Error

Detects odd number of errors

Figure 10.4 *Block code with single parity check error detecting capability.*

Checksum, $C = 11 - \sum\limits_{i=1}^{9}(11 - i)k_i \ (\text{mod-}11)$

i	1	2	3	4	5	6	7	8	9	C
ISBN	0	1	9	8	5	3	8	0	4	9

Figure 10.5 *ISBN codeword and checksum calculated to satisfy $\sum_{i=1}^{10}(11 - i)k_i = 0 \ (mod\text{-}11)$.*

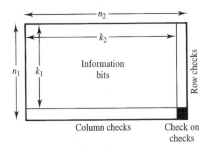

Figure 10.6 *Two-dimensional row–column array code.*

although in language contextual information goes beyond isolated words whilst in a block code each codeword is decoded in isolation.

Modulo-$(2^n - 1)$ checksums are in widespread use for performing error detection on byte-serial network connections. They are usually computed by software during the data block (or packet) construction. One such error detection code is the Internet checksum for protocol messages [Comer], Chapter 18. If a message of length w (16 bit) bytes, $m_{w-1}, \cdots, m_0$, is to be checksummed, the 1 byte checksum is just the complement of:

$$\sum_{i=0}^{w-1} m_i \ (\text{mod-65535})$$

The main operation required for this is summation modulo-65535 (i.e. summation using one's complement word addition). The checksum is included with the transmitted message allowing a recipient to check for transmission errors by performing a similar summation over the received data. If this sum is not zero then a channel error has occurred. Another error detection code is the ISO 2 byte checksum [Fletcher]. The checksum is again included within the transmitted message and a recipient can perform a summation over the received data to confirm that both checksum bytes are zero; if not, a channel error has occurred.

EXAMPLE 10.2

Figure 10.7 illustrates a seven-digit codeword with four information digits (I_1 to I_4) and three parity check digits (P_1 to P_3), commonly referred to as a (7,4) block code. The circles indicate how the information bits contribute to the calculation of each of the parity check bits. Assuming even parity, show the realisation of this encoder using 3-input modulo-2 adders. Calculate the individual parity check bits and encoding of P_1, P_2 and P_3 for the information digits 1011.

Figure 10.7 shows P_1 represented by the modulo-2 sum of I_1, I_3 and I_4. P_2 is the sum of I_1, I_2 and I_4, etc. (Modulo-2 arithmetic was used previously in Table 8.1.) The parity check digits are generated by the circuit in Figure 10.8. For the data sequence 1011, the 3-input modulo-2 adders count the total number of ones which are present at the inputs, and output the least significant bit as the binary coded sum. Thus $P_1 = 1$, $P_2 = 0$ and $P_3 = 0$, giving a coder output of 1011100.

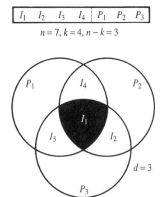

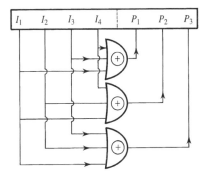

Figure 10.7 *Representation of relationship between parity check and data bits.*

Figure 10.8 *(7,4) block code hardware generation of three parity check digits.*

Figure 10.9 shows how parity check equations for P_1, P_2 and P_3 in the above example may be written using $\oplus$ to represent the modulo-2 or exclusive-or arithmetic operation. Figure 10.9 also shows how these equations can be reduced to matrix form in a *parity check* matrix **H**. The coefficients of the information digits I_1, I_2, I_3 and I_4 are to the left of the dotted partition in the parity check matrix. The top row of the matrix contains the information about parity check P_1, the second row about parity check P_2 and the third row about parity check P_3.

$$P_1 = 1 \times I_1 \oplus 0 \times I_2 \oplus 1 \times I_3 \oplus 1 \times I_4$$
$$P_2 = 1 \times I_1 \oplus 1 \times I_2 \oplus 0 \times I_3 \oplus 1 \times I_4$$
$$P_3 = 1 \times I_1 \oplus 1 \times I_2 \oplus 1 \times I_3 \oplus 0 \times I_4$$

$$\mathbf{H} = \begin{bmatrix} 1 & 0 & 1 & 1 & \vdots & 1 & 0 & 0 \\ 1 & 1 & 0 & 1 & \vdots & 0 & 1 & 0 \\ 1 & 1 & 1 & 0 & \vdots & 0 & 0 & 1 \end{bmatrix}$$

Figure 10.9 *Representation of the code in Figure 10.7 by parity check equations and an **H** matrix.*

Consider the top left hand part of the matrix (1011). The coefficients 1011 correspond to the information digits I_1, I_3 and I_4 in the equation for P_1. Similarly, the second row is 1101 to the left of the partition because I_3 is not involved in calculating parity check P_2 and the corresponding part of the bottom row is 1110 because I_4 is not involved in calculating the parity check P_3, Figure 10.8. To the right of the dotted partition there is a 3×3 diagonal matrix of ones. Each column in this diagonal matrix corresponds to a particular parity check digit. The first column (100) indicates parity check P_1. The second column indicates parity check P_2 and the third column parity check P_3. Later, in section 10.7, a *generator* matrix will be used to obtain the codeword directly from the information vector.

10.4 Probability of error in n-digit codewords

What is the probability of having more than R' errors in an n-digit codeword? First consider the case of exactly j errors in n digits with a probability of error per digit of P_e. From Chapter 3, equation (3.8):

$$P(j \text{ errors}) = (P_e)^j \, (1 - P_e)^{n-j} \times {}^nC_j \qquad (10.1)$$

The probability of having more than R' errors can be written as:

$$P(> R' \text{ errors}) = 1 - \sum_{j=0}^{R'} P(j) \qquad (10.2)$$

Statistical stability controls the usefulness of this equation, statistical convergence occurring for long codewords or blocks (see Figure 3.2). A long block effectively embodies a large number of trials, to determine whether or not an error will occur, and the number of errors in such a block will therefore be close to $P_e n$. Furthermore, the fraction of blocks containing a number of errors that deviates significantly from this value becomes smaller as the block length, n, becomes larger, Figure 3.2. Choosing a code that can correct $P_e n$ errors in a block will ensure that there are very few cases in which the coding system will fail. This is the rationale for long block codes. The attraction of block codes is that they are amenable to precise performance analysis. By far the most important, and most amenable, set of block codes is the linear group codes.

10.5 Linear group codes

The codewords in a linear group code have a one-to-one correspondence with the elements of a mathematical group. Group codes contain the all-zeros codeword and have the property referred to as closure. Thus, taking any two codewords C_i and C_j, then $C_i \oplus C_j = C_k$. (For the all-zeros codeword, when $i = 0$, then $k = j$.) Adding, modulo-2, corresponding pairs of digits in each of the codewords produces another codeword C_k. The presence of the all-zeros codeword and the closure property together make performance calculations with linear group codes particularly easy, see later. Figure 10.10 illustrates a simple group code which will be used here as an example. It is first used to illustrate the property of closure. Figure 10.10 depicts a source alphabet with four members: a, b, c and d (i.e. the number of information digits is $k = 2$). Each symbol is coded into an n-digit codeword (where $n = 5$). This is therefore a (5,2) code. (By the less strict definition, this is also a systematic code where the information digits are in columns 1/2 and 4/5.) Consider the codewords corresponding to c and b. Modulo-2 summing c and b gives the codeword d, illustrating closure.

10.5.1 Members of the group code family

Group codes can be divided into two types: those which are 'polynomial generated' in simple feedback shift registers; and others. The simplicity of the former has rendered the rest irrelevant. The polynomial generated codes can be further divided into subgroups, the main ones being the binary Bose–Chaudhuri–Hocquenghem (BCH) codes and their important, non-binary counterpart, the Reed–Solomon codes. BCH codes are widely tabulated up to $n = 255$ with an error correcting capability of up to 30 digits [Blahut, 1983]. Generally speaking for the same error correcting capability a larger (e.g. $n = 255$) block size offers a higher rate than a shorter (e.g. $n = 63$) block size. Reed–Solomon, non-binary, byte organised codes are used extensively in compact disc players and computer memories.

10.5.2 Performance prediction

Normally all possible codeword pairs would have to be examined, and their Hamming distances measured, to determine the overall performance of a block code. For the case of group codes, however, consideration of each of the codewords with the all-zeros codeword is sufficient. This is a significant advantage of linear group codes and one reason why these codes are so important in relation to other block codes. (Analysis for large n becomes much simpler for group codes, since the number of combinations of codewords, which would otherwise have to be searched, is very large.)

The important quantity, as far as code performance prediction is concerned, is the minimum Hamming distance between any pair of codewords. For the four five-digit codewords in Figure 10.10 – 0 0 0 0 0, 0 0 1 1 1, 1 1 1 0 0, 1 1 0 1 1 – inspection reveals a minimum Hamming distance of 3, i.e. $D_{min} = 3$ for this (5,2) code. Other (n, k) block codes with this minimum distance of 3 are (3,1), (15,11), (31,26), etc.

The weight *structure* of a set of codewords is just a list of the weights of all the codewords. Consider the previous example with four codewords: the weights of these are

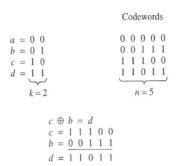

Codewords

$$a = 0\ 0 \qquad 0\ 0\ 0\ 0\ 0$$
$$b = 0\ 1 \qquad 0\ 0\ 1\ 1\ 1$$
$$c = 1\ 0 \qquad 1\ 1\ 1\ 0\ 0$$
$$d = 1\ 1 \qquad 1\ 1\ 0\ 1\ 1$$

$$k = 2 \qquad\qquad n = 5$$

$$c \oplus b = d$$
$$c = 1\ 1\ 1\ 0\ 0$$
$$\underline{b = 0\ 0\ 1\ 1\ 1}$$
$$d = 1\ 1\ 0\ 1\ 1$$

Figure 10.10 *Illustration of the closure property of a group code.*

0, 3, 3 and 4. Ignoring the all-zeros word (as interest is concentrated in the distances from this codeword), the minimum weight in the weight structure (3) is equal to D_{min}, the minimum Hamming distance for the code.

Consider the probability of the *i*th codeword (C_i) being misinterpreted as the *j*th codeword (C_j). This probability depends on the distance between these two codewords (D_{ij}). Since this is a linear group code, this distance D_{ij} is equal to the weight of a third codeword C_k which is actually the modulo-2 sum of C_i and C_j. The probability of C_i being mistaken for C_j is therefore equal to the probability of C_k being mistaken for C_0. Furthermore, the probability of C_k being mistaken for the all-zeros codeword (C_0) is equal to the probability of C_0 being misinterpreted as C_k (by symmetry). The probability of C_0 being misinterpreted as C_k depends only on the weight of C_k.

This reasoning reveals the importance of a linear group code's weight structure since the performance of such a code can be determined completely by consideration of C_0 and the weight structure alone.

10.5.3 Error detection and correction capability

The maximum possible error correcting power, *t*, of a code is defined by its ability to correct *all* patterns of *t* or less errors. It is related to the code's minimum Hamming distance by:

$$t = int\left(\frac{D_{min} - 1}{2}\right)$$

(10.3(a))

where:

$$D_{min} - 1 = e + t$$

(10.3(b))

Here *int*( ) indicates 'the integer part of', *e* is the total number of *detectable* errors (including the correctable *t* errors) and $t \le e$. Taking the case where D_{min} is 3, then there are at least two possible binary words which lie between each pair of valid codewords. In Example 10.1 these could be the binary words 11000 and 11001. If any single error occurs in one of the codewords it can therefore be corrected. Alternatively, if there is no error correction $D_{min} - 1$ errors can be detected (two in this case as both 11000 and 11001 are detectable as errors). Note that the code *cannot* work in both these detection and correction modes simultaneously (i.e. detect two errors and correct one of them).

Longer codes with larger Hamming distances offer greater detection and correction capability by selecting different *t* and *e* values in equation (10.3). $D_{min} = 7$ can offer $t = 1$ bit correction combined with $e = 5$ bit error detection. If *t* is increased to 2 then *e* must decrease to 4. The UK Post Office Code Standards Advisory Group (POCSAG) code with $k = 21$ and $n = 32$ is an $R \approx 2/3$ code with $D_{min} = 6$. This provides a 3 bit *detection* or a 2 bit *correction* capability, for a codeword which is widely used in pager systems, see Chapter 15. The $n = 63$, $k = 57$, BCH code gives $R = 0.9$ with $t = 1$ bit, while reducing *k* to 45 gives $R = 0.7$ with $t = 3$ bits. Further reducing *k* to 24 reduces *R* to below 0.4 but achieves a $t = 7$ bit correction capability. This illustrates the trade-off between rate and error correction power. BCH codes can correct burst and random errors.

10.6 Nearest neighbour decoding of block codes

Encoding is achieved by use of a feedback shift register and is relatively simple as will be shown later. The two most important strategies for decoding are nearest neighbour and maximum likelihood decoding. These are equivalent if the probability of t errors is much greater than that of $t + 1$ errors etc., as in Example 3.4. Using a decoding table based on nearest neighbours, therefore, implies the maximum likelihood decoding strategy, as discussed in the context of decision theory in Chapter 9. This is illustrated with a simple example.

Figure 10.11 is a nearest neighbour decoding table for the previous four-symbol example of Figure 10.10. The codewords are listed along the top of this table starting with the all-zeros codeword in the top left hand corner. Below each codeword all possible received sequences are listed which are at a Hamming distance of 1 from this codeword. (In the case of the all-zeros codeword these are the sequences 10000 to 00001.) If this were a t error correcting code this list would continue with all the patterns of 2 errors, 3 errors, etc., up to all patterns of t errors. Any detected bit pattern appearing in the table is interpreted as representing the codeword at the top of the relevant column, thus allowing the bit errors to be corrected. Below the table in Figure 10.11 there are eight 5 bit words which lie outside the table. These received sequences are equidistant from two possible codewords, so these sequences lie on a decision boundary. It is not possible, therefore, to decide which of the two original codewords they came from, and consequently the errors cannot be corrected. These sequences were referred to previously as *detectable* error sequences.

10.6.1 Hamming bound

Consider the possibility of a code with codewords of length n, comprising k information digits and having error correcting power t. There is an upper bound on the performance of block codes which is given by:

$$2^k \le \frac{2^n}{1 + n + {}^nC_2 + {}^nC_3 + \cdots + {}^nC_t} \tag{10.4}$$

The simplest way to derive equation (10.4) is to inspect the nearest neighbour decoding table for the (n, k), t-error correcting code. Figure 10.12 develops Figure 10.11 into the general case of a t-error correcting code with 2^k codewords. There are, thus, 2^k columns

Codewords	00000	11100	00111	11011
Single-bit error correctable patterns	10000	01100	10111	01011
	01000	10100	01111	10011
	00100	11000	00011	11111
	00010	11110	00101	11001
	00001	11101	00110	11010
Double-bit error detectable patterns	10001	01101	10110	01010
	10010	01110	10101	01001

Figure 10.11 *Nearest neighbour decoding table for the group code of Figure 10.9.*

Figure 10.12 on the page:

1	$0\,0\dots0$	C_2	....	C_{2^k}
n	$100\dots0$ $010\dots0$ $\vdots\quad\ \vdots$ $000\dots1$		Single errors	
$^nC_2$			Double errors	
$\vdots$	$\vdots$	$\vdots$	$\vdots$	$\vdots$
$^nC_1$			t errors	

2^k columns

Rows $= [1 + n + {^nC_2} + \dots + {^nC_1}]$

Figure 10.12 *Decoding table for a t-error correcting (n, k) block code.*

in the decoding table. Consider the left hand column. The all-zeros codeword itself is, obviously, one possible correctly received sequence or valid codeword. Also there are n single error patterns associated with that all-zeros codeword. Further, there are $^nC_2$ patterns of 2 errors, etc., down to ^nC_t patterns of t errors. Totalling the number of entries in this column reveals the total number of rows in the table and the value of the denominator in equation (10.4). Taking this number of rows and dividing into 2^n (which is the total number of possible received sequences), as in equation (10.4), gives the maximum possible number of columns and hence the maximum number of codewords in the given code. If the left hand side of equation (10.4) is greater than the right hand side then no such code exists and n must be increased, k decreased, or t decreased, until equation (10.4) is satisfied. For a *perfect* code, equation (10.4) is an equality. This implies that there are no bit patterns which lie outside the decoding table, avoiding the problem of equidistant errors which occurred in the code of Figure 10.11.

10.7 Syndrome decoding

The difficulty with decoding of block codes using the nearest neighbour decoding table of Figure 10.12 is the physical size of the table for large n. The syndrome decoding technique described here provides a solution to this problem.

10.7.1 The generator matrix

The generator matrix is a matrix of basis vectors. The rows of the generator matrix **G** are used to derive the actual transmitted codewords. This is in contrast with the **H** (or parity check) matrix, Figure 10.9, which does not contain any codewords. The generator matrix **G** for an (n, k) block code can be used to generate the appropriate n-digit codeword from any given k-digit data sequence. The **H** and corresponding **G** matrices for the example (7,4) block code of Figure 10.8 are shown below:

$$\mathbf{H} = \begin{bmatrix} 1\,0\,1\,1:1\,0\,0 \\ 1\,1\,0\,1:0\,1\,0 \\ 1\,1\,1\,0:0\,0\,1 \end{bmatrix} \tag{10.5}$$

$$G = \begin{bmatrix} 1\,0\,0\,0 : 1\,1\,1 \\ 0\,1\,0\,0 : 0\,1\,1 \\ 0\,0\,1\,0 : 1\,0\,1 \\ 0\,0\,0\,1 : 1\,1\,0 \end{bmatrix} \qquad (10.6)$$

Study of **G** shows that on the left of the dotted partition there is a 4×4 unit diagonal matrix and on the right of the partition there is a parity check section. This part of **G** is the transpose of the left hand portion of **H**. As this code has a single error correcting capability then D_{min}, and the weight of the codeword, must be 3. As the identity matrix has a single one in each row then the parity check section must contain at least two ones. In addition to this constraint, rows cannot be identical.

Continuing the (7,4) example, we now show how **G** can be used to construct a codeword, using the matrix equation (4.13) [Spiegel]. Assume the data sequence is 1001. To generate the codeword associated with this data sequence the data vector 1001 is multiplied by **G** using modulo-2 arithmetic:

$$[1\,0\,0\,1] \begin{bmatrix} 1\,0\,0\,0\,1\,1\,1 \\ 0\,1\,0\,0\,0\,1\,1 \\ 0\,0\,1\,0\,1\,0\,1 \\ 0\,0\,0\,1\,1\,1\,0 \end{bmatrix} = [1\,0\,0\,1\,0\,0\,1] \qquad (10.7)$$

The 4×4 unit diagonal matrix in the left hand portion of **G** results in the data sequence 1001 being repeated as the first four digits of the codeword and the right hand (parity check) portion results in the three parity check digits P_1, P_2 and P_3 (in this case 001) being calculated. (This generator matrix could, therefore, be applied to solve the second part of Example 10.2.)

It is now possible to see why the columns to the right of the partition in **G** are the rows of **H** to the left of its partition. From another standpoint the construction of a codeword is viewed as a weighted sum of the rows of **G**. The digits of the data sequence perform the weighting. With digits 1001 in this example, the top row of **G** is weighted by 1, the second row by 0, the third row by 0 and the fourth row by 1. After weighting the corresponding digits from each row are added modulo-2 to obtain the required codeword.

10.7.2 Syndrome table for error correction

Recall the strong inequality that the probability of t errors is much greater than the probability of $t + 1$ errors. This situation always holds in the P_e regime where FECC systems normally operate. Thus nearest neighbour decoding is equivalent to maximum likelihood decoding. Unfortunately, the nearest neighbour decoding table is normally too large for practical implementation which requires a different technique, involving a smaller table, to be used instead. This table, referred to as the syndrome decoding table, is smaller than the nearest neighbour table by a factor equal to the number of codewords in the code set (2^k). This is because the syndrome is independent of the transmitted codeword and only depends on the error sequence as is demonstrated below.

When **d** is a message vector of k digits, **G** is the $k \times n$ generator matrix and **c** is the n-digit codeword corresponding to the message **d**, equation (10.7) can be written as:

$$\mathbf{d}\,\mathbf{G} = \mathbf{c} \qquad (10.8)$$

Furthermore:

$$\mathbf{H}\,\mathbf{c} \;=\; \mathbf{0} \tag{10.9}$$

where $\mathbf{H}$ is the (even) parity check matrix corresponding to $\mathbf{G}$ in equation (10.8). Also:

$$\mathbf{r} \;=\; \mathbf{c} \oplus \mathbf{e} \tag{10.10}$$

where $\mathbf{r}$ is the sequence received after transmitting $\mathbf{c}$, and $\mathbf{e}$ is an error vector representing the location of the errors which occur in the received sequence $\mathbf{r}$. Consider the product $\mathbf{Hr}$ which is referred to as the syndrome vector $\mathbf{s}$:

$$\mathbf{s} \;=\; \mathbf{H}\,\mathbf{r} = \mathbf{H}\,(\mathbf{c} \oplus \mathbf{e})$$

$$= \mathbf{H}\,\mathbf{c} \oplus \mathbf{H}\,\mathbf{e} = \mathbf{0} \oplus \mathbf{H}\,\mathbf{e} \tag{10.11}$$

Thus $\mathbf{s}$ is easily calculated and, if there are no received errors, the syndrome will be the all-zero vector $\mathbf{0}$. Calculating the vector $\mathbf{s}$ provides immediate access to the vector $\mathbf{e}$ and hence the position of the errors. A syndrome table is constructed by assuming transmission of the all-zeros codeword and calculating the syndrome vector associated with each correctable error pattern:

$$\begin{bmatrix} 1\,0\,1\,1\,1\,0\,0 \\ 1\,1\,0\,1\,0\,1\,0 \\ 1\,1\,1\,0\,0\,0\,1 \end{bmatrix} \begin{bmatrix} 0 \\ 0 \\ 0 \\ 0 \\ 0 \\ 0 \\ 0 \end{bmatrix} = \begin{bmatrix} 0 \\ 0 \\ 0 \end{bmatrix} \tag{10.12}$$

Equation (10.12) illustrates the case of no errors in the received sequence leading to the all-zeros syndrome for the earlier (7,4) code example. Figure 10.13 shows the full syndrome table for this (7,4) code. In this case only single errors are correctable and the syndrome table closely resembles the transposed matrix $\mathbf{H}^T$. If a double error occurs then it will normally give the same syndrome as some single error and, since single errors are much more likely than double errors, a single error will be assumed and the wrong codeword will be output from the decoder resulting in a 'sequence' error. This syndrome decoding technique is still a nearest neighbour (maximum likelihood) decoding strategy.

$$\mathbf{H}\,\mathbf{e} = \mathbf{s}$$

Error pattern	Syndrome
0 0 0 0 0 0 0	0 0 0
1 0 0 0 0 0 0	1 1 1
0 1 0 0 0 0 0	0 1 1
0 0 1 0 0 0 0	1 0 1
0 0 0 1 0 0 0	1 1 0
0 0 0 0 1 0 0	1 0 0
0 0 0 0 0 1 0	0 1 0
0 0 0 0 0 0 1	0 0 1

Figure 10.13 *The complete syndrome table for all possible single error patterns.*

EXAMPLE 10.3

As an example of the syndrome decoding technique, assume that the received vector for the (7,4) code is $\mathbf{r} = 1001101$ and find the correct transmitted codeword.

$$
\begin{bmatrix} 1\,0\,1\,1\,1\,0\,0 \\ 1\,1\,0\,1\,0\,1\,0 \\ 1\,1\,1\,0\,0\,0\,1 \end{bmatrix}
\begin{bmatrix} 1 \\ 0 \\ 0 \\ 1 \\ 1 \\ 0 \\ 1 \end{bmatrix}
=
\begin{bmatrix} 1 \\ 0 \\ 0 \end{bmatrix}
$$

The above matrix equation illustrates calculation of the corresponding syndrome (100). Reference to the syndrome table (Figure 10.13) reveals the corresponding error pattern as (0000100). Finally $\mathbf{c} = \mathbf{r} \oplus \mathbf{e}$:

$$
\begin{aligned}
\mathbf{r} &= 1001101 \\
\mathbf{e} &= 0000100 \\
\mathbf{c} &= 1001001
\end{aligned}
$$

to give the corrected transmitted codeword $\mathbf{c}$ as 1001001.

EXAMPLE 10.4

For a (6, 3) systematic linear block code, the codeword comprises $I_1\ I_2\ I_3\ P_1\ P_2\ P_3$ where the three parity check bits P_1, P_2 and P_3 are formed from the information bits as follows:

$$
\begin{aligned}
P_1 &= I_1 \oplus I_2 \\
P_2 &= I_1 \oplus I_3 \\
P_3 &= I_2 \oplus I_3
\end{aligned}
$$

Find: (a) the parity check matrix; (b) the generator matrix; (c) all possible codewords. Determine (d) the minimum weight; (e) the minimum distance; and (f) the error detecting and correcting capability of this code. (g) If the received sequence is 101000, calculate the syndrome and decode the received sequence.

(a)

$$
\mathbf{H} = \begin{bmatrix} 1\,1\,0\,1\,0\,0 \\ 1\,0\,1\,0\,1\,0 \\ 0\,1\,1\,0\,0\,1 \end{bmatrix}
$$

(b)

$$
\mathbf{G} = \begin{bmatrix} 1\,0\,0\,1\,1\,0 \\ 0\,1\,0\,1\,0\,1 \\ 0\,0\,1\,0\,1\,1 \end{bmatrix}
$$

(c) and (d)

Message × **G**		= Codeword	Weight
[0 0 0] × **G**	=	0 0 0 0 0 0	0
[0 0 1] × **G**	=	0 0 1 0 1 1	3
[0 1 0] × **G**	=	0 1 0 1 0 1	3
[1 0 0] × **G**	=	1 0 0 1 1 0	3
[0 1 1] × **G**	=	0 1 1 1 1 0	4
[1 0 1] × **G**	=	1 0 1 1 0 1	4
[1 1 0] × **G**	=	1 1 0 0 1 1	4
[1 1 1] × **G**	=	1 1 1 0 0 0	3

Minimum weight = 3.

(e) D_{min} = minimum weight = 3.

(f) The code is thus single error correcting (or double error detecting if no correction is required).

(g) Now using **H r** = **s**, equation (10.11) becomes:

$$\mathbf{s} = \begin{bmatrix} 1 & 1 & 0 & 1 & 0 & 0 \\ 1 & 0 & 1 & 0 & 1 & 0 \\ 0 & 1 & 1 & 0 & 0 & 1 \end{bmatrix} \begin{bmatrix} 1 \\ 0 \\ 1 \\ 0 \\ 0 \\ 0 \end{bmatrix} = \begin{bmatrix} 1 \\ 0 \\ 1 \end{bmatrix}$$

The decoded codeword could be found by constructing the syndromes for all possible error patterns and modulo-2 adding the appropriate error pattern to the received bit pattern. It is clear, however, that the decoded codeword is [1 1 1 0 0 0] as this is the closest valid codeword to the received bit pattern.

10.8 Cyclic codes

Cyclic codes are a subclass of group codes which do not possess the all-zeros codeword. The Hamming code is an example of a cyclic code. Their properties and advantages are as follows:

- Their mathematical structure permits higher order correcting codes.
- Their code structures can be easily implemented in hardware by using simple shift registers and exclusive-or gates.
- Cyclic code members are all lateral, or cyclical, shifts of one another.
- Cyclic codes can be represented as, and derived using, polynomials.

The third property, listed above, can be expressed as follows. If:

$$C = (I_1, I_2, I_3, \cdots, I_n) \tag{10.13}$$

then:

$$C_i = (I_{i+1}, \cdots, I_n, I_1, I_2, \cdots, I_i) \tag{10.14}$$

(C_i is an i bit cyclic shift of C, and $I_1, \cdots, I_n$ now represent both parity and information bits.) The following three codewords provide an example:

1 0 0 1 0 1 1

0 1 0 1 1 1 0

0 0 1 0 1 1 1

Full description of these codes requires a detailed discussion of group and field theory and takes us into a discussion of prime numbers and Galois fields. Non-systematic cyclic codes are obtained by multiplying the data vector by a generator polynomial with modulo arithmetic. In general cyclical codes are generated from parity check matrices which have cyclically related rows.

Cyclic codes encompass BCH codes and the Reed–Solomon non-binary codes. The Reed–Solomon (RS) code is made up of n symbols where each symbol is m bits long. m can be any length depending on the application; for example, if $m = 8$ bits then each symbol would represent a byte. Thus RS codes operate on a multiple bit symbol principle and not a bit principle as other cyclic codes do. An important property of the RS codes is their burst error correction property. Their error correcting power is:

$$t = \frac{n - k}{2} \tag{10.15}$$

where n and k here relate to encoded symbols, not bits. For example, the (31,15) RS code has 31 5 bit encoded symbols which represent 15 symbols of input information or 75 input information bits. This can correct eight independent bit errors or four bursts of length equal to or less than the 5 bit symbol duration.

10.8.1 Polynomial codeword generation

Systematic cyclic redundancy checks (CRCs) are in widespread use for performing error detection, but not correction, on bit-serial channels. The operation of CRCs can be considered as an algebraic system in which 0 and 1 are the only values and where addition and subtraction involve no carry operations (i.e. arithmetic is modulo-2). (Addition and subtraction are both, therefore, the same as the logical exclusive-or operation.) If a message of length k bits, $m_{k-1}, \cdots, m_1, m_0$ (from the most to least significant bit), is to be transmitted over a channel then, for coding purposes, it may be considered to represent a polynomial of order $k - 1$:

$$M(x) = m_{k-1}x^{k-1} + \cdots + m_1 x + m_0 \tag{10.16}$$

The message $M(x)$ is modified by the generator polynomial $P(x)$ to form the channel coded version of $M(x)$. This is accomplished by multiplying, or bit shifting, $M(x)$ by the order of $P(x)$. $P(x)$ is then divided into the bit shifted or extended version of $M(x)$ and the remainder is then appended to $M(x)$ replacing the zeros which were previously added by the bit shifting operation. Note that the quotient is discarded.

EXAMPLE 10.5

Generate a polynomial codeword from the data sequences (a) 1100, (b) 0101, (c) 1011, where the leading bit in each codeword represents the first sequence bit to enter the coder, for the generator polynomial $x^3 + x + 1$.

For (a) $M(x) = 1100$ (i.e. $x^3 + x^2$) the bit shifted sequence $kM(x)$ is thus 1100000. This is now divided by the polynomial $P(x) = 1011$ (i.e. $x^3 + x + 1$):

$$1011 \overline{\smash{\big)}\,1100000}$$
$$\underline{1011}$$
$$111000$$
$$\underline{1011}$$
$$10100$$
$$\underline{1011}$$
$$010$$

to obtain the remainder 010. The three appended zeros in $kM(x)$ are then replaced by the remainder 010 to realise the transmitted codeword as 1100010.

For case (b)

$$1011 \overline{\smash{\big)}\,0101000}$$
$$\underline{1011}$$
$$100$$

and the corresponding transmitted codeword is 0101100.

For (c) the corresponding transmitted codeword is 1011000.

The validity of codeword (a) can be further checked by dividing it by the corresponding $P(x)$ value:

$$1011 \overline{\smash{\big)}\,1100010}$$
$$\underline{1011}$$
$$111010$$
$$\underline{1011}$$
$$10110$$
$$\underline{1011}$$
$$000$$

to confirm that the remainder is indeed 000.

The required division process can be conveniently implemented in hardware. An encoding circuit, which is equivalent to the above long division operations, is shown in Figure 10.14. The encoding circuit works as follows for the sequence (a) or $M(x) = 1100$. The information bits are transmitted as part of the cyclic code but are also fed back

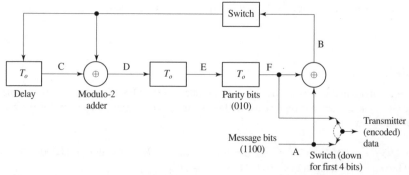

Figure 10.14 *Encoder for a (7,4) cyclic code generated by $P(x) = x^3 + x + 1$.*

via the feedback loop. During this step, the feedback switch is kept closed. As we clock the shift register and input the message codeword the states $A, \cdots, E$ in the shift register follow this bit pattern:

$$\text{Input } A = 1100$$
$$F \oplus A = B = 1110$$
$$\text{previous } B, \text{i.e. } C = 0111$$
$$C \oplus B = D = 1001$$
$$\text{previous } D, \text{i.e. } E = 0100$$
$$\text{previous } E, \text{i.e. } F = 0010$$

$$\text{Parity } bits \text{ } are \text{ } thus \text{ } E \text{ } C \text{ } B = 010$$

Following the transmission of the final bit of information and the clocking of the B data bit into the leading shift register the feedback loop switch is opened leaving a logic 0 at the shift register input. The information remaining at shift register locations E, C, B (at the final time instant in the above example) is thus appended to the data for transmission by the output switch. These remaining parity check bits, in example (a), would be 010, resulting in a (7,4) Hamming code.

On reception the received data is again divided by $P(x)$ and, if the remainder is zero, then no errors have been introduced. Further examination of a non-zero remainder in a syndrome table can allow the bit error positions to be determined and the errors corrected by adding in the error pattern in the decoder [Blahut, 1983], as previously shown in Example 10.3. The syndrome table can be found either by mathematical manipulation or by successive division for each error location.

If the generator polynomial had been $x^3 + x^2 + 1$ the encoder would have used the structure shown in Figure 10.15. The hardware decoding scheme, see Figure 10.16, is basically an inverse version of the encoding scheme of Figure 10.15. If the decoder receives an error it is capable of identifying the position of the error digit via the remainder and the syndrome table.

Thus, the polynomial coded message consists of the original k message bits, followed by $n - k$ additional bits. The generator polynomial is carefully chosen so that almost all errors will be detected. Using a generator of degree k allows the detection of all burst

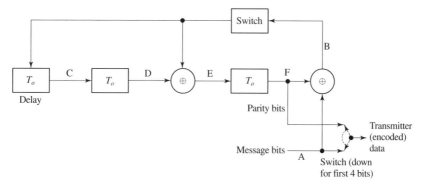

Figure 10.15 *Corresponding encoder for a (7,4) cyclic code generated by* $P(x) = x^3 + x^2 + 1$.

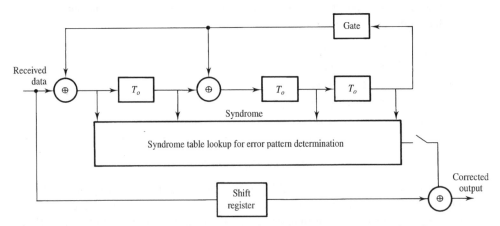

Figure 10.16 *Syndrome calculator and decoder for a (7,4) cyclic code $P(x) = x^3 + x^2 + 1$.*

errors affecting up to k consecutive bits. The generator chosen by ITU-T for the V.41 standard, which is the same as that used extensively on wide area networks, is:

$$M(x) = x^{16} + x^{12} + x^5 + 1 \qquad (10.17)$$

Equation (20.20) provides an example of an ATM CRC polymonial. The generator chosen by IEEE, used extensively on local area networks (see Chapters 18 and 20), is:

$$M(x) = x^{32} + x^{26} + x^{23} + x^{22} + x^{16} + x^{12} + x^{11} + x^{10} + x^8 + x^7 + x^5 + x^4 + x^2 + x + 1 \qquad (10.18)$$

CRC 6 bit codewords are transmitted within the plesiochronous multiplex, Chapter 20, to improve the robustness of frame alignment words. The error correcting power of the CRC code is low and it is mainly used when ARQ retransmission is deployed, rather than for error correction itself.

10.8.2 Interleaving

The largest application area of block codes is in compact disc players which employ a powerful concatenated and cross-interleaved RS coding scheme to handle random and burst errors. Partitioning data into blocks and then splitting the blocks and interleaving them means that a burst transmission error usually degrades only part of each original block. Thus, using FECC, it is possible to correct for a long error burst, which might have destroyed all the information in the original block, at the expense of the delay required for the interleaver encoder/decoder function. In other applications bit-by-bit interleaving is employed to spread burst errors across a data block prior to decoding. Figure 10.17 shows how an input data stream is read, column by column, into a temporary array and then read out, row by row, to achieve the bit interleaving operation. Now, for example, a burst of three errors in the consecutive transmitted data bits, I_1, I_5, I_9, is converted into isolated single errors in the de-interleaved data.

10.9 Encoding of convolutional codes

Convolutional codes are generated by passing a data sequence through a shift register which has two or more sets of register taps (effectively representing two or more different filters) each set terminating in a modulo-2 adder. The code output is then produced by sampling the output of all the modulo-2 adders once per shift register clock period. The coder output is obtained by the convolution of the input sequence with the impulse response of the coder, hence the name convolutional code. Convolution applies even though there are exclusive-or and switch operations rather than multiplies. Figure 10.18 illustrates this with an exceptionally simple example where the output encoder operation can be defined by two generator polynomials. The first and second encoded outputs, $P_1(x)$, $P_2(x)$, can be defined by $P_1(x) = 1 + x^2$ and $P_2(x) = 1 + x$, as in Example 10.5. This shows a three-stage encoder giving a constraint length $n = 3$. The error correcting power is related to the constraint length, increasing with longer lengths of shift register. Figure 10.18 is drawn purely to illustrate the principle of the convolutional coder and to permit the later inclusion of corresponding tree and trellis diagrams of reasonable size and complexity. In any practical coder design the shift register would be longer and the generator polynomials more sophisticated to achieve the required error correction capability.

Assume the data sequence 1101 is input to the three-stage shift register which is initiated or flushed with zeros prior to clocking the sequence through. This example depicts a rate ½ coder ($R = ½$) since there are two output digits for every input information digit. The coder is non-systematic since the data digits are not explicitly present in the transmitted data stream. The first output following a given input is obtained with the switch in its first position and the second output is obtained with the switch in its second position, etc.

This encoder may be regarded as a finite state machine. The first stage of the shift register holds the next input sample and its contents determine the transition to the next state. The final two stages of the shift register hold past inputs and may be regarded as

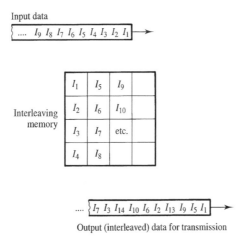

Figure 10.17 *Data block interleaving to overcome burst errors.*

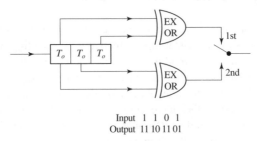

Input 1 1 0 1
Output 11 10 11 01

Figure 10.18 *A simple example of a rate ½ convolutional encoder.*

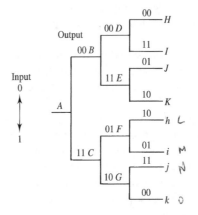

Figure 10.19 *Tree diagram representation of the coder in Figure 10.18.*

determining the 'memory' of the machine. In this example there are two 'memory stages' and hence four possible states. In general an n-stage register would have $2^{(n-1)}$ states. For the $n = 3$ stage coder the four states correspond to the data bit pairs 00, 10, 01, 11 (from prior input data). The convolutional encoder operation may be represented by a tree diagram.

10.9.1 Tree diagram representation

Figure 10.19 depicts the tree diagram corresponding to the example of Figure 10.18. Assume that the encoder is 'flushed' with zeros prior to the first input of data and that it is in an initial state which is labelled A. Conventionally the tree diagram is drawn so that inputting a zero results in exiting the present state by the upper path, while inputting a one causes it to exit by the lower path. Assuming a zero is input, the machine will move to state B and output 00. Outputs are shown on the corresponding branches of the diagram. Alternatively if the machine is in state A and a one is input then it proceeds to state C via the lower branch and 11 is output. Figure 10.19 depicts the first three stages of the tree diagram, after which there appear to be eight possible states. This is at variance with the previous statement, that the state machine in this example has only four states.

There are, in fact, only four distinct states here (00, 10, 01, 11), but each state appears twice. Thus *H* is equivalent to *h*, for example. This duplication of states results from identical prior data bits being stored in the shift register of the encoder, The path *B*, *D* to *H* represents the input of two zeros, as does the path *C*, *F* to *h*, resulting in identical data being stored in the shift register. After the fourth stage each state would appear four times, etc. Two states are identical if, on receiving the same input, they respond with the same output. Following through input data in Figure 10.19 by the path which this data generates allows the states (00, 10, 01, 11) to be identified and the figure annotated accordingly to identify the redundancy. The apparent exponential growth rate in the number of states can be contained by identifying the identical states and overlaying them. This leads to a trellis diagram.

10.9.2 Trellis diagram

Figure 10.20 shows the trellis diagram corresponding to the tree diagram of Figure 10.19. The horizontal axis represents time while the states are arranged vertically. On the arrival of each new bit the tree diagram is extended to the right. Here five stages are shown with the folding of corresponding tree diagram states being evident at the fourth and fifth stages (states *HIJK*, *LMNO*) by the presence of two entry paths to each state. There are still too many states here and inspection will show that *H* and *L*, for example, are equivalent. Thus four unique states may be identified. These are labelled *a*, *b*, *c* and *d* on this diagram again corresponding to the binary data 00, 10, 01 and 11 being stored in the final two stages of the shift register of Figure 10.18.

The performance of the convolutional coder is basically dependent on the Hamming distance between the valid paths through the trellis, corresponding to all the possible, valid, data bit patterns which can occur. The final step in compacting the graphical representation of the convolutional encoder is to reduce this trellis diagram to a state transition diagram.

10.9.3 State transition diagram

Here the input to the encoder is shown on the appropriate branch and the corresponding outputs are shown in brackets beside the input, Figure 10.21. For example, if the encoder is in state *a* (the starting state) and a zero is input then the transition is along the self loop returning to state *a*. The corresponding output is 00, as shown inside the brackets (and

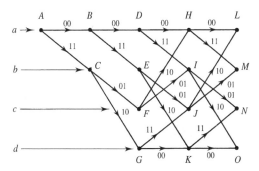

Figure 10.20 *Trellis diagram representation of the coder in Figure 10.18.*

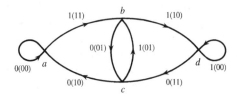

Figure 10.21 *State transition diagram representation of the coder in Figure 10.18.*

along the top line of Figure 10.20). If, on the other hand, a one is input while in state *a*, then 11 is output and the state transition is along the branch from *a* to *b* etc.

10.10 Viterbi decoding of convolutional codes

There are three main types of decoder. These are based on sequential, threshold (majority logic) and Viterbi decoding techniques. The Viterbi technique is by far the most popular.

Data encoded by modern convolutional coders are usually divided into message blocks for decoding, but unlike the block coded messages, where $n < 255$, the convolutional coded message typically ranges from 500 to >10000 bits, depending on the application. This makes decoding of convolutional codes potentially onerous. (The decoder memory requirements grow with message length.) The coder operation is illustrated here by the processing of short fixed length blocks, which are fed through the encoder after it has been 'flushed' with zeros to bring it into state *a*. The block of data is followed with trailing zeros to return the encoder back to state *a* at the end of the coding cycle. This simplifies decoding and 'flushes' the encoder ready for the next block. The zeros do not, however, carry any information and the efficiency, or rate, of the code is consequently reduced.

Secondly, the Viterbi decoding algorithm is used at each stage of progression through the decoding trellis, retaining only the most likely path to a given node and rejecting all other possible paths on the grounds that their Hamming distance is larger and that they are, thus, less likely events than that represented by the shorter distance path. This leads to a linear increase in storage requirement with block length as opposed to an exponential increase.

The Viterbi decoding algorithm implements a nearest neighbour decoding strategy. It picks the path through the decoding trellis, which assumes the minimum number of errors (the probability of t errors being much greater than that of $t+1$ errors, etc.). Conceptually a decoding trellis, similar to the corresponding encoding trellis, is used for decoding.

EXAMPLE 10.6 – Decoding trellis construction

Assume a received data sequence 1010001010 for the encoder operation of Figure 10.18. Identify the errors and derive the corresponding transmitted data sequence.

The ten transmitted binary digits correspond to five information digits. We assume that the first three of these digits are unknown data and the last two are flushing zeros.

Decoding begins by building a decoding trellis corresponding to the encoding trellis starting at state *A* as shown in Figure 10.22. We assume that the first input to the encoder is a zero. Reference to the encoding trellis indicates that on entering a zero with the encoder completely flushed 00 would be output, but 10 has been received. This means that the received sequence is a Hamming distance of 1 from the possible transmitted sequence (with an error in the first output bit). This distance metric is noted along the upper branch from *A* to *B*.

The possibility that the input data may have been a one is now investigated. Again, reference to the encoding trellis indicates that if a one is input to the encoder in state *A*, the encoder will output 11 and follow the lower path to state *C*. In fact, 10 was received, so, again, the actual received sequence is a Hamming distance of 1 from this possible transmitted sequence. (Here the error would be in the second bit.) The distance metric is thus noted as 1 along the branch from *A* to *C*.

Now we return to state *B* and assume that the input was zero followed by another zero. If this were the case, the encoder would have gone from state *B* to state *D* and output 00 (Figure 10.23). The third and fourth digits received, however, were 10 and again there is a Hamming distance of 1 between the received sequence and this possible transmitted sequence. This distance metric is noted on the branch *B* to *D* and a similar operation is performed on branch *B* to *E* where the distance is also 1. Next we consider inputting zero while in state *C*. This would create 01 whilst, in fact, 10 was received. The Hamming distance here is 2. This metric is noted on branch *C* to *F* and attention is turned finally to branch *C* to *G*. Starting in state *C* and inputting a one would have output 10 and, in fact, 10 was received, so at last there is a received pair of digits which does not imply any errors. The *cumulative* distances along the various paths, i.e. the path metrics, are now entered in square brackets above the final states in Figure 10.23.

Figure 10.24 illustrates a further problem. Decoding is now at stage 3 in the decoding trellis and the possibility of being in state *J* is being considered. From this stage on, each of the four states in this example has two entry, and two exit, paths. Conventionally on reaching a state like *J* with two input paths, the cumulative Hamming distance or path metric of the upper route (*ABEJ*) is shown first and the cumulative Hamming distance for the lower path (*ACGJ*) is shown second in the square brackets adjacent to state *J*. The real power of the Viterbi algorithm lies in its rejection of one of those two paths, retaining one path which is referred to as the 'survivor'. If the two paths have different Hamming distances then, since this is a nearest neighbour (maximum likelihood) decoding strategy, the path with the larger Hamming distance is rejected and the path with the smaller Hamming distance, or path metric, is carried forward as the survivor.

In the case illustrated in Figure 10.24 the distances are identical and the decoder must flag an uncorrectable error sequence, if using incomplete decoding. If using complete decoding a random choice is made between the two paths (bearing in mind there is a 50% probability of being wrong). Fortunately, this situation is rare in practice. (The probability of error has been deliberately increased here for illustrative purposes.)

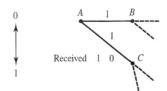

Figure 10.22 *First stage in constructing the decoding trellis for a received sequence from the encoder of Figure 10.18 after receiving two encoded data bits.*

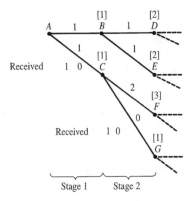

Figure 10.23 *Second stage of trellis of Figure 10.18 after decoding four data bits.*

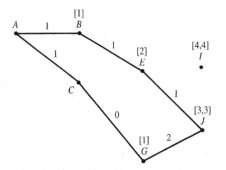

Figure 10.24 *Illustration of a sequence containing a detectable but uncorrectable error pattern.*

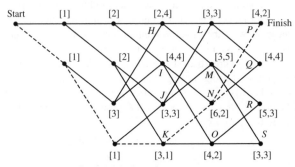

Figure 10.25 *Complete decoding trellis for Example 10.6 with the dashed preferred path and resulting decoded sequence 11100.*

There are two paths to state *H* with path metrics [2,4], Figure 10.25. The path of distance 4 may thus be rejected as being less likely than the path of distance 2, etc. To state *I* there are also two paths of equal length [4,4]. In the final stage state *P* has been labelled as being the finishing point of the decoding process since in this example only three unknown data digits are being transmitted followed by 00 to flush the encoder and bring the decoder back to state *a*. Only the more likely of the two paths to state *P* is retained. This is the lower path with a Hamming distance of 2. Note that although state O also has distances of 2 we cannot progress from O to P to complete the decoding operation. Figure 10.25 shows the complete decoding trellis. Tracing back along the most likely (dashed) path provides the corresponding decoded sequence as (11100) and the implied correct received data as 1110001110. Although state Q also has a cumulative distance of 2 this cannot terminate the correct path as this decoder must be flushed with zeros ready for the next block of data.

10.10.1 Decoding window

In a practical convolutional decoder the block length would usually be very much larger than in the simple example above. The data from a complete frame of a video coded image, Chapter 16, may be sent as a single message block, for example. In such a case the overhead requirement, for accommodating the flushing zeros, becomes negligible. There is a constraint on the length of data which can be retained in the decoder memory, however, when performing the Viterbi decoding operation. The practical limitation is known as the decoding window. In a practical decoder the new distance metrics are added to the previous path metrics to obtain updated path metrics. Details of the paths, which correspond to these various distances, are carried forward in the decoding process. In Figure 10.25 the decoding was performed over a window length of 5 input data bits. (With long block lengths the same procedure would be followed.) Figure 10.26 shows an example of decoding with a window length of only 4. This demonstrates how the decoding of Figures 10.22 to 10.25 moves through the window (sequence (a) to (d)) with only the most likely paths being retained.

When the window length is restricted to 4 then, on the arrival of the final received bit pair, it is no longer possible to continue to examine the start bit as this would have propagated out of the restricted length of the decoding window (through which the trellis is viewed). The window length should be long enough to cover all bursts of decoding

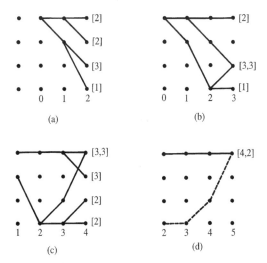

Figure 10.26 *Viterbi decoding within a finite length decoding window.*

errors but, since longer lengths involve more computation, a compromise must be made and the decoder's performance verified by computer simulation. (For practical coders a good rule of thumb is for the decoding window to be set at five times the constraint length of the encoder and the constraint length in practical coders is always much longer than in the simple example shown in Figure 10.18.)

10.10.2 Sequential decoding

Viterbi's algorithm requires all the surviving sequences to be followed throughout the decoding process and leads to excessive memory requirements for long constraint lengths. Complexity can be reduced by sequential decoding, which directly constructs the sequence of states by performing a distance measure at each step. Sequential decoding proceeds forwards until complete decoding is accomplished or the cumulative distance exceeds a preset threshold. When this occurs the algorithm backtracks and selects an alternative path until a satisfactory overall distance is maintained. This works well at low error rates but, when the error rate is high, the number of backward steps can become very large.

The Viterbi decoder has three main components, Figure 10.27. The first is the branch metric value (BMV) calculation unit which finds the Hamming distances for each new branch in the trellis. The add–compare–select unit is the second which calculates and updates the overall path history or path metric values (PMVs) for each path arriving at each node in the trellis. In the third component, the output determination unit, only the surviving paths are retained (i.e. selected) as these have the smallest distances, the paths with higher distances being discarded. When working through the complete trellis, only the most likely overall path is finally retained as the ultimate survivor. Trellis decoding is not restricted to convolutional codes as it is also used for soft decision decoding of block codes [Honary and Markarian]. (A similar decoding trellis is also used later for trellis coded modulation, section 11.4.8 [Biglieri *et al.*, Ungerboeck].)

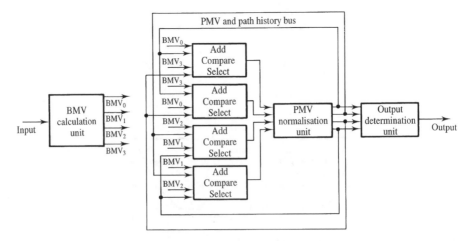

Figure 10.27 *Viterbi decoder circuit for decoding the trellis of Figure 10.20.*

10.11 Practical coders

Examples of practical block codes are the BCH (127,64) which has an error correction capability of $t = 10$ bits; the RS (16,8) or (64,32) codes achieve $t = 4$ symbol capability while the shorter block length of the Golay (23,12) code has a $t = 3$ bit capability. The error rate performance of these, and some other, codes is compared in Figure 10.28 [Farrell], all for DPSK modulation, in which the horizontal axis is energy per input *information* data bit divided by one-sided noise power spectral density E_b/N_0. Figure 10.28 echoes Figure 10.1, showing clearly the point at which the FECC systems outperform uncoded DPSK transmission (see Chapter 11).

Generally for a $t = 1$ BCH block code, which has a larger block length than the simple Hamming (7,4) example, i.e. a (31,26) or a (63,57) code, then the coding gain over the uncoded system, at a P_b of 10^{-5}, is >2 dB. For the longer $t = 3$ bit BCH (127,106) code the coding gain approaches 4 dB. Linear block codes are usually restricted to $n < 255$ by the decoder complexity. In Figure 10.28 the $t = 4$ symbol RS code is inferior to $t = 3$ bit Golay code at high P_b, but this performance is reversed at lower P_b. The (23,12) Golay code, which corrects triple errors, is a perfect code as equation (10.4) becomes:

$$2^{12} \leq \frac{2^{23}}{{}^{23}C_0 + {}^{23}C_1 + {}^{23}C_2 + {}^{23}C_3} \tag{10.19(a)}$$

i.e.:

$$2^{12} \leq \frac{2^{23}}{1 + 23 + 253 + 1771} \tag{10.19(b)}$$

and hence is satisfied as the equality:

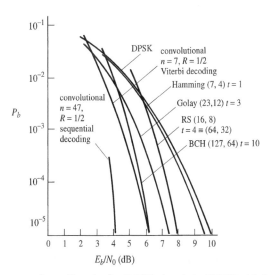

Figure 10.28 *Performance of rate ½ codes for DPSK signals in AWGN with hard decision decoding (source: Farrell, 1989, reproduced with permission of the IEE).*

$$2^{12} = \frac{2^{23}}{2048} = \frac{2^{23}}{2^{11}} \tag{10.20}$$

In general RS codes are attractive for bursty error channels where extremely low P_b values (e.g. 10^{-10}) are required. Convolutional codes are favoured for Gaussian noise channels where more moderate P_b values (e.g. 10^{-6}) are required.

Practical convolution codes often have constraint lengths of $n = 7$ with a rate ½ coder which employs seven delay stages. This requires a decoding window in the trellis of 35 to achieve the theoretical coding gain. With such a window, the encoder message block size can be set appropriate to the specific application. Such a coder then has $D_{min} = 10$ and its performance is equivalent to the BCH (127,64) block code. The convolutional coder is preferred, however, because the encoder is simpler and the decoder can more easily incorporate soft decision techniques (in which a confidence level is retained for the received data). Such soft decisions, which implement maximum likelihood decoding, give approximately 2 dB improvement in coding gain compared to hard decisions. (The Voyager space probes employed soft decision decoders adding 5 to 8 dB coding gain to their link budget, see Chapter 12.)

In general reducing the efficiency, or rate, R, of the coder increases the Hamming distance D_{min} between permissible paths and improves the error correcting power (i.e. the coding gain in dB compared to the uncoded DPSK system). If the performance is normalised by the rate R then there may be little benefit in going to rates of less than ½ as the improvement in error rate is exactly balanced by the reduction in data rate. Whilst additional coding gain improvement is typically 1 dB for an $n = 9$ constraint length, compared to $n = 7$, the improvements beyond this are very small and the decoder complexity is excessively high.

The largest application area for convolutional coders is in rate ¾ compact disc codes which employ a concatenated and cross-interleaved coding scheme. Sequential decoding is a powerful search path technique for use in the decoding trellis, especially with longer constraint length, Figure 10.28. Sophisticated VLSI Viterbi decoders are now available for speeds of 250 kbit/s to 25 Mbit/s with constraint lengths of $n = 7$ at (1996) costs of £10 and above, per FECC decoder. Table 10.2 lists some Qualcom (Q) and Stanford Telecom (ST) convolutional coder chipsets.

Table 10.2 *Examples of commercially available convolutional coder chipsets.*

Data rate (Mbit/s)	25	12	2.5	1	¼
Coder rate, R	$\frac{7}{8}, \frac{3}{4}, \frac{1}{2}, \frac{1}{3}$	$\frac{3}{4}, \frac{2}{3}, \frac{1}{2}$	$\frac{7}{8}, \frac{3}{4}, \frac{1}{2}, \frac{1}{3}$	$\frac{1}{2}, \frac{1}{3}$	$\frac{7}{8}, \frac{3}{4}, \frac{2}{3}, \frac{1}{2}, \frac{1}{3}$
Constraint length	7	7	7	6	6–7
Soft decision capability (bits)	3	3	3	–	3–4
Supplier	Q	Q/ST	Q	ST	Q/ST

10.12 Concatenated coding and turbo codes

Turbo codes, first proposed in [Berrou *et al.*, 1993], have attracted a great deal of recent interest due to their ability to provide very low BER at bit rates and SNRs that closely

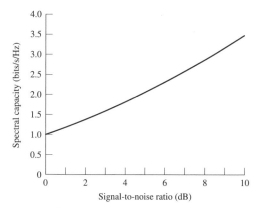

Figure 10.29 *Shannon capacity limit.*

approach the theoretical Shannon limit whilst maintaining a tractable decoding procedure. Figure 10.29 shows the Shannon limit dividing the specific capacity (i.e. maximum information rate per unit bandwidth) E_b/N_0 plane into two regions, one region allowing errorless transmission (in principle) and one not.

Shannon demonstrated that it is possible to transmit information with arbitrarily few errors provided the capacity of the channel is not exceeded, the capacity of the channel depending on bandwidth and SNR, see section 11.4.1. He also showed that one way in which the capacity of a channel could be approached was by mapping a sequence of k information bits randomly onto a code sequence of n code symbols (see section 10.3). This limit is only realised, however, as n and k approach infinity implying unbounded channel latency or delay. Furthermore, with no structure, decoding the random codes becomes problematic, the only solution being an exhaustive search.

Shannon's work motivated a quest to find good practical codes that result in performance as close to the theoretical capacity limit as possible. Traditionally, their design has been tackled by constructing codes with sufficient algebraic structure to facilitate practical decoding schemes. In the 45 years prior to the development of turbo codes many good, traditional, codes were developed. None, however, approached the Shannon limit. It is understandable therefore that the development of turbo codes, and the promise of a coding scheme that could approach the Shannon limit, was initially met with some scepticism.

Despite their performance advantages over more conventional coding schemes, turbo codes [Hanzo *et al.*, 2002] are themselves based on two previously well known techniques, being a combination of concatenated codes and iterative decoding.

10.12.1 Serially concatenated codes

Concatenated coding has been known for many years as a method of combining two or more relatively simple codes to provide a much more powerful code, but with simpler decoding properties than a single larger code of comparable performance. The principle is simple. The output of the first encoder (the outer encoder) is fed into the input of the second encoder (the inner encoder). In the decoder, the second (inner) code is decoded first and its output, the outer code, is then decoded (see Figure 10.30).

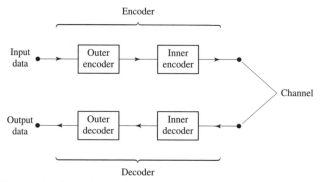

Figure 10.30 *Concatenated encoding and decoding.*

The error correcting performance of a concatenated code is complicated to evaluate since its performance depends sensitively on the distribution of errors. One of the most powerful codes available prior to the discovery of turbo codes was a concatenated code comprising an outer RS block coder followed by an inner, convolutional coder, of constraint length 7. This code has been widely used in applications as diverse as deep-space missions [Miller *et al.*] and digital television broadcasting (ETSI EN 301 DVB standard).

A significant disadvantage of this simple concatenated scheme is error propagation. In the event of a single error at the input of the first decoder, several incorrect bits may be passed to the input of the next decoder, resulting in an even larger number of incorrect bits that may eventually overwhelm the error correcting capacity of the component decoders. Such a situation can be partially alleviated by the use of an interleaver, which simply permutes the order of a sequence of symbols (section 10.8.2). If instead of a large number of bit errors occurring within a single codeword, the errors are redistributed over a larger number of codewords (with correspondingly fewer errors per codeword), then the decoding process is more likely to be successful.

10.12.2 Parallel-concatenated recursive systematic convolutional codes

The term 'turbo code' is something of a misnomer as the word 'turbo' actually refers to the iterative decoding procedure, rather than to the codes themselves. In its most basic form, the encoder of a turbo code comprises two systematic encoders joined together by

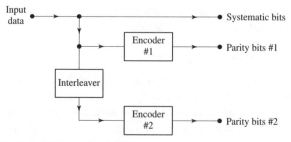

Figure 10.31 *Block diagram of a turbo encoder.*

an interleaver, Figure 10.31. (This form of connection is known as parallel concatenation.) The encoders used by [Berrou *et al.*, 1993] were a form of convolutional code. The convolutional codes described in section 10.9 are non-systematic convolutional (NSC) codes, since the encoders do not generate separately the data and the redundancy bits. NSC encoders have no feedback paths, and in this respect they act like a finite impulse response (FIR) digital filter. Indeed the term 'convolutional' code arises from the fact that the output sequence is the input sequence convolved with the impulse response of the encoder. It is possible to include feedback in the encoder making the structure similar to that of an infinite impulse response (IIR) filter, such encoders being said to be recursive (again borrowing the terminology of digital filters).

NSC codes do not lend themselves to parallel concatenation and a systematic code is therefore required. At high SNRs the BER performance of a classical NSC code is better than that of a classical systematic convolutional code of the same constraint length [Viterbi and Omura]. At low SNRs the reverse is generally true, with the systematic convolutional code having better performance [Berrou and Glavieux]. Recursive systematic convolutional (RSC) codes can be generated from NSC codes by connecting one of the outputs of the encoder directly to the input, whilst other weighted taps from the shift register stages are also fed back to the input. An example of an RSC encoder is shown in Figure 10.32.

Making the codes recursive means that the state of the encoder depends upon past outputs as well as past inputs. Thus a finite length input sequence can generate an infinite length output sequence (unlike an NSC code). This affects the behaviour of the error patterns (a single error in the original message bits producing an infinite number of parity errors), with the result that better overall performance is attained. It can be shown that, like NSC codes, RSC codes are linear.

The operation of the turbo encoder is as follows:

1. The input data sequence is applied directly to encoder 1 and the interleaved version of the same input data sequence is applied to encoder 2, Figure 10.31.
2. The systematic bits (i.e. the original message bits) and the two parity check bit streams (generated by the two encoders) are multiplexed together to form the output of the encoder.

Although the component encoders of the turbo code are convolutional, in practice turbo codes are block codes with the block size being determined by the size of the interleaver.

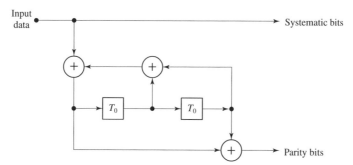

Figure 10.32 *Example of an RSC encoder with n = 3.*

Turbo codes may therefore be described as linear block codes.

The block nature of the turbo encoder presents a number of practical problems: determining precisely the start and finish of an encoded input bit sequence. As with almost all coding schemes it is common practice to initialise the encoder to the all-zeros state before encoding the data. After encoding a number of data bits a number of tail-bits are added to force the encoder back to the all-zeros state at the end of each block of data. (This procedure, also referred to as termination, and the reset sequences subsequently applied, are referred to as terminating sequences.) Termination of turbo codes may be applied to only one of the component encoders or both.

Turbo code interleavers

The novelty of the parallel-concatenated turbo encoder lies in the use of RSC codes and the introduction of an interleaver between the two encoders. The interleaver ensures that two permutations of the same input data are encoded to produce two different parity sequences. The effect of the interleaver is to tie together errors that are easily made in one half of the turbo encoder to errors that are exceptionally unlikely to occur in the other half. This ensures robust performance in the event that the channel characteristics are not known and is the principal reason why turbo codes perform better than traditional codes. The choice of interleaver is, therefore, key to the performance of a turbo coding system. As with any code, turbo code performance can be analysed in terms of the Hamming distance between the codewords. If the applied input sequence happens to terminate one of the encoders it is unlikely that, once interleaved, the sequence will terminate the other leading to a large Hamming distance in at least one of the two encoders. A pseudo-random interleaver is a good choice for a turbo coder since if, by chance, data sequences exist that result in a small overall Hamming distance, there will be relatively few of them because the same sequence elsewhere in the input data block will be interleaved differently. This does have the side effect of giving rise to an 'error floor' (see subsection on performance below).

Puncturing

The turbo code of Figure 10.32 is of (relatively low) rate 1/3. Rather than change the encoder structure, high rate codes are usually generated by a procedure known as puncturing. In the turbo code of Figure 10.32 a change in the code rate to 1/2 could be achieved by puncturing the two parity sequences prior to the multiplexer. One bit might be deleted from each parity output in turn, for example, such that one parity bit remains for each data bit. Similarly deleting other proportions of parity outputs can generate other code rates. At the decoder the deleted bits must be replaced by inserting dummy parity bits, which in the case of a soft decision decoder (as is required for the iterative decoding procedure used for turbo codes) must not bias the decoding procedure.

10.12.3 Turbo decoding

Turbo codes derive their name from the analogy between the iterative decoding process and the 'turbo-engine' principle in which feedback is applied to enhance combustion engine performance.

A key component of iterative (turbo) decoding is the soft-in, soft-out (SISO) decoder. As briefly discussed in Chapter 7 the performance of a demodulator–decoder combination can be substantially improved if soft decisions (providing confidence or reliability information) from the demodulator are available to the decoder, which finally makes hard decisions. To use 'soft information' in an iterative decoding scheme requires that the decoder generate soft information as well as making use of it. Soft information is usually represented quantitatively by a log-likelihood ratio (the logarithm of the likelihood ratio), LLR, for each bit. The absolute magnitude of the LLR is a measure of certainty about a decision. The idea of iterative decoding is that subsequent decoders make use of this information. Note that the LLR mirrors Shannon's quantitative measure of information (see section 9.2).

Before considering the decoding of turbo codes it is instructive to consider the decoding of array codes (see section 10.3.1, Figure 10.6) using the iterative turbo principle. The conventional decoding technique for array codes is to decode the rows and columns separately. This may not always yield the best results. Errors made in the decoding of the rows may result in many more errors when the columns are decoded that might not have occurred if the columns were decoded first followed by the decoding of the rows. (Decoding array codes is similar to solving a crossword puzzle in which there are constraints that have to be met by the words answering the 'across' and 'down' clues.) The optimum decoder would find the maximum likelihood solution for the whole array (thereby giving the minimum number of errors). The penalty, of course, is considerably increased decoder complexity and latency. The decoding of concatenated interleaved codes is made difficult since, conventionally, only one decoder has access to soft information from the demodulator. The solution is to apply soft information to both decoders by de-interleaving the input signal and then applying this sequence to the second decoder. Using soft information from the output of the first decoder and applying it to the input of the second decoder can make further improvements. A final improvement is to apply the output of the second decoder to the first; in effect iterating.

Figure 10.33 shows an iterative decoder for array codes. The information available about various decoding decisions comes from several sources. Some of the information

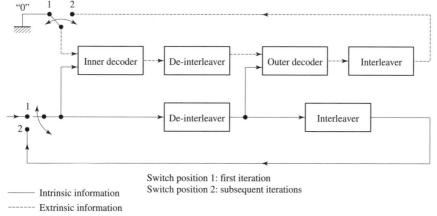

Switch position 1: first iteration
Switch position 2: subsequent iterations

——— Intrinsic information

------ Extrinsic information

Figure 10.33 *Basic iterative decoder.*

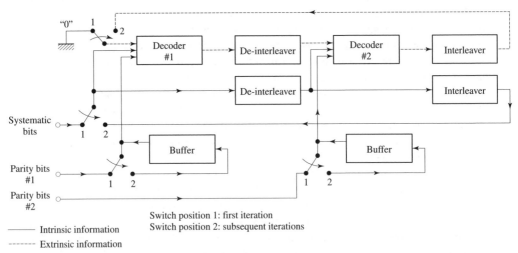

Figure 10.34 *Block diagram of a turbo decoder.*

comes directly from the received data bit itself: this is referred to as intrinsic information. When decoding a concatenated or product code, the information from one decoder that is passed to the next is called extrinsic information. Only the extrinsic information is passed from decoder to decoder, as the intrinsic information is made directly available. After both the inner and outer codes have been decoded once the data is re-interleaved as appropriate and the process repeated as many times as required, until finally a hard decision is made to yield the decoded bits. It is this iterative procedure from which the turbo decoding name is derived. Figure 10.32 shows the basic turbo decoder structure.

Each of the two decoding stages uses a BCJR (Bahl, Cocke, Jelinek, Raviv) algorithm, which was invented to solve a maximum a posteriori probability (MAP) detection problem [Burr]. The BCJR algorithm differs from the Viterbi decoding algorithm in two key respects. Firstly, the BCJR algorithm is a SISO algorithm with two recursions, one forward and one backward, both of which involve soft decisions. In contrast the Viterbi algorithm is a soft input, hard output, decoding algorithm. Secondly, the BCJR algorithm is a MAP decoder in that it minimises the bit errors by estimating the a posteriori probabilities of the individual bits in a codeword; to reconstruct the original data sequence the soft outputs of the BCJR algorithm are hard-limited. On the other hand the Viterbi algorithm is a maximum likelihood sequence estimator in that it maximises the likelihood function for the whole sequence, not for each bit. A variant of the Viterbi algorithm suitable for turbo decoding has been developed. This algorithm is known as the soft output Viterbi algorithm (SOVA) [Hagenauer].

For practical implementations, the nature of the turbo encoder implies that the decoders must operate much faster than the rate at which incoming data is received such that several iterations can be made. Alternatively a single iterating decoder may be replaced by series of interconnected decoders (a decoding pipeline), with the output being available at the final stage. In either case, once a preset number of iterations have been completed the combination of intrinsic and extrinsic information can be used to find the decoded data sequence. Typically, a fixed number of iterations (in the range 4 to 10) is used depending on the type of code and its length.

10.12.4 Turbo code performance

The error performance of a rate 1/2 turbo code with $P_1(x) = 1 + x + x^2 + x^3 + x^4$ and $P_2(x) = 1 + x^4$ having a 256×256 interleaver is shown in Figure 10.35. It can be seen that beyond approximately six iterations the improvement is small. For a decoder with $K = 3$ and 18 iterations, a BER of 10^{-5} is achieved for $E_b/N_0 = 0.9$ dB. Putting this in context Figure 10.28 shows that a conventional convolutional ($n = 47$, rate ½) code requires $E_b/N_0 = 4.0$ dB for the same BER.

Note, however, that impressive coding gain comes at the cost of a large interleaver size. Additionally there may be latency or speed restrictions due to the nature of the decoding. At the time of writing some of the fastest ASIC decoder devices can attain speeds of 90 Mbit/s [see – Advanced hardware architectures WWW site]. The fact that a turbo code appears essentially random to the channel by virtue of the pseudo-random interleaver, and yet at the same time has a physically realisable decoding structure, is central to its impressive performance.

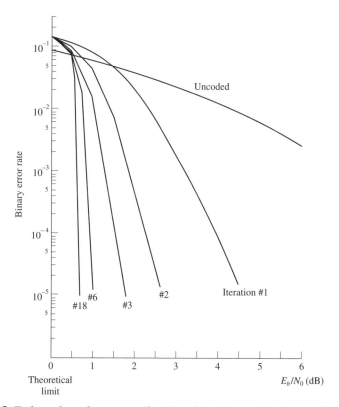

Figure 10.35 *Turbo code performance as function of iteration number (source: Berrou and Glavieux, 1996, reproduced with the permission of the IEEE © 2003 IEEE).*

10.12.5 Other applications of the turbo principle

Since the conception of turbo codes, the turbo principle has been widely applied to other problems. One of these is the decoding of the serially concatenated block codes or array codes described earlier. These are decoded using an iterative decoding technique, which offers a performance not much poorer than turbo codes [Pyndiah]. Serially concatenated block codes or array codes decoded in this way are called turbo product codes (TPCs). Due to their simpler decoding scheme ASIC and FPGA cores for TPCs are available which can operate at rates in excess of 500 Mbit/s. An obvious generalisation to array codes is to consider an array of more than two dimensions. Such codes are a special case of a family of low density parity check codes [Gallagher]. Although these codes do not approach the Shannon limit quite as closely as turbo codes, they do not exhibit the error floor characteristic of turbo codes.

Other applications (in which decoding can be considered part of the problem) that use the turbo principle have been proposed. These include modulation (e.g. turbo trellis coded modulation, T-TCM), carrier recovery, symbol timing, channel estimation and channel equalisation). Although the initial work on turbo codes used only two constituent coders, more can be used. These are known as multiple turbo codes.

10.13 Summary

Error rate control is necessary for many systems to ensure that the probability of bit error is acceptably low. Bit error rates may be reduced by increasing transmitted power, applying various forms of diversity, using echo back and retransmission, employing ARQ, or incorporating FECC. In this chapter the focus has been on channel coding for error detection and correction which are prerequisites, respectively, for ARQ and FECC error control systems.

Channel codes may be systematic or unsystematic. Systematic codes use codewords which contain the information digits, from which the codeword is derived, explicitly. The rate, R, of a code is the ratio of information bits transmitted to total bits transmitted. The Hamming distance between a pair of binary codewords, which is given by their modulo-2 sum, is a measure of how easily one codeword can be transformed into the other. The weight of a binary codeword is equal to the number of binary ones which it contains.

Block codes divide the precoded data into k bit lengths and add $(n - k)$ parity check bits to create a post-coded block, with length n bits. An (n, k) block code, therefore, has an efficiency, or rate, of $R = k/n$. Single parity check codes are block codes which append a single digital 1, or 0, to each codeword in order to ensure that all codewords have either an even (for even parity) or an odd (for odd parity) weight. This allows single error detection. Data can also be arranged in 2-dimensional, rectangular, arrays allowing parity check digits to be added to the ends of rows, and the bottoms of columns. This achieves single error correction.

Group codes (also called linear block codes) are block codes which contain the all-zeros codeword and have the closed set, or linear group, property, i.e. the modulo-2 sum of any pair of valid codewords in the set is another valid codeword. The error correcting power, t, of a linear group code is given by $int((D_{min} - 1)/2)$ where D_{min} is the minimum Hamming distance between any pair of codewords. The error detecting power, e, of a linear group code is given by $D_{min} - t - 1$. Group codes are the most important block codes due to the ease with which their performance can be predicted. (D_{min} is given by the weight of the minimum weight codeword, excluding the all-zeros codeword, in the group.) Block codes can be generated using a generator matrix. Complete codewords are generated by the product of the precoded data vector with the generator matrix.

Block codes are most easily decoded using a nearest neighbour strategy. This is equivalent to maximum likelihood decoding providing that $P(t$ errors) $\gg P(t + 1$ errors), which is always the case in practice. Nearest neighbour decoding can be implemented using a nearest neighbour table or a syndrome table. (The syndrome vector for a given received codeword is the product of the parity check matrix and the received codeword vector. It is also the product of the parity check matrix and the error pattern vector of the received codeword allowing error patterns to be determined from a table of syndromes.) Syndrome decoding is advantageous when block size is large since the syndrome table for all (single) error patterns is much smaller than the nearest neighbour decoding table.

Cyclic codes are linear group codes in which the codeword set consists of only, and all, cyclical shifts of any one member of the codeword set. They are particularly easily generated using shift registers with appropriate feedback connections. Their syndromes are also easily calculated using shift register hardware. CRC codes are systematic cyclic codes which are potentially capable of error correction but are often used for error detection only. A polynomial representation of a precoded block of information bits is multiplied (i.e. bit shifted) by the order of a generator polynomial and then divided by the generator polynomial. The remainder of the division process is appended to the block of information bits to form the complete codeword. Division of received codeword polynomials by the generator polynomial leaves zero remainder in the absence of errors.

Convolution codes are unsystematic and operate on long data blocks. The encoding operation can be described (in increasing order of economy) using tree, trellis or state diagrams. The Viterbi algorithm, which implements a nearest neighbour decoding strategy, is usually used to decode convolution codes. The error correction capability of convolutional coders is not inherently in excess of that for block coders but their decoder and, particularly, encoder designs are simpler.

FECC combinations of block and convolutional codes are widely applied to accommodate random and burst errors which may both arise in communication channels.

Turbo coders represent a new development in high performance error control coding. They comprise parallel-concatenated recursive systematic convolutional coders incorporating interleaving between coders. Their turbo designation arises from the iterative method used in the decoder that represents a form of feedback analogous to feedback in turbo-engines. Turbo codes are particularly attractive as they are potentially capable of realising performance (with practical coding/decoding latency) close to the Shannon limit of errorless transmission at a bit rate corresponding to the channel capacity.

10.14 Problems

10.1 Assume a binary channel with independent errors and $P_e = 0.05$. Assume k-digit symbols from the source alphabet are encoded using an (n, k) block code which can correct all patterns of three or fewer errors. Assume $n = 20$. (a) What is the average number of errors in a block? [1] (b) Assuming binary transmission at 20000 binary digits per second, derive the symbol error rate at the decoder output. [15.8 symbol/s]

10.2 A binary signal is transmitted through a channel which adds zero mean, white, Gaussian noise. The probability of bit error is 0.001. What is the probability of error in a block of four data bits? If the bandwidth is expanded to accommodate a (7,4) block code, what would be the probability of an error in a block of four data bits? [0.0040, 0.0019]

10.3 Assume a systematic (n, k) block code where $n = 4$, $k = 2$ and the four codewords are 0000, 0101, 1011, 1110. (a) Construct a maximum likelihood decoding table for this code. (b) How many errors will the code correct? Are there any errors which are detectable but not correctable? [eight corrected and eight detected but not corrected] (c) Assume this code is used on a channel with $P_e = 0.01$. What is the probability of having a detectable error sequence? What is the probability of having an undetectable error sequence? [0.0388, 0.0006]

10.4 For a (6,3) systematic linear block code, the three parity check digits are:

$$P_1 = 1 \times I_1 \oplus 1 \times I_2 \oplus 1 \times I_3$$

$$P_2 = 1 \times I_1 \oplus 1 \times I_2 \oplus 0 \times I_3$$

$$P_3 = 0 \times I_1 \oplus 1 \times I_2 \oplus 1 \times I_3$$

(a) Construct the generator matrix $\mathbf{G}$ for this code. (b) Construct all the possible codewords generated by this matrix. (c) Determine the error correcting capabilities for this code. [single] (d) Prepare a suitable decoding table. (e) Decode the received words 101100, 000110 and 101010. [111100, 100110, 101011]

10.5 Given a code with the parity check matrix:

$$\mathbf{H} = \begin{bmatrix} 1110\ 100 \\ 1101\ 010 \\ 1011\ 001 \end{bmatrix}$$

(a) Write down the generator matrix showing clearly how you derive it from $\mathbf{H}$. (b) Derive the complete weight structure for the above code and find its minimum Hamming distance. How many errors can this code correct? How many errors can this code detect? Can it be used in correction and detection modes simultaneously? [3, 1, 2, No] (c) Write down the syndrome table for this code showing how the table may be derived by consideration of the all-zeros codeword. Also comment on the absence of an all-zeros column from the $\mathbf{H}$ matrix. (d) Decode the received sequence 1001110, indicate the most likely error pattern associated with this sequence and give the correct codeword. Explain the statement 'most likely error pattern'. [0000010, 1001100]

10.6 When generating a (7,4) cyclic block code using the polynomial $x^3 + x^2 + 1$: (a) What would the generated codewords be for the data sequences 1000 and 1010? [1000110, 1010001] (b) Check that these codewords would produce a zero syndrome if received without error. (c) Draw a circuit to generate this code and show how it generates the parity bits 110 and 001 respectively for the two data sequences in part (a). (d) If the codeword 1000110 is corrupted to 1001110, i.e. an error occurs in the fourth bit, what is the syndrome at the receiver? Check this is the same syndrome as for the codeword 1010001 being corrupted to 1011001. [101]

10.7 A (7,4) block code has the parity check matrix as:

$$H = \begin{bmatrix} 1 & 1 & 0 & 1 & 1 & 0 & 0 \\ 1 & 1 & 1 & 0 & 0 & 1 & 0 \\ 0 & 0 & 1 & 1 & 0 & 0 & 1 \end{bmatrix}$$

This code can correct a single error. (a) Derive the generator matrix for this code and encode the data 1110. (b) Derive a syndrome decoding for the code as described above and decode the received data 1101110. (c) Calculate the maximum number of errors a (15,11) block code can correct. [1110010, 1]

10.8 Given the ½ rate convolutional encoder defined by $P_1(x) = 1 + x + x^2$ and $P_2(x) = 1 + x^2$, and assuming data is fed into the shift register 1 bit at a time, draw the encoder: (a) tree diagram; (b) trellis diagram; (c) state transition diagram. (d) State what the rate of the encoder is. (e) Use the Viterbi decoding algorithm to decode the received block of data, 10001000.

Note: there may be errors in this received vector. Assume that the encoder starts in state a of the decoding trellis in Figure 10.20 and, after the unknown data digits have been input, the encoder is driven back to state a with two 'flushing' zeros. [0000]

10.9 For a convolutional encoder defined by: $P_1(x) = 1 + x + x^2$ and $P_2(x) = 1 + x^2$: (a) State the constraint length of the encoder and the coding rate. (b) The coder is used to encode two data bits followed by two flushing zeros. Encode the data sequences: (i) 10 (ii) 11. Assume that the encoder initially contains all zeros in the shift register and the left hand bit is the first bit to enter the encoder. (c) Take the two encoded bit sequences from parts (b)(i) and (b)(ii) above and invert the second and fifth bit to create received codewords with two errors in each. Decode the altered sequences using a trellis diagram and show whether or not the code can correct the errors you have introduced. [3, 1/3, b(i) 101111110000, b(ii) 101010001110, both decode correctly]

CHAPTER 11

Bandpass modulation of a carrier signal

11.1 Introduction

Modulation, at its most general, refers to the modification of one signal's characteristics in sympathy with another signal. The signal being modified is called the carrier and, in the context of communications, the signal doing the modifying is called the information signal. Chapter 5 described pulse modulation in which the carrier is a rectangular pulse train and the characteristics adjusted are, for example, pulse amplitude, or, in PCM, the coded waveform. In this chapter, intermediate or radio frequency (IF or RF) bandpass modulation is described, in which the carrier is a sinusoid and the characteristics adjusted are amplitude, frequency or phase.

The principal reason for employing IF modulation is to transform information signals (which are usually generated at baseband) into signals with more convenient (bandpass) spectra. This allows:

1. Signals to be matched to the characteristics of transmission lines or channels.
2. Signals to be combined using frequency division multiplexing and subsequently transmitted using a common physical transmission medium.
3. Efficient antennas of reasonable physical size to be constructed for radio communication systems.
4. Radio spectrum to be allocated to services on a rational basis and regulated so that interference between systems is kept to acceptable levels.

11.2 Spectral and power efficiency

Different modulation schemes can be compared on the basis of their spectral[1] and power efficiencies. Spectral efficiency is a measure of information transmission rate per Hz of bandwidth used. A frequent objective of the communications engineer is to transmit a maximum information rate in a minimum possible bandwidth. This is especially true for radio communications in which the radio spectrum is a scarce, and therefore valuable, resource. The appropriate units for spectral efficiency are clearly bit/s/Hz. It would be elegant if an analogous quantity, i.e. power efficiency, could be defined as the information transmission rate per W of received power. This quantity, however, is not useful since the information received (correctly) depends not only on received signal power but also on noise power.

It is thus carrier-to-noise ratio (CNR) or equivalently the ratio of symbol energy to noise power spectral density (NPSD), E/N_0, which must be used to compare the power efficiencies of different schemes. It is legitimate, however, to make comparisons between different digital communications systems on the basis of the relative signal power needed to support a given received information rate assuming identical noise environments. In practice this usually means comparing the signal power required by different modulation schemes to sustain identical BERs for identical transmitted information rates.

11.3 Binary IF modulation

Binary IF modulation schemes represent the simplest type of bandpass modulation. They are easy to analyse and occur commonly in practice. Each of the basic binary schemes is therefore examined below in detail. The ideal BER performance of these modulation schemes could be found directly using a general result, equation (8.61), from Chapter 8. It is instructive, however, to obtain the probability of error, P_e, formula for each scheme by considering matched filtering or correlation detection (see section 8.3) as an ideal demodulation process followed by an ideal, baseband, sampling and decision process.

11.3.1 Binary amplitude shift keying (and on–off keying)

In binary amplitude shift keying (BASK) systems the two digital symbols, zero and one, are represented by pulses of a sinusoidal carrier (frequency, f_c) with two different amplitudes A_0 and A_1. In practice, one of the amplitudes, A_0, is invariably chosen to be zero resulting in on–off keying (OOK) IF modulation, i.e.:

$$f(t) = \begin{cases} A_1 \, \Pi(t/T_o) \cos 2\pi f_c t, & \text{for a digital 1} \\ 0, & \text{for a digital 0} \end{cases} \quad (11.1)$$

where T_o is the symbol duration (as used in Chapters 6 and 8) and Π is the rectangular pulse function.

An OOK modulator can be implemented either as a simple switch, which keys a carrier on and off, or as a double balanced modulator (or mixer) which is used to multiply

---

[1] See footnote on first page of Chapter 8.

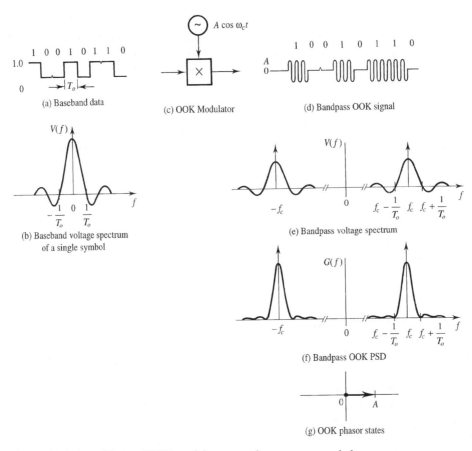

Figure 11.1 *On–off keying (OOK) modulator, waveforms, spectra and phasor states.*

the carrier by a baseband unipolar OOK signal. A schematic diagram of the latter type of modulator is shown with rectangular pulse input and output waveforms, spectra and allowed phasor states in Figure 11.1. The modulated signal has a DSB spectrum centred on $\pm f_c$ (Figures 11.1(e) and (f)) and, since a constant carrier waveform is being keyed, the OOK signal has two phasor states, 0 and $A = A_1$ (Figure 11.1(g)). Detection of IF OOK signals can be coherent or incoherent. In the former case a matched filter or correlator is used prior to sampling and decision thresholding (Figure 11.2(a),(b)). In the latter (more common) case envelope detection is used to recover the baseband digital signal followed by centre point sampling or integrate and dump (I + D) detection (Figure 11.2(d)). (The envelope detector would normally be preceded by a bandpass filter to improve the CNR.) Alternatively, an incoherent detector can be constructed using two correlation channels configured to detect inphase (I) and quadrature (Q) components of the signal followed by I and Q squaring operations and a summing device (Figure 11.2(e)). This arrangement overcomes the requirement for precise carrier phase synchronisation (cf. Figure 11.2(b)). Whilst such a dual channel incoherent detector may seem unnecessarily complicated compared with the other incoherent detectors, recent

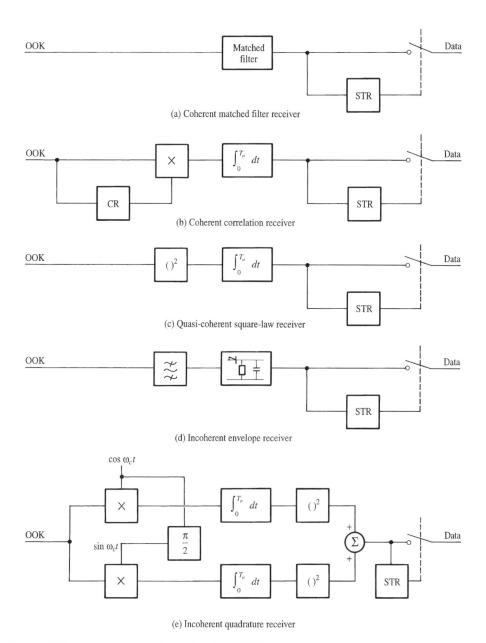

Figure 11.2 *Coherent and incoherent bandpass OOK receivers.*

advances in VLSI technology mean that they can often be implemented (digitally) as smaller, lighter and cheaper components than the filters and envelope detectors used in more traditional designs. (Symbol timing recovery is discussed in section 6.7 and carrier recovery is discussed later in this chapter.)

The decision instant voltage, $f(nT_o)$, at the output of an OOK matched filter or correlation detector (see equation (8.37)) is:

$$f(nT_o) = \begin{cases} kE_1, & \text{digital 1} \\ 0, & \text{digital 0} \end{cases} \tag{11.2}$$

where E_1 (V^2 s) is the normalised energy contained in symbol 1 and k has units of Hz/V. The normalised noise power, σ^2 (V^2), at the detector output (see equation (8.49)) is:

$$\sigma^2 = k^2 E_1 N_0/2 \tag{11.3}$$

where N_0 (V^2/Hz) is the normalised one-sided noise power spectral density at the matched filter, or correlator, input. (Note that the constant k is not shown in equations (8.37) and (8.49) since it is assumed, there, to be 1.0 Hz/V.) The post-filtered decision process is identical to the baseband binary decision process described in Chapter 6. Equation (6.8) can therefore be used with $\Delta V = k(E_1 - 0)$ and $\sigma^2 = k^2 E_1 N_0/2$ to give the probability of symbol error:

$$P_e = \frac{1}{2}\left[1 - \text{erf}\,\frac{1}{2}\left(\frac{E_1}{N_0} \right)^{1/2} \right] \tag{11.4}$$

Equation (11.4) can be expressed in terms of the time averaged energy per symbol, $\langle E \rangle = \frac{1}{2}(E_1 + E_0)$, where, for OOK, $E_0 = 0$, i.e.:

$$P_e = \frac{1}{2}\left[1 - \text{erf}\,\frac{1}{\sqrt{2}}\left(\frac{\langle E \rangle}{N_0} \right)^{1/2} \right] \tag{11.5}$$

Finally equations (11.4) and (11.5) can be expressed in terms of received carrier-to-noise ratios (C/N) using the following relationships:

$$C = \langle E \rangle/T_o \quad (\text{V}^2) \tag{11.6}$$

$$N = N_0 B \quad (\text{V}^2) \tag{11.7}$$

$$\langle E \rangle/N_0 = T_o B\, C/N \tag{11.8}$$

where C is the received carrier power averaged over all symbol periods and N is the normalised noise power in a bandwidth B Hz.[2] This gives:

$$P_e = \frac{1}{2}\left[1 - \text{erf}\,\frac{(T_o B)^{1/2}}{\sqrt{2}}\left(\frac{C}{N} \right)^{1/2} \right] \tag{11.9}$$

For minimum bandwidth (i.e. Nyquist) pulses $T_o B = 1.0$ and $\langle E \rangle/N_0 = C/N$. Bandlimited signals will result in symbol energy being spread over more than one symbol period, however, resulting (potentially) in ISI. This will degrade P_e with respect to that given in equations (11.4), (11.5) and (11.9) unless proper steps are taken to ensure ISI-free sampling instants at the output of the receiver matched filter (as in section 8.4).

Incoherent detection of OOK is, by definition, insensitive to the phase information

---

[2] If pulse shaping, or filtering, has been employed to bandlimit the transmitted signal, the obvious interpretation of B is the predetection bandwidth, equal to the signal bandwidth. Here C/N in equation (11.9) will be the actual CNR measured after a predetection filter with bandwidth just sufficient to pass the signal intact. In the case of rectangular pulse transmission (or any other non-bandlimited transmission scheme), B must be interpreted simply as a convenient bandwidth within which the noise power is measured or specified (typically chosen to be the double sided Nyquist bandwidth, $B = 1/T_o$). C/N in equation (11.9) does not then correspond to the CNR at the input to the matched filter or correlation receiver since, strictly speaking, this quantity will be zero.

contained in the received symbols. This lost information degrades the detector's performance over that given above. The degradation incurred is typically equivalent to a 1 dB penalty in receiver CNR at a P_e level of 10^{-4}. The modest size of this CNR penalty means that incoherent detection of OOK signals is almost always used in practice.

EXAMPLE 11.1

An OOK IF modulated signal is detected by an ideal matched filter receiver. The non-zero symbol at the matched filter input is a rectangular pulse with an amplitude 100 mV and a duration of 10 ms. The noise at this point is known to be white and Gaussian, and has an RMS value of 140 mV when measured in a noise bandwidth of 10 kHz. Calculate the probability of bit error.

Energy per non-zero symbol:

$$E_1 = v_{RMS}^2 \, T_o$$

$$= \left(\frac{100 \times 10^{-3}}{\sqrt{2}} \right)^2 10 \times 10^{-3} = 5.0 \times 10^{-5} \ (\text{V}^2 \text{ s})$$

Average energy per symbol:

$$\langle E \rangle = \frac{E_1 + 0}{2} = 2.5 \times 10^{-5} \ (\text{V}^2 \text{ s})$$

Noise power spectral density (from equation (11.7)):

$$N_0 = \frac{N}{B_N} = \frac{n_{RMS}^2}{B_N}$$

$$= \frac{(140 \times 10^{-3})^2}{10 \times 10^3} = 1.96 \times 10^{-6} \ (\text{V}^2/\text{Hz})$$

From equation (11.5):

$$P_e = \frac{1}{2} \left[1 - \text{erf} \ \frac{1}{\sqrt{2}} \left(\frac{\langle E \rangle}{N_0} \right)^{1/2} \right]$$

$$= \frac{1}{2} \left[1 - \text{erf} \ \frac{1}{\sqrt{2}} \left(\frac{2.5 \times 10^{-5}}{1.96 \times 10^{-6}} \right)^{1/2} \right]$$

$$= \frac{1}{2} \left[1 - \text{erf} \, (2.525) \right]$$

Using error function tables:

$$P_e = \frac{1}{2} \left[1 - 0.999645 \right] = 1.778 \times 10^{-4}$$

11.3.2 Binary phase shift keying (and phase reversal keying)

Binary phase shift keying (BPSK) impresses baseband information onto a carrier, by changing the carrier's phase in sympathy with the baseband digital data, i.e.:

$$f(t) = \begin{cases} A \ \prod(t/T_o) \cos 2\pi f_c t, & \text{for a digital 1} \\ A \ \prod(t/T_o) \cos(2\pi f_c t + \phi), & \text{for a digital 0} \end{cases} \tag{11.10}$$

In principle any two phasor states can be used to represent the binary symbols but usually antipodal states are chosen (i.e. states separated by $\phi = 180°$). For obvious reasons this type of modulation is sometimes referred to as phase reversal keying (PRK). A PRK transmitter with typical baseband and IF waveforms, spectra and allowed phasor states is shown in Figure 11.3. PRK systems must obviously employ coherent detectors which can be implemented as either matched filters or correlators. Since the zero and one symbols are antipodal, Figure 11.3(g), only one receiver channel is needed, Figure 11.4. The post-filtered decision instant voltages are $\pm kE$ (V) where E (V^2 s) is the normalised energy residing in either symbol. The normalised noise power, σ^2 (V^2), at the filter output is the same as in the BASK case. Substituting $\Delta V = 2kE$ and $\sigma^2 = k^2 EN_0/2$ into equation (6.8) gives:

$$P_e = \frac{1}{2}\left[1 - \mathrm{erf}\left(\frac{E}{N_0}\right)^{1/2}\right] \qquad (11.11)$$

(Note that in this case $E = \langle E \rangle$.) Using equation (11.8) the PRK probability of symbol error can be expressed as:

$$P_e = \frac{1}{2}\left[1 - \mathrm{erf}\,(T_o B)^{1/2}\left(\frac{C}{N}\right)^{1/2}\right] \qquad (11.12)$$

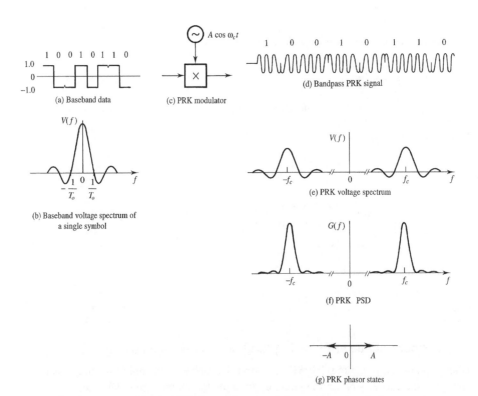

(a) Baseband data

(c) PRK modulator

(d) Bandpass PRK signal

(b) Baseband voltage spectrum of a single symbol

(e) PRK voltage spectrum

(f) PRK PSD

(g) PRK phasor states

Figure 11.3 *Phase reversal keying (PRK) modulator waveforms, spectra and phasor states.*

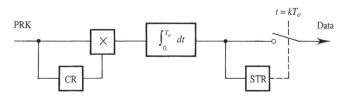

Figure 11.4 *PRK correlation detector.*

For more general BPSK modulation where the difference between phasor states is less than 180° the P_e performance is most easily derived by resolving the allowed phasor states into a residual carrier and a reduced amplitude PRK signal, Figure 11.5. The residual carrier, which contributes nothing to symbol detection, can be employed as a pilot transmission and used for carrier recovery purposes at the receiver. If the difference between phasor states is $\Delta\theta$ (Figure 11.5) and a BPSK 'modulation index', m, is defined by:

$$m = \sin\left(\frac{\Delta\theta}{2}\right) \tag{11.13}$$

then m is the proportion of the transmitted signal voltage which conveys information and the corresponding proportion of total symbol energy is:

$$m^2 = \sin^2\left(\frac{\Delta\theta}{2}\right) = \frac{1}{2}(1 - \cos\Delta\theta) \tag{11.14}$$

$\cos\Delta\theta$ is the scalar product of the two (unit amplitude) symbol phasors. Denoting this quantity by the normalised correlation coefficient ρ, then:

$$m = \sqrt{\tfrac{1}{2}(1 - \rho)} \tag{11.15}$$

The BPSK probability of symbol error is found by replacing E in equation (11.11) with $m^2 E$, i.e.:

$$P_e = \frac{1}{2}\left[1 - \mathrm{erf}\, m\left(\frac{E}{N_0}\right)^{1/2}\right] \tag{11.16}$$

and equation (11.16) can be rewritten as:

$$P_e = \frac{1}{2}\left[1 - \mathrm{erf}\,\frac{\sqrt{1-\rho}}{\sqrt{2}}\left(\frac{E}{N_0}\right)^{1/2}\right] \tag{11.17}$$

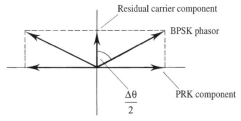

Figure 11.5 *Resolution of BPSK signal into PRK signal plus residual carrier.*

It is, of course, no coincidence that equation (11.17) is identical to equation (8.61) since, as pointed out in both Chapter 8 and section 11.3, this is a general result for all binary systems having equal energy, equiprobable, symbols. The same caution must be exercised when using equations (11.11), (11.12), (11.16) and (11.17) as when using equations (11.4), (11.5) and (11.9) since all those equations assume ISI-free reception. Filtering of PRK signals to limit their bandwidth may, therefore, result in a P_e which is higher than these equations imply.

EXAMPLE 11.2

A 140 Mbit/s ISI-free PRK signalling system uses pulse shaping to constrain its transmission to the double sideband Nyquist bandwidth. The received signal power is 10 mW and the one-sided noise power spectral density is 6.0 pW/Hz. Find the BER expected at the output of an ideal matched filter receiver. If the phase angle between symbols is reduced to 165° in order to provide a residual carrier, find the received power in the residual carrier and the new BER.

The double sided Nyquist bandwidth is given by:

$$B = \frac{1}{T_o}$$

i.e.:

$$T_o B = 1.0$$

$$N = N_0 B = N_0/T_o = N_0 R_s$$

$$= 6.0 \times 10^{-12} \times 140 \times 10^6 = 8.4 \times 10^{-4} \text{ W}$$

Now using equation (11.12):

$$P_e = \frac{1}{2}\left[1 - \text{erf}\,(T_o B)^{1/2}\left(\frac{C}{N}\right)^{1/2}\right] = \frac{1}{2}\left[1 - \text{erf}\left(\frac{10 \times 10^{-3}}{8.4 \times 10^{-4}}\right)^{1/2}\right]$$

$$= \tfrac{1}{2}[1 - \text{erf}\,(3.450)] = \tfrac{1}{2}[1 - 0.999998934] = 5.33 \times 10^{-7}$$

$$\text{BER} = P_e R_s$$

$$= 5.33 \times 10^{-7} \times 140 \times 10^6 = 74.6 \text{ error/s}$$

If $\Delta\theta$ is reduced to 165° then:

$$m = \sin\left(\frac{\Delta\theta}{2}\right) = \sin\left(\frac{165}{2}\right) = 0.9914$$

Proportion of signal power in residual carrier is:

$$1 - m^2 = 1 - 0.9914^2 = 0.0171$$

Power received in residual carrier is therefore:

$$C(1 - m^2) = 10 \times 10^{-3} \times 0.0171 = 1.71 \times 10^{-4} \text{ W}$$

Information bearing carrier power is:

$$C\,m^2 = 10 \times 10^{-3}\,(0.9914)^2 = 9.829 \times 10^{-3} \text{ W}$$

Now from equation (11.12):

$$P_e = \frac{1}{2}\left[1 - \operatorname{erf}(T_o B)^{1/2}\left(\frac{C\,m^2}{N}\right)^{1/2}\right] = \frac{1}{2}\left[1 - \operatorname{erf}\left(\frac{9.829 \times 10^{-3}}{8.4 \times 10^{-4}}\right)^{1/2}\right]$$

$$= \tfrac{1}{2}\,[1 - \operatorname{erf}(3.421)] = \tfrac{1}{2}\,[1 - 0.999998688] = 6.55 \times 10^{-7}$$

$$\text{BER} = P_e R_s$$

$$= 6.55 \times 10^{-7} \times 140 \times 10^6 = 91.7 \text{ error/s}$$

11.3.3 Binary frequency shift keying

Binary frequency shift keying (BFSK) represents digital ones and zeros by carrier pulses with two distinct frequencies, f_1 and f_2, i.e.:

$$f(t) = \begin{cases} A\,\prod(t/T_o)\cos 2\pi f_1 t, & \text{for a digital 1} \\ A\,\prod(t/T_o)\cos 2\pi f_2 t, & \text{for a digital 0} \end{cases} \tag{11.18}$$

Figure 11.6 shows a schematic diagram of a BFSK modulator, signal waveforms and signal spectra. (In practice the BFSK modulator would normally be implemented as a numerically controlled oscillator.) The voltage spectrum of the BFSK signal is the superposition of the two OOK spectra, one representing the baseband data stream modulated onto a carrier with frequency f_1 and one representing the inverse data stream

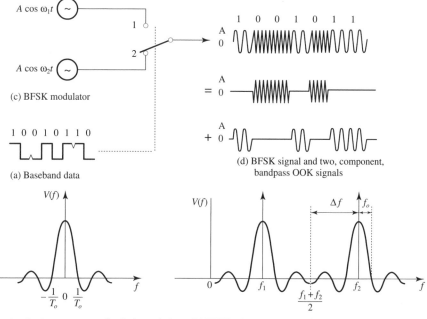

(c) BFSK modulator

(a) Baseband data

(b) Baseband voltage spectrum of a single symbol

(d) BFSK signal and two, component, bandpass OOK signals

(e) BFSK voltage spectrum (of two symbols, 0 and 1). Note overlapping OOK spectra

Figure 11.6 *Binary frequency shift keying (BFSK) modulators, waveforms and spectra.*

modulated onto a carrier with frequency f_2 (Figure 11.6(e)). It is important to realise, however, that the PSD of a BFSK signal is not the superposition of two OOK PSDs. This is because in the region where the two OOK spectra overlap the spectral lines must be added with due regard to their phases. (In practice, when the separation of f_1 and f_2 is large, and the overlap correspondingly small, then the BFSK PSD is *approximately* the superposition of two OOK PSDs.)

Detection of BFSK can be coherent or incoherent although the latter is more common. Incoherent detection suffers the same CNR penalty, compared with coherent detection, as is the case for OOK systems. Figure 11.7 shows coherent and incoherent BFSK receivers. The receiver in Figure 11.7(c) is an FSK version of the quadrature receiver shown in Figure 11.2. The coherent receiver (Figure 11.7(a)) is an FSK version of the PSK receiver in Figure 11.4.

FSK does not provide the noise reduction of wideband analogue FM transmissions. If, however, a BFSK carrier frequency, f_c, Figure 11.6(e), is defined by:

$$f_c = \frac{f_1 + f_2}{2} \quad \text{(Hz)} \tag{11.19}$$

and a BFSK frequency deviation, Δf, is defined (Figure 11.6(e)) as:

$$\Delta f = \frac{f_2 - f_1}{2} \quad \text{(Hz)} \tag{11.20}$$

then using the first zero crossing points in the BFSK voltage spectrum to define its bandwidth, B, gives:

$$B = 2\Delta f + 2f_o \tag{11.21}$$

Here $f_o = 1/T_o$ is both the nominal bandwidth and the baud rate of the baseband data stream. Equation (11.21) is strongly reminiscent of Carson's rule for the bandwidth of an FM signal [Stremler]. If the binary symbols of a BFSK system are orthogonal (see section 2.5.3), i.e.:

$$\int_0^{T_o} \cos(2\pi f_1 t) \cos(2\pi f_2 t) \, dt = 0 \tag{11.22}$$

then, when the output of one channel of a coherent BFSK receiver is a maximum, the output of the other channel will be zero. After subtracting the post-filtered signals arising from each receiver channel the orthogonal BFSK decision instant voltage is:

$$f(nT_o) = \begin{cases} kE, & \text{for a digital 1} \\ -kE, & \text{for a digital 0} \end{cases} \tag{11.23}$$

If the one-sided NPSD at the BFSK receiver input is N_0 (V^2/Hz) then the noise power, $\sigma_1^2 = k^2 E N_0/2$, received via channel 1 and the noise power, $\sigma_2^2 = k^2 E N_0/2$, received via channel 2 will add power-wise (since the noise processes will be uncorrelated). The total noise power at the receiver output will therefore be:

$$\sigma^2 = \sigma_1^2 + \sigma_2^2 = k^2 E N_0 \quad (V^2) \tag{11.24}$$

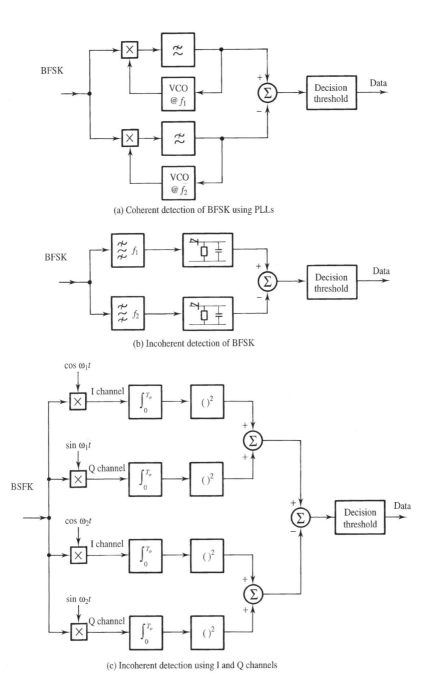

(a) Coherent detection of BFSK using PLLs

(b) Incoherent detection of BFSK

(c) Incoherent detection using I and Q channels

Figure 11.7 *Coherent and incoherent BFSK receivers.*

Table 11.1 P_e *formulae for ideal coherent detection of baseband and IF modulated binary signals.*

		P_e	
Baseband signalling	Unipolar (OOK)	$\frac{1}{2}\,\mathrm{erfc}\,\sqrt{\dfrac{1}{2}\dfrac{E}{N_0}}$ (8.61)	$\frac{1}{2}\,\mathrm{erfc}\,\sqrt{\dfrac{1}{4}\dfrac{S}{N}}$ (6.10(b))
	Polar	$\frac{1}{2}\,\mathrm{erfc}\,\sqrt{\dfrac{E}{N_0}}$ (8.61)	$\frac{1}{2}\,\mathrm{erfc}\,\sqrt{\dfrac{1}{2}\dfrac{S}{N}}$ (6.12)
IF/RF signalling	OOK	$\frac{1}{2}\,\mathrm{erfc}\,\sqrt{\dfrac{1}{2}\dfrac{E}{N_0}}$	$\frac{1}{2}\,\mathrm{erfc}\,\sqrt{\dfrac{T_o B}{2}\dfrac{C}{N}}$
	BFSK (orthogonal)	$\frac{1}{2}\,\mathrm{erfc}\,\sqrt{\dfrac{1}{2}\dfrac{E}{N_0}}$	$\frac{1}{2}\,\mathrm{erfc}\,\sqrt{\dfrac{T_o B}{2}\dfrac{C}{N}}$
	PRK	$\frac{1}{2}\,\mathrm{erfc}\,\sqrt{\dfrac{E}{N_0}}$	$\frac{1}{2}\,\mathrm{erfc}\,\sqrt{T_o B\,\dfrac{C}{N}}$

Substituting $\Delta V = 2kE$ from equation (11.23) and $\sigma = k\sqrt{EN_0}$ from equation (11.24) into equation (6.8) gives the probability of error for coherently detected orthogonal BFSK:

$$P_e = \frac{1}{2}\left[1 - \mathrm{erf}\,\frac{1}{\sqrt{2}}\left(\frac{E}{N_0}\right)^{\!\frac{1}{2}}\right] \tag{11.25}$$

Notice that here, as for BPSK signalling, $E = \langle E \rangle$. Since equation (11.25) is identical to equation (11.5) for OOK modulation the BFSK expression for P_e in terms of CNR is identical to equation (11.9).

Table 11.1 compares the various formulae for P_e obtained in section 11.3, Chapter 6 and Chapter 8.

11.3.4 BFSK symbol correlation and Sunde's FSK

Equation (11.25) applies to the coherent detection of orthogonal BFSK. It is, of course, possible to choose symbol frequencies (f_1 and f_2) and a symbol duration (T_o) such that the symbols are not orthogonal, i.e.:

$$\int_0^{T_o} \cos(2\pi f_1 t)\cos(2\pi f_2 t)\,dt \neq 0 \tag{11.26}$$

If this is the case then there will be non-zero sampling instant outputs from both BFSK receiver channels when either symbol is present at the receiver input. The key to understanding how this affects the probability of symbol error is to recognise that the common fraction of the two symbols (i.e. the fraction of each symbol which is common to the other) can carry no information. The common fraction is the normalised correlation coefficient, ρ, of the two symbols. Denoting a general pair of symbols by $f_1(t)$ and $f_2(t)$ the normalised correlation coefficient is defined by equation (2.89):

$$\rho = \frac{\langle f_1(t)f_2(t)\rangle}{\sqrt{\langle |f_1(t)|^2\rangle}\,\sqrt{\langle |f_2(t)|^2\rangle}} \tag{11.27}$$

(The normalisation process can be thought of as scaling each symbol to an RMS value of 1.0 before cross correlating.) For practical calculations equation (11.27) can be rewritten as:

$$\rho = \frac{1}{f_{1RMS}\,f_{2RMS}}\,\frac{1}{T_o}\int_0^{T_o} f_1(t)f_2(t)\,dt$$

$$= \frac{1}{\sqrt{E_1 E_2}}\int_0^{T_o} f_1(t)f_2(t)\,dt \tag{11.28}$$

and for BFSK systems, in which $E_1 = E_2 = E$, this becomes:

$$\rho = \frac{2}{T_o}\int_0^{T_o}\cos(2\pi f_1 t)\cos(2\pi f_2 t)\,dt \tag{11.29}$$

Figure 11.8 shows BFSK symbol correlation, ρ, plotted against the difference in the number of carrier half cycles contained in the symbols. The zero crossing points represent orthogonal signalling systems and $2(f_2 - f_1)T_o = 1.43$ represents a signalling system somewhere between orthogonal and antipodal, see section 6.4.2. (This is the optimum operating point for BFSK in terms of power efficiency and yields a 0.8 dB CNR saving over orthogonal BFSK. It corresponds to symbols which contain a difference of approximately 3/4 of a carrier cycle.)

Whilst any zero crossing point on the ρ–T_o diagram of Figure 11.8 corresponds to orthogonal BFSK it is not possible to use the first zero (i.e. $2(f_2 - f_1)T_o = 1$) if detection is incoherent. (This can be appreciated if it is remembered that for BFSK operated at this point the two symbols are different by only half a carrier cycle. A difference this small can be detected as a change in phase (of 180°) over the symbol duration but not measured reliably in this time as a change in frequency.) The minimum frequency separation for successful incoherent detection of orthogonal BFSK is therefore given by the second zero crossing point (i.e. $2(f_2 - f_1)T_o = 2$) in Figure 11.8 which corresponds to one carrier cycle difference between the two symbols or $\Delta f = f_o/2$ in Figure 11.6. BFSK operated at the second zero of the ρ–T_o diagram is called Sunde's FSK. The voltage and power

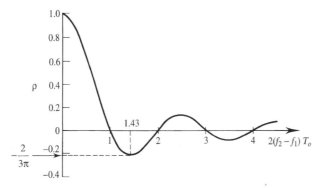

Figure 11.8 *BFSK ρ–T_o diagram, $\rho = \{\sin[\pi(f_2 - f_1)T_o]\cos[\pi(f_2 - f_1)T_o]\}/[\pi(f_2 - f_1)T_o]$.*

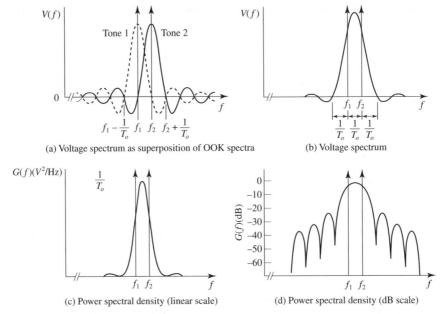

(a) Voltage spectrum as superposition of OOK spectra

(b) Voltage spectrum

(c) Power spectral density (linear scale)

(d) Power spectral density (dB scale)

Figure 11.9 *Comparison of spectra for Sunde's FSK.*

spectra for this scheme are shown in Figure 11.9. If the first spectral zero definition of bandwidth (or equivalently Carson's rule) is applied to Sunde's FSK the bandwidth would be given by:

$$B = (f_2 - f_1) + \frac{2}{T_o} = \frac{3}{T_o} \quad \text{(Hz)} \tag{11.30}$$

The overlapping spectral lines of the component OOK signals, however, result in cancellation giving a $1/f^4$ roll-off in the envelope of the power spectral density. This rapid roll-off allows the practical bandwidth of Sunde's FSK to be taken, for most applications, to be:

$$B = f_2 - f_1 = \frac{1}{T_o} \quad \text{(Hz)} \tag{11.31}$$

EXAMPLE 11.3

A rectangular pulse BFSK system operates at the third zero crossing point of the ρ–T_o diagram of Figure 11.8. The maximum available transmitter power results in a carrier power at the receiver input of 60 mW. The one-sided NPSD referred to the same point is 0.1 nW/Hz. What is the maximum bit rate which the system can support if the probability of bit error is not to fall below 10^{-6}? What is the nominal bandwidth of the BFSK signal if the lower frequency symbol has a frequency of 80 MHz?

$$P_e = \frac{1}{2} \left[1 - \text{erf} \frac{1}{\sqrt{2}} \left(\frac{E}{N_0} \right)^{\frac{1}{2}} \right]$$

i.e.:

$$\frac{E}{N_0} = 2 \left[\text{erf}^{-1} (1 - 2P_e) \right]^2 = 2 \left[\text{erf}^{-1} \left(1 - 2 \times 10^{-6} \right) \right]^2$$

$$= 2 \left[\text{erf}^{-1} (0.999998) \right]^2 = 2 [3.361]^2 = 22.59$$

Therefore:

$$E = 22.59 \, N_0 = 22.59 \times 0.1 \times 10^{-9} = 2.259 \times 10^{-9} \quad \text{J}$$

Now from equation (11.6):

$$T_o = \frac{E}{C} = \frac{2.259 \times 10^{-9}}{60 \times 10^{-3}} = 3.765 \times 10^{-8} \text{ s}$$

$$R_b = R_s = \frac{1}{T_o} = \frac{1}{3.765 \times 10^{-8}} = 2.656 \times 10^7 \text{ bit/s}$$

Since operation on the diagram of Figure 11.8 is at $n = 3$ then:

$$3 = 2 (f_2 - f_1) T_o$$

$$f_2 = \frac{3}{2T_o} + f_1$$

$$= \frac{3}{2 \times 3.765 \times 10^{-8}} + 80 \times 10^6 = 1.198 \times 10^8 \text{ Hz}$$

$$\Delta f = \frac{f_2 - f_1}{2} = \frac{119.8 - 80}{2} = 19.9 \text{ MHz}$$

and from equation (11.21):

$$B = 2 (\Delta f + f_o)$$

$$= 2 (19.9 + 26.6) = 93.0 \text{ MHz}$$

11.3.5 Comparison of binary shift keying techniques

It is no coincidence that OOK and orthogonal FSK systems are equally power efficient, i.e. have the same probability of symbol error for the same $\langle E \rangle / N_0$, Table 11.1. This is because they are both orthogonal signalling schemes. PRK signalling is antipodal and, therefore, more power efficient, i.e. it has a better P_e performance or, alternatively, a power saving of 3 dB. If comparisons are made on a peak power basis then orthogonal FSK requires 3 dB more power than PRK, as expected, but also requires 3 dB less power

Table 11.2 *Relative power efficiencies of binary bandpass modulation schemes.*

	Bandpass OOK	*Orthogonal BFSK*	*PRK*
$\dfrac{E_1}{N_0}$	4	2	1
$\dfrac{\langle E \rangle}{N_0}$	2	2	1

than OOK since all the energy of the OOK transmission is squeezed into only one type of symbol. These relative power efficiencies are summarised in Table 11.2.

Although the spectral efficiency for all unfiltered (i.e. rectangular pulse) signalling systems is, strictly, 0 bit/s/Hz (see section 8.2) it is possible to define a nominal spectral efficiency using the signal pulse's main lobe bandwidth, $B = 2/T_o$. (This does not correspond to the theoretical minimum channel bandwidth which is $B = 1/T_o$.) Table 11.3 summarises the nominal spectral efficiencies of unfiltered bandpass BASK, BFSK and BPSK systems and compares them with the nominal efficiency of unfiltered baseband OOK. It is interesting to note that Sunde's FSK could be regarded as being more spectrally efficient than BASK and BPSK due to the more rapid roll-off of its spectral envelope. (If the null-to-null definition of bandwidth is adhered to strictly then Sunde's FSK has a nominal efficiency of 0.33 bit/s/Hz which is less than BASK and BPSK.) The probability of symbol error against $\langle E \rangle / N_0$ for orthogonal (OOK and orthogonal BFSK) and antipodal (PRK) signalling is shown in Figure 11.10.

It is useful to develop the generalised expression for P_e which can be applied to all binary (coherently detected, ideal) modulation schemes. For any two-channel coherent receiver the decision instant voltages for equal energy symbols will be:

$$f(nT_o) = \begin{cases} k \left[E - \int_0^{T_o} f_1(t) f_2(t) \, dt \right], & \text{symbol 0 present} \\[3mm] k \left[-E + \int_0^{T_o} f_1(t) f_2(t) \, dt \right], & \text{symbol 1 present} \end{cases}$$

$$= \begin{cases} k \left[E - \rho E \right], & \text{symbol 0 present} \\[2mm] k \left[-E + \rho E \right], & \text{symbol 1 present} \end{cases} \tag{11.32}$$

Table 11.3 *Relative spectral efficiencies of binary bandpass modulation schemes.*

	Baseband BASK	Bandpass BASK	Orthogonal BFSK $(n \geq 3)$*	Sunde's BFSK $(n = 2)$*	BPSK
Data rate (bit/s)	$1/T_o$	$1/T_o$	$1/T_o$	$1/T_o$	$1/T_o$
Nominal bandwidth (Hz)	$1/T_o$	$2/T_o$	$(n+4)/2T_o$	$1/T_o$† or $3/T_o$	$2/T_o$
Nominal spectral efficiency (bit/s/Hz)	1	1/2	$2/(n+4)$	1† or 1/3	1/2
Minimum ‡ (ISI-free) bandwidth	$1/2T_o$	$1/T_o$	$(n+2)/2T_o$	$2/T_o$	$1/T_o$
Maximum ‡ spectral efficiency	2	1	$2/(n+2)$	1/2	1

* n = zero crossing operating point on ρ–T_o diagram of Figure 11.8
† Depends on definition of bandwidth
‡ Based on absolute bandwidth

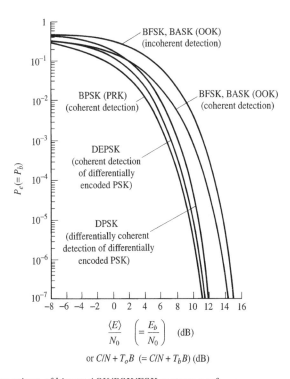

Figure 11.10 *Comparison of binary ASK/PSK/FSK systems performance.*

The decision instant voltage difference between symbols is:

$$\Delta V = k\left[2E - 2\rho E\right] = k2E(1 - \rho) \tag{11.33(a)}$$

The two channels result in a total noise power of:

$$\sigma^2 = k^2 E N_0 (1 - \rho) \tag{11.33(b)}$$

which using equation (6.8) gives a probability of error of:

$$P_e = \frac{1}{2}\left[1 - \text{erf} \sqrt{\frac{1-\rho}{2}\left(\frac{E}{N_0}\right)^{\frac{1}{2}}}\right] \tag{11.34(a)}$$

where the correlation coefficient $\rho = 0$ for OOK, $\rho = 0$ for orthogonal BFSK, $\rho = -2/3\pi$ for optimum BFSK and $\rho = -1$ for PRK. (It can now be seen why ρ was chosen to represent $\cos \Delta\theta$ in equation (11.15).) The equivalent expression in terms of C/N is:

$$P_e = \frac{1}{2}\left[1 - \text{erf} \sqrt{\frac{1-\rho}{2}(T_oB)^{\frac{1}{2}}\left(\frac{C}{N}\right)^{\frac{1}{2}}}\right] \tag{11.34(b)}$$

11.3.6 Carrier recovery, phase ambiguity and DPSK

Coherent detection of IF modulated signals is required to achieve the lowest error rate, Figure 11.10. This usually requires a reference signal which replicates the phase of the signal carrier. A residual or pilot carrier, if present in the received signal, can be extracted using a filter or phase locked loop or both, then subsequently amplified and employed as a coherent reference. This is possible (though not often implemented) for ASK and FSK signals since both contain discrete lines in their spectra. BPSK signals with phasor states separated by less than 180° also contain a spectral line at the carrier frequency and can therefore be demodulated in the same way. PRK signals, however, contain no such line and carrier recovery must therefore be achieved using alternative methods.

One technique which can be used is to square the received PRK transmission creating a double frequency carrier (with no phase transitions). This occurs because $\sin^2 2\pi f_c t$ and $\sin^2(2\pi f_c t + \pi)$ are both equal to $\sin 2\pi 2 f_c t$. A phase locked loop can then be used as a frequency divider to generate the coherent reference. Such a carrier recovery circuit is shown in Figure 11.11. (Note that the phase locked loop locks to a reference 90° out of phase with the input signal. Thus a 90° phase shifting network is required either within the loop, or between the loop and the demodulator, to obtain the correctly phased reference signal for demodulation.) A second technique is to use a Costas loop [Dunlop and Smith, Lindsey and Simon] in place of a conventional PLL. The Costas loop, Figure 11.12, consists essentially of two PLLs operated in phase quadrature and has the property that it will lock to the suppressed carrier of a PRK signal.

Both the squaring loop and Costas loop solutions to PRK carrier recovery suffer from a 180° phase ambiguity, i.e. the recovered carrier may be either inphase or in antiphase with the transmitted (suppressed) carrier. This ambiguity can lead to symbol inversion of the demodulated data. Whilst data inversion might be tolerated by some applications, in others it would clearly be disastrous. There are two distinct approaches to resolving phase ambiguities in the recovered carrier.

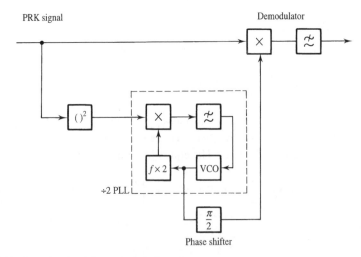

Figure 11.11 *Squaring loop for suppressed carrier recovery.*

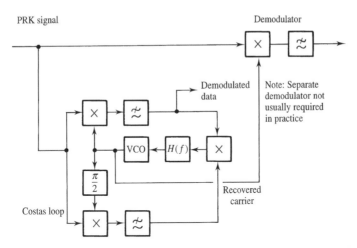

Figure 11.12 *Costas loop for suppressed carrier recovery.*

The first involves periodic transmission of a known data sequence. (This sequence is normally one part of a larger 'preamble' sequence transmitted prior to a block of data, see Chapters 18 and 20.) If the received 'training' sequence is inverted this is detected and a second inversion introduced to correct the data. The second approach is to employ differential encoding of the data before PRK modulation (Figure 11.13). This results, for example, in digital ones being represented by a phase transition and digital zeros being represented by no phase transition. The phase ambiguity of the recovered carrier then becomes irrelevant. Demodulation of such encoded PSK signals can be implemented using conventional carrier recovery and detection followed by baseband differential decoding (Figure 11.14). (STR techniques are described in section 6.7.) Systems using this detection scheme are called differentially encoded PSK (DEPSK) and have a probability of symbol error, P'_e, given by:

$$P'_e = P_e(1 - P_e) + P_e(1 - P_e)$$
$$= 2(P_e - P_e^2) \tag{11.35}$$

where P_e is the probability of error for uncoded PSK.

The first term, $P_e(1 - P_e)$, in equation (11.35) is the probability that the current symbol in the decoder is in error and the previous symbol is correct. The second (identical) term is the probability that the current symbol is correct and the previous symbol is in error. (If both current and previous symbols are detected in error the

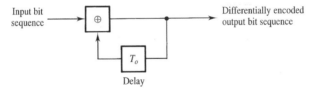

Figure 11.13 *Differential encoding.*

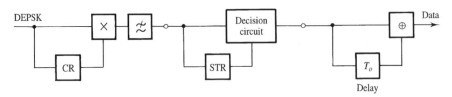

Figure 11.14 *Differentially encoded PSK (DEPSK) detection.*

decoded symbol will, of course, be correct.)

An alternative method of demodulating PSK signals with differential coding is to use one symbol as the coherent reference for the next symbol (Figure 11.15). Systems using this detection scheme are called differential PSK (DPSK). They have the advantage of simpler, and therefore cheaper, receivers compared to DEPSK systems. They have the disadvantage, however, of having a noisy reference signal and therefore degraded P'_e performance compared to DEPSK and coherent PSK, Figure 11.10. It might appear that since reference and signal are equally noisy the P'_e degradation in DPSK systems would correspond to a 3 dB penalty in CNR. In practice the degradation is significantly less than this (typically 1 dB) since the noise in the signal and reference channels is correlated. DEPSK and DPSK techniques are easily extended from biphase to multiphase signalling (see section 11.4.2).

11.4 Modulation techniques with increased spectral efficiency

Spectral efficiency, η_s, as defined in equation (8.1), depends on symbol (or baud) rate, R_s, signal bandwidth, B, and entropy, H (as defined in equation (9.3)), i.e.:

$$\eta_s = \frac{R_s H}{B} \quad \text{(bit/s/Hz)} \tag{11.36(a)}$$

Since $R_s = 1/T_o$ and $H = \log_2 M$, for statistically independent, equiprobable symbols (section 9.2.4) then η_s can be expressed as:

$$\eta_s = \frac{\log_2 M}{T_o B} \quad \text{(bit/s/Hz)} \tag{11.36(b)}$$

It is apparent from equation (11.36(b)) that spectral efficiency is maximised by making the symbol alphabet size, M, large and the $T_o B$ product small. This is exactly the strategy employed by spectrally efficient modulation techniques.

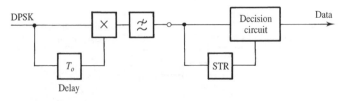

Figure 11.15 *Differential PSK (DPSK) detection.*

Pulse shaping (or filtering) to decrease B for a given baud rate, $R_s = 1/T_o$, has already been discussed in Chapter 8. There, however, the discussion was in the context of baseband signals where the minimum $T_o B$ product (avoiding ISI) was limited by:

$$T_o B \geq 0.5 \qquad\qquad (11.37(\text{a}))$$

and the minimum bandwidth $B = 1/(2T_o)$ was called the single sided Nyquist bandwidth. In the context of IF modulation (i.e. bandpass signals) the modulation process results in a double sideband (DSB) signal. The minimum ISI-free $T_o B$ product is then given by:

$$T_o B \geq 1.0 \qquad\qquad (11.37(\text{b}))$$

and the minimum bandwidth is now $B = 1/T_o$, sometimes called the double sided Nyquist bandwidth. Dramatic increases in η_s, however, must usually come from increased alphabet size. Operational systems currently exist with $M = 64, 128, 256$ and 1024. In principle such multisymbol signalling can lead to increased spectral efficiency of MASK, MPSK and MFSK systems (the first letter in each case represents M-symbol). Only MASK, MPSK and hybrid combinations of these two are used in practice, however. This is because MFSK signals are normally designed to retain orthogonality between all symbol pairs. In this case increasing M results in an approximately proportional increase in B which actually results in a decrease in spectral efficiency. MASK and MPSK signals, however, are limited to alphabets of two (OOK) and four (4-PSK) symbols respectively if orthogonality is to be retained. MASK and MPSK therefore must sacrifice orthogonality to achieve values of M greater than four.

Sub-orthogonal signalling requires greater transmitted power than orthogonal signalling for the same P_e. MASK and MPSK systems are therefore spectrally efficient at the expense of increased transmitted power. Conversely, (orthogonal) MFSK systems are power efficient at the expense of increased bandwidth. This is one manifestation of the general trade-off which can be made between power and bandwidth represented at its most fundamental by the Shannon–Hartley law.

11.4.1 Channel capacity

The Shannon–Hartley channel capacity theorem states that the maximum rate of information transmission, $R_{\max}$, over a channel with bandwidth B and signal-to-noise ratio S/N is given by:

$$R_{\max} = B \log_2 \left(1 + \frac{S}{N}\right) \quad \text{bit/s} \qquad\qquad (11.38(\text{a}))$$

Thus as B increases S/N can be decreased to compensate [Hartley]. Note that in a 3.2 kHz wide audio channel with an SNR of 1000 (30 dB) the theoretical maximum bit rate, $R_{\max}$, is slightly in excess of 30 kbit/s. The corresponding maximum spectral efficiency, $R_{\max}/B$, for a channel with 30 dB SNR is 10 bit/s/Hz. We can determine the value that the channel capacity approaches as the channel bandwidth B tends to infinity:

$$R_\infty = \lim_{B \to \infty} B \log_2 \left(1 + \frac{S}{N_0 B}\right)$$

$$= \lim_{B \to \infty} \frac{S}{N_0} \log_2 \left(1 + \frac{S}{N_0 B}\right)^{N_0 B/S} \qquad\qquad (11.38(\text{b}))$$

$$= \frac{S}{N_0} \log_2 e = 1.44 \frac{S}{N_0}$$

This gives the maximum possible channel capacity as a function of signal power and noise power spectral density. In an actual system design, the channel capacity might be compared to this value to decide whether a further increase in bandwidth is worth while. In a practical system it is realistic to attempt to achieve a transmission rate of one-half R_∞. Some popular modulation schemes which trade an increase in SNR for a decrease in bandwidth, to achieve improved spectral efficiency, are described below.

11.4.2 *M*-symbol phase shift keying

In the context of PSK signalling M-symbol PSK (i.e. MPSK) implies the extension of the number of allowed phasor states from 2 to 4, 8, 16, $\cdots$ (i.e. 2^n). The phasor diagram (also called the constellation diagram) for 16-PSK is shown in Figure 11.16, where there are now 16 distinct states. The constellation diagram, equivalent to an Argand diagram, plots the complex baseband representation of the MPSK modulated signal vector $f(t)$ in terms of its real and imaginary baseband components, as an extension of the BPSK representation in Figure 11.3(g). Note that as the signal amplitude is constant these states all lie on a circle in the complex plane. Four-phase modulation (4-PSK) with phase states $\pi/4$, $3\pi/4$, $5\pi/4$ and $7\pi/4$ can be considered to be a superposition of two PRK signals using quadrature ($\cos 2\pi f_c t$ and $\sin 2\pi f_c t$) carriers. This type of modulation (usually called quadrature or quaternary phase shift keying, QPSK) is discussed separately in section 11.4.4.

The probability of symbol error for MPSK systems is found by integrating the 2-dimensional pdf of the noise centred on the tip of each signal phasor in turn over the corresponding error region and averaging the results. The error region for phasor state 0 is shown as the unhatched region in Figure 11.17. It is possible to derive an exact, but complicated, integral expression for the probability of symbol error in the presence of Gaussian noise. Accurate P_e curves found by numerical evaluation of this expression are shown in Figure 11.18(a). A simple but good approximation [Stein and Jones] which is useful for $M \geq 4$ is:

$$P_e \approx 1 - \mathrm{erf}\left[\sin\left(\frac{\pi}{M}\right)\left(\frac{E}{N_0}\right)^{1/2} \right] \tag{11.39(a)}$$

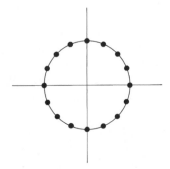

Figure 11.16 *Phasor states (i.e. constellation diagram) for 16-PSK.*

(Note that when $M = 2$ this expression gives a P_e which is twice the correct result.) The approximation in equation (11.39(a)) becomes better as both M and E/N_0 increase. Equation (11.39(a)) can be rewritten in terms of CNR using $C = E/T_o$ and $N = N_0B$ from equations (11.6) and (11.7) as usual, i.e.:

$$P_e \approx 1 - \text{erf}\left[(T_oB)^{1/2} \sin\left(\frac{\pi}{M} \right)\left(\frac{C}{N} \right)^{1/2} \right]$$ (11.39(b))

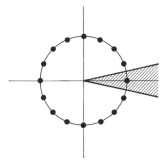

Figure 11.17 *Error region (unhatched) for $\phi = 0°$ state of a 16-PSK signal.*

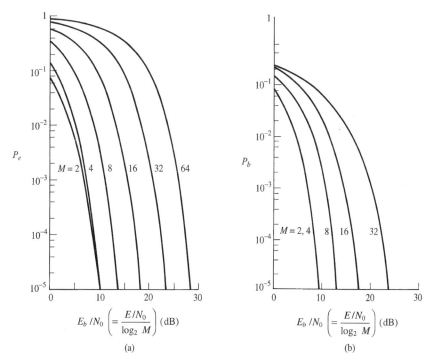

Figure 11.18 *MPSK symbol and bit error probabilities: (a) probability of symbol error, P_e, against E_b/N_0; and (b) probability of bit error, P_b, against E_b/N_0.*

Multisymbol signalling can be thought of as a coding or bit mapping process in which n binary symbols (bits) are mapped into a single M-ary symbol (as discussed in section 6.4.7), except that here each symbol is an IF pulse. A detection error in a single symbol can therefore translate into several errors in the corresponding decoded bit sequence. The probability of bit error, P_b, therefore depends not only on the probability of symbol error, P_e, and the symbol entropy, $H = \log_2 M$, but also on the code or bit mapping used and the types of error which occur. For example, in the 16-PSK scheme shown in Figure 11.17 the most probable type of error involves a given phasor state being detected as an adjacent state. If a Gray code is used to map binary symbols to phasor states this type of error results in only a single decoded bit error. In this case, providing the probability of errors other than this type is negligible, then the bit error probability is:

$$P_b = \frac{P_e}{\log_2 M} \tag{11.40(a)}$$

In order to compare the performance of different modulation schemes on an equitable basis it is useful to express performance in terms of P_b as a function of average energy per bit, E_b. Since the energy, E, of all symbols in an MPSK system is identical the average energy per bit is:

$$E_b = \frac{E}{\log_2 M} \tag{11.40(b)}$$

Substituting equations (11.40) into (11.39):

$$P_b = \frac{1}{\log_2 M} \left\{ 1 - \mathrm{erf}\left[\sin\left(\frac{\pi}{M}\right) \sqrt{\log_2 M} \left(\frac{E_b}{N_0}\right)^{\!\frac{1}{2}} \right] \right\} \tag{11.41(a)}$$

In terms of CNR this becomes:

$$P_b = \frac{1}{\log_2 M} \left\{ 1 - \mathrm{erf}\left[(T_o B)^{\frac{1}{2}} \sin\left(\frac{\pi}{M}\right)\left(\frac{C}{N}\right)^{\!\frac{1}{2}} \right] \right\} \tag{11.41(b)}$$

More accurate closed form expressions for Gray coded and natural binary bit mapped MPSK P_b performance have been derived [Irshid and Salous]. Figure 11.18(a) shows the probability of symbol error, P_e, against E_b/N_0, and Figure 11.18(b) shows P_b against E_b/N_0. (Figure 13.41 clarifies the distinction between P_e and P_b.)

Differential MPSK (DMPSK) can be used to simplify MPSK receiver design and circumvent the phase ambiguity normally present in the recovered carrier. Binary digits are mapped to phase difference between adjacent symbols and each symbol is detected at the receiver using the previous symbol as a coherent reference. The P_e performance of DMPSK is degraded over that for MPSK since the reference and received symbol are now equally noisy. As for DPSK the degradation is not equivalent to a CNR reduction of 3 dB because the noise present in the signal and reference channels is correlated and therefore, to some extent, cancels. Phase noise correlation makes increasingly little difference, however, to the performance degradation as M increases and the phasor states become crowded together. In the limit of large M the degradation approaches the expected 3 dB limit.

Since each symbol of an MPSK signal has an identical amplitude spectrum to all the other symbols the spectral occupancy of MPSK depends only on baud rate and pulse shaping and is independent of M. For unfiltered MPSK (i.e. MPSK with rectangular pulses) the nominal (i.e. main lobe null-to-null) bandwidth is $2/T_o$ Hz. In this case the spectral efficiency (given by equation (11.36(b))), is:

$$\eta_s = 0.5 \log_2 M \quad (\text{bit/s/Hz}) \tag{11.42(a)}$$

The maximum possible, ISI-free, spectral efficiency occurs when pulse shaping is such that signalling takes place in the double sided Nyquist bandwidth $B = 1/T_o$ Hz, i.e.:

$$\eta_s = \log_2 M \quad (\text{bit/s/Hz}) \tag{11.42(b)}$$

Thus we usually say BPSK has an efficiency of 1 bit/s/Hz and 16-PSK 4 bit/s/Hz. Table 11.4 compares the performance of several PSK systems.

Table 11.4 *Comparison of several PSK modulation techniques.*

	Required E_b/N_0 for $P_b = 10^{-6}$	*Minimum channel bandwidth for ISI-free signalling (R_b = bit rate)*	*Max spectral efficiency (bit/s/Hz)*	*Required CNR in minimum channel bandwidth*
PRK	10.6 dB	R_b	1	10.6 dB
QPSK	10.6 dB	$0.5R_b$	2	13.6 dB
8-PSK	14.0 dB	$0.33R_b$	3	18.8 dB
16-PSK	18.3 dB	$0.25R_b$	4	24.3 dB

EXAMPLE 11.4

An MPSK, ISI-free, system is to operate with 2^N PSK symbols over a 120 kHz channel. The minimum required bit rate is 900 kbit/s. What minimum CNR is required to maintain reception with a P_b no worse than 10^{-6}?

Maximum (ISI-free) baud rate:

$$R_s = 1/T_o = B$$

Therefore $R_s \leq 120$ kbaud (or k symbol/s). Minimum required entropy is therefore given by:

$$H \geq \frac{R_b}{R_s} = \frac{900 \times 10^3}{120 \times 10^3} = 7.5 \quad \text{bit/symbol}$$

Minimum number of symbols required is given by:

$$H \leq \log_2 M$$

$$M \geq 2^H = 2^{7.5}$$

Since M must be an integer power of 2:

$$M = 2^8 = 256 \quad \text{and} \quad H = \log_2 M = 8$$

For Gray coding of bits to PSK symbols:

$$P_e = P_b \log_2 M = 10^{-6} \log_2 256 = 8 \times 10^{-6}$$

$$P_e = 1 - \text{erf}\left[(T_o B)^{\frac{1}{2}} \sin\left(\frac{\pi}{M}\right)\left(\frac{C}{N}\right)^{\frac{1}{2}} \right]$$

Now:

$$R_s = \frac{R_b}{\log_2 M} = \frac{900 \times 10^3}{8} = 112.5 \times 10^3 \text{ baud}$$

$$T_o = \frac{1}{R_s} = 8.889 \times 10^{-6} \text{ s}$$

$$T_o B = 8.889 \times 10^{-6} \times 120 \times 10^3 = 1.067$$

Thus:

$$\frac{C}{N} = \left[\frac{\text{erf}^{-1}(1 - P_e)}{(T_o B)^{\frac{1}{2}} \sin\left(\dfrac{\pi}{M}\right)} \right]^2 = \left[\frac{\text{erf}^{-1}(1 - 8 \times 10^{-6})}{(1.067)^{\frac{1}{2}} \sin\left(\dfrac{\pi}{256}\right)} \right]^2$$

$$= \left[\frac{\text{erf}^{-1}(0.999992)}{1.033 \sin\left(\dfrac{\pi}{256}\right)} \right]^2 = \left(\frac{3.157}{0.01268} \right)^2 = 61988 = 47.9 \text{ dB}$$

11.4.3 Amplitude/phase keying and quadrature amplitude modulation

For an unsaturated transmitter operating over a linear channel it is possible to introduce amplitude as well as phase modulation to give an improved distribution of signal states in the signal constellation. The first such proposal (*c.* 1960) introduced a constellation with two amplitude rings and eight phase states on each ring, Figure 11.19(a). For obvious reasons modulation schemes of this type are called amplitude/phase keying (APK). Subsequently it was observed that with half the number of points on the inner ring, Figure 11.19(b), a 3 dB performance improvement could be gained, as the constellation points are more evenly spaced over 12 distinct phases. The square constellation (Figure 11.19(c)), introduced *c.* 1962, is easier to implement and has a slightly better P_e performance yet. Since a square constellation APK signal can be interpreted as a pair of multilevel ASK (MASK) signals modulated onto quadrature carriers it is normally called quadrature amplitude modulation (QAM) [Hanzo, Webb and Keller]. (The terms APK and QAM are sometimes used interchangeably – here, however, we always use the term

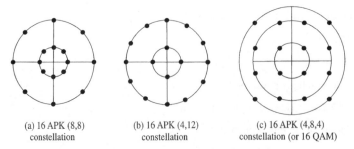

| (a) 16 APK (8,8) | (b) 16 APK (4,12) | (c) 16 APK (4,8,4) |
| constellation | constellation | constellation (or 16 QAM) |

Figure 11.19 *Three possible 16-state QAM signal constellations.*

QAM to represent the square constellation subset of APK.)

Figure 11.20 shows the time domain waveforms for a 16-state QAM constellation which are obtained by encoding binary data in 4 bit sequences; 2 bit sequences are each encoded in the I and Q channels into four-level signals, as shown in Figures 11.20(a),(c) and (b),(d). These are combined to yield the full 16-state QAM, complex, signal, Figure 11.20(e). Ideally all of the 16 states are equiprobable and statistically independent. Note that at the symbol sampling times there are four possible amplitudes in the inphase and quadrature channels reflecting the constellation's 4×4 structure, Figure 11.19(c). In fact the complex signal has three possible amplitude levels with the intermediate value being the most probable.

The 16-QAM signals can be generated as shown in Figure 11.20(g). A 16-QAM signal has two amplitude levels and two distinct phase states in both the real and imaginary channels, see Figures 11.20(c) and (d). Two bits at a time are taken by each of the I and Q channels which use low resolution (2 bit) digital to analogue converters to generate the appropriate bipolar drive signals, Figures 11.20(a) and (b), for the four-quadrant multipliers, Figure 11.20(g). The final summation results in the three-level,

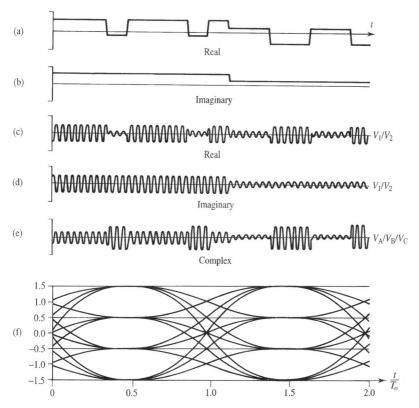

Figure 11.20 *The 16-state QAM signal: (a) four-level baseband signals in the inphase and (b) quadrature branches; (c), (d) corresponding four-level modulated complex signals; (e) resulting combined complex (three-level) QAM signal; (f) demodulated signal eye diagram over two symbols (in the I or Q channels).*

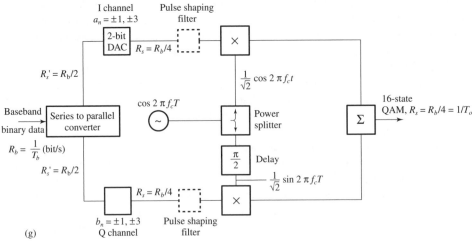

Figure 11.20(g) *Schematic for a 16-QAM modilator.*

12-phase signal of Figure 11.20(e), i.e. the constellation of Figure 11.19(c). As there are separate I and Q channels, and 2 bit converters in each of these, then 4 input bits define each of the 16 symbols giving a spectral efficiency of 4 bit/s/Hz if ideal filtering (or pulse shaping) is employed prior to the I and Q channel multipliers.

A simple approximation for the probability of symbol error for MQAM (M even) signalling in Gaussian noise is [Carlson]:

$$P_e = 2\left\{\frac{M^{1/2}-1}{M^{1/2}}\right\}\left[1 - \text{erf}\sqrt{\frac{3}{2(M-1)}\left(\frac{\langle E \rangle}{N_0}\right)^{1/2}}\right] \tag{11.43}$$

where $\langle E \rangle$ is the average energy per QAM symbol. For equiprobable rectangular pulse symbols $\langle E \rangle$ is given by:

$$\langle E \rangle = \frac{1}{3}\left(\frac{\Delta V}{2}\right)^2 (M-1)\, T_o \tag{11.44}$$

where ΔV is the voltage separation between adjacent inphase or quadrature MASK levels and T_o is the symbol duration. Using $C = \langle E \rangle / T_o$ and $N = N_0 B$, equation (11.43) can be rewritten as:

$$P_e = 2\left\{\frac{M^{1/2}-1}{M^{1/2}}\right\}\left[1 - \text{erf}\sqrt{\frac{3\,T_o B}{2(M-1)}\left(\frac{C}{N}\right)^{1/2}}\right] \tag{11.45}$$

For Gray code mapping of bits along the inphase and quadrature axes of the QAM constellation the probability of bit error P_b is given approximately by equation (11.40(a)). Denoting the average energy per bit by $E_b = \langle E \rangle / \log_2 M$, equation (11.43) can be written as:

$$P_b = \frac{2}{\log_2 M}\left\{\frac{M^{1/2}-1}{M^{1/2}}\right\}\left[1 - \text{erf}\sqrt{\frac{3\log_2 M}{2(M-1)}\left(\frac{E_b}{N_0}\right)^{1/2}}\right] \tag{11.46(a)}$$

and equation (11.45) as:

$$P_b = \frac{2}{\log_2 M} \left\{ \frac{M^{\frac{1}{2}} - 1}{M^{\frac{1}{2}}} \right\} \left[1 - \text{erf} \sqrt{\frac{3\,T_o B}{2(M-1)}} \left(\frac{C}{N} \right)^{\frac{1}{2}} \right] \qquad (11.46(b))$$

Like MPSK all the symbols in a QAM (or APK) signal occupy the same spectral space. The spectral efficiency is therefore identical to MPSK and is given (for statistically

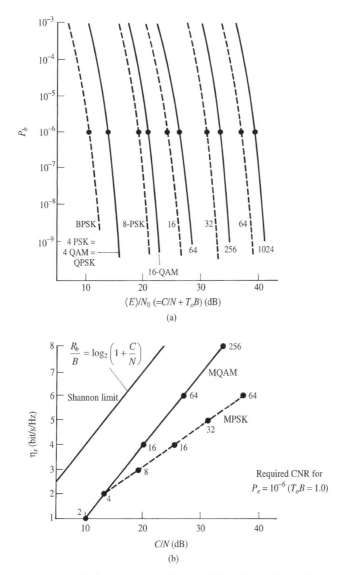

Figure 11.21 *P_e and spectral efficiency for multiphase PSK and M-QAM modulation: (a) bit error probability against CNR with PSK shown as dashed and QAM as solid curves; (b) comparison of the spectral efficiency of these modulation schemes.*

independent equiprobable symbols) by equation (11.36b). Unfiltered and Nyquist filtered APK signals therefore have (nominal and maximum) spectral efficiencies given by equations (11.42(a)) and (11.42(b)) respectively. Figure 11.21(a) compares the bit error probability of MPSK and MQAM systems. Note that the superior constellation packing in QAM over MPSK gives a lower required E_b/N_0 for the same P_b value. Figure 11.21(b) shows the spectral efficiencies of Nyquist filtered ($T_oB = 1$) MQAM and MPSK systems plotted against the CNR required for a bit error probability of 10^{-6}. Table 11.5 compares the C/N and E_b/N_0 required for a selection of MPSK and QAM schemes to yield a probability of bit error of 10^{-6}.

Table 11.5 *Comparison of various digital modulation schemes ($P_b = 10^{-6}$, $T_oB = 1.0$).*

Modulation	C/N ratio (dB)	E_b/N_0 (dB)
PRK	10.6	10.6
QPSK	13.6	10.6
4-QAM	13.6	10.6
8-PSK	18.8	14.0
16-PSK	24.3	18.3
16-QAM	20.5	14.5
32-QAM	24.4	17.4
64-QAM	26.6	18.8

EXAMPLE 11.5

Find the maximum spectral efficiency of ISI-free 16-QAM. What is the noise induced probability of symbol error in this scheme for a received CNR of 24.0 dB if the maximum spectral efficiency requirement is retained? What is the Gray coded probability of bit error?

$$\eta_s = \frac{R_s H}{B} = \frac{H}{T_oB} \quad \text{(bit/s/Hz)}$$

For maximum (ISI-free) spectral efficiency $T_oB = 1$ and $H = \log_2 M$, i.e.:

$$\eta_s = \log_2 M = \log_2 16 = 4 \quad \text{bit/s/Hz}$$

Probability of symbol error is given by equation (11.45) with $T_oB = 1$ and $C/N = 10^{2.4} = 251.2$:

$$P_e = 2\left\{\frac{4-1}{4}\right\}\left[1 - \text{erf}\sqrt{\frac{3}{2(16-1)}(251.2)^{\frac{1}{2}}}\right]$$

$$= 1.5[1 - \text{erf}(5.012)]$$

For $x \geq 4.0$ the error function can be approximated by:

$$\text{erf}(x) = 1 - \frac{e^{-x^2}}{(\sqrt{\pi} \, x)}$$

Therefore:

$$P_e = \frac{1.5 \, e^{-5.012^2}}{(\sqrt{\pi} \, 5.012)} = 2.08 \times 10^{-12}$$

For Gray coding:

$$P_b = \frac{P_e}{\log_2 M} = \frac{2.08 \times 10^{-12}}{\log_2 16} = 5.2 \times 10^{-13}$$

EXAMPLE 11.6

Repeat Example 11.4 as an MQAM system, recalculating the new value for the minimum CNR required to maintain reception with a P_b no worse than 10^{-6} and comment on this new result.

Minimum number of symbols required is still:

$$M \geq 2^H = 2^{7.5}$$

and since M must be an integer power of 2, $M = 2^8 = 256$. Repeating equation (11.46(b)):

$$P_b = \frac{2}{\log_2 M} \left\{ \frac{M^{\frac{1}{2}} - 1}{M^{\frac{1}{2}}} \right\} \left[1 - \operatorname{erf} \sqrt{\frac{3\,T_o B}{2(M-1)} \left(\frac{C}{N} \right)^{\frac{1}{2}}} \right].$$

where $T_o B = 1.067$ as before. Thus:

$$\left(\frac{C}{N} \right)^{\frac{1}{2}} = \frac{\operatorname{erf}^{-1}\left[1 - \left(\dfrac{P_b}{\dfrac{2(M^{\frac{1}{2}}-1)}{M^{\frac{1}{2}} \log_2 M}} \right) \right]}{\sqrt{\dfrac{3 T_o B}{2(M-1)}}} = \frac{\operatorname{erf}^{-1}\left[1 - \left(\dfrac{10^{-6}}{\dfrac{2(16-1)}{16 \times 8}} \right) \right]}{\sqrt{\dfrac{3 \times 1.067}{2 \times 255}}}$$

$$\left(\frac{C}{N} \right) = \frac{\left[\operatorname{erf}^{-1}\left(1 - 10^{-6} \times 4.2667 \right) \right]^2}{(\sqrt{0.0062764})^2} = \frac{(3.251)^2}{0.0062764} = 32.3 \ \text{dB}$$

The required SNR value for 256-QAM is much lower than for the 256-PSK result in Example 11.4. This is consistent with Figure 11.21 noting that the figure is drawn for $P_e = 10^{-6}$ rather than $P_b = 10^{-6}$.

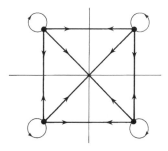

Figure 11.22 *Quadrature phase shift keying (QPSK) signal constellation showing allowed state transitions.*

11.4.4 Quadrature phase shift keying (QPSK) and offset QPSK

Quadrature phase shift keying (QPSK) [Aghvami] can be interpreted either as 4-PSK with carrier amplitude A (i.e. quaternary PSK) or as a superposition of two (polar) BASK signals with identical 'amplitudes' $\pm A/\sqrt{2}$ and quadrature carriers $\cos 2\pi f_c t$ and $\sin 2\pi f_c t$, i.e. 4-QAM. The constellation diagram of a QPSK signal is shown in Figure 11.22. The accompanying arrows show that all transitions between the four states are possible. Figure 11.23 shows a schematic diagram of a QPSK transmitter and Figure 11.24 shows the receiver.

The transmitter and receiver are effectively two PRK transmitters and receivers arranged in phase quadrature, the inphase (I) and quadrature (Q) channels each operating at half the bit rate of the QPSK system as a whole. If pulse shaping and filtering are

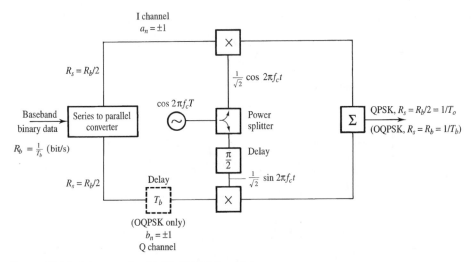

Figure 11.23 *Schematic for QPSK (OQPSK) modulator.*

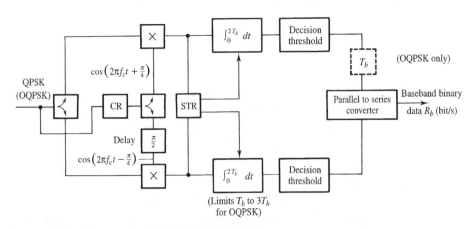

Figure 11.24 *Schematic for QPSK (OQPSK) demodulator.*

absent the signal is said to be unfiltered, or rectangular pulse, QPSK. Figure 11.25 shows an example sequence of unfiltered QPSK data and Figures 11.26(a) and (b) show the corresponding power spectral density with $T_o = 2T_b$.

The spectral efficiency of QPSK is twice that for BPSK. This is because the symbols in each quadrature channel occupy the same spectral space and have half the spectral width of a BPSK signal with the same data rate as the QPSK signal. This is illustrated in

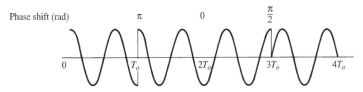

Figure 11.25 *Unfiltered (constant envelope) QPSK signal ($T_o = 2T_b$).*

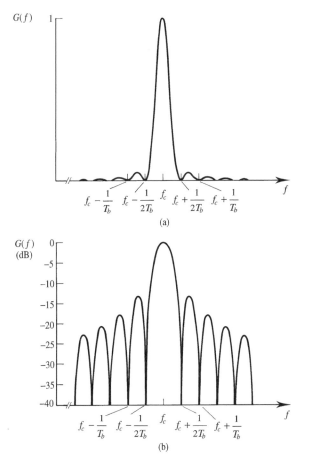

Figure 11.26 *Power spectral density of QPSK/OQPSK signals. $G(f) \propto [(\sin(4\pi fT_b))/(4\pi fT_b)]^2$ on (a) linear; and (b) dB scales.*

the form of orthogonal voltage spectra in Figure 11.27.

The P_e performance of QPSK systems will clearly be worse than that of PRK systems, Figure 11.18, since the decision regions on the constellation diagram, Figure 11.22, are reduced from half spaces to quadrants. The P_b performances of these modulation schemes are, however, identical. This is easily appreciated by recognising that each channel (I and Q) of the QPSK system is independent of (orthogonal to) the other. In principle, therefore, the I channel (binary) signal could be transmitted first followed by the Q channel signal. The total message would take twice as long to transmit in this form but, because each QPSK I or Q channel symbol is twice the duration and half the power of the equivalent PRK symbols, the total message energy (and therefore the energy per bit, E_b) is the same in both the QPSK and PRK cases, Figure 11.28. The P_b performance of ideal QPSK signalling is therefore given by equation (11.11) with E interpreted as the energy per bit, E_b, i.e.:

$$P_b = \frac{1}{2}\left[1 - \mathrm{erf}\left(\frac{E_b}{N_0}\right)^{\frac{1}{2}}\right] \tag{11.47(a)}$$

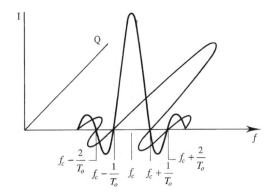

Figure 11.27 *QPSK orthogonal inphase (I) and quadrature (Q) voltage spectra ($T_o = 2T_b$).*

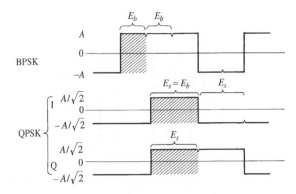

Figure 11.28 *Envelopes of (unfiltered) BPSK and equivalent QPSK signals showing distribution of bit energy in I and Q channels.*

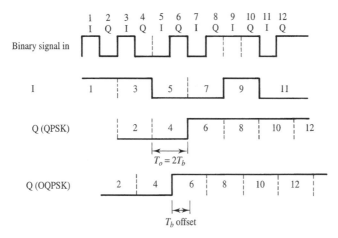

Figure 11.29 *Input bit stream and I and Q channel bit streams for QPSK and OQPSK systems.*

In terms of CNR, therefore:

$$P_b = \frac{1}{2}\left[1 - \text{erf} (T_b B)^{\frac{1}{2}} \left(\frac{C}{N}\right)^{\frac{1}{2}}\right] \tag{11.47(b)}$$

where the bit period T_b is half the QPSK symbol period, T_o. This is the origin of the statement that the BER performances of QPSK and PRK systems are identical. (Note that the minimum $T_o B$ product for ISI-free QPSK signalling is 1.0, corresponding to $T_b B = 0.5$.) The probability of symbol error, P_e, is given by the probability that the I channel bit is detected in error, or the Q channel bit is detected in error, or both channel bits are detected in error, i.e.:

$$P_e = P_b(1 - P_b) + (1 - P_b)P_b + P_b P_b = 2P_b - P_b^2 \tag{11.48}$$

Offset QPSK (OQPSK), also sometimes called staggered QPSK, is identical to QPSK except that, immediately prior to multiplication by the carrier, the Q channel symbol stream is offset with respect to the I channel symbol stream by half a QPSK symbol period (i.e. one bit period), Figure 11.23. The OQPSK relationships between the input binary data stream and the I and Q channel bit streams are shown in Figure 11.29, and the OQPSK constellation and state transition diagram is shown in Figure 11.30. The lines on Figure 11.30 represent the allowed types of symbol transition and show that transitions across the origin (i.e. phase changes of 180°) are, unlike QPSK, prohibited. The offset between I and Q channels means that OQPSK symbol transitions occur, potentially, every T_b seconds (i.e. $T_o = T_b$).

The spectrum, and therefore spectral efficiency, of OQPSK is thus identical to that of QPSK. (Transitions occur more often but, on average, are less severe.) OQPSK has a potential advantage over QPSK, however, when transmitted over a non-linear channel. This is because envelope variations in the signal at the input of a non-linear device produce distortion of the signal (and therefore the signal spectrum) at the output of the device. This effect leads to spectral spreading which may give rise, ultimately, to adjacent channel interference. To minimise spectral spreading (and some other

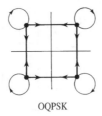

OQPSK

Figure 11.30 *Offset QPSK (OQPSK) constellation diagram showing allowed state transitions.*

undesirable effects, see section 14.3.3) in non-linear channels, constant envelope modulation schemes are desirable. Since some degree of filtering (i.e. pulse shaping) is always present in a modulated RF signal (even if this is unintentional, and due only to the finite bandwidth of the devices used in the modulator) then the 180° phase transitions present in a QPSK signal will result in severe (∞ dB) envelope fluctuation. In contrast the envelope fluctuation of filtered OQPSK is limited to 3 dB since only one quadrature component of the signal can reverse at any transition instant, Figure 11.31. The spectral spreading suffered by an OQPSK signal will therefore be less than that suffered by the equivalent QPSK signal for the same non-linear channel.

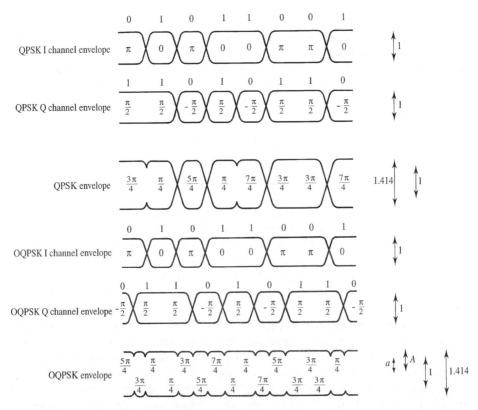

Figure 11.31 *Origin and comparison of QPSK and OQPSK envelope fluctuations ($a/A = 1/\sqrt{2}$).*

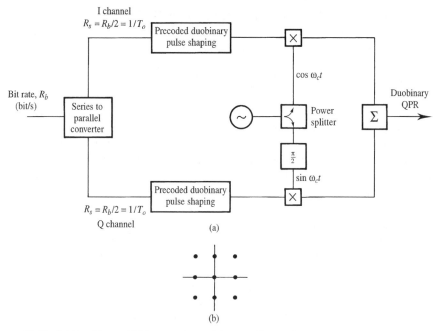

Figure 11.32 *(a) Duobinary QPR transmitter; (b) constellation diagram.*

A further derivative of OQPSK modulation is $\pi/4$ QPSK [Burr] where the transmitted signal is derived by alternating between two separate $\pi/4$ shifted QPSK constellations. This, like OQPSK, has the effect of prohibiting transitions across the origin of the constellation diagram and thus limits amplitude modulation (though not to such low levels as OQPSK). $\pi/4$ QPSK offers simpler STR than in OQPSK but carrier recovery becomes a more complicated operation. The $\pi/4$ QPSK constellation is used in the TETRA, IS-54 and IS-136 mobile radio standards, see Chapter 15.

Quadrature partial response systems

Quadrature partial response (QPR) modulation is a quadrature modulated form of partial response signalling (sections 8.2.7 and 8.2.8). A block diagram showing the structure of a duobinary QPR transmitter is shown in Figure 11.32(a) and the resulting duobinary constellation diagram is shown in Figure 11.32(b). The (absolute) bandwidth of the (DSB) QPR signal is $1/T_o$ (Hz) or $1/2T_b$ (Hz) and the spectral efficiency is 2 bit/s/Hz. This is the same spectral efficiency as baseband duobinary signalling since a factor of 2 is gained by using quadrature carriers but a factor of 2 is lost by using DSB modulation. The duobinary QPR probability of bit error in the presence of white Gaussian noise [Taub and Schilling] corresponds to a CNR of approximately 3 dB more than that required for the same P_e performance in QPSK and 2 dB less than that required in 8-PSK.

11.4.5 Minimum shift keying

Minimum shift keying (MSK) is a modified form of OQPSK in that I and Q channel sinusoidal pulse shaping is employed prior to multiplication by the carrier, Figure 11.33 [Pasapathy]. The transmitted MSK signal can be represented by:

$$f(t) = a_n \sin\left(\frac{2\pi t}{4T_b}\right)\cos 2\pi f_c t + b_n \cos\left(\frac{2\pi t}{4T_b}\right)\sin 2\pi f_c t \tag{11.49}$$

where a_n and b_n are the nth I and Q channel symbols. Figures 11.34(a) to (c) show an example sequence of binary data, the corresponding, sinusoidally shaped, I and Q symbols and the resulting MSK signal. MSK signalling is an example from a class of modulation techniques called continuous phase modulation (CPM). It has a symbol constellation which must now be interpreted as a time varying phasor diagram, Figure 11.34(d). The phasor rotates at a constant angular velocity from one constellation point to an adjacent point over the duration of one MSK symbol. (Like OQPSK the MSK symbol period is the same as the bit period of the unmodulated binary data.) When $a_n = b_n$ the phasor rotates clockwise and when $a_n \neq b_n$ the phasor rotates anticlockwise. A consequence of MSK modulation is that one-to-one correspondence between constellation points and the original binary data is lost.

An alternative interpretation of MSK signalling is possible in that it can be viewed as a special case of BFSK modulation (and is therefore sometimes called continuous phase FSK). This is obvious, in that when the phasor in Figure 11.34(d) is rotating anticlockwise the MSK symbol has a constant frequency of $f_c + 1/(4T_b)$ Hz and when rotating clockwise it has a frequency of $f_c - 1/(4T_b)$. This corresponds to BFSK operated at the first zero crossing point of the ρ–T_o diagram (see section 11.3.4). When viewed this way MSK can be seen to be identical to BFSK with inherent differential coding (since frequency f_1 is transmitted when $a_n = b_n$ and frequency f_2 is transmitted when

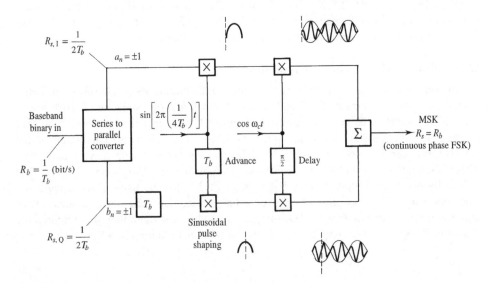

Figure 11.33 *Minimum shift keying (MSK) transmitter.*

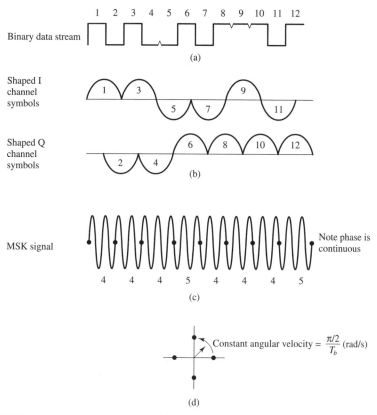

Figure 11.34 *MSK transmitter waveforms and phasor diagrams: (a) binary data; (b) sinusoidally shaped channel symbols; (c) transmitted MSK signal; (d) MSK phasor diagram.*

$a_n \neq b_n$). A one-to-one relationship between bits and frequencies can be re-established, however, by differentially precoding the serial input bit stream prior to MSK modulation. The appropriate precoder is identical to that used for differential BPSK, Figure 11.13. When precoding is used the modulation is called fast frequency shift keying (FFSK) although this term is sometimes also used indiscriminately for MSK without precoding.

Since MSK operates at the first zero on the ρ–T_o diagram of Figure 11.8 one of the BFSK signalling frequencies has an integer number of cycles in the symbol period and the other has either one half cycle less or one half cycle more.

The normalised power spectral density of MSK/FFSK is shown in Figure 11.35. The spectra of MSK/FFSK and QPSK/OQPSK are compared, on a dB scale, in Figure 11.36. As can be seen the MSK spectrum has a broader main lobe but more rapidly decaying sidelobes, which is particularly attractive for FDM systems in order to achieve reduced adjacent channel interference. The probability of bit error for ideal MSK detection, Figure 11.37, is identical to that for QPSK (equations (11.47)) systems since orthogonality between I and Q channels is preserved. (MSK with differentially precoded data, i.e. FFSK, suffers the same degradation in P_b performance as differentially encoded PSK.)

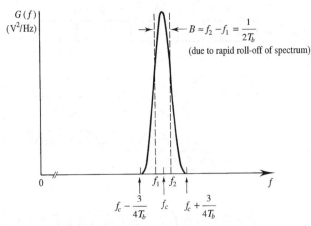

Figure 11.35 *Power spectral density of MSK/FFSK signal. $G(f) \propto [(\cos(2\pi fT_b))/(1 - 16f^2T_b^2)]^2$.*
(Note: no spectral lines are present.)

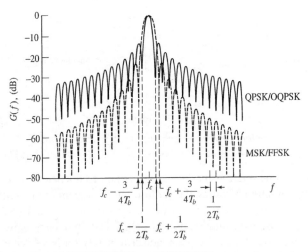

Figure 11.36 *Comparison of QPSK/OQPSK and MSK/FFSK spectra (spectral envelopes' roll-off with $(f - f_c)^{-2}$, i.e. −6 dB/octave, and $(f - f_c)^{-4}$, i.e. −12 dB/octave).*

EXAMPLE 11.7

Find the probability of bit error for a 1.0 Mbit/s MSK transmission with a received carrier power of −130 dBW and an NPSD, measured at the same point, of −200 dBW/Hz.

$$C = 10^{\frac{-130}{10}} = 10^{-13} \text{ W}$$

$$E_b = CT_b = 10^{-13} \times 10^{-6} = 10^{-19} \text{ J}$$

$$N_0 = 10^{\frac{-200}{10}} = 10^{-20} \text{ W/Hz}$$

$$P_b = \frac{1}{2}\left[1 - \mathrm{erf}\left(\frac{E_b}{N_0}\right)^{1/2}\right] = \frac{1}{2}\left[1 - \mathrm{erf}\left(\frac{10^{-19}}{10^{-20}}\right)^{1/2}\right]$$

$$= \frac{1}{2}[1 - \mathrm{erf}\,(3.162)] = \frac{1}{2}[1 - 0.99999924] = 3.88 \times 10^{-6}$$

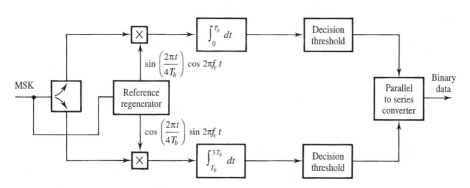

Figure 11.37 *Schematic of MSK demodulator.*

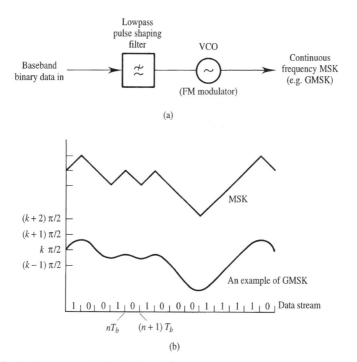

Figure 11.38 *(a) Generation of GMSK; (b) MSK and GMSK phase trajectories for typical bit sequences with the MSK trajectory displaced vertically for clarity (source: Hirade and Murota, 1979, reproduced with the permission of the IEEE © 2003 IEEE).*

11.4.6 Gaussian MSK

Although the phase of an MSK signal is continuous, its first derivative (i.e. frequency) is discontinuous. A smoother modulated signal (with correspondingly narrower spectrum) can be generated by reducing this frequency discontinuity. The conceptually simplest way of achieving this is to generate the MSK signal directly as a BFSK signal (with $(f_2 - f_1)T_o = \frac{1}{2}$) using a voltage controlled oscillator, and to employ premodulation pulse shaping of the baseband binary data, Figure 11.38(a). If the pulse shape adopted is Gaussian then the resulting modulation is called Gaussian MSK (GMSK). Figure 11.38(b) compares the phase trajectory of MSK and GMSK signals and Figure 11.39 shows the corresponding power spectra. The parameter $B_b T_o$ in Figure 11.39 is the normalised bandwidth of the premodulation low pass filter in Figure 11.38(a). (B_b is the baseband bandwidth and T_o is the symbol, or bit, period.) $B_b T_o = \infty$ corresponds to no filtering and therefore MSK signalling. Figure 11.40 shows the impulse response of the premodulation filter for various values of $B_b T_o$. The baseband Gaussian pulses essentially occupy $1/(B_b T_o)$ bit periods which results in significant sampling instant ISI for $B_b T_o < 0.5$. The improvement in spectral efficiency for $B_b T_o < 0.5$ outweighs the resulting degradation in BER performance, however, in some applications. Figure 11.41 shows measured P_e performance curves for GMSK modulation. GMSK with $B_b T_o = 0.3$ is the modulation scheme adopted by the CEPT's Groupe Spéciale Mobile (Global System for Mobile), GSM, for the implementation of digital cellular radio systems in Europe, see Chapter 15.

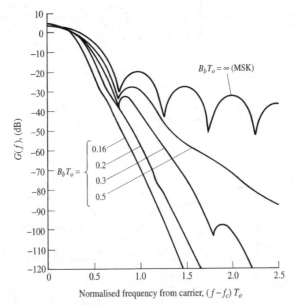

Figure 11.39 *Comparison of GMSK spectra with various values of $B_b T_o$ (B_b is the bandwidth of the premodulation pulse shaping filter, $T_o (= T_b)$ is the symbol period (source: Parsons and Gardiner, 1989).*

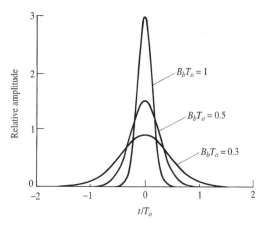

Figure 11.40 *Gaussian filter impulse response for different B_bT_o products.*

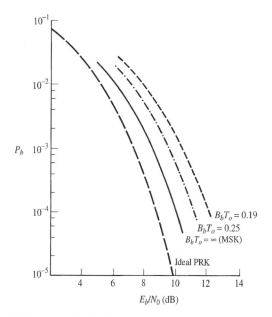

Figure 11.41 *Measured P_b against E_b/N_0 for MSK with various values of B_bT_o after bandlimiting by an ideal bandpass channel with bandwidth $B = 0.75/T_o$. ($T_o = T_b$.) (Source: Hirade and Murota, 1979, reproduced with the permission of the IEEE © 2003 IEEE).*

Although Figure 11.38(a) implies a simple implementation for GMSK, in practice the implementation is significantly more complex.

11.4.7 Trellis coded modulation

Trellis coded modulation (TCM) [Ungerboeck] is a combined coding and modulation technique for digital transmission over bandlimited channels. In TCM there is a restriction on which states may occur so that for a given received symbol sequence the number of possible transmitted symbol sequences is reduced. This increases the decision space across the constellation. Its main attraction is that it allows significant coding gains over the previously described (uncoded) multilevel modulation schemes, without compromising the spectral efficiency.

TCM employs redundant non-binary modulation in combination with a finite state encoder which governs the selection of transmitted symbols to generate the coded symbol sequences. In the receiver, the noisy symbols are decoded by a soft decision, maximum likelihood sequence (Viterbi) decoder as described for FEC decoding in Chapter 10. Simple four-state TCM schemes can improve the robustness of digital transmission against additive noise by 3 dB, compared to conventional uncoded modulation. With more complex TCM schemes, the coding gain can reach 6 dB or more. These gains are obtained without bandwidth expansion, or reduction of the effective information rate, as required by traditional error correction schemes. Figure 11.42 depicts symbol sets and Figure 11.43 the state transition (trellis) diagrams for uncoded 4-PSK modulation and coded 8-PSK modulation with four trellis states. The trivial one-state diagram in Figure 11.43(a) is shown only to illustrate uncoded 4-PSK from the viewpoint of TCM. Each one of the four connected paths labelled 0, 2, 4, 6 through the trellis in Figure 11.43(a) represents an allowed symbol sequence based on two binary bits (since $2^2 = 4$). In both systems, starting from any state, four transitions can occur, as required to encode two information bits per modulation interval (and obtain a spectral efficiency of 2 bit/s/Hz).

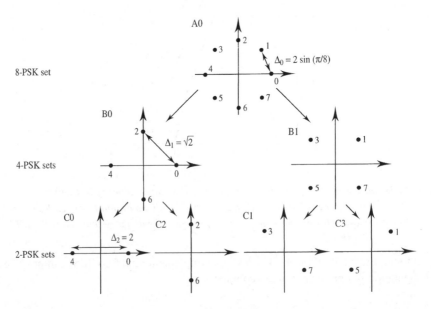

Figure 11.42 *Partitioning of 8-PSK into four-phase (B0/B1) and two-phase (C0 ⋯ C3) subsets.*

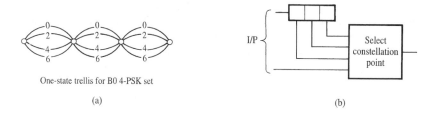

One-state trellis for B0 4-PSK set

(a)

I/P

Select constellation point

(b)

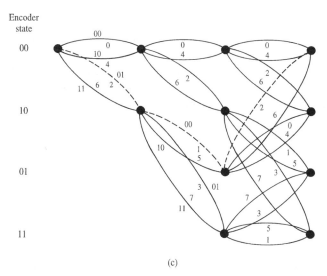

(c)

Figure 11.43 *Trellis coded modulation (TCM): (a) trellis diagram for 4-PSK uncoded transmission where there is no restriction in the choice of constellation point; (b) four-state TCM encoder; (c) resulting four-state TCM 8-PSK trellis diagram.*

The four 'parallel' transition paths in the one-state trellis diagram of Figure 11.43(a) for uncoded 4-PSK do not restrict the sequence of 4-PSK symbols that can be transmitted as there is no sequence coding. For unit transmitted amplitude the smallest distance between the 4-PSK signals is $\sqrt{2}$, denoted by Δ_1 in Figure 11.42; 4-PSK at 2 bit/s/Hz has identical spectral efficiency to rate 2/3 FECC data transmitted over an uncoded 8-PSK system.

In order to understand the ideas behind TCM the 8-PSK signal set can be partitioned as shown in Figure 11.42 into B and C subsets with increasing distance or separation between the constellation states [Burr]. The encoder, like the convolutional coder of Chapter 10, consists of a shift register which contains the current data and a number of previous data bits, so that the state of the encoder depends on the previous data samples, Figure 11.43(b). The nodes in Figure 11.43(c) represent the state of the encoder at a given time, while the branches leaving each node represent the different input data which arrives in that symbol period. Each branch is labelled with the input data bit pairs and the number of the constellation point which is transmitted for that encoder state.

Any pair of paths between the same two nodes represents a pair of codewords which could be confused. In TCM we choose a symbol from the different subsets of the constellation for each branch in the trellis to maximise the distance between such pairs and, therefore, minimise the error probability. This means using points from the same C subset, Figure 11.42, for these parallel branches. Other branches which diverge from, or converge to, the same node require a smaller minimum distance and are chosen from the same B subset.

Consider the trellis diagram of Figure 11.43(c). The closest paths (in terms of Euclidean distance) to the all-zeros symbol path, which is used as a reference, are shown dashed. There are four branches in the trellis from each node, corresponding to 2 bits per transmitted constellation point (or channel symbol). Thus the scheme can be compared directly with uncoded QPSK, which also transmits 2 bits per symbol. But in uncoded QPSK the minimum Euclidean distance between transmitted signals is that between two adjacent constellation points, i.e. $\Delta_1 = \sqrt{2}$, Figure 11.42. The TCM scheme achieves $\sqrt{2}$ times this distance, i.e. $\Delta_2 = 2$. As squared distances correspond to signal powers, the squared Euclidean distance between two points determines the noise tolerance between them. The TCM code has twice the square distance of uncoded QPSK, tolerating twice, or 3 dB, more noise power and coders are readily available, Table 11.6.

Table 11.6 *Commercially available TCM coder chipset examples.*

Data rate (Mbit/s)		75
Rate, R	2/3	3/4
Modulation	8-PSK	16-PSK
Coding gain (dB) at 10^{-5}	3.2	3.0

In general TCM offers coding gains of 3 to 6 dB at spectral efficiencies of 2 bit/s/Hz or greater. An 8-state coder (with a 16-state constellation) gives 4 dB of gain while 128 states are required for 6 dB gain. TCM is thus a variation on FECC, Chapter 10, where the redundancy is obtained by increasing the number of vectors in the signal constellation rather than by explicitly adding redundant digits, thus increasing bandwidth. Further design of the TCM signal constellation to cluster the points into *clouds* of closely spaced points with larger gaps between the individual clouds can obtain error control where there are varying degrees of protection on different information bits. Such systems, which offer unequal error protection, are being actively considered for video signal coding where the priority bits reconstruct the basic picture and the less well protected bits add in more detail, to avoid picture loss in poor channels.

EXAMPLE 11.8

Compare the carrier-to-noise ratio required for a P_b of 10^{-6} with: (a) 4-PSK; and (b) TCM derived 8-PSK, using Figure 11.43. How does the performance of the TCM system compare with PRK operation? Is it possible to make any enhancements to the TCM coder to achieve further reductions in the CNR requirement?

For uncoded 4-PSK the distance between states of amplitude 1 is $\sqrt{2}$ and the CNR for $P_b = 10^{-6}$ is given in Figure 11.21 and Table 11.5. The CNR requirement (for $T_o B = 1$) is 13.6 dB.

For TCM based 8-PSK the minimum distance is increased from $\sqrt{2}$ to 2 and the CNR requirement reduces by 3 dB. Thus the minimum CNR will be 10.6 dB.

TCM 8-PSK therefore operates at the same CNR as PRK but achieves twice the spectral efficiency of 2 bit/s/Hz.

The TCM encoder performance can be improved further by extending the constraint length, but this increases the encoder and decoder complexity. [Blahut, 1990] shows, in his Table 6.1, that increasing the constraint length to 8 increases the coding gain to 5.7 dB.)

11.5 Power efficient modulation techniques

Some communications systems operate in environments where large bandwidths are available but signal power is limited. Such power limited systems rely on power efficient modulation schemes to achieve acceptable bit error, and data, rates. In general data rate can be improved by increasing the number of symbols (i.e. the alphabet size) at the transmitter. If this is to be done without degrading P_e then the enlarged alphabet of symbols must remain at least as widely spaced in the constellation as the original symbol set. This can be achieved without increasing transmitted power by adding orthogonal axes to the constellation space – a technique which results in multidimensional signalling. Power can also be conserved by carefully optimising the arrangement of points in the constellation space. The most significant power saving comes from optimising the lattice pattern of constellation points. This is called symbol packing and results in an increased constellation point density without decreasing point separation. An additional (small) power saving can be obtained by optimising the boundary of the symbol constellation.

11.5.1 Multidimensional signalling and MFSK

Multiple frequency shift keying (MFSK) is a good example of a power efficient, multidimensional, modulation scheme if, as is usually the case, its symbols are designed to be mutually orthogonal, section 2.5.3. Figure 11.44(a), developed from Figure 11.9, shows the voltage spectrum of an orthogonal MFSK signal as a superposition of OOK signals and Figure 11.44(b) shows the power spectrum plotted in dB. The increased data rate realised by MFSK signalling is achieved entirely at the expense of increased bandwidth. Since each symbol (for equiprobable, independent, symbol systems) conveys $H = \log_2 M$ bits of information then the nominal spectral efficiency of orthogonal MFSK is given by:

$$\eta_s = \frac{\log_2 M}{(n/2)(M-1)+2} \quad \text{(bit/s/Hz)} \tag{11.50}$$

where n is the selected zero crossing point on the ρ–T_o diagram of Figure 11.8 and the (nominal) signal bandwidth is defined by the first spectral nulls above and below the highest and lowest frequency symbols respectively. For incoherent detection ($n \geq 2$) the maximum spectral efficiency ($n = 2$) in Figure 11.44(b) is given by:

$$\eta_s = \frac{\log_2 M}{M+1} \quad \text{(bit/s/Hz)} \tag{11.51}$$

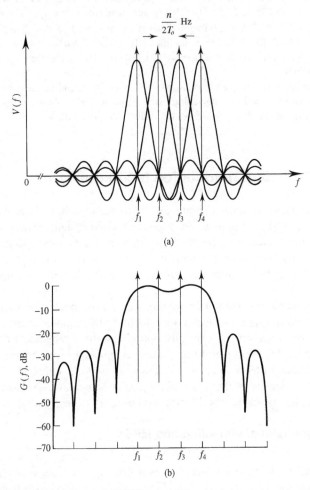

Figure 11.44 *Spectrum of orthogonal MFSK signal (M = 4): (a) as a superposition of OOK signal voltage spectra and (b) combined power spectrum on a dB scale. Tone spacing corresponds to second zero crossing point (n = 2) on ρ–T$_o$ diagram.*

EXAMPLE 11.9

Find the maximum spectral efficiency of an 80 kbaud, 8-FSK, orthogonal signalling system which operates with a frequency separation between adjacent tones corresponding to the third zero crossing point of the ρ–T$_o$ diagram of Figure 11.8. Compare this with the nominal spectral efficiency as given by equation (11.50).

The frequency separation between orthogonal tones satisfies:

$$n = 2 \, (f_i - f_{i-1}) \, T_o$$

In this case, therefore:

$$(f_i - f_{i-1}) = \frac{3}{2\,T_o}$$

$$= \frac{3 \times 80 \times 10^3}{2} = 120 \times 10^3 \quad \text{Hz}$$

The maximum (ISI-free) spectral efficiency is realised when $T_o B_{symbol} = 1.0$ and $H = \log_2 M$. The bandwidth of each FSK symbol is therefore $B_{symbol} = 1/T_o$ Hz and for the MFSK signal is:

$$B_{opt\ MFSK} = (M-1)(f_i - f_{i-1}) + 2\,(\tfrac{1}{2} \times B_{symbol})$$

$$= (8-1)\,120 \times 10^3 + 80 \times 10^3 = 920 \times 10^3 \quad \text{Hz}$$

The maximum spectral efficiency is therefore given by:

$$\eta_s = \frac{\log_2 M}{T_o B}$$

$$= \frac{(\log_2 8) \times 80 \times 10^3}{920 \times 10^3} = 0.261 \quad \text{bit/s/Hz}$$

The nominal spectral efficiency given by equation (11.50) is:

$$\eta_s = \frac{\log_2 8}{(3/2)\,(8-1) + 2} = 0.240 \quad \text{bit/s/Hz}$$

It can be seen that the difference between the nominal and maximum spectral efficiencies (and signal bandwidths) becomes small for MFSK signals as M (and/or tone separation) becomes large.

Multidimensional signalling can also be achieved using sets of orthogonally coded bit patterns, Figure 11.45 [developed from Sklar, p. 253]. (See also equations (15.17) to

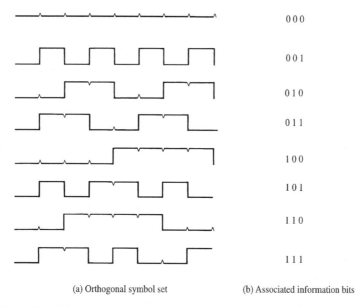

(a) Orthogonal symbol set   (b) Associated information bits

Figure 11.45 *Orthogonal code set comprising eight symbols.*

(15.20).) For an M-symbol alphabet each coded symbol (ideally) carries $H = \log_2 M$ bits of information but, to be mutually orthogonal, symbols must be M binary digits (or 'chips') long, Figure 11.45. The nominal bandwidth of such symbols is M/T_o Hz for baseband signalling and $2M/T_o$ Hz for bandpass signalling, where T_o/M is the duration of the binary chips which make up the orthogonal symbols. For equiprobable symbols the nominal spectral efficiency for bandpass orthogonal code signalling is therefore:

$$\eta_s = \frac{\log_2 M}{2M} \quad \text{(bit/s/Hz)} \tag{11.52}$$

and the maximum spectral efficiency (with $T_o B = M$) is twice this. For optimum (coherent) detection of any equal energy, equiprobable, M-ary orthogonal symbol set (including MFSK), in the presence of white Gaussian noise, the probability of symbol error is traditionally bounded by the (union bound) formula:

$$P_e \leq \frac{1}{2} (M - 1) \left[1 - \text{erf} \sqrt{\frac{E}{2N_0}} \right] \tag{11.53(a)}$$

A tighter bound, however, given by [Hughes] is:

$$P_e \leq 1 - \left\{ \frac{1}{2} \left[1 + \text{erf} \sqrt{\frac{E}{2N_0}} \right] \right\}^{M-1} \tag{11.53(b)}$$

where E, as usual, is the symbol energy. The energy per information bit is $E_b = E/\log_2 M$. Figure 11.46 shows P_e against E_b/N_0. The parameter $H = \log_2 M$ is the number of information bits associated with each (equiprobable) symbol. Since the symbols are mutually orthogonal the distance between any pair is a constant. It follows that all types of symbol error are equiprobable, i.e.:

$$P(i_{RX}|j_{TX}) = P_e \quad \text{for all } i \neq j \tag{11.54}$$

From Figure 11.45(b) it can be seen that for any particular information bit associated with any given symbol only $M/2$ symbol errors out of a total of $M - 1$ possible symbol errors will result in that bit being in error. The probability of bit error for orthogonal signalling (including MFSK and orthogonal code signalling) is therefore:

$$P_b = \frac{(M/2)}{M - 1} P_e \tag{11.55}$$

Figure 11.47 shows the probability of bit error against E_b/N_0 for orthogonal M-ary systems.

   If antipodal symbols are added to an orthogonal symbol set the resulting set is said to be *biorthogonal*. The P_e (and therefore P_b) performance is improved since the average symbol separation in the constellation space is increased for a given average signal power. Figure 11.48 shows the biorthogonal P_b performance. The spectral efficiency of biorthogonal systems is improved by a factor of two over orthogonal systems since pairs of symbols now occupy the same spectral space. Biorthogonal signals must, of course, be detected coherently.

   Inspection of Figure 11.45 reveals that for an orthogonal codeword set one binary digit (or chip) is identical for all codewords. This binary digit therefore carries no information and need not be transmitted. An alphabet of codewords formed by deleting

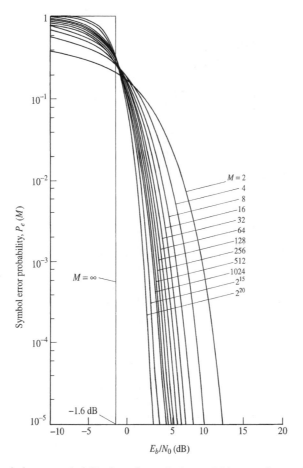

Figure 11.46 *Symbol error probability for coherently detected M-ary orthogonal signalling (source: Lindsey and Simon, 1973).*

the redundant chip from an orthogonal set is said to be *transorthogonal*. It has a correlation between symbol (i.e. codeword) pairs given by:

$$\rho = -\frac{1}{M-1} \tag{11.56}$$

and therefore has slightly better P_e (and P_b) performance than orthogonal signalling. Since one binary digit from a total of M has been deleted it has a spectral efficiency which is $M/(M-1)$ times better than the parent orthogonal scheme.

The major application of MFSK-type signalling systems is the telegraphy system used for teleprinter messages. This system, called Piccolo [Ralphs], has 32 tones and sends data at only 10 symbol/s with 100 ms duration symbols. (This was chosen to overcome the 0 to 8 ms multipath effects in over-the-horizon HF radio links.)

Another system is currently being developed for multicarrier broadcast to mobiles. It uses *simultaneous* MFSK or orthogonal FDM (OFDM) transmissions [Alard and

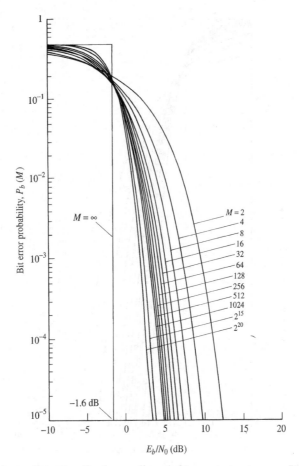

Figure 11.47 *Coherent detection of orthogonally coded transmission (source: Lindsey and Simon, 1973).*

Lassalle] to reduce a single high data rate signal into many parallel low data rate signals, again with lengthened symbol periods. By suitable FEC coding (Chapter 10) of the 'orthogonal carrier' channels, errors introduced by multipath effects can be tolerated in some of the channels without degrading significantly the overall system error rate. This system is now being developed separately for audio and TV broadcast applications.

11.5.2 Orthogonal frequency division multiplex (OFDM)

The concept behind OFDM is that the modulator comprises a serial to parallel data converter followed by a set of mutually orthogonal parallel modulators as defined in equation (11.22). In the conversion k sequential high speed (short T_b duration) data bits are mapped into k parallel low speed data bits, Figure 11.49(a). In the serial to parallel conversion process k bits are transferred into the holding register and held there for k symbol periods. The OFDM symbol duration is thus extended to $k \times T_b$ s. The individual carriers are spaced by $\Delta f = 1/(k \times T_b)$ to ensure orthogonality between the

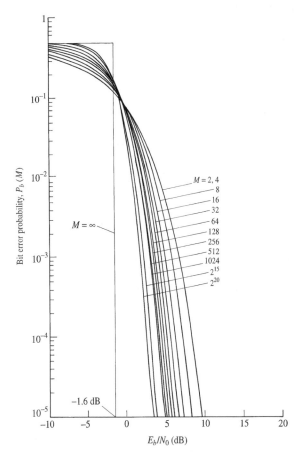

Figure 11.48 *Coherent detection of biorthogonally coded transmission (source: Lindsey and Simon, 1973).*

individual frequencies, i.e. to realise an OFDM signal.

In place of receiving a single wideband modulated signal there are now k simultaneous narrowband FDM signals. Due to multipath effects, see section 12.4.5, some of the FDM channels may suffer fading and loss of received data. This is mitigated by wrapping the OFDM modulator and demodulator in a convolutional coder/decoder, Chapter 10, to form a COFDM system.

In many practical implementations of COFDM [Forrest] the number of parallel channels ranges from 1000 to 8000 and so the discrete modulators in the transmitter are replaced by an inverse fast Fourier transform (IFFT) processor, Figure 11.49(b), and the receiver demodulator by a forward FFT processor as these provide exactly the required time to frequency and frequency to time mapping functions. Further sophistication is achieved by selecting appropriate QPSK, 16-QAM or 64-QAM modulators on each individual channel to increase the overall data throughput rate.

COFDM is being adopted widely in audio and video broadcast applications, see section 16.8, and is being proposed for wireless LAN transmission. OFDM is now the

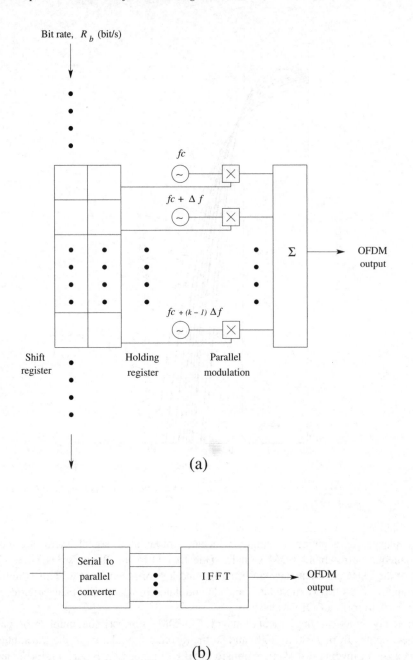

(a)

(b)

Figure 11.49 *(a) Schematic of an OFDM transmitter design based on k parallel low data rate modulators and (b) practical realisation with an inverse fast Fourier transform (IFFT) processor for implementing large numbers of discrete modulators (i.e. >1000 of parallel channels).*

preferred coding format for broadband wireless local access, see section 20.11 (ADSL and IEEE 802.16 wireless MAN) [and also WWW for OFDM Forum].

11.5.3 Optimum constellation point packing

For a given alphabet of symbols distributed over a fixed number of orthogonal axes there are many possible configurations of constellation points. In general these alternative configurations will each result in a different transmitted power. Power efficient systems may minimise transmitted power for a given P_e performance by optimising the packing of constellation points. This is well illustrated by the alternative constellation patterns possible for the (2-dimensional) 16-APK system shown in Figure 11.19. These patterns are sometimes referred to as (8,8), (4,12) and (4,8,4) APK where each digit represents the number of constellation points on each amplitude ring.

The power efficiency of these constellation patterns increases moving from (8,8) to (4,8,4). The QAM rectangular lattice (represented in Figure 11.19(c) by (4,8,4) APK) is already quite power efficient and as such is a popular practical choice for M-ary signalling. A further saving of 0.6 dB is, however, possible by replacing the rectangular lattice pattern with a hexagonal lattice, Figure 11.50, which is related to the cross constellation [Burr]. This represents the densest possible packing of constellation points (in two dimensions) for fixed separation of adjacent points.

The power saved by optimising symbol packing increases as the number of constellation dimensions is increased. For a 2-dimensional constellation the saving gained by replacing a rectangular lattice with a hexagonal lattice is modest. The optimum lattice for a 64-dimensional constellation, however, gives a saving over a rectangular lattice of 8 dB [Forney *et al.*].

11.5.4 Optimum constellation point boundaries

Further modest savings in power can be made by optimising constellation boundaries. Consider, for example, Figure 11.51(a) which represents conventional 64-QAM signalling. If the points at the corners of the constellation are moved to the positions shown in Figure 11.51(b) then the voltage representing each of the moved symbols is reduced by a factor of (9.055/9.899) and their energy reduced by a factor of 0.84. Since the energy residing in each of the other symbols is unaffected the average transmitted power is reduced.

Not surprisingly the most efficient QAM constellation point boundary is that which most closely approximates a circle, and for a 3-dimensional signalling scheme the most efficient boundary would be a sphere. It is clear that this idea extends to N-dimensional signalling where the most efficient boundary is an N-dimensional sphere. The power

Figure 11.50 *Optimum 2-dimensional lattice pattern.*

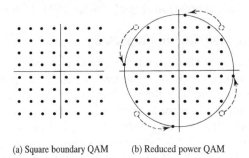

(a) Square boundary QAM   (b) Reduced power QAM

Figure 11.51 *64-QAM constellations.*

savings obtained in moving from N-dimensional cubic boundaries to N-dimensional spherical boundaries increase with N. For $N = 2$ (i.e. QAM) the saving is only 0.2 dB whilst for $N = 64$ the saving has increased to 1.3 dB [Forney *et al.*].

If efficient constellation point boundaries are used with efficient constellation point packing then the power savings from both are, of course, realised. Figure 11.52 shows such a (2-dimensional) 64-APK constellation.

11.6 Data modems

Voiceband data modems are sophisticated electronic subsystems which permit digital data to be transmitted and received without error over analogue voiceband (3.2 kHz bandwidth) telephone circuits. These modems convert the data stream into modulated signals using multi-amplitude, multiphase transmission schemes. Early modem designs used simple FSK schemes to signal at 1200 and 2400 bit/s over four-wire circuits, the key to efficient reception being to equalise the channel in the receiver, to overcome the effects of bandlimiting and dispersion.

The protocols associated with data modem signal transmissions are covered in Chapter 18.

The many different techniques used in voiceband data modems are covered by the ITU-T V series of recommendations. The simple 16-QAM constellation of Figure

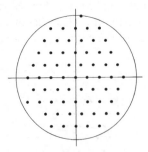

Figure 11.52 *Lattice and boundary efficient 64-APK.*

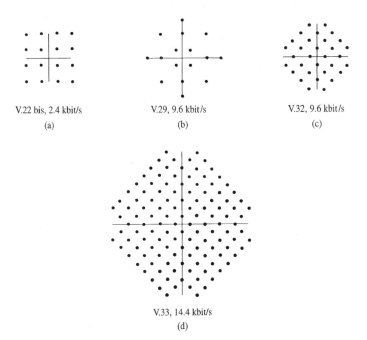

V.22 bis, 2.4 kbit/s (a)

V.29, 9.6 kbit/s (b)

V.32, 9.6 kbit/s (c)

V.33, 14.4 kbit/s (d)

Figure 11.53 *Examples of signal constellations used in speech band data modems: (a) and (c) as used on switched lines; and (b) on leased lines.*

11.53(a) is used in V.22 data modems. Table 20.4 gives greater technical detail on the V. series of data modems. The V.32 recommendation covers three different modulation options, 4800 bit/s and 9600 bit/s using conventional modulation techniques and 9600 bit/s TCM, see section 11.4.7. At these low data rates the complexity of TCM is not a problem as current receiver processors have sufficient time to make all the required computations during the symbol period. A standard 9600 bit/s modem (V.29, Figure 11.53(b)) needs a four-wire connection to operate. Leased circuits are available with four-wire connections, but each public switched telephone network (PSTN) connection provides only two wires. Two PSTN lines are thus needed to back up a single leased circuit. V.32 modems, however, provide major cost savings and other benefits. Instead of a four-wire leased circuit, only a two-wire circuit is required, i.e. a single PSTN connection is adequate.

In the 1980s several manufacturers produced modems with data rates from 9600 bit/s to 19.2 kbit/s for operation over dial-up networks. With V.33 modems operating at 14.4 kbit/s TCM (having full two-way duplex capability, Chapter 15) and the V.fast standards, one can now readily achieve data rates of 24 kbit/s and 28.8 kbit/s over voiceband circuits. Given that the Shannon–Hartley law, equation (11.38(a)), allows 32 kbit/s for a 3.2 kHz bandwidth voiceband circuit with 30 dB SNR, then these sophisticated modems appear to be approaching the theoretical limits of performance[3].

---

[3] The fact that systems (e.g. xDSL) are possible which appear to exceed this limit simply reflects the fact that the channel is not strictly limited in bandwidth to 3.2 kHz.

The major recent advances have been in the use of digital signal processing (DSP) to implement the equaliser combined with personal computer use spurring the widespread modem use on two-wire normal domestic telephone circuits. Today's modems are implemented on a single printed circuit card and have combined the better SNR and reduced impulsive noise of telephone circuits with developments in VLSI circuit density to achieve rates of 56 kbit/s over the standard twisted pair two-wire telephone connection. This represents over a 40-fold increase in rate from early 1200 bit/s designs and more than a 40-fold reduction in size illustrating the power of advanced DSP techniques. Higher rates beyond 56 kbit/s are achieved over shorter connection lengths by the digital subscriber line (xDSL) techniques described in section 6.9.

11.7 Summary

IF or RF modulation is used principally to shift the spectrum of a digital information signal into a convenient frequency band. This may be to match the spectral band occupied by a signal to the passband of a transmission line, to allow frequency division multiplexing of signals, or to enable signals to be radiated by antennas of practical size.

Two performance measures are commonly used to compare different IF modulation techniques – spectral efficiency and power efficiency. The former has units of bit/s/Hz and the latter relates to the required value of E_b/N_0 for a given probability of bit error.

There are three generic IF modulation techniques for digital data. These are ASK, PSK and FSK. Binary ASK and FSK are usually operated as orthogonal signalling schemes. Binary PSK is often operated as an antipodal scheme and therefore requires 3 dB less (average) power. ASK and FSK modulated signals can be detected incoherently which simplifies transmitter and receiver design. A small P_e penalty is incurred in this case over matched filter, or correlation, detection. PSK systems may suffer from phase ambiguity. This can be avoided, however, using either a known phase training sequence, transmitted in a preamble to the information data blocks, or differential coding.

MPSK, APK and QAM are spectrally efficient modulation schemes. QPSK and its variants are also reasonably spectrally efficient. Sophisticated 64-QAM and 256-QAM systems are now being applied in digital microwave long haul radio communication systems (see Chapter 14), but the power amplifier linearity requirements are severe for high level QAM systems. TCM is a technique which increases power efficiency without compromising spectral efficiency by combining modulation with error control coding.

MFSK is usually operated as an orthogonal modulation scheme and is therefore power efficient (as is orthogonal code signalling). Non-orthogonal M-ary schemes can be made more power efficient by distributing the symbols (constellation points) over an increased number of orthogonal axes. Further improvements in power efficiency can be made by choosing the optimum symbol packing arrangement for the number of dimensions used and adopting an (N-dimensional) spherical boundary for the set of constellation points.

Many of these techniques are widely applied in speech band data modems (described in section 11.6) and also in wideband high speed microwave communication systems (described in Chapter 14).

11.8 Problems

11.1 A rectangular pulse OOK signal has an average carrier power, at the input to an ideal correlation receiver, of 8.0 nW. The (one-sided) noise power spectral density, measured at the same point, is 2.0×10^{-14} W/Hz. What maximum bit rate can this system support whilst maintaining a P_e of 10^{-6}? [17.7 kbit/s]

11.2 A BPSK, 1.0 Mbaud, communication system is to operate with a P_e of 8×10^{-7}. The signal amplitude at the input to the ideal correlation receiver is 150 mV and the one-sided noise power spectral density at the same point is 15 pW/Hz. The impedance level at the correlation receiver input is 50 Ω. What minimum phase shift between phasor states must the system employ and what residual carrier power is therefore available for carrier recovery purposes? [122°, 5.23×10^{-5} W]

11.3 Define on–off keying (OOK), frequency shift keying (FSK) and phase shift keying (PSK) as used in binary signalling. Compare their respective advantages and disadvantages.

In a Datel 600 modem FSK tone frequencies of 1300 Hz and 1700 Hz are used for a signalling rate of 600 bit/s. Comment on the consistency between these tone frequencies, the ρ–T_o diagram, and the signalling rate. What value would the upper tone frequency need to have, to handle a 900 bit/s signalling rate? [1900 Hz]

11.4 A binary, rectangular pulse, BFSK modulation system employs two signalling frequencies f_1 and f_2 and operates at the second zero of the ρ–T_o diagram. The lower frequency f_1 is 1200 Hz and the signalling rate is 500 baud. (a) Calculate f_2. (b) Sketch the PSD of the FSK signal. (c) Calculate the channel bandwidth which would be required to transmit the FSK signal without 'significant' distortion. Indicate in terms of the PSD those frequencies which you assume should be passed to keep the distortion 'insignificant'. [1700 Hz, $B = 1500$ Hz]

11.5 If the CNR of the system described in problem 11.4 is 4 dB at the input to an ideal (matched filter) receiver and the channel has a low pass rectangular frequency characteristic with a cut-off frequency of 3.6 kHz, estimate the probability of bit error. [1.06×10^{-5}]

11.6 (a) Define the correlation coefficient of two siganls: $s_1(t)$ and $s_2(t)$. (b) Sketch the correlation coefficient of two FSK rectangular RF signalling pulses, such as might be used in an FSK system, as a function of tone spacing. (c) Find the P_e of an orthogonal FSK binary signalling system using ideal matched filter detection if the signalling pulses at the matched filter input have rectangular envelopes of 10 μs duration and 1.0 V peak-to-peak amplitude. The one-sided NPSD at the input to the filter is 5.5 nW/Hz and the impedance level at the filter input is 50 Ω. [1.653×10^{-2}]

11.7 If the spectrum of each signalling pulse described in problem 11.6(c) is assumed to have significant spectral components only within the points defined by its first zeros, what is the minimum bandwidth of the FSK signal? [0.25 MHz]

11.8 A receiver has a mean input power of 25 pW and is used to receive binary FSK data. The carrier frequencies used are 5 MHz and 5.015 MHz. The noise spectral density at the receiver has a value of 2.0×10^{-16} W/Hz. If the error probability is fixed at 2×10^{-4} find the maximum data rate possible, justifying all assumptions. [10 kbits/s] If the transmitter is switched to PRK what new bit rate can be accommodated for the same error rate? [20 kbit/s]

11.9 What is meant by orthogonal and antipodal signalling? Can the error performance of an FSK system ever be better than that given by the orthogonal case and, if so, how is the performance improved, and at what cost?

11.10 A DEPSK, rectangular pulse, RZ signal, with 50% duty cycle, has a separation of 165° between phasor states. The baud rate is 50 kHz and the peak received power (excluding noise) at the input of an ideal correlation receiver is 100 μW. The (one-sided) noise power spectral density measured at the same point is 160 pW/Hz. What is the probability of symbol error? [4.56×10^{-4}]

11.11 Find the maximum spectral efficiency of ISI-free 16-PSK. What is the probability of symbol error of this scheme for a received CNR of 24 dB if the maximum spectral efficiency requirement is retained? What is the Gray coded probability of bit error?
[4 bit/s/Hz, 1.227×10^{-5}, 3.068×10^{-6}]

11.12 Sketch the constellation diagram of 64-QAM. What is the spectral efficiency of this scheme if pulse shaping is employed such that $BT_o = 2$? What is the best possible spectral efficency whilst maintaining zero ISI? Why is QAM (with the exception of QPSK) usually restricted to use in linear (or approximately linear) channel applications? [3 bit/s/Hz, 6 bit/s/Hz]

11.13 (a) Draw a block diagram of a QPSK modulator. Show clearly on this diagram or elsewhere the relationship in time between the binary signals in I and Q channels at the points immediately preceding multiplication by the I and Q carriers. How does this relationship differ in OQPSK systems? What advantage do OQPSK systems have over QPSK systems? Sketch the phasor diagrams for both systems showing which signal state transitions are allowed in each.
(b) Calculate the probability of bit error for an ideal OQPSK system if the single sided NPSD and bit energy are 1.0 pW/Hz and 10.0 pJ respectively. [3.88×10^{-6}]

11.14 Justify the statement: 'The performances of BPSK and QPSK systems are identical'.

11.15 Two engineers are arguing about MSK modulation. One says it is a special case of BFSK and the other asserts it is a form of OQPSK. Resolve their argument. Is incoherent detection of MSK possible? (Explain your answer.) Draw a block diagram of an MSK transmitter and sketch a typical segment of its output.

11.16 A power limited digital communications system is to employ orthogonal MFSK modulation and ideal matched filter detection. P_b is not to exceed 10^{-5} and the available receiver carrier power is limited to 1.6×10^{-8} V^2. Find the minimum required size of the MFSK alphabet which will achieve a bit rate of 1.0 Mbit/s at the specified P_b if the two-sided noise power spectral density at the receiver input is 10^{-15} V^2/Hz. (The use of a computer running an interactive maths package such as MATLAB, or a good programmable calculator, will take some of the tedium out of this problem.) [$M = 6$]

11.17 A 3-APK modulation scheme contains the constellation points 0, $e^{j\pi/3}$, $e^{j2\pi/3}$. What is the power saving (in dB) if this scheme is replaced with a minimum power 3-PSK scheme having the same P_e performance? [3 dB]

11.18 Repeat Example 11.4 for the same bandwidth and bit rate but this time replace MPSK by MQAM modulation and derive the minimum CNR required for reception with P_b no worse than 10^{-7}.

11.19 Sketch the constellation diagram for an ideal $\pi/4$ QPSK signal and find the theoretical limit of the signal's amplitude fluctuations (in dB).

11.20 Sixteen-state quadrature amplitude modulation (QAM), with cosinusoidal roll-off spectrum shaping, is to be used to transmit data at 60Mbit/s over a microwave channel that is restricted in its −3dB bandwidth to only 18MHz.
(a) State what will be the symbol rate of the system and determine the roll-off factor required for the spectrum shaping. If the average bit energy to one-sided noise spectral density ratio is $E_b/N_0 = 15$ dB for a bit error ratio (BER) of $P_b = 10^{-6}$, determine the required carrier-to-noise ratio.
(b) What is the disadvantage of this modulation system when high power amplifiers, such as those deployed in satellite repeaters, are used? [14.21 dB]

CHAPTER 12

System noise and communications link budgets

12.1 Introduction

Signal-to-noise ratio (SNR) is a fundamental limiting factor on the performance of communication systems. It can often be improved by a receiving system using appropriate demodulation and signal processing techniques. It is always necessary, however, to know the carrier-to-noise ratio (CNR) present at the input of a communications receiver to enable its performance to be adequately characterised. (The term CNR was used widely in place of SNR in Chapter 11, and is used here, since for most modulation schemes the carrier power is not synonymous with the impressed information signal power.) This chapter reviews some important noise concepts and illustrates how system noise power can be calculated. It also shows how received carrier power can be calculated (at least to first order accuracy) and therefore how a CNR can be estimated.

12.2 Physical aspects of noise

Whilst this text is principally concerned with systems engineering, it is interesting and useful to establish the physical origin of the important noise processes and the physical basis of their observed characteristics (especially noise power spectral density).

Consideration here is restricted to those noise processes arising in the individual components (resistors, transistors, etc.) of a subsystem, often referred to collectively as circuit noise. Noise which arises from external sources and is coupled into a communication system by a receiving antenna is discussed in section 12.4.3.

12.2.1 Thermal noise

Thermal noise is produced by the random motion of free charge carriers (usually electrons) in a resistive medium. These random motions represent small random currents which together produce a random voltage, via Ohm's law, across, for example, the terminals of a resistor. (Such noise does not occur in ideal capacitors as there are no free electrons in a perfect dielectric material.) Thermally excited motion takes place at all temperatures above absolute zero (0 K) since, by definition, temperature is a measure of average kinetic energy per particle. Thermal noise is a limiting factor in the design of many, but not all, radio communications receivers at UHF (0.3 to 3 GHz), Table 1.4, and above.

The power spectral density of thermal noise (at least up to about 10^{12} Hz) can be predicted from classical statistical mechanics. This theory indicates that the average kinetic (i.e. thermal) energy of a molecule in a gas which is in thermal equilibrium and at a temperature T, is $1.5kT$ joules. k is therefore a constant which relates a natural, or atomic, temperature scale to a man-made scale. It is called Boltzmann's constant and has a value of 1.381×10^{-23} J/K.

The *principle of equipartition* says that a molecule's energy is equally divided between the three dimensions or *degrees of freedom* of space. More generally for a system with any number of degrees of freedom the principle states that:

the average thermal energy per degree of freedom is ½ kT joules.

Equipartition can be applied to the transmission line shown in Figure 12.1. On closing the switches thermal energy will be trapped in the line as standing electromagnetic waves. Since each standing wave mode on the line has two degrees of freedom (its pulsations can exist as sinusoidal or cosinusoidal functions of time) then each standing wave will contain kT joules of energy. The wavelength of the nth mode (Figure 12.2) is given by:

$$\lambda_n = \frac{2l}{n} \quad \text{(m)} \tag{12.1}$$

where l is the length of the transmission line. Therefore the mode number, n, is given by:

$$n = \frac{2lf_n}{c} \tag{12.2}$$

where c is the electromagnetic velocity of propagation on the line. The frequency difference between adjacent modes is:

$$f_n - f_{n-1} = \frac{nc}{2l} - \frac{(n-1)\,c}{2l}$$

$$= \frac{c}{2l} \quad \text{(Hz)} \tag{12.3}$$

Figure 12.1 *Transmission line for trapping thermal energy.*

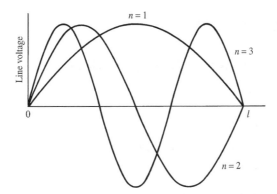

Figure 12.2 *Standing waves of trapped thermal energy on short-circuited transmission line.*

Therefore the number of modes, v, in a bandwidth B is:

$$v = \frac{B}{f_n - f_{n-1}} = \frac{2lB}{c} \tag{12.4}$$

The spatial (line) energy density of each standing wave mode is kT/l (J/m). Since each standing wave is composed of two travelling waves (with opposite directions of travel) then the spatial energy density in each travelling wave is $\frac{1}{2}kT/l$ (J/m). The energy per second travelling past a given point in a given direction (i.e. the available power) per oscillating mode is therefore:

$$P_{mode} = \frac{\frac{1}{2} kT}{\tau} \quad \text{(J/s)} \tag{12.5}$$

where τ is the energy transit time from one end of the line to the other. Using $\tau = l/c$:

$$P_{mode} = \frac{\frac{1}{2} kTc}{l} \quad \text{(J/s)} \tag{12.6}$$

The available power, N, in a bandwidth B (i.e. v modes) is therefore given by:

$$N = P_{mode} v \tag{12.7}$$

$$= \frac{\frac{1}{2} kTc}{l} \frac{2lB}{c} \quad \text{(J/s)}$$

i.e.:

$$N = kTB \quad \text{(W)} \tag{12.8}$$

Equation (12.8) is called Nyquist's formula and is accurate so long as classical mechanics can be assumed to hold. The constant behaviour with frequency, however, of the available (one-sided) noise power spectral density, $G_n(f) = kT$, incorrectly predicts infinite power when integrated over all frequencies. This paradox, known as the ultraviolet catastrophe, is resolved when quantum mechanical effects are accounted for. The complete version of equation (12.8) then becomes:

$$N = \int_0^B G_n(f)\, df \quad \text{(W)} \tag{12.9(a)}$$

where:

$$G_n(f) = \frac{hf}{e^{\left(\frac{hf}{kT}\right)} - 1} \quad \text{(W/Hz)} \qquad (12.9(b))$$

and h in equation (12.9(b)) is Planck's constant which has a value of 6.626×10^{-34} (J s). For $hf \ll kT$ then:

$$e^{\left(\frac{hf}{kT}\right)} \approx 1 + \frac{hf}{kT} \qquad (12.10)$$

and equation (12.9(b)) reduces to the Nyquist form:

$$G_n(f) \approx kT \quad \text{(W/Hz)} \qquad (12.11)$$

The power spectral densities of thermal noise at four temperatures are compared in Figure 12.3.

The Thévenin equivalent circuit of a thermal noise source is shown in Figure 12.4. Since the available noise power for almost all practical temperatures and frequencies (excluding optical applications) is kTB W then the maximum available RMS noise voltage V_{na} across a (conjugately) matched load is:

$$V_{na} = \sqrt{NR} = \sqrt{kTBR} \quad \text{(V)} \qquad (12.12)$$

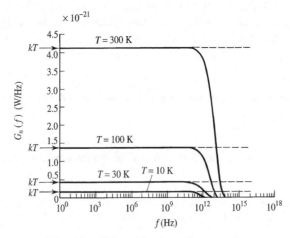

Figure 12.3 *Power spectral densities of thermal noise at four temperatures.*

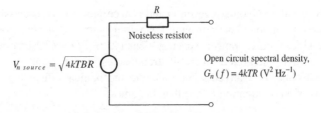

Figure 12.4 *Thévenin equivalent circuit of a thermal noise source.*

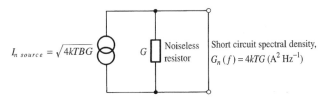

Figure 12.5 *Norton equivalent circuit of a thermal noise source.*

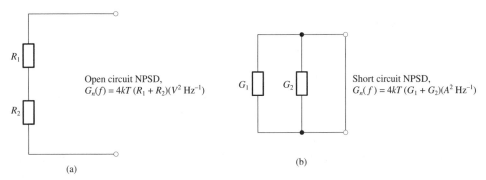

Figure 12.6 *NPSD for resistors in (a) series and (b) parallel.*

where R is the load (and source) resistance. Since the same voltage is dropped across the source and load resistances the equivalent RMS noise voltage of the source (i.e. the open circuit voltage of the Thévenin equivalent circuit) is given by:

$$V_{n\ source} = 2\sqrt{kTBR} = \sqrt{4kTBR} \quad (\text{V}) \tag{12.13}$$

An equivalent random current source can also be defined using the Norton equivalent circuit in Figure 12.5. In this case the current source has an RMS value given by:

$$I_{n\ source} = \sqrt{4kTBG} \quad (\text{A}) \tag{12.14}$$

where $G = 1/R$ is the load (and source) conductance. For passive components (such as resistors) connected in series and parallel their equivalent mean square noise voltages and currents add in an obvious way. For example, the series combination of resistances R_1 and R_2 in Figure 12.6(a) results in:

$$V_{n\ source}^2 = V_{n\ source\ 1}^2 + V_{n\ source\ 2}^2 \quad (\text{V}^2) \tag{12.15(a)}$$

i.e.:

$$V_{n\ source} = \sqrt{4kTB(R_1 + R_2)} \quad (\text{V}) \tag{12.15(b)}$$

and the parallel combination of conductances G_1 and G_2 in Figure 12.6(b) results in:

$$I_{n\ source}^2 = I_{n\ source\ 1}^2 + I_{n\ source\ 2}^2 \quad (\text{A}^2) \tag{12.16(a)}$$

i.e.:

$$I_{n\ source} = \sqrt{4kTB(G_1 + G_2)} \quad (\text{A}) \tag{12.16(b)}$$

Alternatively, and entirely unsurprisingly, the equivalent voltage source of the parallel combination is given by:

$$V_{n\ source} = \frac{I_{n\ source}}{G}$$

$$= \sqrt{4kTB \frac{R_1 R_2}{R_1 + R_2}} \quad (V) \tag{12.16(c)}$$

i.e. the two parallel resistors together behave as a single resistor with the appropriate parallel value.

12.2.2 Non-thermal noise

Although the time averaged current flowing in a device may be constant, statistical fluctuations will be present if individual charge carriers have to pass through a potential barrier. The potential barrier may, for example, be the junction of a PN diode, the cathode of a vacuum tube or the emitter–base junction of a bipolar transistor. Such statistical fluctuations constitute shot noise. The traditional device used to illustrate the origin of shot noise is a vacuum diode, Figure 12.7. Electrons are emitted thermally from the cathode. Assuming there is no significant space charge close to the cathode surface due to previously emitted electrons (i.e. assuming that the diode current is temperature limited), then electrons are emitted according to a Poisson statistical process (see Chapter 19).

The spatial (line) charge density, σ, between cathode and anode is given by:

$$\sigma = \frac{nq_e}{l} \quad (C/m) \tag{12.17}$$

where n is the average number of electrons in flight, l is the distance between cathode and anode and q_e is the charge on an electron. If the average electron velocity is v then the average current, I_{DC}, in (any part of) the circuit is:

$$I_{DC} = \sigma v \quad (C/s) \tag{12.18}$$

and the average contribution to I_{DC} from any individual electron is:

$$i_e = \frac{I_{DC}}{n} = q_e \frac{v}{l} \quad (C/s) \tag{12.19}$$

The current associated with any given electron can therefore be considered to flow only for that time the electron is in flight. Since l/v is the average cathode–anode transit time,

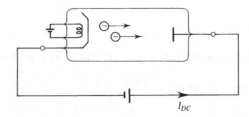

Figure 12.7 *Schematic diagram for a vacuum diode.*

τ, for the electrons, equation (12.19) can be written as:

$$i_e = \frac{q_e}{\tau} \quad \text{(A)} \qquad (12.20)$$

Thus each electron gives rise to a current pulse of duration τ as it moves from cathode to anode. (The precise shape of the current pulse will depend on the way in which the electron's velocity varies during flight. For the purpose of this discussion, however, the pulse can be assumed to be rectangular, Figure 12.8.)

The total current will be a superposition of such current pulses occurring randomly in time with Poisson statistics. Since each pulse has a spectrum with (nominal) bandwidth $1/\tau$ Hz (in the ideal rectangular pulse case the energy spectrum will have a sinc^2 shape, Figure 12.9) then for frequencies very much less than $1/\tau$ (Hz) shot noise has an essentially white spectrum.

The (two-sided) current spectral density of the pulse centred on $t = 0$ due to a single electron is:

$$I(f) = \int_{-\infty}^{\infty} \frac{q_e}{\tau} \Pi\left(\frac{t}{\tau}\right) e^{-j2\pi ft} \, dt$$

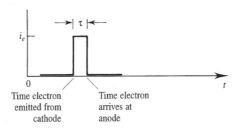

Figure 12.8 *Current pulse due to a single electron in diode circuit of Figure 12.7.*

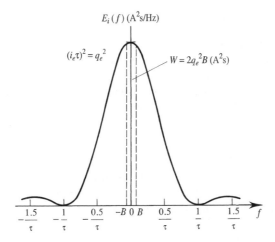

Figure 12.9 *Energy spectral density of current pulse shown in Figure 12.8.*

$$= q_e \, \text{sinc}(\tau f) \quad \text{(A/Hz)} \tag{12.21}$$

The (normalised, two-sided) energy spectral density of this current pulse is therefore:

$$E_i(f) = q_e^2 \, \text{sinc}^2(\tau f) \quad \text{(A}^2 \text{ s/Hz)} \tag{12.22}$$

For frequencies much less than $1/\tau$ Hz, $E_i(f)$ is a constant, namely q_e^2.

Using this model the energy, W, contained within a bandwidth B ($\ll 1/\tau$) is given by:

$$W = 2q_e^2 \, B \quad \text{(A}^2 \text{ s)} \tag{12.23}$$

For m electrons arriving at the anode in a time T the total shot noise (normalised) power will therefore be:

$$I_n^2 = \frac{2mq_e^2 B}{T} \quad \text{(A}^2) \tag{12.24}$$

mq_e/T, however, can be identified as the total DC current flowing in the circuit, i.e.:

$$I_n = \sqrt{2I_{DC}q_e B} \quad \text{(A)} \tag{12.25}$$

Equation (12.25) shows that shot noise RMS current is proportional to both the square root of the DC current and the square root of the measurement bandwidth. Although this result has been derived for a vacuum diode it is also correct for a PN junction diode and for the base and collector (with $I_{DC} = I_B$ and $I_{DC} = I_C$ as appropriate) of a bipolar junction transistor. Figure 12.10 shows a bipolar transistor along with its equivalent thermal and shot noise sources. $G_{V_{nB}}$ is the power spectral density of the thermal noise voltage generated in the transistor's base spreading resistance. $G_{I_{nB}}$ is the power spectral density of the shot noise current associated with those carriers flowing across the emitter–base junction which subsequently arrive at the base terminal of the transistor. (The fluctuation in base current caused by $G_{V_{nB}}$ and $G_{I_{nB}}$ will be amplified, like any other base current changes, at the transistor collector.) $G_{I_{nC}}$ is the power spectral density of the shot noise current associated with the carriers flowing across the emitter–base junction which subsequently arrive at the transistor collector. For FETs the shot noise is principally associated with gate current and can be modelled by equation (12.25) with $I_{DC} = I_G$.

Flicker, or $1/f$, noise also occurs in most active (and some passive) devices. It is rather device specific and is therefore not easily modelled. Its power spectral density has approximately a $1/f$ characteristic for frequencies below a few kilohertz. Above a few kilohertz flicker noise has an essentially flat spectrum but in any case is weak and usually neglected. Because flicker noise power is concentrated at low frequencies it is sometimes called pink noise.

12.2.3 Combining white noise sources

In many applications it is sufficient to account for thermal noise and shot noise only. This also has the practical advantage that circuit noise calculations can be made initially on a power spectral density basis, the total power then being found using the circuit noise bandwidth.

Consider Figure 12.11(a). There are two principal sources of shot noise and, effectively, three sources of thermal noise in this circuit. The former arise from the base–emitter junction of the transistor and the latter arise from the source resistance, R_s

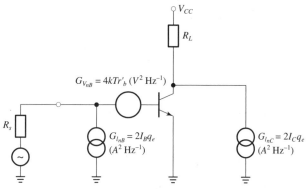

Figure 12.10 *Power spectral densities of thermal (G_{V_n}) and shot (G_{I_n}) noise processes in a bipolar transistor.*

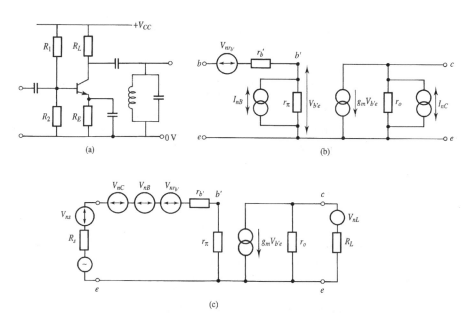

Figure 12.11 *(a) Single stage transistor amplifier; (b) transistor equivalent circuit with noise sources; and (c) equivalent circuit including driving, biasing and load components with most noise sources referred to input.*

(comprising the ouput resistance of the preceding stage in parallel with the biasing resistors R_1 and R_2), the base spreading resistance, $r_{b'}$, and the load resistance, R_L. Figure 12.11(b) shows an equivalent circuit of the transistor, incorporating its noise sources, and Figure 12.11(c) shows an equivalent circuit of Figure 12.11(a) with all noise sources (excluding that due to the load) transferred to the transistor's input and represented as RMS noise voltages. The noise sources in Figure 12.11(c) are given by:

$$V_{nr_{b'}}^2 = 4\,kTr_{b'}B \qquad\qquad (12.26(a))$$

$$V_{nB}^2 = I_{nB}^2 (R_s + r_{b'})^2 = 2q_e I_B B (R_s + r_{b'})^2 \qquad (12.26(b))$$

$$V_{nC}^2 = (I_{nC}/h_{fe})^2 (R_s + r_{b'} + r_\pi)^2$$

$$= (2q_e I_C B)(R_s + r_{b'} + r_\pi)^2/(g_m r_\pi)^2 \qquad (12.26(c))$$

$$V_{ns}^2 = 4\,kTR_s B \qquad (12.26(d))$$

$$V_{nL}^2 = 4\,kTR_L B \qquad (12.26(e))$$

where B is the noise bandwidth of the circuit, q_e is the electronic charge, I_B and I_C are the quiescent base and collector currents respectively, $h_{fe} = g_m r_\pi$ is the small signal forward current gain of the transistor and g_m is the transistor's transconductance.

The, effective, transistor input noise (which excludes that due to the source resistance) is therefore given by:

$$V_{n\,tran}^2 = V_{nC}^2 + V_{nB}^2 + V_{nr_{b'}}^2$$

$$= (2q_e I_C B)(R_s + r_{b'} + r_\pi)^2/(g_m r_\pi)^2 + 2q_e I_B B(R_s + r_{b'})^2 + 4\,kTr_{b'}B$$

$$= 4\,kT \left[0.5(R_s + r_{b'} + r_\pi)^2/(g_m r_\pi^2) + 0.5g_m(R_s + r_{b'})^2/h_{FE} + r_{b'} \right] B$$

$$= 4\,kTR_{tran} B \qquad (12.27)$$

where h_{FE} is the transistor's DC forward current gain and use has been made of $g_m = q_e I_C/(kT) \approx 40 I_C$ at room temperature. R_{tran}, defined by the square bracket in equation (12.27), represents a hypothetical resistor located at the (noiseless) transistor's input which would account (thermally) for the actual noise introduced by the real (noisy) transistor. The total noise at the transistor output (including that, V_{nL}^2, generated by the load resistor, R_L) is therefore:

$$V_{n\,olp}^2 = (V_{ns}^2 + V_{n\,tran}^2) \left(\frac{r_\pi}{R_s + r_{b'} + r_\pi} \right)^2 g_m^2 \left(\frac{r_o R_L}{r_o + R_L} \right)^2 + V_{nL}^2 \qquad (12.28)$$

Since the source resistance and load resistance depend, at least partly, on the output impedance of the preceding stage and input impedance of the following stage the importance of impedance optimisation in low noise circuit design can be appreciated. The following example illustrates a circuit noise calculation and also shows, explicitly, how such a calculation can be related to the conventional figure of merit (i.e. noise factor or noise figure, see section 12.3.3) commonly used to summarise the noise performance of a circuit, system or subsystem. (In practice communications circuits usually operate at high frequency where s-parameter transistor descriptions are appropriate. For a full treatment of low noise, high frequency, circuit design techniques the reader is referred to specialist texts, e.g. [Smith, Yip, Liao].)

EXAMPLE 12.1

The transistor in Figure 12.11(a)) has the following parameters: $r_{b'} = 150\ \Omega$, $r_o = 40\ k\Omega$, $h_{fe} = 100$, $h_{FE} = 80$. The circuit in which the transistor is embedded has the component values: $R_1 = 10\ k\Omega$, $R_2 = 3.3\ k\Omega$, $R_E = 1\ k\Omega$, $R_L = 3.3\ k\Omega$. The output resistance of the signal source which drives this circuit is 1.7 kΩ. Find: (a) the transistor's equivalent thermal input noise resistance, R_{tran}; (b) the noise power spectral density at the transistor outputs; and (c) the ratio of source plus transistor noise to transistor noise only. (For part (b) assume that, apart from thermal noise arising from the signal source's output resistance, the source is noiseless.) In part (c) the ratio which you have been asked to calculate is a noise factor, see section 12.3.3.

(a) The transistor transconductance, g_m, depends on the collector current and is found by:

$$R_{in} \approx (h_{FE} + 1) \, R_E = (80 + 1) \, 1 \times 10^3 = 81 \text{ k}\Omega$$

$$V_B = \frac{R_2 R_{in} / (R_2 + R_{in})}{[R_2 R_{in} / (R_2 + R_{in})] + R_1} \, V_{CC}$$

$$= \frac{(3.3 \times 81) / (3.3 + 81)}{[(3.3 \times 81) / (3.3 + 81)] + 10} \times 10 = 2.4 \text{ V}$$

$$V_E = V_B - V_{BE} = 2.4 - 0.6 = 1.8 \text{ V}$$

$$I_C \approx I_E = \frac{V_E}{R_E} = \frac{1.8}{1 \times 10^3} = 1.8 \text{ mA}$$

The transconductance is the reciprocal of the intrinsic emitter resistance, r_e, i.e.:

$$g_m = \frac{1}{r_e} = \frac{I_C \, (\text{mA})}{25} = \frac{1.8}{25} = 0.072 \text{ S}$$

$$r_\pi = \frac{h_{fe}}{g_m} = \frac{100}{0.072} = 1400 \, \Omega$$

The source resistance, R_s, is the parallel combination of the signal source's output resistance, R_o, and the transistor's biasing resistors:

$$R_s = \left[\frac{1}{R_o} + \frac{1}{R_1} + \frac{1}{R_2} \right]^{-1}$$

$$= \left[\frac{1}{1700} + \frac{1}{10000} + \frac{1}{3300} \right]^{-1} = 1000 \, \Omega$$

From equation (12.27):

$$R_{tran} = \frac{0.5 \, (R_s + r_{b'} + r_\pi)^2}{g_m \, r_\pi^2} + \frac{0.5 \, g_m \, (R_s + r_{b'})^2}{h_{FE}} + r_{b'}$$

$$= \frac{0.5 \, (1000 + 150 + 1400)^2}{(0.072) \, (1400^2)} + \frac{0.5 \, (0.072) \, (1000 + 150)^2}{80} + 150$$

$$= 23 + 600 + 150 = 770 \, \Omega$$

(b) The total noise power expected at the transistor output is the sum of contributions from the source resistance, the transistor and the load. Thus, using equation (12.28):

$$V_{n \, o/p}^2 = (V_{ns}^2 + V_{n \, tran}^2) \left(\frac{r_\pi}{R_s + r_{b'} + r_\pi} \right)^2 g_m^2 \left(\frac{r_o \, R_L}{r_o + R_L} \right)^2 + V_{nL}^2$$

where $V_{ns}^2 = 4 \, kTR_s B$ and $V_{nL}^2 = 4 \, kTR_L B$. Therefore:

$$V_{n \, o/p}^2 = 4 \, kT \left[(R_s + R_{tran}) \left(\frac{r_\pi}{R_s + r_{b'} + r_\pi} \right)^2 g_m^2 \left(\frac{r_o \, R_L}{r_o + R_L} \right)^2 + R_L \right] B \, (\text{V}^2)$$

$$= 4 \times 1.38 \times 10^{-23} \times 290 \left[(1000 + 770) \left(\frac{1400}{1150 + 1400} \right)^2 0.072^2 \left(\frac{40000 \times 3300}{40000 + 3300} \right)^2 + 3300 \right] B$$

$$G_{n\,o/p} = V_{n\,o/p}^2 / B$$

$$= 4 \times 4 \times 10^{-21} \; [(1000 + 770)\,15000 + 3300]$$

$$= 16 \times 10^{-21} \left[27 \times 10^6 + 3.3 \times 10^3 \right] = 4.3 \times 10^{-13} \; \text{V}^2/\text{Hz}$$

(c) Noise factor, f (see section 12.3.3), is given by:

$$f = \frac{R_s + R_{tran}}{R_s} = \frac{1000 + 770}{1000} = 1.77$$

12.3 System noise calculations

The gain, G, of a device is often expressed in decibels, as the ratio of the output to input voltages or powers, i.e. $G_{dB} = 20 \log_{10}(V_o/V_i)$ or $G_{dB} = 10 \log_{10}(P_o/P_i)$. Being a ratio gain is not related to any particular power level. There is a logarithmic unit of power, however, which defines power in dB above a specified reference level. If the reference power is 1 mW the units are denoted by dBm (dB with respect to a 1 mW reference) and if the reference level is 1 W the units are denoted by dBW. Thus 10 mW would correspond to +10 dBm and 100 mW would be +20 dBm, etc.; dBm can be converted to dBW using +30 dBm $\equiv$ 0 dBW. Table 12.1 illustrates these relationships.

Table 12.1 *Signal power measures.*

dBW	Power level (W)	dBm
30 dBW	1000	60 dBm
20 dBW	100	50 dBm
10 dBW	10	40 dBm
0 dBW	1	30 dBm
−10 dBW	1/10	20 dBm
−20 dBW	1/100	10 dBm
−30 dBW	1/1000	0 dBm
−40 dBW	1/10000	−10 dBm

The ways in which the noise performance of individual subsystems is specified, and the way these specifications are used to calculate the overall noise performance of a complete system, are now discussed. The overall noise characteristics of individual subsystems are usually either specified by the manufacturer or measured using special instruments (e.g. noise figure meters). If $T = 290$ K (ambient or room temperature) and B = 1 Hz, then the available noise power is $1.38 \times 10^{-23} \times 290 \times 1 = 4 \times 10^{-21}$ W/Hz. Expressing this power (in a 1 Hz bandwidth) in dBW gives a noise power spectral density of −204 dBW/Hz or −174 dBm/Hz. Power, in dBm or dBW, can be scaled by simply adding or subtracting the gain of amplifiers or attenuators, measured in dB, to give directly the output power, again in dBm or dBW.

12.3.1 Noise temperature

Consider again equation (12.8) which defines available noise power. The actual noise power delivered is less than this if the source and load impedances are not matched (in the maximum power transfer, i.e. conjugate impedance, sense). A convenient way to specify noise power is via an equivalent thermal noise temperature, T_e, given by:

$$T_e = \frac{N}{kB} \ (\text{K}) \tag{12.29}$$

The noise temperature of a subsystem (say an amplifier) is *not* the temperature of the room it is in *nor* even the temperature inside its case. It is the (hypothetical) temperature which an ideal resistor matched to the input of the subsystem would need to be at in order to account for the extra available noise observed at the device's output *over and above that which is due to the actual input noise*. This idea is illustrated schematically in Figure 12.12. The total available output noise of a device is therefore given by:

$$N = (kT_sB + kT_eB)G \tag{12.30}$$

where the first term in the brackets is the contribution from the input (or source) noise and the second term is the contribution from the subsystem itself. G is the (power) gain (expressed as a ratio) of the subsystem which can be greater or less than 1.0. For example, an amplifier would have a gain exceeding unity while the mixer in a downconverter, Figure 1.4, would typically have a gain of approximately 0.25. Whilst the contribution of the subsystem to the total output noise, and therefore the noise temperature, will generally be *influenced* by its physical temperature, it will also depend on its design and the quality of its component parts, particularly with respect to the sections at, or close to, its 'front' (i.e. input) end.

Some non-thermal noise processes, such as $1/f$ or flicker noise, have non-white spectra which if included in equation (12.29) would make T_e a function of bandwidth. In practice, however, for most systems work, white noise is assumed to dominate, making T_e bandwidth independent. If thermal noise is dominant (as is often the case in low noise systems) then physical cooling of the device will improve its noise performance.

It has been stated above that the equivalent noise temperature of a device has a dependence on its physical temperature. The strength of this dependence is determined by the relative proportions of thermal to non-thermal noise which the device generates. Strictly speaking, therefore, the noise temperature quoted for a device or subsystem relates to a specific device physical temperature. This temperature is invariably the

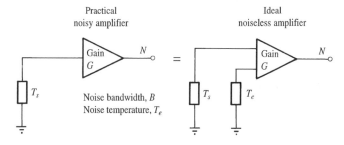

Figure 12.12 *Equivalent noise temperature (T_e) of an amplifier.*

equilibrium temperature which the device attains (under normal operating conditions which may incorporate heat sinks, etc.) when the ambient temperature around it is 290 K. The operating environments of electronic devices normally have ambient temperatures which are sufficiently close to this to make errors in noise calculations based on an assumed temperature of 290 K negligible for most practical purposes.

EXAMPLE 12.2

Calculate the output noise of the amplifier, shown in Figure 12.12, assuming that the amplifier gain is 20 dB, its noise bandwidth is 1 MHz, its equivalent noise temperature is 580 K and the noise temperature of its source is 290 K.

The available NPSD from the matched source at temperature T_0 is given in equation (12.11) by:

$$G_{n,s}(f) = kT_0 \quad \text{(W/Hz)}$$

(T_0 is a widely used symbol denoting 290 K.) The available NPSD from the equivalent resistor at temperature T_e (modelling the internally generated noise) is:

$$G_{n,e}(f) = kT_e \quad \text{(W/Hz)}$$

The total equivalent noise power at the input is:

$$N_{in} = k(T_0 + T_e)B \quad \text{(W)}$$

The total noise power at the amplifier output is therefore:

$$N_{out} = Gk(T_0 + T_e)B \quad \text{(W)}$$
$$= 10^{20/10} \times 1.38 \times 10^{-23} \times (290 + 580) \times 1 \times 10^6$$
$$= 1.2 \times 10^{-12} \quad \text{W} = -119.2 \text{ dBW} \quad \text{or} \quad -89.2 \text{ dBm}$$

12.3.2 Noise temperature of cascaded subsystems

The total system noise temperature, at the output of a device or subsystem, $T_{syst\ out}$, can be found by dividing equation (12.30) by kB, i.e.:

$$T_{syst\ out} = (T_s + T_e)G \tag{12.31}$$

If several subsystems are cascaded, as shown in Figure 12.13, the noise temperatures at the output of each subsystem are given by:

$$T_{out\ 1} = (T_s + T_{e1})G_1 \tag{12.32(a)}$$

$$T_{out\ 2} = (T_{out\ 1} + T_{e2})G_2 \tag{12.32(b)}$$

$$T_{out\ 3} = (T_{out\ 2} + T_{e3})G_3 \tag{12.32(c)}$$

The noise temperature at the output of subsystem 3 is therefore:

$$T_{out\ 3} = \{[(T_s + T_{e1})G_1 + T_{e2}]G_2 + T_{e3}\}G_3 \tag{12.33}$$
$$= T_s G_1 G_2 G_3 + T_{e1} G_1 G_2 G_3 + T_{e2} G_2 G_3 + T_{e3} G_3$$

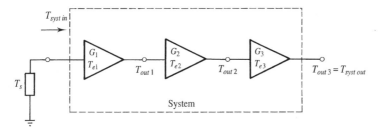

Figure 12.13 *Partitioning of system into cascaded amplifier subsystems.*

This temperature can be referred to the input of subsystem 1 by dividing equation (12.33) by the total gain, $G_1 G_2 G_3$, i.e.:

$$T_{syst\ in} = T_s + T_{e1} + \frac{T_{e2}}{G_1} + \frac{T_{e3}}{G_1 G_2} \tag{12.34}$$

or:

$$T_{syst\ in} = T_s + T_e \tag{12.35}$$

where T_e is the equivalent (input) noise temperature of the system excluding the source and $T_{syst\ in}$ is the overall input noise temperature *including the source*. The total noise temperature output of a set of cascaded subsystems is then simply given by:

$$T_{syst\ out} = T_{syst\ in} G_1 G_2 G_3 \tag{12.36}$$

It is important to remember that the 'gain' of the individual subsystems can be greater *or less* than 1.0. In the latter case the subsystem is lossy. (A transmission line, for instance, with 10% power loss would have a gain of 0.9 or −0.46 dB.) The equivalent (input) noise temperature, $T_{e,l}$, of a lossy device at a physical temperature T_{ph} can be found from its 'gain' using:

$$T_{e,l} = \frac{T_{ph}(1 - G_l)}{G_l} \tag{12.37}$$

(The subscript *l* here reminds us that a lossy device is being considered and that the gain G_l is therefore less than 1.0.) Equation (12.37) is easily verified by considering a transmission line terminated in matched loads as shown in Figure 12.14. Provided the loads, R_s and R_L, and the transmission line are in thermal equilibrium there can be no net

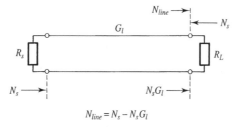

Figure 12.14 *Source, transmission line and load in thermal equilibrium.*

flow of noise power across the terminals at either end of the transmission line. From Nyquist's noise formula, equation (12.8), the power available to the transmission line from R_s is:

$$N_s = kT_{ph}B \tag{12.38}$$

If the transmission line has (power) gain G_l (< 1.0), then the power available to R_L from R_s is:

$$N'_s = kT_{ph}BG_l \tag{12.39}$$

Since R_L supplies:

$$N_L = kT_{ph}B \tag{12.40}$$

of power to the transmission line then the transmission line itself must supply the balance of power to R_L, i.e.:

$$N_{line} = kT_{ph}B - kT_{ph}BG_l \tag{12.41}$$

Dividing equation (12.41) by kB gives:

$$T_{line} = T_{ph}(1 - G_l) \tag{12.42}$$

Referring the line temperature to the source terminals then gives equation (12.37).

EXAMPLE 12.3

Figure 12.15 shows a simple superheterodyne receiver consisting of a front end low noise amplifier (LNA), a mixer and two stages of IF amplification. The source temperature of the receiver is 100 K and the characteristics of the individual receiver subsystems are:

Device	Gain (or conversion loss)	T_e
LNA	12 dB	50 K
Mixer	−6 dB	
IF Amp 1	20 dB	1000 K
IF Amp 2	30 dB	1000 K

Calculate the noise power at the output of the second IF amplifier in a 5.0 MHz bandwidth.

Using equation (12.34):

$$T_{syst\ in} = T_s + T_{e1} + \frac{T_{e2}}{G_1} + \frac{T_{e3}}{G_1G_2} + \frac{T_{e4}}{G_1G_2G_3}$$

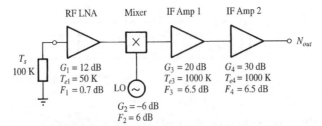

Figure 12.15 *Superheterodyne receiver.*

where:

$G_1 = 12$ dB $= 15.8$, $G_2 = -6$ dB $= 0.25 \, (= G_l)$, $G_3 = 20$ dB $= 100$, $T_s = 100$ K,
$T_{e1} = 50$ K, $T_{e2} = T_{ph}(1 - G_l)/G_l = 290(1 - 0.25)/0.25 = 870$ K, $T_{e3} = 1000$ K, $T_{e4} = 1000$ K.
(Notice that the physical temperature of the mixer has been assumed to be 290 K. This is the normal assumption unless information to the contrary is given. Notice also that the gain of the final stage is irrelevant as far as the system equivalent input temperature is concerned.)
Therefore:

$$T_{syst\ in} = 100 + 50 + \frac{870}{15.8} + \frac{1000}{15.8 \times 0.25} + \frac{1000}{15.8 \times 0.25 \times 100}$$

$$= 100 + 50 + 55 + 253 + 3 = 461 \text{ K}$$

If the effective noise bandwidth of the system (often determined by the final IF amplifier) is 5.0 MHz then the total noise output power will be:

$$N_{out} = kT_{syst\ in} BG_1 G_2 G_3 G_4$$

$$= 1.38 \times 10^{-23} \times 461 \times 5 \times 10^6 \times 15.8 \times 0.25 \times 100 \times 1000$$

$$= 1.26 \times 10^{-8} \text{ W} = -49.0 \text{ dBm}$$

It is interesting to repeat this calculation with the positions of the LNA and mixer reversed. In this case:

$$T_{syst\ in} = 100 + 870 + \frac{50}{0.25} + \frac{1000}{0.25 \times 15.8} + \frac{1000}{0.25 \times 15.8 \times 100}$$

$$= 100 + 870 + 200 + 253 + 3 = 1426 \text{ K}$$

The equivalent system input noise temperature, and therefore the total system output noise, has been degraded by a factor of 3. The reason for the presence of a low noise amplifier at the front end of a receiver thus becomes obvious.

12.3.3 Noise factor and noise figure

The noise factor, f, of an amplifier is defined as the ratio of SNR at the system input to SNR at the system output *when the input noise corresponds to a temperature of 290 K,* i.e.:

$$f = \frac{(S/N)_i}{(S/N)_o} \bigg|_{N_i = k\ 290\ B} \tag{12.43}$$

where:

$$(S/N)_i = \frac{\text{signal power at input, } S_i}{k\ 290\ B} \tag{12.44(a)}$$

and:

$$(S/N)_o = \frac{\text{signal power at output, } S_o}{k(290 + T_e)B\ G} \tag{12.44(b)}$$

The specification of the input noise temperature as 290 K allows all devices and systems to be compared fairly, on the basis of their quoted noise factor. f is therefore a figure of merit for comparing the noise performance of different devices and systems. Substituting equations (12.44) into (12.43):

$$f = \frac{S_i / (k\ 290\ B)}{GS_i / [G\ k(290 + T_e)B]} = \frac{290 + T_e}{290} \qquad (12.45(\text{a}))$$

i.e.:

$$f = 1 + \frac{T_e}{290} \qquad (12.45(\text{b}))$$

It is important to remember that, strictly speaking, f is only the ratio of input to output SNR if:

1. The device is operating at its equilibrium temperature in an ambient (290 K) environment. (This is necessary for the quoted T_e to be reliable.)
2. The source temperature at the device input is 290 K.

In practice it is the second condition which is most likely to be unfulfilled. Despite the arbitrary nature of the assumed source temperature in the definition of f, it is still possible to make accurate calculations of overall system noise temperature (and therefore overall SNRs) even when $T_s \neq 290$ K. This is because the noise factor of any device can be converted to an equivalent noise temperature using:

$$T_e = (f - 1)290 \quad (\text{K}) \qquad (12.46)$$

and equivalent noise temperature makes no assumption at all about source temperature. If preferred, however, the noise effects due to several subsystems can be cascaded before converting to noise temperatures. Comparing equations (12.34) and (12.35) the equivalent noise temperature of the cascaded subsystems is:

$$T_e = T_{e1} + \frac{T_{e2}}{G_1} + \frac{T_{e3}}{G_1 G_2} \quad (\text{K}) \qquad (12.47)$$

This can be rewritten in terms of noise factors as:

$$(f - 1)290 = (f_1 - 1)290 + \frac{(f_2 - 1)290}{G_1} + \frac{(f_3 - 1)290}{G_1 G_2} \qquad (12.48)$$

Dividing by 290 and adding 1:

$$f = f_1 + \frac{f_2 - 1}{G_1} + \frac{f_3 - 1}{G_1 G_2} \qquad (12.49)$$

This is called the Friis noise formula. The final system noise temperature (referred to the input of device 1) is then calculated from equations (12.46) and (12.35).

Traditionally noise factor is quoted in decibels, i.e.:

$$F = 10 \log_{10} f \quad (\text{dB}) \qquad (12.50)$$

and in this form is called the *noise figure*. It is therefore essential to remember to convert F to a ratio before using it in the calculations described above. (Care is required since the terms noise figure and noise factor and the symbols f and F are often used interchangeably in practice.) Table 12.2 gives some noise figures, noise factors and their equivalent noise temperatures.

The noise figure of lossy devices (such as transmission lines or passive mixers) is related to their 'gain', G_l, by:

$$f = 1/G_l \tag{12.51}$$

where f and G_l are expressed as ratios, or alternatively by:

$$F = - G_l \quad \text{(dB)} \tag{12.52}$$

where F and G_l are expressed in decibels. A transmission line with 10% power loss (i.e. $G_l = 0.9$) therefore has a noise factor given by:

$$f = 1/G_l = 1/0.9 = 1.11 \tag{12.53}$$

(or 0.46 dB as a noise figure). A mixer with a conversion loss of 6 dB (i.e. a conversion gain of –6 dB) has a noise figure given by:

$$F = - G_l = 6 \text{ dB} \tag{12.54}$$

(or 4.0 as a noise factor). Strictly speaking a mixer would have slightly greater noise figure than this due to the contribution of non-thermal noise by the diodes in the mixer circuit. For passive mixers such non-thermal effects can usually be neglected. For active mixers this might not be so (although in this case there would probably be a conversion gain rather than loss).

Table 12.2 *Comparison of noise performance measures.*

T_e	f	F	Comments
0 K	1.00	0 dB	Perfect (i.e. noiseless) device
15 K	1.05	0.2 dB	Excellent LNA
60 K	1.20	0.8 dB	Good LNA
120 K	1.40	1.5 dB	Typical LNA
580 K	2.00	3.0 dB	Typical amplifier
900 K	4.00	6.0 dB	Poor quality amplifier
10000 K	35.50	15.5 dB	Temperature of a noise source

EXAMPLE 12.4

The output noise of the system shown in Figure 12.15 is now recalculated using noise factors instead of noise temperatures.

The noise factor of the entire system is given in equation (12.49) by:

$$
\begin{aligned}
f &= f_1 + \frac{(f_2 - 1)}{G_1} + \frac{(f_3 - 1)}{G_1 G_2} + \frac{(f_4 - 1)}{G_1 G_2 G_3} \\
&= 10^{0.7/10} + \frac{10^{6/10} - 1}{15.8} + \frac{10^{6.5/10} - 1}{15.8 \times 0.25} + \frac{10^{6.5/10} - 1}{15.8 \times 0.25 \times 100} \\
&= 1.17 + \frac{2.98}{15.8} + \frac{3.47}{15.8 \times 0.25} + \frac{3.47}{15.8 \times 0.25 \times 100} \\
&= 1.17 + 0.19 + 0.88 + 0.01 = 2.25
\end{aligned}
$$

(or $F = 3.5$ dB)

The equivalent system noise temperature at the input to the low noise amplifier (LNA) is:

$$T_e = (f - 1)290$$

$$= (2.25 - 1)290 = 362 \text{ K}$$

and the total noise power at the system output in a bandwidth of 5.0 MHz is therefore:

$$N = k(T_s + T_e)BG_{syst}$$

$$= 1.38 \times 10^{-23} \times (100 + 362) \times 5 \times 10^6 \times 10^{56/10}$$

$$= 1.27 \times 10^{-8} \text{ W} = -49.0 \text{ dBm}$$

12.4 Radio communication link budgets

A communication system link budget refers to the calculation of received SNR given a specification of transmitted power, transmission medium attenuation and/or gain, and all sources of noise.

The calculation is often set out systematically (in a similar way to a financial budget) accounting explicitly for the various sources of gain, attenuation and noise. For single section line communications the essential elements are transmitted power, cable attenuation, receiver gain and noise figure. The calculation is then simply a matter of applying the Friis formula, or its equivalent, as described in sections 12.3–12.3.3. For radio systems the situation is different in that signal energy is lost not only as a result of attenuation (i.e. energy which is dissipated as heat) but also due to its being radiated in directions other than directly towards the receiving antenna. For multisection communication links the effects of analogue, amplifying, repeaters can be accounted for using their gains and noise figures in the Friis formula, and the effects of digital regenerative repeaters can be accounted for by summing the BERs of each section (providing the BER is small, see section 6.3). Before the details of a radio communication link budget are described some important antenna concepts are reviewed.

12.4.1 Antenna gain, effective area and efficiency

An isotropic antenna (i.e. one which radiates electromagnetic energy equally well in all directions), radiating P_{rad} W of power, supports a power density at a distance R (Figure 12.16) given by:

$$W_{isotrope} = \frac{P_{rad}}{4\pi R^2} \quad (\text{W/m}^2) \tag{12.55}$$

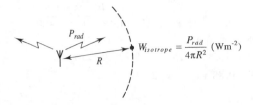

Figure 12.16 *Power density radiated by an isotropic antenna.*

The observation point is assumed, here, to be in the far field of the antenna, i.e. $R \geq 2D^2/\lambda$ where D is the largest dimension (often the diameter) of the antenna. In this region the reactive and radiating antenna near fields are negligible, and the radiating far field can be assumed to be a transverse electromagnetic (TEM) wave. The radiating far field pattern is independent of distance and, in any small region of space, the wavefront can be considered approximately plane. The field strength in the antenna far field is related to power density by:

$$\frac{E^2_{RMS}}{Z_o} = W \quad (W/m^2) \tag{12.56}$$

where $Z_o = E/H = 377 \, \Omega$ is the plane wave impedance of free space. Since $1/(4\pi R^2)$ in equation (12.55) represents a purely geometrical dilution of power density as the spherical wave expands, this factor is often referred to as the *spreading loss*. The *radiation intensity*, I, in an isotropic antenna's far field is given by:

$$I_{isotrope} = \frac{P_{rad}}{4\pi} \quad (W/\text{steradian or } W/rad^2) \tag{12.57}$$

The mutually orthogonal requirement on $\mathbf{E}, \mathbf{H}$ and $\mathbf{k}$ in the far field (where $\mathbf{E}$ is electric field strength, $\mathbf{H}$ is magnetic field strength and $\mathbf{k}$ is vector wave number pointing in the direction of propagation) excludes the possibility of realising isotropic radiators. Thus all practical antennas radiate preferentially in some directions over others. If radiation intensity $I(\theta, \phi)$ is plotted against spherical coordinates θ and ϕ the resulting surface (i.e. radiation pattern) will be spherical for a (hypothetical) isotrope and non-spherical for any realisable antenna, Figure 12.17. The gain, $G(\theta, \phi)$, of a transmitting antenna can be defined in azmuth θ and elevation ϕ as the ratio of radiation intensity in the direction θ, ϕ to the radiation intensity observed by replacing the antenna with a *lossless* isotrope:

$$\begin{aligned} G(\theta, \phi) &= \frac{I(\theta, \phi)}{I_{lossless \, isotrope}} \\ &= \frac{I(\theta, \phi)}{P_T/(4\pi)} \end{aligned} \tag{12.58(a)}$$

where P_T is the transmitter output power and the antenna is assumed to be well matched to its transmission line feed. A related quantity, antenna directivity, $D(\theta, \phi)$, can be defined by:

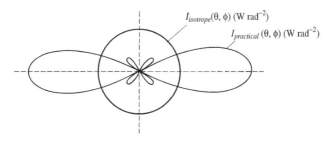

Figure 12.17 *Two-dimensional polar plots of antenna radiation intensity for isotropic and practical antenna.*

$$D(\theta, \phi) = \frac{I(\theta, \phi)}{I_{isotrope}}$$

$$= \frac{I(\theta, \phi)}{P_{rad}/(4\pi)} \tag{12.58(b)}$$

where the hypothetical isotropic antenna is now assumed to have a loss equal to that of the actual antenna. Directivity and gain therefore have identical shape and are related by:

$$G(\theta, \phi) = \eta_{\Omega} D(\theta, \phi) \tag{12.58(c)}$$

where $\eta_{\Omega} = P_{rad}/P_T$ is called the ohmic efficiency of the antenna. The ohmic losses relate physically to the electromagnetic energy dissipated by induced conduction currents flowing in the metallic conductors of the antenna and induced displacement currents flowing in the dielectric of the antenna. (The former usually dominate the latter and are often referred to as I^2R losses.) As an alternative both antenna gain and directivity can be defined as ratios of far-field power densities, i.e.:

$$G(\theta, \phi) = \frac{W(\theta, \phi, R)}{W_{lossless\ isotrope}(R)} \tag{12.59(a)}$$

and

$$D(\theta, \phi) = \frac{W(\theta, \phi, R)}{W_{isotrope}(R)} \tag{12.59(b)}$$

Using equations (12.55), (12.58), (12.59) and $P_{rad} = \eta_{\Omega} P_T$, the power density at a distance R from a transmitting antenna can be found from either of the following formulae:

$$W(\theta, \phi, R) = \frac{P_{rad}}{4\pi R^2} D(\theta, \phi) \quad (\text{W/m}^2) \tag{12.60(a)}$$

$$W(\theta, \phi, R) = \frac{P_T}{4\pi R^2} G(\theta, \phi) \quad (\text{W/m}^2) \tag{12.60(b)}$$

Figure 12.18 shows a typical, Cartesian coordinate, directivity pattern for a microwave reflector antenna. The effective area, a_e, of a receiving antenna, is defined as the ratio of carrier power C received at the antenna terminals, when the antenna is illuminated by a plane wave from the direction θ, ϕ, to the power density in the plane wave (Figure 12.19), i.e.:

$$a_e(\theta, \phi) = \frac{C(\theta, \phi)}{W} \quad (\text{m}^2) \tag{12.61}$$

(The plane wave is assumed to be polarisation matched to the antenna.) Unsurprisingly, there is an intimate connection between an antenna's gain and its effective area which can be expressed by:

$$G(\theta, \phi) = \frac{4\pi a_e(\theta, \phi)}{\lambda^2} \tag{12.62}$$

Equation (12.62) is a widely used expression of antenna reciprocity which can be derived from the Lorentz reciprocity theorem [Collin]. Usually interest is focused on the gain and effective area corresponding to the direction of maximum radiation intensity. This

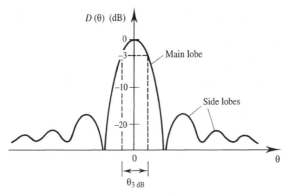

Figure 12.18 *Two-dimensional Cartesian plot of directivity (in dB) for microwave antenna (3 dB beamwidth of axisymmetric reflectors may be estimated using $\theta_{3dB} = 1.2\lambda/D$ rad).*

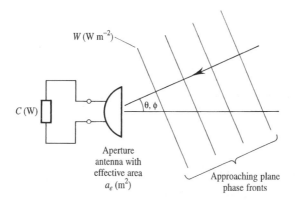

Figure 12.19 *Aperture antenna receiving plane wave from θ, ϕ direction.*

direction is called antenna boresight and is always implied if θ and ϕ are not specified.

For aperture antennas, such as paraboloidal reflectors, it is intuitively reasonable to associate a_e, at least approximately, with the physical antenna aperture. Unfortunately, they cannot be assumed to be identical since not all parts of the antenna aperture are fully utilised. There are several reasons for this including non-uniform illumination of the aperture by the feed, non-zero illumination of the region outside the aperture by the feed, aperture blockage caused by the presence of the feed and feed struts, and random errors in the definition of the antenna's reflecting surface. These and other effects can be accounted for by reducing the physical area, A_{ph}, of an aperture antenna by an aperture efficiency factor, η_{ap}. This gives the antenna's *effective* aperture, A_e, i.e.:

$$A_e = \eta_{ap} A_{ph} \quad (\text{m}^2) \tag{12.63(a)}$$

These aperture losses are in addition to the ohmic losses referred to earlier. Ohmic loss is accounted for by the ohmic efficiency, which when applied to the effective aperture gives the antenna's effective area as defined in equation (12.61), i.e.:

$$a_e = \eta_\Omega A_e \quad (\text{m}^2) \tag{12.63(b)}$$

The effective area of the antenna is therefore related to its physical area by:

$$a_e = \eta_{ap} \eta_\Omega A_{ph} \quad (\text{m}^2) \tag{12.64}$$

η_{ap} and η_Ω can be further broken down into more specific sources of loss. Any more detailed accounting for losses, however, is usually important only for antenna designers and is of rather academic interest to communications systems engineers. (Care should be taken in interpreting antenna efficiencies, however, since the terms effective aperture and effective area are often used interchangeably, as are the terms aperture efficiency and antenna efficiency.)

It is important to realise that antenna aperture efficiency is certain to be less than unity only for antennas with a well defined aperture of significant size (in terms of wavelengths). Wire antennas, such as dipoles, for example, have aperture efficiencies greater than unity if A_{ph} is taken to be the area presented by the wire to the incident wavefront.

EXAMPLE 12.5

The power density, radiation intensity and electric field strength are now calculated at a distance of 20 km from a microwave antenna having a directivity of 42.0 dB, an ohmic efficiency of 95% and a well matched 4 GHz transmitter with 25 dBm of output power.

The gain is given by:

$$\begin{aligned}
G &= \eta_\Omega D \\
&= 10 \log_{10} \eta_\Omega + D_{dB} \quad (\text{dB}) \\
&= 10 \log_{10} 0.95 + 42.0 = -0.2 + 42.0 = 41.8 \quad \text{dB}
\end{aligned}$$

and, from equation (12.60(b)), the received power density is given by:

$$\begin{aligned}
W &= \frac{P_T}{4\pi R^2} G_T \\
&= P_T + G_T - 10 \log_{10} (4\pi R^2) \quad \text{dBm/m}^2 \\
&= 25 + 41.8 - 10 \log_{10} (4\pi \times 20000^2) \\
&= -30.2 \quad \text{dBm/m}^2 = -60.2 \quad \text{dBW/m}^2 \\
&= 9.52 \times 10^{-7} \quad \text{W/m}^2
\end{aligned}$$

The radiation intensity is given by equation (12.58(b)) as:

$$\begin{aligned}
I &= \frac{P_{rad}}{4\pi} D = \frac{\eta_\Omega P_T}{4\pi} D \\
&= \frac{0.95 \times 10^{\frac{25}{10}}}{4\pi} \times 10^{\frac{42}{10}} \quad \text{mW/rad}^2 \\
&= 3.789 \times 10^5 \quad \text{mW/rad}^2
\end{aligned}$$

and from equation (12.56):

$$E_{RMS} = \sqrt{W\ Z_o} = \sqrt{9.52 \times 10^{-7} \times 377} \quad \text{V/m} = 18.9 \quad \text{mV/m}$$

If an identical receiving antenna is located 20 km from the first, the available carrier power, C, at its terminals could be calculated as follows:

$$\lambda = \frac{c}{f} = \frac{3 \times 10^8}{4 \times 10^9} = 0.075 \text{ m}$$

The antenna effective area, equation (12.62), is:

$$a_e = G\frac{\lambda^2}{4\pi} = 10^{\frac{41.8}{10}}\left[\frac{(0.075)^2}{4\pi}\right]$$

$$= 6.775 \text{ m}^2$$

and from equation (12.61):

$$C = W\ a_e$$

$$= 9.52 \times 10^{-7} \times 6.775 = 6.45 \times 10^{-6} \text{ W} = -21.9 \text{ dBm}$$

12.4.2 Free space and plane earth signal budgets

Consider the free space radio communication link shown in Figure 12.20. The power density at the receiver radiated by a lossless isotrope would be:

$$W_{lossless\ isotrope} = \frac{P_T}{4\pi R^2} \quad \text{(W/m}^2) \tag{12.65}$$

For a practical transmitting antenna with gain G_T the power density at the receiver is actually:

$$W = \frac{P_T}{4\pi R^2}\ G_T \quad \text{(W/m}^2) \tag{12.66}$$

The carrier power available at the receive antenna terminals is given by:

$$C = W\ a_e \quad \text{(W)} \tag{12.67}$$

which, on substituting equation (12.66) for W, gives:

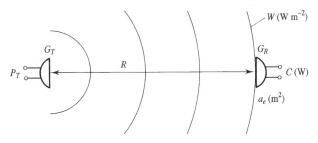

Figure 12.20 *Free space propagation.*

$$C = \frac{P_T}{4\pi R^2} G_T a_e \quad \text{(W)} \tag{12.68}$$

Using antenna reciprocity (equation (12.62)):

$$C = \frac{P_T}{4\pi R^2} G_T G_R \frac{\lambda^2}{4\pi} \quad \text{(W)} \tag{12.69}$$

where G_R is the gain of the receiving antenna. (Note that $\lambda^2/4\pi$ can be identified as the effective area, a_e, of a lossless isotrope.) Equation (12.69) can be rewritten as:

$$C = P_T G_T \left(\frac{\lambda}{4\pi R} \right)^2 G_R \quad \text{(W)} \tag{12.70}$$

This is the basic free space transmission loss formula for radio systems. The quantity $P_T G_T$ is called the effective isotropic radiated power (EIRP) and the quantity $[\lambda/(4\pi R)]^2$ is called the free space path loss (FSPL). Notice that FSPL is a function of wavelength. This is because it contains a factor to convert the receiving antenna effective area to gain in addition to the (geometrical) spreading loss.

Equation (12.70) is traditionally expressed using decibel quantities:

$$C = \text{EIRP} - \text{FSPL} + G_R \quad \text{(dBW)} \tag{12.71(a)}$$

where:

$$\text{EIRP} = 10 \log_{10} P_T + 10 \log_{10} G_T \quad \text{(dBW)} \tag{12.71(b)}$$

and:

$$\text{FSPL} = 20 \log_{10} \left(\frac{4\pi R}{\lambda} \right) \quad \text{(dB)} \tag{12.71(c)}$$

(The same symbol is used here for powers and gains whether they are measured in natural units or decibels. The context, however, leaves no doubt as to which is intended.)

Figure 12.21 shows a radio link operating above a plane earth. In this case there are two possible propagation paths between the transmitter and receiver, one direct and the other reflected via the ground. Assuming a complex (voltage) reflection coefficient at the ground, $\rho e^{j\phi}$, then the field strength, E, at the receiver will be changed by a (complex) factor, F, given by:

$$F = 1 + \rho e^{j\phi} e^{-j2\pi(d_2 - d_1)/\lambda} \tag{12.72}$$

where d_1 and d_2 are the lengths of the direct and reflected paths respectively and $2\pi(d_2 - d_1)/\lambda = \theta$ is the resulting excess phase shift of the reflected, over the direct, path.

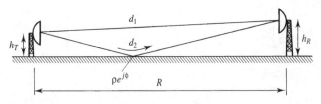

Figure 12.21 *Propagation over a plane earth.*

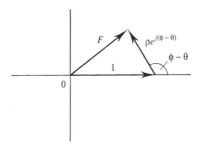

Figure 12.22 *Phasor addition of (normalised) fields from direct and reflected paths in Figure 12.21.*

Figure 12.22 illustrates the (normalised) phasor addition of the direct and reflected fields. Assuming perfect reflection at the ground (i.e. $\rho e^{j\phi} = -1$) and using $e^{-j\theta} = \cos\theta - j\sin\theta$ the magnitude of the field strength gain factor can be written, for practical multipath geometries (see Problem 12.8), as:

$$|F| = 2\sin\left(\frac{2\pi h_T h_R}{\lambda R}\right) \tag{12.73}$$

where h_T and h_R are transmit and receive antenna heights respectively and R is the horizontal distance between transmitter and receiver, Figure 12.21. The power density at the receiving antenna aperture, and therefore the received power, is increased over that for free space by a factor $|F|^2$. Thus, using natural (not decibel) quantities:

$$C = \text{EIRP} \times \text{FSPL} \times |F|^2 \times G_R$$

$$= P_T G_T \left(\frac{\lambda}{4\pi R}\right)^2 4\sin^2\left(\frac{2\pi h_T h_R}{\lambda R}\right) G_R \quad \text{(W)} \tag{12.74}$$

A typical interference pattern resulting from the two paths is illustrated as a function of horizontal range and height in Figure 12.23(a) and Figure 12.23(b) respectively. Equation (12.74) can be expressed in decibels as:

$$C = P_T + G_T - \text{FSPL} + G_R + 6.0 + 20\log_{10}\left|\sin\left(\frac{2\pi h_T h_R}{\lambda R}\right)\right| \quad \text{(dBW)} \tag{12.75}$$

Figure 12.24 shows the variation of received power on a dB scale with respect to transmitter–receiver distance. (A plot of power density in dBW/m^2 or electric field in dBμV/m has identical shape.) Notice that the peaks, corresponding to points of constructive interference, are 6 dB above the free space value and the troughs, corresponding to points of destructive interference, are, in principle, ∞ dB below the free space value. (In practice the minima do not represent zero power density or field strength because $\rho < 1.0$.) The furthest point of constructive interference occurs for the smallest argument of $\sin[(2\pi h_T h_R)/(\lambda R)]$ which gives 1.0, i.e. when:

$$\frac{2\pi h_T h_R}{\lambda R} = \frac{\pi}{2} \tag{12.76(a)}$$

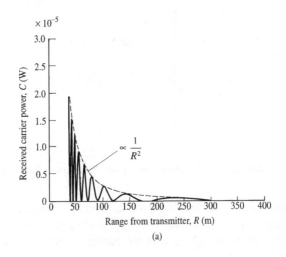

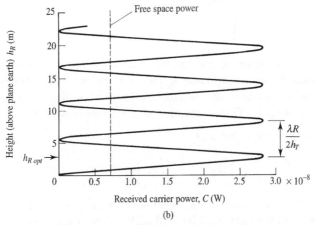

Figure 12.23 *Variation of received power with: (a) distance at a height of 2 m; (b) height at a range of 1000 m from transmitter. (Parameters as used in equation (12.74): $P_T G_T = 10\ W,\ \lambda = 0.333\ m,\ h_T = 30\ m,\ G_R = 0\ dB.$)*

The distance R_{max}, to which this corresponds, is:

$$R_{\text{max}} = \frac{4h_T h_R}{\lambda} \quad (\text{m}) \tag{12.76(b)}$$

For $R > R_{\text{max}}$ the total field decays monotonically. When $2\pi h_T h_R / \lambda R$ is small the approximation $\sin x \approx x$ can be used and the received power becomes:

$$C = P_T G_T \left(\frac{h_T h_R}{R^2} \right)^2 G_R \quad (\text{W}) \tag{12.77}$$

or, in decibels:

$$C = P_T + G_T - \text{PEPL} + G_R \quad (\text{dBW}) \tag{12.78(a)}$$

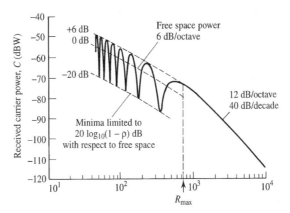

Figure 12.24 *Variation of received power with range for plane earth propagation. (Parameters used are the same as for Figure 12.23.)*

where PEPL, the plane earth path loss, is given by:

$$\text{PEPL} = 20 \log_{10} \left(\frac{R^2}{h_T h_R} \right) \text{ (dB)} \tag{12.78(b)}$$

Equation (12.77) shows that over a plane conducting earth the received power, and therefore the power density at the receiving antenna aperture, decays as $1/R^4$. The corresponding field strength therefore decays as $1/R^2$.

The optimum receive antenna height, $h_{R\ opt}$, for fixed λ, R and h_T is found by differentiating $|F|$ in equation (12.73) with respect to h_R and setting the result to zero, i.e.:

$$\frac{d|F|}{dh_R} = 2 \cos \left(\frac{2\pi h_T h_R}{\lambda R} \right) \frac{2\pi h_T}{\lambda R} = 0 \tag{12.79}$$

This requires that:

$$\frac{2\pi h_T h_{R\ opt}}{\lambda R} = \frac{n\pi}{2}, \quad n = 1, 3, 5, \cdots \tag{12.80(a)}$$

i.e.:

$$h_{R\ opt} = n \frac{\lambda R}{4h_T} \text{ (m)} \tag{12.80(b)}$$

where n is an odd integer. (A second derivative $(d^2|F|/dh_R^2)$ confirms that this is a condition for maximum, rather than minimum, received power.) Clearly antenna heights would not normally be chosen to be higher than necessary and so, in the absence of other considerations, n would be chosen to be equal to 1. (It would actually be useful to optimise an antenna height only if the ground reflection was reasonably stable in both magnitude *and phase* – a condition not often encountered.)

EXAMPLE 12.6

A 6 GHz, 40 km, LOS link uses 2.0 m axisymmetric paraboloidal reflectors for both transmitting and receiving antennas. The ohmic and aperture efficiencies of the antennas are 99% and 70% respectively and both antennas are mounted at a height of 25 m. The transmitter power is 0 dBW. Find the received power for both free space and plane earth conditions. Which condition is most likely to prevail if the path is over water?

$$\lambda = \frac{c}{f} = \frac{3 \times 10^8}{6 \times 10^9} = 0.05 \text{ m}$$

From equation (12.64) we have:

$$a_e = \eta_\Omega \eta_{ap} A_{ph}$$

$$= 0.99 \times 0.70 \times \pi \times 1.0^2 = 2.177 \text{ m}^2$$

and using equation (12.62):

$$G = \frac{4\pi a_e}{\lambda^2} = \frac{4\pi \times 2.177}{0.05^2} = 1.094 \times 10^4 = 40.4 \text{ dB}$$

$$\text{FSPL} = 20 \log_{10}\left(\frac{4\pi R}{\lambda}\right)$$

$$= 20 \log_{10}\left(\frac{4\pi 40 \times 10^3}{0.05}\right) = 140.0 \text{ dB}$$

For free space conditions equation (12.71(a)) gives:

$$C = P_T + G_T - \text{FSPL} + G_R$$

$$= 0 + 40.4 - 140.0 + 40.4 = -59.2 \text{ dBW}$$

For plane earth conditions (assuming perfect reflection) equation (12.76(b)) gives:

$$R_{max} = \frac{4h_T h_R}{\lambda} = \frac{4 \times 25^2}{0.05} = 50000 \text{ m}$$

The receive antenna is thus closer to the transmit antenna than the furthest point of constructive interference and equation (12.75) rather than (12.78) must therefore be used.

$$C = C|_{free\ space} + 6.0 + 20 \log_{10}\left|\sin\left(\frac{2\pi h_T h_R}{\lambda R}\right)\right|$$

$$= -59.2 + 6.0 + 20 \log_{10}\left|\sin\left(\frac{2\pi \times 25^2}{0.05 \times 40 \times 10^3}\right)\right|$$

$$= -59.2 + 6.0 - 0.7 = -53.9 \text{ dBW}$$

Note that this is only 0.7 dB below the maximum possible received power so the antenna heights must be close to optimum for plane earth propagation. Equation (12.80) shows that optimum (equal) antenna heights would be given by:

$$h_R h_T = \frac{\lambda R}{4} = \frac{0.05 \times 40 \times 10^3}{4} = 500 \text{ m}^2$$

i.e.: $h_R = h_T = \sqrt{500} = 22.36$ m

As has been said in the text, however, it is unlikely that such precise positioning of antennas would be of benefit in practice. To assess whether free space or plane earth propagation is likely to occur, a first order calculation based on antenna beamwidth can be carried out as follows:

$$\text{Antenna beamwidth} \approx 1.2 \, \frac{\lambda}{D} \text{ rad}$$

$$= 1.2 \left(\frac{0.05}{2.0} \right) = 0.03 \text{ rad } (= 1.7°)$$

Under normal atmospheric conditions for a horizontal path over water, with equal height transmit and receive antennas, the point of specular reflection will be half way along the link. The vertical width of the antenna beam (between −3 dB points) is given by:

$$\text{Vertical width} \approx \text{beamwidth} \times \tfrac{1}{2} \text{ path length}$$

$$= 0.03 \times \tfrac{1}{2} \times 40 \times 10^3 = 600 \text{ m}$$

Since half this is much greater than the path clearance and since water is a good reflector at microwave frequencies then specular reflection (for a calm water surface) is likely to be strong and a plane earth calculation is appropriate.

12.4.3 Antenna temperature and radio noise budgets

The overall noise power in a radio communications receiver depends not only on internally generated receiver noise but also on the electromagnetic noise collected by the receiver's antenna. Just as the equivalent thermal noise of a circuit or receiver subsystem can be represented by a noise temperature, so too can the noise received by an antenna. The equivalent noise temperature of a receiving antenna, T_{ant}, is defined by:

$$T_{ant} = \frac{\text{available NPSD at antenna terminals}}{\text{Boltzmann's constant}} \qquad (12.81)$$

$$= \frac{G_N(f)}{k} \text{ (K)}$$

where $G_N(f)$ is assumed to be white and the noise power spectral density is one sided. Antenna noise originates from several different sources. Below about 30 MHz it is dominated by the broadband radiation produced in lightning discharges associated with thunderstorms. This radiation is trapped by the ionosphere and so propagates world-wide. Such noise is sometimes referred to as atmospherics. The ionosphere is essentially transparent above about 30 MHz and between this frequency and 1 GHz the dominant noise is galactic. This has a steeply falling spectral density with increasing frequency (the slope is about −25 dBK/decade). It arises principally due to synchrotron radiation produced by fast electrons moving through the galactic magnetic field. Because the galaxy is very oblate in shape, and also because the earth is not located at its centre, galactic noise is markedly anisotropic and is much greater when the receiving antenna is pointed towards the galactic centre than when it is pointed to the galactic pole.

Above 1 GHz galactic noise is relatively weak. This leaves atmospheric thermal radiation and ground noise as the dominant noise processes. Atmospheric and ground noise is approximately flat with frequency up to about 10 GHz, its spectral density depending sensitively on antenna elevation angle. As elevation increases from 0° to 90° the thickness of atmosphere through which the antenna beam passes decreases as does the influence of the ground, both effects leading to a decrease in received thermal noise. In this frequency range a zenith-pointed antenna during clear sky conditions may have a noise temperature close to the cosmic background temperature of 3 K. Above 10 GHz resonance effects (of water vapour molecules at 22 GHz and oxygen molecules at 60 GHz) lead to increasing atmospheric attenuation and, therefore, thermal noise emission. The typical 'clear sky' noise temperature, as would be measured by a lossless narrow beam antenna, is illustrated over a band of frequencies from HF to SHF in Figure 12.25.

In addition to noise received as electromagnetic radiation by the antenna, thermal noise will also be generated in the antenna itself. A simple equivalent circuit of an antenna is shown in Figure 12.26. The (total) antenna noise temperature, T_{ant}, is the sum of its aperture temperature, T_A (originating from external sources and reduced by ohmic losses), and its equivalent internal, thermal, temperature, i.e.:

$$T_{ant} = T_A \eta_\Omega + T_{ph}(1 - \eta_\Omega) \quad (K) \tag{12.82}$$

where η_Ω is the antenna ohmic efficiency and T_{ph} is the physical temperature of the

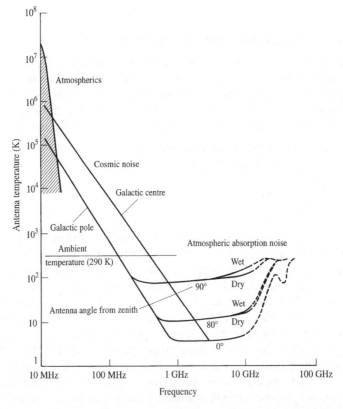

Figure 12.25 *Antenna sky noise temperature as a function of frequency and antenna elevation angle (source: from J.D. Kraus, 'Radio Astronomy' Cygnus-Quasar books, 1986, with permission).*

antenna. For noise purposes, then, the radiation resistance, R_r, in Figure 12.26 can be assumed to be at the equivalent antenna aperture temperature and the ohmic resistance, R_Ω, can be assumed to be at the antenna physical temperature. The ohmic efficiency of the antenna is related to R_r and R_Ω by:

$$\eta_\Omega = \frac{R_r}{R_r + R_\Omega} \tag{12.83}$$

The calculation of T_A can be complicated but may be estimated using:

$$T_A = \frac{1}{4\pi} \int\int_{4\pi} D(\theta, \phi)\varepsilon(\theta, \phi)T_{ph}(\theta, \phi) \ d\Omega \quad (\text{K}) \tag{12.84}$$

where $\varepsilon(\theta, \phi)$ and $T_{ph}(\theta, \phi)$ are the emissivity and physical temperature, respectively, of the material lying in the direction θ, ϕ, and $d\Omega$ is an element of solid angle. (The emissivity of a surface is related to its voltage reflection coefficient, ρ, by $\varepsilon = 1 - \rho^2$ and the product $\varepsilon(\theta, \phi)T_{ph}(\theta, \phi)$ is sometimes called brightness temperature, $T_B(\theta, \phi)$.)

The quantity $D(\theta, \phi)\varepsilon(\theta, \phi)d\Omega$ in equation (12.84) can be interpreted as the fraction of power radiated by the antenna which is *absorbed* by the material lying in the direction between θ, ϕ and $\theta + d\theta$, $\phi + d\phi$. If the brightness temperature of the environment around the antenna changes discretely (assuming, for example, different, but individually uniform, temperatures for the ground, clear sky and sun) then equation (12.84) can be written in the simpler form:

$$T_A = \sum_i \alpha_i T_{ph,i} \tag{12.85}$$

where α_i is the fraction of power radiated by the antenna which is absorbed by the ith body and $T_{ph,i}$ is the physical temperature of the ith body.

For more precise calculations still, scattering of radiation from one source by another (e.g. scattering of ground noise into the antenna by the atmosphere) must be accounted for. This level of detail, however, is usually the concern of radio remote sensing engineers rather than communications engineers.

Having established an antenna noise temperature the overall noise temperature of a radio receiving system is:

$$T_{syst \ in} = T_{ant} + T_e \tag{12.86}$$

where T_e is the equivalent input noise temperature of the receiver.

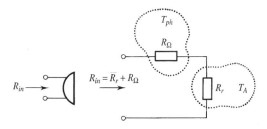

Figure 12.26 *Equivalent circuit of an antenna for noise calculations (R_r is radiation resistance, R_Ω is ohmic resistance and R_{in} is input resistance).*

12.4.4 Receiver equivalent input CNR

The equivalent input CNR of a radio receiver is given by combining equations (12.70) or (12.74) and (12.86) in the receiver bandwidth, i.e.:

$$\frac{C}{N} = \frac{C}{kT_{syst\ in}B} \tag{12.87}$$

The baseband BER and/or SNR of the demodulated signal are then found by applying any CNR detection gain provided by the demodulator and/or decoder, and finding the probability of bit error for the particular digital modulation scheme being used (including the mitigating effects of error control coding). If required the PCM decoded SNR of a baseband analogue signal can then be found by applying an equation such as (5.29).

It should be remembered that although the emphasis here has been on naturally occurring noise some systems are interference limited. Such interference may be random, quasi-periodic, intelligible (in which case it is usually called crosstalk) or a combination of these. Cellular radio is an example of a system which usually operates in an interference limited environment.

EXAMPLE 12.7

A 10 GHz terrestrial line-of-sight link has good clearance over rough terrain such that free space propagation can be assumed. The free space signal power at the receiving antenna terminals is -40.0 dBm. The overall noise figure of the receiver is 5.0 dB and the noise bandwidth is 20 MHz. Estimate the actual and effective clear sky CNRs at the antenna terminals assuming that the antenna has an ohmic efficiency of 95% and is at a physical temperature of 280 K. Also make a first order estimate of the effective CNR during a rain fade of 2 dB, assuming the rain is localised and occurs close to the receiving antenna.

From Figure 12.25 the clear sky aperture temperature, T_A, at 10 GHz for a horizontal link is about 100 K. The antenna temperature is given in equation (12.82) by:

$$T_{ant} = T_A\eta_\Omega + T_{ph}(1 - \eta_\Omega)$$
$$= 100 \times 0.95 + 280\ (1 - 0.95) = 95 + 14 = 109\ \text{K}$$

(Notice that the contribution from the physical temperature of the antenna is, in this case, probably within the uncertainty of the estimate of aperture temperature.) Now from equation (12.8):

$$N = kTB = 1.38 \times 10^{-23} \times 109 \times 20 \times 10^6 = 3.01 \times 10^{-14}\ \text{W} = -135.2\ \text{dBW}$$

The actual clear sky CNR at the antenna terminals with a received power of -70 dBW is given by:

$$\frac{C}{N} = -70.0 - (-135.2) = 65.2\ \text{dB}$$

The equivalent noise temperature of the receiver is given by equation (12.46) as:

$$T_e = (f - 1)290 = (10^{\frac{5}{10}} - 1)290 = 627\ \text{K}$$

The system noise temperature (referred to the antenna output) is:

$$T_{syst\ in} = T_{ant} + T_e = 109 + 627 = 736\ \text{K}$$

The effective system noise power (referred to the antenna output) is:

$$N = kT_{syst\ in}B = 1.38 \times 10^{-23} \times 736 \times 20 \times 10^6$$

$$= 2.03 \times 10^{-13}\ \text{W} = -126.9\ \text{dBW}$$

The effective CNR is:

$$\left.\frac{C}{N}\right|_{eff} = C - N\ \text{(dB)}$$

$$= -70.0 - (-126.9) = 56.9\ \text{dB}$$

During a 2 dB rain fade carrier power will be reduced by 2 dB to –42 dBm and noise will be increased. A first order estimate of noise power during a fade can be found as follows.

Assuming the physical temperature of the rain (T_{rain}) is the same as that of the antenna then the aperture temperature may be recalculated using the transmission line of equation (12.42) as:

$$T_A = T_{sky} \times \text{fade} + T_{rain}(1 - \text{fade})$$

$$= 100 \times 10^{\frac{-2}{10}} + 280\ (1 - 10^{\frac{-2}{10}})$$

$$= 63 + 103 = 166\ \text{K}$$

$$T_{ant} = T_A\ \eta_\Omega + T_{ph}(1 - \eta_\Omega)$$

$$= 166 \times 0.95 + 280(1 - 0.95)$$

$$= 158 + 14 = 172\ \text{K}$$

$$N = k(T_{ant} + T_e)B$$

$$= 1.38 \times 10^{-23}\ (172 + 627) \times 20 \times 10^6$$

$$= 2.21 \times 10^{-13}\ \text{W} = -126.6\ \text{dBW}$$

The effective CNR during the 2 dB fade is therefore:

$$\left.\frac{C}{N}\right|_{eff} = C - N$$

$$= -72.0 - (-126.6) = 54.6\ \text{dB}$$

The example shown above is intended to illustrate the concepts discussed in the text and probably contains spurious precision. The uncertainties associated with real systems mean that in practice a first order estimate of CNR would probably be based on a worst case antenna noise temperature of, say, 290 K. (The difference between this assumption and the above effective CNR calculation is only $10 \log_{10} [(290 + 627)/(172 + 627)] = 0.6$ dB.)

12.4.5 Multipath fading and diversity reception

Multipath fading occurs to varying extents in many different radio applications [Rummler, 1986]. It is caused whenever radio energy arrives at the receiver by more than one path. Figure 12.27 illustrates how multipath propagation may occur on a point-to-point line-of-sight microwave link. In this case multiple paths may occur due to ground

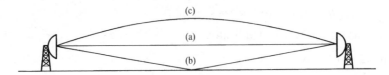

Figure 12.27 *Multipath in line-of-sight terrestial link due to: (a) direct path plus (b) ground reflection and/or (c) reflection from (or refraction through) a tropospheric layer.*

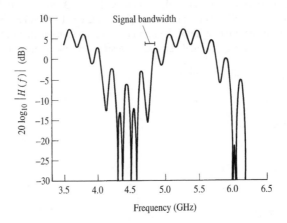

Figure 12.28 *Amplitude response of a frequency selective channel for three-ray multipath propagation with ray amplitudes and delays of: 1.0, 0 ns; 0.9, 0.56 ns; 0.1, 4.7 ns.*

reflections, reflections from stable tropospheric layers (with different refractive index) and refraction by tropospheric layers with extreme refractive index gradients. Other systems suffer multipath propagation due to the presence of scattering obstacles. This is the case for urban cellular radio systems for example, Figure 15.1. These multipath signals must be summed by phasor addition, Figure 12.22, to obtain the received signal power level.

There are two principal effects of multipath propagation on systems, their relative severity depending essentially on the relative bandwidth of the resulting channel compared with that of the signal being transmitted. For fixed point systems such as the microwave radio relay network the fading process is governed by changes in atmospheric conditions. Often, but not always, the spread of path delays is sufficiently short for the frequency response of the channel to be essentially constant over its operating bandwidth. In this case fading is said to be flat since all frequency components of a signal are subjected to the same fade at any given instant. When several or more propagation paths exist the fading of signal *amplitude* obeys Rayleigh statistics (due to the central limit theorem). If the spread of path delays is longer then the frequency response of the channel may change rapidly on a frequency scale comparable to signal bandwidth, Figure 12.28. In this case the fading is said to be frequency selective and the received signal is subject to severe amplitude and phase distortion. Adaptive equalisers may then be required to flatten and linearise the overall channel characteristics. The effects of flat fading can be combated by increasing transmitter power whilst the effects of frequency

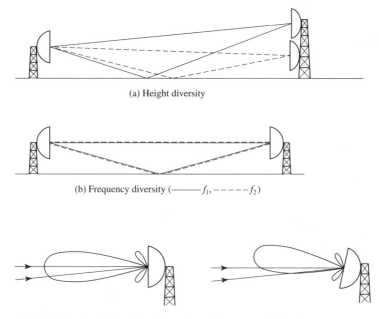

(a) Height diversity

(b) Frequency diversity ($\longrightarrow f_1, ----f_2$)

(c) Angle diversity (main and diversity, co-located, antennas unlikely to both respond well to multipath ray(s))

Figure 12.29 *Three types of diversity arrangements to combat multipath fading.*

selective fading cannot. For microwave links which are subject to flat fading a fade margin is usually designed into the link budget to offset the expected multipath (and rain induced) fades. The magnitude of this margin depends, of course, on the required availability of the link.

To reduce the necessary fade margin to acceptable levels diversity reception is sometimes employed. Figures 12.29(a) to (c) illustrate the principles of three types of diversity system, namely space (also called height), frequency and angle diversity. In all cases the essential assumption is that it is unlikely that both the main and diversity channels will suffer severe fades at the same instant. Selecting the channel with largest CNR, or combining channels with weightings in proportion to their CNRs, will clearly result in improved overall CNR.

12.5 Fibre optic transmission links

Optical fibres comprise a core, cladding and protective cover and are much lighter than metallic cables [Gower]. This advantage, coupled with the rapid reduction in propagation loss to its current value of 0.2 dB/km or less, Figure 1.6(e), and the enormous potential bandwidth available, make optical fibre now the only serious contender for the majority of long haul trunk transmission links. The potential capacity of optical fibres is such that all the radar, navigation and communication signals in the microwave and millimetre wave region, which now exist as free space signals, could be accommodated within 1% of the potential operational bandwidth of a single fibre. Current commercial systems can now accommodate 0.1 to 10 million simultaneous telephone calls in a single fibre.

12.5.1 Fibre types

In the fibre a circular core of refractive index n_1 is surrounded by a cladding layer of refractive index n_2, where $n_2 < n_1$, Figure 12.30. This results in optical energy being guided along the core with minimal losses.

The size of the core and the nature of the refractive index change from n_1 to n_2 determine the three basic types of optical fibre [Gower], namely:

- multimode step index;
- multimode graded index;
- monomode step index.

The refractive index profiles, typical core and cladding layer diameters and a schematic ray diagram representing the distinct optical modes in these three fibre types are shown in Figure 12.30. The optical wave propagates down the fibre via reflections at the refractive index boundary or refraction in the core. In multimode fibres there was, originally, a 20% difference between the refractive indices of the core and the cladding. In the more recently developed monomode fibres, the difference is much smaller, typically 0.5%. The early multimode step index fibres were cheap to fabricate but they had limited bandwidth and a limited section length between repeaters. In a 1 km length

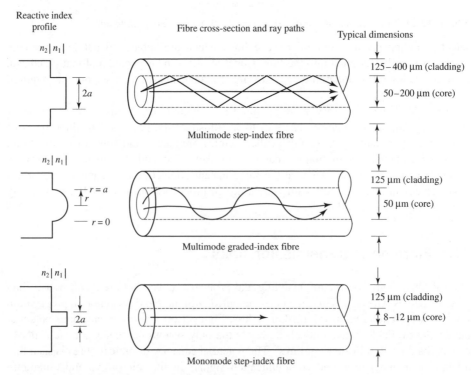

Figure 12.30 *Three distinct types of optical fibre.*

of modern multimode fibre with a 1% difference in refractive index between core and cladding, pulse broadening or dispersion, caused by the difference in propagation velocity between the different electromagnetic modes, limits the maximum data rate to, typically, 10 Mbit/s. Graded index fibres suffer less from mode dispersion because the ray paths representing differing modes encounter material with differing refractive index. Since propagation velocity is higher in material with lower refractive index, the propagation delay for all the modes can, with careful design of the graded-n profile, be made approximately the same. For obvious reasons *modal* dispersion is absent from monomode fibres altogether and in these fibre types *material* dispersion is normally the dominant pulse spreading mechanism. Modal dispersion becomes a problem in systems with bit rates exceeding 40 Gbit/s. Material dispersion occurs because refractive index is, generally, a function of wavelength, and different frequency components therefore propagate with different velocities. (Material dispersion is exacerbated due to the fact that practical sources often emit light with a narrow, but not monochromatic, spectrum.) The rate of change of propagation velocity with frequency (dv/df), and therefore dispersion, in silica fibres changes sign at around 1.3 μm, Figure 12.31, resulting in zero material dispersion at this wavelength. (Fortuitously, this wavelength also corresponds to a local minimum in optical attenuation, see Figure 1.6(e).)

If both modal and material dispersion are zero, or very small, *waveguide* dispersion, which is generally the weakest of the dispersion mechanisms, may become significant. Waveguide dispersion arises because the velocity of a waveguide mode depends on the normalised dimensions (d/λ) of the waveguide supporting it. Since the different frequency components in the transmitted pulse have different wavelengths these components will travel at different velocities even though they exist as the same electromagnetic mode. Because both material and waveguide dispersion relate to changes in propagation velocity with wavelength they are sometimes referred to collectively as chromatic dispersion. The various fibre types are further defined in ITU-T recommendation G.652.

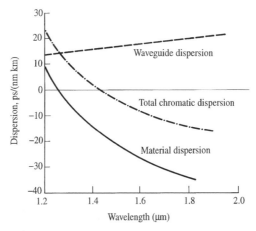

Figure 12.31 *Variation of material dispersion and waveguide dispersion, giving zero total dispersion near $\lambda = 1.5$ μm (source, Flood and Cochrane, 1991, reproduced with the permission of Peter Peregrinus).*

12.5.2 Fibre transmission systems

There have been three generations of optical fibre systems operating at 0.85 μm, 1.3 μm and 1.5 μm wavelengths to progressively exploit lower optical attenuation, Figure 1.6(e), and permit longer distances to be achieved between repeaters. Monomode fibres, with core diameters in the range 8 to 12 μm, have been designed at the two longer wavelengths for second and third generation systems. Since material dispersion is zero at around 1.3 μm, where the optical attenuation in silica is also a local minimum, this was the wavelength chosen for second generation systems, which typically operated at 280 Mbit/s.

Third generation systems operate at wavelengths around 1.5 μm and bit rates of 622 Mbit/s to exploit the lowest practical optical attenuation value of 0.2 dB/km, and tolerate the resulting increased chromatic dispersion which in practice may be 15 to 20 ps/(nm km). Thus, the choice at present between 1.3 and 1.5 μm wavelength depends on whether one wants to maximise link repeater spacing or signalling bandwidth.

Figure 12.31 shows that if core dimensions, and core cladding refractive indices, are chosen correctly, however, material dispersion can be cancelled by waveguide dispersion at about 1.5 μm resulting in very low total chromatic dispersion in this lowest attenuation band.

The impact of fibre developments is clearly seen in Figures 1.7 and 12.32. The latter illustrates the evolution of transmission technology for a 100 km wideband link. The coaxial cable used in the 1970s with its associated 50 repeaters had a mean time between failures (MTBF) of 0.4 years, which was much lower than the 2 year MTBF of the plesiochronous multiplex equipment, Chapter 20, at each terminal station. Multimode fibre (MMF) still needed a repeater every 2 km but was a more reliable transmission medium. The real breakthrough came with single mode fibre (SMF) and, with the low optical attenuation at 1.5 μm, there is now no need for any repeaters on a 100 km link, in which the fibre path loss is typically 10 to 28 dB. (This is *very* much lower than the microwave systems of section 12.4.2.) In the early 1990s there were 1.5 million km of

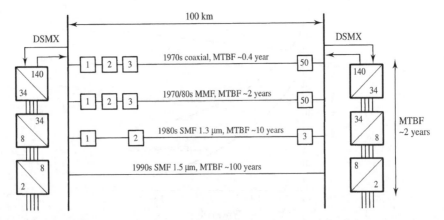

Figure 12.32 *Evolution of a 100 km link from coaxial to non-WDM optical transmission (source: Cochrane, 1990, reproduced with the permission of British Telecommunications plc.).*

installed optical fibre carrying 80% of the UK telephone traffic. This UK investment represented 20% of the world transmission capability installed in optical fibre at that time. Optical fibre transmission capacity doubles each year with an exponentially reducing cost. State of the art transmission systems are limited by the bandwidth of the modulation in the fibre rather than the rate of the modulating source.

12.5.3 Optical sources

Two devices are commonly used to generate light for fibre optic communications systems: light emitting diodes (LEDs) and injection laser diodes (ILDs). The edge emitting LED is a PN junction diode made from a semiconductor material such as aluminium–gallium arsenide (AlGaAs) or gallium arsenide–phosphide (GaAsP). The wavelength of light emitted is typically 0.94 μm and output power is approximately 3 mW at 100 mA of forward diode current. The primary disadvantage of this type of LED is the non-directionality of its light emission which makes it a poor choice as a light source for fibre optic systems. The planar heterojunction LED generates a more brilliant light spot which is easier to couple into the fibre. It can also be switched at higher speeds to accommodate wider signal bandwidth.

The injection laser diode (ILD) is similar to the LED but, above the threshold current, an ILD oscillates and lasing occurs. The construction of the ILD is similar to that of an LED, except that the ends are highly reflective. The mirror-like ends trap photons in the active region which, as they are reflected back and forth, stimulate free electrons to recombine with holes at a higher-than-normal energy level to achieve the lasing process. ILDs are particularly effective because the optical radiation is easy to couple into the fibre. Also the ILD is powerful, typically giving 10 dB more output power than the equivalent LED, thus permitting operation over longer distances. Finally, ILDs generate close to monochromatic light, which is especially desirable for single mode fibres operating at high bit rates.

12.5.4 Optical detectors

There are two devices that are commonly used to detect light energy in fibre optic systems: PIN (positive–intrinsic–negative) diodes and APDs (avalanche photodiodes). In the PIN diode, the most common device, light falls on the intrinsic material and photons are absorbed by electrons, generating electron–hole pairs which are swept out of the device by the applied electric field. The APD is a positive–intrinsic–positive–negative structure, which operates just below its avalanche breakdown voltage to achieve an internal gain. Consequently, APDs are more sensitive than PIN diodes, each photon typically producing 100 electrons, their outputs therefore requiring less additional amplification.

12.5.5 Optical amplifiers

In many optical systems it is necessary to amplify the light signal to compensate for fibre losses. Light can be detected, converted to an electrical signal and then amplified conventionally before remodulating the semiconductor source for the next stage of the communications link. Optical amplifiers, based on semiconductor or fibre elements employing both linear and non-linear devices, are much more attractive and reliable; they permit a range of optical signals (at different wavelengths) to be amplified simultaneously and are especially significant for submarine cable systems.

Basic travelling wave semiconductor laser amplifier (TWSLA) gains are now typically in the range 15 to 20 dB, i.e. slightly higher than that shown in Figure 12.33. These Fabry–Perot lasers are multimode in operation and are used in medium distance systems. Single mode operation is possible with distributed feedback (DFB) lasers for longer distance, high bit rate, systems. In common with all optical amplifiers, the TWSLA generates spontaneous emissions which results in an optical (noise) output in the absence of an input signal. For a system with cascaded TWSLAs the noise terms can accumulate.

Further research has resulted in fibre amplifiers consisting of 10 m to 50 km of doped or undoped fibre. These amplifiers use either a linear, rare earth (erbium), doping mechanism or the non-linear Raman/Brillouin mechanism [Cochrane *et al.*]. The erbium doped fibre amplifier (EDFA) uses a relatively short section (1 to 100 m) of silica fibre pumped with optical, rather than electrical, energy. EDFAs offer efficient coupling of fibre-to-fibre splices, and thus high gains (20 dB) are achievable, Figure 12.33, over a 30 to 50 nm optical bandwidth. Practical amplifier designs generally have gains of 10 to 15 dB. The key attraction of this amplifier is the excellent end-to-end link SNR which is achievable and the enormous 4 to 7 THz (Hz $\times 10^{12}$) of optical bandwidth. (This far exceeds the 300 GHz of the entire radio, microwave and millimetre-wave spectrum.) Two interesting features of these amplifiers are the precise definition of the operating wavelength via the erbium doping and their relative immunity to signal dependent effects. They can therefore be engineered to maintain wide bandwidths when cascaded, and can operate equally well with OOK, FSK or PSK signals.

Injecting a high power laser beam into an optical fibre results in Raman scattering. Introducing a lower intensity signal bearing beam into the same fibre with the pump energy results in its amplification with gains of approximately 15 dB per W of pump power, coupled with bandwidths that are slightly smaller than the erbium amplifiers, Figure 12.33. Unlike EDFAs Raman amplifiers can also be tuned to a desired system wavelength by appropriate choice of pump wavelength.

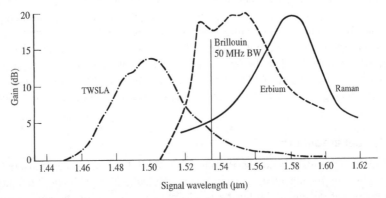

Figure 12.33 *Comparison of the gain of four distinct optical amplifier types (source: Cochrane, 1990, reproduced with the permission of British Telecommunications plc.).*

Brillouin scattering is a very efficient non-linear amplification mechanism that can realise high gains with modest optical pump powers (approximately 1 mW). However, the bandwidth of only 50 MHz, Figure 12.33, is very limited in pure silica which makes such devices more applicable as narrowband tunable filters. This limited bandwidth fundamentally restricts the Brillouin amplifier to relatively low bit rate communication systems.

Comparing the features of optical amplifiers shows that long haul optical transmission systems (10000 km) are now readily realisable, with fibre amplifier based repeaters, and bit rates of 2.5 to 10 Gbit/s. At the present time TWSLA and erbium amplifiers generally require similar electrical pump power but TWSLAs achieve less gain, Figure 12.33, due to coupling losses. Erbium fibre amplifiers have the lowest noise and WDM channel crosstalk performance of all the amplifier types reported but high splice reflectivities can cause these amplifiers to enter the lasing condition. Optical isolators are often included to alleviate lasing. With the exception of Brillouin amplifiers, these amplifiers can be expected to be used across a broad range of system applications including transmitter power amplifiers, receiver preamplifiers, in-line repeater amplifiers and switches. The key advantage of EDFAs is that the lack of conversion to, and from, electrical signals for amplification gives rise to the 'dark fibre' – a highly reliable data super highway operating at tens of Gbit/s.

12.5.6 Optical repeater and link budgets

The electro-optic repeater is similar to the metallic line regenerator of Figures 6.15 and 6.30. For monomode fibre systems the light emitter is a laser diode and the detector uses an APD. The symbol timing recovery circuit uses zero crossing detection of the equalised received signal followed by pulse regeneration and filtering to generate the necessary sampling signals. Due to the high data rates, the filters often use surface acoustic wave devices [Matthews] exploiting their high frequency operation combined with acceptable Q value. The received SNR is given by:

$$\frac{S}{N} = \frac{I_p^2}{2q_e B(I_p + I_D) + I_n^2} \tag{12.88}$$

where I_p is the photodetector current, I_D is the leakage current, q_e is the charge on an electron, B is bandwidth and I_n is the RMS thermal noise current given by:

$$I_n^2 = \frac{4kTB}{R_L} \tag{12.89}$$

For received power levels of −30 dBm, $I_n^2 \gg 2q_e B(I_p + I_D)$ giving, typically, a 15 to 20 dB SNR and hence an OOK BER in the range 10^{-7} to $< 10^{-10}$ [Alexander].

There are many current developments concerned with realising monolithic integrated electro-optic receivers. Integrated receivers can operate with sensitivities of −20 to −30 dBm and a BER of 10^{-10} at 155, 625 and 2488 Mbit/s, with a 20 dB optical overload capability. They are optimised for low crosstalk with other multiplex channels.

The power budget for a typical link in a fibre transmission system might have a transmitted power of 3 dBm, a 60 km path loss of 28 dB, a 1 dB path dispersion allowance and 4 dB system margin to give a −30 dBm received signal level which is consistent with a low cost 155 Mbit/s transmission rate. The higher rate of 2.5 Gbit/s, with a similar receiver sensitivity of −30 dBm, would necessitate superior optical interfaces on such a 60 km link.

Current systems achieve bit rates of 10 Gbit/s using up to 20 cascaded amplifiers spanning approximately 10000 km. BERs of 10^{-4} to 10^{-8} were measured in the early 1990s on a 500 km, five-amplifier system, operating at 565 Mbit/s, in which the individual amplifier gains were 7 to 12 dB [Cochrane *et al.*]. Such systems are not limited by electrical noise in the receiver but rather by the accumulated amplified spontaneous emission noise of the EDFAs.

EXAMPLE 12.8

A monomode, 1.3 μm, optical fibre communications system has the following specification:

Optical output of transmitter, P_T	0.0 dBm
Connector loss at transmitter, L_T	2.0 dB
Fibre specific attenuation, γ	0.6 dB/km
Average fibre splice (joint) loss, L_S	0.2 dB
Fibre lengths, d	2.0 km
Connector loss at receiver, L_R	1.0 dB
Design margin (including dispersion allowance), M	5.0 dB
Required optical carrier power at receiver, C	−30 dBm

Find the maximum loss limited link length which can be operated without repeaters.

Let estimated loss limited link length be D' km and assume, initially, that the splice loss is distributed over the entire fibre length.

$$\text{Total loss} = L_T + (D' \times \gamma) + \left(\frac{D'}{d} \times L_S\right) + L_R + M$$

$$= 2.0 + 0.6D' + \left(\frac{0.2}{2.0} \times D'\right) + 1.0 + 5.0$$

$$= 8.0 + 0.7D' \text{ dB}$$

$$\text{Allowed loss} = P_T - C = 0.0 - (-30) = 30.0 \text{ dB}$$

Therefore:

$$8.0 + 0.7D' = 30$$

$$D' = \frac{30.0 - 8.0}{0.7} = 31.4 \text{ km}$$

The assumption of distributed splice loss means that this loss has been overestimated by:

$$\Delta L_S = L_s [D' - int (D'/d)d] / 2$$

$$= 0.2 [31.4 - int (31.4/2)2] / 2$$

$$= 0.14 \text{ dB}$$

This excess loss can be reallocated to fibre specific attenuation allowing the link length to be extended by:

$$\Delta D = \Delta L_S/\gamma = 0.14/0.6 = 0.2 \text{ km}$$

The maximum link length, D, therefore becomes:

$$D = D' + \Delta D = 31.4 + 0.2 = 31.6 \ \text{km}$$

12.5.7 Optical FDM

With the theoretical 50 THz of available bandwidth in an optical fibre transmission system, and with the modest linewidth of modern optical sources, it is now possible to implement optical FDM and transmit multiple optical carriers along a single fibre. The optical carriers might typically be spaced by 1 nm wavelengths. With the aid of optical filters these signals can be separated in the receiver to realise wavelength division multiplex (WDM) communications [Oliphant *et al.*]. In excess of 100 separate channels can be accommodated using this technique, with 2.5 to 10 Gbit/s on each channel, and practical systems with 200 channel capacity are commercially available. WDM signals are thus amplified using a single fibre amplifier without having to demultiplex to obtain the individual optical carriers. WDM thus increases by a hundredfold the information carrying capacity of fibre based systems now that the necessary components for modulators and demodulators are fully developed. WDM is of vital importance in delivering the required high bandwidth, long haul data links for internet and other applications.

Soliton transmission uses pulses that retain their shape for path lengths of thousands of kilometres due to the reciprocal effects of chromatic dispersion and a refractive index which is a function of intensity. Such systems have been constructed for 1000 km paths with bit rates of 10 to 50 Gbit/s. In the laboratory, 10^6 km recirculating links have been demonstrated, corresponding to many circulations of the earth before the received SNR is unacceptable [Cochrane *et al.*].

12.5.8 Optical signal routers

WDM is only one of the current developments in optical technology for wideband or high data rate transmission. Another active research area is the implementation of all-optical add and drop multiplexers, Figure 20.24. This offers the opportunity of applying photonic rather than electronic techniques to access one or more of the wavelengths to directly drop out or add in wideband digital communications payload traffic. The creation of all optical multiplexers is an extremely powerful technique (for egress and ingress) of traffic. For full flexibility it is necessary to be able to incorporate wavelength conversion to route traffic from any one of the hundred incoming wavelengths to a different outgoing λ value. Technologies such as micromechanical systems (MEMS) with silicon mirrors acting as space switches and tunable lasers for wavelength conversion are possible new components under active development for all-optical signal routers.

12.6 Summary

Noise is present in all communications systems and, if interpreted to include interference, is always a limiting factor on their performance. Both thermal and shot noise have white power spectral densities and are therefore easy to quantify in a specified measurement bandwidth. In many systems calculations, non-white noise is often neglected and white noise is represented by an equivalent thermal spectral density or equivalent thermal noise temperature. The noise factor of a system is the ratio of its input SNR to output SNR when the system's source temperature is 290 K. The total noise available from several cascaded subsystems can be found using the noise factors, and gains, of each subsystem in the Friis formula.

Signal budgets are used to determine the carrier power present at the input to a communications receiver. In a radio system this involves antenna gains and path loss in addition to transmitter power. Two formulae for path loss are commonly used, namely the free space and plane earth formulae. Which is used for a particular application depends on frequency, antenna beamwidths, path clearance and the surface reflection coefficient of the earth beneath the path. Free space propagation results in a power density which decays as $1/R^2$ and a field strength which decays as $1/R$. Plane earth propagation results in a power density which decays as $1/R^4$ and a field strength which decays as $1/R^2$. For paths with particularly stable ground reflections it may be possible to mount antennas at an optimum height in order to take advantage of the 6 dB signal enhancement available at points of constructive interference.

Noise budgets in radio systems must account not only for the noise generated within the receiver itself but also for noise received by the antenna. The latter (assumed white) can be represented by an antenna temperature which is added to the equivalent noise temperature of the receiver to get the effective system input temperature. The effective system input noise power is then found using the Nyquist noise formula. The link budget includes both signal and noise budgets and establishes the received carrier-to-noise ratio.

Fibres currently operate, in decreasing order of attenuation, at wavelengths of 0.85, 1.3 and 1.5 μm. They can be divided into three types depending on the profile of their refractive index variations. Multimode step index fibres have relatively large core dimensions and suffer from modal dispersion which limits their useful bandwidth. Multimode graded index fibres also have large core dimensions but use their variation in refractive index to offset the difference in propagation velocity between modes resulting in larger available bandwidth. Monomode (step index) fibres suffer only chromatic dispersion which arises partly from the frequency dependence of the refractive index (material dispersion) and partly from the frequency dependence of a propagating electromagnetic mode velocity in a waveguide of fixed dimensions. These two contributions can be made to cancel at an operating wavelength around 1.5 μm, however, which also corresponds to a minimum in optical attenuation of about 0.2 dB/km. This wavelength is therefore an excellent choice for long, low dispersion, high bit rate, links. Repeaterless links, hundreds of kilometres long, operating at Gbit/s data rates are now widely deployed. Wavelength division multiplex (WDM) increases by a hundred times the communications capacity of a single fibre and is of vital importance in long haul data links.

12.7 Problems

12.1   A preamplifier with 20 dB gain has an effective noise temperature of 200 K. What is its noise figure in dB? If this preamplifier is followed by a power amplifier of 30 dB gain with a noise figure F of 4 dB what is the degradation in dB for the overall noise figure of the two-amplifier cascade compared to the preamplifier alone? [2.28 dB, 0.04 dB]

12.2 A microwave radio receiver has the following specification:

Antenna noise temperature	= 80 K
Antenna feeder physical temperature	= 290 K, Loss = 2.0 dB
Low noise amplifier noise figure	= 2.2 dB, Gain = 12.0 dB
Frequency downconverter conversion loss	= 6.0 dB
First IF amplifier noise temperature	= 870 K, Gain 30.0 dB
Second IF amplifier noise figure	= 25.0 dB, Gain 30.0 dB
Receiver bandwidth	= 6.0 MHz

Assuming that the bandwidth of the receiver is determined by the second IF amplifier find the noise figure of the receiver, the system noise temperature of the receiver, and the noise power which you would expect to measure at the receiver output. [6.3 dB, 1025 K, –36.7 dBm]

12.3 A digital communications system requires a CNR of 18.0 dB at the receiver's detection circuit input in order to achieve a satisfactory BER. The actual CNR at the receiver input is 23.0 dB and the present (unsatisfactory) CNR at the detection circuit input is 16.0 dB. The source temperature of the receiver is 290 K. What noise figure specification must be met by an additional amplifier (to be placed at the front end of the receiver) for a satisfactory BER to be realised if the gain of the amplifier is: (a) 6.0 dB; (b) 10.0 dB; and (c) 14.0 dB? [3.3 dB, 4.4 dB, 4.8 dB]

12.4 A 36.2 GHz, 15.5 km, single hop microwave link has a transmitter power of 0 dBm. The transmit and receiver antennas are circularly symmetric paraboloidal reflectors with diameters of 0.7 m and 0.5 m, respectively. Assuming both antennas have ohmic efficiencies of 97% and aperture efficiencies of 67% calculate: (a) the transmitted boresight radiation intensity; (b) the spreading loss; (c) the free space path loss; (d) the effective isotropic radiated power; (e) the power density at the receiving antenna's aperture; and (f) the received power. [3.64 W/steradian, 94.8 dB/m^2, 147.4 dB, 46.6 dBm, –78.2 dBW/m^2, –57.1 dBm]

12.5 If the effective noise bandwidth of the receiver in Problem 12.4 is 100 kHz, the antenna's aperture temperature is 200 K, its physical temperature is 290 K and the noise figure of the entire receiver (from antenna output to detector input) is 4.0 dB calculate the CNR at the detector input. [63.4 dB]

12.6 Derive the fundamental free space transmission loss equation using the concepts of spreading loss and antenna reciprocity. Identify the free space path loss in the equation you derive and explain why it is a function of frequency.

12.7 Explain the concept of path gain in the context of plane earth propagation. How does this lead to a $1/R^4$ dependence of power density? What is the equivalent distance dependence of field strength?

12.8 Show that equation (12.73) follows from equation (12.72) when $\rho e^{j\phi} = -1$.

12.9 A VHF communications link operates over a large lake at a carrier frequency of 52 MHz. The path length between transmitter and receiver is 18.6 km and the heights of the sites chosen for the location of the transmitter and receiver towers are 72.6 m and 95.2 m respectively. Assuming that the cost of building a tower increases at a rate greater than linearly with tower height find the most economic tower heights which will take full advantage of the 'ground' reflected ray. Estimate by what height the water level in the lake would have to rise before the 'ground' reflection advantage is lost completely. [80.3 m, 80.3 m, 69.0 m]

12.10 Digital MPSK transmissions carried on a 6 GHz terrestial link require a bandwidth of 24 MHz. The transmitter carrier power level is 10 W and the hop distance is 40 km. The antennas used each have 40 dB gain and filter, isolator and feeder losses of 4 dB. The receiver has a noise figure of 10 dB in the specified frequency band. Calculate the carrier to thermal noise power ratio at the receiver output. [62 dB]

Comment on any effects which might be expected to seriously degrade this carrier-to-noise ratio in a practical link. If a minimum carrier-to-noise ratio of 30 dB is required for the

MPSK modulation what is the fade margin of the above link? [32 dB]

12.11 A QAM 4 GHz link requires a bandwidth of 15.8 MHz. The radiated carrier power is 10 W and the hop distance 80 km. It uses antennas each having 40 dB gain, has filter, isolator and antenna feeder losses totalling 7 dB at each end and a receiver noise figure of 10 dB. If the Rayleigh fade margin is 30 dB, calculate the carrier to thermal noise power ratio (CNR) at the receiver. What complexity of QAM modulation will this CNR support for an error rate of 10^{-6}? [32.5 dB, 128 state]

12.12 A small automatic laboratory is established on the moon. It contains robotic equipment which receives instructions from earth via a 4 GHz communications link with a bandwidth of 30 kHz. The EIRP of the earth station is 10 kW and the diameter of the laboratory station's receiving antenna is 3.0 m. If the laboratory's receiving antenna were to transmit, 2% of its radiated power would illuminate the lunar surface and 50% would illuminate the earth. (The rest would be radiated into space which has the cosmic background brightness temperature of 3 K.) The noise figure of the laboratory's receiver is 2.0 dB. If the daytime physical temperature of the lunar surface is 375 K and the brightness temperature of the earth is 280 K estimate the daytime CNR at the laboratory receiver's detection circuit input. (Assume that the laboratory antenna has an ohmic efficiency of 98%, an aperture efficiency of 72%, and a physical temperature equal to that of the lunar surface. Also assume that the earth and moon both behave as black bodies, i.e. have emissivities of 1.0.) The distance from the earth to the moon is 3.844×10^8 m. [23.0 dB]

12.13 If the nighttime temperature of the lunar surface in Problem 12.12 is 125 K find the nighttime improvement in CNR over the daytime CNR. Is this improvement: (a) of real engineering importance; (b) measurable but not significant; or (c) undetectably small? [0.14 dB]

CHAPTER 13

Communication systems simulation

13.1 Introduction

Most of the material in the preceding chapters has been concerned with the development of equations which can be used to predict the performance of digital communications systems. An obvious example is the error function formulae used to find the probability of bit error for ideal systems assuming zero ISI, matched filter receivers and additive, white, Gaussian noise. Such models are important in that they are simple, give instant results, and provide a series of 'reference points' in terms of the relative (or perhaps potential) performance of quite different types of system. In addition they allow engineers to develop a quantitative feel for how the performance of systems will vary as their parameters are changed. They also often provide theoretical limits on performance guarding the design engineer against pursuit of the unobtainable.

The principal limitation of such equations arises from the sometimes unrealistic assumptions on which they are based. Filters, for example, do not have rectangular amplitude responses, oscillators are subject to phase noise and frequency drift, carrier recovery circuits do not operate with zero phase error, sampling circuits are prone to timing jitter, etc.

Hardware prototyping during the design of systems avoids the limitation of idealised models in so far as real-world imperfections are present in the prototype. Designing, implementing and testing hardware, however, is expensive and time consuming, and is becoming more so as communication systems increase in sophistication and complexity. This is true to the extent that it would now usually be impossible to prototype all credible solutions to a given communications problem. Computer simulation of communication systems [Jeruchim et al., 2000] falls into the middle ground between idealised modelling using simple formulae and hardware prototyping. It occupies an intermediate location along all the following axes:

crude – accurate
simple – complex
cheap – expensive
quick – time consuming

In addition computer simulations are often able to account for system non-linearities which are notoriously difficult to model analytically.

Computer simulations of communication systems usually work as follows. The system is broken down into functional blocks each of which can be modelled mathematically by an equation, a rule, an input/output lookup table or in some other way. These subsystems are connected together such that the outputs of some blocks form the input of others and vice versa. An information source is then modelled as a random or pseudo-random sequence of bits (see section 13.4). The signal is sampled and the samples fed into the input of the first subsystem. This subsystem then operates on these samples according to its system model and provides modified samples at its output. These samples then become the inputs for the next subsystem, and so on, typically until the samples represent the received, demodulated, information bit stream. Random samples representing noise and/or interference are usually added at various points in the system. Finally the received information bits are compared with the original information source bits to estimate the BER of the entire communication system. Intermediate results, such as the spectrum of the transmitted signal and the pdf of signal plus noise in the receiver, can also be obtained.

The functional definition of some subsystems, e.g. modulators, is easier in the time domain whilst the definition of others (e.g. filters) is easier in the frequency domain. Conversion between time and frequency domains is an operation which may be performed many times when simulations are run. Convolution, multiplication and discrete Fourier transforms are therefore important operations in communications simulation.

The accuracy of a well designed simulation, in the sense of how closely it matches a hardware prototype, depends essentially on the level of detail at which function blocks are defined. The more detailed the model the more accurate might be expected its results. The penalty paid, of course, is in the effort required to develop the model and the computer power needed to simulate the results in a reasonable time. Only an overview of the central issues involved in simulation is presented here. A detailed and comprehensive exposition of communication system simulation is given in [Jeruchim *et al.*, 1984]. In essence, however, the normal simulation process can be summarised as:

1. Derivation of adequate models for all input signals (including noise) and subsystems.
2. Conversion, where possible, of signals and subsystem models to their equivalent baseband form.
3. Sampling of all input signals at an adequately high rate.
4. Running of simulations, converting between time and frequency domains as necessary.
5. Use of Monte Carlo or quasi-analytic methods to estimate BERs.
6. Conversion of output signals back to passband form if necessary.
7. Display of intermediate signals, spectra, eye diagrams, etc., as required.

Much of the work in the preceding chapters (especially Chapters 2 to 4) has been directed at modelling the signals, noise and subsystems which commonly occur in digital communications systems. This chapter therefore concentrates on steps 2 to 7 above.

13.2 Equivalent complex baseband representations

Consider a microwave LOS communication system operating with a carrier frequency of 6 GHz and a bandwidth of 100 MHz. To simulate this system it might superficially appear that a sampling rate of 2×6050 MHz would be necessary. This conclusion is, of course, incorrect as should be apparent from the bandpass sampling theorem discussed in Chapter 5. In fact, the most convenient way of representing this narrowband system for simulation purposes is to work with equivalent baseband quantities.

13.2.1 Equivalent baseband signals

A (real) passband signal can be expressed in polar (i.e. amplitude and phase) form by:

$$x(t) = a(t) \cos [2\pi f_c t + \phi(t)]$$

$$= \Re\left\{ a(t)\, e^{j2\pi f_c t}\, e^{j\phi(t)} \right\} \tag{13.1(a)}$$

or, alternatively, in Cartesian (i.e. inphase and quadrature) form by:

$$x(t) = x_I(t) \cos 2\pi f_c t - x_Q(t) \sin 2\pi f_c t \tag{13.1(b)}$$

The corresponding complex baseband (or low pass) signal is defined as:

$$x_{LP}(t) = a(t)\, e^{j\phi(t)} \tag{13.2(a)}$$

or:

$$x_{LP}(t) = x_I(t) + j\, x_Q(t) \tag{13.2(b)}$$

(Multiplying equation (13.2(b)) by $e^{j2\pi f_c t}$ and taking the real part demonstrates the correctness of the + sign here.) $x_{LP}(t)$ is sometimes called the *complex envelope* of $x(t)$. The baseband nature of $x_{LP}(t)$ is now obvious and the modest sampling rate required to satisfy Nyquist's theorem is correspondingly obvious. Notice that the transformation from the real passband signal of equation (13.1(a)) to the complex baseband signal of equation (13.2(a)) can be viewed as a two-step process, i.e.:

1. $x(t)$ is made cisoidal (or *analytic*) by adding $j\, a(t) \sin[2\pi f_c t + \phi(t)]$.
2. The carrier is suppressed by dividing by $e^{j2\pi f_c t}$.

Since $a(t) \sin[2\pi f_c t + \phi(t)]$ is derived from $x(t)$ by shifting all positive frequency components by $+90°$ and all negative frequency components $-90°$, step 1 corresponds to adding $j\hat{x}(t)$ to $x(t)$ where ^ denotes the (time domain) Hilbert transform (see section 4.5). The relationship between $x(t)$ and $x_{LP}(t)$ can therefore be summarised as:

$$x_{LP}(t) = [x(t) + j\, \hat{x}(t)]\, e^{-j2\pi f_c t} \tag{13.3}$$

$x(t)$ and $x_{LP}(t)$ are shown schematically, for an APK signal, in Figure 13.1.

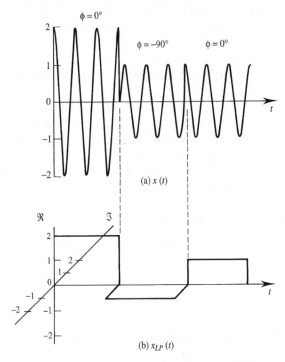

Figure 13.1 *Real passband signal x(t) and its complex envelope $x_{LP}(t)$.*

The spectrum, $X_{LP}(f)$, of $x_{LP}(t)$ can be found by applying the corresponding frequency domain steps to the spectrum, $X(f)$, of $x(t)$, i.e.:

1. $X(f)$ has its negative frequency components suppressed and its positive frequency components doubled. This is demonstrated using phasor diagrams, for a sinusoidal signal, in Figure 13.2.
2. The (doubled) positive frequency components are shifted to the left by f_c Hz.

Step 1 is more formally expressed as the addition to $X(f)$ of $j\hat{X}(f) = j[-j\text{sgn}(f)X(f)] = \text{sgn}(f)X(f)$. Step 2 follows from the Fourier transform frequency translation theorem. Figure 13.3 shows the relationship between $X(f)$ and $X_{LP}(f)$. Notice that the spectrum of the complex envelope does not have the Hermitian symmetry characteristic of real signals. Steps 1 and 2 together can be summarised by:

$$X_{LP}(f) = 2X(f + f_c) u(f + f_c) \qquad (13.4)$$

The first factor on the right hand side of equation (13.4) doubles the spectral components, the second moves the entire spectrum to the left by f_c Hz and the third factor suppresses all spectral components to the left of $-f_c$ Hz, Figure 13.4.

13.2.2 Equivalent baseband systems

Equivalent baseband representations of passband systems can be found in the same way as for signals. A filter with a passband (Hermitian) frequency response $H(f)$ (Figures 13.5(a) to (c)), and (real) impulse response $h(t)$, has an equivalent baseband frequency response (Figures 13.5(d) to (f)):

$$H_{LP}(f) = H(f + f_c) \, u(f + f_c) \tag{13.5(a)}$$

and baseband impulse response:

$$h_{LP}(t) = \tfrac{1}{2} \, [h(t) + j \, \hat{h}(t)] \, e^{-j2\pi f_c t} \tag{13.5(b)}$$

Alternatively the complex baseband impulse response can be expressed in terms of its inphase and quadrature components, i.e.:

$$h_{LP}(t) = \tfrac{1}{2} \, [h_I(t) + j \, h_Q(t)] \tag{13.6}$$

where the (real) passband response is:

$$h(t) = h_I(t) \cos 2\pi f_c t - h_Q(t) \sin 2\pi f_c t \tag{13.7}$$

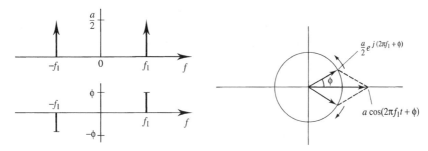

(a) Amplitude/phase spectrum and phasor diagram for a single spectral component

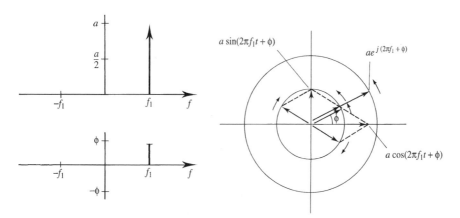

(b) Cancellation of negative frequency phasors, reinforcement of positive frequency phasors

Figure 13.2 *Phasor diagram demonstration of the equivalence between (a) the addition of an imaginary quadrature version of a signal and (b) the suppression of the negative frequency components plus a doubling of the positive frequency components.*

13.2.3 Equivalent baseband system output

The output of an equivalent low pass linear system when excited by an equivalent low pass signal is found, in the time domain, using convolution in the usual way (but taking care to convolve both real and imaginary components) and represents the equivalent low pass output of the system, $y_{LP}(t)$, i.e.:

$$y_{LP}(t) = h_{LP}(t) * x_{LP}(t)$$

$$= \tfrac{1}{2} [h_I(t) + j\, h_Q(t)] * [x_I(t) + j\, x_Q(t)] \tag{13.8}$$

Notice that when compared with the definition of equivalent baseband signals (equations (13.4), (13.3) and (13.2(b))) the equivalent baseband system definitions (equations (13.5(a)), (13.5(b)) and (13.6)) are smaller by a factor of ½. This is to avoid, for example, the baseband equivalent frequency response of a lossless passband filter having a voltage gain of 2.0 in its passband. (Many authors make no such distinction between the definitions of baseband equivalent signals and systems, in which case the factor of ½ usually appears in the *definition* of equivalent baseband convolution.) Recognising that the equivalent baseband system output signal, $y_{LP}(t)$, can be expressed as inphase and quadrature components of the passband output signal, $y(t)$, i.e.:

$$y_{LP}(t) = y_I(t) + j\, y_Q(t) \tag{13.9}$$

where:

$$y(t) = y_I(t) \cos 2\pi f_c t - y_Q(t) \sin 2\pi f_c t \tag{13.10}$$

and equating real and imaginary parts in equations (13.8) and (13.9) gives:

$$y_I(t) = \tfrac{1}{2} [h_I(t) * x_I(t) - h_Q(t) * x_Q(t)] \tag{13.11(a)}$$

$$y_Q(t) = \tfrac{1}{2} [h_Q(t) * x_I(t) + h_I(t) * x_Q(t)] \tag{13.11(b)}$$

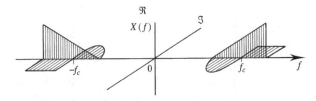

(a) Hermitian spectrum of real passband signal

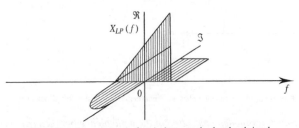

(b) Non-Hermitian spectrum of equivalent complex baseband signal

Figure 13.3 *Spectra of real passband signal and its complex envelope.*

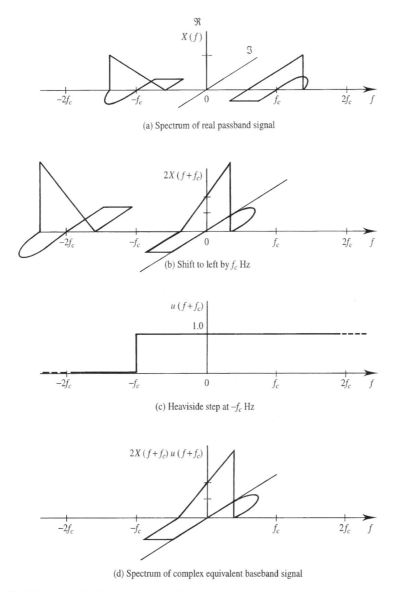

(a) Spectrum of real passband signal

(b) Shift to left by f_c Hz

(c) Heaviside step at $-f_c$ Hz

(d) Spectrum of complex equivalent baseband signal

Figure 13.4 *Relationship between passband and equivalent low pass spectra.*

These operations are illustrated schematically in Figure 13.6. Equations (13.10) and (13.11) thus give the passband output of a system directly in terms of the inphase and quadrature baseband components of its passband input and impulse response. $y(t)$ can also be found from $y_{LP}(t)$ by reversing the steps used in going from equations (13.1(a)) to (13.2(a)), i.e.:

$$y(t) = \Re\left\{ y_{LP}(t)\, e^{j2\pi f_c t} \right\}$$

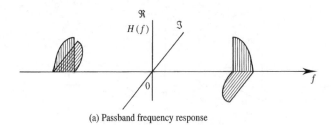

(a) Passband frequency response

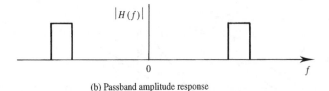

(b) Passband amplitude response

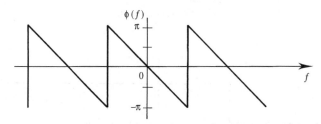

(c) Phase response of passband filter

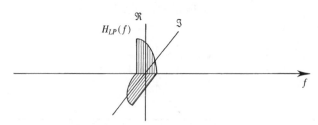

(d) Equivalent baseband filter frequency response

Figure 13.5 *Passband and equivalent low pass frequency, amplitude and phase responses.*

$$= \tfrac{1}{2} \left[y_{LP}(t) \, e^{j2\pi f_c t} + y_{LP}^*(t) \, e^{-j2\pi f_c t} \right] \tag{13.12(a)}$$

Alternatively, if $Y_{LP}(f)$ has been found from:

$$Y_{LP}(f) = H_{LP}(f) \, X_{LP}(f)$$

then $Y(f)$ can be obtained using the equivalent frequency domain quantities and operations of equation (13.12(a)) (see Figure 13.7), i.e.:

$$Y(f) = \text{Hermitian}\{Y_{LP} \, (f - f_c)\} \tag{13.12(b)}$$

$$= \frac{Y_{LP} \, (f - f_c) + Y_{LP}^* \, (-f - f_c)}{2}$$

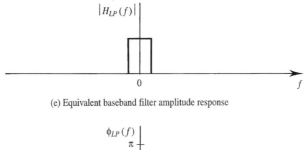

(e) Equivalent baseband filter amplitude response

(f) Equivalent baseband filter phase response

Figure 13.5-ctd. *Equivalent low pass amplitude and phase responses.*

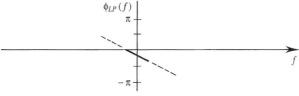

(a) Real bandpass model

(b) Equivalent complex envelope model

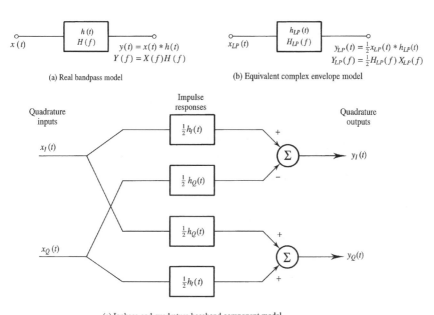

(c) Inphase and quadrature baseband component model

Figure 13.6 *Convolution of inputs and impulse responses with equivalent baseband operations.*

(That any function, in this case $Y_{LP}(f - f_c)$, can be split into Hermitian and anti-Hermitian parts is easily demonstrated as follows:

$$X(f) = \frac{X(f) + X^*(-f)}{2} + \frac{X(f) - X^*(-f)}{2} \tag{13.13}$$

The first term has even real part and odd imaginary part and is therefore Hermitian. The

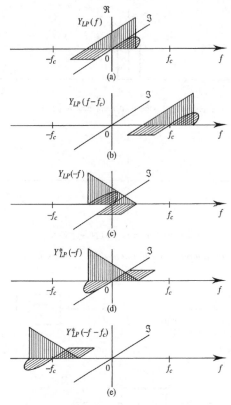

Figure 13.7 *Obtaining the spectrum of real passband signal, (b) plus (e), from the spectrum of an equivalent complex baseband signal, (a).*

second term has odd real part and even imaginary part and is therefore anti-Hermitian.)

13.2.4 Equivalent baseband noise

Equivalent baseband noise, $n_{LP}(t)$, can be modelled in exactly the same way as an equivalent baseband signal, i.e.:

$$n_{LP}(t) = r(t)\, e^{j\theta(t)}$$
$$= n_I(t) + j\, n_Q(t) \qquad (13.14)$$

where the passband noise process is:

$$n(t) = r(t)\, e^{j\theta(t)}\, e^{j2\pi f_c t}$$
$$= n_I(t) \cos 2\pi f_c t - n_Q(t) \sin 2\pi f_c t \qquad (13.15)$$

Figure 13.8 illustrates the relationship between the passband and equivalent baseband processes in the time and phase domains. For the special, but very important, case of a strict-sense, zero mean, Gaussian, narrowband noise process the properties of the baseband processes $n_I(t)$ and $n_Q(t)$ are summarised in Table 13.1.

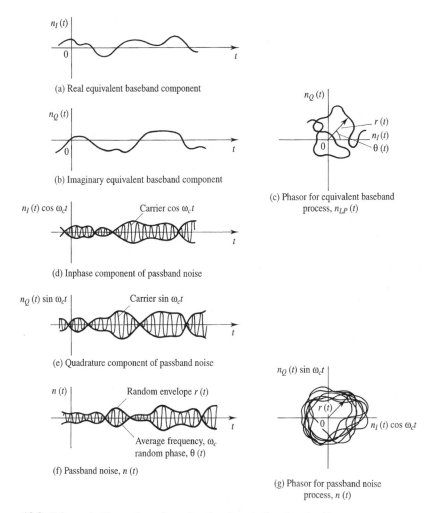

$n_I(t)$

0

t

(a) Real equivalent baseband component

$n_Q(t)$

0

t

(b) Imaginary equivalent baseband component

$n_I(t) \cos \omega_c t$ Carrier $\cos \omega_c t$

t

(d) Inphase component of passband noise

$n_Q(t) \sin \omega_c t$ Carrier $\sin \omega_c t$

t

(e) Quadrature component of passband noise

$n(t)$ Random envelope $r(t)$

t

Average frequency, ω_c
random phase, $\theta(t)$

(f) Passband noise, $n(t)$

$n_Q(t)$

$r(t)$
$n_I(t)$
0
$\theta(t)$

(c) Phasor for equivalent baseband
process, $n_{LP}(t)$

$n_Q(t) \sin \omega_c t$

$r(t)$
0
$n_I(t) \cos \omega_c t$

(g) Phasor for passband noise
process, $n(t)$

Figure 13.8 *Schematic illustration of passband and equivalent baseband noise processes with corresponding phasor trajectories.*

Most of these properties are intuitively reasonable and are, therefore, not proved here. (Proofs can be found in [Taub and Schilling].) One anti-intuitive aspect of the equal variance property, however, is that each of the quadrature baseband processes alone contains the same power (i.e. has the same variance) as the passband noise process. (To the authors, at least, intuition would suggest that each of the baseband processes would contain half the power of the passband process.) This problem is easily resolved, however, by considering the power represented by equation (13.15), i.e.:

$$\langle n^2(t) \rangle = \langle [n_I(t) \cos 2\pi f_c t - n_Q(t) \sin 2\pi f_c t]^2 \rangle$$
$$= \tfrac{1}{2}\langle n_I^2(t) \rangle + \tfrac{1}{2}\langle n_Q^2(t) \rangle \qquad (13.16(a))$$

And since $\langle n_I^2(t) \rangle = \langle n_Q^2(t) \rangle$ (acceptable on intuitive grounds) then:

$$\langle n^2(t) \rangle = \langle n_I^2(t) \rangle = \langle n_Q^2(t) \rangle \tag{13.16(b)}$$

Table 13.1 *Properties of equivalent baseband Gaussian noise quadrature processes.*

Property	Definition
Zero mean	$\langle n_I(t) \rangle = \langle n_Q(t) \rangle = 0$
Equal variance	$\langle n_I^2(t) \rangle = \langle n_Q^2(t) \rangle = \langle n^2(t) \rangle = \sigma^2$
Zero correlation	$\langle n_I(t)\, n_Q(t) \rangle = 0$
Gaussian quad. components	$p(n_I) = p(n_Q) = [1/(\sqrt{2\pi}\,\sigma)]\, e^{-\frac{n_x}{2\sigma^2}}$
Rayleigh amplitude	$p(r) = (r/\sigma^2)\, e^{-r^2/2\sigma^2}, \quad r \geq 0$
Uniform phase	$p(\theta) = 1/(2\pi), \qquad\qquad |\theta| < \pi$

Figure 13.9 shows, in a systems context, how the passband process $n(t)$ could be generated from the baseband processes $n_I(t)$ and $n_Q(t)$. Notice that the power in each quadrature leg is halved after multiplication with the carrier.

Since noise processes do not have a well defined voltage spectrum (preventing equation (13.4) from being used to find an equivalent baseband spectrum) the (power) spectral (density) description of $n_I(t)$ and $n_Q(t)$ is found by translating the positive frequency components of $G_n(f)$ down by f_c Hz, translating the negative frequencies up by f_c Hz, and adding, i.e.:

$$G_{n_I}(f) = G_{n_Q}(f)$$

$$= G_n(f + f_c)\, u(f + f_c) + G_n(f - f_c)\, u(-f + f_c) \tag{13.16(c)}$$

The relationship between the PSD of $n(t)$ and that of $n_I(t)$ and $n_Q(t)$ is illustrated in Figure 13.10.

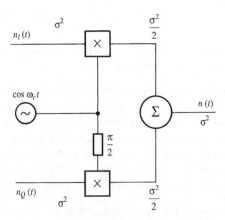

Figure 13.9 *Relationship between baseband quadrature noise components and the passband Gaussian noise process.*

13.3 Sampling and quantisation

Sampling and quantisation, as they affect communication systems generally, have been discussed in Chapter 5. Here these topics are re-examined in the particular context of simulation.

13.3.1 Sampling equivalent baseband signals

Nyquist's sampling theorem, if correctly interpreted, can be applied to any process, including an equivalent, complex, baseband process. In this case the baseband signal has a conventionally defined bandwidth which is only half the bandwidth, B, of the real passband signal, Figure 13.11. Thus a straightforward application of Nyquist's theorem gives a minimum sampling rate:

$$f_s \geq 2 \frac{B}{2} = B \text{ Hz} \tag{13.17}$$

Superficially equation (13.17) looks wrong in that it suggests all the information present in a passband signal with bandwidth B Hz is preserved in only half the number of samples expected. This paradox is resolved by remembering that for a complex baseband signal there will be two real sample values for each sampling instant (i.e. an inphase, or real sample, and a quadrature, or imaginary, sample). The total number of real numbers characterising a given passband signal is therefore the same, whether or not an equivalent baseband representation is used.

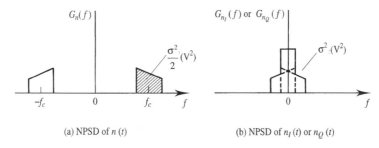

(a) NPSD of $n(t)$ (b) NPSD of $n_I(t)$ or $n_Q(t)$

Figure 13.10 *Relationship between PSD of each equivalent baseband quadrature component of $n(t)$ and PSD of $n(t)$.*

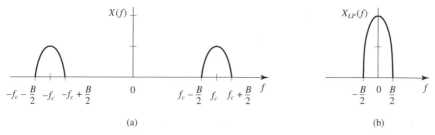

(a)   (b)

Figure 13.11 *(a) Passband and (b) equivalent baseband frequency spectra.*

Table 13.2 *Worst case SDR for various numbers of samples per symbol. (After Jeruchim et al., 1984)*

f_s/R_s (samples/symbol)	SDR (dB)
4	15.8
8	18.7
10	19.8
16	21.9
20	22.9

Sampling at a rate of f_s Hz defines a simulation bandwidth of $f_s/2$ Hz in the sense that any spectral components which lie within this band will be properly simulated whilst spectral components outside this band will be aliased. The selection of f_s is therefore a compromise between the requirement to keep f_s low enough so that simulation can be carried out in a reasonable time with modest computer resources, and the requirement to keep f_s high enough for aliasing errors to be acceptably low. The aliasing errors are quantified in section 5.3.4 by a signal to distortion ratio (SDR) defined as the ratio of unaliased signal power to aliased signal power, Figure 13.12. SDR is clearly a function of the number of samples per symbol (i.e. f_s/R_s). Table 13.2 shows several corresponding pairs of SDR and f_s/R_s for the (worst) case of an unfiltered (i.e. rectangular pulse) symbol stream.

f_s/R_s is typically selected such that SDR is 10 dB greater than the best SNR to be simulated. For signals with significant pulse shaping the SDR for a given value of f_s/R_s is higher than that shown in Table 13.2. Eight samples per symbol, therefore, may often represent sufficiently rapid sampling for the simulation of realistic systems.

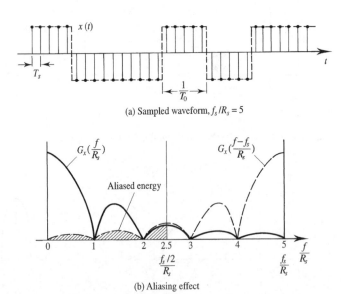

(a) Sampled waveform, $f_s/R_s = 5$

(b) Aliasing effect

Figure 13.12 *Sampling of random binary waveform (source: Jeruchim et al., 1984, reproduced with permission of the IEEE © 2003 IEEE).*

13.3.2 Quantisation

Simulation quantisation errors can be equated with the limited precision with which a computer can represent numbers. The rounding errors due to this limited precision effectively add noise to the waveform being simulated. In digital communications the system's ADC itself introduces quantisation error (sections 5.5 and 5.6) quantified by a signal to quantisation noise ratio (SN_qR) and given approximately (for linearly quantised voice signals) by $6(n-1)$ dB where n is the number of bits (typically 8) representing each level, equation (5.23). The simulation induced SN_qR will depend in a similar way on the binary word size which the computer uses to represent numbers. This word size would normally be at least 16 bits, typically 32 bits. Whilst this suggests that simulation induced quantisation error will be negligible with respect to system induced quantisation error it may sometimes be the case that simulation quantisation errors accumulate in calculations. For long simulations (perhaps millions of symbols) accumulated errors may become significant. Using double precision arithmetic, where possible, will obviously help in this respect.

13.4 Modelling of signals, noise and systems

One of the strengths of simulation is that it can often be interfaced to real signals and real systems hardware. A real voice signal, for example, might be rapidly sampled and those samples used as the information source in a simulation. Such a simulation might also include measured frequency responses of actual filters and measured input/output characteristics of non-linear amplifiers which are to be incorporated in the final system hardware. (In this respect the distinction between hardware prototyping and software simulation can become blurred.) Nevertheless, it is still a common requirement to model signals, noise and systems using simple (and rather idealised) assumptions. A variety of such models are discussed below.

13.4.1 Random numbers

Statistically independent, random, numbers with a uniform pdf are easily generated by computers. Many algorithms have been proposed, but one [Park and Miller] which appears to have gained wide acceptance for use on machines using 32 bit integer arithmetic is:

$$x(k) = 7^5 \, x(k-1) \, (2^{31} - 1) \tag{13.18}$$

$x(k)$ represents a sequence of integer numbers drawn from a uniform pdf with minimum and maximum values of 1 and $2^{31} - 1$ respectively. To generate a sequence of numbers, $x_u(k)$, with uniform pdf between 0 and 1, equation (13.18) is simply divided by $2^{31} - 1$. ($x_u(k)$ never takes on a value of exactly zero since the algorithm would then produce zero values indefinitely.) The initial value of the sequence $x(0) \neq 0$ is called the generator *seed* and is chosen by the user. Strictly speaking only the choice of seed is random since thereafter the sequence is deterministic, repeating periodically. The sequences arising from well designed algorithms such as equation (13.18), however, have many properties in common with truly random sequences and, from an engineering point of view, need not usually be distinguished from them. They are therefore referred to as pseudo-random (or pseudo-noise) sequences. A sequence $y(k)$ with a pdf $p_Y(y)$ can be derived from the sequence $x_u(k)$ using the target cumulative distribution of Y, Figure 13.13. The CD of Y (i.e. $P(Y \leq y)$) is first found (by integrating $p_Y(y)$ if necessary). The values of $x_u(k)$ are then mapped to $y(k)$ according to this curve as shown in Figure 13.13. For simple pdfs this transformation can sometimes be accomplished analytically resulting in a simple formula relating $y(k)$ and $x_u(k)$. Otherwise $P(Y \leq y)$ can be defined by tabulated values and the individual numbers transformed by interpolation.

The method described above is general and could be used to generate numbers with a Gaussian pdf. It is often easier, for this special case, however, to take advantage of the central limit theorem and add several independent, uniformly distributed, random sequences, i.e.:

$$y(k) = \sum_{i=1}^{N} x_{u,i}(k) - \frac{N}{2} \tag{13.19}$$

Using equation (3.16) we see that $\bar{X}_u$ is 0.5, and subtracting $N/2$ in equation (13.19) ensures that $\bar{Y} = 0$. Furthermore the variance of X_u is 1/12. Choosing $N = 12$ therefore ensures that $\sigma_y^2 = 1.0$ without the need for any additional scaling. (Higher values of N would, of course, improve the accuracy of the resulting Gaussian pdf.)

Correlated random numbers are easily generated from statistically independent random numbers by forming linear combinations of sequence pairs. For example, a random sequence, $y(k)$, with autocorrelation properties, $R_y(\kappa)$ (as defined in section 3.3.3) specified by:

$$R_y(0) = \sigma_Y^2 \tag{13.20(a)}$$

$$R_y(\pm 1) = \alpha \, \sigma_Y^2, \quad \alpha < 1 \tag{13.20(b)}$$

$$R_y(\pm 2) = \beta \, \sigma_Y^2, \quad \beta < 1 \tag{13.20(c)}$$

$$R_y(\pm \kappa) = 0, \quad \kappa > 2 \tag{13.20(d)}$$

$$\overline{y(k)} = 0 \tag{13.20(e)}$$

can be formed from a zero mean, unit variance, statistically independent sequence, $s(k)$, using the linear transform:

$$y(k) = w_1 s(k) + w_2 s(k-1) + w_3 s(k-2) \tag{13.21}$$

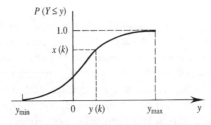

Figure 13.13 *Cumulative distribution used to transform a uniformly distributed random variable to a random variable Y with pdf $p_Y(y)$.*

Substituting equation (13.21) into equation (13.20(a)) gives:

$$\sigma_Y^2 = \overline{y^2(k)} - \overline{[y(k)]}^2 \tag{13.22(a)}$$

$$= \overline{[w_1 s(k) + w_2 s(k-1) + w_3 s(k-2)]^2} - 0$$

Since $s(k)$ is a sequence of uncorrelated numbers only terms, in the expansion of equation (13.22(a)), having factors with equal arguments of s give non-zero results. Therefore:

$$\sigma_Y^2 = w_1^2 \overline{s^2(k)} + w_2^2 \overline{s^2(k-1)} + w_3^2 \overline{s^2(k-2)} \tag{13.22(b)}$$

And since the variance of $s(k)$ is 1.0 then:

$$\sigma_Y^2 = w_1^2 + w_2^2 + w_3^2 \tag{13.22(c)}$$

Similarly:

$$\alpha \, \sigma_Y^2 = \overline{[w_1 s(k) + w_2 s(k-1) + w_3 s(k-2)] \, [w_1 s(k-1) + w_2 s(k-2) + w_3 s(k-3)]}$$

$$= w_2 w_1 \overline{s^2(k-1)} + w_3 w_2 \overline{s^2(k-2)} \tag{13.23(a)}$$

all other terms being zero. Thus:

$$\alpha \, \sigma_Y^2 = w_2 w_1 + w_3 w_2 \tag{13.23(b)}$$

and:

$$\beta \, \sigma_Y^2 = \overline{[w_1 s(k) + w_2 s(k-1) + w_3 s(k-2)] \, [w_1 s(k-2) + w_2 s(k-3) + w_3 s(k-4)]}$$

$$= w_3 w_1 \overline{s^2(k-2)} \tag{13.24(a)}$$

i.e.:

$$\beta \, \sigma_Y^2 = w_3 w_1 \tag{13.24(b)}$$

Equations (13.22) to (13.24) generalise for a series with non-zero correlation over an N-term window to:

$$R_y(\kappa) = \sum_{i=1}^{N-\kappa} w_i w_{\kappa+i} \tag{13.25}$$

Equation (13.20(e)) is automatically satisfied since $\overline{s(k)} = 0$. Equation (13.25) provides N independent equations which are solved simultaneously to give the appropriate weighing factors, $w_1, w_2, \cdots, w_N$.

The general problem of simultaneously obtaining a specified pdf and specified PSD is a difficult one. This is because the linear system represented by equation (13.21) generally changes the pdf in an unpredictable way. The exception to this, of course, is for random sequences with Gaussian pdf which can be filtered to realise a specified PSD without altering its Gaussian characteristic (see section 4.6.3).

13.4.2 Random digital symbol streams

Random digital symbol streams can be easily generated from a set of random numbers as follows. Consider a sequence of independent random numbers, $y(k)$, with the pdf $p_Y(y)$ shown in Figure 13.14. If three symbols represented by three voltage levels $v = 0, 1, 2$ V are required with probabilities of 0.5, 0.25, 0.25 respectively then $p_Y(y)$ is divided into three areas corresponding to those probabilities. This defines thresholds y_1 and y_2 (Figure 13.14). Each random variable sample can then be mapped to a random symbol using the rule:

$$v(k) = \begin{cases} 0, & y(k) \leq y_1 \\ 1, & y_1 < y(k) \leq y_2 \\ 2, & y(k) > y_2 \end{cases} \tag{13.26}$$

As an alternative to using a random number algorithm (such as equation (13.18)) followed by equation (13.26), pseudo-random digital symbol streams can be generated using shift registers with appropriate feedback connections. An example generator for a binary sequence is shown in Figure 13.15. Such generators are simple and easily implemented in hardware as well as software. (This makes them useful as signal sources for field measurements of BER when no secure, i.e. errorless, reference channel is available.) The properties of the pseudo-random bit sequences (PRBSs) generated in this way can be summarised as follows:

1. The sequence is periodic.
2. In each period the number of binary ones is one more than the number of binary zeros.
3. Among runs of consecutive ones and zeros one-half of the runs of each kind are of length 1, one-quarter are of length 2, one-eighth are of length 3, etc., as long as these fractions give meaningful numbers.
4. If a PRBS is compared, term by term, with any cyclical shift of itself the number of agreements differs from the number of disagreements by one.

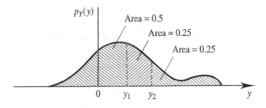

Figure 13.14 *Pdf for a random variable used to generate a random digital symbol stream.*

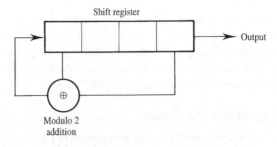

Figure 13.15 *Use of a shift register to generate pseudo-random bit sequences.*

The autocorrelation function of a K bit, periodic, PRBS, $z(k)$, is defined by:

$$R_z(\tau) = \frac{1}{K} \sum_{k=1}^{K} z(k)z(k - \tau) \tag{13.27}$$

and is shown in Figure 13.16(b) for a polar sequence with amplitude ± 1, Figure 13.16(a). Its voltage and power spectra are shown in Figures 13.16(c) and (d).

The general algorithm implemented by an n-element shift register with modulo-2 feedback, Figure 13.17, can be expressed mathematically as:

$$v(k) = w_{n-1}v(k - 1) \oplus w_{n-2}v(k - 2) \oplus \cdots \oplus w_0 v(k - n) \tag{13.28}$$

where $\oplus$ denotes modulo-2 addition. w_i denotes the feedback weighting (1 or 0) associated with the register's $(i + 1)$th element and corresponds in hardware to the presence or absence of a connection.

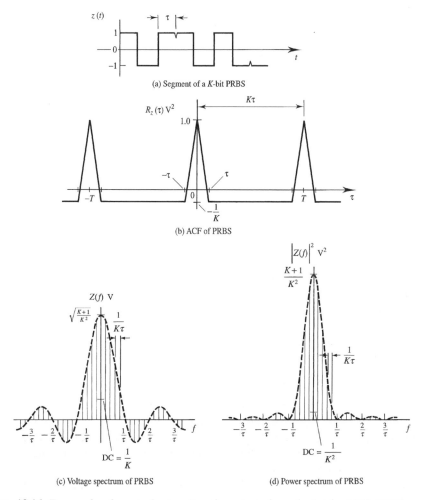

(a) Segment of a K-bit PRBS

(b) ACF of PRBS

(c) Voltage spectrum of PRBS

(d) Power spectrum of PRBS

Figure 13.16 *Temporal and spectral properties of a rectangular pulsed polar, NRZ, PRBS.*

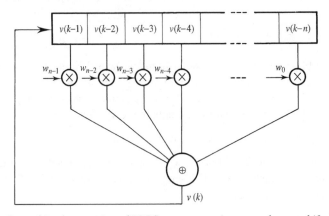

Figure 13.17 *General implementation of PRBS generator using an n-element shift register.*

Table 13.3 *PRBS sequence feedback connections.*

Number of shift register elements, n	Non-zero feedback taps in addition to w_0	Sequence length $K = 2^n - 1$
4	w_3	15
5	w_3	31
6	w_5	63
7	w_6	127
8	w_6, w_5, w_4	255
9	w_5	511
10	w_7	1023
12	w_{11}, w_8, w_6	4095
14	w_{13}, w_8, w_4	16383
16	w_{15}, w_{13}, w_4	65535

The all-zeros state is prohibited in the PRBS, as generated above, since this would result in an endless stream of zeros thereafter. A maximal length PRBS is one in which all possible n bit patterns (except the all-zeros pattern) occur once and once only in each period. The length of such a sequence is therefore $K = 2^n - 1$ bits. Not all arrangements of feedback connection give maximal length sequences. To establish whether a given set of connections will yield a maximal length sequence the polynomial:

$$f(x) = w_0 + w_1 x + w_2 x^2 + \cdots + w_{n-1} x^{n-1} + x^n \qquad (13.29)$$

is formed and then checked to see if it is *irreducible*. (Note that $w_0 = 1$, otherwise the final element of the register is redundant.) An irreducible polynomial of degree n is one which cannot be factored as a product of polynomials with degree lower than n. A test for irreducibility is that such a polynomial will not divide exactly (i.e. leaving zero remainder) into $x^P + 1$ for all $P < 2^n + 1$.

Table 13.3 [adapted from Jeruchim *et al.*, 1984] gives a selection of shift register feedback connections which yield maximal length PRBSs.

13.4.3 Noise and interference

Noise is modelled, essentially, in the same way as the signals described in section 13.4.2. The following interesting point arises, however, in the modelling of white Gaussian noise. Because a simulation deals only with noise samples then, providing these noise samples are uncorrelated (and being Gaussian, statistically independent), no distinction can be made between any underlying (continuous) noise processes which satisfy:

$$R_n(kT_s) = 0 \qquad\qquad (13.30)$$

where $T_s = 1/f_s$ is the simulation sampling period. The simulation is identical, then, whether the underlying noise process is strictly white with an impulsive autocorrelation function, or bandlimited to $f_s/2$ Hz with the sinc shaped autocorrelation function shown in Figure 13.18. Provided $f_s/2$ (sometimes called the simulation bandwidth) is large with respect to the bandwidth of the system being simulated then the results of the simulation will be unaffected by this ambiguity. White Gaussian noise is therefore, effectively, simulated by generating independent random samples from a Gaussian pdf with variance (i.e. normalised power), σ_n^2, given by:

$$\sigma_n^2 = \frac{N_0 f_s}{2} = \frac{N_0}{2 T_s} \quad (\text{V}^2) \qquad\qquad (13.31)$$

where N_0 (V^2/Hz) is the required *one-sided* NPSD.

Impulsive noise is characterised by a transient waveform which may occur with random amplitude at random times, or with fixed amplitude periodically, or with some combination of the two, Figure 13.19. The noise may be generated at baseband or passband. In radio systems, however, the noise will be filtered by the receiver's front end and IF strip after which even baseband generated impulses will have bandpass characteristics, often resembling the receiver's impulse response. (Impulsive noise does not imply a sequence of impulses in the Dirac delta sense but usually does imply transient pulses with an effective duration which is short compared to the average interpulse spacing.) A useful way of modelling impulsive noise is to separate the statistical aspects from the deterministic pulse shape. This can be done by using a (complex) random number generator to model the amplitude and phase of the pulses, a Poisson counting process to model the arrival times, T_i, of the pulses and a deterministic function, $I(t)$, to

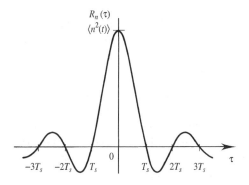

Figure 13.18 *ACF of bandlimited white noise with bandwidth B = f_s/2 Hz.*

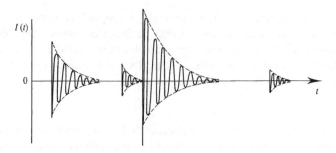

Figure 13.19 *Impulsive noise with random interpulse spacing and random pulse amplitude.*

model the pulse shape (e.g. $e^{-at} \cos \omega_c t\, u(t)$). Such a model would be specified at passband by:

$$I(t) \;=\; \sum_{i=1}^{N} A_i\, e^{-\frac{t-T_i}{\tau}} \cos(\omega_c t + \theta_i)\, u(t - T_i) \tag{13.32}$$

The equivalent baseband representation would be:

$$I_{LP}(t) \;=\; \sum_{i=1}^{N} A_i\, e^{j\theta_i}\, e^{-\frac{t-T_i}{\tau}}\, u(t - T_i) \tag{13.33}$$

Rewriting equation (13.33) as a convolution and expanding the exponential, i.e.:

$$I_{LP}(t) \;=\; e^{-\frac{t}{\tau}}\, u(t) * \sum_{i=1}^{N} A_i\, (\cos \theta_i + j \sin \theta_i)\, \delta(t - T_i) \tag{13.34}$$

emphasises the separation of pulse shape from pulse statistics. θ_i would normally be assumed to have a uniform pdf and a typical pdf for A_i might be log-normal. If the impulse noise is a Poisson process then the pdf of interarrival time between pulses is:

$$p_{\Delta T_i}(\Delta T_i) \;=\; \lambda\, e^{-\lambda\, \Delta T_i} \tag{13.35}$$

where $\Delta T_i = T_i - T_{i-1}$ and λ is the average pulse arrival rate (see Chapter 19).

Interference usually implies either an unwanted periodic waveform or an unwanted (information bearing) signal. In the former case pulse trains and sinusoids, for example, are easily generated in both passband and equivalent baseband form. In the latter case the interfering signal(s) can be generated in the same manner as wanted signals.

13.4.4 Time invariant linear systems

Linear subsystems, such as pulse shaping filters in a transmitter and matched filters in a receiver, are usually specified by their frequency response (both amplitude and phase). There is a choice to be made in terms of the appropriate level of idealisation when specifying such subsystems. An IF filter in a receiver may, for example, be represented by a rectangular frequency response at one extreme or a set of tabulated amplitude and phase values, obtained, across the frequency band, from measurements on a hardware prototype, at the other extreme. (In this case, unless the frequency domain points have the same frequency resolution as the simulation Fourier transform algorithm, interpolation and/or resampling with the correct resolution will be required.) In between these extreme cases analytical or tabulated models of classical filter responses (e.g. Butterworth, Chebyshev, Bessel, elliptic) may be used. Digital filter structures, typically implemented using tapped delay lines [Mulgrew *et al.*], are especially easy to simulate, at least in principle. The effect of a frequency response on an input signal can be found by convolving the impulse response of the filter (which will be complex if equivalent baseband representations are being used) with its input. (The impulse response is obtained from the frequency response by applying an inverse FFT, see section 13.5.) Alternatively, block filtering can be applied in which the input time series is divided into many equal length segments, Fourier transformed using an FFT algorithm, multiplied by the (discrete) frequency response and then inverse transformed back to a time series. The implementation of block filtering, including important aspects such as appropriate zero padding, is described in [Strum and Kirk].

13.4.5 Non-linear and time varying systems

Amplitude compression, as used in companding (section 5.7.3), is a good example of baseband, memoryless, non-linear, signal processing. This type of non-linearity can be modelled either analytically (using, for example, equation (5.31)) or as a set of tabulated points relating instantaneous values of input and output. (Interpolation may be necessary in the latter case.)

For memoryless, non-linear, systems operating on narrowband signals the useful equivalent baseband model shown in Figure 13.20 [developed from Tranter and Kosbar] can be used. This model works because, for a sinusoidal input, memoryless non-linearities produce outputs only at the frequency of the input and its harmonics. The harmonics can be filtered out at the non-linearity's output. The overall effect is of a 'linear' system with amplitude dependent gain (AM/AM conversion).

If a non-linearity has memory of intermediate length (i.e. memory which is significant with respect to, and may be many times, the carrier period but which is nevertheless short compared to changes in the carrier's complex envelope) then this too can be simulated as an equivalent baseband system, Figure 13.21. The amplitude of the signal (being related to the envelope) is changed in an essentially memoryless way. The phase of the signal, however, may now be affected by the non-linearity in a way which depends on the signal amplitude (AM/PM conversion). The power amplifier in satellite transponders can often be modelled in this way.

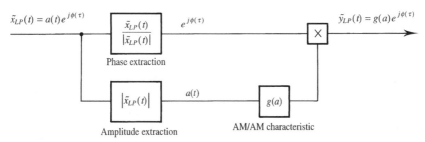

Figure 13.20 *Equivalent baseband model for a non-linear, memoryless, bandpass system.*

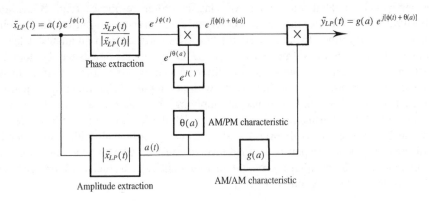

Figure 13.21 *Equivalent baseband model for a non-linear bandpass system with memory of intermediate length.*

There are other types of non-linear bandpass system which fall into neither of the above special categories. In general it is not possible to obtain equivalent baseband models for these processes. Simulation of these non-linearities must normally, therefore, be executed at passband.

Adaptive equalisers (usually implemented as tapped delay lines with variable tap weights) and adaptive delta modulation (section 5.8.5) are examples of time varying, linear, subsystems. The simulation aspects of such subsystems are essentially straight-forward, the design of algorithms to produce the required adaptive behaviour being the principal challenge. It is worth observing, however, that an equaliser operating on the demodulated I and Q components of a quadrature modulated signal generally comprises four separate tapped delay lines. Two lines operate in I and Q channels, individually controlling I and Q channel ISI, whilst two operate across I and Q channels controlling crosstalk, as in Figure 13.6(c). These four, real, tapped lines can be replaced by two lines, in which the weighting factors applied to each tap are complex, and which operate on the complex envelope of the I and Q channel signals.

13.5 Transformation between time and frequency domains

Both time domain and spectral quantities are of interest in the design and evaluation of communication systems. This alone would be sufficient reason to require transformation between domains when simulating systems. In fact transformation between domains is also desirable since some simulation operations can be implemented more efficiently in the frequency domain than in the time domain.

Computer simulations work with discrete samples of time waveforms (i.e. time series) and discrete samples of frequency spectra. The discrete Fourier transform (DFT) and its relative, the discrete Fourier series (DFS), are the sampled signal equivalent of the Fourier transform and Fourier series described in Chapter 2. From a mathematical point of view the DFT, DFS and their inverses can simply be defined as a set of consistent formulae without any reference to their continuous function counterparts. Almost always, however, in communications engineering the discrete time series and frequency spectra on which these transforms operate represent sampled values of underlying continuous functions. Their intimate and precise connection with continuous Fourier operations is therefore emphasised here.

13.5.1 DFT

Consider the N-sample time domain signal (or time series), $v_s(t)$, shown in Figure 13.22(a). The sample values can be represented by a series of weighted impulses, Figure 13.22(b), and expressed mathematically by:

$$v_\delta(t) = v_0\, \delta(t) + v_1\, \delta(t - \Delta t) + v_2\, \delta(t - 2\Delta t) + \cdots \quad (\text{V/s}) \tag{13.36}$$

Notice that $v_\delta(t)$ has units of V/s. This is easily demonstrated by using the sampling property of impulse functions under integration to recover the original time series which has units of V, i.e.:

$$v_s(t) = \left\{ \int_{-\infty}^{\infty} v_0\delta(t)\, dt, \; \int_{-\infty}^{\infty} v_1\delta(t - \Delta t)\, dt, \; \int_{-\infty}^{\infty} v_2\delta(t - 2\Delta t)\, dt, \cdots \right\}$$

$$= \left\{ v_0 \int_{-\infty}^{\infty} \delta(t)\, dt, \; v_1 \int_{-\infty}^{\infty} \delta(t - \Delta t)\, dt, \; v_2 \int_{-\infty}^{\infty} \delta(t - 2\Delta t)\, dt, \cdots \right\}$$

$$= \{v_0, \; v_1, \; v_2, \cdots, \; v_{N-1}\} \quad (\text{V}) \tag{13.37}$$

(Since the sample values, v_i, have units of V this means that the impulses, $\delta(t - \Delta t)$, have units of s^{-1}, i.e. their area, or strength, is dimensionless.) The voltage spectrum of $v_\delta(t)$ is given by:

$$V_\delta(f) = \text{FT}\,\{v_\delta(t)\}$$

$$= v_0 \int_{-\infty}^{\infty} \delta(t)e^{-j2\pi ft}\, dt \; + \; v_1 \int_{-\infty}^{\infty} \delta(t - \Delta t)e^{-j2\pi ft}\, dt + \cdots \quad (\text{V}) \tag{13.38}$$

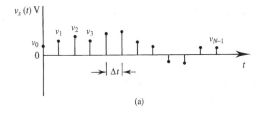

(a)

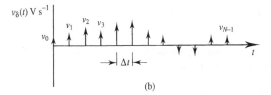

(b)

Figure 13.22 *(a) Sampled continuous signal or time series, and (b) its representation as a sum of impulses or delta functions.*

(Notice that because equation (13.38) is the transform of a sequence of weighted impulses with units of V/s the voltage spectrum has units of V only, not V/Hz as usual.) Using the sampling property of $\delta(t)$ under integration equation (13.38) becomes:

$$V_\delta(f) = v_0 e^{-j0} + v_1 e^{-j2\pi f \Delta t} + v_2 e^{-j2\pi f 2\Delta t} + \cdots \quad \text{(V)} \tag{13.39}$$

$V_\delta(f)$ in equation (13.39) is a *continuous* function, i.e. it is defined for all values of f. Using the summation notation equation (13.39) can be written more succinctly as:

$$V_\delta(f) = \sum_{\tau=0}^{N-1} v_\tau \, e^{-j2\pi f(\tau \Delta t)} \quad \text{(V)} \tag{13.40}$$

The values of the spectrum, $V_\delta(f)$, at the discrete frequencies f_0, f_1, f_2, etc., are:

$$V_\delta(f_\nu) = \sum_{\tau=0}^{N-1} v_\tau \, e^{-j2\pi f_\nu(\tau \Delta t)} \quad \text{(V)} \tag{13.41}$$

and if the frequencies of interest are equally spaced by Δf Hz then:

$$V_\delta(f_\nu) = \sum_{\tau=0}^{N-1} v_\tau \, e^{-j2\pi(\nu \Delta f)(\tau \Delta t)} \quad \text{(V)} \tag{13.42}$$

where $\nu = 0, 1, 2, 3, \cdots$, etc. Since the time series contains N samples it represents a signal with duration, T, given by:

$$T = N\Delta t \quad \text{(s)} \tag{13.43}$$

Nyquist's sampling theorem (Chapter 5) asserts that the lowest observable frequency (excluding DC) in the time series is:

$$f_1 = \frac{1}{N\Delta t} = \Delta f \quad \text{(Hz)} \tag{13.44}$$

Thus:

$$\Delta f \, \Delta t = \frac{1}{N} \tag{13.45}$$

and equation (13.42) can be written as:

$$V_\delta(f_\nu) = \sum_{\tau=0}^{N-1} v_\tau \, e^{-j2\pi \frac{\nu}{N} \tau} \quad \text{(V)} \tag{13.46}$$

Equation (13.46) is the definition, adopted here, for the *forward* DFT. Comparing this with the conventional (i.e. continuous) FT:

$$V(f) = \int_{-\infty}^{\infty} v(t) \, e^{-j2\pi ft} \, dt \quad \text{(V/Hz)} \tag{13.47}$$

the difference is seen, essentially, to be the absence of a factor corresponding to dt. $V_\delta(f_\nu)$ is therefore related to $V(f)$ by:

$$V(f)|_{f=f_\nu} \approx V_\delta(f_\nu) \, \Delta t \quad \text{(V/Hz)} \tag{13.48}$$

The reason why an approximation sign is used in equation (13.48) will become clear later. A note of caution is appropriate at this point. Equation (13.46) as a definition for the DFT is not universal. Sometimes a factor of $1/N$ and sometimes a factor of $1/\sqrt{N}$ is included in the formula. If the absolute magnitude of a voltage spectrum is important it is essential, therefore, to know the definition being used. Furthermore proprietary DFT software may not include an implementation of equation (13.48). Care is therefore needed in correctly interpreting the results given by DFT software.

13.5.2 DFS

If $v(t)$ is periodic then its FT should represent a *discrete* voltage spectrum (in contrast to discrete values taken from a continuous spectrum). Comparing the DFT (equation (13.46)) with the formula for a set of Fourier series coefficients (Chapter 2):

$$\tilde{C}_v = \frac{1}{T} \int_0^T v(t)\, e^{-j2\pi f t}\, dt \quad \text{(V)} \tag{13.49}$$

and remembering that the length of the time series is given by $T = N\Delta t$ then, to make $V_\delta(f_v)$ reflect $\tilde{C}_v$ properly, an extra factor $1/N$ is required, i.e.:

$$\tilde{C}_v = \frac{1}{N\Delta t} V_\delta(f_v)\, \Delta t = \frac{1}{N} V_\delta(f_v) \quad \text{(V)} \tag{13.50}$$

The (forward) *discrete Fourier series* (DFS) is therefore defined by:

$$\tilde{C}(f_v) = \frac{1}{N} \sum_{\tau=0}^{N-1} v_\tau\, e^{-j2\pi \frac{v}{N} \tau} \quad \text{(V)} \tag{13.51}$$

The need for the factor $1/N$ in equation (13.51) is seen most easily for the DC ($v = 0$) value which is simply the time series average.

13.5.3 DFS spectrum and rearrangement of spectral lines

Providing all the samples in the time series, v_τ, are real then the spectrum defined by equation (13.51) has the following properties:

1. Spectral lines occurring at $v = 0$ and $v = N/2$ are real. All others are potentially complex.
2. $\tilde{C}(f_{N-v}) = \tilde{C}^*(f_v)$, i.e. the DFS amplitude spectrum is even and the DFS phase spectrum is odd.

Figure 13.23 illustrates these properties for a 16-sample time series. The harmonic number, v, along the x-axis of Figure 13.23 is converted to conventional frequency f (in Hz) using:

$$f_v = v f_1 = \frac{v}{N\Delta t} \quad \text{(Hz)} \tag{13.52}$$

This leads, superficially, to a paradox in that f_{N-1} appears to correspond to a frequency of (almost) $1/\Delta t$ Hz yet Nyquist's sampling theorem asserts that frequencies no higher than $1/(2\Delta t)$ Hz can be observed. The paradox is resolved by recognising that no additional information is contained in the harmonics $N/2 < v < N$ since these are conjugates of the harmonics $0 < v < N/2$. A satisfying interpretation of the 'redundant' harmonics ($v > N/2$) is as the negative frequency components of a double sided spectrum. The

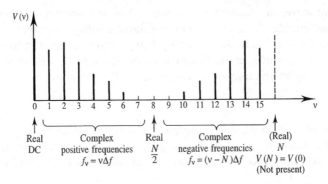

Figure 13.23 *Schematic illustration of discrete Fourier series for a 16-sample time series.*

conventional frequency spectrum is therefore constructed from the DFS (or DFT) by shifting all the lines from the top half of the DFS spectrum down in frequency by $N\Delta f$ Hz as shown in Figure 13.24. (Half the component at $v = N/2$ can also be shifted by $-N\Delta f$ Hz to retain overall symmetry, if desired.) The highest observable frequency is then given by:

$$f_{N/2} = \frac{1}{2\,\Delta t} \quad \text{(Hz)} \tag{13.53}$$

as expected.

13.5.4 Conservation of information

In the time domain an N-sample real time series is represented by N real numbers. In the frequency domain the same signal is represented by $N - 2$ complex numbers and 2 real numbers (i.e. $2(N - 2) + 2$ real numbers). Half of the complex numbers, however, are complex conjugates of the other half. Thus in the frequency domain, as in the time domain, the signal is represented by N *independent* real numbers.

13.5.5 Phasor interpretation of DFS

In equation (13.51) each (complex) spectral line, $\tilde{C}(f_3)$, for example, is a sum of N phasors (one arising from each time sample), all with identical frequency, f_3, but each with a different phase, $\theta_\tau = -2\pi(3/N)\tau$. Figures 13.25(a) and (b) illustrate the phasor diagrams for $\tilde{C}(f_1)$ and $\tilde{C}(f_2)$ respectively corresponding to an eight-sample time series.

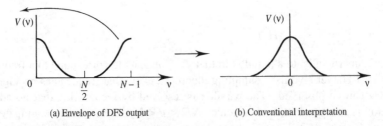

(a) Envelope of DFS output   (b) Conventional interpretation

Figure 13.24 *Interpretation of DFS (or DFT) components $v > N/2$ as negative frequencies.*

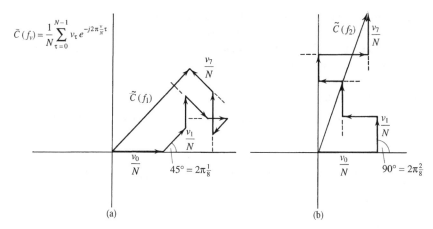

$$\tilde{C}(f_v) = \frac{1}{N}\sum_{\tau=0}^{N-1} v_\tau\, e^{-j2\pi\frac{v}{N}\tau}$$

Figure 13.25 *Phasor interpretation for DFS showing composition of: (a) fundamental ($v = 1$) and (b) second harmonic ($v = 2$) components of an eight-sample time series.*

In Figure 13.25(a) (where $v = 1$) the phasors advance by $2\pi(1/8)$ rad $= 45°$ each time τ is incremented. In Figure 13.25(b) (where $v = 2$) the phasors advance by $2\pi(2/8)$ rad $= 90°$. (There is an additional π rad phase change when the time series sample is negative.)

13.5.6 Inverse DFS and DFT

Consider the forward DFS:

$$\tilde{C}(f_v) = \frac{1}{N}\sum_{\tau=0}^{N-1} v_\tau\, e^{-j2\pi\frac{v}{N}\tau} \quad \text{(V)} \tag{13.54}$$

After rearranging, this represents N cisoids (or phasors) each rotating with a different frequency, i.e.:

$$\tfrac{1}{2}\,\tilde{C}(f_{-N/2}),\cdots,\tilde{C}(f_{-2}),\tilde{C}(f_{-1}),\tilde{C}(f_0),\tilde{C}(f_1),\tilde{C}(f_2),\cdots,\tfrac{1}{2}\,\tilde{C}(f_{N/2})$$

Each time sample, v_τ, is the sum of these cisoids evaluated at time $t = \tau\Delta t$. Figure 13.26, for example, shows the phasor diagram for $N = 8$ at the instant $t = 3\Delta t$ (i.e. $\tau = 3$). The resultant on this diagram corresponds to the third sample of the time series. (The diagram is referred to as a phasor diagram, here, even though different phasor pairs are rotating at different frequencies.) By inspection of Figure 13.26 the *inverse* DFS can be seen to be:

$$v_\tau = \sum_{v=0}^{N-1} \tilde{C}(f_v)\, e^{j\frac{2\pi}{N}v\tau} \quad \text{(V)} \tag{13.55}$$

Similarly the inverse DFT is:

$$v_\tau = \frac{1}{N}\sum_{v=0}^{N-1} V_\delta(f_v)\, e^{j\frac{2\pi}{N}v\tau} \quad \text{(V)} \tag{13.56}$$

(Notice that the factors of $1/N$ are arranged in the DFS, DFT and their inverses, such that applying a given discrete transform, followed by its inverse, results in no change.)

13.5.7 DFT accuracy

Having calculated a spectrum using a DFT, and rearranged the spectral lines as described in section 13.5.3, the question might be asked – how well do the resulting values represent the continuous FT of the continuous time function underlying the time series? This question is now addressed.

Sampling and truncation errors

There are two problems which have the potential to degrade accuracy. These are:

1. sampling;
2. truncation.

Aliasing, due to sampling, is avoided providing sampling is at the Nyquist rate or higher, i.e.:

$$\frac{1}{\Delta t} \geq 2f_H \quad (\text{Hz}) \tag{13.57}$$

The truncation problem arises because it is only possible to work with time series of finite length. A series with N samples corresponds to an infinite series multiplied by a rectangular window, $\Pi(t/T)$, of width $T = N\Delta t$ s, Figure 13.27. This is equivalent to convolving the FT of $v_\delta(t)$ with the function $T\text{sinc}(Tf)$, Figure 13.28. The convolution smears, or smooths, $V(f)$ on a frequency scale of approximately $2/T$ Hz (i.e. the width of the smoothing function's main lobe). It might be argued that a time series for which the

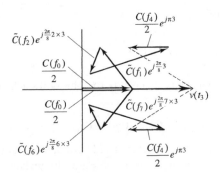

Figure 13.26 *'Phasor' diagram at the instant $t = 3\Delta t$ (i.e. $\tau = 3$) for a real, eight-sample, time series.*

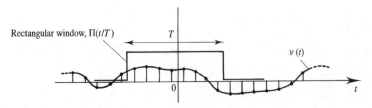

Figure 13.27 *A finite (eight-sample) time series interpreted as the product of an infinite series and a rectangular window.*

underlying function is strictly time limited to $N\Delta t$ s does not suffer from truncation error. In this case, however, the function cannot be bandlimited and the Nyquist sampling rate would be ∞ Hz making it impossible to avoid aliasing errors. Conversely if the underlying function is strictly bandlimited (allowing the possibility of zero aliasing error) then it cannot be time limited, making truncation errors unavoidable. (It is, of course, possible to reduce both types of error by sampling at a faster rate and increasing the length of the time series. This means increasing the overall number of samples, however, requiring greater computer resources.)

Frequency sampling, smoothing, leakage and windowing

Discreteness and periodicity are the corresponding properties of a function expressed in the frequency and time domains. This is summarised in Table 13.4.

Table 13.4 *Relationship between periodic and discrete signals.*

Time domain	Frequency domain
Periodic	Discrete
Discrete	Periodic

Since the DFT gives discrete samples of $V(f)$ (at least approximately) then the function of which it is the precise FT (or FS) is periodic. Furthermore, since the time series, $v_\delta(t)$, is discrete it should have a periodic spectrum. This implies that a replicated version of a DFT, such as that shown in Figure 13.29(b), is the exact FT (or FS) of a function such as that shown in Figure 13.29(a).

Figure 13.30 illustrates how the implicit periodicity of both the underlying time series and the exact spectrum of this time series impacts on truncation errors. It can be seen that in addition to smoothing of the baseband spectrum, energy leaks into the DFT from the higher frequency spectral replicas via the window spectrum with which the exact spectrum is convolved. This type of error is called leakage and can be reduced by decreasing the sidelobes of the window function's spectrum. Shaping of the time domain window function to realise low sidelobes in its spectrum tends, however, to increase the width of the spectral main lobe. There is, consequently, a trade-off to be made between leakage and smoothing errors. The optimum shape for a time domain window depends on the particular application, the data being transformed and engineering judgement. The commonly encountered window functions are discussed in [Brigham].

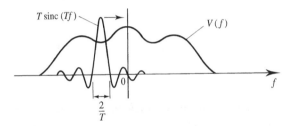

Figure 13.28 *Convolution in frequency domain of FT $\{v(t)\}$ with FT $\{\Pi(t/T)\}$, corresponding to rectangular windowing in Figure 13.27.*

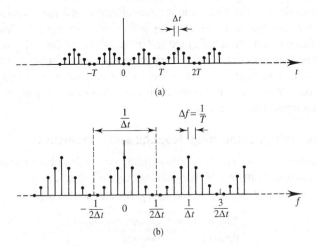

Figure 13.29 *Sampled and replicated versions of a signal in (a) time and (b) frequency domain forming a precise FT or FS pair.*

The effect of crude windowing, with a rectangular shape for example, does not, *necessarily*, lead to poor accuracy in the calculated spectrum of a time series. If the spectrum of the unwindowed time series is slowly changing on the frequency scale of the oscillations of the window's sidelobes then cancellation tends to occur between the leakage from adjacent sidelobes, Figure 13.31. A corollary of this is a frequency domain manifestation of Gibb's phenomenon, Figure 13.32. If the time series spectrum changes rapidly on the frequency scale of window spectrum oscillations, then as the window spectrum slides across the time series spectrum the convolution process results in window spectrum oscillations being reproduced in the spectrum of the windowed time series. This problem is at its worst in the region of time series spectral discontinuities where it can lead to significant errors. The same effect occurs in the region of any impulses present in the time series spectrum. In particular this means that the DC level (i.e. average value) of a time series should be removed before a DFT is applied. If this is overlooked it is possible that the window spectrum reproduced by convolution with the 0 Hz impulse may obscure, and be mistaken for, the spectrum of the time series data, Figure 13.33. (If impulses in the spectrum of the time series are important they should be removed, before application of a DFT, and then reinserted into the calculated spectrum.)

Trailing zeros

The frequency spacing of spectral values obtained from a DFT is given by:

$$\Delta f = f_1 = \frac{1}{T} \text{ (Hz)} \tag{13.58}$$

where $T = N\Delta t$ is the (possibly windowed) length of the time series. Δf is sometimes called the resolution of the DFT. (This does not imply that all spectral features on a scale of Δf are necessarily resolved since resolution in this sense may be limited by windowing effects.) Δf can only be made smaller by increasing the length of the time series data record, i.e. increasing T (or N). The alternative is to artificially extend the

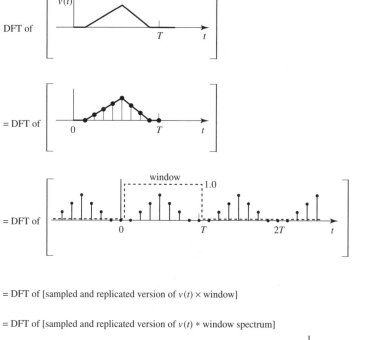

= DFT of [sampled and replicated version of $v(t)$ × window]

= DFT of [sampled and replicated version of $v(t)$ * window spectrum]

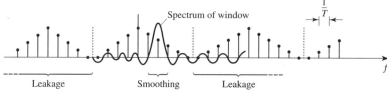

Figure 13.30 *Origin of smoothing and leakage errors in a DFT.*

data record with additional trailing zeros. This is illustrated in Figure 13.34 and is called zero padding. Notice that the highest observable frequency, $f_{N/2}$, determined by the sampling period, Δt, is unaffected by zero padding but, since more samples are included in the DFT operation, there are more samples in the output display, giving finer sampling in the frequency domain. Since the genuine data record has not been extended, however, the underlying resolution is unaltered, the zero padding samples having simply allowed the DFT output to be more finely interpolated [Mulgrew *et al.*].

Random data and spectral estimates

Consider the stationary random process, $v_s(t)$, illustrated in Figure 13.35(a). If several segments of this random time series are windowed and transformed using a DFT then the result is a number of voltage spectra, each with essentially meaningless phase information. If power spectra, $G_i(f_v)$, are obtained using:

$$G_i(f_v) = \frac{1}{T} \mid \text{DFT } [v_\tau \, w_i(\tau)]\Delta t \mid^2 \quad (\text{V}^2/\text{Hz}) \tag{13.59}$$

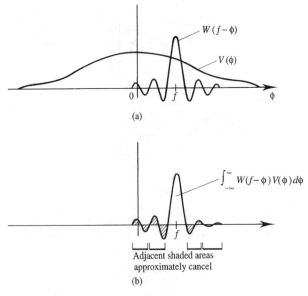

(a)

(b)

Adjacent shaded areas
approximately cancel

Figure 13.31 *Approximate cancellation of leakage errors for a windowed signal with slowly changing spectrum: (a) convolution of underlying spectrum with oscillating window spectrum; (b) result of convolution at frequency shift $\phi = f$.*

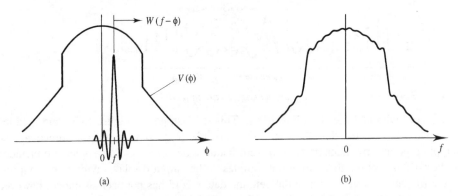

(a)   (b)

Figure 13.32 *Frequency manifestation of Gibb's phenomenon: (a) convolution of a spectrum containing rapidly changing region with oscillating spectrum of a windowing function; (b) resulting oscillations in spectrum of windowed function.*

where $w_i(\tau)$ is the window applied to the ith segment of the data and $G_i(f_v)$ is the power spectrum of the ith segment, then the size of a spectral line at a given frequency fluctuates randomly from spectrum to spectrum, Figure 13.35(b). Surprisingly, the random error in the calculated samples of the power spectral density does not decrease if the length of time series segments is increased. This is because the frequency 'resolution' improves proportionately. For example, if the length of a time series segment is doubled the spacing of its spectral lines is decreased by a factor of 2. Twice the number of spectral lines are generated by the DFT and thus the *information per spectral line remains constant.* The random errors in such a spectrum can be decreased by:

1. Averaging the corresponding values (lines) over the spectra found from independent time segments, i.e. using:

$$G(f_v) = \frac{1}{M} \sum_{i=1}^{M} G_i(f_v) \quad (\text{V}^2/\text{Hz}) \tag{13.60}$$

2. Averaging adjacent values in a single spectrum.

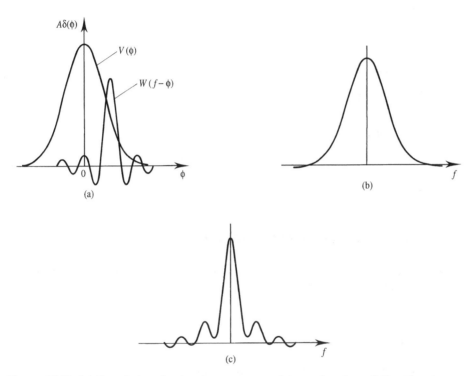

Figure 13.33 *(a) Convolution of a signal spectrum containing an impulse at 0 Hz with a sinc function reflecting rectangular windowing of a signal with a large DC value, (b) result of convolution excluding DC impulse, (c) convolution including DC impulse.*

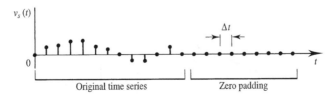

Figure 13.34 *Illustration of zero padding a time series with trailing zeros.*

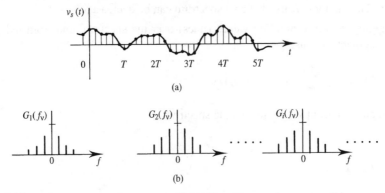

Figure 13.35 *(a) Stationary random process divided into T-second segments, (b) power spectra calculated from each time series segment.*

An alternative approach is to use the Wiener–Kintchine theorem and transform the autocorrelation function (ACF) of the time series, i.e.:

$$G(f_v) = \text{DFT}\,\{\text{ACF}\,[v_\tau\,w(\tau)]\}\,\Delta t \quad (\text{V}^2/\text{Hz}) \tag{13.61}$$

where the units of the discrete ACF are V^2. The number of spectral lines generated by equation (13.61) now depends on the maximum temporal displacement (or lag) used in the ACF rather than the length of the time series used.

13.6 Discrete and cyclical convolution

For sampled data such as that used in computer simulations the (discrete) convolution between two time series $x_0, x_1, x_2, \cdots, x_{N-1} = \{x_\tau\}$ and $y_0, y_1, y_2, \cdots, y_{M-1} = \{y_\tau\}$ is defined by:

$$z_n = \sum_{\tau=0}^{n} x_\tau\, y_{n-\tau} \quad (\text{V}^2) \tag{13.62}$$

where τ is the (integer) sample number of (both) time series (i.e. $x_\tau = x(\tau\Delta t)$), Δt is the sampling period and $n = 0, 1, 2, \cdots, N + M - 2$ is the shift in sample numbers between x_τ and the time reversed version of y_τ. (If the discrete convolution is being used to evaluate the convolution of a pair of underlying continuous signals, then equation (13.62) must be

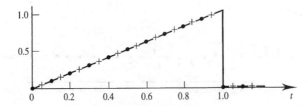

Figure 13.36 *Sampled signal showing alternative sampling instants.*

multiplied by the sampling period, Δt. The result will then have units of V^2 s as expected.) The question may be asked – does the precise timing of the sampling instants affect the result of the numerical (or discrete) convolution defined in equation (13.62)? The slightly surprising answer is generally yes. Figure 13.36 shows a signal sampled at 10 Hz. Dots represent samples which start at the origin (defined by the start of the continuous signal) and crosses represent samples which start half a sample period after the origin. Since convolution involves integration it is usually the case that the set of sampling points which best represent the area of the underlying function will give the best result in the sense that application of equation (13.62) will give an answer closest to the analytical convolution of the underlying functions. This implies that the crosses in Figure 13.36 represent superior sampling instants, which is also consistent with an intuitive feeling that a sample should be at the centre of the function segment which it represents.

The crosses in Figure 13.36 also have the advantage that none fall on a point of discontinuity. If sampling at such points cannot be avoided then an improvement in terms of area represented is obtained by assigning a value to that sample equal to the mean of the function value on either side of the discontinuity. In Figure 13.36, therefore, the sample (dot) for the point $t = 1.0$ would be better placed at 0.5 V than 0 V. (This is consistent with the fact that physical signals do not contain discontinuities and that bandlimited signals, corresponding to truncated Fourier series, converge at points of discontinuity to the mean of the signal value on either side of that discontinuity, see section 2.2.3.)

In practice convolution is often implemented by taking the inverse DFT of the product of the DFTs of the individual time series, i.e.:

$$\{x_\tau\} * \{y_\tau\} \equiv \text{DFT}^{-1}\left[\text{DFT}\{x_\tau\}\,\text{DFT}\{y_\tau\}\right] \quad (V^2) \tag{13.63}$$

For equation (13.63) to yield sensible results the sampling period and length of both time series must be the same. (Zero padding can be used to equalise series lengths if necessary.) Writing out equation (13.63) explicitly (using equations (13.56) and (13.46)), and using primes and double primes to keep track of sample numbers in the different time series:

$$z(\tau) = \{x_{\tau'}\} * \{y_{\tau''}\} = \frac{1}{N}\sum_{v=0}^{N-1}\left[\sum_{\tau'=0}^{N-1} x_{\tau'}\, e^{-j2\pi\frac{v}{N}\tau'}\sum_{\tau''=0}^{N-1} y_{\tau''}\, e^{-j2\pi\frac{v}{N}\tau''}\right]e^{j\frac{2\pi}{N}v\tau}$$

$$= \frac{1}{N}\sum_{v=0}^{N-1}\left[\sum_{\tau'=0}^{N-1}\sum_{\tau''=0}^{N-1} x_{\tau'}\, y_{\tau''}\, e^{-j2\pi\frac{v}{N}(\tau'+\tau'')}\right]e^{j\frac{2\pi}{N}v\tau}$$

$$= \frac{1}{N}\sum_{\tau'=0}^{N-1}\sum_{\tau''=0}^{N-1} x_{\tau'}\, y_{\tau''}\left[\sum_{v=0}^{N-1} e^{j2\pi\frac{v}{N}(\tau-\tau'-\tau'')}\right] \tag{13.64}$$

The square bracket on the last line of equation (13.64) is zero unless:

$$\tau - \tau' - \tau'' = nN \quad \text{(for any integer } n\text{)} \tag{13.65}$$

in which case it is equal to N. Equation (13.64) can therefore be rewritten as:

$$z(\tau) = \sum_{\tau'=0}^{N-1}\sum_{\tau''=0}^{N-1} x_{\tau'}\, y_{\tau''} \quad (\tau'' = \tau - \tau' - nN \text{ only})$$

$$= \sum_{\tau'=0}^{N-1} x_{\tau'} \, y_{\tau - \tau' - nN} \tag{13.66}$$

The implication of equation (13.66) is that the τth sample in the convolution result is the same for any n, i.e.:

$$z(\tau) = \dots$$

$$= \sum_{\tau'=0}^{N-1} x_{\tau'} y_{\tau - \tau' + N}$$

$$= \sum_{\tau'=0}^{N-1} x_{\tau'} y_{\tau - \tau'}$$

$$= \sum_{\tau'=0}^{N-1} x_{\tau'} y_{\tau - \tau' - N}$$

$$= \dots \tag{13.67}$$

This means that $y_{\tau - \tau'}$ (i.e. $y_{\tau''}$) is cyclical with period N as shown in Figure 13.37. Alternative interpretations are that as the time shift variable, τ, changes, elements of the

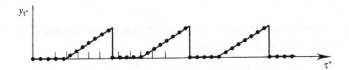

Figure 13.37 *Periodic interpretation of $y_{\tau''}$ resulting in cyclic convolution when using $x_\tau * y_\tau = \mathrm{DFT}^{-1}[\mathrm{DFT}(x_\tau)\mathrm{DFT}(y_\tau)]$ to implement discrete convolution.*

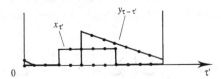

Figure 13.38 *Sample recycling interpretation of y_τ for cyclic convolution.*

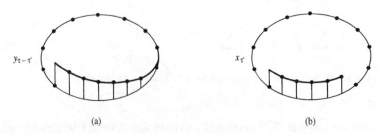

(a)   (b)

Figure 13.39 *Closed loop interpretation of x_τ and y_τ for cyclic convolution.*

time series, $y_{\tau''}$, lying outside the (possibly windowed) time series, $x_{\tau'}$, are recycled as shown in Figure 13.38, or that the time series are arranged in closed loops as shown in Figure 13.39. Figures 13.38 and 13.39 both define the cyclical convolution operation which is the equivalent time series operation to the multiplication of DFTs.

The cyclic (or periodic) convolution of two N-element time series is contained in a series which has, itself, N elements. (For normal, or aperiodic, convolution the series has $2N - 1$ elements.) A consequence is that if two functions are convolved using equation (13.63) (thus implying cyclical convolution) then enough zero padding must be used to ensure that one period of the result is visible, isolated by zeros on either side. If insufficient leading and/or trailing zeros are present in the original time series then the convolved sequences will have overlapping ends, Figure 13.40, making normal interpretation difficult. (In general, to avoid this, the number of leading plus trailing zeros required prior to cyclical convolution is equal to the number of elements in the functions to be convolved from the first non-zero element to the last non-zero element.)

13.7 Estimation of BER

For digital communications the quantity most frequently used as an objective measure of performance is symbol error rate, SER, or equivalently probability of symbol error, P_e. Sometimes more detailed information is desirable: for example, it might be important to know whether errors occur independently of each other or whether they tend to occur in bursts. Here we give a brief outline of two of the methods by which SER (and, if necessary, second order performance information) can be estimated using simulation. These are the Monte Carlo method and the quasi-analytical method. To some extent these methods represent examples of SER estimation techniques located at opposite ends of a spectrum of techniques. Other methods exist which have potential advantages under particular circumstances. A detailed discussion of many of these methods is given in [Jeruchim *et al.*, 1984].

13.7.1 Monte Carlo simulation

This is the conceptually simplest and most general method of estimating SERs. The detected symbol sequence at the receiver is compared symbol by symbol with the (error-free) transmitted sequence and the errors are counted, Figure 13.41. The estimated P_e is then given by:

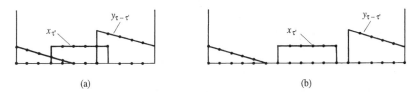

Figure 13.40 *Use of extra zeros to pad x_τ and y_τ in order to avoid overlapping of discrete convolution replicas in cyclical convolution result: (a) insufficient leading/trailing zeros leading to overlapping of replicas; (b) padded functions avoiding overlapping.*

$$P_e = \frac{\text{error count}}{\text{total symbol count}}$$

(13.68)

and the SER is given in equation (6.13) as:

$$\text{SER} = P_e R_s \quad (\text{error/s})$$

(13.69)

where R_s is the baud rate. Direct Monte Carlo estimation of BER is effected in a similar way, Figure 13.41. The penalty paid for the simplicity and generality of Monte Carlo simulation is that if SER is to be measured with both precision and confidence for low error rate systems, then large numbers of symbols must be simulated. This implies large computing power and/or long simulation times. If errors occur independently of each other, and a simulation is sufficiently long to count at least 10 errors, then the width of the P_e interval in which we can be 90%, 95% and 99% confident that the true P_e lies may be

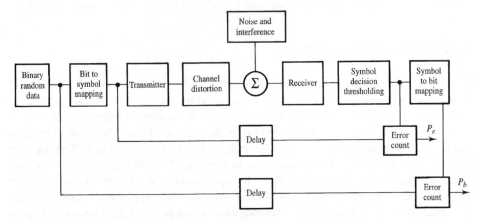

Figure 13.41 *Principles of Monte Carlo evaluation of SER and BER.*

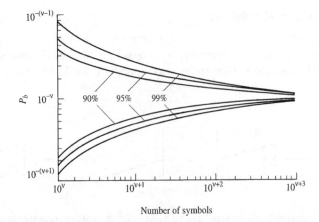

Figure 13.42 *Confidence bands on P_b when observed value is 10^{-v} for the Monte Carlo technique based on the normal approximation (source: Jeruchim et al., 1984, reproduced with permission of the IEEE © 2003 IEEE).*

found from Figure 13.42. If errors are not independent (e.g. they may occur in bursts due to the presence of impulsive noise) then each error in a given burst clearly yields less information, on average, about the error statistics than in the independent error case. (At its most obvious each subsequent error in a burst is less surprising, and therefore less informative, than the first error.) It follows that a greater number of errors would need to be counted for a given P_e confidence interval in this case than in the independent error case.

13.7.2 Quasi-analytic simulation

Quasi-analytic (QA) simulation can dramatically reduce the required computer power and/or run time compared with Monte Carlo methods. This is because the QA method simulates only the effect of system induced distortion occurring in the signal rather than including the effects of additive noise. It does depend, however, on a knowledge of the total noise pdf at the decision circuit input. The essence of QA simulation is best illustrated by a binary signalling example. Figure 13.43(a) shows an (unfiltered) baseband binary signal, $v_s(t)$, representing the output of a binary information source. Figure 13.43(b) shows the demodulated, but distorted, signal, $v_d(t)$, at the decision circuit

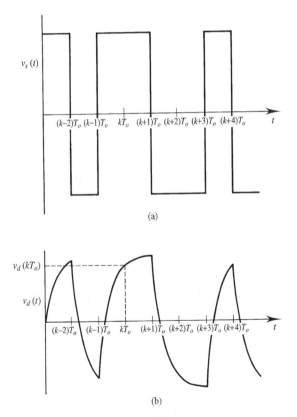

(a)

(b)

Figure 13.43 *Signals at: (a) transmitter binary source output; (b) receiver decision circuit input after transmission through a distorting but noiseless channel.*

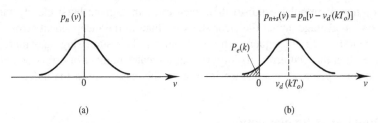

Figure 13.44 *Pdfs of: (a) noise only; (b) signal plus noise at kth sampling instant.*

input. If purely additive noise with pdf $p_n(v)$ is present at the decision circuit input the total pdf of signal plus noise at the kth sampling instant is given by:

$$p_{s+n}(v) = p_n[v - v_d(kT_0)] \tag{13.70}$$

where $v_d(kT_0)$ is the decision instant signal. $v_d(kT_0)$ will depend on the history of the bit sequence via the ISI due to the impulse response (i.e. distorting effect) of the entire system.

The probability, $P_e(k)$, that the noise (if present) would have produced an error at the kth bit sampling instant is:

$$P_e(k) = \int_{-\infty}^{0} p_n[v - v_d(kT_0)]\, dv = \int_{-\infty}^{-v_d(kT_0)} p_n(v)\, dv \tag{13.71(a)}$$

if $v_s(kT_0) > 0$ (i.e. v_s represents a transmitted digital 1), and is:

$$P_e(k) = \int_{0}^{\infty} p_n[v - v_d(kT_0)]\, dv = \int_{-v_d(kT_0)}^{\infty} p_n(v)\, dv \tag{13.71(b)}$$

if $v_s(kT_0) < 0$ (i.e. v_s represents a transmitted digital 0). $p_n(v)$, $p_n[v - v_d(kT_0)]$ and $v_d(kT_0)$ are shown in Figure 13.44. If $p_n(v)$ is Gaussian then the evaluation of the integrals is particularly easy using error function lookup tables or series approximations.

The overall probability of error is found by averaging equations (13.71) over many (N, say) bits, i.e.:

$$P_e = \frac{1}{N} \sum_{k=1}^{N} P_e(k) \tag{13.72}$$

N must be sufficiently large to allow essentially all possible combinations of bits, in a time window determined by the memory of the system, to occur. This ensures that all possible distorted signal patterns will be accounted for in the averaging process.

EXAMPLE 13.1

As an example of the power and utility of proprietary simulation packages a simplified model of a satellite communication system is analysed here using Signal Processing WorkSystem (SPW)™ marketed by Cadence Design Systems, Inc. SPW is immensely powerful as a simulation tool and

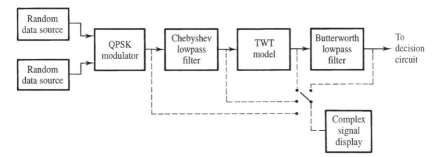

Figure 13.45 *Block diagram for example simulation.*

only a fraction of its facilities are illustrated in this example. The system is shown in Figure 13.45 which has been redrawn to simplify that produced using SPW's block diagram editor. It consists of a pair of (independent) random binary data sources which provide the input to a QPSK modulator, Figure 11.23. The symbol rate of each source is 1.0 baud.

The output of the modulator is an equivalent baseband signal and is therefore complex. The transmitted, uplink, signal is filtered by a 6-pole Chebyshev filter. The bandwidth of this (equivalent baseband) filter is 1.1 Hz. A satellite channel is modelled analytically using a non-linear input/output characteristic typical of a travelling wave tube (TWT) amplifier operating with 3.0 dB of input back-off which accounts for both AM/AM and AM/PM distortion (see section 14.3.3). A parameter of the TWT model is its average input power which is set to 2.0 W. The received, downlink, signal is filtered by a 6-pole Butterworth filter. Signal sink blocks are used to record the time series data generated as the simulation progresses. It is these signal records which are analysed to produce the required simulation results.

The simulation sampling rate used is 16 Hz which, for a baud rate of 1.0, corresponds to 16 sample/symbol. The number of samples simulated (called iterations in SPW) is 3000. Unrealistic parameters in simulations (e.g. unit baud rate and filter bandwidths of the order of 1 Hz) are typical. It is, usually, necessary only for parameters to be correct relative to each other for the simulation to give useful results. (The interpretation of those results must, of course, be made in the proper context of the parameters used.) Figures 13.46(a), (b), (c) and (d) show the simulated (complex) time series data at the input to the QPSK modulator, and the outputs from the Chebyshev filter, the TWT and the Butterworth filter. Note that the 'random' component of the signals in Figures 13.46(b), (c) and (d) is not due to noise but is due entirely to the distortion introduced by the channel.

Figures 13.47(a) and (b) show the magnitude and phase spectra of the pre- and post-filtered transmitted uplink signals respectively, calculated using a 1024 (complex) point FFT. (The time series were windowed with a Hamming function prior to the FFT in this case.) Figure 13.47(c) shows the spectrum of the TWT output. It is interesting to note the regenerative effect that the TWT non-linearity has on the sidelobes of the QPSK signal, previously suppressed by the Chebyshev filter. (This effect was briefly discussed in section 11.4.4.) The frequency axes of the spectra are easily calibrated remembering that the highest observable frequency (shown as 0.5 in the SPW output) corresponds to half the sampling rate, i.e. $16/2 = 8$ Hz, see section 13.5.3.

A QA simulation routine available in SPW has been run to produce the SER versus E_b/N_0 curve shown in Figure 13.48. QA simulation is applicable if the system is linear between the point at which noise is added and the point at which symbol decisions are made. In this example we assume that noise is added at the satellite (TWT) output. This corresponds to the, often realistic, situation in which the downlink dominates the CNR performance of a satellite communication system, see section 14.3.3.

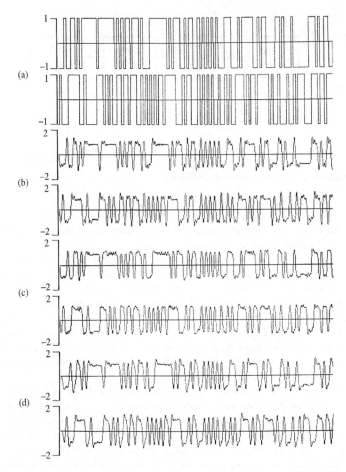

Figure 13.46 *Inphase (upper) and quadrature (lower) time series data at (a) QPSK modulator output and the outputs of: (b) Chebyshev filter; (c) TWT; (d) Butterworth filter.*

The way in which SPW implements QA simulation is as follows:

1. The equivalent noise bandwidth, B_N, of the system segment between the noise injection point and the receiver decision circuit is estimated from a (separate) simulation of this segment's impulse response, $h(t)$, i.e.:

$$B_N = \int_0^\infty |h(t)|^2 \, dt \qquad\qquad (13.73)$$

2. The signal power at the input to the receiver, C (in this example the input to the Butterworth filter), is estimated from the time series contained in the appropriate signal sink file (i.e. iagsig/twtout.sig, see Figure 13.46(c)).

3. The required range and interval values of E_b/N_0 are specified and, for each value of E_b/N_0, the normalised (Gaussian) noise power, $N \, (= \sigma^2)$, is found using:

$$\sigma^2 = N_0 B_N \quad (V^2) \qquad\qquad (13.74(a))$$

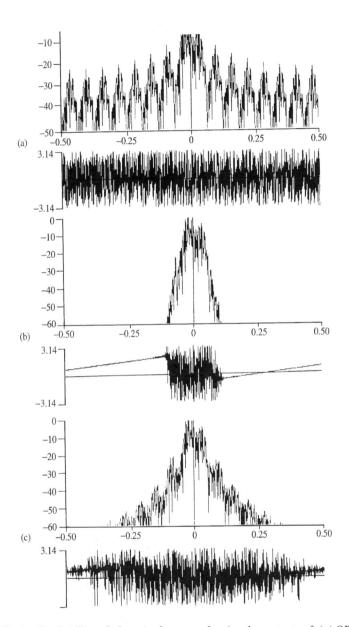

Figure 13.47 *Amplitude (dB) and phase (rad) spectra for signal at outputs of: (a) QPSK modulator; (b) Chebyshev filter; (c) TWT.*

where:

$$N_0 = \frac{CT_b}{(E_b/N_0)} = \frac{C/(R_s H)}{(E_b/N_0)} \quad \text{(V}^2\text{/Hz)} \tag{13.74(b)}$$

(Here H is the number of binary digits/symbol, i.e. it is the entropy assuming zero redundancy. Thus in this example $H = 2$.)

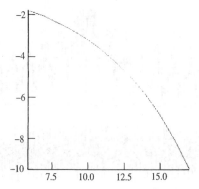

Figure 13.48 *Result of P_e versus E_b/N_0 (dB) for a quasi-analytic (QA) analysis of system shown in Figure 13.45. (P_e is plotted on a logarithmic scale so 10^{-2} is represented by -2.)*

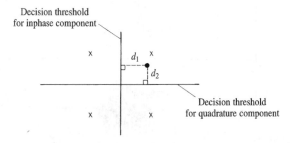

Figure 13.49 *Definition of QPSK received sample distances d_1 and d_2 from the decision thresholds ($\times$ represents transmitted constellation point and $\bullet$ a received sample).*

4. If *centre* point sampling is required then the distances, d_1 and d_2, from the decision thresholds of the (complex) sample at the *centre* of each received symbol is calculated, see Figure 13.49. (For MPSK and MQAM systems SPW assumes constellations which are regular, thus allowing all constellation points to be folded into the first quadrant before this calculation is carried out.) Although centre point sampling is typical, SPW allows the user to specify which sample within the symbol is to be used as the decision sample.
5. The probability of error for the particular sampling point selected, in the particular symbol being considered, is then calculated using d_1, d_2, σ and the error function.
6. The probability of error is averaged over many (N) symbols. (N is typically a few hundred but should be large enough to allow all sequences of symbols, possible in a time window equal to the duration of the impulse response estimated in 1, to occur at least once.)

Many other analysis routines are provided in SPW, e.g. the scatter plot of Figure 13.50 which shows the scatter of simulation points, at the Butterworth filter output (decision circuit input), about the nominal points of the QPSK constellation diagram. Figure 13.51 shows the eye diagram (across two symbols) of the received signal's inphase components at the Butterworth filter output, on a much expanded time scale compared to Figure 13.46.

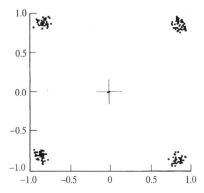

Figure 13.50 *Scatter of simulation QPSK constellation points at Butterworth filter output (decision circuit input) in Figure 13.45.*

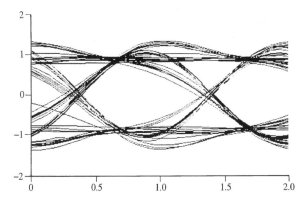

Figure 13.51 *Eye diagram for inphase component (I) of signal at Butterworth filter output.*

13.8 Summary

Simulation is now a vital part of the design process for all but the simplest of communication systems. It enables performance to be assessed in the presence of noise, interference and distortion, and allows alternative design approaches to be compared before an expensive construction phase is implemented. Since simulation is based on discrete samples of underlying continuous signals, it is usual for passband systems (with the exception of those which are non-linear and have long memory) to be simulated as equivalent (complex) baseband processes. The distortion introduced by the sampling and quantisation, required for simulation, needs to be carefully considered to ensure it does not significantly alter the simulation results. Noise modelling using random numbers or pseudo-random bit sequences must also be implemented carefully to ensure that simulated noise samples faithfully represent both the spectral and pdf properties of the actual noise present in the system.

Transformation between time and frequency domains during a simulation is frequently required. This is because some simulation processes are more easily (or more efficiently) implemented in one domain rather than the other, and also because some simulation results are more easily interpreted in one domain rather than the other. Digital signal processing algorithms (principally the FFT) are therefore used to translate between domains. An adequate understanding of effects such as smoothing, leakage, windowing and zero padding is required if these algorithms are to be used to best effect. It is also important that the output of DFT/DFS software is properly interpreted and processed if a numerically accurate representation of a power, or energy, spectral density is required.

Finally, in the assessment of most types of digital communications systems the principal objective measure of performance is BER. For systems which are linear, between the point at which noise is introduced and the point at which symbol decisions are made, quasi-analytic (QA) simulation is extremely efficient in terms of the computer resources required. QA simulation does require the noise pdf at the decision circuit input to be known, however. The most general method of estimating BER uses Monte Carlo simulation in which neither a linearity restriction (on the system through which the noise passes prior to detection), nor any a priori knowledge of the noise characteristics at the decision circuit input, is needed. Monte Carlo simulation can become very expensive in terms of computer power and/or run time, however, if accurate estimates of small error probabilities are required.

Part Three

Applications

Part Three shows how the principles described in Part Two are applied in a selection of fixed and mobile data applications for voice and video transmission.

The link budget analysis presented in Chapter 12 is extended here in Chapter 14 to less idealised fixed service, terrestrial and satellite, microwave communication systems, and includes important propagation effects such as rain fading, multipath fading and signal scintillation. The special problems posed by the exceptionally long range of satellite systems are also discussed as are frequency allocations, multiplexing and multiple accessing schemes. This also includes an introduction to optical fibre transmission systems.

Mobile, cellular and paging applications are described in Chapter 15 using FDMA, TDMA and CDMA techniques. This includes examples of current systems such as personal cordless, GSM 900, DCS 1800, cdmaOne and DECT. These systems all limit their transmissions to small geographical areas, or cells, permitting frequency reuse in close proximity, without incurring intolerable levels of interference. This maximises the number of active users per cell who can be accommodated per MHz of allocated bandwidth. This chapter also includes evolving standards for CDMA spread spectrum cellular radio in third generation and future satellite–mobile systems.

Finally, Chapter 16 discusses the specific requirements of digitisation, transmission and storage for video applications. It includes examples of digital TV development as well as low bit rate video compression coders using transform and model based coding techniques, as employed in JPEG, MPEG and other rapidly evolving video coding standards.

CHAPTER 14

Fixed-point microwave communications

14.1 Introduction

The two most important public service developments in fixed-point radio communications over the last 40 years are the installation of national microwave relay networks and international satellite communications. In both cases these systems originally used analogue modulation and frequency division multiplexing and represented simply an increase in capacity, improvement in quality, or increase in convenience, over the traditional wired PSTN. Their subsequent development, however, has been in favour of digital modulation and time division multiplexing. The proliferation of services which can be offered using digital transmission (via ISDN, Chapter 20) has been at least as important in this respect as improvements in digital technology.

14.2 Terrestrial microwave links

Many wideband point-to-point radio communications links employ microwave carriers in the 1 to 20 GHz frequency range. These are used principally by PTTs to carry national telephone and television signals. Their extensive use can be gauged by the number of antenna masts now seen carrying microwave dishes at the summits of hills located between densely populated urban areas. Figure 14.1 shows the main trunk routes of the UK microwave link network. The following points can be made about microwave links:

1. Microwave energy does not follow the curvature of the earth, or diffract easily over mountainous terrain, in the way that MW and LW transmissions do. Microwave transmissions are, therefore, restricted essentially to line-of-sight (LOS) links.

Figure 14.1 *The UK microwave communications wideband distribution network.*

2. Microwave transmissions are particularly well suited to point-to-point communications since narrow beam, high gain, antennas of reasonable size can be easily designed. (At 2 GHz the wavelength is 0.15 m and a 10-wavelength reflector, i.e. a 1.5 m dish, is still practical. Antenna gains are thus typically 30 to 50 dB.)

3. At about 1 GHz circuit design techniques change from using lumped to distributed elements. Above 20 GHz it becomes difficult and/or expensive to generate reasonable amounts of microwave power.

Antennas are located on high ground to avoid obstacles such as large buildings or hills, and repeaters are used every 40 to 50 km to compensate for path losses. (On a 6 GHz link with a hop distance of 40 km the free space path loss, given by equation (12.71(c)), is 140 dB, see Example 12.6. With transmit and receive antenna gains of 40 dB, however, the basic transmission loss, P_T/C, reduces to 60 dB. This provides sufficient received power for such links to operate effectively.)

Frequencies are allocated in the UK by the Radiocommunication Agency. The principal frequency bands in current use are near 2 GHz, 4 GHz, 6 GHz, 11 GHz and 18 GHz. There are also allocations near 22 GHz and 28 GHz. (The frequency bands listed here are those which are allocated mainly to common carriers. Other allocations for more general use do exist.)

14.2.1 Analogue systems

In the UK microwave links were widely installed during the 1960s for analogue FDM telephony. In these systems each allocated frequency band is subdivided into a number of (approximately) 30 MHz wide radio channels. Figure 14.2 shows how the 500 MHz band, allocated at 4 GHz, is divided into 16 separate channels with 29.65 MHz centre spacings. Each radio channel supports an FDM signal (made up of many individual SSB voice signals) which is frequency modulated onto a carrier. This could be called an SSB/FDM/FM system. Adjacent radio channels use orthogonal antenna polarisations, horizontal (H) and vertical (V), to reduce crosstalk. The allocated band is split into a low and high block containing eight radio channels each, Figure 14.2. One block is used for transmission and the other for reception.

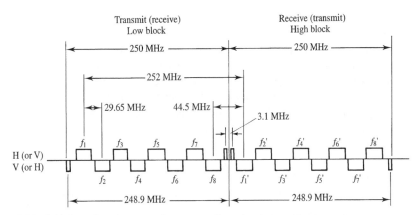

Figure 14.2 *Splitting of a microwave frequency allocation into radio channels.*

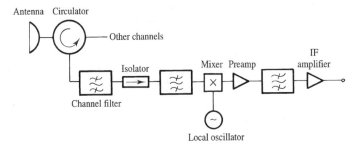

Figure 14.3 *Extraction (dropping) of a single radio channel in a microwave repeater.*

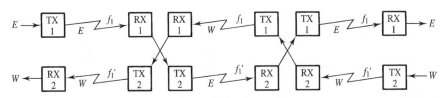

Figure 14.4 *Frequency allocations on adjacent repeaters.*

In the superheterodyne receiver a microwave channel filter, centred on the appropriate radio channel, extracts that channel from the block as shown in Figure 14.3. The signal is mixed down to an intermediate frequency (IF) for additional filtering and amplification. The microwave link can accommodate traffic simultaneously in all 16 channels. It is customary, however, not to transmit a given data signal on the same radio channel over successive hops. Thus in Figure 14.4, for example, f_1 is used for the first hop but f_1', Figure 14.2, is used on the second hop. Under anomalous propagation conditions, temperature inversions can cause ducting to occur with consequent low loss propagation over long distances, Figure 14.5. This may cause overreaching, with signals being received not only at the next repeater but also at the one after that. Moving the traffic to a different radio channel on the next hop thus ensures that interference arising via this mechanism is essentially uncorrelated with the received signal.

14.2.2 Digital systems

The first digital microwave (PSTN) links were installed in the UK in 1982 [Harrison]. They operated with a bit rate of 140 Mbit/s at a carrier frequency of 11 GHz using QPSK modulation. In more recent systems there has been a move towards 16- and 64-QAM (see Chapter 11). Figure 14.6 shows a block diagram of a typical microwave digital radio terminal. (Signal predistortion shown in the transmitter can be used to compensate for distortion introduced by the high power amplifier.)

The practical spectral efficiency of 4 to 5 bit/s/Hz, which 64-QAM systems offer, means that the 30 MHz channel can support a 140 Mbit/s multiplexed telephone traffic signal. Figure 14.7 is a schematic of a digital regenerative repeater, assuming DPSK modulation, for a single 30 MHz channel. Here the circulator and channel filter access the part of the microwave spectrum where the signal is located. With the ever increasing demand for high capacity transmission, digital systems have a major advantage over the older analogue systems in that they can operate satisfactorily at a much lower carrier-to-noise ratio, Figure 11.21. There is also strong interest in even higher level modulation schemes, such as 1024-QAM, to increase the capacity of the traditional 30 MHz channel still further.

Microwave radio links at 2 and 18 GHz are also being applied, at low modulation rates, in place of copper wire connections, in rural communities for implementing the local loop exchange connection [Harrison]. (These are configured as point-to-multipoint links, not unlike the multidrop systems of section 17.7.1. Furthermore the channel access scheme can use derivatives of the digital cellular systems of section 15.4.)

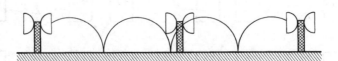

Figure 14.5 *Schematic illustration of ducting causing overreaching.*

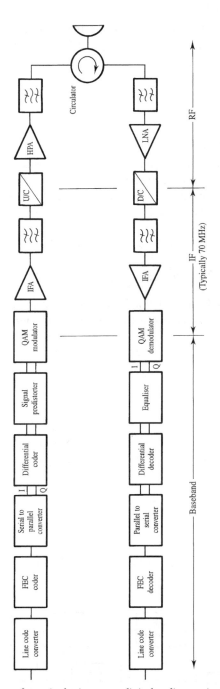

Figure 14.6 *Block diagram of a typical microwave digital radio terminal.*

14.2.3 LOS link design

The first order design problem for a microwave link, whether analogue or digital, is to ensure adequate clearance over the underlying terrain. Path clearance is affected by the following factors:

1. Antenna heights.
2. Terrain cover.
3. Terrain profile.
4. Earth curvature.
5. Tropospheric refraction.

The last of these, tropospheric refraction, occurs because the refractive index, n, of the troposphere depends on its temperature, T, pressure, P, and water vapour partial pressure, e. Since significant changes in P, T and e make only small differences to n, the tropospheric refractive index is usually characterised by the related quantity, refractivity, defined by $N = (n-1)10^6$.

A good model [ITU-R, Rec. 453], relating N to P, T and e, is:

$$N = 77.6 \frac{P}{T} + 3.73 \times 10^5 \frac{e}{T^2} \tag{14.1}$$

where P and e are in millibars and T is in K, as in Chapter 12. Although refractivity is dimensionless the term 'N units' is usually appended to its numerical value.

Equation (14.1) and the variation of P, T and e with altitude result in an approximately exponential decrease of N with height, h, i.e.:

$$N(h) = N_s \, e^{-h/h_0} \tag{14.2}$$

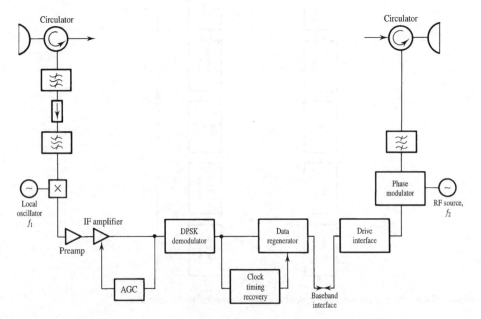

Figure 14.7 *Digital DPSK regenerative repeater for a single 30 MHz radio channel.*

where N_s is surface refractivity and h_0 is a scale height. ITU-R Rec. 369 defines a standard reference troposphere, Figure 14.8, as one for which $N_s = 315$ and $h_0 = 7.35$ km. Over the first kilometre of height this is usually approximated as a linear height dependence with refractivity gradient given by:

$$\left. \frac{dN}{dh} \right|_{first\ km} = N(1) - N(0)$$

$$= -40 \ (N \text{ units/km}) \tag{14.3}$$

The vertical gradient in refractive index causes microwave energy to propagate not in straight lines but along approximately circular arcs [Kerr] with radius of curvature, r, given by:

$$\frac{1}{r} = -\frac{1}{n} \frac{dn}{dh} \cos \alpha \tag{14.4}$$

Here α is the grazing angle of the ray to the local horizontal plane, Figure 14.9. For terrestrial LOS links α is small, i.e. $\cos \alpha \approx 1$, and for a standard troposphere with $n \approx 1$ and $dn/dh = -40 \times 10^{-6}$ the radius of curvature is 25000 km. Microwave energy under these conditions thus bends towards the earth's surface but with a radius of curvature much larger than that of the earth itself, Figure 14.10. (The earth's mean radius, a, may be taken to be 6371 km.) There are two popular ways in which ray curvature is accounted for in path profiling. One subtracts the curvature of the ray (equation (14.4)) from that of the earth giving (for $n = 1$ and $\alpha = 0$):

$$\frac{1}{a_e} = \frac{1}{a} + \frac{dn}{dh} \tag{14.5}$$

This effectively decreases the curvature of both ray and earth until the ray is straight, Figure 14.11, resulting in an effective earth radius, a_e. This may be called the straight ray model. Alternatively the negative of this transformation can be applied, i.e.:

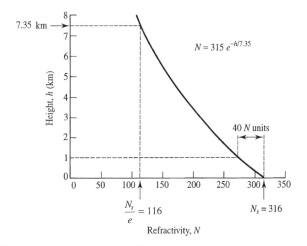

Figure 14.8 *ITU-R standard refractivity profile.*

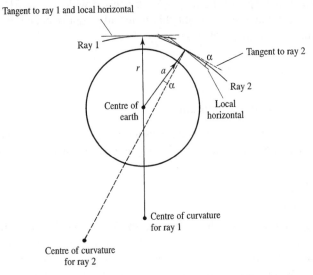

Figure 14.9 *Illustration of circular paths for rays in atmosphere with vertical n-gradient ($\alpha = 0$ for ray 1, $\alpha \neq 0$ for ray 2). Geometry distorted for clarity.*

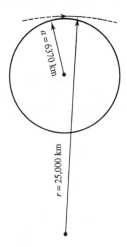

Figure 14.10 *Relative curvatures of earth's surface and ray path in a standard atmosphere.*

$$\frac{1}{r_e} = -\left(\frac{1}{a} + \frac{dn}{dh}\right) \tag{14.6}$$

which decreases the curvature of both ray and earth until the earth is flat, Figure 14.12, resulting in an effective radius of curvature, r_e, for the ray. This may be called the flat earth model. For the straight ray model the ratio of effective earth radius to actual earth radius is called the k factor, i.e.:

$$k = \frac{a_e}{a} \tag{14.7}$$

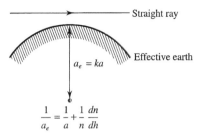

Figure 14.11 *Straight ray model. (Note n ≈ 1.)*

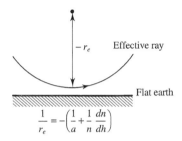

Figure 14.12 *Flat earth model (n≈1 and negative radius indicates ray is concave upwards).*

which, using equation (14.5) and the definition of refractivity, is given by:

$$k = \frac{1}{1 + a\ \dfrac{dN}{dh} \times 10^{-6}} \tag{14.8}$$

k factor thus represents an alternative way of expressing refractivity lapse rate, *dN/dh*. In the lowest kilometre of the standard ITU-R troposphere *k* = 4/3. Table 14.1 shows the relationship between *k* and *dN/dh* and Figure 14.13 shows the characteristic curvatures of rays under standard, sub-refracting, super-refracting and ducting conditions, each drawn (schematically) on a *k* = 4/3 earth profile. For standard meteorological conditions path profiles can be drawn on special *k* = 4/3 earth profile paper.

Table 14.1 *Equivalent values of refractivity lapse rate and k factor.*

$\dfrac{dN}{dh}$ (*N* units/km)	157	78	0	−40	−100	−157	−200	−300
k factor	$\dfrac{1}{2}$	$\dfrac{2}{3}$	1	$\dfrac{4}{3}$	2.75	∞	−3.65	−1.09

Sub-refraction   Super-refraction   Ducting

<----------------------->|<---------------->|<----------------------->

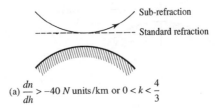

$$(a) \frac{dn}{dh} > -40 \ N \ \text{units/km or } 0 < k < \frac{4}{3}$$

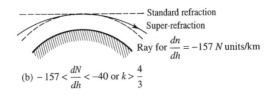

$$(b) -157 < \frac{dN}{dh} < -40 \text{ or } k > \frac{4}{3}$$

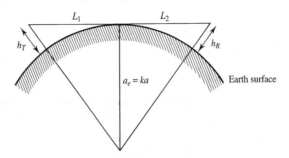

$$(c) \frac{dn}{dh} < -157 \text{ or } k < 0$$

Figure 14.13 *Characteristic ray trajectories drawn with respect to a k = 4/3 earth radius.*

EXAMPLE 14.1

Show that the line-of-sight range over a smooth spherical earth is given by $L = \sqrt{2ka} \ (\sqrt{h_T} + \sqrt{h_R})$ where k is a k-factor, a is earth radius, h_T is transmit antenna height and h_R is receive antenna height.

Consider the geometry shown in Figure 14.14.

$$L_1^2 = (h_T + a_e)^2 - a_e^2$$

Figure 14.14 *Geometry for maximum range LOS link over a smooth, spherical, earth.*

$$= h_T^2 + 2\,h_T\,a_e$$

Similarly:

$$L_2^2 = h_R^2 + 2\,h_R\,a_e$$

Since $h_T \ll a_e$ and $h_R \ll a_e$ then:

$$L_1^2 \approx 2\,h_T\,a_e$$

$$L_2^2 \approx 2\,h_R\,a_e$$

Therefore:

$$L = L_1 + L_2 = \sqrt{2\,a_e}\left(\sqrt{h_T} + \sqrt{h_R}\right)$$

Fresnel zones and path profiling

Fresnel zones are defined by the intersection of Fresnel ellipsoids with a plane perpendicular to the LOS path. The ellipsoids in turn are defined by the loci of points, Figure 14.15, which give an excess path length, ΔL, over the direct path of:

$$\Delta L = n\frac{\lambda}{2} \quad \text{(m)} \tag{14.9}$$

The first, second and third Fresnel radii are illustrated in Figure 14.16. For points not too near either end of the link (i.e. for $r_n \ll d_1$ and $r_n \ll d_2$) the nth Fresnel zone radius, r_n, is given by:

$$r_n \approx \sqrt{n\frac{\lambda\,d_1\,d_2}{d_1 + d_2}} \tag{14.10}$$

where d_1 and d_2 are distance from transmitter and receiver respectively. With personal computers it is now practical to plot ray paths for any k-factor on a 4/3 earth radius profile and, making due allowance for terrain cover, find the minimum clearance along the path as a fraction of r_1. (The required clearance of a link is often specified as a particular fraction of the first Fresnel zone radius under meteorological conditions corresponding to a particular k-factor.) This design exercise is called path profiling. The

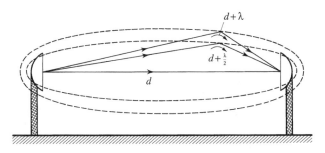

Figure 14.15 *Fresnel ellipsoids.*

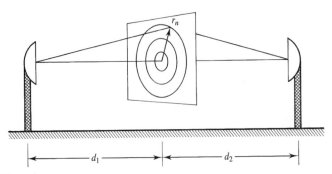

Figure 14.16 *Fresnel zones.*

principles of path profiling, however, are simple and best illustrated by describing the
process as it might be implemented manually:

1. A terrain profile (ignoring earth curvature) is plotted on Cartesian graph paper
   choosing any suitable vertical and horizontal scales.
2. For all points on the path profile likely to give rise to poor clearance (e.g. local
   maxima, path midpoint, etc.) the height of the earth's bulge, h_B, is calculated using:

$$h_B = \frac{d_1 d_2}{2a} \ \text{(m)} \tag{14.11}$$

   (h_B is the height of the, smooth, earth's surface above the straight line connecting the
   points at sea level below transmit and receive antennas.)
3. The effect of tropospheric refraction on clearance is calculated using h_B and k-factor
   to give:

$$h_{TR} = h_B(k^{-1} - 1) \ \text{(m)} \tag{14.12(a)}$$

4. The required Fresnel zone clearance (in metres) is calculated using:

$$h_{FZC} = f\, r_1 \ \text{(m)} \tag{14.12(b)}$$

   where r_1 is given by equation (14.10) and f is typically 1.0 for $k = 4/3$. (In the UK
   $f = 0.6$ for $k = 0.7$ is often used.)
5. An estimate is obtained for the terrain cover height, h_{TC}.
6. $h_{TC} + h_{TR} + h_B + h_{FZC}$ is plotted on the terrain path profile.
7. A straight line passing through the highest of the plotted points then allows
   appropriate antenna heights at each end of the link to be established, Figure 14.17.

There will usually be a trade-off between the height of transmit and receive antennas.
Normally these heights would be chosen to be as equal as possible in order to minimise
the overall cost of towers. If one or more points of strong reflection occur on the path,
however, the antenna heights may be varied to shift these points away from areas of high
reflectivity (such as regions of open water).

If adequate Fresnel zone clearance cannot be guaranteed under all conditions then
diffraction may occur leading to signal fading. If the diffracting obstacle can be modelled
as a knife edge, Figure 14.18, the diffraction loss can be found from Figure 14.19. The
parameter v in this figure is called the Fresnel diffraction parameter and is given by:

$$v = \sqrt{\frac{2}{\lambda} \frac{d_1 + d_2}{d_1 d_2}} \, h \tag{14.13}$$

where d_1 and d_2 locate the knife edge between transmitter and receiver, and h is the minimum clearance measured perpendicularly from the LOS path over the diffracting edge. (If the obstacle blocks the LOS path then h is negative.) If the obstacle cannot be modelled as a knife edge then the diffraction loss will generally be greater. A model for cylindrical edge diffraction, given in [Doughty and Maloney], takes the form of a knife edge loss plus a curvature loss and a correction factor.

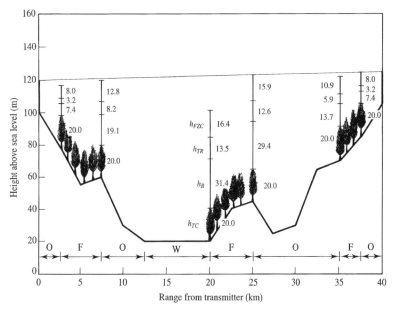

Figure 14.17 *Path profile for hypothetical 4 GHz LOS link designed for 0.6 Fresnel zone (FZ) clearance when k = 0.7 (O = open ground, F = forested region, W = water).*

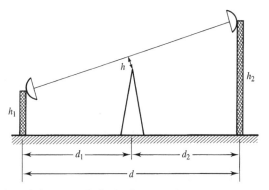

Figure 14.18 *Definition of clearance, h, for knife edge diffraction.*

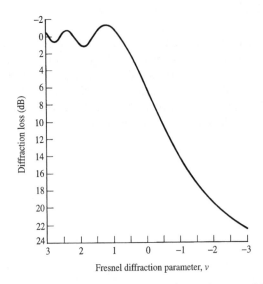

Figure 14.19 *Diffraction loss over a knife edge (negative loss indicates a diffraction gain).*

EXAMPLE 14.2

A 6 GHz microwave link operating over flat ground is 36 km long. At 15 km from the transmitter a long building with a pitched roof is oriented perpendicularly to the LOS path. The height of the building to the vertex of the roof is 30 m. How high must the transmitter and receiver antenna towers be if, under standard refraction conditions ($k = 4/3$): (i) first Fresnel zone clearance is to be maintained and (ii) diffraction fading is not to exceed 8 dB?

(i)

$$\lambda = \frac{3 \times 10^8}{6 \times 10^9} = 0.05 \text{ m}$$

$$h_B = \frac{d_1 \, d_2}{2a} = \frac{15000 \times 21000}{2 \times 6371000} = 24.72 \text{ m}$$

$$h_{TR} = h_B \, (k^{-1} - 1) = 24.72 \left(\frac{3}{4} - 1 \right) = -6.18 \text{ m}$$

$$h_{FZC} = f \, r_1 = 1.0 \sqrt{\frac{\lambda \, d_1 \, d_2}{d_1 + d_2}} = \sqrt{\frac{0.05 \times 15000 \times 21000}{36000}} = 20.92 \text{ m}$$

Required antenna heights are given by:

$$h_T = h_R = h_B + h_{TR} + h_{FZC} = 24.72 - 6.18 + 20.92$$

$$= 39.46 \text{ m}$$

(ii) If diffraction loss is not to exceed 8 dB then, from Figure 14.19, $v = -0.25$. Rearranging equation (14.13):

$$h = \frac{v}{\sqrt{\dfrac{2}{\lambda} \dfrac{d_1 + d_2}{d_1 \, d_2}}} = \frac{-0.25}{\sqrt{\dfrac{2}{0.05} \dfrac{36000}{15000 \times 21000}}} = -3.70 \text{ m}$$

Required antenna heights are now given by:

$$h_T = h_R = h_B + h_{TR} + h = 24.72 - 6.18 - 3.70 = 14.84 \text{ m}$$

14.2.4 Other propagation considerations for terrestrial links

Once an LOS link has been profiled a detailed link budget would normally be prepared. A first order free space signal budget is described in Chapter 12 (section 12.4.2). The actual transmitter power required, however, will be greater than that implied by this calculation since, even in the absence of diffraction, some allowance must be made for atmospheric absorption, signal fading and noise enhancements. For terrestrial microwave links with good clearance the principal mechanisms causing signal reductions are:

1. Background gaseous absorption.
2. Rain fading.
3. Multipath fading.

The principal sources of noise enhancement are:

1. Thermal radiation from rain.
2. Interference caused by precipitation scatter and ducting.
3. Crosstalk caused by cross-polarisation.

In addition to reducing the CNR (and/or carrier to interference ratio) signal fading processes also have the potential to cause distortion of the transmitted signal. For narrowband transmissions this will not usually result in a performance degradation beyond that due to the altered CNR. For sufficiently wideband digital transmissions, however, such distortion may result in extra BER degradation.

Background gaseous absorption

Figure 14.20 shows the nominal specific attenuation, $\gamma(f)$ (dB/km), of the atmosphere due to its gaseous constituents for a horizontal path at sea level. (Separate, and more detailed, curves for the dry air and water vapour constituents are given in [ITU-R, Rec. 676].) The rather peaky nature of this curve is the result of molecular resonance effects (principally due to water vapour and oxygen). For normal applications it is adequate simply to multiply $\gamma(f)$ by the path length, L, and include the resulting attenuation in the detailed link budget.

Rain fading

Significant rain intensity (measured in mm/h) occurs only for small percentages of time. The specific attenuation which is present during such rain is, however, large for frequencies above a few GHz. When fading of this sort occurs the BER can degrade due to a reduction in signal level and an increase in antenna noise temperature.

For signals with normal bandwidths the fading can be considered to be flat and the distortion negligible. In this case the attenuation exceeded for a particular percentage of time is all that is required. This can be estimated as follows:

1. The point rain rate, R_P (mm/h), exceeded for the required time percentage is found either from local meteorological records or by using an appropriate model.

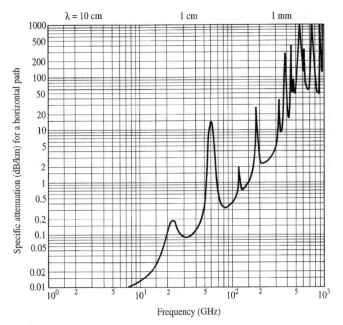

Figure 14.20 *Specific attenuation due to gaseous constituents for transmissions through a standard atmosphere (20°C, pressure one atmosphere, water vapour content 7.5 g/m³). (Source: ITU-R Handbook of Radiometeorology, 1996, with permission of ITU.)*

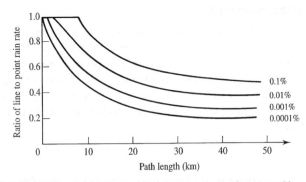

Figure 14.21 *Relationship between point and line rain rates as a function of hop length and percentage time point rain rate is exceeded (source: Hall and Barclay, 1989, reproduced with permission of Peter Peregrinus).*

2. The equivalent uniform (or line) rain rate, R_L, is found using curves such as those in Figure 14.21. This accounts for the non-uniform spatial distribution of rain along the path.

3. The specific attenuation, γ_R, for the line rain rate of interest may be estimated using Figure 14.22, or, more accurately, using a formula of the form:

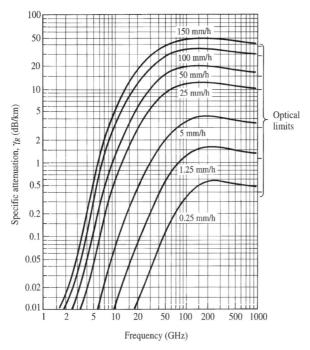

Figure 14.22 *Specific attenuation due to rain (curves derived on the basis of spherical raindrops). (Source: ITU-R Handbook of Radiometeorology, 1996, reproduced with the permission of ITU.)*

$$\gamma_R = k\, R_L^{\alpha} \quad \text{(dB/km)} \tag{14.14}$$

where the regression coefficients k and α are given in [ITU-R, Rec. 838] for a particular polarisation and frequency of interest.

4. The total path attenuation, A (dB), exceeded for the required percentage of time is then found using:

$$A = \gamma_R L \quad \text{(dB)} \tag{14.15}$$

Multipath fading on LOS links

Although broadband measurements have shown that several (identifiably discrete) propagation paths can occur on LOS links the consensus at present is that, in practice, the channel can be adequately modelled with only three paths [Rummler, 1979]. The impulse response of such a three-ray model is (to within a multiplicative constant):

$$h(t) = \delta(t) + \alpha\delta(t - T_1) + \beta\delta(t - T_2) \tag{14.16}$$

where T_1 and T_2 define the relative delays between the rays, and α and β define the relative strengths of the rays. The frequency response, $H(f)$, corresponding to equation (14.16) is:

$$H(f) = 1 + \alpha e^{-j2\pi f T_1} + \beta e^{-j2\pi f T_2} \tag{14.17}$$

If $T_1 \ll 1/B$, where B is the bandwidth of the transmitted signal, then fT_1 is approximately constant over the range $f_c \pm B/2$ (where f_c is the carrier frequency). Equation (14.17) can be written as:

$$H(f) = ae^{-j\theta}\left[1 + \frac{\beta}{a} e^{-j(2\pi fT_2 - \theta)}\right]$$

(14.18)

where:

$$ae^{-j\theta} = 1 + \alpha e^{-j2\pi fT_1}$$

(14.19(a))

$$a = \sqrt{1 + \alpha^2 + 2\alpha \cos(2\pi fT_1)}$$

(14.19(b))

$$\theta = \tan^{-1}\left(\frac{\alpha \sin 2\pi fT_1}{1 + \alpha \cos 2\pi fT_1}\right)$$

(14.19(c))

The factor $e^{-j\theta}$ in equation (14.18) represents an overall phase shift caused by the first delayed ray which can be ignored[1]. Furthermore, since a notch occurs in the channel's amplitude response at:

$$f(= f_o) = (\theta \pm \pi)/2\pi T_2 \quad \text{(Hz)}$$

(14.20)

then equation (14.18) can be rewritten as:

$$H(f) = a\left[1 - be^{-j2\pi(f-f_o)T_2}\right]$$

(14.21)

where $b = \beta/a$. If $b < 1$ then equation (14.21) represents a minimum phase frequency response. If unity and b in equation (14.21) are interchanged then the amplitude response, $|H(f)|$, remains unchanged but the phase response becomes non-minimum. Apart from an extra overall phase shift of $-2\pi T_2 f + 2\pi T_2 f_o + \pi$ (the first term representing pure delay, the second term representing intercept distortion and the third term representing signal inversion) the non-minimum phase frequency response can be modelled by changing the sign of the exponent in equation (14.21) which can then be written as:

$$H(f) = a\left[1 - be^{\pm j2\pi(f - f_o)T_2}\right]$$

(14.22)

to represent both minimum and non-minimum phase conditions [CCIR, Report 718]. Equation (14.22) has four free parameters (a, b, f_o and T_2) which is excessive, in the sense that if $T_2 < (6B)^{-1}$ then normal channel measurements are not accurate enough to determine all four parameters uniquely. One parameter can therefore be fixed without significantly degrading the formula's capacity to match measured channel frequency responses. The parameter normally fixed is T_2 and the value chosen for it is often set using the rule $T_2 = (6B)^{-1}$. Using this rule, joint statistics of the quantities:

$$A = 10\log_{10}\left[a^2(1 + b^2)\right]$$

(14.23(a))

---

[1]   A frequency response with $\phi(f) = \theta$, where θ is a non-zero constant, has zero delay distortion (since $d\phi/df = 0$) but non-zero intercept distortion. Intercept distortion, however, results only in a change in the 'phase' relationship between a wave packet carrier and its envelope, the shape of the envelope remaining unchanged. Providing such distortion changes slowly compared to the time interval between phase training sequences it will have no significant effect on the performance of a PSK communications system.

$$B = 10\log_{10}(2a^2b) \qquad\qquad (14.23(\text{b}))$$

have been determined experimentally for several specific links [CCIR, Report 338]. *A* and *B* appear to be well described by a bivariate Gaussian random variable. The mean, variance and correlation of this distribution for three different links are shown in Table 14.2.

Table 14.2 *Measured statistics of multipath channel parameters (after CCIR, Report 338).*

Path length	Frequency	Bandwidth	$\bar{A}$	$\bar{B}$	σ_A	σ_B	ρ
37 km	11 GHz	55 MHz	−7.25	−5.5	6.5	6.5	0.45
50 km	11 GHz	55 MHz	−8.25	−3.0	9.0	8.5	0.75
42 km	6 GHz	25 MHz	−24.00	−14.5	7.5	7.5	0

Equations (14.23(a) and (b)) can be inverted to allow the joint distribution of *a* and *b* to be derived.

Multipath fading is a potentially severe problem for wideband digital links. Full transversal filter equalisation, which can compensate for both the amplitude and the phase distortions in the transmission medium, is therefore generally desirable. Current digital repeater and receiver equipment has considerable sophistication, often incorporating equalisers which remove distortion arising not only from propagation effects but also from other sources such as non-ideal filters. ([ITU-R, Rec. 530] details methods for predicting fading due to multipath effects in combination with other clear air mechanisms.)

Mechanisms of noise enhancement

The presence of loss on a propagation path will increase the noise temperature of a receiving antenna due to thermal radiation. Rain is the most variable source of such loss along most microwave paths. During a severe fade the aperture temperature (see Chapter 12) of the antenna, T_A, will approach the physical temperature of the rain producing the

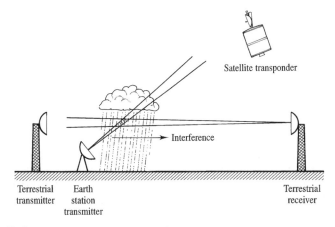

Figure 14.23 *Hydrometeor scatter causing interference between co-frequency systems.*

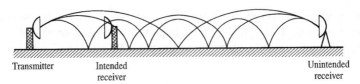

Figure 14.24 *Interference caused by ducting.*

loss. A simple model relating T_A to fade depth is:

$$T_A = T_M(1 - \alpha) \tag{14.24}$$

where α is the fade depth expressed as a fraction of unfaded power (i.e. $\alpha = 10^{-\text{atten(dB)/10}}$) and T_M is the absorption temperature of the medium. T_M is not identical to the medium's physical temperature since, in addition to direct thermal radiation, it accounts for scattering and other effects. T_M has been related empirically to surface air temperature, T_S, by [Freeman]:

$$T_M = 1.12\, T_S - 50 \quad \text{(K)} \tag{14.25}$$

where T_S is expressed in K. Equations (14.24) and (14.25) are only appropriate, of course, for values of attenuation which result in aperture temperatures in excess of the clear sky values indicated by Figure 12.25 (i.e. approximately 100 K for terrestrial links between 1 and 10 GHz). Clear sky aperture temperature increases rapidly with frequency above 10 GHz and, for terrestrial links, can be assumed to be 290 K at 20 GHz and above for most purposes (see later, Figure 14.30).

A quite different mechanism of noise enhancement also occurs due to rain, namely precipitation scatter. Here the existence of a common volume between the transmitting antenna of one system and the receiving antenna of a nominally independent system has the potential to couple energy between the two, Figure 14.23. Such coupled energy represents interference but it is thought that its effects are conservatively modelled by an equal amount of thermal noise power. Meteorological ducting conditions can also cause anomalous, long distance, propagation of microwave signals resulting in interference or crosstalk between independent systems, Figure 14.24. Detailed modelling of precipitation scatter, ducting and methods to reduce the interference caused by them has recently been the subject of intensive study [COST 210].

EXAMPLE 14.3

A 7 GHz terrestrial LOS link is 40 km long and operates with good ground clearance (several Fresnel zones) in a location which experiences a rain rate of 25 mm/h or greater for 0.01% of time. The link has a bandwidth of 1.0 MHz, antenna gains of 30.0 dB and an overall receiver noise figure of 5.0 dB. Assuming the ratio of line to point rain rate, R_L/R_p, is well modelled by Figure 14.21, for this location, estimate the transmitter power required to ensure a CNR of 30.0 dB is achieved or exceeded for 99.99% of time.

From Figure 14.21:

$$R_{L,0.01} = 0.4 \, R_{p,0.01} = 0.4 \times 25 = 10 \text{ mm/h}$$

Using Figure 14.22, specific attenuation, $\gamma_R = 0.2$ dB/km, i.e.:

$$A_{R,0.01} = \gamma_R \, L = 0.2 \times 40 = 8.0 \text{ dB}$$

Assuming a surface temperature of 290 K, equation (14.25) gives:

$$T_M = 1.12 \, T_S - 50 = 1.12 \times 290 - 50 = 275 \text{ K}$$

From equation (14.24):

$$T_A = T_M \left(1 - 10^{-\frac{A}{10}} \right) = 275 \left(1 - 10^{-\frac{8}{10}} \right) = 231 \text{ K}$$

Equivalent noise temperature of receiver:

$$T_e = (f - 1) \, 290 = \left(10^{\frac{5}{10}} - 1 \right) 290 = 627 \text{ K}$$

Total system noise temperature:

$$T_{syst} = T_A + T_e = 231 + 627 = 858 \text{ K}$$

Noise power:

$$N = kTB = 1.38 \times 10^{-23} \times 858 \times 1 \times 10^6 = 1.18 \times 10^{-14} \text{ W} = -139.3 \text{ dBW}$$

Received carrier power:

$$\begin{aligned}
C &= P_T + G_T - \text{FSPL} + G_R - A_R \\[4pt]
&= P_T + 30.0 - 20 \log_{10} \left(\frac{4\pi \, 40 \times 10^3}{0.0429} \right) + 30.0 - 8.0 \\[4pt]
&= P_T - 89.4 \text{ dBW}
\end{aligned}$$

$$\frac{C}{N} = P_T - 89.4 - (-139.3) = P_T + 49.9 \text{ dB}$$

$$P_T = \frac{C}{N} - 49.9 = 30.0 - 49.9 = -19.9 \text{ dBW or } 10.1 \text{ dBm}$$

(Note that the calculation shown above has been designed to illustrate the application of the preceding material and probably contains spurious precision. In particular a simpler estimate for T_A of 290 K results in an increase in P_T of only 0.3 dB.)

Cross-polarisation and frequency reuse

It is possible to double the capacity of a microwave link by using orthogonal polarisations for independent co-frequency channels. For QPSK systems this combines with the advantage of orthogonal inphase and quadrature signalling to give a four-fold increase in transmission capacity over a simple BPSK, single polarisation, system. Unusual propagation conditions along the radio path can give rise to polarisation changes (called cross-polarisation), which results in potential crosstalk at the receiver. The use of corrugated horns and scalar feeds in the design of microwave antennas reduces antenna induced cross-polarisation during refractive bending or multipath conditions. Similarly the use of vertical and horizontal linear polarisations minimises rain induced cross-polarisation which occurs when the angle between the symmetry axis of falling rain drops and the electric field vector of the signal is other than 0° or 90°.

14.3 Fixed-point satellite communications

The use of satellites is one of the three most important developments in telecommunications over the past 40 years. (The other two are cellular radio and the use of optical fibres.) Geostationary satellites, which are essentially motionless with respect to points on the earth's surface and which first made satellite communications commercially feasible, were proposed by the scientist and science fiction writer Arthur C. Clarke. The geostationary orbit lies in the equatorial plane of the earth, is circular and has the same sense of rotation as the earth, Figure 14.25. Its orbital radius is 42164 km and since the earth's mean equatorial radius is 6378 km its altitude is 35786 km. (For simple calculations of satellite range from a given earth station, the earth is assumed to be spherical with radius 6371 km.) There are other classes of satellite orbit which have advantages over the geostationary orbit for certain applications. These include highly inclined highly elliptical (HIHE) orbits, polar orbits and low earth orbits (LEOs), Figure 14.25. For fixed-point communications the geostationary orbit is the most commercially important, for the following reasons:

1. Its high altitude means that a single satellite is visible from a large fraction of the earth's surface (42% for elevation angles > 0° and 38% for elevation angles > 5°). Figure 14.26 shows the coverage area as a function of elevation angle for a geostationary satellite with a global beam antenna. (Elevation angles < 10° are not recommended, and angles < 5° are not used, because of the severe scintillation and fading of the signal, and high antenna noise temperature, which occur due to the

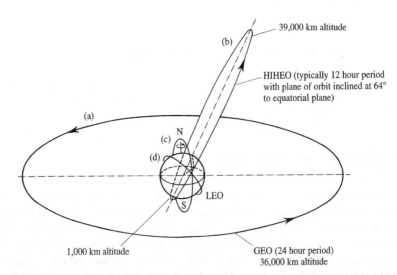

Figure 14.25 *Selection of especially useful satellite orbits: (a) geostationary (GEO); (b) highly inclined highly elliptical (HIHEO); (c) polar orbit; and (d) low earth (LEO).*

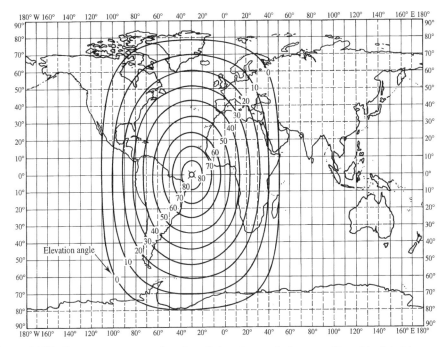

Figure 14.26 *Coverage areas as a function of elevation angle for a satellite with global beam antenna (from CCIR Handbook, 1988, reproduced with the permission of ITU).*

large thickness of atmosphere traversed by the propagation path.)

2. No tracking of the satellite by earth station antennas is necessary.
3. No handover from one satellite to another is necessary since the satellite never sets.
4. Three satellites give almost global coverage, Figure 14.27. (The exception is the polar regions with latitudes > 81° for elevation angles > 0° and latitudes > 77° for elevation angles > 5°.)
5. No Doppler shifts occur in the received carrier.

The following advantages apply to geostationary satellites but may also apply, to a greater or lesser extent, to some communication satellites in non-geostationary orbits:

1. The communications channel can be either broadcast or point-to-point.
2. New communication network connections can be made simply by pointing an antenna at the satellite. (For non-geostationary satellites this is not entirely trivial since tracking and/or handover are usually necessary.)
3. The cost of transmission is independent of distance.
4. Wide bandwidths are available, limited at present only by the speed of the transponder electronics and receiver noise performance.

Despite their very significant advantages, geostationary satellites do suffer some disadvantages. These include:

1. Polar regions are not covered (i.e. latitudes > 77° for elevation angles > 5°).
2. High altitude means large FSPL (typically 200 dB).

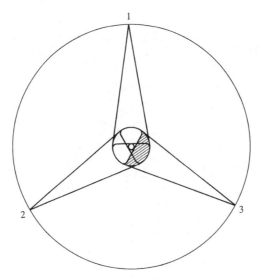

Figure 14.27 *Global coverage (excepting polar regions) from three geostationary satellites. (Approximately to scale, innermost circle represents 81° parallel.)*

3. High altitude results in long propagation delays (approximately 1/8 s for uplink and 1/8 s for downlink).

The last disadvantage means that inadequately suppressed echoes from subscriber receiving equipment arrive at the transmitting subscriber 0.5 s after transmission. Both the 0.25 s delay between transmission and reception and the 0.5 s delay between transmission and echo can be disturbing to telephone users. This means echo suppression or cancellation equipment, which may use techniques [Mulgrew *et al.*] not dissimilar to the equalisers of section 8.5, is almost always required.

14.3.1 Satellite frequency bands and orbital spacing

Figure 14.28 shows the principal European frequency bands allocated to fixed-point satellite services. The 6/4 GHz (C-band) allocation is now fairly congested and new systems are being implemented at 14/11 GHz (Ku-band); 30/20 GHz (Ka-band) systems are currently being investigated. The frequency allocation at 12 GHz is mainly for direct broadcast satellites (DBSs). Intersatellite crosslinks use the higher frequencies as here there is no atmospheric attenuation. The higher of the two frequencies allocated for a satellite communications system is invariably the uplink frequency. This is because the satellite has limited antenna size and a high antenna noise temperature (typically 290 K). The gain of the satellite receiving antenna (and therefore the satellite G/T) is maximised by using the higher frequency on the uplink.

The reason why two frequencies are necessary at all (one for the uplink and one for the downlink) is that the isolation between the satellite transmit and receive antennas is finite. Since the satellite transponder has enormous gain there would be the possibility of positive feedback and oscillation if a frequency offset were not introduced.

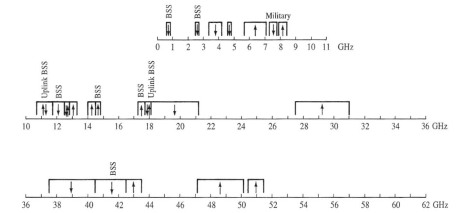

Figure 14.28 *Approximate uplink (↑) and downlink (↓) allocations for region 1 (Europe, Africa, former USSR, Mongolia) fixed satellite, and broadcast satellite (BSS), services.*

Although the circumference of a circle of radius 42000 km is large, the number of satellites which can be accommodated in the geostationary orbit is limited by the need to illuminate only one satellite when transmitting signals from a given earth station. If other satellites are illuminated then interference may result. For practical antenna sizes 4° spacing is required between satellites in the 6/4 GHz bands. Since narrower beamwidths are achievable in the 14/11 GHz band, 3° spacing is permissible here and in the 30/20 GHz band spacing can approach 1°.

14.3.2 Earth station look angles and satellite range

Figure 14.29 shows the geometry of an earth station (E) and geostationary satellite (S). Some careful trigonometry shows that the earth station antenna elevation angle, α, the azimuth angle, β, and the satellite range from the earth station, R_{ES}, are given by:

$$\alpha = \tan^{-1}\left[(\cos\gamma - 0.15127)/\sin\gamma\right] \tag{14.26}$$

$$\beta = \pm\cos^{-1}\left[-\tan\theta_E/\tan\gamma\right] \tag{14.27}$$

$$R_{ES} = 23.188 \times 10^6 \sqrt{3.381 - \cos\gamma} \ (\text{m}) \tag{14.28}$$

where γ, the angle subtended by the satellite and earth station at the centre of the earth, is given by:

$$\gamma = \cos^{-1}\left[\cos\theta_E\cos(\phi_E - \phi_S)\right] \tag{14.29}$$

θ_E is the earth station latitude (positive north, negative south), ϕ_E and ϕ_S are the earth station and satellite longitude respectively (positive east, negative west).

Notice that azimuth, β, is defined clockwise (eastwards) from north. Negative β therefore indicates an angle anticlockwise (westwards) from north. The negative sign is taken in equation (14.27) if the earth station is east of the satellite and the positive sign is taken otherwise.

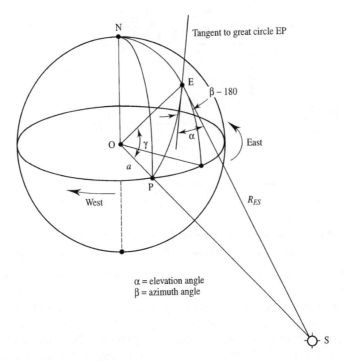

Figure 14.29 *Geostationary satellite geometry.*

EXAMPLE 14.4

Find the look angles and range to a geostationary, Indian Ocean, satellite located at 60.0° E as seen from Edinburgh in the UK (55.95° N, 3.20° W).

$$\gamma = \cos^{-1} [\cos \theta_E \cos (\phi_E - \phi_S)]$$

$$= \cos^{-1} [\cos (55.95°) \cos (-3.20° - 60.0°)]$$

$$= 75.38° \ (= 1.316 \text{ rad})$$

$$\alpha = \tan^{-1} [(\cos \gamma - 0.15127)/\sin \gamma]$$

$$= \tan^{-1} [(\cos (1.316) - 0.15127)/\sin (1.316)]$$

$$= 0.1038 \text{ rad} \ (= 5.95°)$$

$$\beta = \pm \cos^{-1} [- \tan \theta_E/\tan \gamma]$$

$$= \pm \cos^{-1} [- \tan (55.95°)/\tan (75.38°)]$$

$$= \pm 112.7°$$

Since the earth station is west of the satellite the upper (positive) sign is taken, i.e. the satellite azimuth is +112.7° (eastwards from north). Now from equation (14.28):

$$R_{ES} = 23.188 \times 10^6 \sqrt{3.381 - \cos \gamma}$$

$$= 23.\,188 \times 10^6 \, \sqrt{3.\,381 - \cos{(1.\,316)}}$$

$$= 4102 \times 10^4 \, \text{m}$$

$$= 41020 \, \text{km}$$

14.3.3 Satellite link budgets

On the uplink the satellite antenna, which looks at the warm earth, has a noise temperature of approximately 290 K. The noise temperature of the earth station antenna, which looks at the sky, is usually in the range 5 to 100 K for frequencies between 1 and 10 GHz, Figure 12.25. For frequencies above 10 GHz resonance effects of water vapour and oxygen molecules, at 22 GHz and 60 GHz respectively, become important, Figure 14.30. For wideband multiplex telephony, the front end of the earth station receiver is also cooled so that its equivalent noise temperature, T_e, section 12.3.1, is very much smaller than 290 K. This therefore achieves a noise floor that is very much lower than −174 dBm/Hz.

The gain, G, of the cooled low noise amplifier boosts the received signal so that the following amplifiers in the receiver can operate at room temperature. Traditional operating frequencies for satellite communications systems were limited at the low end to >1 GHz by galactic noise and at the high end to <15 GHz by atmospheric thermal noise and rain attenuation, Figure 12.25. However, in recent years a more detailed physical and statistical understanding of rain fading, and other hydrometeor effects, has made operation in higher frequency bands practical.

The satellite transponder is the critical component in a satellite link as its transmitter is invariably power limited by the onboard power supply (i.e. solar cell area and battery capacity). The downlink usually has the worst power budget and this often constrains performance. EIRP is also dependent on satellite antenna design and in particular the

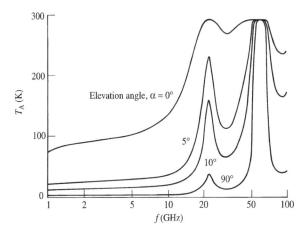

Figure 14.30 *Antenna aperture temperature, T_A, in clear air (pressure one atmosphere, surface temperature 20°C, surface water vapour concentration 10 g/m³). (Source: ITU-R Handbook of Radiometeorology, 1996, reproduced with the permission of the ITU.)*

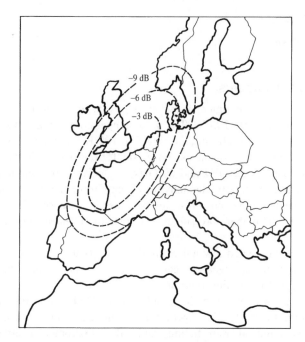

Figure 14.31 *Contours of EIRP with respect to EIRP on antenna boresight.*

earth 'footprint'. Spot beam antennas covering only a small part of the earth, e.g. Figure 14.31, have higher gain giving EIRP values of 30 to 40 dBW or greater.

In addition to being power limited the satellite transponder must be as small and light as possible due to the large launching costs, per kg of weight and m³ of space. This means that the transponder's high power amplifier (HPA), Figure 14.32, must be operated at as high an output power as possible to maintain adequate downlink CNR. As a consequence the transponder is operated in its non-linear region near saturation, resulting in amplitude to amplitude (AM/AM) and amplitude to phase (AM/PM) conversion, Figure 14.33. Intermodulation products (IPs), arising due to mixing of nominally independent signals, simultaneously present in the transponder, can be severe, effectively reducing the overall CNR. Some *back-off* from the transponder saturating input and output power is therefore necessary. (For digital satellite systems the IP problem and resulting need for back-off is usually less serious than for analogue systems, section 14.3.5.) Input and output back-off (BO_i and BO_o respectively) are shown in Figure 14.33. Typical values of back-off are a few dB, BO_i being somewhat greater than BO_o. The transmitted power for the satellite downlink and received power for the satellite uplink during clear sky conditions are therefore given respectively by:

$$P_T = P_{o\,sat} - BO_o \tag{14.30}$$

$$C = P_{i\,sat} - BO_i \tag{14.31}$$

where $P_{o\,sat}$ and $P_{i\,sat}$ are the saturated transponder output and input powers. The uplink CNR can therefore be calculated using:

$$\left(\frac{C}{N} \right)_u = W_s - \text{BO}_i + 10 \log_{10} \left(\frac{\lambda^2}{4\pi} \right) + \left(\frac{G}{T} \right)_s + 228.6 - 10 \log_{10} B \qquad (14.32(\text{a}))$$

where W_s is the power density (dBW/m²) at the satellite receiving antenna required to saturate the transponder, $10 \log_{10}(\lambda^2/4\pi)$ is the effective area of an isotrope (dBm²), $(G/T)_s$ is the satellite G/T (dB/K), 228.6 (= $-10 \log_{10} k$) is Boltzmann's constant (dBW/HzK), and $10 \log_{10} B$ is bandwidth (dB Hz). Note that the transmitting earth station EIRP, EIRP$_e$, and the received power density at the satellite are related by:

$$W_s - \text{BO}_i = \text{EIRP}_e - \text{spreading loss} - L_{Au} \quad (\text{dBW/m}^2) \qquad (14.32(\text{b}))$$

where L_{Au} is uplink atmospheric attenuation. The equivalent formula, to equation (14.32(a)), for the downlink is:

$$\left(\frac{C}{N} \right)_d = \text{EIRP}_s - \text{BO}_o + \left(\frac{G}{T} \right)_e - \text{FSPL}_d + 228.6 - 10 \log_{10} B - L_{Fd} - L_{Ad} \qquad (14.33)$$

where EIRP$_s$ is the saturated satellite EIRP, $(G/T)_e$ is the earth station G/T, L_{Fd}

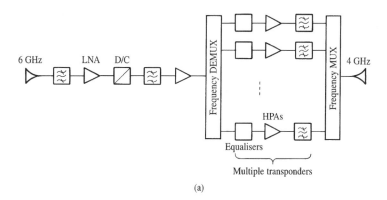

(a)

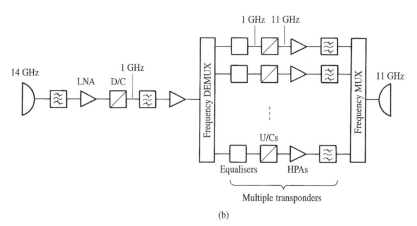

(b)

Figure 14.32 *Simplified block diagram of satellite transponders: (a) single conversion C-band; (b) double conversion Ku-band (redundancy not shown).*

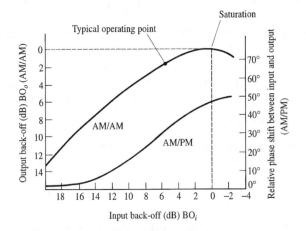

Figure 14.33 *Amplitude and phase characteristic for typical satellite transponder TWT amplifier.*

represents the downlink fixed losses (such as earth station transmission line loss) and L_{Ad} is the downlink atmospheric attenuation.

In satellite multiplex telephony systems, large earth station antennas with kW transmitters give very large EIRP$_e$ (typically 90 dBW) and the uplink exhibits good noise, or E_b/N_0, performance. The transmitter power on the satellite is typically restricted by battery and solar cell capacity to 10 to 100 W (and hence EIRP$_s$ is typically restricted to 30 to 50 dBW). The downlink often, therefore, limits overall CNR and BER performance.

A typical, clear sky, link budget for a 6/4 GHz satellite communications system is shown in Table 14.3.

As illustrated schematically in Figure 14.34 the received downlink carrier power, C_d, for a transparent transponder, is given by:

$$C_d = C_u G L_d \quad \text{(W)} \tag{14.34}$$

where C_u is the *received* uplink power, G is the operating gain of the transponder and L_d represents *all* downlink losses.

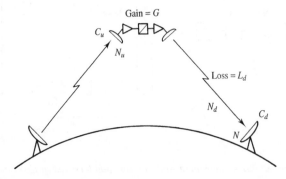

Figure 14.34 *CNRs on uplink and downlink.*

Table 14.3 *Typical 4/6 GHz satellite link budget.*

Uplink (6 GHz)		
Saturation flux density		-72.2 dBW/m^2
Input back-off, BO$_i$		-5.8 dB
Satellite antenna gain, G_R	23.1 dB	
Satellite system noise temperature, T_{syst}	27.6 dBK	
Satellite G/T	-4.5 dB/K	-4.5 dB/K
Effective area of isotrope		-37.0 dB m^2
Minus Boltzmann's constant		228.6 dBW/HzK
Minus transponder bandwidth		-75.6 dB Hz
Resulting clear sky CNR$_u$		33.5 dB

Downlink (4 GHz)		
Saturated transponder output power, P_o	7.0 dBW	
Satellite antenna gain, G_T	22.5 dB	
EIRP$_s$	29.5 dBW	29.5 dBW
Output back-off, BO$_o$		-3.2 dB
FSPL, $20\log_{10}(4\pi R/\lambda)$		-196.1 dB
Earth station antenna gain, G_R	58.3 dB	
Clear sky earth station noise temperature, T_{syst}	18.8 dBK	
Clear sky G/T	39.5 dB/K	39.5 dB/K
Minus Boltzmann's constant		228.6 dBW/HzK
Minus transponder bandwidth		-75.6 dB Hz
Atmospheric attenuation, L_{Ad}		-0.8 dB
Fixed losses, L_{Fd}		-2.0 dB
Resulting clear sky CNR$_d$		19.9 dB

Furthermore, the total received downlink noise power, N, can be expressed by:

$$N = N_u G\, L_d + N_d \quad \text{(W)} \tag{14.35}$$

where N_u is the noise power contributed by the uplink and N_d is the noise power contributed by the downlink. Thus:

$$\frac{N}{C_d} = \frac{N_u G\, L_d + N_d}{C_u G\, L_d}$$

$$= \frac{N_u}{C_u} + \frac{N_d}{C_d} \tag{14.36}$$

The overall CNR for the satellite link can therefore be written as:

$$\frac{C}{N} = \frac{1}{\left(\dfrac{C}{N}\right)_u^{-1} + \left(\dfrac{C}{N}\right)_d^{-1}} \tag{14.37}$$

(dropping the subscript d from the received signal power, C, to conform to convention).

More accurate calculations of satellite systems performance must account for the effects of intermodulation products and interference on uplink and/or downlink. Since noise, intermodulation and interference processes are independent, equation (14.37) can be extended as follows:

$$\frac{C}{N} = \frac{1}{\left(\dfrac{C}{N}\right)_u^{-1} + \left(\dfrac{C}{N}\right)_d^{-1} + \left(\dfrac{C}{N}\right)_{IP}^{-1} + \left(\dfrac{C}{I}\right)_u^{-1} + \left(\dfrac{C}{I}\right)_d^{-1}} \qquad (14.38)$$

where $(C/N)_{IP}$ is the carrier to intermodulation noise ratio and C/I is the carrier to interference ratio. For digital satellite systems (section 14.3.6) in which only a single carrier is present in the transponder at any given time, then intermodulation noise is absent and $(C/N)_{IP}^{-1}$ is zero. For the example link budget in Table 14.3 the overall clear sky CNR, assuming no intermodulation products and neglecting interference, would be:

$$\frac{C}{N} = \frac{1}{(\text{antilog}_{10}\ 3.35)^{-1} + (\text{antilog}_{10}\ 1.99)^{-1}} \qquad (14.39)$$

$$= \frac{1}{(2239)^{-1} + (97.72)^{-1}} = \frac{1}{0.045 \times 10^{-2} + 1.023 \times 10^{-2}}$$

$$= \frac{1}{1.068 \times 10^{-2}}$$

$$= 93.63 = 19.7 \ \text{dB}$$

Note in most cases the CNR performance of the system is dominated by the downlink.

The clear sky CNR calculated above allows only for gaseous background attenuation in the term L_A. In practice a satellite system link budget must also account for fading of the signal and enhancement of the noise (both due, mainly, to the sporadic presence of rain along the propagation path).

Rain induced specific attenuation (in dB/km) varies with frequency, Figure 14.22. The effective length of the propagation path subject to a given rain event depends on elevation angle and climatic factors. *Typical* fade margins included in link budgets to account for 99.99% of meteorological conditions are given in Table 14.4 for the different satellite frequency bands. The detailed calculation of gaseous background attenuation and rain margins is described in section 14.3.4.

Table 14.4 *Typical values for fade margin in different frequency bands.*

Elevation angle	C	Ku	Ka
10°	2.0 dB	8 dB	15 dB
30°	1.0 dB	6 dB	10 dB
90°	0.7 dB	5 dB	8 dB

EXAMPLE 14.5

An 11.7 GHz satellite downlink operates from geosynchronous orbit with 25 W of transmitter output power connected to a 20 dB gain antenna with 2 dB feeder losses. The earth station is at a range of 38000 km from the satellite and uses a 15 m diameter receive antenna, with 55%

efficiency, feeding a low noise (cooled) amplifier which results in a receiver system noise temperature of 100 K. If an E_b/N_0 of 20 dB is required for adequate BER performance what maximum bit rate can be accommodated using BPSK modulation, assuming performance is limited by the downlink and atmospheric attenuation can be neglected?

Transmitter power = 14 dBW
EIRP = 14 + 20 − 2 = 32 dBW = 62 dBm
Free space loss = $20 \log 10 (4\pi 38 \times 10^6)/0.0256 = 205.4$ dB

Receiver antenna $G_R = \dfrac{4\pi}{\lambda^2} \dfrac{\pi d^2}{4} 0.55 = 62.7$ dB

Received power level = 62 − 205.4 + 62.7 dBm = −80.7 dBm
Receiver noise at 100 K noise temperature = −178.6 dBm/Hz
If $BT_o = 1.0$ then $E_b/N_0 = C/N$
Available margin for E_b/N_0 and modulation bandwidth = 178.6 − 80.7 = 97.9 dB Hz
If $E_b/N_0 = 20$ dB then the margin for modulation = 77.9 dB Hz = 61.6 MHz
With BPSK at 1 bit/s/Hz then the modulation rate can be 61.6 Mbit/s.

Alternatively:
If $G_R = 62.7$ dB and $T_{syst} = 100$ K then $G/T = 42.7$ dB/K
Then radiated power at receiver antenna = + 62 − 205.4 dBm = −143.4 dBm
Power at receiver input = − 143.4 + 42.7 dBm/K = −100.7 dBm/K
Boltzmann's constant = − 198.6 dBm/Hz/K
Difference = 97.9 dBHz
Allowing 20 dB for acceptable E_b/N_0 leaves 77.9 dBHz which will support a 61.6 Mbit/s BPSK symbol rate.

14.3.4 Slant path propagation considerations

The discussion of satellite link budgets in section 14.3.3 referred to atmospheric effects which must be accounted for to achieve adequate system availability. The principal effects which contribute to changes in signal level on earth–space paths from that expected for free space propagation are:

1. Background atmospheric absorption.
2. Rain fading.
3. Scintillation.

The principal mechanisms of noise and interference enhancement are:

1. Sun transit.
2. Rain enhancement of antenna temperature.
3. Interference caused by precipitation scatter and ducting.
4. Crosstalk caused by cross-polarisation.

Background gaseous absorption

Gaseous absorption on slant path links can be described by $A = \gamma L$ but with L replaced by effective path length in the atmosphere, L_{eff}. L_{eff} is less than the physical path length in the atmosphere due to the decreasing density of the atmosphere with height. In practice the total attenuation, $A(f)$, is usually calculated using curves of zenith attenuation, Figure 14.35, and a simple geometrical dependence on elevation angle, α, i.e.:

$$A(f) = \frac{A_{zenith}(f)}{\sin \alpha} \qquad (14.40)$$

A_{zenith} is the one-way total zenith attenuation and depends on both frequency and surface pressure (reflecting the height of the earth station above sea level). Curve A is for a dry atmosphere and curve B includes the effect of water vapour at a concentration which is typical of temperate climates. (The scale height of the water vapour concentration is 2 km.) Correction factors have been derived which can be used with Figure 14.35 and equation (14.40) to find slant path gaseous attenuation for other surface pressures and water vapour densities [Freeman].

Rain fading

The same comments can be made for rain fading on slant paths as those which have already been made for terrestrial paths. The slant path geometry, however, means that the calculation of effective path length depends not only on the horizontal structure of the rain but also on its vertical structure.

One model [ITU-R, Rec. 618] for predicting the rain fading exceeded for a given percentage of time therefore incorporates the following formula which derives an effective rain height from earth station latitude, i.e.:

$$h_R \text{ (km)} = \begin{cases} 3.0 + 0.028\phi, & 0 < \phi < 36° \\ 4.0 - 0.075(\phi - 36), & \phi \geq 36° \end{cases} \qquad (14.41)$$

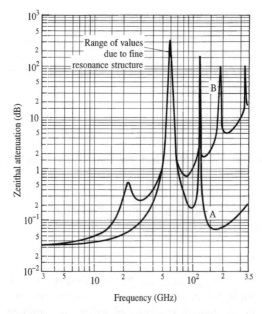

Figure 14.35 *Total ground level zenith attenuation (15°C, 1013 mb) for, A, dry atmosphere and, B, a surface water vapour content of 7.5 g/m³-decaying exponentially with height. (Source: ITU-R Rec. P.676, 1995, reproduced with the permission of the ITU.)*

The slant path length below the rain height is found from Figure 14.36, i.e.:

$$L_s = \frac{h_R - h_E}{\sin \alpha} \quad \text{(km)} \tag{14.42}$$

where h_E is the height of the earth station and α is the slant path elevation angle. (A more accurate formula which takes account of earth curvature is used for $\alpha < 5°$.) The ground projection of L_s is calculated using:

$$L_G = L_s \cos \alpha \quad \text{(km)} \tag{14.43}$$

and a path length reduction formula appropriate for an exceedance value of 0.01% of time is applied, i.e.:

$$r_{0.01} = \frac{1}{1 + L_G(e^{0.015 R_{0.01}})/35} \tag{14.44}$$

(For $R_{0.01} > 100$ mm/h then 100 mm/h is used instead of $R_{0.01}$.) The one-minute rain rate exceeded for 0.01% of time, $R_{0.01}$, is estimated, preferably from local meteorological data, and the corresponding 0.01% specific attenuation is calculated using equation (14.14). (Alternatively a first order estimate of $\gamma_{R\,0.01}$ can be made by interpolating the curves of Figure 14.22.) The total path rain attenuation exceeded for 0.01% of time is then given by:

$$A_{0.01} = \gamma_{R\,0.01} L_s r_{0.01} \quad \text{(dB)} \tag{14.45}$$

The attenuation exceeded for some other time percentage, p (between 0.001% and 1.0%), can be estimated using the empirical scaling law:

$$A_p = A_{0.01} 0.12 p^{-(0.546 + 0.043 \log_{10} p)} \quad \text{(dB)} \tag{14.46}$$

The slant path attenuation prediction method described here is different from that described in section 14.2.4 (for terrestrial links) in that the former uses an actual rain rate and a path length reduction factor whilst the latter uses an actual path length and a rain rate reduction factor. Clearly both types of model are equivalent in that they take account of the non-uniform distribution of rain along an extended path and either approach can be applied to either type of link. In particular the rain rate prediction method described for

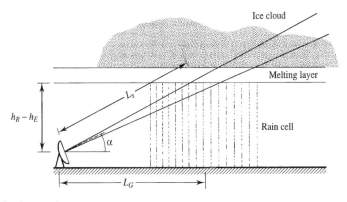

Figure 14.36 *Slant path geometry.*

satellite links can also be applied to terrestial links [ITU-R, Rec. 530] if L_G and L_s are replaced by the actual terrestial path length, L.

Scintillation

Scintillation refers to the relatively small fluctuations (usually less than, or equal to, a few dB peak to peak) of received signal level due to the inhomogeneous and dynamic nature of the atmosphere. Spatial fluctuations of electron density in the ionosphere and fluctuations of temperature and humidity in the troposphere result in non-uniformities in the atmospheric refractive index. As the refractive index structure changes and/or moves across the slant path (with, for example, the mean wind velocity) these spatial variations are translated to time variations in received signal level. The fluctuations occur typically on a time scale of a few seconds to several minutes. Scintillation, unlike rain fading, can result in signal enhancements as well as fades. The CNR is degraded, however, during the fading part of the scintillating signal and as such has the potential to degrade system performance. Whilst severe fading is usually dominated by rain and occurs for only small percentages of time the less severe fading due to scintillation occurs for large percentages of time and may be significant in the performance of low margin, low availability, systems such as VSATs (see section 14.3.9). At very low elevation angles multipath propagation due to reflection from, and/or refraction through, stable atmospheric layers may occur. Distinguishing between severe scintillation and multipath propagation in this situation may, in practice, be difficult, however. Scintillation intensity is sensitively dependent on elevation angle, increasing as elevation angle decreases.

Mechanisms of noise enhancement

Excess thermal noise arising from rain, precipitation scatter, ducting and cross-polarisation may all affect satellite systems in essentially the same way as terrestrial systems (see section 14.2.4). Rain induced cross-polarisation, however, is usually more severe on slant path links since the system designer is not free to choose the earth station's polarisation. Furthermore, since the propagation path continues above the rain height, tropospheric ice crystals may also contribute to cross-polarisation. Earth–space links employing full frequency reuse (i.e. orthogonal polarisations for independent co-frequency carriers) may therefore require adaptive cross-polar cancellation devices to maintain satisfactory isolation between carriers.

Sun transit refers to the passage of the sun through the beam of a receiving earth station antenna. The enormous noise temperature of the sun effectively makes the system unavailable for the duration of this effect. Geostationary satellite systems suffer sun transit for a short period each day around the spring and vernal equinoxes. (Other celestial noise sources can also cause occasional increases in earth station noise temperature.)

System availability constraints

The propagation effects described above will degrade a system's CNR below its clear sky level for a small, but significant, fraction of time. In order to estimate the constraints which propagation effects put on system availability (i.e. the fraction of time that the CNR exceeds its required minimum value) the clear sky CNR must be modified to account for these propagation effects.

In principle, since received signal levels fluctuate due to variations in gaseous absorption and scintillation, these effects must be combined with the statistics of rain fading to produce an overall fading cumulative distribution, in order to estimate the CNR exceeded for a given percentage of time. Gaseous absorption and scintillation give rise to relatively small fade levels compared to rain fading (at least at the large time percentage end of the fading CD) and it is therefore often adequate, for traditional high availability systems, to treat gaseous absorption as constant (as in the link budget of section 14.3.3) and neglect scintillation altogether. (This approach may not be justified in the case of VSATs (see section 14.3.9), for which transmit power, G/T, and consequent availability are all low.) Once the uplink and downlink fade levels for the required percentage of time have been established then the CNRs can be modified as described below.

The uplink CNR exceeded for $100 - p\%$ of time (where typically $100 - p\% = 99.99\%$, i.e. $p = 0.01\%$), $(C/N)_{u, 100-p}$, is simply the clear sky CNR, $(C/N)_u$, reduced by the fade level exceeded for $p\%$ of time, $F_u(p)$, i.e.:

$$\left(\frac{C}{N}\right)_{u, 100-p} = \left(\frac{C}{N}\right)_u - F_u(p) \quad \text{(dB)} \tag{14.47}$$

The uplink noise is not increased by the fade since the attenuating event is localised to a small fraction of the receiving satellite antenna's coverage area. (Even if this were not so the temperature of the earth behind the event is essentially the same as the temperature of the event itself.)

If uplink interference arises from outside the fading region then the uplink carrier to interference ratio exceeded for $100 - p\%$ of time will also be reduced by $F_u(p)$, i.e.:

$$\left(\frac{C}{I}\right)_{u, 100-p} = \left(\frac{C}{I}\right)_u - F_u(p) \quad \text{(dB)} \tag{14.48}$$

In the absence of uplink fading (or the presence of uplink power control to compensate uplink fades) the downlink CNR exceeded for $100 - p\%$ of time is determined by the downlink fade statistics alone. $(C/N)_d$, however, is reduced not only by downlink carrier fading (due to downlink attenuating events) but also by enhanced antenna noise temperature (caused by thermal radiation from the attenuating medium in the earth station's normally cold antenna beam), i.e.:

$$\left(\frac{C}{N}\right)_{d, 100-p} = \left(\frac{C}{N}\right)_d - F_d(p) - \left(\frac{N_{faded}}{N_{clear\ sky}}\right)_{dB} \quad \text{(dB)} \tag{14.49}$$

where $N_{faded} = k(T_{ant} + T_e)B$, see Chapter 12.

A simple, but conservative, estimate of $(C/N)_{d, 100-p}$ can be made by assuming, in the 'clear sky' link budget, a (worst case) antenna noise temperature which is equal to the physical temperature of the lossy medium (typically 290 K) and ignoring $(N_{faded}/N_{clear\ sky})_{dB}$ in equation (14.49).

Downlink carrier to interference ratio is usually unaffected by a downlink fade since both the wanted and interfering signals are equally attenuated, i.e.:

$$\left(\frac{C}{I}\right)_{d, 100-p} = \left(\frac{C}{I}\right)_d \quad \text{(dB)} \tag{14.50}$$

From a system design point of view fade *margins* can be incorporated into the satellite uplink and downlink budgets such that under clear sky conditions the system operates with the correct back-off but with excess uplink and downlink CNR (over those required for adequate overall CNR) of $F_u(p)$ and $F_d(p)$ respectively. Assuming fading does not occur simultaneously on uplink and downlink this ensures that an adequate overall CNR will be available for $100 - 2p\%$ of time. More accurate estimates of the system performance limits imposed by fading would require joint statistics of uplink and downlink attenuation, consideration of changes in back-off produced by uplink fades (including consequent *improvement* in intermodulation noise), allowance for possible cross-polarisation induced crosstalk, hydrometeor scatter and other noise and interference enhancement effects.

EXAMPLE 14.6

A 30 GHz receiving earth station is located near Bradford (54° N, 2° W). It has an overall receiver noise figure of 5.0 dB and a free space link budget which yields a CNR of 35.0 dB. (The free space budget does not account for background gaseous attenuation of the carrier but does allow for normal atmospheric noise.) The station is at a height of 440 m above sea level and the elevation angle to the satellite is 29°. If the one-minute rain rate exceeded for 0.01% of time at the earth station is 28 mm/h estimate the CNR exceeded for 99.9% of time assuming that the CNR is downlink limited and uplink power control is used to compensate all uplink fades.

From Figure 14.35 $A_{zenith} = 0.23$ dB. (This is strictly the value for an earth station at sea level, but since Figure 14.20 shows a specific attenuation *at sea level* of only 0.09 dB/km then the error introduced is less than $0.09 \times 0.44 = 0.04$ dB and can therefore be neglected.)

Clear sky slant path attenuation from equation (14.40):

$$A = \frac{A_{zenith}}{\sin \alpha} = \frac{0.23}{\sin 29°} = 0.5 \text{ dB}$$

$$\left. \frac{C}{N} \right|_{clear\ sky} = \left. \frac{C}{N} \right|_{free\ space} - A$$

$$= 35.0 - 0.5 = 34.5 \text{ dB}$$

From equation (14.41) rain height is:

$$h_R = 4.0 - 0.075 (54 - 36) = 2.65 \text{ km}$$

Slant path length in rain (equation (14.42)) is:

$$L_s = \frac{2.65 - 0.44}{\sin 29°} = 4.56 \text{ km}$$

Ground projection of path in rain (equation (14.43)):

$$L_G = 4.56 \cos 29° = 3.99 \text{ km}$$

Path length reduction factor (equation (14.44)):

$$r_{0.01} = \left[1 + \frac{3.99 \, e^{(0.015 \times 28)}}{35} \right]^{-1} = 0.852$$

Using Figure 14.22 specific attenuation, γ_R, for 30 GHz at 28 mm/h can be estimated to be 5.3

dB/km. (A more accurate ITU-R model using the formula aR^b gives 5.6 dB/km for horizontal polarisation and 4.7 dB/km for vertical polarisation.)

From equation (14.45):

$$A_{0.01} = 5.3 \times 4.56 \times 0.852 = 20.6 \text{ dB}$$

Using equation (14.46):

$$A_{0.1} = 20.6 \times 0.12 \times 0.1^{-(0.546 + 0.043 \log_{10} 0.1)} = 7.9 \text{ dB}$$

Using equations (14.24) and (14.25) and assuming a surface temperature of 290 K:

$$T_A = [(1.12 \times 290) - 50] \, [1 - 10^{-\frac{7.9}{10}}] = 230 \text{ K}$$

Ratio of effective noise powers under faded and clear sky conditions is:

$$
\begin{aligned}
\frac{N_{faded}}{N_{clear\,sky}} &= \frac{T_{syst\,faded}}{T_{syst\,clear\,sky}} \\[2mm]
&= \frac{T_{A\,faded} + (f - 1)\,290}{T_{A\,clear\,sky} + (f - 1)\,290} \\[2mm]
&= \frac{230 + (10^{\frac{5}{10}} - 1)\,290}{50 + (10^{\frac{5}{10}} - 1)\,290} \\[2mm]
&= \frac{230 + 627}{50 + 627} = 1.266 = 1.0 \text{ dB}
\end{aligned}
$$

(Note that the noise enhancement in this case is within the uncertainty introduced by estimating γ_R from Figure 14.22.) From equation (14.49):

$$\left(\frac{C}{N}\right)_{d,\,99.9} = 34.5 - 7.9 - 1.0 = 25.6 \text{ dB}$$

14.3.5 Analogue FDM/FM/FDMA trunk systems

Figure 14.37 shows a schematic diagram of a large, traditional, earth station. Such an earth station would be used mainly for fixed point-to-point international PSTN communications. The available transponder bandwidth (typically 36 MHz) is subdivided into several transmission bands (typically 3 MHz wide) each allocated to one of the participating earth stations, Figure 14.38. All the signals transmitted by a given earth station, irrespective of their destination, occupy that earth station's allocated transmission band. Individual SSB voice signals arriving from the PSTN at an earth station are frequency division multiplexed (see Figure 5.12) into a position in the earth station's transmission band which depends on the voice signal's destination. Thus all the signals arriving for transmission at earth station 2 and destined for earth station 6 are multiplexed into sub-band 6 of transmission band 2. The FDM signal, consisting of all sub-bands, is then frequency modulated onto the earth station's IF carrier. The FDM/FM signal is subsequently upconverted (U/C) to the 6 GHz RF carrier, amplified (to attain the required EIRP) and transmitted.

A receiving earth station demodulates the carriers from *all* the other earth stations in the network. (Each earth station therefore requires $N - 1$ receivers where N is the number of participating earth stations.) It then filters out the sub-band of each

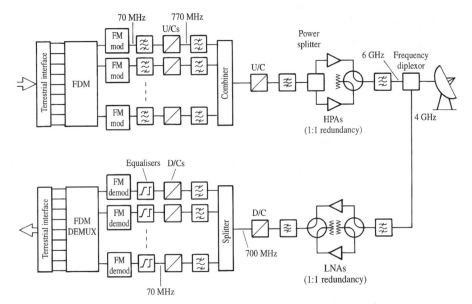

Figure 14.37 *Simplified block diagram of a traditional FDM/FM/FDMA earth station (only HPA/LNA redundancies shown).*

transmission band designated to itself and discards all the other sub-bands. The sub-band signals are then demultiplexed, the resulting SSB voice signals demodulated if necessary (i.e. translated back to baseband) and interfaced once again with the PSTN. This method of transponder resource sharing between earth stations is called frequency division multiple access (FDMA).

When assessing the SNR performance of FDM/FM/FDMA voice systems the detection gain of the FM demodulator must be included. Assuming that operation is at a CNR above threshold, this gain is given by [Pratt and Bostian]:

$$\frac{(S/N_b)_{wc}}{C/N} = \left(\frac{\Delta f_{RMS}}{f_M} \right)^2 \frac{B}{b} \tag{14.51}$$

where:

$(S/N_b)_{wc}$ is the SNR of the worst case voice channel (see Figure 14.39),
Δf_{RMS} is the RMS frequency deviation of the FM signal,
f_M is the maximum frequency of the modulating (FDM) signal,
B is the bandwidth of the modulated (FM) signal,
b is the bandwidth of a single voice channel (typically 3.1 kHz).

The quantities needed to apply equation (14.51) can be estimated using:

$$f_M = 4.2 \times 10^3 N \quad \text{(Hz)} \tag{14.52}$$

where N is the number of voice channels in the FDM signal,

$$B = 2(\Delta f + f_M) \quad \text{(Hz)} \tag{14.53}$$

where Δf is the peak frequency deviation of the FM signal, and

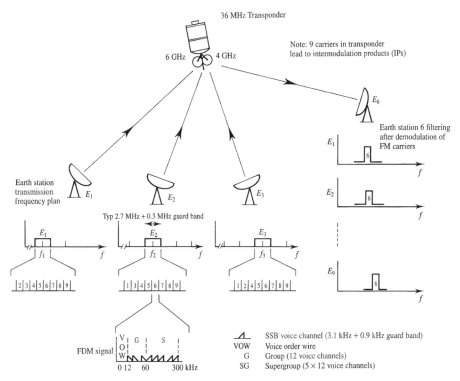

Figure 14.38 *Illustration of MCPC FDM/FM/FDMA single transponder satellite network and frequency plan for the transponder (with nine participating earth stations).*

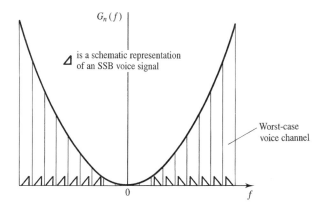

Figure 14.39 *Parabolic noise power spectral density after FM demodulation.*

$$\Delta f = \begin{cases} 3.\,16\,\Delta f_{RMS}, & \text{for } N > 24 \\ 6.\,5\,\Delta f_{RMS}, & \text{for } N \le 24 \end{cases} \qquad (14.54)$$

(The FDM signal is usually amplitude limited to ensure the peak to RMS frequency deviation ratio is ≤ 3.16. Since the FDM signal is essentially a sum of many independent voice signals its pdf will be Gaussian, $\Delta f / \Delta f_{RMS} = 3.16$, corresponding to the 0.1% extreme value, Figure 4.22. Clipping will therefore take place for approximately 0.2% of time.) The RMS frequency deviation is set by adjusting the FM modulator constant, K (Hz/V), of the earth station transmitter, i.e.:

$$\Delta f_{RMS} = K \sqrt{\langle g^2(t) \rangle} \tag{14.55}$$

where $g(t)$ is the modulating (FDM) signal. In practice the modulator constant, K, is set using a 1 kHz, 0 dBm test tone as the modulating signal. The test tone RMS frequency deviation, Δf_{RMS}^T, can be related to the required FDM signal frequency deviation, Δf_{RMS}, by:

$$\Delta f_{RMS} = l \, \Delta f_{RMS}^T \tag{14.56}$$

where:

$$20 \log_{10} l = \begin{cases} -1 + 4 \log_{10} N, & 12 \leq N \leq 240 \\ -15 + 10 \log_{10} N, & N > 240 \end{cases} \tag{14.57}$$

In addition to the FM detection gain given by equation (14.51) an extra SNR gain can be obtained by using a pre-emphasis network prior to modulation and a de-emphasis network after demodulation. Using the ITU pre-emphasis/de-emphasis standards the pre-emphasis SNR gain is 4 dB. Finally, the combined frequency response of a telephone earpiece and the subscriber's ear matches the spectrum of the voice signal better than the spectrum of the noise. This results in a further (if partly subjective) improvement in SNR. This improvement is accounted for by what is called the *psophometric weighting* and has a numerical value of 2.5 dB. Since many voice channels are modulated (as a single FDM signal) onto a single carrier, FDM/FM/FDMA is often referred to as a multiple channel per carrier (MCPC) system. MCPC is efficient providing each earth station is heavily loaded with traffic.

For lightly loaded earth stations MCPC suffers the following disadvantages:

1. Expensive FDM equipment is necessary.
2. Channels cannot be reconfigured easily and must therefore be assigned on essentially a fixed basis.
3. Each earth station carrier is transmitted irrespective of traffic load. This means that full transponder power is consumed even if little or no traffic is present.
4. Even under full traffic load, since an individual user speaks for only about 40% of time, significant transponder resource is wasted.

An alternative to MCPC for lightly loaded earth stations is a single channel per carrier (SCPC) system. In this scheme each voice signal is modulated onto its own individual carrier and each voice carrier is transmitted only as required. This saves on transponder power at the expense of a slightly increased bandwidth requirement. This scheme might be called FM/FDM/FDMA in contrast to the FDM/FM/FDMA process used by MCPC systems. The increased bandwidth per channel requirement over MCPC makes it an uneconomical scheme for traditional point-to-point international trunk applications. The fact that the channels can be demand assigned (DA) as traffic volumes fluctuate, and that the carrier can be switched on (i.e. voice activated) during the 35–40% of active speech time typical of voice signals (thus saving 4 dB of transponder power), makes SCPC superior to MCPC for systems with light, or highly variable, traffic.

Another type of SCPC system dispenses with FM entirely. Compatible SSB systems simply translate the FDM signal (comprising many SSB voice signals) directly to the RF transmission band (using amplitude modulation). This is the most bandwidth efficient system of all and is not subject to a threshold effect as FM systems are. Compatible SSB does not, however, have the large SNR detection gain that both FDM/FM/FDMA and FM/FDM/FDMA systems have.

14.3.6 Digital TDM/PSK/TDMA trunk systems

Time division multiple access (TDMA) is an alternative to FDMA for transponder resource sharing between earth stations. Figure 14.40 illustrates the essential TDMA principle. Each earth station is allocated a time slot (in contrast to an FDMA frequency slot) within which it has sole access to the entire transponder bandwidth. The earth station time slots, or bursts, are interleaved on the uplink, frequency shifted, amplified and retransmitted by the satellite to all participating earth stations. One earth station periodically transmits a reference burst in addition to its information burst in order to synchronise the bursts of all the other earth stations in the TDMA system. Time division multiplexing and digital modulation are obvious techniques to use in conjunction with TDMA. In order to minimise AM/PM conversion in the non-linear transponder, Figure 14.33, constant envelope PM is attractive. MPSK is therefore used in preference to MQAM, for example (see Chapter 11). Since some filtering of the PSK signal prior to transmission is necessary (for spectrum management purposes) even MPSK envelopes are not, in fact, precisely constant. QPSK signals, for instance, have envelopes which fall to zero when both inphase and quadrature symbols change simultaneously, Figure 11.31. Offset QPSK (OQPSK) reduces the maximum envelope fluctuation to 3 dB by offsetting

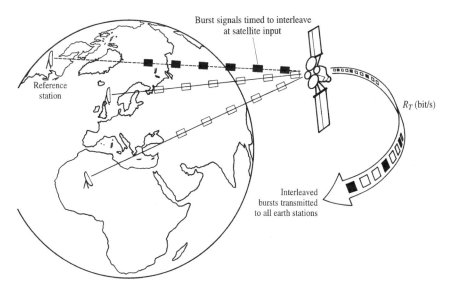

Figure 14.40 *Principle of time division multiplex access (TDMA).*

inphase and quadrature symbols by half a symbol period (i.e. one information bit period), Figure 11.31. Chapter 11 discusses bandpass modulation (including OQPSK) in detail. Figure 14.41 shows a schematic diagram of a TDM/PSK/TDMA earth station.

For digital satellite systems having only a single carrier present in the transponder at any one time then intermodulation products are absent and $(C/N)_{IP}^{-1}$ in equation (14.38) is zero. Recall (Chapter 11) that the quantity E_s/N_0 is related to C/N by:

$$\frac{\langle E_s \rangle}{N_0} = \frac{C}{N} BT_o \tag{14.58}$$

where T_o is the symbol period (i.e. the reciprocal of the baud rate, R_s) and BT_o depends on the particular digital modulation scheme and filtering employed. Equation (14.58) can also be expressed in terms of bit energy, E_b, and bit duration, T_b, i.e.:

$$\frac{E_b}{N_0} = \frac{C}{N} BT_b \tag{14.59}$$

For QPSK modulation, which is the primary TDM/PSK/TDMA modulation standard currently used by INTELSAT, the E_b/N_0 required to support a given P_b performance is found using equation (11.47):

$$P_b = \frac{1}{2} \operatorname{erfc} \left(\frac{E_b}{N_0} \right)^{1/2} \tag{14.60}$$

and the required CNR is then found using equations (11.6) and (11.7):

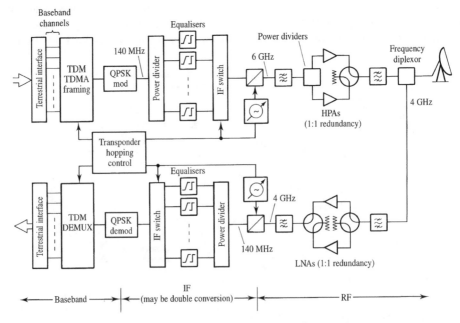

Figure 14.41 *Simplified block diagram of traditional TDM/QPSK/TDMA earth station. (Only HPA/LNA redundancies are shown.)*

$$\frac{C}{N} = \frac{E_b/T_b}{N_0 B} = \frac{E_b R_b}{N_0 B} \tag{14.61}$$

or in decibels:

$$\left(\frac{C}{N}\right)_{dB} = \left(\frac{E_b}{N_0}\right)_{dB} - (BT_b)_{dB} \tag{14.62}$$

(For minimum bandwidth, ISI-free, filtering such that the transmitted signal occupies the DSB Nyquist bandwidth then $BT_b = 1.0$ (or 0 dB) and $C/N = E_b/N_0$.) In practice an implementation margin of a few decibels would be added after using equations (14.60) and (14.61) to allow for imperfect modulation, demodulation, etc.

A typical frame structure showing the TDMA slots allocated to different earth stations is shown in Figure 14.42. Two reference bursts are often included (provided by different earth stations) so that the system can continue to function in the event of losing one reference station due, for example, to equipment failure. Typically the frame period T_F is of the order of 5 ms.

The traffic bursts each consist of a preamble followed by the subscriber traffic. The preamble, which might typically be 280 QPSK symbols long, is used for carrier recovery, symbol timing, frame synchronisation and station identification. In addition it supports voice and data channels (voice order wires) to enable operations and maintenance staff at different earth stations in the network to communicate without using traffic slots. Reference bursts have the same preamble as the traffic bursts followed by control and delay signals (typically eight symbols in duration) which ensure that the TDMA bursts from participating earth stations are timed to interleave correctly at the transponder input.

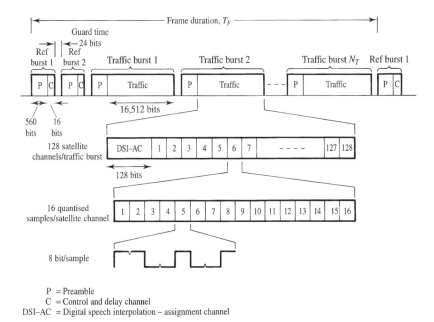

P = Preamble
C = Control and delay channel
DSI-AC = Digital speech interpolation – assignment channel

Figure 14.42 *Typical TDMA frame structure. (DSI-AC time slot is discussed in section 14.3.7.)*

Subscriber traffic is subdivided into a number (typically 128) of *satellite channels*, Figure 14.42. In conventional preassigned (PA) systems the satellite channels are pre-divided into groups, each group being assigned to a given destination earth station.

Each satellite channel carries (typically) 128 bits (64-QPSK symbols) representing 16 consecutive 8 bit PCM samples from a single voice channel. For a conventional 8 kHz PCM sampling rate this corresponds to $16 \times 125 \mu s = 2$ ms of voice information. In this case the frame duration would therefore be limited in length to 2 ms so that the next frame could convey the next 2 ms of each voice channel. Channels carrying non-voice, high data rate, information are composed of multiple voice channels. Thus, for example, a 320 kbit/s signal would require five 64 kbit/s voice channels.

Frames may be assembled into master frames as shown in Figure 14.43. The relative lengths of information bursts from each earth station can then be varied from master frame to master frame depending on the relative traffic loads at each station. This would represent a simple demand assigned system.

The frame efficiency, η_F, of a TDM/PSK/TDMA system is equal to the proportion of frame bits which carry revenue earning traffic, i.e.:

$$\eta_F = \frac{b_F - b_o}{b_F} \tag{14.63}$$

where b_F is the total number of frame bits and b_o is the number of overhead (i.e. non-revenue earning) bits. The total number of frame bits is given by:

$$b_F = R_T T_F \tag{14.64}$$

where R_T is the TDMA bit rate and T_F is the frame duration. The number of overhead bits per frame can be calculated using:

$$b_o = N_R b_R + N_T (b_p + b_{AC}) + (N_R + N_T) b_G \tag{14.65}$$

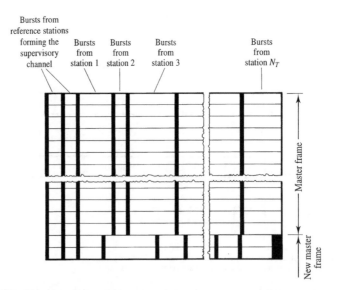

Figure 14.43 *TDMA master frame structure.*

where N_R is the number of participating reference stations, N_T is the number of participating traffic stations, b_R is the number of bits in a reference burst, b_P is the number of preamble bits (i.e. all bits excluding traffic bits) in a traffic burst, b_{AC} is the number of digital speech interpolation–assignment channel bits per traffic burst and b_G is the number of guard bits per reference or traffic burst.

Typically η_F is about 90% for a TDMA system with 15 to 20 participating earth stations and a frame period of several milliseconds.

The number of voice channels which a TDMA system can support is called its voice channel capacity, χ. This can be calculated from:

$$\chi = \frac{R_i}{R_v} \tag{14.66}$$

where R_i is the information bit rate and R_v is the bit rate of a single voice channel.

The information bit rate is given by:

$$R_i = \frac{b_F - b_o}{T_F}$$

$$= \eta_F R_T \quad \text{(bit/s)} \tag{14.67}$$

and the bit rate of a single voice channel is given by:

$$R_v = f_s n \quad \text{(bit/s)} \tag{14.68}$$

where f_s is the sampling rate (usually 8 kHz) and n is the number of PCM bits per sample (typically 8), Chapter 6. (For ADPCM, R_v is 32 kbit/s, Table 9.3.) Finally, the average number of voice channels *per earth station access* $\bar{\chi}_A$ is:

$$\bar{\chi}_A = \frac{\chi}{N} \tag{14.69}$$

where N is the number of accesses per frame. (If all earth stations only access once per frame then N is also, of course, the number of participating earth stations.)

The primary TDM/PSK/TDMA modulation standard used by INTELSAT is summarised in Table 14.5.

Table 14.5 *INTELSAT TDM/PSK/TDMA modulation standard.*

Modulation	QPSK
Nominal symbol rate	60.416 Mbaud
Nominal bit rate	120.832 Mbit/s
Encoding	Absolute (i.e. not differential phase)
Phase ambiguity resolution	Using unique word in preambles

EXAMPLE 14.7

The frame length of the pure TDMA system illustrated in Figure 14.42 is 2.0 ms. If the QPSK symbol rate is 60.136 Mbaud and all traffic bursts are of equal length determine: (i) the maximum number of earth stations which the system can serve and (ii) the frame efficiency.

$$R_T = 2 \times \text{QPSK baud rate} = 2 \times 60.136 \times 10^6 = 1.20272 \times 10^8 \text{ bit/s}$$

$$b_F = R_T T_F = 1.20272 \times 10^8 \times 2.0 \times 10^{-3} = 240544 \text{ bits}$$

Overhead calculation:

$$b_o = N_R b_R + N_T (b_p + b_{AC}) + (N_R + N_T) b_G$$

Now for a 560 + 16 bit reference, b_R, 560 bit preamble, b_p, 24 bit guard interval, b_G, and 128 bit DSI assignment channel:

$$b_o = 2 (560 + 16) + (560 + 128) N_T + (2 + N_T) 24$$

$$= 1200 + 712 N_T$$

$$b_F - b_o = 240544 - (1200 + 712 N_T) = 239344 - 712 N_T$$

(i)

$$N_T = \frac{b_F - b_o}{16512 - 128} = \frac{239344 - 712 N_T}{16384}$$

i.e.:

$$N_T (16384 + 712) = 239344$$

Therefore: $N_T = 14$, i.e. a maximum of 14 earth stations may participate.

(ii)

$$\eta_F = \frac{b_F - b_o}{b_F} = \frac{240544 - (1200 + 712 \times 14)}{240544} = 0.954$$

i.e. frame efficiency is 95.4%.

14.3.7 DA-TDMA, DSI and random access systems

Preassigned TDMA (PA-TDMA) risks the situation where, at a certain earth station, all the satellite channels assigned to a given destination station are occupied whilst free capacity exists in channels assigned to other destination stations. Demand assigned TDMA (DA-TDMA) allows the reallocation of satellite channels in the traffic burst as the relative demand between earth stations varies. In addition to demand assignment of satellite channels within the earth station's traffic burst DA-TDMA may also allow the number of traffic bursts per frame, and/or the duration of the traffic bursts, allocated to a given earth station to be varied.

Digital speech interpolation (DSI) is another technique employed to maximise the use made of available transponder capacity. An average speaker engaged in conversation actually talks for only about 35% of the time. This is because for 50% of the time he, or she, is passively listening to the other speaker and for 30% of the remaining 50% of the time there is silence due to pauses and gaps between phrases and words. DSI systems automatically detect when speech is present in the channel, and during speech absences reallocate the channel to another user. The inevitable clipping at the beginning of speech which occurs as the channel is being allocated is sufficiently short for it to go unnoticed.

Demand assigned systems require extra overhead in the TDMA frame structure to control the allocation of satellite channels and the relative number per frame, and lengths, of each earth station's traffic bursts. For systems with large numbers of earth stations each contributing short, bursty, traffic at random times then random access (RA) systems may use transponder resources more efficiently than DA systems. The earth stations of

RA systems attempt to access the transponder (i.e. in the TDMA context, transmit bursts) essentially at will. There is the possibility, of course, that the traffic bursts (usually called packets in RA systems) from more than one earth station will collide in the transponder causing many errors in the received data. Such collisions are easily detected, however, by both transmitting and receiving earth stations. After a collision all the transmitting earth stations wait for a random period of time before retransmitting their packets.

Many variations and hybrids of the multiple access techniques described here have been used, are being used, or have been proposed, for satellite communications systems. A more detailed and quantitative discussion of these techniques and their associated protocols can be found in [Ha].

14.3.8 Economics of satellite communications

The cost of a long distance point-to-point terrestrial voice circuit is about 2000 dollars p.a. Lease of a private, high quality, landline from New York to Los Angeles for FM broadcast use costs 13000 dollars p.a. Satellite communications systems can provide the equivalent services at lower cost. The monthly lease for a video bandwidth satellite transponder is typically 50000 to 200000 dollars whilst the hourly rate for satellite TV programme transmission can be as low as 200 dollars. Access to digital audio programmes can be as low as 2500 dollars per month for high quality satellite broadcasts. The advent of the space shuttle has greatly reduced satellite launch costs below the 30000 dollars/kg of conventional rocket techniques making them much more competitive. (The figures given here reflect 1990 costs.)

14.3.9 VSAT systems

Satellite communication is predominantly used for international, point-to-point, multiplex telephony, fax and data traffic, making high EIRP transmissions necessary due to the wideband nature of the signal multiplex. For low data rates (2.4 to 64 kbit/s) with narrow signal bandwidth the reduced noise in the link budget allows the use of satellites with very small aperture (typically 1 m diameter antenna) terminals (VSATs) and modest power (0.1 to 10 W) earth station transmitters [Everett]. This has given rise to the development of VSAT low data rate networks, in which many remote terminals can, for example, access a central computer database. These systems use roof, or garden, located antennas which permanently face a geostationary satellite. VSAT systems are widely deployed in the USA and may have up to 10000 VSAT earth station terminals in a single network. They are used for retail point-of-sale credit authorisation, cash transactions, reservations, stock control and other data transfer tasks. They are also now being expanded to give worldwide coverage in order to meet the telecommunications requirements of large international companies.

The use of non-geostationary satellites for mobile and personal communications with even lower antenna gains is discussed in Chapter 15.

14.3.10 Satellite switched TDMA and onboard signal processing

Satellites operating with small spot beams have high antenna gains. This implies either a low onboard power requirement or a large bandwidth and therefore high potential bit rate. If many spot beams with good mutual isolation are used, frequency bands can be reused

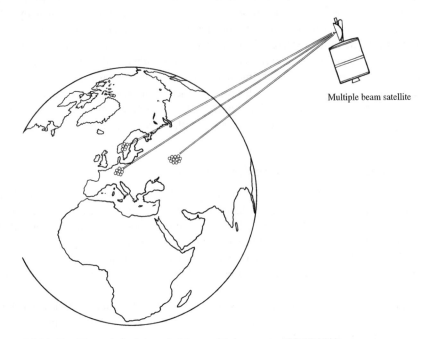

Multiple beam satellite

Figure 14.44 *Satellite switched time division multiplex access (SS-TDMA).*

thus increasing spectrum utilisation efficiency, Figure 14.44. Connectivity between a system's participating earth stations is potentially decreased, however, since a pair of earth stations in different spot beams can communicate only if their beams are connected. Satellite switched TDMA has the potential to re-establish complete connectivity between earth stations using a switching matrix onboard the satellite, Figure 14.45, see also section 20.6. The various sub-bursts (destined for different receiving stations) of a transmitting station's traffic burst can be directed by the matrix switch to the correct downlink spot beams. Furthermore, for areas with a sparse population of users, such that many fixed spot beams are uneconomic, the beams may be hopped from area to area and the uplink bursts from each earth station demodulated and stored. Onboard signal processing is then used to reconfigure the uplink bursts into appropriately framed downlink bursts before the signals are remodulated and transmitted to the appropriate earth stations as the downlink spot beam is hopped. Onboard demodulation and remodulation also has the normal advantage of digital communications, i.e. the uplink and downlink noise is decoupled. The NASA advanced communications satellite was used in the middle 1990s to evaluate these types of system.

14.4 Summary

Point-to-point, terrestrial, microwave links now play a major part in the arterial trunk routes of the PSTN. Originally carrying SSB/FDM/FM telephony they now carry mixed services, principally using PCM/TDM/QAM. Typically, repeaters are spaced at 40 to 50

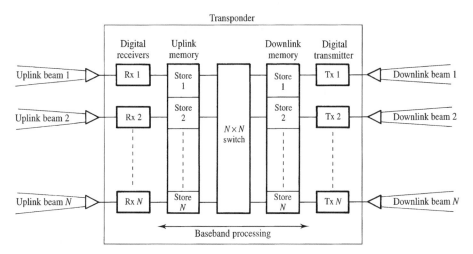

Figure 14.45 *SS-TDMA transponder.*

km intervals and the total microwave link bandwidth is split into high and low frequency blocks, each of which contains eight 30 MHz channels. Adjacent channel isolation is improved by alternating vertical and horizontal polarisations across the transmission band. Correlation between the expected signal and interference due to overreaching caused by ducting is avoided by alternating transmit and receive channels between high and low channel blocks on adjacent hops.

Proper path profiling of LOS microwave links ensures adequate clearance under specified propagation conditions. If clearance is less than the first Fresnel zone then serious diffraction losses may occur. Propagation paths may be conveniently plotted as straight rays over an earth with appropriately modified curvature, or as rays with modified curvature over a plane earth. Bulk refractive index conditions are described by a k-factor which is the ratio of effective earth radius to actual earth radius in the straight ray model. Standard propagation conditions correspond to a k-factor of 4/3 (which represents a refractivity lapse rate of $40N$ units/km). Positive $k < 4/3$ represents sub-refractive conditions and $k > 4/3$ represents super-refraction. Negative k represents ducting.

For accurate clear sky link budgets, background gaseous absorption must be accounted for in addition to free space path loss. Gaseous absorption is small below 10 GHz but rises rapidly with frequency due primarily to water vapour and oxygen resonance lines which occur at 22 GHz and 60 GHz respectively. Rain attenuation also increases rapidly with frequency and must be accounted for if links are to meet a specified availability. Multipath propagation may, like rain, result in deep signal fades but, unlike rain fading which is essentially flat, can also cause amplitude and phase distortion to wideband signals. Microwave paths which suffer from severe frequency selective fading can benefit significantly from the application of diversity techniques and/or the use of adaptive equalisation.

Hydrometeor scatter may result in interference between nominally independent co-frequency systems. Hydrometeor induced cross-polarisation may also result in crosstalk between the channels of systems using polarisation, in whole or in part, as the channel isolating mechanism.

Fixed service satellite communications make use of the geostationary orbit. Satellites in this orbit are located above the earth's equator at a height of 36000 km. They are stationary with respect to a point on the earth's surface and tracking is therefore unnecessary. A single geostationary satellite is visible from 42% of the earth's surface. With the exception of the polar regions three such satellites can give global coverage. The combined propagation delay of uplink and downlink is about 0.25 s making echo cancellation essential. FSPL is of the order of 200 dB making large transmitter powers, large antennas and sensitive receivers necessary for high data rate transmissions. Like terrestrial microwave paths gaseous absorption and rain fading must be considered when engineering an earth–space communications link. Scintillation on earth–space paths might also be important for low margin, low availability systems. Fixed-point satellite services currently operate in the C- and Ku-bands although services using the Ka-band will probably follow shortly.

The predominant multiplexing/modulation/multiple accessing technique used currently for international PSTN satellite telephony is FDM/FM/FDMA. Since satellite HPAs are operated near saturation the use of FDMA, which results in multiple FM carriers in the transponder, gives rise to large intermodulation products which must normally be accounted for in the overall CNR calculation. As with terrestrial LOS relay links, the trend in satellite communications is towards digital transmission. FDM/FM/FDMA systems will therefore gradually be replaced by TDM/PSK/TDMA systems. Intermodulation products will be of less concern since in (pure) TDMA systems only one carrier is present in the transponder at any particular time. TDMA systems require less back-off, therefore, than the equivalent FDMA systems. This means that constant envelope modulation is even more desirable to minimise AM/AM and AM/PM distortion. QPSK and OQPSK are attractive modulation techniques in these circumstances.

The bits transmitted by an earth station in a TDMA network are grouped into frames (and frames into master frames). Each frame carries overhead bits which serve various synchronisation, monitoring and control purposes. The frame efficiency is defined as the proportion of frame bits which carry (revenue earning) traffic.

For systems which have fluctuating traffic loads demand assignment may be used to reallocate capacity to those earth stations with highest demand. For systems with large numbers of lightly loaded earth stations, random access techniques may be appropriate.

VSAT systems can use modest transmitter powers and small antennas to transmit low bit rate (and therefore narrow bandwidth) data. Such technology is relatively inexpensive allowing commercial (and other) organisations to operate their own networks of earth stations. Switched spot beam satellites with onboard signal processing, and satellite based mobile communications systems using both high and low earth orbits (HEOs and LEOs), are currently being developed.

14.5 Problems

14.1 Show that the effective height of the (smooth) earth's surface above a cord connecting two points on the surface, for a straight ray profile, is given (for practical microwave link geometries) by $d_1 d_2/(2ka)$, where k is the k-factor and a is the earth radius. Hence confirm the correctness of equations (14.11) and (14.12(a)).

14.2 A 10.3 GHz microwave link with good ground clearance is 60 km long. The link's paraboloidal reflector antennas have a diameter of 2.0 m and are aligned to be correctly pointed during standard ($k = 4/3$) atmospheric conditions. What limiting k-factors can the link tolerate before 6 dB of power is lost from its signal budget due to refractive decoupling of the antennas? To what refractivity lapse rates do these k-factors correspond and in what regime (sub-refraction/super-refraction/ducting) do these lapse rates reside? (Assume that the antennas are mounted at precisely the same heights above sea level and that their 3 dB beamwidths are given by $1.2\lambda/D$ rad where λ is wavelength and D is antenna diameter.)

14.3 If the transmitting antenna in Problem 14.2 is 50 m high and is aligned under standard refractive index conditions such that its boresight is precisely horizontal, what must be the difference between transmit and receive antenna heights for the receive antenna to be located at the centre of the transmitted beam? [158.9 m]

14.4 Show that the radius of the nth Fresnel zone is given by equation (14.10).

14.5 A microwave LOS link operates at 10 GHz over a path length of 32 km. A minimum ground clearance of 50 m occurs at the centre of the path for a k-factor of 0.7. Find the Fresnel zone clearance of the path in terms of (a) the number of Fresnel zones cleared and (b) the fraction (or multiple) of first Fresnel zone radius. [10, 3.2]

14.6 A 6 GHz LOS link has a path length of 13 km. The height of both antennas above sea level at each site is 20 m. The only deviation of the path profile from a smooth earth is an ancient earth works forming a ridge 30 m high running perpendicular to the path axis 3 km from the receiver. If the transmitted power is 0 dBW, the transmit antenna has a gain of 30 dB and the receive antenna has an effective area of 4.0 m^2 estimate the power at the receive antenna terminals assuming knife edge diffraction. By how much would both antennas need to be raised if 0.6 first Fresnel zone clearance were required? (Assume knife edge diffraction and no atmospheric losses other than gaseous background loss.) [−74 dBW]

14.7 A terrestrial 20 GHz microwave link has the following specification:

Path length	35.0 km
Transmitter power	−3.0 dBW
Antenna gains	25.0 dB
Ground clearance	Good (no ground reflections or diffraction)
Receiver noise figure	3.0 dB
Receiver noise bandwidth	100 kHz

Estimate the effective CNR exceeded for 99.9% of time if the one-minute rain rate exceeded for 0.1% of time is 10 mm/h. (Assume that Figure 14.21 applies to the climate in which the link operates and that the surface temperature is 290 K.) [30.4 dB]

14.8 A geostationary satellite is located at 35° W. What are its look angles and range from: (a) Bradford, UK (54° N, 2° W); (b) Blacksburg, USA (37° N, 80° W); (c) Cape Town, South Africa (34° S, 18° E)? [22°, −140°, 39406 km; 27°, 121°, 38914 km, 22°, 292°, 39365 km]

14.9 A geostationary satellite sits 38400 km above the earth's surface. The zenith pointing uplink is at 6 GHz and has an earth station power of 2 kW with $G_T = 61$ dB. If satellite antenna gain $G_R = 13$ dB, calculate the power in mW at the satellite receiver. Calculate the effective carrier-to-noise ratio, given that the receiver bandwidth is 36 MHz, its noise figure F is 3 dB and the thermal noise level is −174 dBm/Hz. [0.54 ×10^{-6} mW, 32.7 dB]

14.10 A receiver for geostationary satellite transmission at 2 GHz has an equivalent noise temperature of 160 K and a bandwidth of 1 MHz. The receiving antenna gain is 35 dB and the antenna noise temperature is 50 K. If the satellite antenna gain is 10 dB and expected total path losses are 195 dB, what is the minimum required satellite transmitter power to achieve a 20 dB CNR at the output of the receiver? [24.6 dBW]

14.11 Ship–shore voice communications are conducted via an INTELSAT IV satellite in geostationary orbit at an equidistant range from ship and shore stations of 38400 km. One 28 kbit/s voice channel is to be carried in each direction. The shipboard terminal contains a

2.24 m diameter parabolic antenna of 65% efficiency. An uncooled low noise receiver tuned to 6040 MHz, having a bandwidth of 28 kHz and a system noise temperature of 150 K, is mounted directly behind the antenna. The satellite EIRP in the direction of the ship is 41.8 dBm. The shore based terminal has an antenna of gain 46 dB at the satellite–shore link frequency of 6414.6 MHz. The satellite EIRP in the direction of the shore is 31.8 dBm. The shore's receiver has a bandwidth of 28 kHz and a noise temperature of 142 K. Calculate the carrier-to-noise ratios in dB for reception at the ship and the shore from the two satellite downlinks. [15 dB, 10.1 dB]

14.12   A 14/11 GHz digital satellite, transparent transponder, communications link has the following specification:

Uplink saturating flux density	-83.2 dBW/m^2
Input back-off	8.0 dB
Satellite G/T	1.8 dB/K
Transponder bandwidth	36.0 MHz
Uplink range	41000 km
Saturated satellite EIRP	45.0 dBW
Output back-off	2.8 dB
Downlink range	39000 km
Earth station G/T	31.0 dB/K
Downlink earth station fixed losses	3.5 dB

If the uplink and downlink earth station elevation angles are 5.6° and 20.3° respectively estimate the overall link, clear sky, CNR. (Assume pure TDMA operation such that there is no intermodulation noise, and assume that interference is negligible.) What clear sky earth station EIRP is required on the uplink? [15.3 dB, 72.8 dBW]

14.13   If uplink power control is used to precisely compensate uplink fading in the system described in Problem 14.12 and the downlink earth station is located in Bradford (54° N, 2° W) at a height of 440 m, where the one-minute rain rate exceeded for 0.01% of time is 25 mm/h, estimate the overall CNR exceeded for the following time percentages: (a) 0.1%; (b) 0.01%; (c) 0.001%. (Assume that under clear sky conditions one-third of the total system noise can be attributed to the aperture temperature of the antenna and two-thirds originates in the receiver.) What would be the maximum possible ISI-free bit rate if the modulation is QPSK and what values of BER would you expect to be exceeded for 0.1, 0.01 and 0.001% of time? [12.8 dB; 7.1 dB; −3.9 dB, 459 error/s; 8.5 kerror/s; 35.7 Merror/s]

14.14   A particular satellite communications system has the following TDMA frame structure:

Single reference burst containing 88 bits.
Preamble to each traffic burst containing 144 bits.
Frame duration of 750 μs.
Guard time after each burst of 24 bits duration.
Overall TDMA bit rate of 90.389 Mbit/s.

Ten earth stations are each allocated two traffic bursts per frame, and one station provides (in addition) the single reference burst. What is the frame efficiency assuming DSI is not employed? If the satellite were used purely for standard (64 kbit/s) PCM voice transmission, what would be the TDMA voice channel capacity of the system? How many consecutive samples from each voice channel must be transmitted per frame? [94.9%, 1340, 6]

14.15   A 14 GHz satellite downlink operates from synchronous orbit with 20 W output power connected to a 26 dB gain antenna with 3 dB feeder and TWT losses. The receiver uses a 12 m diameter antenna with 94% efficiency combined with a cooled amplifier with 100 K receiver noise temperature. If the energy per bit, divided by noise power spectral density, E_b/N_0, must be 18 dB and assuming the usual range to the synchronous orbit, what maximum bit rate of signal can be accommodated over this link? Comment on the practicality of this value. [446.6 Mbit/s].

CHAPTER 15

Mobile and cellular radio

15.1 Introduction

In comparison to the relative stability and modest technical developments which are occurring in long haul wideband microwave communication systems there is rapid development and expanding deployment of new mobile personal communication systems. These range from wide coverage area pagers, for simple data message transmission, through to sophisticated cellular systems for voice and data transmission, which employ common standards and hence achieve contiguous coverage over large geographical areas, such as all the major urban centres and transport routes in Europe, Asia or the continental USA. This chapter discusses the specific channel characteristics of mobile systems and examines the typical cellular clusters adopted to achieve continuous communication with the mobile user. It then highlights the important properties of current, and emerging, TDMA and code division multiple access (CDMA), mobile digital cellular communication systems.

15.1.1 Private mobile radio

Terrestrial mobile radio works best at around 250 MHz as lower frequencies than this suffer from noise and interference while higher frequencies experience multipath propagation from buildings, etc., section 15.2. In practice modest frequency bands are allocated between 60 MHz and 2 GHz. Private mobile radio (PMR) is the system which is used by taxi companies, county councils, health authorities, ambulance services, fire services, the utility industries, etc., for mobile communications.

PMR has three spectral allocations at VHF, one just below the 88 to 108 MHz FM broadcast band and one just above this band with another allocation at approximately 170

MHz. There are also two allocations at UHF around 450 MHz. All these spectral allocations provide a total of just over 1000 radio channels with the channels placed at 12½ kHz channel spacings or centre frequency offsets. Within the 12½ kHz wide channel the analogue modulation in PMR typically allows 7 kHz of bandwidth for the signal transmission. When further allowance is made for the frequency drift in the oscillators of these systems a peak deviation of only 2 to 3 kHz is available for the speech traffic. Traffic is normally impressed on these systems by amplitude modulation or frequency modulation and again the receiver is of the ubiquitous superheterodyne design, Figure 1.4. A double conversion receiver with two separate local oscillator stages is usually required to achieve the required gain and rejection of adjacent channel signals.

One of the problems with PMR receivers is that they are required to detect very small signals, typically -120 dBm at the antenna output, corresponding to 0.2 μV (RMS into 50 Ω), and, after demodulating this signal, produce an output with perhaps 1 W of audio power. It is this stringent gain requirement which demands double conversion. In this type of equipment the first IF is normally at 10.7 MHz and the second IF is very often at 455 kHz. Unfortunately, with just over 1000 available channels for the whole of the UK and between 20000 and 30000 issued licences for these systems, it is inevitable that the average business user will have to share the allocated channel with other companies in his/her same geographical area.

There are various modes of operation for mobile radio communications networks, the simplest of which is single frequency simplex. In simplex communication, traffic is broadcast, or one way. PMR uses half duplex (see later Table 15.3) where, at the end of each transmission period, there is a handover of the single channel to the user previously receiving, in order to permit him/her to reply over the same channel. This is efficient in that it requires only one frequency allocation for the communications link but it has the disadvantage that all units can hear all transmissions provided they are within range of the mobile and base station. An improvement on half-duplex is full-duplex operation where two possible frequencies are allocated for the transmissions. One frequency is used for the forward or downlink, namely base-to-mobile communications, while a second frequency is used for the reverse or uplink channel, namely mobile-to-base communications. This permits simultaneous two-way communication and greatly reduces the level of interference, but it halves the overall capacity. One possible disadvantage is that mobiles are now unable to hear each other's transmissions, which can lead to contention with two mobiles attempting to initiate a call, at the same time, on the uplink in a busy system.

Although PMR employs relatively simple techniques with analogue speech transmission there have been many enhancements to these systems over the years. Data transmission is now in widespread use in PMR systems using FSK modulation (see section 11.3.3). Data transmission also allows the possibility of hard copy graphics output and it gives direct access to computer services such as databases, etc. Data preambles can also be used, in a *selective calling* mode, when initiating a transmission to address a specific receiver and thus obtain more privacy within the system.

The problems in PMR are basically two-fold. One is the very restricted number of channels which are available. The second concerns the fact that mobile equipment will only operate when it is close to the base station transmitter which is owned by the company or organisation using the system. It has been the desire to design a wider coverage system, which also overcomes the restrictions of the limited number of channels, that has given rise to cellular radio and the new trans European trunked radio (TETRA) standards [Dunlop *et al.*, WWW] for paramilitary, i.e. police, fire service, etc., use. Following the development of national and international coverage systems, PMR is now used mainly by taxi operators. TETRA has developed a memorandum of understanding [WWW] for coordination of the adoption of the standard.

15.1.2 Radio paging systems

Another simple communication system, which is similar to broadcast, is one-way paging. These started as on-site private systems employing 1 W transmitters with, approximately, 1 μV sensitivity superheterodyne receivers. The early systems used sequential tone (FSK) transmission in the UHF band [Macario] and simply alerted a specific receiver that it had been called. Digital pagers were then further developed for wide area public paging, with the POCSAG coded service opening in London in the early 1980s. POCSAG had a capacity of 2 million active pagers, a message rate of 1 per second and a message length of 40 characters. This was achieved with a data rate of 512 bit/s using NRZ FSK modulation and a tone separation of 9 kHz.

A key feature, necessary to achieve long battery life, is that these simple receivers switch to standby when no message is being transmitted. Paging messages are batched and preceded by a preamble to ensure that all pager receivers are primed to receive the signals. The alphanumeric message, which is displayed on the paged receiver, is sent as sets of (32,21) $R = 2/3$ POCSAG coded data vectors, Chapter 10 and [Macario]. The European radio message system (ERMES) was standardised within ETSI for *international* paging. At 9.6 kbit/s ERMES supports four of the current 1.2 kbit/s POCSAG messages in simultaneous transmission. In 1994, INMARSAT launched a worldwide satellite based pocket pager with a £300 receiver cost. Text messaging in cellular systems has in many ways now minimised the attraction of pagers.

15.2 Mobile radio link budget and channel characteristics

The mobile communications channel suffers from several, potentially serious, disadvantages with respect to static, line-of-sight (LOS) links. These are:

1. Doppler shifts in the carrier due to relative motion between the terminals.
2. Slow spatial fading due, principally, to topographical shadowing effects along the propagation path.
3. Rapid spatial fading due to regions of constructive and destructive interference between signals arriving along different propagation paths. (Such multipath fading may also occur on fixed point-to-point systems but is usually less severe and more easily mitigated.)
4. Temporal fading due, principally, to the mobile terminal's motion through this spatially varying field.
5. Frequency selective fading when the signals are broadband.
6. Time dispersion due to multipath propagation.
7. Time variation of channel characteristics due, principally, to movement of the mobile terminal.

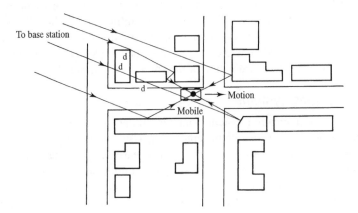

Figure 15.1 *Multipath origin of Doppler shift, fading and dispersion in a mobile radio channel (d indicates possible points of diffraction).*

Some of these effects are, of course, intimately connected and represent different manifestations of the same physical processes. Figure 15.1 illustrates the origin of these effects, with the signals reflecting off neighbouring buildings (or vehicles) before being summed at the mobile terminal receiving antenna. Note that Figure 15.1 is shadowed as the direct path is not available at the receiver due to the obstruction (i.e. shadowing) of the buildings.

15.2.1 Prediction of median signal strength

The fading processes referred to above [Patzold] mean that, in general, only statistical statements can be made about the signal strength in a particular place at a particular time[1]. Usually these statements are made in the form of cumulative distributions of signal strength. Ideally such distributions could be measured for each possible location of mobile and base station during the design of a mobile communications system. This, however, is usually neither practical nor economic and simplified models are used to predict these cumulative distributions. A common assumption used in these models is that propagation is essentially governed by equation (12.78) with a correction factor, incorporated to account for departures from a perfectly reflecting plane earth and the resulting random fading. A semi-empirical model [Ibrahim and Parsons] uses exactly this approach. The received median carrier strength in the model is derived from equation (12.78) as:

$$C = P_T + G_T - \text{PEPL} + G_R - \beta \quad \text{(dBW)} \tag{15.1(a)}$$

where PEPL is the plane earth path loss and:

$$\beta = 20 + \frac{f_{MHz}}{40} + 0.18L - 0.34H + K \quad \text{(dB)} \tag{15.1(b)}$$

---

[1] Recent spectacular improvements in the memory size, and speed, of computers combined with a similar reduction in hardware costs have made possible the simulation of mobile radio channels using detailed topographical databases combined with geometrical optic methods. Much effort is currently being invested in producing practical systems design tools from programs using such 'deterministic' modelling techniques.

and:

$$K = \begin{cases} 0.094U - 5.9, & \text{for inner city areas} \\ 0, & \text{elsewhere} \end{cases} \quad (15.1(c))$$

β is called a clutter factor, L is called the land usage factor and U the degree of urbanisation. L and U are defined by the percentage of land covered by buildings, and the percentage of land covered by buildings with four or more storeys, respectively, in a given 0.5 km square. These quantities were chosen as parameters for the model because they are collected, and used, by UK local authorities in their databases of land usage. H is the height difference in metres between the 0.5 km squares containing the transmitter and receiver. The RMS error between predicted and measured path loss using this model in London is about 2 dB at 168 MHz and about 6 dB at 900 MHz.

More sophisticated propagation loss models also take into account the height of the base station, h_{BS}, and the height of the mobile handset or station, h_{MS}, both defined in m when calculating the path loss. One widely used model is that developed by Hata from Japanese measurement data [Okamura *et al.*]. The path loss in the Hata urban model, defined as L_{Hu}, is:

$$L_{Hu} = 69.55 + 26.16 \log_{10} f_{MHz} - 13.82 \log_{10} h_{BS} - a(h_{MS})$$
$$+ (44.9 - 6.55 \log_{10} h_{BS}) \log_{10} R_{km} \quad \text{(dB)} \quad (15.2)$$

where f_{MHz} is the propagation frequency in MHz, $a(h_{MS})$ is a terrain dependent correction factor, while R_{km} is the path loss distance in km. The correction factor $a(h_{MS})$ for small and medium sized cities was found to be:

$$a(h_{MS}) = (1.1 \log_{10} f_{MHz} - 0.7) h_{MS} - (1.56 \log_{10} f_{MHz} - 0.8) \quad (15.3(a))$$

while for larger cities it is frequency dependent:

$$a(h_{MS}) = \begin{cases} 8.29[\log_{10}(1.54h_{MS})]^2 - 1.1, & \text{if } f_{MHz} \leq 200 \\ 3.2[\log_{10}(11.75h_{MS})]^2 - 4.97, & \text{if } f_{MHz} \geq 400 \end{cases} \quad (15.3(b))$$

The *typical suburban Hata model* applies a correction factor to the urban model yielding:

$$L_{H\,suburban} = L_{Hu} - 2[\log_{10}(f_{MHz}/28)]^2 - 5.4 \quad \text{(dB)} \quad (15.4)$$

The rural Hata model also modifies the urban formula:

$$L_{H\,rural} = L_{Hu} - 4.78(\log_{10} f_{MHz})^2 + 18.33 \log_{10} f_{MHz} - 40.94 \quad \text{(dB)} \quad (15.5)$$

The fundamental limitations of these model parameters are:

$$\begin{array}{ll} f_{MHz}: & 150\text{--}1500 \text{ (MHz)} \\ h_{BS}: & 30\text{--}200 \text{ (m)} \\ h_{MS}: & 1\text{--}10 \text{ (m)} \\ R_{km}: & 1\text{--}20 \text{ (km)} \end{array}$$

Thus for a 900 MHz GSM/CDMA system these conditions can be usually satisfied but for a 1.8 GHz urban microcell the limits must be stretched.

Steele and Hanzo have provided a detailed analysis of the Hata model at different antenna heights and they concluded that, when compared with real propagation data, in the range $R_{km} = 0.1$ to 1 km the Hata model is 10 dB more pessimistic than fitting regression lines to the collected data [Steele and Hanzo]. However, the Hata model provides a very good and widely used approximation for deriving the median signal strength for cellular mobile users.

In many cellular systems the propagation loss within the cell is modelled by a log-normal distribution following two distinct slopes or power laws. Square law propagation, as in equation (12.70), is used close to the base station with typically a fourth power law response closer to the mobile handset [Shepherd *et al.*].

15.2.2 Slow and fast fading

Slow fading [Patzold], due to topographic diffraction along the propagation path, tends to obey log-normal statistics and occurs, in urban areas, typically on a scale of tens of metres. Its statistics can be explained by the cascading (i.e. multiplicative) effects of independent shadowing processes and the central limit theorem (see section 3.2.9). Fast fading, due to multipath propagation where the receiver experiences several time delayed signal replicas, tends to obey Rayleigh statistics, Figure 4.23. This is explained by the additive effects of independently faded and phased signals and the central limit theorem (see sections 3.2.9 and 4.7.1). The spatial scale of Rayleigh fading is typically half a wavelength. On a spatial scale of up to a few tens of metres fading can, therefore, usually be assumed to be a purely Rayleigh process given by:

$$p(r) = \begin{cases} (r/\sigma^2)e^{\frac{-r^2}{2\sigma^2}}, & r \geq 0 \\ 0, & r < 0 \end{cases} \tag{15.6}$$

where r is the signal amplitude and σ is the standard deviation of the parent Gaussian distribution. The corresponding exceednce is:

$$P(r > r_{ref}) = e^{\frac{-r_{ref}^2}{2\sigma^2}} \tag{15.7}$$

And since the median value of a Rayleigh distributed quantity is related to the standard deviation of its parent Gaussian distribution by:

$$r_{median} = 1.1774\sigma \tag{15.8}$$

then the exceednce can be rewritten as:

$$P(r > r_{ref}) = e^{-\left(\frac{r_{ref}}{1.2\,r_{median}}\right)^2} \tag{15.9}$$

Equation (15.9) allows the probability that the signal amplitude exceeds any particular reference value for a given median signal level to be found if slow, i.e. log-normal, fading can be neglected. If log-normal fading cannot be neglected (which is usually the case in practice) then the signal level follows a more complicated (Suzuki) distribution [Parsons and Gardiner].

15.2.3 Dispersion, frequency selective fading and coherence bandwidth

Multipath propagation results in a received signal that is dispersed in time. For digital signalling, this time dispersion leads to a form of ISI whereby a given received data sample is corrupted by the responses of neighbouring data symbols. The severity of this ISI depends on the degree of multipath induced time dispersion (quantified by the multipath RMS delay spread of the radio channel) relative to the data symbol period. It is generally agreed that if the ratio of the RMS delay spread to symbol period is greater than about 0.3, then multipath induced ISI must be corrected if the system's performance is to be acceptable.

In the 900 MHz frequency band used for cellular mobile radio, wideband propagation measurements have shown that worst case RMS delay spreads are usually less than 12 μs. More particularly, urban areas tend to have RMS delay spreads of about 2 to 3 μs, with significant echo power up to about 5 μs, Figure 15.2, while rural and hilly areas have RMS delay spreads of about 5 to 7 μs [Parsons]. The Doppler shift, f_d, in Hz is given by:

$$f_d = \frac{v}{\lambda} = \frac{v}{c} f_{Hz} \tag{15.10}$$

where v is the relative velocity in the direction of propagation and c is the velocity of light, i.e. 3×10^8 m/s. Figure 15.2 also illustrates Doppler shift due to relative motion of the transmitter, receiver or reflectors. At 900 MHz a relative velocity of approximately 1.2 km/h thus produces a 1 Hz Doppler shift.

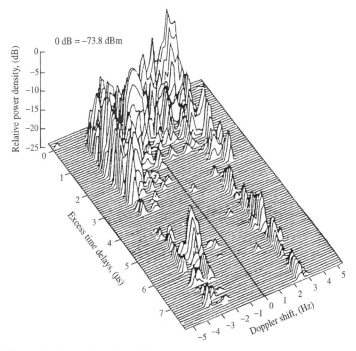

Figure 15.2 *Typical urban multipath profile (source: Parsons, 1991, reproduced with the permission of Peter Peregrinus).*

Multipath propagation results in a time varying fading signal as the mobile moves position, Figure 15.3. As the mobile moves then the magnitude and phase (i.e. delay) of the multipath components in Figures 15.1 and 15.2 alter. Now a single antenna receiver can only sum these components resulting in constructive and destructive interference depending on the precise phasing of the signal components. (Figure 15.3 shows the instantaneous received power for a mobile, with respect to the local spatial average or mean power level, measured when travelling round a path, at approximately fixed range from the base station.) This figure shows spatial fading due to mobile motion. If the reflectors are moving objects such as vehicles then temporal fading will result for a stationary mobile user.

Multipath delay spread can be, equivalently, quantified in the frequency domain by the channel's coherence bandwidth, which is roughly the reciprocal of the RMS delay spread. The coherence bandwidth gives a measure of spectral flatness in a multipath channel. If two signal components are separated by greater than the coherence bandwidth of a channel, then they tend to fade independently and the overall signal is said to experience frequency-selective fading. On the other hand, if a signal's bandwidth is less than the coherence bandwidth of the channel, then the channel effectively exhibits spectrally flat fading in which all the signal components tend to fade simultaneously.

15.2.4 Multipath modelling and simulation

Multipath simulation parameters were provided for GSM 900 MHz system evaluation by a European collaboration [COST 207]. The COST 207 group decided that the multipath should be modelled in terms of time delays with Doppler shifts associated with each delay to give a received signal, $z(t)$:

$$z(t) = \int\int_{R^2} y(t - \tau)\, S(\tau_i, f)\, e^{2\pi j f \tau}\, df d\tau \qquad (15.11)$$

where the right hand terms represent the delayed signals, their amplitudes and associated Doppler spectra [COST 207].

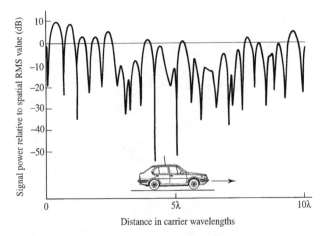

Figure 15.3 *Received power profile for a moving vehicle, as in Figure 12.28.*

The model is based on a wide sense stationary uncorrelated scattering model where the following information is used to describe or characterise the channel:

- discrete number of taps with associated delay and average power level,
- Rayleigh distributed instantaneous amplitude for each tap varying according to the Doppler spectrum $S(\tau_i, f)$ for tap i.

Doppler spectra are typically modelled by the classical distribution for short ($< \frac{1}{2}$ μs) delays:

$$S(\tau_i, f) = \frac{A}{\sqrt{\left[1 - \left(\dfrac{f}{f_d}\right)^2\right]}}, \qquad -f_d < f < f_d \qquad (15.12)$$

The shortest path is often modelled by a Ricean distribution [Rice], comprising the sum of a classical spectrum plus one direct path, and Gaussian distributions, equation (4.59), are used to model delays over $\frac{1}{2}$ μs.

For rural or non-hilly areas (RAs) then the COST model has 100 ns delay or tap spacing with 4 or 6 taps, Table 15.1:

Table 15.1 *The 6-tap rural area (RA) COST 207 multipath model.*

Tap no.	Delay (μs)	Average power level (dB)
1	0	0
2	0.1	−4
3	0.2	−8
4	0.3	−12
5	0.4	−16
6	0.5	−20

while typically urban (TU) uses a 12- or 6-tap model with much longer overall delay, Table 15.2:

Table 15.2 *The 6-tap simplified typical urban (TU) COST 207 multipath model.*

Tap no.	Delay (μs)	Average power level (dB)
1	0	−3
2	0.2	0
3	0.6	−2
4	1.6	−6
5	2.4	−8
6	5.0	−10

These tables give the average power level at each tap or delay value after applying the Doppler equation (15.12). These delays are considered to be appropriate for modelling 900 MHz GSM and CDMA transmission systems. Note that the TU model allows for shadowing by placing the largest signal at a 0.2 μs relative delay with respect to the direct transmission path which has been obscured (i.e. shadowed), as in Figure 15.1.

[Greenstein *et al.*] also provides further models for use at 900 MHz.

Path loss models at 1.8–1.9 GHz for wideband channels favour using the COST 231–Hata model as this gives a good approximation to practice for flat terrain with high base station height. In moderate or hilly terrain then other models [Erceg *et al.*] are often used.

15.3 Nationwide cellular radio communications

15.3.1 Introduction

The growth in the demand for mobile radio services soon exceeded the capacity of, or possible spectral allocations for, PMR-type systems. In the 1970s the concept therefore evolved of using base stations with modest power transmitters, serving all mobile subscribers in a restricted area, or cell, with adjacent cells using different operating frequencies. The key facet in cellular systems, introduced progressively through the 1980s, is that the power level is restricted so that the same frequency allocations can be reused in the adjacent cluster of cells, Figure 15.4. This constitutes a type of FDMA. Note that only cluster sizes of 3, 4, 7, 9, 12, etc., tesselate (i.e. lead to regular repeat patterns without gaps).

The protection ratio for these systems is the ratio of signal power from the desired transmitter (located at the centre of a cell) to the power received from a co-channel cell using the same operating frequency. Small clusters of 3 or 4 cells give 12 to 15 dB protection ratios which will only permit the use of a robust digital modulation method such as BPSK, Figures 11.10 and 11.21. As the cluster size increases so does the protection ratio. This arrangement is complicated as it involves handover as the mobile roams through the cells, but the ability to reuse the frequencies in close geographic proximity is a considerable advantage and this cellular concept has now therefore been adopted worldwide for mobile telephone systems.

Cellular radio communication has experienced rapid growth particularly in Asia and Europe. Hitherto, this unprecedented increase in user demand has been spurred on primarily by the business sector and contractors such as plumbers and carpenters but now cellular equipment is widely used by students and young people as well as business executives. It was the achievement of personal convenience and freedom at relatively low cost in the mid 1990s which made cellular radio really attractive to the general public. In

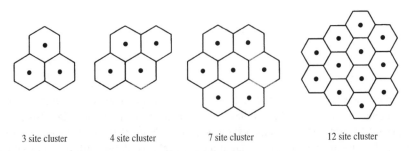

　3 site cluster　　　4 site cluster　　　7 site cluster　　　　　12 site cluster

Figure 15.4 *Examples of possible cluster patterns for cellular systems.*

1994 there were only 3 million registered users of the UK cellular systems, by 1996 this grew to almost 7 million and, in 2000, the penetration was 30 million users or 40% of the total population. (In 2000 Scandinavia had the highest market penetration worldwide (60–70%) for cellular telephony, the USA was 30% and, surprisingly, Germany was only 20%.)

It is predicted that, worldwide, user demand will continue to increase. In fact, mobile cellular telephone systems in some large cities are currently congested to near capacity. Thus, it will be necessary to continuously increase the capacity of mobile cellular systems for the foreseeable future and to increase the bandwidth as voice transmission moves to data transmission for file transfer, multimedia and other mobile computing applications.

15.3.2 Personal cordless communications

The simplest of these systems was the 12-channel domestic, analogue, cordless (CT1) telephone system which operated within 50 to 100 m of the base unit. The second generation 40-channel digital telepoint equipment (CT2) with a 50 to 200 m range allowed more mobility but, due to its still restricted range, needed many base stations. These telepoint FDMA systems, Tables 15.3 and 15.4, used simple 32 kbit/s ADPCM speech coders (section 5.8.3), with 10 mW of output power, to achieve two-way communication on one 72 kbit/s circuit with only a 5 ms round trip delay. Channels were allocated at call set-up, from the 40 available channels in the public base stations. The base stations are all connected to the PSTN giving telephony coverage similar to wired handsets.

The replacement system, DECT, was originally conceived as a cordless private branch exchange. However, DECT is more advanced than CT2 in that it operates at 1900 MHz with TDMA, Tables 15.3 and 15.4, has more advanced signalling and handover, and supports basic rate ISDN access (see Chapter 20). DECT uses TDMA with a 10 ms frame time which is split into 5 ms for the uplink and 5 ms for the downlink transmissions, using time division duplex (TDD). Within this frame it supports 12 separate TDMA time-slots and, with 12 RF channels in the 1880 to 1900 MHz spectral allocation, the base station capacity is approximately 140 simultaneous mobile users.

Table 15.3 *Modulation and access techniques for analogue and digital mobile systems.*

System	Mode		Access	Modulation
Paging	Simplex	-	-	FM/FSK
PMR	Half duplex	-	-	AM/FM
TACS	Full duplex	FDD	FDMA	FM
CT2	Full duplex	TDD	FDMA	FSK
DECT	Full duplex	TDD	MF-TDMA	Gaussian FSK
GSM/DCS	Full duplex	FDD	TDMA	GMSK
CDMA	Full duplex	FDD	CDMA	BPSK/QPSK

DECT (see also section 21.7.2) is designed primarily for indoor operation where the multipath delay spread is less than 50 ns as the bit period is 870 ns. Thus it cannot be extended for use in outdoor cellular systems unless they are of the smaller coverage (microcellular) system design. Due to the simplicity of DECT, it cannot handle significant Doppler shift due to handset motion and hence can only be used for mobiles travelling at walking pace. Also, due to the small range and hence cell sizes, DECT systems can only operate within localised areas as it is uneconomic to construct the 25000 100 m diameter cells, for example, which would be required to achieve full coverage of a large city centre such as London. However, these microcells are significant in that they use low power transmissions of 1 to 100 mW, which is attractive for battery powered personal mobile handsets and, with the base station antenna below roof height, there is very little energy that is radiated into adjacent cells.

Table 15.4 *Comparison of European digital cordless and cellular telephony systems.*

	DECT	GSM 900	DCS 1800
Operating band (MHz)	1880–1900	890–960	1710–1880
Bandwidth (MHz)	20	2×35	2×75
Access method	MF-TDMA	TDMA	TDMA
Peak data rate (kbit/s)	1152	270	270
Carrier separation (kHz)	1728	200	200
Channels per carrier	12	8	8
Speech coding	32 kbit/s	22.8 kbit/s	22.8 kbit/s
Coding/equalisation	no	yes	yes
Modulation	Gaussian FSK	GMSK	GMSK
Traffic channels/MHz	7	19	19
Mobile power output (W)	0.25	0.8–2.0	0.25–1
Typical cell size	50–200 m	0.3–35 km	0.02–8 km
Operation in motion	walking pace	> 250 km/h	> 130 km/h
Capacity (Erlangs/km^2)	10000+	1000	2000

15.3.3 Analogue cellular radio communication

First generation cellular systems employed analogue narrowband FM techniques [Black, Lee 1993]. For example, the North American advanced mobile phone system (AMPS) provided full-duplex voice communications with 30 kHz channel spacing in the 800 to 900 MHz band, while the UK total access communication system, TACS – and extended TACS (ETACS) – operated with 25 kHz channel spacing in the 900 MHz band.

In a narrowband FM/FDMA cellular system the total available radio spectrum is divided into disjoint frequency channels, each of which is assigned to one user. The total available number of channels is then divided amongst the cells in each cluster and reused in every cell cluster. Each cell in a given cluster is assigned a different set of frequency channels to minimise the adjacent channel interference while the cell size/cluster spacing is chosen to minimise co-channel interference. Analogue systems in general need larger received signal to interference ratio than digital systems and thus they use the bigger clusters in Figures 15.4 and 15.5 to give the required physically greater spacing between cells operating at the same centre frequency, in the adjacent clusters.

15.3.4 Cell sizes

Cell size is dependent on expected call requirements. From knowledge of the total number of subscribers within an area, the probability of their requiring access and the mean duration of the calls, the traffic intensity in Erlangs [Dunlop and Smith] can be calculated. Erlang tables can then be used to ascertain the number of required channels for a given blocking, or lost call, probability.

For the cellular geometries of Figure 15.5 it is possible to calculate the carrier to interference ratio (C/I) close to the edge of the cell, when receiving signals from all the adjacent co-frequency cells. In an n-cell cluster, using a fourth power propagation law, the carrier to interference ratio is:

$$\frac{C}{I} = \frac{r^{-4}}{(n-1)d^{-4}} \tag{15.13}$$

where r is the cell radius and d the reuse distance between cell centres operating at the same centre frequency. (This is an approximate formula, arrived at by inspection, which is close, but not equal, to the worst case. A more accurate formula is derived in [Lee 1995].) Figure 15.5 shows the reuse distance between adjacent 7-cell clusters. The cluster size required to achieve a given d/r (and therefore the corresponding C/I) can be found using:

$$\frac{d}{r} = \sqrt{3n} \tag{15.14}$$

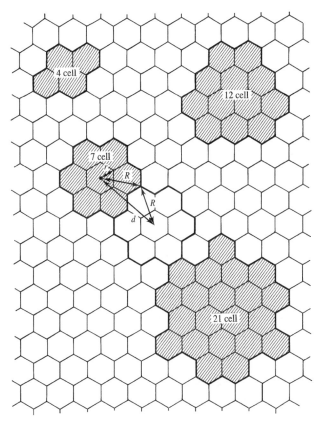

Figure 15.5 *Methods of grouping cells in order to cover a given area by repeating pattern.*

Although essentially general, equation (15.14) is particularly easily proved for the 7-cell cluster (Figure 15.5) as follows:

$$n = \frac{\text{cluster area}}{\text{cell area}} \tag{15.15}$$

The ratio of cluster to cell radii is therefore given by:

$$\frac{R}{r} = \frac{\sqrt{\text{cluster area}}}{\sqrt{\text{cell area}}} = \sqrt{n} \tag{15.16}$$

The angle between R and d shown in Figure 15.5 is 30°, therefore:

$$\frac{d}{2} = R \cos 30° \tag{15.17}$$

i.e.:

$$d = \sqrt{3}R \tag{15.18}$$

Substituting $R = \sqrt{n}r$ from equation (15.16) into (15.18) gives:

$$\frac{d}{r} = \sqrt{3n} \tag{15.19}$$

It is now possible to refer to Table 15.4 to obtain the available radio channel capacity within each cell. The total number of channels which are required to accommodate the expected traffic within a geographical area defines the required number of cells, or clusters. Division of the area covered by the total number of cells in use then provides the required area per individual cell. To accommodate higher traffic capacity, required in dense urban areas such as central London, necessitates smaller cell sizes, Figure 15.6, compared to more rural areas.

At the frequencies presently used for cellular radio, propagation usually follows (approximately) an inverse square law for field strength (i.e. an inverse fourth power law for power density). The median signal strength can be estimated using the propagation models in section 15.2.

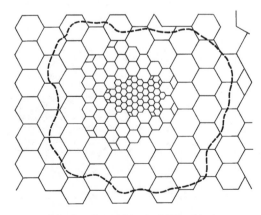

Figure 15.6 *Cell patterns used for London within the M25 orbital motorway, showing smaller cell patterns where most users are expected.*

As all cells use the same base station transmitter power, to minimise interference, the power level is set to ensure that there is adequate received signal at the edges of each cell compared to the thermal noise level. Transmission at a higher power level is not desirable as it simply wastes power. In the mobile the transmitter output power is controlled by feedback from the base station to minimise both interference between users and the mobile handset and battery drain (to maximise the time between recharges).

15.3.5 System configuration

In the UK cellular telephone network the base stations are connected to switching centres which hold information on the mobile transceivers which are active, including their location. Thus on calling a mobile transceiver from the wired PSTN one is connected first to the nearest switching centre and then on to the appropriate switching centre to route the call to the current mobile location, Figure 15.7. Packet data routing in mobile systems is described in section 20.5.2. Within the area covered by a switching centre the mobile experiences handover as it roams from cell to cell. These digital switching centres, which form the interface between the wired PSTN and the mobile network, each have the capacity to handle 20 million Erlangs of traffic and up to 250000 to 500000 subscribers in 1995. There were typically 30 switching centres for the dense London mobile traffic. The location of individual subscribers within the cells is tracked and this positional information within the overall mobile network is constantly updated in the home subscriber databases, Figure 15.7. More advanced positioning will be implemented progressively for mobile location estimation.

15.4 Digital TDMA terrestrial cellular systems

Although there are still some analogue cellular handsets in use in the UK, most of the 30 million users are connected to digital cellular networks which offer superior speech quality and improved privacy, and provide a natural extension to data transmission. Most importantly, digital cellular systems have the potential to provide significantly higher capacity than analogue cellular systems, while utilising the same available bandwidth, and can achieve worldwide interoperability.

15.4.1 TDMA systems

The digital cellular systems for Europe, North America and Japan are all based on TDMA. The pan-European global system for mobile communications (GSM), previously called Groupe Spéciale Mobile [Lee 1995, Black], which was the world's first TDMA cellular system, transmits an overall bit rate of 271 kbit/s in a bandwidth of 200 kHz, with eight TDMA users per carrier operating at 900 MHz, Figure 15.8. Thus, GSM 900 [WWW] effectively supports one TDMA user per 25 kHz. On the other hand, the North American narrowband IS-54 TDMA system accommodates three TDMA users per carrier by transmitting an overall bit rate of about 48 kbit/s over the same 30 kHz bandwidth of the existing AMPS system. The Japanese system is similar to the North American system, supporting three TDMA users per carrier with an overall bit rate of about 40 kbit/s transmitted over a 25 kHz bandwidth.

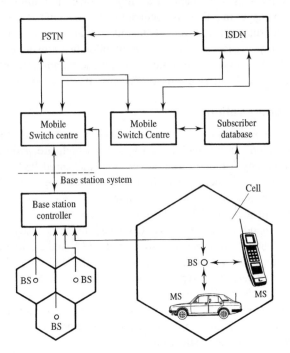

Figure 15.7 *Mobile system operation with mobile and base stations (MS) & (BS).*

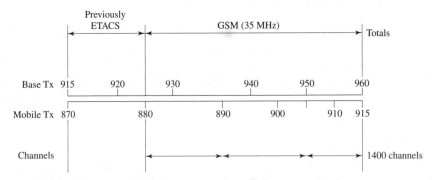

Figure 15.8 *UK allocation of 900 MHz spectrum for cellular communications.*

Such narrowband TDMA digital cellular systems operate at lower CNR than analogue systems and improve system capacity by improving frequency reuse through cell size reduction. Further capacity gains can also be realised by these digital systems through low bit rate speech coding, higher trunking efficiency and more effective frequency reuse due to increased robustness to co-channel interference. However, TDMA digital cellular systems are susceptible to multipath induced ISI which must be mitigated with relatively complex adaptive equalisation techniques, Table 15.4.

When it commenced operation in mid 1991, the GSM system provided duplex communications with 45 MHz separation between the frequency bands of the 890 to 915

MHz uplink, and the 935 to 960 MHz downlink, Figure 15.8. Over time, with the withdrawal of analogue systems, GSM has been extended to the 35 MHz bandwidth of Figure 15.8 with approximately 1400-channel capacity. As a vehicle mounted system, the mobile power output ranges from 800 mW to 20 W, permitting up to 30 km cell sizes in low traffic density rural areas. GSM in the UK has two operators, Vodafone and O_2.

In addition, GSM has been extended, via the DCS 1800 standard, to provide increased capacity for the large, hand-held portable market. DCS 1800 operates in the 1800 MHz band, with twice the available bandwidth and capacity of GSM, and accommodates low power (0.25 to 1 W), small size portable handsets. At the higher power levels this corresponds to DCS 1800 cell sizes ranging from 0.5 km (urban) to 8 km (rural) with high grade coverage. Table 15.4 provides a comparison of these systems. DCS 1800 is thus optimised more for high density, high capacity situations (not necessarily with the facility to handle fast moving mobile subscribers), to achieve a personal communications network (PCN) meeting the current communication expectations of subscribers.

The DCS 1800 system was initially introduced within the M25 orbital motorway area in London in 1993 as the PCN service before being rolled out to provide national coverage. DCS 1800 in the UK is licensed predominantly to T-Mobile and Orange, offers twice the number of radio channels and hence twice the capacity of GSM, Table 15.4. As in GSM, handsets are controlled by smart cards to allow several users access to the same handset with separate billing. The GSM/DCS standard was, in 1996, used in 270 networks in 98 countries and in 2000 GSM based systems carried 80% of all the worldwide digital mobile traffic. In 2001 GSM serviced 380M, i.e. 60% of the total 640M worldwide mobile telephone handset users.

15.4.2 TDMA data format and modulation

In GSM 900 and DCS 1800, eight TDMA users communicate over one carrier and share one base station transceiver, thereby reducing equipment costs. At the lowest level, the GSM TDMA format consists of eight user time-slots, each of 0.577 ms duration, within a frame of 4.615 ms duration (see Figure 15.9). This frame duration is similar to DECT but, in GSM, the transmissions are only one way and the uplink and downlink have *separate* frequency allocations as they use frequency division duplex, Table 15.3. TDMA users communicate by transmitting a burst of data symbols within their allotted time-slot in each frame. Each TDMA data burst contains 156.25 bits consisting of 116 coded speech bits, 26 training bits and 14.25 start, stop and guard bits (see Figure 15.9). This results in the overall transmitted bit rate of 270.8 kbit/s. GSM is relatively complex with significant overhead for control signalling and channel coding. However, because of the resulting redundancy, signal quality can be maintained even if one out of five frames is badly corrupted due to multipath, interference or noise induced errors.

GSM uses Gaussian minimum shift keying (GMSK) modulation with a bandwidth–time product of 0.3, section 11.4.6. With its Gaussian prefiltering, GMSK generates very low adjacent channel interference (greater than 40 dB below the in-band signal). The GMSK signal is transmitted at the bit rate of 270.8 kbit/s within a bandwidth of 200 kHz, resulting in a spectral (or bandwidth) efficiency of 1.35 bit/s/Hz. With eight TDMA users sharing a 200 kHz channel, Figure 15.9, GSM has essentially the same spectral capacity (in terms of number of users per unit bandwidth) as the previous analogue TACS system. However, GSM achieves a higher overall capacity (in

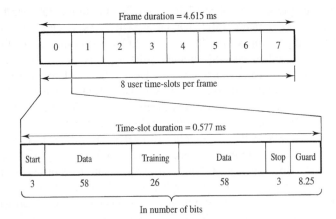

Figure 15.9 *Frame and time-slot details for data bit allocation in GSM.*

bit/s/Hz/km^2) by exploiting the digital signal's greater robustness to interference. Specifically, GSM operates at a signal to interference ratio of 9 dB while TACS required about 18 dB. Consequently, a GSM system achieves a higher frequency reuse efficiency with smaller cells and typical cell cluster sizes of 9 to 12, Figure 15.4.

15.4.3 Speech and channel coding

In order to achieve spectrally efficient coding with toll quality speech, GSM employs enhancements of LPC, section 9.7.1, with residual pulse excitation and a long term prediction [Gray and Markel], which compresses the speech traffic down to 13 kbit/s. There is also an enhanced version of this full rate coder which uses algebraic code excited linear prediction, also at the same 13 kbit/s rate.

Because of the low bit rates in these coders, speech quality is sensitive to bit errors. GSM overcomes this with channel coding in the form of 3 bit CRC error detection (Chapter 10) on 50 of the 182 most significant bits and half-rate convolutional coding with Viterbi decoding (Chapter 10) to protect further the 182 bits, Figure 15.10. The remaining 78 out of 260 bits are less important in achieving high speech quality and are left unprotected. The redundancy introduced by this channel coding results in an overall bit rate of 22.8 kbit/s (i.e. 456 bits transmitted in 20 ms), Table 15.4. Furthermore, these 456 coded bits are interleaved (see Chapter 10) over eight TDMA time-slots to protect

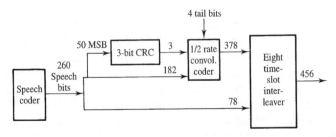

Figure 15.10 *Initial GSM channel coding technique to assemble a 20 ms time-slot.*

against burst errors resulting from Rayleigh fading. In addition to this channel coding, data encryption can also be used to achieve secure communications. It is this coding of digital data that gives GSM increased resilience to interference and hence more intense frequency reuse for high subscriber capacity.

Subsequent developments of enhanced performance half-rate speech coders increase further the active user capacity per unit of allocated bandwidth. The half-rate GSM speech coder uses CELP–vector sum excited linear prediction to reduce the rate to only 7 kbit/s.

15.4.4 Other operational constraints

Channel equalisation is required to overcome multipath induced ISI. Moreover, due to the time varying nature of the multipath fading channel, a user encounters a different channel impulse response at every TDMA burst. Thus, it is necessary to incur the overhead of transmitting a known training sequence in each time-slot, Figure 15.9, to allow the receiver to learn the channel impulse response and to adapt its filter coefficients accordingly. Furthermore, because the fading is potentially rapid in a mobile radio system, the channel response can change appreciably during a single TDMA burst. Consequently, the equaliser must be adaptively updated by tracking the channel variations. The GSM recommendations do not specify a particular equalisation approach, but the signal structure does lend itself to maximum likelihood sequence estimation with Viterbi decoding.

There are two methods of overcoming the deep fades of Figure 15.3. The GSM system combines slow frequency hopping with convolutional coding to ensure that a stationary mobile handset is not continuously in a deep null as altering the frequency of transmission changes the fading pattern. The use of slow frequency hopping (FSK) amongst carrier channels is a useful diversity method against Rayleigh fading. Since fading is independent for frequency components separated by greater than the coherence bandwidth of the channel, then FSK frequency agility reduces the probability that the received signal will fall into a deep fade for a long duration. This is especially effective for a slowly moving or stationary mobile which might otherwise experience a deep fade for a relatively long time.

Alternatively by using widely spaced and hence uncorrelated base station antennas then switching or combining the two antenna outputs also minimises the effect of these deep nulls.

For a given cellular layout and frequency plan, interference is reduced by proper use of adaptive power control, handover, discontinuous transmission and slow frequency hopping. For GSM mobile, adaptive power control is mandatory. The vehicular mobile has a transmit power dynamic range of about 30 dB (i.e. 20 mW to 20 W) adjusted at a rate of 2 dB every 60 ms, to achieve constant received signal power at the base station. Personal handsets have transmit power levels of 10 mW to 1 W.

The GSM handover strategy [Lee 1995] is based on finding a base station with equal or higher received signal strength, regardless of the received signal level from the current base station. Due to the variability in cell sizes and distances, propagation times from transmitter to receiver can range from about 3 to 100 μs. To ensure that adjacent TDMA time-slots from different mobile transmitters do not overlap at the base station receiver, each mobile must transmit its TDMA bursts with timing advances consistent with its distance from the base station. This timing information is intermittently measured at the base station and updates are sent to each mobile to ensure that this is achieved.

15.4.5 Trunked radio for paramilitary use

Another related TDMA mobile radio standard is the trans European trunked radio (TETRA) network which has been developed as part of the public safety radio communications service (PSRCS) for use by police, utilities, customs officers, etc. TETRA [Dunlop *et al.*] in fact is part of wider international collaborations for paramilitary radio use.

In these portable radios there is a need for frequency hopping (FH) to give an anti-eavesdropping capability and encryption for security of transmission to extend military mobile radio capabilities to paramilitary use, i.e. for police, customs and excise officers, etc. These capabilities are included in the multiband interteam radio (MBITR) for the associated public safety comms office (APCO) in the USA while Europe has adopted the TETRA standard.

TETRA is essentially the digital TDMA replacement of the analogue PMR systems of section 15.1.1. The TETRA standard has spectrum allocations of 380 to 400 and 410 to 430 MHz, with the lower band used for mobile transmissions and the upper band for base station use. TETRA mobiles have 1 W output power and the base stations 25 W using $\pi/4$ QPSK modulation. The TETRA network is designed to give a guaranteed probability of bit error with the data throughput rate varying, to meet the required quality of service. TETRA can accommodate up to four users each with a basic speech or data rate of 7.2 kbit/s. With coding and signalling overheads the final transmission rate for the four-user slot is 36 kbit/s. This equipment are larger and more sophisticated than a commercial cell phone, and they sell for a very much higher price than cell phones because the production runs are much smaller. However, their advanced capabilities are essential for achieving paramilitary communications which are secure from eavesdropping.

15.5 Code division multiple access (CDMA)

Analogue communication systems predominantly adopt frequency division multiple access (FDMA), where each subscriber is allocated a narrow frequency slot within the available channel, Figure 5.12. The alternative TDMA (GSM) technique allocates the entire channel bandwidth to a subscriber but constrains him/her to transmit only regular short bursts of wideband signal. Both these accessing techniques are well established for long haul terrestrial, satellite and mobile communications as they offer very good utilisation of the available bandwidth.

15.5.1 The CDMA concept

The inflexibility of these coordinated accessing techniques has resulted in the development of new systems based on the uncoordinated spread spectrum concept [Dixon, Lee 1993, Viterbi]. In these systems the bits of slow speed data traffic from each subscriber are deliberately multiplied by a high chip rate spreading code, $s_k(n)$, forcing the low rate (narrowband data signal) to fill a wide channel bandwidth, Figure 15.11.

Spread spectrum multiplexing thus has the following characteristics:

- channel bandwidth many times data bandwidth,
- bandspread narrowband data traffic into common wideband channel,
- employment of coding and modulation techniques at the transmitter,
- requirement for correlator or matched filter receiver to detect and demodulate the signals.

Spreading ratios, i.e. ratios of the transmitted (chip) bandwidth to data (bit) bandwidth, are typically between 100 and 10000. (This is in contrast to the longer duration M-ary symbols of Chapter 11. In CDMA, the symbol rate is represented by the chips of spreading code which are of much shorter duration than the information digits.) Many subscribers can then be accessed by allocating a unique, orthogonal, spreading code, $s_k(n)$, to each user, Figure 15.11. This constitutes a CDMA system. The signals, which are summed in the channel, have a flat, noise-like, spectrum allowing each individual transmission to be effectively hidden within the multiple access interference.

The key benefits of spread spectrum or CDMA are:

- the receiver only detects correctly coded data sequences,
- there is a signal-to-noise ratio (SNR) gain on reception,
- other user interference is reduced,
- it achieves an (uncoordinated) multiple access capability,
- the wide transmission bandwidth can resolve individual multipath responses,
- the receiver can recover precise received signal timing.

In the receiver, detection of the desired signal is achieved by correlation, section 2.6, against a local reference code, which is identical to the particular spread spectrum code employed prior to transmission, Figure 15.11 (i.e. $s_k(n)$ is used to decode the subscriber k transmissions). The orthogonal property of the allocated spreading codes means that the output of the correlator is essentially zero for all except the desired transmission. Thus correlation detection gives a processing gain or SNR improvement, G_p, equal to the spreading ratio:

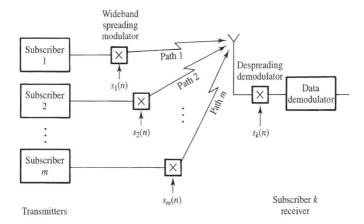

Figure 15.11 *Principle of (direct sequence) code division multiple access (CDMA).*

$$G_p = 10 \log_{10} \left(\frac{\text{transmitted signal bandwidth}}{\text{original data bandwidth}} \right) \text{ (dB)}$$

$$= 10 \log_{10} \left(\frac{R_c}{R_b} \right) \text{ (dB)} \tag{15.20(a)}$$

where R_c is the chip rate and R_b the binary digit rate. Equivalently this can be expressed as:

$$G_p = 10 \log_{10} (T_b B) \text{ (dB)} \tag{15.20(b)}$$

where B is the bandwidth of the spreading code and T_b is the, relatively long, bit duration of the information signal.

The following (trivial) example shows how a short 3-chip spreading code, $s(n)$, can be added modulo-2 to the slow speed data. Each data bit has a duration equal to the entire 3-chip spreading code, and is synchronised to it, to obtain the transmitted product sequence, $f(n)$.

Data		+1	+1	+1	−1	−1	−1	+1	+1	+1
Spreading code	$s_k(n)$	+1	−1	+1	+1	−1	+1	+1	−1	+1
Product sequence	$f(n)$	+1	−1	+1	−1	+1	−1	+1	−1	+1

15.5.2 CDMA receiver design

Both coherent receiver architectures (i.e. the active correlator and matched filter, of Figure 15.12) can be used for decoding or despreading the CDMA signals. The matched filter receiver [Turin, 1976] implements convolution (section 4.3.3) using a finite impulse response filter, Figure 15.12(a), whose coefficients are the time reverse of the expected sequence, to decode the transmitted data.

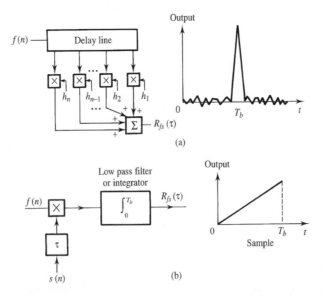

Figure 15.12 *Spread spectrum receiver typical outputs: (a) matched filter; (b) active correlator.*

For a discrete or sampled input signal, $f(n)$, with corresponding frequency spectrum, $F(f)$, the receiver filter frequency response, $H(f)$, which maximises the output SNR is given by:

$$H(f) = k \ F^*(f)e^{(-j2\pi f t_m)} \qquad (15.21)$$

and its corresponding time domain impulse response is:

$$h(n) = k \ f^* \ (n_o - n) \qquad (15.22)$$

where the superscript * denotes the complex conjugate operation.

This implies that the impulse response of the matched filter receiver in Figure 15.12(a) is a *time reversed* ($-n$) or mirror image version of the expected signal $f(n)$, *delayed* by n_o samples where n_o is the duration of the input signal, i.e. t_m. (See sections 8.3.1 to 8.3.3 for a more detailed discussion of matched filtering.) This condition ensures that $|H(f)| = |F(f)|$, i.e. the frequency response of the filter is carefully matched to the frequency spectrum of the expected signal. The complex conjugate operation ensures that all the different signal components add in phase at the (optimum) sampling instant. Thus the filter coefficients or weights in Figure 15.12(a) (which essentially repeats Figure 8.34) would be $h_1 = +1$, $h_2 = -1$ and $h_3 = +1$ for the previous 3-chip code. (Note that in this simple example the code is symmetric and the time reversal is not therefore explicitly evident.) The output is given by the convolution of the received signal with the stored weight values, as shown previously in Chapters 4 and 8.

Time reversal occurs because h_1 must be the final tap weight in Figure 15.12(a) to match with the first chip of the spreading sequence as it propagates through the delay line. However, the *impulse response* of the Figure 15.12(a) filter has h_n as the first output sample and h_1 as the final sample, hence the time reversal operation.

The signal to interference ratio at the filter input is given by:

$$\left(\frac{S}{I}\right)_{in} = \frac{S}{I_o B} \qquad (15.23)$$

where I_o is the power spectral density of all unwanted signals, S is the desired signal power and B is the bandwidth of the spread signal, $f(n)$. The output SNR, after adding the processing gain arising from matched filter detection, equation (15.19), is then:

$$\left(\frac{S}{N}\right)_{out} = \frac{S}{I_o f_m} \qquad (15.24)$$

where f_m ($= R_b$) is the bandwidth of the desired information signal, i.e. $1/T_b$ in equation (15.20(b)).

If the receiver is not synchronised then the received signal will propagate through the matched filter which outputs the complete correlation function. The large peak confirms that the correct code is indeed being received and provides accurate timing information for the received signal, Figure 15.12(a). Note that with binary stored weight values the filter design is especially simple. Semiconductor suppliers market single chip 'correlators' with up to 64 binary weighted taps at 20 to 30 Mchip/s input rates.

When timing information is already available then the simpler active correlator receiver, Figure 15.12(b), can be used (also shown in Figure 15.11). This receiver only operates correctly when the local receiver reference $s_k(n)$ is accurately matched, and correctly timed, with respect to the spreading code within the received signal $f(n)$. Here the mixer performs the multiplication to give an all-ones output and the integrator sums the ones over the data bit period, T_b. Strictly speaking the input is the spreading sequence $f(n)$ plus interference from all the other users and multipath responses, as in Figure 8.32. Thus, in practical systems, the idealised outputs in Figure 15.12 will be degraded by noise and other user interference. Synchronisation can be obtained by sliding the reference signal through the received signal. This can be an extremely slow process, however, for large $T_b B$ spreading waveforms. It is generally accepted that serial search with the Figure 15.12(b) correlator takes a time equal to the product of the timing uncertainty in the received code and the processing gain!

The pseudo-random bit sequence (PRBS) or pseudo-noise (PN) sequence, as obtained using a linear feedback shift register (Figure 13.15), can be used as the spreading code. For example, one 7-chip PN sequence is $1, 1, 1, -1, -1, 1, -1$ and a 15-chip PN sequence is $1, 1, 1, 1, -1, 1, -1, 1, 1, -1, -1, 1, -1, -1, -1$ [Golomb *et al.*]. As the spreading code is often phase modulated onto the carrier and coherently demodulated $+1/-1$ is a more appropriate representation for the 1/0 binary digits. These codes are repeated cyclically to form continuous spreading codes. For spread spectrum applications the most important properties of PN sequences are that the *cyclic* autocorrelation sidelobe levels are small. For a sequence of K chips the sidelobes are equal to -1, Figure 15.13(a), with a peak amplitude of K. (Figure 15.13 repeats Figure 13.16 but without normalisation.) The aperiodic autocorrelation, for an isolated PN sequence transmission, has peak sidelobes of $< \sqrt{K}$, Figure 15.13(b). The low level of cyclic autocorrelation sidelobes (-1) in Figure 15.13(a) only occurs for the zero Doppler shift case, while the plot with Doppler offset is more like the aperiodic performance of Figure 15.13(b).

The PN sequence produces an approximate balance in the number of 1 and -1 digits and its two-valued autocorrelation function in Figure 15.13(a) closely resembles that of a white noise waveform, Figure 3.28. The origin of the name pseudo-noise is that the

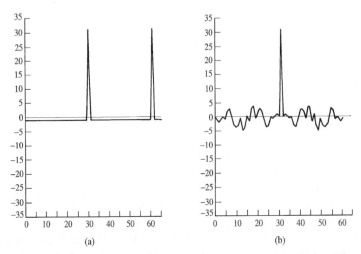

(a)   (b)

Figure 15.13 *PN code autocorrelation function: (a) for continuous $K = 31$ chip PN coded transmission; and (b) burst or aperiodic transmission again for $K = 31$.*

digital signal has an autocorrelation function which is very similar to that of a white noise signal. Note that the rise time and fall time for the peak of the function in Figure 15.13 is one chip period in duration and hence its bandwidth is directly related to the clock, or chip, rate of the spreading sequence.

The key facet of the CDMA system is that there is discrimination against narrowband and wideband interference, such as signals spread with a code sequence which is different to that used in the receiver, Figure 15.14. The level of the suppression, which applies to CW as well as code modulated waveforms, is given by the processing gain of equations (15.20).

15.5.3 Spreading sequence design

Maximal length sequences

Maximal length shift register sequences or m-sequences are so called due to their property that all possible shift register states except the all-zero state occur in a single, K length, cycle of the generated sequence [Golomb *et al.*]. Therefore for a shift register with n elements, the longest or maximum (m) length sequence which can be generated is $K = 2^n - 1$, section 13.4.2.

Due to the occurrence of all shift register states (except the all-zero state) each m-sequence will consist of 2^{n-1} ones and $(2^{n-1} - 1)$ zeros. There will be one occurrence of a run of n ones, and one run of $(n - 1)$ zeros. The number of runs of consecutive ones and zeros of length $(n - 2)$ and under will double for each unit reduction in run length and be divided equally between ones and zeros, see section 13.4.2 on PRBSs. Typical m-sequence generator feedback connections were shown in Table 13.3 and the overall size of the m-sequence set for different sequence lengths is given here in Table 15.5.

The correlation properties of the m-sequence family are interesting due to their flat periodic autocorrelation sidelobes, Figure 15.13(a), which provide a good approximation to an impulsive autocorrelation function. The cross correlation profile for a pair of m-sequences (forward and time reversed) is more typical of the generalised cross correlation function.

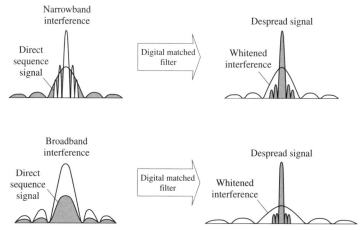

Figure 15.14 *Narrowband and wideband interference reduction in spread spectrum signals.*

It can be seen from Table 15.5 that the available set size is very much smaller than the sequence length, K, particularly in the case of the even shift register orders. Therefore, as a multiple access code set, the PN sequence does not provide adequate capability for system subscribers who each require their own, unique, code assignment. An analytical expression for bounds on the maximum cross correlation level has become available [Pursley and Roefs] for the aperiodic sequence cross correlation function. These sequences are not really appropriate for CDMA applications as the cross correlation levels are too large. m-sequence sets in isolation are not, therefore, favoured for practical, high traffic capacity CDMA systems, hence the development of other sequence sets.

Table 15.5 *Set sizes and periodic cross correlation peak levels for m-sequences and Gold codes.*

n	K	m-sequences		Gold codes	
		Set size	Peak level	Set size	Peak level
3	7	2	5	9	5
4	15	2	9	17	9
5	31	6	11	33	9
6	63	6	23	65	17
7	127	18	41	129	17
8	255	16	95	257	33
9	511	48	113	513	33
10	1023	60	383	1025	65
11	2047	176	287	2049	65
12	4095	144	1407	4097	129

Gold codes

Selected pairs of m-sequences exhibit a three-valued periodic cross correlation function, with a reduced upper bound on the correlation levels as compared with the rest of the m-sequence set. This m-sequence family subset is referred to as the preferred pair and one such unique subset exists for each sequence length. For the preferred pair of m-sequences of order n, the periodic cross correlation and autocorrelation sidelobe levels are restricted to the values given by $(-t(n), -1, t(n) - 2)$ where:

$$t(n) = \begin{cases} 2^{(n+1)/2} + 1, & n \text{ odd} \\ 2^{(n+2)/2} + 1, & n \text{ even} \end{cases} \qquad (15.25)$$

The enhanced correlation properties of the preferred pair can be passed on to other sequences derived from the original pair. By a process of modulo-2 addition of the preferred pair, Figure 15.15, the resulting derivative sequence shares the same features and can be grouped with the preferred pair as a member of the newly created family. This process can be repeated for all possible cyclically shifted modulo-2 additions of the preferred pair of sequences, producing new family members at each successive shift. A zero-shift 31-chip code can be generated by modulo-2 addition of two parent Gold sequences:

$$+1+1+1+1+1-1-1-1+1+1-1+1+1+1-1+1-1 \quad \cdots \quad -1+1+1-1-1$$
$$+1+1+1+1+1-1-1+1+1-1-1+1+1-1-1-1-1+1 \quad \cdots \quad -1+1+1+1-1-1$$
$$\overline{-1-1-1-1-1-1+1+1+1+1-1+1+1-1+1+1 \quad \cdots \quad -1-1-1+1-1}$$

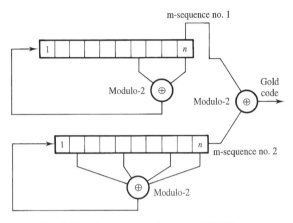

m-sequence no. 1

Figure 15.15 *Generation of a Gold code sequence for use in CDMA.*

and the corresponding 5-chip shifted Gold code for the second sequence would be:

$$+1+1+1+1+1-1-1-1+1+1+1-1+1+1+1-1+1+1-1 \quad \cdots \quad -1+1+1-1-1$$
$$-1-1+1-1-1+1+1+1-1-1-1-1+1+1-1+1+1+1-1+1 \quad \cdots \quad +1+1+1+1+1+1$$
$$\overline{+1+1-1+1+1+1+1+1-1+1+1+1-1-1+1+1-1+1+1+1+1 \quad \cdots \quad +1-1-1+1+1+1}$$

For this case, i.e. with an $n = 5$ generating register, any shift in initial condition from zero to 30 chip can be used (a 31-chip shift is the same as the zero shift). Thus, from this Gold sequence generator, 33 maximal length codes are available when the preferred (original) pair are added as valid Gold codes. Extending this, any appropriate two-register generator of length n can generate $2^n + 1$ maximal length sequences (of $2^n - 1$ code length) including the preferred pair of sequences. The family of sequences derived from and including the original preferred pair are known collectively as Gold codes.

Since there are K possible cyclic shifts between the preferred pairs of m-sequences of length K, the available set size for the Gold code family is $K + 2$. The available sequence numbers for the Gold code set are compared with the available number of m-sequences in Table 15.5. It is apparent that, not only do Gold codes have improved periodic cross correlation properties, but they are also available in greater abundance than the m-sequence set from which they are derived. However, as with the periodic m-sequence case, this attractive correlation property is lost as soon as the data traffic modulation is added.

Gold code, shift register based designs exist, the direct approach consisting of two m-sequence generators employing the necessary preferred pair of feedback polynomials. This type of generator is shown in Figure 15.15 and can be used to synthesise all the preferred pair derived sequences in the Gold code by altering the relative cyclic shift through control of the shift register initial states. Switching control is also required over the output 'exclusive-or' operation in order that the pure m-sequences can also be made available.

Gold codes are widely used in spread spectrum systems such as the international Navstar global positioning system (GPS), which uses 1023-chip Gold codes for the civilian clear access (C/A) part of the positioning service.

Walsh sequences

Walsh sequences [Beauchamp, 1990] have the attractive property that all codes in a set are precisely orthogonal. A series of codes $w_k(t)$, for $k = 0, 1, 2, \cdots, K$, are orthogonal with weight K over the interval $0 \le t \le T$, section 2.5.3 (where T usually equals T_b), when:

$$\int_0^T w_n(t)w_m(t)\, dt = \begin{cases} K, & \text{for } n = m \\ 0, & \text{for } n \neq m \end{cases} \tag{15.26}$$

where n and m have integer values and K is a non-negative constant which does not depend on the indices m and n but only on the code length K. This means that, in a *fully synchronised* communication system where each user is uniquely identified by a different Walsh sequence from a set, the different users will not interfere with each other *at the proper correlation instant*, when using the same channel.

Walsh sequence systems are limited to code lengths of $K = 2^n$ where n is an integer. When Walsh sequences are used in communication systems the code length K enables K orthogonal codes to be obtained. This means that the communication system can serve as many users per cell as the length of the Walsh sequence. This is broadly similar to the CDMA capability of Gold codes, Table 15.5.

There are several ways to generate Walsh sequences, but the easiest involves manipulations with Hadamard matrices. The orders of Hadamard matrices are restricted to the powers of two, the lowest order (two) Hadamard matrix being defined by:

$$H_2 = \begin{bmatrix} 1 & 1 \\ 1 & -1 \end{bmatrix} \tag{15.27}$$

Higher order matrices are generated recursively from the relationship:

$$H_K = H_{K/2} \otimes H_2 \tag{15.28}$$

where $\otimes$ denotes the Kronecker product. The Kronecker product is obtained by multiplying each element in the matrix $H_{K/2}$ by the matrix H_2. This generates K codes of length K, for $K = 2, 4, 8, 16, 32$, etc. Thus for $K = 4$:

$$H_4 = \begin{bmatrix} 1 & 1 & 1 & 1 \\ 1 & -1 & 1 & -1 \\ 1 & 1 & -1 & -1 \\ 1 & -1 & -1 & 1 \end{bmatrix} \tag{15.29(a)}$$

The difference between the Hadamard matrix and the Walsh sequence is only the order in which the codes appear; the codes themselves are the same. The reordering to Walsh sequences is done by numbering the Walsh sequences after the number of zero crossings the individual codes possess, to put them into 'sequency' order. The Walsh sequences thus result when the H_4 rows are reordered:

$$W_4 = \begin{bmatrix} 1 & 1 & 1 & 1 \\ 1 & 1 & -1 & -1 \\ 1 & -1 & -1 & 1 \\ 1 & -1 & 1 & -1 \end{bmatrix} \tag{15.29(b)}$$

Due to orthogonality, when the autocorrelation response peaks the cross correlations for all other time synchronised Walsh sequenced transmissions are zero. The time sidelobes, however, in autocorrelated and cross correlated Walsh sequences have considerably *larger* magnitudes than those for Gold or PN sequences. Also the orthogonal performance no longer applies in the imperfect (multipath degraded) channel.

15.5.4 Data modulation

Typically high chip rate, short PN, Gold or Walsh coded CDMA spreading sequences are multiplied by the slower data traffic and this signal is used to phase-shift-key a carrier as described in Chapter 11. When the received signal is fed into a matched filter receiver then the baseband output, as shown in Figure 15.16, results. Figure 15.16 shows a 31-chip spreading code which is keyed by the low rate data traffic sequence $-1, -1, +1, -1, +1, +1, -1, -1$, before reception in the 31-tap matched filter of Figure 15.12(a).

The matched filter or correlation receiver only provides a sharply peaked output if the correct codeword is received. CDMA systems thus rely on the auto- and cross correlation properties of codes such as those described above to minimise multiple access interference between subscribers.

15.5.5 CDMA multipath processing

The channel multipath of Figures 15.1 and 15.3 destroys the orthogonal properties of the CDMA spreading codes. If a CDMA signal has sufficiently wide bandwidth (i.e. greater than the coherence bandwidth of the channel) it is able to resolve the individual multiple paths or taps which were defined in Tables 15.1 and 15.2, producing several peaks at the code matched filter output at different time instants.

For efficient operation the receiver must collect and use all the received multipath signals in a coherent manner. It does this by using a channel equaliser which

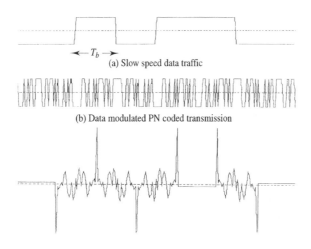

(a) Slow speed data traffic

(b) Data modulated PN coded transmission

(c) Receiver matched filter response to above encoded sequence

Figure 15.16 *Data modulation and demodulation on direct sequence coded signals.*

continuously adapts to the time varying mobile channel characteristic, effectively implementing a filter which is matched to this characteristic. The RAKE filter [Turin, 1980], in which the weights in the sum bus of a tapped delay line are derived by measuring the channel impulse response, is one example. The RAKE filter is a channel matched (FIR) filter where the filter weights are set as the complex conjugate of the magnitudes of the delayed multipath components, i.e. they are the time reverse of the channel response. The RAKE filter typically provides 3 dB saving in power for a given error rate in a typical urban channel.

A CDMA RAKE receiver can also be configured to talk simultaneously to two separate base stations from different, but closely spaced, cell sites. This provides superior performance at the cell boundary, where the signal strength is weakest, and can also be used to achieve a soft handover capability between cells.

15.5.6 The cdmaOne system

Second generation mobile CDMA cellular systems [Viterbi, Lee and Miller] use a spreading ratio, or K value, of approximately 127. In the US cdmaOne CDMA standard [Steele, *et al.,* IS-95 and 3gpp2 WWW sites], the basic user transmission (speech coder output) rate is 9.6 kbit/s. This is spread with a channel chip rate of 1.2288 Mchip/s (implying a total spreading factor of 128). Detailed analysis has shown that this $T_b B$ product is sufficient to support all the expected users within a typical cell; 1.2288 Mbit/s is also an acceptable speed for the VLSI modem electronics. The spreading process deployed is different on the downlink and uplink paths.

The downlink information, transmitted to the mobile, is split into four channels. These are the pilot, synchronisation, paging and traffic channels. The receiver must demodulate these channels, to acquire the necessary systems information, etc., before data transmission can start. This is needed to synchronise the mobile handset to the receiver's local basestation, permit him/her to measure and compensate for the multipath on the communication channel and alert him/her to incoming calls. The CDMA spreading is done by a combination of short Walsh and 32767-chip short PN spreading codes. The PN spreader uses a complex (inphase and quadrature) modulation technique. Although the overall spreading ratio is 127 the spreader uses 64 Walsh codes. These are partitioned into one synchronisation and one pilot channel and 62 remaining channels where up to 7 of the 62 traffic channels are used for paging the remote handset. Walsh codes are thus used to label the channels in a true CDMA scheme. The all-ones Walsh code, W_0, representing the top row in equation (15.26), is used for the pilot channel and the alternating 1 −1 1 −1 code, W_{32}, is used for the 1200 bit/s synchronisation channel.

The pilot channel is thus a continuous coherent transmission with each base station using a unique offset from the same long $2^{42} - 1$ PN code which permits the mobile transceiver to note the presence of the strongest base station and to measure or estimate channel impulse response parameters such as delay, phase and magnitude of the three expected multipath components for use in the RAKE antimultipath receiver processor. The pilot is often transmitted at 4 to 6 dB higher power level than the traffic channels to facilitate this measurement. (All base stations transmit the same PN code but a different relative delay, or code phase, is allocated to each individual station.)

The synchronisation channel, which is convolutionally encoded, interleaved and then coded with the W_{32} Walsh sequence, provides the Walsh sequence allocation information to identify the coded mobile transmissions. The synchronisation channel permits the mobile to note the precise timing of the base station spreading codes, particularly time of day and long $2^{42} - 1$ PN code synchronisation (timing) information for the mobile station decoder.

With this access completed the mobile station then listens on a paging channel for system information and it enters an idle state to save battery power if it is not in active communication. The paging channel is convolutionally encoded, interleaved, encrypted with the long PN sequence and then coded with the user-specific Walsh sequence. This provides access to the system overhead information and specific messages for the mobile handset.

The speech traffic data rates in cdmaOne are 9.6/4.8/2.4/1.2 kbit/s depending on speaker activity. To ensure a constant rate of 19.2 ksymbol/s, after half-rate convolutional coding (section 10.9), the lower data rates of 1.2 to 4.8 kbit/s are simply repeated. The traffic data is then interleaved, Figure 10.23, and scrambled by the long $2^{42} - 1$ chip PN code generator before Walsh coding with its own unique CDMA code and complex PN spreading by the final 32767-chip code. The traffic channel also contains the embedded power control signals which adjust the transmitted power level every 1.25 ms.

Each mobile in a given cell is assigned a different Walsh spreading sequence, providing perfect separation among the signals from different users, at least for a single path channel. To reduce interference between mobiles that use the same spreading sequence in different cells, and to provide the desired wideband spectral characteristics (from Walsh functions which have variable power spectral characteristics), all signals in a particular cell are subsequently scrambled using the short $2^{15} - 1 = 32767$ chip PN sequence which is generated at the *same* rate as the Walsh sequence (1.2288 Mchip/s). Orthogonality among users within a cell is preserved because all signals are scrambled identically and the transmissions are synchronous.

The CDMA uplink uses a somewhat different scheme. Here the speech is rate 1/3 convolutional coded (Chapter 10) with a constraint length of 9 with data repetition again depending on speech activity. The uplink does not transmit directly the bit symbols as in the downlink but, following interleaving, it groups 6 bits into unique symbols. This mapping to one of the $2^6 = 64$ encoded orthogonal Walsh sequences (i.e. using a 64-ary orthogonal signalling system) was implied previously in Figure 11.45 where three symbols were effectively mapped into one of eight orthogonal sequences. Each of these $2^6 = 64$ symbols generates the appropriate 64-chip Walsh code which is then combined with both the long $2^{42} - 1$ PN code and the 32767-chip quadrature spreading code prior to transmission.

Thus on the uplink the Walsh codes are used to convey data to the base station. A base station can recognise the signals intended for it by using the appropriate offset in the $2^{42} - 1$ chip long PN spreading code. The base station uses 64 correlators for the 64 Walsh codes and the resulting streams of recovered 6 bit symbols are reordered to represent each mobile station transmission before employing convolution decoding. The rate 1/3 coding and the mapping onto Walsh sequences achieves a greater tolerance to interference than would be realised from traditional PN spreading and offers the possibility of employing more sophisticated receiver (array) processing in the base station.

The mobile receiver in cdmaOne operates with matched filter detection of the coded waveform plus a RAKE antimultipath filter. It is generally accepted for CDMA matched filter detection in a perfect channel without multipath that the error ratio will follow the characteristic of Figure 15.17 which approximates to the expected downlink performance. Figure 15.17 is derived from [Mowbray *et al.*]. This implies approximately 12 to 15% user capacity with respect to the spreading ratio which is employed, i.e. 15 to 17 users with a typical spreading gain of 127. However, as CDMA can reuse frequencies in each cell, then a 7-cell CDMA system will have the same 100% capacity of a 7-cell TDMA cluster. Also if one exploits the silences in telephone speech signals, see DSI in section 14.3.7 or talkspurts in section 19.4.1, then, when the transmitter is switched off in the silence periods, interference is reduced and CDMA active user capacity can be more flexibly increased than in TDMA.

Current research is investigating more sophisticated multiuser detection techniques [Verdu] where the demodulator either explicitly or implicitly assumes that other CDMA users are present and uses this information to reduce the error ratio well below that of matched filter detection in Figure 15.17. This achieves a BER of 10^{-3} to 10^{-4} which is relatively independent of the number of active users [Verdu].

The pilot channel is typically transmitted with 20% of the total base station power. This high level signal permits the receiver to measure the strength and delay of all the nearby base station transmissions in order to select the closest ones yielding the greatest received power. The pilot signal also permits the measurement, using a correlator, of the channel multipath response at the mobile, Figure 15.2, this response being used to continuously adjust the timing delay and gain of the taps in the RAKE receiver.

For CDMA systems to work effectively all signals must be received with comparable power level at the base station. If this is not so then the cross correlation levels of

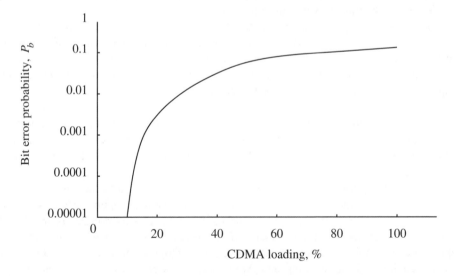

Figure 15.17 *Typical bit error ratio for CDMA systems for differing numbers of active users or loading as a percentage of the spreading gain. This is plotted for matched filter detection with perfect power control.*

unwanted stronger, nearby, signals may swamp the weaker wanted, more distant, signal. In addition to the 'near–far' problem, which arises from very different path lengths, different fading and shadowing effects are experienced by different transceivers at the same distance from the base station.

If the receiver input power from one CDMA user is η_1 then for k equal power, active multiple access users, the total receiver input power is $k\eta_1$. The receiver has a processing gain, given by G_p in equation (15.19), to discriminate against other user multiple access interference. Thus the receiver output SNR is given by:

$$\left(\frac{S}{N}\right)_{out} = \frac{\eta_1}{k\eta_1} G_p = \frac{G_p}{k} \tag{15.30}$$

or, for a (given) receiver output SNR, required to reliably detect the data, the multiple access capacity can be estimated from:

$$k = \frac{G_p}{\left(\dfrac{S}{N}\right)_{out}} \tag{15.31}$$

CDMA cellular systems must thus deploy accurate power control on the uplink to adjust mobile transmitter power levels, dependent on their location in the cell, in order to ensure equal received power level at the base station for all users.

Fast closed loop control is used, commands being transmitted at a rate of 800 bit/s, within the speech frames. The base station sends ± 0.5 dB step power control instructions every 1.25 ms to each mobile station to cover the 85 dB dynamic range in the available range of mobile output power level, to accommodate the received fading signal. It is essential to control the individual received power levels to within ± 1 dB if the system is to be able to handle the desired number of users as the limiting factor in CDMA is the other user interference, as shown in Figure 15.17. This is much more accurate CDMA power control than that used on a TDMA, i.e. in the GSM, system.

The reduction in transmitter power away from the cell boundary, i.e. closer to the base station, also saves battery drain further prolonging the mobile's talk time. These CDMA systems reuse identical code sets in each adjacent cell and so form a single cell cluster system.

At both the base station and the mobile, RAKE receivers are used to resolve and combine multipath components, significantly reducing fading. This receiver architecture is also used to provide base station diversity during 'soft' handoffs, whereby a mobile making the transition between cells maintains simultaneous links with both base stations during the transition. Deployment of cdmaOne systems [IS-95] in the Los Angeles, California area started in 1995 and they are now widely deployed in USA and Asia.

Benefits of CDMA

Although CDMA represents a sophisticated system which can only operate with accurate uplink power control, it offers some unique attractions for cellular mobile communications. The voice activity factor in a normal conversation, where each person only speaks for 3/8 of the time, can also be exploited to switch off transmission in the quiet periods further reducing the battery power requirement and reducing CDMA interference.

Besides the potential of offering higher capacity and flexibility, there are several attributes of CDMA that are of particular benefit to cellular systems. These are summarised below:

- **No frequency planning needed** – In FDMA and TDMA, the coordinated frequency planning for a service region is critical. Because the frequency bands allocated to CDMA can be reused in every CDMA cell, no frequency planning is required. As CDMA deployment grows, there is no need to continually redesign the existing system to coordinate frequency planning.
- **Soft capacity** – For FDMA or TDMA, when a base station's frequency bands or time-slots are fully occupied, no additional calls can be accommodated by that cell. Instead of facing a hard limit on the number of available channels a CDMA system can handle more users at the expense of introducing a gradual degradation of C/I, and therefore link quality. CDMA service providers can thus trade off capacity against link quality.
- **Simpler accessing** – Unlike FDMA and TDMA, coordination of signals into prespecified frequency allocations or time-slots is no longer required in CDMA.
- **Micro-diversity** – CDMA mitigates multipath propagation using a RAKE receiver. This combines all the delayed signals within the cell to improve the received SNR over other cellular systems.
- **Soft handover** – Soft handover refers to a CDMA call being processed simultaneously by two base stations while a mobile is in the boundary area between two cells. CDMA is able to perform soft handover because all users in the system share the same carrier frequency, so no retuning is involved in switching from one base station to another as is required by FDMA and TDMA systems.
- **Dynamic power control** – Stringent power control is essential to maximise CDMA system capacity. This is exercised by embedding power control commands into the voice traffic. The average transmitted power of a CDMA mobile is thus significantly less than that of FDMA or TDMA systems.
- **Lower battery dissipation rates** – As the transmitted power in a handset is reduced, so is the battery dissipation, prolonging the available talk time.

All these benefits are incorporated in the cdmaOne specification [IS-95]. CdmaOne has now evolved into CDMA2000. This has a 1× or single band (cdmaOne-type) system and a higher capacity 3× multicarrier system, which is equivalent to three cdmaOne systems operating in parallel or separate carriers (i.e. FDMA). It is thus proposed to use CDMA in most third generation systems.

Other systems which use CDMA, or spread spectrum techniques, are the NAVSTAR-GPS satellite based navigation system. This is a ranging system in which a correlation peak determines the time of arrival of a signal and hence the range to one of the many satellites orbiting overhead, thus establishing precise timing and position information.

15.5.7 Frequency hopped transmission

An alternative to PSK modulation with the spreading sequence is to alter, hop or shift the transmitter frequency in discrete steps to achieve a wide spread bandwidth. Figure 15.18 shows in (a) a hopping pattern where the selected frequency increases linearly with time and in (b) a random hopping pattern. In contrast to slow frequency hopping (FH) to achieve diversity and alleviate multipath effects, section 15.4.4, these fast FH techniques

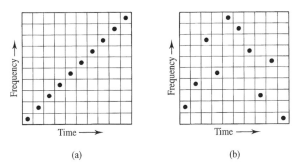

(a)   (b)

Figure 15.18 *(a) Representation for linear step frequency hopped waveform with $K = 10$; and (b) example of a 10×10 random frequency hop pattern.*

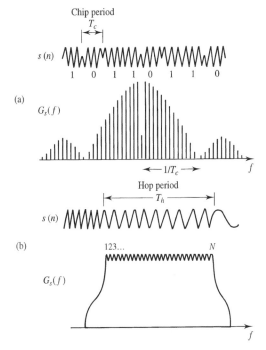

Figure 15.19 *Time domain and frequency domain representations of spread spectrum waveforms: (a) phase modulated transmissions; (b) frequency hopped transmissions.*

permit a spread spectrum communication system to be designed, in principle, with any desired number of frequency slots, several hundred being typical. Individual frequencies are selected in a frequency synthesiser by detecting several adjacent chips of the spreading code and decoding the word to identify the transmit frequency. Orthogonality between adjacent frequency slots is ensured if the dwell time on each frequency, T_h, equals the reciprocal of the slot separation, see section 11.5.1 on MFSK. FH offers a much flatter transmitted spectrum than spread spectrum PSK, Figure 15.19. Also, for a given RF bandwidth, the FH dwell time, T_h, is much longer than the equivalent chip time, T_c, when direct sequence PSK techniques are used, easing synchronisation acquisition in the FH system. For slow FH systems, such as those sometimes used by the military, coherence between hops is not required as the hopping rate is much less than the bit rate. Here the spread spectrum technique is used primarily to achieve protection against hostile jamming signals, to overcome slow fading and to achieve a level of privacy.

15.6 Mobile satellite based systems

A current development in satellite communications is to extend communication to the mobile subscriber anywhere on the globe. This requires the system to handle large numbers of small capacity earth stations which have relatively low gain antennas compared to the stationary antennas which are employed in the INTELSAT and VSAT systems. One mobile communications example is the INMARSAT ship-to-shore telephone and data service which is used on large cruise ships, by construction workers, and by UN peacekeepers. INMARSAT A stations, whose model M (briefcase sized) terminals cost £20000, provide 64 kbit/s data rate links for speech on which call costs are £3 per minute (1996 prices). The cheaper C terminals have only 600 bit/s capability through the MARISAT geostationary satellites using L-band UHF transmissions. In these systems the mobile has only limited motion and hence a modest gain (10 to 25 dB) steered dish is still employed as the receiving/transmitting antenna.

Another mobile example is the Skyphone system which routes digitally coded speech traffic from commercial aircraft to a satellite ground station using the INMARSAT satellites as the space-borne repeaters [Schoenenberger]. Here the antenna gain is lower as it has an omnidirectional coverage pattern but with sophisticated speech coding the data rate is reduced to approximately 12 kbit/s to compensate for the low power link budget.

For personal communications, with individual subscribers in mind, the 1992 World Radio Administrative Conference (WRAC) made frequency allocations at L-band (1.6 GHz) for mobile-to-satellite, and S-band (2.5 GHz) for satellite-to-mobile, channels. Here mobile receivers will again use antennas with very modest gain (0 to 6 dB). The satellites will require large antennas to provide the required spot beam coverage, particularly when the link budget is marginal. For this reason alternative systems employing inclined highly elliptical orbit (HEO) or low earth orbit (LEO) satellites are proposed, Figure 14.25.

The most promising inclined HEO is the Molniya orbit, Figure 14.25. It is asynchronous with a period of 12 hours. Its inclination angle (i.e. the angle between the orbital plane and the earth's equatorial plane) is 63.4° and its apogee (orbital point furthest from earth) and perigee (orbital point closest to earth) are about 39000 km and 1000 km respectively. The Molniya orbit's essential advantage is that, from densely populated northern latitudes, a satellite near apogee has a high elevation angle and a small transverse velocity relative to an earth station. Its high elevation angle reduces the potential impact on the link budget of shadowing by local terrain, vegetation and tall buildings. (The last can be especially problematic for mobile terminals in urban areas.) Its small angular velocity near apogee makes a satellite in this orbit appear almost stationary for a significant fraction of the orbital period. Three satellites, suitably phased around a Molniya orbit, can therefore provide continuous coverage at a high elevation angle. Modest gain, zenith pointed, antennas can be used for the mobile terminals, the quasi-stationary nature of the satellite making tracking unnecessary. On either side of apogee satellites in HEOs have the disadvantage of producing large Doppler shifts in the carrier frequency due to their relatively large radial velocity with respect to the earth station. Handover mechanisms must also be employed to allow uninterrupted service as one satellite leaves the apogee region and another enters it.

There are now a number of evolving mobile satellite systems using LEOs [Sheriff and Hu]. With multiple spot beams these form cells which move across the earth's surface with the LEO orbit parameters. The Iridium system, launched in 1999, had 66 satellites based on TDMA/FDMA interconnected to 12 gateways with intersatellite links, but the system folded due to revenue problems. The Globalstar system uses 48 satellites and a modified cdmaOne access but, as it has no intersatellite links, it needs more gateways to the earth centred telecommunications network. These techniques could provide a worldwide space based personal cellular system for use with simple handsets. It is likely that future subscriber mobile handsets will then be reprogrammable for use with either satellite or terrestrial mobile systems.

Broadband systems are also being considered such as Teledesic [WWW] which proposes 840 LEO satellites and Skybridge [WWW] both of which aim to give 2 Mbit/s data access, see Table 20.12. Other approaches are looking at balloons or airships to achieve lower platform altitudes and hence reduced path loss to provide higher data rate MAN-type service in the millimetre wave band, Table 1.1, to cover major cities.

It will not be possible in the systems described above to use the fixed FDM and TDM access schemes employed in the INTELSAT systems, Chapter 14. Mobile systems may have to employ different modulation and multiple accessing formats on the uplink and on the downlink. In order to allow access from low power narrowband mobile transmissions FDMA will probably be used on the uplink with a single narrowband channel allocated to each carrier or discrete transmission. On the downlink it will be preferable to employ

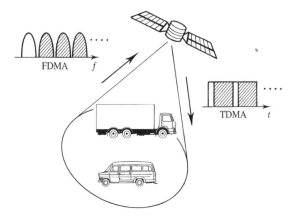

Figure 15.20 *Satellite communication to small mobile transceivers where the uplink and downlink typically use different multiple access techniques.*

TDMA as this is less affected by non-linearities in the satellite high power amplifiers than FDMA. Such a scheme involves extra complexity, as an onboard processor will be required on the satellite to demodulate the individual FDMA signals, regenerate them and then remodulate them in TDMA format, Figure 15.20. Also with small antenna diameters at mobile (microterminal) ground stations the bit rates must be low, but commensurate with the bandwidth of encoded speech, Table 9.3, to give sufficient SNR for reception at the satellite. Satellite links are widely used in Europe for distributing radio pager messages to VSATs for subsequent terrestial transmission.

15.7 Third generation mobile cellular standards

15.7.1 Mobile data transmission

TDMA mobile handsets, such as GSM, have always offered limited data transmission rates for text messaging, etc., using the short message service (SMS). In 2001 two new data services were launched to extend second generation system capabilities. These are the high speed circuit switched data (HSCSD) at 28.8 kbit/s and general packet radio system (GPRS) [Halonen *et al.*] with rates from 9.05 to 21.4 kbit/s, in a single GSM time-slot, and up to 171.2 kbit/s if all the eight slots in Figure 15.9 are used for data traffic. HSCSD overcame the relatively low GSM data rate capability of 9.6 kbit/s, in what is basically designed as a speech transmission system. As a circuit switched facility HSCSD is the preferred data access technique for video telephony or video conference and, for example, to transfer a large data file over a busy network.

The attraction of the internet protocol (IP) packet switched GPRS [WWW] is that there is no requirement for dial-up, with consequent call time charges, as the IP connection is readily available when a handset is powered up and thus connected to the network. GPRS thus offers the capability to intermittently transfer large and small data files to implement efficiently a wireless application protocol (WAP) based mobile network [WWW]. The enhanced data rate for GPRS evolution (EDGE) [Halonen *et al.*] further extends GPRS to 384 kbit/s by using eight-state coded modulation in the radio channel.

These data services permit the mobile handset user to access internet services using WAP. In late 2001 there were 50 million WAP enabled handsets worldwide and 10000 WAP sites in 95 countries. The Japanese i-mode was one popular example of these services. As the third generation system rolls out we expect, by 2005, to have 10^9 mobile internet users. These 3G systems ultimately promise 2 Mbit/s data rate capability which is well matched to the image sequence transmission rates of sections 16.7.2 and 16.7.3.

As a general statement lower data rates of 20 to 30 kbit/s will have full mobility to users in cars and trains, 384 kbit/s will offer reduced mobility, while 2 Mbit/s will require a stationary or possibly a very slow moving (pedestrian) terminal. Higher order modulation, e.g. QAM, as used in EDGE, can increase further the data rate but this is less robust to noise, interference and fading, Chapter 11. It can only be contemplated close to a base station for a very limited number of active users.

In addition to these full system data services back into the network, the Bluetooth standard [WWW] offers a short range 10 to 300 m data link to connect peripheral devices such as keyboards, printers, etc., to a mobile communications device and create a personal area network (PAN). Bluetooth operates in the unlicensed 2.4 GHz industrial scientific and measurement (ISM) band and provides low cost, low power, data networking at rates up to 721 kbit/s.

15.7.2 3G developments

The evolution of the third generation (3G) systems began in the 1980s when the ITU produced the initial recommendations for a new universal mobile telecommunication system (UMTS) [WWW], Figure 15.21. The 3G mobile radio service aims to provide low to high data rate services, with a maximum data rate of 2Mbit/s. Multimedia applications encompass services such as voice, audio/video, graphics, data, Internet access and e-mail. These packet and circuit switched services have to be supported by the radio interface and the network subsystem.

In 1998 15 possible radio transmission technologies (RTT) were submitted initially to ITU [WWW IMT-2000] for evaluation and adoption for use in the new standards which became known as IMT-2000. In January 1998, the European standardisation body for 3G mobile radio systems, the ETSI Special Mobile Group, agreed on a radio access scheme for 3G UMTS terrestrial radio access (UTRA). UTRA consists of two modes: frequency division duplex (FDD) where the uplink and downlink are transmitted on different frequencies; and time division duplex (TDD) where the uplink and downlink are transmitted on the same carrier frequency, but multiplexed in time. The agreement assigns wideband CDMA (WCDMA) to the paired bands (i.e. for UTRA FDD) and TD-CDMA to the unpaired bands (i.e. for UTRA TDD). TD-CDMA is based on a combination of TDMA frame access with CDMA in the time-slots, while WCDMA is a pure CDMA based system. Both modes of UTRA have been harmonised with respect to basic system parameters such as carrier spacing, chip rate and frame length. The interworking of UTRA with GSM is thus assured.

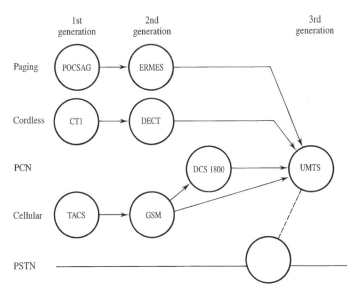

Figure 15.21 *Evolution of mobile and personal communication systems and equipment.*

In parallel with the European activities extensive work on 3G mobile radio has been performed simultaneously in Japan. The Japanese standardisation body also chose WCDMA, so that the Japanese and European proposals for the FDD mode were already aligned closely at an early stage. Very similar concepts have also been adopted by the North American standardisation body.

In order to work towards a global 3G mobile radio standard, the third generation partnership project (3GPP) was formed in December 1998. The 3GPP consists of members of the standardisation bodies in Europe, the USA, Japan, Korea and China. It has merged the already well harmonised proposals of the regional standardisation bodies and continues to work on a common 3G *international* mobile radio standard, still called UTRA. UTRA is based on the evolved GSM core network incorporating both the FDD and TDD modes. But the 3GPP Project 2 (3GPP2), on the other hand, works towards a 3G mobile radio standard based on an cdmaOne/IS-95 evolution that was originally called cdma2000.

In June 1999, a harmonised global third generation (G3G) concept evolved, which has been accepted by 3GPP and 3GPP2. The harmonised concept is a single standard with the following three modes of operation:

- Direct spread CDMA (DS-CDMA), based on UTRA FDD as specified by 3GPP.
- Multicarrier CDMA (MC-CDMA), based on cdma2000 using FDD, i.e. 3GPP2.
- TDD (CDMA TDD), based on UTRA TDD as specified by 3GPP.

The harmonised concept for the CDMA based proposals was achieved by aligning the radio parameters as much as possible and defining a model of a generic protocol stack to simplify global roaming and enable direct connection to an evolved GSM media access protocol.

The specifications elaborated by 3GPP and 3GPP2 are now, among others, part of the ITU Recommendations for International Mobile Telecommunications 2000 (IMT-2000). IMT-2000 aims to cover indoor, outdoor and vehicle mounted scenarios with circuit and packet switched traffic, see Chapter 20, at a quality of service (QoS) comparable with wired telecomms networks. See, for example, the IMT-2000 and UMTS RTT standards.

UTRA thus provides efficient access to mobile multimedia services with seamless roaming and service capabilities. It offers high speech quality and mobile data services with data rates of up to 2 Mbit/s. UTRA consists of two modes that complement each other:

- UTRA FDD is well suited to applications in public macro and microcell environments with typical data rates up to 384 kbit/s and high mobility.
- UTRA TDD on the other hand is preferred for micro and picocell environments as well as for licensed and unlicensed cordless and wireless local loop [Stavroulakis] applications. It facilitates the use of the unpaired spectrum for asymmetric services with higher rates, up to 2 Mbit/s.

15.7.3 Wideband CDMA

The 3G proposals were predominantly based on wideband CDMA (WCDMA) [Holma and Toskala] and a mix of FDD and TDD access techniques [Jabbari, Ojanpera and Prasad]. WCDMA is favoured to meet the 3G aims in poor propagation environments with a mix of high and modest speed data traffic. It is generally accepted that CDMA is the preferred access technique and, with the increase in the data rate, then the spreading modulation needs to increase to wideband transmission.

The lower FDD uplink band from 1920 to 1980 MHz will be paired with a 2110 to 2170 MHz FDD downlink. In addition the unpaired bands at 1900 to 1920 MHz and 2010 to 2025 MHz are preferred for the asymmetric TDD internet access service. TDD is more flexible as time-slots can be dynamically reassigned to uplink and downlink functions, as required for asymmetric transfer of large files or video on demand traffic.

There are also a number of evolving mobile satellite systems using LEOs for narrowband and broadband systems. However, as satellite systems evolve they are clearly a major component of the worldwide communications system as they are the only method of delivery service to the remote and thinly inhabited parts of the earth. Thus fully functional IMT transceivers will often have to be multistandard to accommodate a satellite service.

15.8 Summary

Private mobile radio utilises VHF and UHFs. It is not usually interfaced with the PSTN and is now principally used by the haulage industry and taxi companies. The number of PMR channels available in the UK is a little over 1000. Each channel has an allocated bandwidth of 12½ kHz. Narrowband FM is normally used for speech traffic while FSK modulation can be used for data transmission.

Like satellite communications, first generation cellular radio systems used analogue modulation. In the UK this generation was represented by TACS which operated using narrowband FM/FDMA with 25 kHz channels and a carrier frequency of 900 MHz. The essential feature of cellular systems is the division of the coverage area into cell clusters. The allocated frequencies are divided between the cells of a cluster. The same frequencies are also used (or reused) in the cells of the adjacent clusters. A base station at the (nominal) centre of each cluster uses modest transmit power so that interference in adjacent cells is kept to acceptable levels. Base stations are connected to the wired PSTN via switching centres. The received signal strength is subject to slow (log-normal) fading due to (multiplicative) shadowing effects and fast (Rayleigh) fading due to (additive) multipath effects.

Second generation cellular systems are digital. In Europe this generation is represented by GSM. In this system eight TDMA users share each 200 kHz wide, 271 kbit/s, channel. The bandwidth per user is therefore 25 kHz, as in TACS, but overall spectrum is conserved due to the lower C/I tolerated by digital systems allowing the use of smaller cells and, consequently, more intensive frequency reuse. GSM uses GMSK modulation with $BT_b = 0.3$, LPC derivatives for speech bandwidth compression and CRC error correction. For large power delay spread the channel is frequency selective and time dispersion results in serious ISI. Adaptive equalisation is therefore required. Adaptive power control at the mobile is also used to maintain adequate received power at the base station whilst minimising interference. CDMA is more sophisticated than TDMA and it is destined for progressively increasing deployment in future cellular systems.

Digital systems have been extended to provide data access with HSCSD, GPRS and EDGE at rates from a few kbit/s up to 384 kbit/s. Further developments of the digital TDMA concept for the paramilitary emergency services and public utilities are covered by the TETRA standard, which achieves secure communications by incorporating frequency hopping and encryption techniques.

Third generation communication systems will integrate further the digital systems described here (and others) to provide an increased range of services to each user with data rates up to 2 Mbit/s. These services will be based on an ISDN infrastructure and will be available to interface to all terminal equipment including those which are mobile and hand-held. This is the medium range objective at the core of the PCN concept.

This chapter has brought together many of the techniques covered in earlier chapters. It has shown how source and channel coding techniques combined with VHF, UHF or microwave transmission frequencies, and advanced receiver processing, can combat ISI and multipath effects. This now permits the design and realisation of advanced communication systems which are well matched to the needs of the mobile user. Such is the attraction of lightweight handsets that, in 2000, there were 550 million users worldwide and it is predicted confidently that in 2005 mobile users will exceed the fixed wireline telephone subscribers.

15.9 Problems

15.1 For the GSM cellular system of Table 15.1, with allocations as in Figure 15.8, what is the number of simultaneous calls which can be handled in a 7-cell cluster as in Figures 15.4 or 15.5? [1400 users].

15.2 Compare the answer in Problem 15.1 above with an analogue TACS scheme for a 12-cell cluster. [1400 users, 40 users per MHz in both systems]

15.3 If the analogue and digital (GSM) systems have the same capacity what is the advantage of the GSM system?

15.4 What is the capacity per base station and total system capacity for a DECT standard system? [144 users].

15.5 What are the fundamental differences between the two CDMA receiver implementations in Figure 15.12?

15.6 Compare the capacity of GSM TDMA and cdmaOne systems in terms of active users per MHz of available bandwidth for a 10-cell cluster. [both systems 40 users per MHz]

CHAPTER 16

Video transmission and storage

16.1 Introduction

This chapter considers the encoding, transmitting, storing and displaying of video pictures by both traditional (analogue) and modern (digital) techniques. Here we aim to give only an introduction to the subject; for more advanced and detailed information, the reader is referred to one of the many specialist textbooks, e.g. [Grob, Lenk].

The fundamental requirement of any video display system is the ability to convey, in a usable form, a stream of information relating to the different instances of a picture. This must contain two basic elements, namely some description of the section of the picture being represented, e.g. brightness, and an indication of the *location* (in space and time) of that section. This implies that some *encoding* of the picture is required.

There are many different approaches to the encoding problem. We shall look at some of the common solutions which have been adopted, although other solutions also exist. Initially, we consider analogue based solutions, which transfer and display video information in real time with no direct mechanism for short term storage. Later digitally based solutions, which allow for storage of the video image sequence, are considered. To date, however, most digital video sequences are currently displayed using analogue techniques similar to those employed in current domestic television receivers.

The problem of representing a small section of an image (often called a picture element or pixel) is solved by different encoding mechanisms in different countries. All essentially separate the information contained in a pixel into black and white (intensity) and colour components. The pixel, although normally associated with digital images, is relevant to analogue images as it represents the smallest independent section of an image. The size of the pixel limits *resolution* and, therefore, the quality of the image. It also affects the amount of information needed to represent the image. For real time transmission this in turn determines the bandwidth of the signal conveying the video stream (see Table 1.2 and Chapter 9 on information theory).

The problem of identifying the location of a pixel is addressed by allowing the image to be represented by a series of lines one pixel wide, scanned in a predetermined manner across the image, Figure 16.1. The pixels are then transmitted serially, special pulse sequences being used to indicate the start of both a new line and a new picture or image frame according to the encoding scheme used (see later Figure 16.4). Note that these systems rely upon the transmitter and receiver remaining synchronised. In the UK, the encoding mechanism used for TV broadcasting is known as PAL (phase alternate line), in the USA NTSC (National Television Standards Committee) is used, while in France SECAM (système en couleurs à mémoire) has been adopted.

16.2 Colour representation

A pixel from a colour image may be represented in a number of ways. The usual representations are:

1. An independent intensity (or luminance) signal, and two colour (or chrominance) signals normally known as hue and saturation.
2. Three colour signals, typically the intensity values of red, green and blue, each of which contains part of the luminance information.

In the second technique a white pixel is obtained by mixing the three primary colours in appropriate proportions. The colour triangle of Figure 16.2 shows how the various colours are obtained by mixing. Figure 16.2 also interprets, geometrically, the hue and saturation chrominance information. The hue is an angular measure on the colour triangle whilst the proportion of saturated (pure) colour to white represents radial distance. In practice we also need to take account of the response of the human eye which varies with colour or wavelength as shown in Figure 16.3. Thus for light to be interpreted as white we actually need to add 59% of green light with 30% of red and 11% of blue light. The luminance component Y is thus related to the intensity values of the

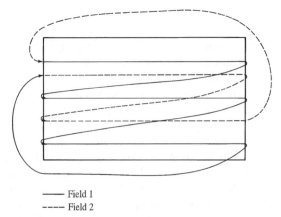

—— Field 1
---- Field 2

Figure 16.1 *Line scanning TV format with odd and even fields.*

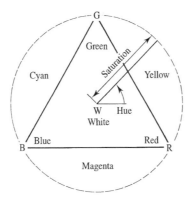

Figure 16.2 *Colour triangle showing hue and saturation.*

red (R), green (G) and blue (B) contributions by the following approximate formula:

$$Y = 0.3R + 0.59G + 0.11B \qquad (16.1)$$

In practice, luminance and colour information are mathematically linked by empirical relationships. The principal benefit of separating the luminance and chrominance signals is that the luminance component only may then be used to reproduce a monochrome version of the image. This approach is adopted in colour TV transmission for compatibility with the earlier black and white TV transmission system.

The theory governing the production of a range of colours from a combination of three primary colours is known as additive mixing. (This should not be confused with subtractive mixing as used in colour photography.) It is possible for full colour information to be retrieved if the luminance and two colours are transmitted. In practice, colour/luminance difference signals (e.g. R − Y) are transmitted, and these are modified to fit within certain amplitude constraints. The colour difference signals, or colour separated video components, U and V, are:

$$U = 0.88(R - Y) \qquad (16.2)$$

and

$$V = 0.49(B - Y) \qquad (16.3)$$

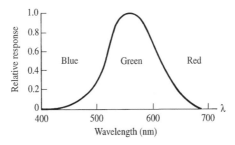

Figure 16.3 *Response of the human eye to colour.*

Most video cameras and cathode-ray tube (CRT) displays produce an image in the so-called RGB format, i.e. using three signals of mixed intensity and colour information. This is transformed into YUV format before video transmission and reformed into an RGB representation for colour display.

16.3 Conventional TV transmission systems

16.3.1 PAL encoding

The PAL encoding mechanism, as employed in the UK, provides for both colour information transmission and regeneration of a scanned image after display synchronisation. The latter is achieved by transmitting the image as a sequence of lines (625 in total) which are displayed to form a sequence of frames (25 per second), Figure 16.1. The system is basically analogue in operation, the start of line and start of frame being represented by pulses with predefined amplitudes and durations.

Line scanning to form the image is illustrated in Figure 16.1. In PAL systems, a complete image, or frame, is composed of two fields (even and odd) each of which scans through the whole image area, but includes only alternate lines. The even field contains the even lines of the frame and the odd field contains the odd lines. The field rate is 50 fields/s with 312½ lines per field and the frame rate is 25 frames/s.

The line structure is shown in Figure 16.4. Figure 16.4(a) shows the interval between fields and Figure 16.4(b) shows details of the signal for one scan line. This consists of a line synchronisation pulse preceded by a short duration (1.5 μs) period known as the front porch and followed by a period known as the back porch which contains a 'colour burst' used for chrominance synchronisation, as described shortly. The total duration of this section of the signal (12 μs) corresponds to a period of non-display at the receiver and is known as 'line blanking'. A period of active video information then follows where the signal amplitude is proportional to luminous intensity across the display and represents, in sequence, all the pixel intensities in one scan line. Each displayed line at the receiver is therefore of 52 μs duration and is repeated with a period of 64 μs to form the field display.

In all, 625 lines are transmitted over the two fields of one frame. However, only 575 of these contain active video. The non-active lines contain field synchronisation pulses, and also data for teletext-type services. The video information carried by a PAL encoded signal is contained within a number of frequency bands, although to save bandwidth, the chrominance information is inserted into a portion of the high frequency section of the luminance signal, Figure 16.4(c). This band is chosen to avoid harmonics of the line scan frequency which contain most of the luminance energy. Fortunately the colour resolution of the human eye is less than its resolution for black and white images and hence we can conveniently transmit chrominance information as a reduced bandwidth signal within the luminance spectrum.

The chrominance information is carried as a quadrature amplitude modulated signal, using two 4.43 MHz (suppressed) carriers separated by a 90° phase shift, to carry the colour difference signals of equations (16.2) and (16.3). This is similar to the QAM described in Chapter 11, but with analogue information signals. If the carrier frequency is f_c (referred to as the colour subcarrier frequency) then the resulting colour signal S_c is:

$$S_c = U\cos(2\pi f_c t) + V\sin(2\pi f_c t) \tag{16.4}$$

The resulting chrominance signal bandwidth is 2 MHz. This is sometimes called the YIQ signal. Figure 16.4(c) also shows an additional audio signal whose subcarrier frequency is 6 MHz.

In order to demodulate the QAM chrominance component, a phase locked local oscillator running at the colour subcarrier frequency f_c is required at the receiver. This is achieved by synchronising the receiver's oscillator to the transmitted colour subcarrier, using the received 'colour burst' signal, Figure 16.4(b).

Since the chrominance component is phase sensitive, it will be seriously degraded by any relative phase distortion within its frequency band due, for example, to multipath propagation of the transmitted RF signal. This could lead to serious colour errors but the effect is reduced by reversing the phase of one of the colour difference components on

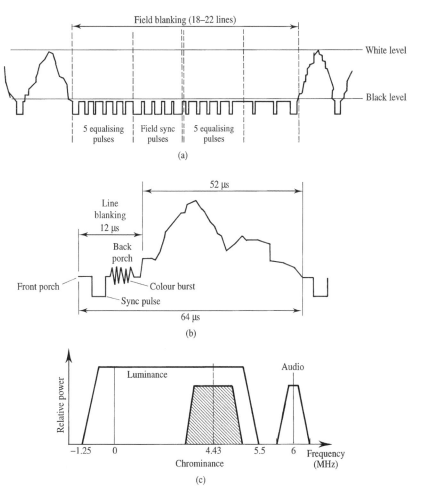

Figure 16.4 *TV waveform details: (a) field blanking information at the end of a frame; (b) detail for one line of video signal; (c) spectrum of the video signal.*

alternate line scans. Thus the R – Y channel is reversed in polarity on alternating line transmissions to alleviate the effects of differential phase in the transmission medium.

The PAL signal spectrum shown in Figure 16.4(c) is the baseband TV signal and if RF TV transmission is required, this must be modulated on to a suitable carrier and amplified. Typical UHF carrier frequencies extend to hundreds of MHz with power levels up to hundreds of kW (even MW in some cases). Terrestrial TV channels 21 to 34 lie between 471.25 and 581.25 MHz and channels 39 to 68 fall between 615.25 and 853.25 MHz, Table 1.4. Satellite TV occupies a band at 11 GHz with 16 MHz interchannel spacings.

16.3.2 PAL television receiver

Figure 16.5 shows a simplified block diagram of the main functional elements of a colour TV receiver. The RF signal from the antenna or other source is selected and amplified by the tuner and IF (intermediate frequency) stages and finally demodulated to form a baseband PAL signal. Four information signals are then extracted: the audio, luminance, chrominance and synchronisation signals.

The chrominance signal is demodulated by mixing with locally generated inphase and quadrature versions of the colour subcarrier. The resulting colour difference and luminance signals are then summed in appropriate proportions and the R, G and B output signals amplified to sufficient levels to drive the CRT video display.

The x and y deflections of the electron beams are produced by magnetic fields provided by coils situated on the outside of the CRT. These coils are driven from ramp generators, synchronised to the incoming video lines such that the beam is deflected horizontally across the display face of the tube from left to right during the active line period, and vertically from top to bottom during the active field period.

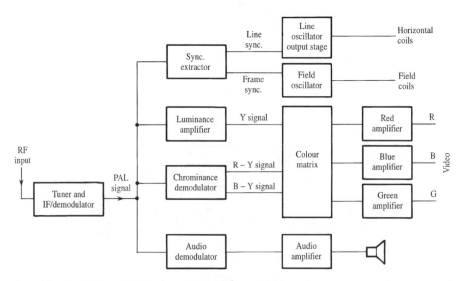

Figure 16.5 *Simplified block diagram of colour TV receiver.*

16.3.3 Other encoding schemes

The NTSC system has many fundamental similarities to PAL, but uses different transmission rates. For example, the frame rate is 30 Hz, and the number of lines per frame is 525. PAL is, in fact, an enhancement of the basic principles of the NTSC system. The main difference between the two is in the phase distortion correction provided by PAL. This is not present in the NTSC system, and hence NTSC can be subject to serious colour errors under conditions of poor reception. Any attempted phase correction is controlled entirely by the receiver, and most receivers allow the viewer to adjust the phase via a manual control. The overall spectral width of NTSC is somewhat smaller than PAL.

The SECAM system uses a similar frame and line rate to PAL and, like PAL, was developed in an attempt to reduce the phase distortion sensitivity of NTSC. Essentially, it transmits only one of the two chrominance components per line, switching to the second chrominance component for the next line.

Transmission of digitally encoded video images is currently being developed and implemented. Some of the basic digital techniques are discussed in later sections.

16.4 High definition TV

16.4.1 What is HDTV?

High definition television (HDTV) first came to public attention in 1981, when NHK, the Japanese broadcasting authority, demonstrated it in the USA. HDTV [Prentiss] is defined by the ITU-R study group as:

> *A system designed to allow viewing at about three times the picture height, such that the system is virtually transparent to the quality or portrayal that would have been perceived in the original scene ... by a discerning viewer with normal visual acuity.*

Although, in principle, HDTV could be delivered using either analogue or digital transmission technologies, the technical and economic advantages of the latter mean that the terms HDTV and Digital TV are now practically synonymous.

HDTV proposals are for a screen which is wider than the conventional TV image by about 33%. It is generally agreed that the HDTV aspect ratio will be 16:9, as opposed to the 4:3 ratio of conventional TV systems. This ratio has been chosen because psychological tests have shown that it best matches the human visual field. It also enables use of existing cinema film formats as additional source material, since this is the same aspect ratio used in normal 35 mm film. Figure 16.6(a) shows how the aspect ratio of HDTV compares with that of conventional TV, using the same resolution, or the same surface area as the comparison metric.

To achieve the improved resolution the video image used in HDTV must contain over 1000 lines, as opposed to the 525 and 625 provided by the existing NTSC and PAL systems. The exact value is chosen to be a simple multiple of one or both of the vertical resolutions used in conventional TV. However, due to the higher scan rates the bandwidth requirement for analogue HDTV would be approximately 12 MHz, compared to the nominal 6 MHz of conventional TV, Table 1.2.

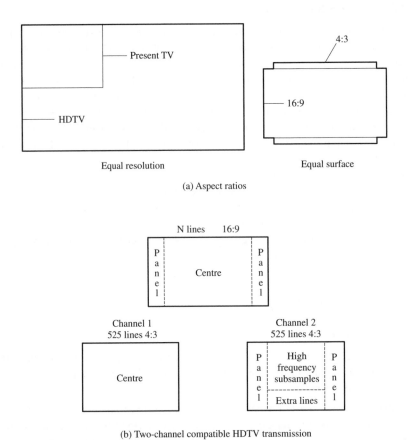

(a) Aspect ratios

(b) Two-channel compatible HDTV transmission

Figure 16.6 *(a) Comparison between conventional TV and HDTV, (b) two-channel transmission.*

The introduction of a non-compatible TV transmission format for HDTV would require the viewer either to buy a new receiver, or to buy a converter. The initial thrust in Japan was towards a two-channel HDTV format illustrated in Figure 16.6(b) which was compatible with conventional TV standards, and thus could be received by conventional receivers, with conventional quality, assuming analogue transmission. However, to get the full benefit of HDTV, a new wide screen, high resolution receiver has to be purchased.

16.4.2 Studio standards

Initially there were two main proposals for a worldwide HDTV *studio* system, with characteristics as shown in Table 16.1. The Japanese broadcasting company, NHK, has proposed one of the systems while the joint European project, Eureka, proposed an alternative standard.

The European standard uses a 50 Hz field rate to provide relatively easy conversion to both 60 and 50 Hz conventional, and 60 Hz HDTV, systems. It is also well suited to transfer from film; 1250 lines were chosen as this is exactly double the number of lines of the European conventional standards. A conversion to the American 525 lines standard is slightly more difficult, involving a ratio of 50/21.

Table 16.1 *Proposed HDTV studio production standards.*

	Europe	*North America, Japan*
Total lines/picture	1250	1125
Active lines/picture	1192	1035
Scanning method	1:1	2:1 Interlaced
Aspect ratio	16:9	16:9
Field frequency (Hz)	50	60
Line frequency (kHz)	62.5	33.75
Samples/active line	1920	1920

16.4.3 Transmissions

In order to achieve compatibility with conventional TV early proposals were to split the HDTV information and transmit it in two separate channels. When decomposing a 1250-line studio standard, for US transmission, the line interpolator would extract the centre portion of every second line of the top 1050 lines, and send it in channel 1 as the reduced resolution 525-line image, lower left part of Figure 16.6(b). The panel extractor then removes the side panels, which make up the extra width of the picture, adds in the bottom 200 lines which were not sent with channel 1 plus the missing alternate lines and sends this as channel 2. Transmission of both channels would be by conventional analogue methods.

Recent European developments have swung in favour of a digital HDTV transmission standard based on digital video broadcast (DVB). The move to digitised video permits the use of compression coding techniques, such as MPEG, see later section 16.7.3, to minimise the transmission bit rate and hence fit high definition digital TV into a broadly similar bandwidth allocation as analogue TV. The other major development is the move to COFDM transmissions for broadcast audio and video, see section 11.5.2, as it is more tolerant of the multipath fading channel conditions which are encountered in broadcasting to mobile or moving receivers. This is discussed later in section 16.8.

16.5 Digital video

Video in either RGB or YUV format can be produced in digital form [Forrest]. In this case discrete samples of the analogue video signal are digitised, to give a series of PCM words which represent the pixels. The words are divided into three fields representing each of the three signals RGB or YUV, section 16.2. Pixel word sizes range from 8 bits to 24 bits. Typical configurations include:

24 bits – where R = G = B = 8 bits and
16 bits – where Y = 8 bits and U = V = 4 bits.

In the latter case the chrominance signals are transmitted with less accuracy than the luminance but the eye is not able to discern the degradation. It is then possible either to store the samples in a memory device (e.g. compact disc (CD)), or to transmit them as a digital signal. Figure 16.7 shows the entropy (see section 9.2.4) of the luminance and chrominance signals in a video sequence comprising a fast moving sports sequence. Cb is the chrominance signal B − Y and Cr is R − Y. Scenes with less movement, such as a head and shoulders newsreader, have correspondingly lower entropy values which can be taken full advantage of if frame to frame differential coding is employed. Note, in Figure 16.7, that the luminance information (Y) requires a higher accuracy in the quantisation operation than does the chrominance information.

It would initially seem ideal to use a computer network, or computer, modem and telephone network, for video delivery. Unfortunately, the amount of data which would need to be transmitted is usually excessive. For a conventional TV system, the equivalent digital bit rate is around 140 Mbit/s which is not compatible with a standard modem. Even if the quality of the received image is reduced, data rates are still high. For example, if an image with a reduced resolution of 256×256 pixels is considered where each pixel consists of 16 bits (YUV) and a standard video frame rate of 25 frames per second is used, the bit rate R_b is given by:

$$R_b = 256 \times 256 \times 16 \times 25 = 26.2 \text{ Mbit/s} \tag{16.5}$$

Even a single frame requires 1 Mbit, or 132 kbyte, of storage. For full resolution (720 pixel by 480 pixel) ITU-R 601 PAL TV with 8 bit quantisation for each of the three RGB colour components, the rate grows to 207 Mbit/s and HDTV, with 1920 pixel by 1250 pixel resolution, requires almost 2 Gbit/s. A further problem exists in storage and display as the access rate of CD-ROM is 120 kbyte/s while fast hard disk is only 500 kbyte/s. Fortunately, a general solution to the problem of data volume exists. In our sports sequence example, 16 bits were needed to represent each YUV pixel of the image (Figure 16.7). However, the long term average entropy of each pixel is considerably smaller than this and so data compression can be applied.

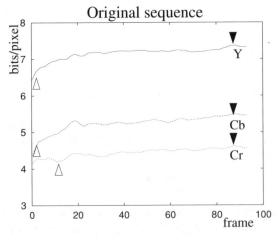

Figure 16.7 *Typical entropy measures in a colour video image sequence.*

16.6 Video data compression

Data rate reduction is achieved by exploiting the *redundancy* of natural image sequences which arises from the fact that much of a frame is constant or predictable, because most of the time changes *between* frames are small. This is demonstrated in Figure 16.8 which shows two sequences of frames with a uniform background. The pixels representing the background are identical. The top sequence of frames is very similar in many respects, expect for some movement of the subject. The lower sequence has more movement which increases the entropy. It is thus inefficient to code every image frame into 1 Mbit of data and ignore the predictive properties possessed by the previous frames. Also there are few transmission channels for which we can obtain the 25 to 2000 MHz bandwidth needed for uncompressed, digital, TV. For example, the ISDN digital telephone channel has 128 kbit/s capacity, while ISDN primary access rate is 2 Mbit/s and the full STM-1 rate is only 45 Mbit/s (see Chapter 20).

Removal of redundancy is achieved by image sequence coding. Compression operations (or algorithms) may work within a single frame (intra-frame), or between frames of a sequence (inter-frame), or a combination of the two. Practical compression systems tend to be hybrid in that they combine a number of different compression mechanisms. For example, the output of an image compression algorithm may be Huffman coded (Chapter 9) to further reduce the final output data rate.

We will look firstly at a few of the basic compression principles, before progressing to review practical systems. General coding techniques such as DPCM discussed in sections 5.8.2 and 5.8.3 are also applicable to image compression. Figure 16.9 shows the inter-frame difference between two images, displayed as a grey scale image. DPCM would achieve 2 to 3 times compression compared to conventional PCM quantisation. However, video compression requires 20:1 to 200:1 compression ratios. Video compression and expansion equipment is often referred to as a 'video CODEC' inferring the ability to both transmit and receive images. In practice, not all equipment is able to do this, and the term is sometimes used to refer to the transmitter (coder) or receiver (decoder) only.

⟶ frame(t) ⟶ frame(t+1) ⟶ frame(t+2) ⟶ frame(t+3) ⟶

Figure 16.8 *Upper sequence has similarity between adjacent frames implying high redundancy in the video image sequence but this is not evident in the lower sports sequence.*

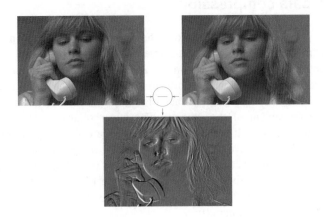

Figure 16.9 *Difference image between two frames in a video image sequence.*

16.6.1 Run length coding

This intra-frame compression algorithm is best suited to graphic images, or video images with large sections made up of identical pixels. The algorithm simply detects the presence of a sequence of identical pixel values (usually operating on the luminance component only), and notes the start point, and number of pixels in the run (the 'run length'). This information is then transmitted in place of the original pixel values. (The technique was described as applied in facsimile transmission in Chapter 9.) To achieve a significant coding gain the run lengths must be large enough to provide a saving when considering the additional overhead of the addressing and control information which is required.

16.6.2 Conditional replenishment

This is an intra-frame coding algorithm which requires a reference frame to be held at the transmitter. The algorithm first divides the frame into small elements, called blocks, although lines can be used. Each pixel element is compared with the same location in the reference frame, and some measure of the difference between the pixel elements is calculated. If this is greater than a decision threshold, the pixel is deemed to have changed and the new value is sent to the receiver and updated in the reference frame. If the difference is not greater than the threshold, no data is sent. The frame displayed is therefore a mixture of old and new pixel element values.

16.6.3 Transform coding

As its name suggests, transform coding attempts to convert or transform the input samples of the image from one domain to another [R.J. Clarke, 1985]. It is usual to apply a 2-dimensional (2-D) transform to the 2-dimensional image and then quantise the transformed output samples. Note that the transform operation does not in itself provide compression; often one transform coefficient is output for each video amplitude sample input, as in the Fourier transform (Chapters 2 and 13). However, many grey scale patterns in a 2-dimensional image transform into a much smaller number of output samples which have *significant* magnitude. (Those output samples with insignificant magnitude need not be retained.) Furthermore, the quantisation resolution of a particular output sample which *has* significant magnitude can be chosen to reflect the importance of that sample to the overall image.

To reconstruct the image, the coefficients are input to an inverse quantiser and then inverse transformed such that the original video samples are reconstructed, Figure 16.10. Much work has been undertaken to discover the optimum transforms for these operations, and the Karhunen–Loeve transform [R.J. Clarke, 1985] has been identified as one which minimises the overall mean square error between the original and reconstructed images. Unfortunately, this is complex to implement in practice and alternative, sub-optimum transforms are normally used.

These alternative transforms, which include sine, cosine and Fourier transforms amongst others, must still possess the property of translating the input sample energy into another domain. The cosine transform has been shown to produce image qualities similar to the Karhunen–Loeve transform for practical images (where there is a high degree of inter-frame pixel correlation) and is now specified in many standardised compression systems.

The 2-D discrete cosine transform (DCT) [R.J. Clarke, 1985] can be defined as a matrix with elements:

$$
C_{nk} = \begin{cases} \sqrt{\dfrac{1}{N}} \cos\left[\dfrac{n(2k+1)\pi}{2N}\right], & n = 0, \quad 0 \le k \le N-1 \\[2em] \sqrt{\dfrac{2}{N}} \cos\left[\dfrac{n(2k+1)\pi}{2N}\right], & 1 \le n \le N-1, 0 \le k \le N-1 \end{cases} \tag{16.6}
$$

where N is the number of samples in one of the dimensions of the normally square transform data block. The fact that the video data comprises real (pixel intensity) rather than complex sample values favours the use of the DCT over the DFT.

The next stage in a transform based video compression algorithm is to quantise the output frequency samples. It is at this stage that compression can occur. Recall that the output samples now represent the spatial frequency components of the input 'signal' [Clarke, 1998, Mulgrew *et al.*]. The first output sample represents the DC, or average, value of the 'signal' and is referred to as the DC coefficient. Subsequent samples are AC coefficients of increasing spatial frequency. Low spatial frequency components imply slowly changing features in the image while sharp edges or boundaries demand high spatial frequencies.

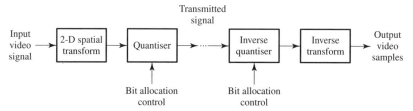

Figure 16.10 *Simplified block diagram illustrating transform coder operation.*

Much of the image information registered by the human visual process is contained in the lower frequency coefficients, with the DC coefficient being the most prominent. The quantisation process therefore involves allocating a different number of bits to each coefficient. The DC coefficient is allocated the largest number of bits (i.e. it is given the highest resolution), with increasingly fewer bits being allocated to the AC coefficients representing increasing frequencies. Indeed, many of the higher frequency coefficients are allocated 0 bits (i.e. they are simply not required to reconstruct an acceptable quality image).

The combined operations of transformation, quantisation and their inverses are shown in Figure 16.10. Note that the quantisation and inverse quantisation operations may be adaptive in the number of bits allocated to each of the transformed values. Modification of the bit allocations allows the coder to transmit over a channel of increased or decreased capacity as required, perhaps due to changes in error rates or a change of application. With the advent of ATM (section 20.10), such a variable bit rate (bursty) signal can now be accommodated.

Practical transform coders include other elements in addition to the transformation and quantisation operations described above. Further compression can be obtained by using variable length (e.g. Huffman) type coding operating on the quantiser output. In addition, where coding systems operate on continuous image sequences, the transformation may be preceded by inter-frame redundancy removal (e.g. movement detection) ensuring that only changed areas of the new frame are coded. Some of these supplementary techniques are utilised in the practical standard compression mechanisms described below.

16.7 Compression standards

16.7.1 COST 211

This CODEC specification was developed in the UK in the 1980s by British Telecom, and the CODEC has been used extensively for professional video conference applications. Its output bit rate is high compared with more recent developments based on transform techniques. A communication channel of at least 340 kbit/s is required for the COST 211 CODECs, the preferred operation being at the 2 Mbit/s ISDN primary access rate (Chapter 20). The COST 211 CODEC provides full frame rate (25 frame/s) video assuming a conference scene without excessive changes between frames. The CODEC has a resolution of 286 lines, each of 255 pixels. The compression uses intra-frame algorithms, initially utilising conditional replenishment where the changed or altered pixel information is transmitted, followed by DPCM coding (section 5.8.2) which is applied to the transmitted pixels. Finally, the output is Huffman encoded.

Further compression modes are available, including subsampling (skipping over sections of the frame, or even missing complete frames) to enable the CODEC output to be maintained at a constant rate if replenishment threshold adjustment is not sufficient.

The conditional replenishment, or movement detector, part of the CODEC is shown in Figure 16.11. This operates on a pixel-by-pixel basis and calculates the weighted sum of five consecutive pixel differences. The difference sum is then compared with a variable threshold value to determine if the pixel has changed. The threshold is adjusted to keep

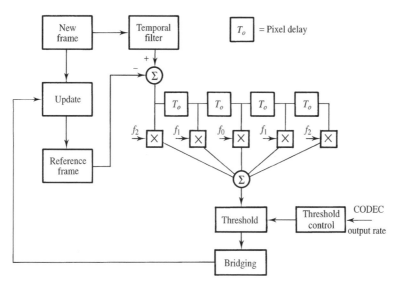

Figure 16.11 *COST 211 CODEC movement detector.*

the CODEC output rate constant. Changed pixels are both transmitted and stored in the local reference frame at the transmitter.

16.7.2 JPEG

JPEG is an international standard for the compression and expansion of single frame monochrome and colour images. It was developed by the Joint Photographic Experts Group (JPEG) and is actually a set of general purpose techniques which can be selected to fulfil a wide range of different requirements. However, a common core to all modes of operation, known as the baseline system, is included within JPEG. JPEG is a transform based coder and some, but not all, operating models use the DCT applied to blocks of 8×8 image pixels yielding 64 output coefficients. Figure 16.12 shows a 352×240 pixel image, partitioned into 330 macroblocks. The 16×16 pixel macroblocks are processed slice by slice in the DCT coder.

The coefficients are then quantised by a user defined quantisation table which specifies a quantiser step size for each coefficient in the range 1 to 255. The DC coefficient is coded as a difference value from that in the previous block, and the sequence of AC coefficients is reordered as shown in Figure 16.13(a) by zig-zag scanning in ascending order of spatial frequency, and hence decreasing magnitude, progressing from the more significant to the least significant components. Thus the quantiser step size increases and becomes coarser with increasing spatial frequency in the image to aid the final data compression stage, Figure 16.13(b).

The final stage consists of encoding the quantised coefficients according to their statistical probabilities or entropy (Chapter 9 and Figure 16.7). Huffman encoding is used in the baseline JPEG systems. JPEG also uses a predictor, measuring the values from three adjacent transformed pixels, to estimate the value of the pixel which is about to be encoded. The predicted pixel value is subtracted from the actual value and the

Figure 16.12   *Image partitioning into macroblocks (MB).*

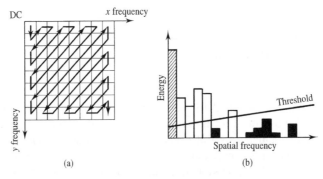

Figure 16.13   *(a) Zig-zag scanning to reorder transformed data into increasing spatial frequency components (b) and permit variable length coding of the scanned components.*

difference signal is sent to the Huffman coder as in DPCM (Chapter 5).

JPEG compression ratios vary from 2:1 to 20:1. Low compression ratios achieve lossless DPCM coding where the reconstructed image is indistinguishable from the original while high compression ratios can reduce the storage requirements to only ¼ bit per pixel and the transmission rate, for image sequences, to 2 Mbit/s. VLSI JPEG chips existed in 1993 which could process data at 8 Mbyte/s to handle 352×288 pixel (¼ full TV frame) images at 30 frame/s.

In recent years wavelet based coders have gained popularity, over the DCT of the baseline JPEG system, because the wavelet coder operates over the entire image rather than the restricted size DCT blocks. JPEG 2000 is one example of a wavelet based coder. As JPEG is a single frame coder it is not optimised to exploit the inter-frame correlation in image sequence coding. The following schemes are therefore preferred for compression of video data.

16.7.3 MPEG-1 and MPEG-2

This video coding specification was developed by the Motion Picture Experts Group (MPEG) as a standard for coding image *sequences* to a bit rate of about 1.5 Mbit/s for MPEG-1 and 2 to 8 Mbit/s for MPEG-2. MPEG-1 applies to non-interlaced video while MPEG-2 was ratified in 1995 for broadcast (interlaced) TV transmissions. The lower rate was developed, initially, for 352×288 pixel images because it is compatible with digital storage devices such as hard disk drives, compact disks, digital audio tapes, etc. The algorithm is deliberately flexible in operation, allowing different image resolutions, compression ratios and bit rates to be achieved. The basic blocks are:

- Motion compensation
- DCT
- Variable length coding

A simplified diagram of the encoding operation is shown in Figure 16.14. As the algorithm is intended primarily for the storage of image *sequences* it incorporates motion compensation which is not included in JPEG. A compromise still exists between the need for high compression and easy regeneration of randomly selected frame sequences. MPEG therefore codes the frames in one of three ways, Figure 16.15.

Intra-frames (I-frames), which are coded independently of other frames, allow random access, but provide limited compression. They form the start points for replay sequences. Unidirectional predictive coded frames (P-frames) can achieve motion prediction from *previous* reference frames and hence, with the addition of motion compensation, the bit rate can be reduced. Bidirectionally predictive coded frames (B-frames) provide greatest compression but require two reference frames, one previous

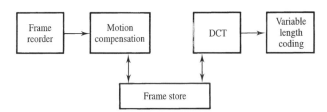

Figure 16.14 *Block diagram for simplified MPEG encoder.*

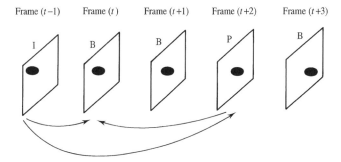

Figure 16.15 *I, P, B image frames, as used in the MPEG coder.*

frame and one future frame, in order to be regenerated. Figure 16.15 shows such an I, P, B picture sequence. The precise combination of I-, P- and B-frames which are used depends upon the specific application.

The motion prediction and compensation uses a block based approach on 16×16 pixel block sizes as in Figure 16.12. The concept underlying picture motion compensation is to estimate the motion vector, Figure 16.16, and then use this to enable the information in the previous frame to be used in reconstructing the current frame, minimising the need for new picture information. MPEG is a continuously evolving standard. MPEG-2 is also now used for coding digital TV where partial images are combined into the final image. Here the processing is dependent on the image content to enhance the compression capability and permit use with ATM at a P_b of 10^{-4}. MPEG-2 data rates vary between VHS video cassette player quality at 1.5 Mbit/s through to 30 Mbit/s for digital TV images. It is generally recognised that fast moving sports scenes require the higher bit rates of 6 to 8 Mbit/s and, by reducing these to 1.5 Mbit/s, then the quality degrades, to much like that of a VHS video cassette player. The MPEG-2 standard is used within digital video disc (DVD) and digital video recorder (DVR) commercial products. MPEG-3 was launched for HDTV but it was found that MPEG-2 techniques were generally adequate.

16.7.4 MPEG-4

MPEG-4, initiated in 1993, is a standard for interactive multimedia applications. Key objectives of MPEG-4 video coding are to be tolerant of or robust to transmission network errors, to have high interactive functionality (e.g. for audio and video manipulation) to allow accessing or addressing of the stored data by content. Thus it is able to accept both natural (pixel based) and synthetic data and, at the same time, achieve a high compression efficiency. It also facilitates transmission over mobile telephone networks and the Internet at rates of 20 kbit/s to 1 Mbit/s.

MPEG-4 uses content based coding where the video images are separated or partitioned into objects such as background, moving person, text overlay, etc. Video data representing each of these video objects (VOs) is then separated out and encoded as a separate layer or video object plane (VOP) bit stream which includes shape, transparency, spatial coordinates, i.e. location data, etc., relevant to the video object. Objects are selected from a video sequence using, for example, edge detection techniques, Figure

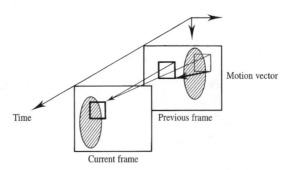

Figure 16.16 *Estimation of the motion vector, between consecutive frames, with object movement.*

(a)

Figure 16.17 *Segmentation of full image (a) in (b) to locate the edges or boundaries of an object and removal of selected object for subsequent editing into other images (source: Ghanbari, 1999, reproduced with the permission of Peter Peregrinus).*

16.17, combined with motion vectors to identify a moving object such as a person or vehicle. Edges with similar motion can be connected to form a contour around the moving object or VOP in the overall image. A user can then decode all the individual object layers to reconstruct the original image sequence or, alternatively, the image can be manipulated to remove objects, zoom scale or translate objects by simple bit stream editing operations. Thus backgrounds can be easily changed, Figure 16.17, by editing the bit stream to accommodate natural or synthetic images.

The basic encoder functionality is similar to MPEG-1 and MPEG-2 coders which employ a DCT to measure image intensity coupled with motion vector estimation , but in MPEG-4 the image objects can have arbitrary shape or contour in contrast to the rectangular blocks encoded by MPEG-1 and MPEG-2.

Many examples of MPEG-4 scene coding have been presented to illustrate the benefits of the technique. One commonly demonstrated is a tennis player in a court surrounded by spectators. The scene is partitioned into the flexible tennis player with coherent motion in the foreground and the court, crowd and line judges which are modelled as a static background. Thus the background object is built up progressively and only transmitted as a single frame while the contour coordinates and motion vectors of the moving tennis player are allocated the majority of the transmitted data bits.

Coding with MPEG-4 gives much better subjective picture quality than MPEG-1 for a modest bit rate of say 1 Mbit/s, i.e. it offers superior compression gain for multimedia database applications. Generally, broadcast TV requires the on-line processing and coding of MPEG-2 to faithfully reproduce the motion in the entire image and this necessitates a higher data rate. However, it is now recognised that MPEG-4 offers VHS quality video at 0.7 to 0.8 Mbit/s transmission rates in place of the 1.2 to 1.5 Mbit/s required for MPEG-1.

16.7.5 MPEG-7

Work started in 1999 on MPEG-7 as a new standard [Manjunath *et al.*] to characterise audio and video sequences so that they can be browsed and archived in databases. This is based on filtering the information via a content description to create personal access to the sequences or data of interest. The characterisation is based on both signal properties (colour, number of objects, human presence, etc.) and text descriptions (authors, names of persons in scenes, types of scene, etc.). The work on defining MPEG-7 promises to assist in accessing the video sequences people are interested in viewing from a large stored database.

16.7.6 H.261 and H.263

The H.261 algorithm was developed prior to MPEG in the 1980s for the purpose of image transmission rather than image storage. It is designed to produce a constant output of $p \times 64$ kbit/s, where p is an integer in the range 1 to 30. This allows transmission over a digital network or data link of varying capacity. It also allows transmission over a single 64 kbit/s digital telephone channel for low quality videotelephony, or at higher bit rates for improved picture quality. The basic coding algorithm is similar to that of MPEG in that it is a hybrid of motion compensation, DCT and straightforward DPCM (intra-frame coding mode), Figure 16.18, without the MPEG I-, P-, B-frames. The DCT operation is performed at a low level on 8×8 blocks of error samples from the predicted

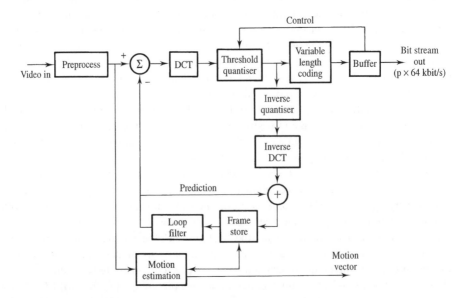

Figure 16.18 *Schematic representation of the H.261 encoder standard.*

luminance pixel values, with subsampled blocks of chrominance data. The motion compensation is performed on the macroblocks of Figure 16.12, comprising four of the previous luminance and two of the previous chrominance blocks.

H.261 is widely used on 176×144 pixel images. The ability to select a range of output rates for the algorithm allows it to be used in different applications. Low output rates ($p = 1$ or 2) are only suitable for face-to-face (videophone) communication. H.261 is thus the standard used in many commercial videophone systems. Video-conferencing would require a greater output data rate ($p > 6$) and might go as high as 2 Mbit/s for high quality transmission with larger image sizes.

A further development of H.261 is H.263 for lower fixed transmission rates. This deploys arithmetic coding in place of the variable length coding in Figure 16.17 and, with other modifications, the data rate is reduced to only 20 kbit/s. The key advances in H.263 are that it employs hybrid interpicture prediction and distinct methods of coding the transform coefficients and motion vectors.

16.7.7 Model based coding

At the very low bit rates (20 kbit/s or less) associated with videotelephony, the requirements for image transmission stretch the compression techniques described earlier to their limits. In order to achieve the necessary degree of compression they often require reduction in spatial resolution or even the elimination of frames from the sequence. Model based coding (MBC) attempts to exploit a greater degree of redundancy in images than current techniques, in order to achieve significant image compression but without adversely degrading the image content information. It relies upon the fact that the image quality is largely subjective. Providing that the appearance of scenes within an observed image is kept at a visually acceptable level, it may not matter that the observed image is not a precise reproduction of reality.

One MBC method for producing an artificial image of a head sequence utilises a feature codebook where a range of facial expressions, sufficient to create an animation, are generated from sub-images or templates which are joined together to form a complete face. The most important areas of a face, for conveying an expression, are the eyes and mouth, hence the objective is to create an image in which the movement of the eyes and mouth is a convincing approximation to the movements of the original subject. When forming the synthetic image, the feature template vectors which form the closest match to those of the original moving sequence are selected from the codebook and then transmitted as low bit rate coded addresses.

By using only 10 eye and 10 mouth templates, for instance, a total of 100 combinations exists implying that only a 6 bit codebook address need be transmitted. It has been found that there are only 13 visually distinct mouth shapes for vowel and consonant formation during speech. However, the number of mouth sub-images is usually increased, to include intermediate expressions and hence avoid step changes in the image.

Another common way of representing objects in 3-dimensional computer graphics is by a net of interconnecting polygons. A model is stored as a set of linked arrays which specify the coordinates of each polygon vertex, with the lines connecting the vertices together forming each side of a polygon. To make realistic models, the polygon net can be shaded to reflect the presence of light sources.

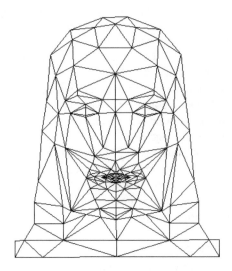

Figure 16.19 *Image representation by wire frame model.*

The wire frame model [Welch, 1991] can be modified to fit the shape of a person's head and shoulders. The wire frame, composed of over 100 interconnecting triangles, can produce subjectively acceptable synthetic images, providing that the frame is not rotated by more than 30° from the full-face position. The model, shown in Figure 16.19, uses smaller triangles in areas associated with high degrees of curvature where significant movement is required. Large flat areas, such as the forehead, contain fewer triangles. A second wire frame is used to model the mouth interior.

A synthetic image is created by texture mapping detail from an initial full-face source image, over the wire frame. Facial movement can be achieved by manipulation of the vertices of the wire frame. Head rotation requires the use of simple matrix operations upon the coordinate array. Facial expression requires the manipulation of the features controlling the vertices.

This model based feature codebook approach suffers from the drawback of codebook formation. This has to be done off-line and, consequently, the image is required to be prerecorded, with a consequent delay. However, the actual image sequence can be sent at a very low data rate. For a codebook with 128 entries where 7 bits are required to code each mouth, a 25 frame/s sequence requires less than 200 bit/s to code the mouth movements. When it is finally implemented, rates as low as 1 kbit/s are confidently expected from MBC systems, but they can only transmit image sequences which match the stored model, e.g. head and shoulders displays.

16.8 Digital video broadcast

Digital video broadcast (DVB) [Collins] for digital TV has been the subject of considerable research in Europe. Here there are now complementary DVB standards for terrestrial TV (DVB-T), satellite TV (DVB-S) and cable systems (DVB-C) [Forrest].

These systems are all aimed at transmission in a radio channel which is compatible with analogue TV, Figure 16.4, and a bandwidth of 8 MHz has been selected for DVB-T. The preferred modulation technique is COFDM, see section 11.5.2, and transform sizes of 2048 and 8196 (2k and 8k) have been adopted in the standard [DVB].

For DVB the COFDM technique is married with MPEG-2 video coding and at rates of > 6 Mbit/s there are essentially no visible imperfections or artifacts in the received signal. In DVB two layers of error control coding are applied in the form of a Reed–Solomon 204,188 outer coder and an inner convolutional encoder with a rate between 1/2 and 7/8, see Chapter 10. Also a short guard interval is introduced between the COFDM symbols into which the subcarrier transmissions are extended. Providing the guard interval is longer than the delay spread of the channel this ensures that, over the nominal symbol duration, the subchannels remain orthogonal (preventing interchannel interference) and adjacent symbols on the same subchannel don't overlap (preventing intersymbol interference).

The available capacity of the channel depends on the QPSK/QAM signal constellation used on the individual OFDM carriers. In the 8 MHz channel 64-QAM offers 38 Mbit/s while the 36 MHz wideband satellite repeater must use the simpler QPSK modulation due to power and linearity constraints, Chapter 14, but achieves a similar transmission rate of 30 to 40 Mbit/s. Precise payload rate depends on FECC coder rate. MPEG-2 coding thus permits 6 to 8 digital video TV channels to be transmitted in one satellite transponder and, as compression standards improve, so does the transponder capacity.

For wireless and packet based transmission special error protection techniques such as unequal error protection give a more robust transmission capability. Here, for example, image line synchronisation information can be better preserved to permit superior picture decoding in channels which are degraded by bursty noise or where packet loss might result.

16.9 Packet video

Packet switched (section 20.7) video is the term used to describe low bit rate encoded video, from one of the above CODECs, which is then transmitted over the ISDN as packet data similar to packet speech [ITU-T Rec. H.323]. Due to the operation of the CODEC and the nature of the video signal, the resulting data rate varies considerably between low data rates when the image sequence contains similar frames, as in Figure 16.8, to high data rates when there is a sudden change in the sequence and almost the entire image frame must be encoded and transmitted. This is called variable bit rate (VBR) traffic, Figures 16.20 and 20.73, which is accommodated by packet transmission on an ATM network, section 20.10. Figure 16.20 shows the regular variation in bit rate in an MPEG coder arising from quantising the I-, B-, P-frames, Figure 16.15, plus the major alterations in rate due to motion or dramatic scene changes in the video sequence. At 30 frame/s this represents an overall variation in the required transmission rate between 600 kbit/s and 6 Mbit/s.

VBR systems rely on the fact that their average transmission rate has a modest bandwidth requirement. Hence if many VBR sources are statistically multiplexed then

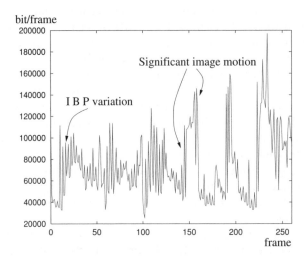

Figure 16.20   *Variable rate bursty traffic from a colour video image CODEC. Note the regular short term variation in the MPEG coded data and the major scene changes.*

their individual bit rate variations will be averaged and they can be carried over a system with a transmission bandwidth which is compatible with the modest *average* bandwidth requirement of each source. Tariffs will then be levied which use this modest data rate requirement, rather than providing each user with access to a peak data rate channel and incurring correspondingly high costs.

The principles underlying packet video [Hanzo *et al.*] are similar to those described in Chapter 19. We must inevitably accept some loss of video data when the overall demand for transmission bandwidth exceeds the available channel allocation since arbitrarily large amounts of data cannot be held within the finite length of the buffer in Figure 16.18 (and Figure 20.22). Much current research is aimed at investigating this problem as the ISDN system expands and ATM techniques start to become more widely used. Some of this work is aimed at the design of the video encoding algorithm to control the statistics of the output VBR traffic; other work is aimed at investigating coding methods which split the video data into high and low priority traffic.

The overall protocol [Falconer and Adams], which corresponds to the lowest two layers of the OSI reference model (see section 17.12), represents one potential method of achieving different priority of traffic. By careful control of the video encoding technique we can ensure that high priority bits are allocated to the major video image features while low priority bits add more detail to this basic image. Techniques such as this allow speech and video data to be transmitted, with some packet loss, while still providing a quality of service which is acceptable to the user, for a transmission cost which is much lower than that of providing the full bandwidth requirement for 100% of the time.

16.10 Summary

The smallest elemental area of a 2-dimensional image which can take on characteristics independent of other areas is called a picture cell or pixel. The fundamental requirement of an image transmission system is the ability to convey a description of each pixel's characteristics: intensity, colour and location in space. For video images, information giving the pixel's location in time is also required.

A pixel from a colour image may be represented by three colour intensity signals (red, green and blue) or by a white (red + green + blue) intensity signal (luminance) and two colour difference (chrominance) signals (white − red and white − blue). The colour difference signals may be transformed geometrically, using the colour triangle, into saturation and hue. The luminance plus chrominance representation makes colour image transmission compatible with monochrome receivers and is therefore used for conventional TV broadcasts.

In the UK the two chrominance signals of a TV broadcast are transmitted using quadrature double sideband suppressed carrier modulation of a subcarrier. Since this requires a phase coherent reference signal for demodulation, the reconstructed image is potentially susceptible to colour errors caused by phase distortion, within the chrominance signal frequency band, arising from multipath propagation. These errors can be reduced by reversing the carrier phase of one of the quadrature modulated chrominance signals on alternate picture lines which gives the system its name − 'phase alternate line' or PAL. The PAL signal transmits 25 frames per second, each frame being made up of 625 lines. The frame is divided into two fields, one containing even lines only and the other containing odd lines only. The field rate is therefore 50 Hz. The total bandwidth of a PAL picture signal is approximately 6 MHz.

HDTV may be either analogue (for compatibility with traditional broadcast receivers), digital or mixed. The picture resolution of HDTV is about twice that of conventional TV as is the required (analogue) signal bandwidth. Compatibility with conventional TV receivers can be maintained but it is much more likely that HDTV will ultimately adopt an all-digital transmission system.

Digital video can be implemented by sampling and digitising an analogue video signal. Typically, the luminance signal will be subject to a finer quantisation process than the chrominance signals (e.g. 8 bit versus 4 bit quantisation). Unprocessed digital video signals have a bandwidth which is too large for general purpose telecommunications transmission channels. To utilise digital video for services such as videophone or teleconferencing the digital signal therefore requires data compression. Several coding techniques for redundancy removal, including run length coding, conditional replenishment, transform coding, Huffman coding and DPCM, may be used, either in isolation or in combination, to achieve this.

Several international standards exist for the transmission and/or storage of images. These include JPEG which is principally for single images (i.e. stills) and MPEG which is principally for moving pictures (i.e. video).

Low bit rate transmission of video pictures can be achieved, using aggressive coding, if only modest image reproduction quality is required. Thus the H.261 standard CODEC allows videophone services to be provided using a single 64 kbit/s digital telephone channel. The concepts used within the H.261 standard have been refined and extended to many more video applications in the MPEG standards and a complete set of MPEG video standards have evolved which are used for digital video and TV transmission. Extremely low bit rate video services can be provided using model based coding schemes which are analogous to the low bit rate speech vocoder techniques described in Chapter 9.

Variable bit rate transmissions, which may arise as a result of video coding, can be accommodated by packet transmission over ATM networks. Predicting the overall performance of packet systems requires queueing theory which is discussed in Chapter 19. Networks, protocols and the foundations of ATM data transmission are discussed in Chapters 17, 18 and 21.

16.11 Problems

16.1 A 625-line black and white television picture may be considered to be composed of 550 picture elements (pixels) per line. (Assume that each pixel is equiprobable among 64 distinguishable brightness levels.) If this is to be transmitted by raster scanning at a 25 Hz frame rate calculate, using the Shannon–Hartley theorem of Chapter 11, the minimum bandwidth required to transmit the video signal, assuming a 35 dB signal-to-noise ratio on reception. [4.44 MHz]

16.2 What is the minimum time required to transmit one of the picture frames described in Problem 16.1 over a standard 3.2 kHz bandwidth telephone channel, as defined in Figure 5.12, with 30 dB SNR? [65 s]

16.3 A colour screen for a US computer aided design product may be considered to be composed of a 1000×1000 pixel array. Assume that each pixel is coded with straightforward 24 bit colour information. If this is to be refreshed through a 64 kbit/s ISDN line, calculate the time required to update the screen with a new picture. [6.25 min]

16.4 Derive an expression for the bandwidth or effective data rate for a television signal for the following scan parameters: frame rate = P frame/s; number of lines per frame = M; horizontal blanking time = B; horizontal resolution = x pixels/line where each pixel comprises one of k discrete levels. Hence calculate the bandwidth of a television system which employs 819 lines, a 50 Hz field rate, a horizontal blanking time of 18% of the line period and 100 levels. The horizontal resolution is 540 pixels/line. [75.5 Mbit/s]

16.5 Use the Walsh transform matrix of equation (15.29(b)) to transform the data vector $x = 5, 4, 8, 6$ and calculate the values for the four transformed output samples. For an orthonormal function, section 2.5.3, the energy content of the input must equal the energy content of the transformed output. Thus introduce a scalar weight into equation (15.29(b)) to ensure that this property is achieved and calculate the required value for the scalar weight. Using the orthonormal values calculate the percentage error if the two smallest transformed output values are deleted. [23, −5, −1, 3, weight 0.5, 1.77%]

Part Four

Networks

Part Four is devoted to communication networks which now exist on all scales from geographically small wireless personal LANs to the global ISDN.

Chapter 17 describes the topologies and protocols employed by networks to ensure the reliable, accurate and timely delivery of information packets between network terminals. Stars, rings, buses and their associated medium access protocols are discussed, and international standards such as ISO OSI, X.25, Ethernet and FDDI are described. This chapter includes circuit, message and packet switching techniques. Chapter 18 covers network protocols, synchronisation, flow and congestion control, routing, end-to-end error control and quality of service issues as used in the seven layers of the ISO OSI model. This includes a detailed discussion of the X.21 and X.25 interfaces.

Chapter 19 covers queueing theory which is used to predict the delay suffered by digital information packets as they propagate through a data network. This includes various queue types, waiting times, Markovian queue theory, Erlang distributions and extends queue theory from fixed to mobile applications. This last topic covers the speech source model and equilibrium probability.

Chapter 20 examines public networks. A general introduction is provided to graph theory, as applied to the cost and capacity matrices, which define the capability of networks. The current plesiochronous digital hierarchy (PDH) is then reviewed before introducing a more detailed discussion of the synchronous digital heirarchy (SDH) which is progressively replacing PDH. The chapter then discusses space and time switching, circuit and packet switched technologies and modem standards. It concludes with a brief discussion of PSTN/PDN data access techniques including the ISDN standard, ATM and the future development of the local loop into DSL, and the new, emerging, fixed wireless access standards.

Part Four ends in Chapter 21 with broadcast networks and LANs. It discusses accessing, polling, token passing and contention in both wired LANs, e.g. Ethernet variants, and wireless LANs. The concluding discussion of personal and home area networks covers Bluetooth, and other wireless as well as wired home networks and residential gateways.

CHAPTER 17

Network applications, topologies and architecture

17.1 Introduction

A telecommunications network consists of a set of communication links that interconnect a number of user devices. The links may form a (fixed) and common transmission medium between all the connected devices (broadcast networks) or they may be re-configurable, intersecting at switches that allow point-to-point paths to be established through the network between any two connected devices (switched networks). In 1969, the first major packet switched communications network, the ARPANET, began operation. The network was originally conceived by the Advanced Research Projects Agency (ARPA) of the US Department of Defense for the interconnection of dissimilar computers, each with a specialised capability. Today systems range from small networks interconnecting microcomputers, hard disks and laser printers in a single room (e.g. Appletalk), through terminals and computers within a single building or campus (e.g. Ethemet), to large geographically distributed networks spanning the globe (e.g. the Internet).

This chapter describes a variety of communications network applications and outlines the network's functional role. A network classification scheme is given that allows similarities and differences to be recognised between the practical instances of what can be a confusing array of network technologies. The common network topologies (i.e. the connection patterns) used to link network terminals and nodes are reviewed and, finally, the layered approach to architectures that is necessary to reduce the design and implementation of a complex network to a tractable problem is described.

17.2 **Network applications**

The number of current communications network applications is enormous. In addition to conventional voice communication these applications include:

1. **Travel sales and administration** – One of the first large scale uses of a data network was the coordination of airline seat booking. The availability of seats on a given flight is known only at one location, but there is a large number of booking offices where this information is required at short notice. This is the classic example of a network application where a large number of geographically dispersed terminals need to interrogate a central database.

2. **Banking** – The organisation of a high street bank requires a central database containing information about each customer's account, and a large number of branch offices spread throughout the country for serving customers. Transactions at the branches must be recorded in the central database and information about the status of customers' accounts must be made available at the branch offices. Initially this information transfer was made over the 'plain old telephone' (POT) network with data often being transmitted overnight at the end of each day's transactions to take advantage of the cheap nighttime tariffs offered by the telephone service providers. Customer identity must sometimes be checked, particularly when an automatic teller machine is being used. This of course requires immediate communication and cannot wait until a more suitable transmission time.

3. **Scientific and technical computing** – The first networks grew up around scarce computing resources, allowing a number of users to access them without being physically close to the machine. Although the first networks represented a great technical advance they were inconvenient to use since the transmission rate was very low (typically 10 character/s) resulting in significant delay and thus limiting 'real time' user interaction. Today such computing networks cover geographical areas of every size from campus local area networks (LANs) to worldwide global area networks (GANs). Transmission speeds can be very high offering almost instantaneous access and genuine interaction supporting, for example, remote virtual reality applications and the rapid transfer of large databases and high resolution images.

4. **Electronic funds transfer (EFT)** – Whereas the banking network described above signals the state of various accounts and the money deposited/withdrawn at a given branch, etc., EFT is an electronic means of actually transferring money. Although using the same principles as an ordinary data transfer network, security must be of a higher order both to prevent errors from corrupting the transferred information and to prevent fraudulent transfers from taking place. Data encryption and message authentication techniques have been developed that are making EFT a reliable and increasingly common method of transferring money.

5. **Point-of-sale (POS) transfer** – In situations where large volume sales occur, there is great advantage to recording the transaction directly. Product data can be read, optically, from a barcode on product packaging and customer identity can be read, magnetically, from a personalised credit or debit card. A network is required to link each of the POS terminals to a local stock control database and to interrogate the customer's bank account. The purchaser's account status can therefore be checked

and funds transferred immediately from this account to the vendor's account. Stock levels are monitored directly and reordering of items from the vendor's supplier can be automated. Such tight control of stock levels can be vitally important for fast turnaround goods, particularly perishable items.

6. **Telecontrol** – Any large distribution or collection network (e.g. for electricity, gas, oil, water, sewage) functions better if it can be monitored and controlled. All these utilities therefore employ large telecontrol networks that enable a centrally placed controller to examine the flow of the product over the network and respond immediately to sudden changes in demand or faults. By their nature such networks are often in electrically hostile environments but the integrity of the information transfer must be maintained since the consequences of acting on incorrect information are potentially serious. Telecontrol networks must often be available continuously, so standby routes and equipment diversity may be essential.

7. **Internet** – The rapid growth of this global network, initially to serve the communication needs of the international academic community, now provides a general medium for relaxation, entertainment and commerce as well as the interchange of scientific and technical information. It links tens of thousands of national and local networks into a single global entity. Web browsers allow information to be found and extracted from locations worldwide without the user even needing to be aware of the information's geographical location.

The applications described above and summarised schematically in Figure 17.1 represent a small subset of those currently possible. The number of future applications of communications networks is, for all practical purposes, limitless.

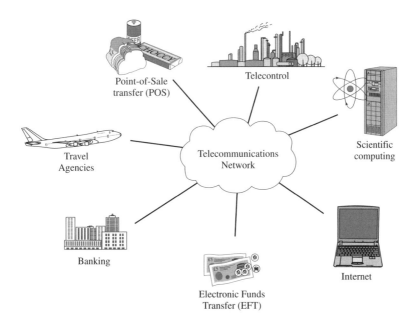

Point-of-Sale transfer (POS)

Telecontrol

Travel Agencies

Telecommunications Network

Scientific computing

Banking

Electronic Funds Transfer (EFT)

Internet

Figure 17.1 *Example network applications.*

17.3 Network function

The essential function of a network is to emulate as far as possible the fully interconnected mesh illustrated in Figure 17.2 in which a dedicated transmission link is provided between all terminal (or user) pairs. It is easy to see that the number, N, of (bidirectional) transmission links required for this type of ideal network is given by:

$$N = \frac{n(n-1)}{2} \qquad (17.1)$$

This is perfectly reasonable for small values of n and small geographical separations between terminals. Such a transmission infrastructure clearly becomes prohibitively expensive, however, for large values of n and large geographical separations. Twenty million telephone handsets, for example, in the UK public switched telephone network (PSTN) would need approximately 200000 billion transmission links for a fully interconnected national network alone. Furthermore, adding only a single node to a fully interconnected mesh necessitates a further $n-1$ new links, making expansion of such networks practically impossible. A network that looks to the user like a fully interconnected mesh but operates with a small fraction of these transmission links is therefore highly desirable.

17.4 Network classification

There are two basic approaches to emulating a fully interconnected mesh with realistic transmission resources. The first is to connect all the users to a common transmission medium and the second is to use links that can be reconfigured between users as required. The former approach results in a *broadcast* network whilst the latter results in a *switched* network. This could be called the great network divide since it represents two quite different network philosophies. In broadcast networks all users share the whole of the common transmission medium, any message transmitted from one terminal consequently being received by all terminals. In switched networks users are connected by a (much) less than fully interconnected mesh of mutually isolated links with switches at mesh

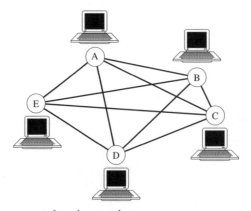

Figure 17.2 *Fully interconnected mesh network.*

nodes that are able to create a route through the network between any two terminals. A message transmitted over a switched network by one terminal is only, therefore, received at the terminal intended to be its destination. The fundamental problem in switched networks is the fair, but efficient, allocation of links to provide an effective path between user pairs on demand, and these networks can therefore be subclassified according to the switching method they employ. The fundamental problem to be solved for broadcast networks is the fair, but efficient, sharing of the common transmission medium between users, and broadcast networks can therefore be subclassified according to their rules for medium access control, referred to as their medium access control (MAC) protocol. Figure 17.3 shows a taxonomy of network types constructed along these lines.

Networks may also be classified in terms of their geographical size depending on whether their coverage is local, metropolitan or wide. A local area network (LAN) typically extends over about 1 km (e.g. an educational or industrial campus), a metropolitan area network (MAN) typically extends over about 10 km (e.g. a city) and a wide area network (WAN) typically extends over 100 km or greater (typically a country). Global area networks (GANs) are also sometimes referred to which may represent a single large WAN or a set of interconnected WANs. Some typical features of LANs, MANs and WANs are summarised in Table 17.1 [after Smythe, 1991].

Table 17.1   *Comparison of LAN/MAN/WAN characteristics.*

Issue	*WANs*	*MANs*	*LANs*
Geographical size	1000s km	1–100 km	0–5 km
No. of nodes	10000s	1–500	1–200
R_b	0.1–100 kbit/s	1–100 Mbit/s	1–100s Mbit/s
P_b	10^{-3}–10^{-6}	$<10^{-9}$	$<10^{-9}$
Delays	>0.5 s	100–100s ms	1–100 ms
Routing	sophisticated	simple	none
Linkages	gateways	bridges	bridges

The distinction between networks based on coverage can be mapped onto the great network divide rather closely. LANs are almost always broadcast networks whilst MANs and WANs are invariably switched networks.

An alternative distinction between LANs, MANs and WANs depends on what might loosely be called electrical size rather than geographical size. The electrical size, in this context, is defined by the ratio of network delay to average duration of a single transmission (i.e. data packet duration). This ratio is much smaller than unity for a LAN, much larger than unity for a WAN, and of the order of unity for a MAN [Smythe, 1991]. A distinction based on electrical size has the advantage of being well defined, but is also rather academic, since the looser definitions relating to geographical size are almost universally used in practice.

A third distinction between LANs, MANs and WANs can be made in terms of the technologies, data rates and protocols employed. LANs are almost always high data rate broadcast networks, each transmission being received by all connected terminals. MANs are usually high data rate switched networks and WANs are lower data rate switched networks. Figure 17.4 shows the relationships between LANs, MANs, WANs, and a selection of other network types and terminologies.

17.5 Switched network topologies and representation

Any network [Hioki] must fundamentally be based on some interconnection topology, to link its constituent terminals. The principal network topologies are reviewed here in the context of the great network divide (switched and broadcast) that has already been identified. Switched network topologies consist of a set of point-to-point links

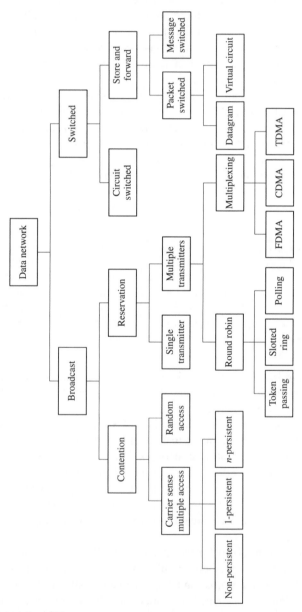

Figure 17.3 *A data network taxonomy.*

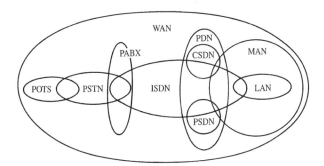

Figure 17.4 *Relationships between network architectures. Key - POTS: Plain old telephone system, PSTN: Public switched telephone system, PABX: Private automatic branch exchange, ISDN: Integrated services digital network, PDN: Public data network, CSDN: Circuit switched data network, PSDN: Packet switched data network, LAN: Local area network, MAN: Metropolitan area network, WAN: Wide area network.*

interconnected by switches at network nodes that allow the link connections to be reconfigured so that paths can be created between users as required. Three varieties of switched topology are described below.

17.5.1 Star or hub

Switched star network configurations have now existed for over a century in the access network of the PSTN and, for this reason, represent perhaps the best understood class of network. In the star configuration the nodes or devices of the network are connected by point-to-point links to a central node, computer or exchange, Figure 17.5. Network control in this type of topology is usually centralised at the hub of the star.

The star network has two major limitations:

1. The remote devices are unable to communicate directly and must do so via the central node or hub, which is required to switch these transmissions as well as carrying out its primary processing function.

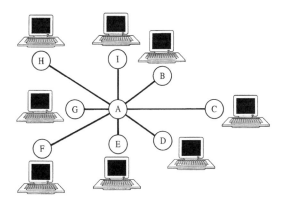

Figure 17.5 *Star network.*

2. Such a network is very vulnerable to failure, either of the central node, causing a complete suspension of operation, or of a transmission link. Reliability and/or redundancy are therefore particularly important considerations for this topology.

Despite these limitations, the star configuration is important since it has been used extensively for telephone exchange connections and in fibre optic systems. (Until relatively recently it has been difficult to realise cheap, passive, optical couplers for implementation of an optical fibre bus system, for example.) It also has the advantage that expansion of a star network by a single node requires only a single extra connection. The star configuration is best suited to one-to-many types of application (e.g. a host computer to slave terminals network).

17.5.2 Tree

A tree topology is illustrated in Figure 17.6. The defining property of a tree network is that there is only one possible path between any pair of nodes which makes the network routing algorithm essentially trivial. The structure is hierarchical in that network nodes at higher levels of the structure have a role in connecting a greater proportion of the set of all possible connections than network nodes at lower levels and in this sense are more important. (The hierarchical level of a node refers, here, to the number of nodes in the path connecting it with the tree's apex, or root node – the number counted including the node itself and the root node. This may or may not coincide with some other definition of hierarchical node importance in terms of the equipment installed at a particular location.) It is a property of tree networks that each node is directly connected to one, and only one, higher level node. Like the star network, therefore, expansion of a tree network by one node requires only one additional link. This is not coincidental since a star network represents the special case of a tree network with only two hierarchical levels.

17.5.3 Mesh

If a link is added to a tree network such that a node is connected to more than one higher level node or to a peer level node then more than one path will exist between at least one pair of terminals and the network topology becomes a mesh. If all nodes are connected to all other nodes then the network becomes the special (and ideal) case of the fully interconnected mesh described in section 17.3.

Mesh networks, Figure 17.7, are primarily used in many-to-many types of application (e.g. peer device communications as typically provided for by a WAN).

Fully interconnected mesh networks, for more than a small number of nodes, are generally expensive or impractical for the reasons outlined in section 17.3. They are very resilient to failure, however, since alternative routes are available if any link becomes unavailable. Where links are long or data volumes low, a public packet switched service may offer a significant cost advantage over a private mesh network.

17.5.4 Matrix representation

The topology of a network with N nodes can be described by an $N \times N$ matrix. The element m_{ij} is unity if a connection between nodes i and j exists and is zero if a connection does not exist. Since nodes do not have links connecting them to themselves, the diagonal elements of the matrix are zero, i.e. $m_{ii} = 0$.

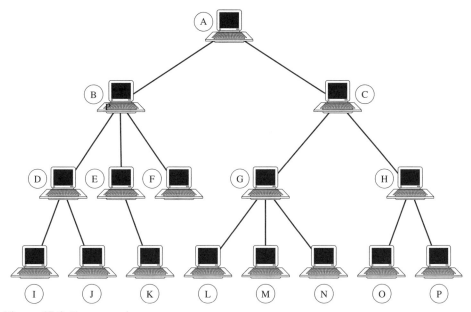

Figure 17.6 *Tree network.*

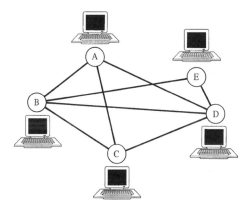

Figure 17.7 *Mesh network.*

The matrices describing the star, tree and mesh connected networks of Figures 17.5, 17.6 and 17.7 (using the convention nodes A, B, C, $\cdots$ = 1, 2, 3, $\cdots$) are therefore given, respectively, by:

$$
M_{star} = \begin{bmatrix}
0 & 1 & 1 & 1 & 1 & 1 & 1 & 1 & 1 \\
1 & 0 & 0 & 0 & 0 & 0 & 0 & 0 & 0 \\
1 & 0 & 0 & 0 & 0 & 0 & 0 & 0 & 0 \\
1 & 0 & 0 & 0 & 0 & 0 & 0 & 0 & 0 \\
1 & 0 & 0 & 0 & 0 & 0 & 0 & 0 & 0 \\
1 & 0 & 0 & 0 & 0 & 0 & 0 & 0 & 0 \\
1 & 0 & 0 & 0 & 0 & 0 & 0 & 0 & 0 \\
1 & 0 & 0 & 0 & 0 & 0 & 0 & 0 & 0
\end{bmatrix}
$$

$$
M_{tree} = \begin{bmatrix}
0 & 1 & 1 & 0 & 0 & 0 & 0 & 0 & 0 & 0 & 0 & 0 & 0 & 0 & 0 & 0 \\
1 & 0 & 0 & 1 & 1 & 1 & 0 & 0 & 0 & 0 & 0 & 0 & 0 & 0 & 0 & 0 \\
1 & 0 & 0 & 0 & 0 & 0 & 1 & 1 & 0 & 0 & 0 & 0 & 0 & 0 & 0 & 0 \\
0 & 1 & 0 & 0 & 0 & 0 & 0 & 0 & 1 & 1 & 0 & 0 & 0 & 0 & 0 & 0 \\
0 & 1 & 0 & 0 & 0 & 0 & 0 & 0 & 0 & 0 & 1 & 0 & 0 & 0 & 0 & 0 \\
0 & 1 & 0 & 0 & 0 & 0 & 0 & 0 & 0 & 0 & 0 & 0 & 0 & 0 & 0 & 0 \\
0 & 0 & 1 & 0 & 0 & 0 & 0 & 0 & 0 & 0 & 0 & 1 & 1 & 1 & 0 & 0 \\
0 & 0 & 1 & 0 & 0 & 0 & 0 & 0 & 0 & 0 & 0 & 0 & 0 & 0 & 1 & 1 \\
0 & 0 & 0 & 1 & 0 & 0 & 0 & 0 & 0 & 0 & 0 & 0 & 0 & 0 & 0 & 0 \\
0 & 0 & 0 & 1 & 0 & 0 & 0 & 0 & 0 & 0 & 0 & 0 & 0 & 0 & 0 & 0 \\
0 & 0 & 0 & 0 & 1 & 0 & 0 & 0 & 0 & 0 & 0 & 0 & 0 & 0 & 0 & 0 \\
0 & 0 & 0 & 0 & 0 & 0 & 1 & 0 & 0 & 0 & 0 & 0 & 0 & 0 & 0 & 0 \\
0 & 0 & 0 & 0 & 0 & 0 & 1 & 0 & 0 & 0 & 0 & 0 & 0 & 0 & 0 & 0 \\
0 & 0 & 0 & 0 & 0 & 0 & 1 & 0 & 0 & 0 & 0 & 0 & 0 & 0 & 0 & 0 \\
0 & 0 & 0 & 0 & 0 & 0 & 0 & 1 & 0 & 0 & 0 & 0 & 0 & 0 & 0 & 0 \\
0 & 0 & 0 & 0 & 0 & 0 & 0 & 1 & 0 & 0 & 0 & 0 & 0 & 0 & 0 & 0
\end{bmatrix}
$$

$$
M_{mesh} = \begin{bmatrix}
0 & 1 & 1 & 1 & 0 \\
1 & 0 & 1 & 1 & 1 \\
1 & 1 & 0 & 1 & 0 \\
1 & 1 & 1 & 0 & 1 \\
0 & 1 & 0 & 1 & 0
\end{bmatrix}
$$

17.6 Generic network switching philosophies

There are three generic approaches to creating an information path between two communicating terminals using a switched network. In order of historical development these are circuit switching, message switching and packet switching. They each have their own advantages and disadvantages that make them appropriate for a particular application and type of traffic. It is true to say, however, that the triumph of digital over analogue communications and the changing mix of traffic types have resulted in a general trend away from circuit switching and towards packet switching. A brief description of the most important aspects of these switching philosophies is given below.

17.6.1 Circuit switching

In circuit switched systems a continuous physical link is established between the pair of communicating data terminals for the entire duration of the communications session. (The term physical link here includes the possibility of radio links.) Circuit switching commits the overall route through a network for the entire duration of a call whether the connection is carrying useful signals or not. It is also inefficient for variable bit rate transmission since the circuit must always support the highest data rate expected. Furthermore, congestion or call blocking (the inability of a connection to be made) can occur, even when many parts of the network are idle.

Although there is usually a significant delay in setting up a circuit switched route (typically of the order of a second) the user has dedicated use of the route for the duration of the call with essentially instantaneous access.

For bursty data communication, such as interrogation of a remote database, the channel lies idle for a significant fraction of the time, making poor use of the network plant. In the case of voice transmission over an expensive facility (e.g. a transoceanic cable or satellite link), the quiet periods when a speaker pauses can be utilised for an alternative conversation, and thus more efficient use made of the limited transmission resource. This generic technique, often called time assigned speech interpolation (TASI), is well illustrated in the context of satellite systems, section 14.3.7, where it is referred to as digital speech interpolation (DSI).

Switching at a network node can be accomplished by a space division switching matrix using metallic contacts as was widely applied in old analogue telephone exchanges. Newer digital exchanges employ time division switching in which each incoming data stream is assigned to a particular time-slot in a multiplex frame and the outgoing time-slot is selected electronically, leading to a flexible program controlled switch (see section 20.6.2).

17.6.2 Message switching

In message switched systems the complete message (of any reasonable length) is stored at each network node and forwarded when the ongoing route is clear. Physical connections between node pairs are made only for the duration of the message transfer between those node pairs and are broken as soon as the message transfer is complete. No complete physical path need therefore exist between communicating terminals at any time. Since the message is always stored somewhere in the network the message can be retained until the receiver is ready to accept it. Call blocking, therefore, does not occur (although this and the avoidance of call set-up time are at the cost of potentially large and variable message delay).

Whilst significant storage capacity is required at each node, where the switch takes the form of a general purpose minicomputer, there are important advantages to using message switching over circuit switching for short messages. These are:

1. The transmission capacity of the network is better utilised.
2. Multiple destinations can be specified for any particular message.
3. Priority can be allocated to an urgent message since each message is examined at each node as it passes through the network.
4. Errors can be corrected on a link-by-link basis (retransmitting the message over each offending link) since end-to-end communication is not in real time.
5. Since the message is of finite length, its transmission speed or its format can be changed at any node within the network. The two communicating user stations can therefore use different clock rates and different representational codes for the data.

Message switching is most useful for short messages such as telegrams, electronic mail, short files, transaction queries and responses, etc.

17.6.3 Packet switching

In packet switched networks each message is divided into many packets (typically of a few thousand bits) that are then routed individually through the network. Each packet is stored and forwarded at each network node. Messages are reassembled from their constituent packets at the receiving terminal.

Two distinct varieties of packet switching exist: *virtual circuit* packet switching in which all packets follow the same route through the network, and *datagram* packet switching in which different packets (within the same message) contain their own address information and are routed entirely independently. Virtual circuit systems ensure that packets are received in their correct chronological order but require a route to be established through the network before transmission of data takes place. A set-up procedure, similar to that needed for a switched circuit, is therefore necessary. The virtual circuit represents only a logical connection (rather than a physical connection), however, and does not commit transmission capacity. The information required to route later packets through the network is lodged at the network nodes during the set-up procedure. After this packets need only carry information associating them with a particular logical channel (e.g. a logical channel number).

Datagram systems avoid the need for any path set-up procedure (which wastes time especially for short messages) but must include full destination addresses and packet sequencing information in each data packet to enable correct message delivery and re-assembly. The absence of a path set-up procedure qualifies the datagram approach as a connectionless mode of transmission whilst circuit switching, message switching and virtual circuit packet switching are all connection oriented transmission modes.

Packet switching has all the advantages of message switching over circuit switching but suffers lower overall delay. This advantage also extends to smaller packet systems over larger packet systems as shown by the following simplified analysis.

Consider a message of M bytes, transmitted at a rate of R bit/s, divided up into K packets. If it is necessary to include an overhead of H bytes to identify and control packets irrespective of packet length then the time occupied by a single packet, T_P, is given by:

$$T_P = \frac{8}{R}\left(\frac{M}{K} + H\right) \text{ (s)} \tag{17.2}$$

If there are N nodes in a network path and each node receives and stores each incoming packet before transmitting it onwards then $(N + K)$ packet durations elapse before the complete message is received. Figure 17.8 illustrates this for the case of $N = 3$ and $K = 4$.

(Note that with a complete message, only one link is occupied at any one time whereas with a packetised message each link contains one packet, the network thereby being more efficiently utilised.) The overall transmission time, T_{tot}, is therefore given by:

$$T_{tot} = \frac{8}{R}\left(\frac{M}{K} + H\right)(N + K) \text{ (s)} \tag{17.3}$$

The following example based on the above analysis shows the relationship between network latency and packet length. This (crudely derived) relationship is illustrated in Figure 17.9.

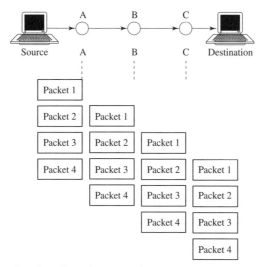

Figure 17.8 *Progress of packets through a network.*

EXAMPLE 17.1

A message of 1024 bytes, which is transmitted over a 2.048 Mbit/s network, encounters five network nodes en route. Assuming that 10 bytes of overhead per packet are required, estimate the overall transmission time when the message is divided into 1, 2, 3, 4, 5, 7 and 10 packets.

Using the simple model of equation (17.3) we have $M = 1024$, $N = 5$ and $R = 2.048$ Mbit/s.

Choosing $K = 4$, for example, the packets each contain 256 bytes of information and 10 bytes of overhead making a total of 266 bytes. For this case we have:

$$T_{tot} = \frac{8}{2048} \left(\frac{1024}{4} + 10 \right)(5 + 4)$$

$$= 9.35 \text{ ms}$$

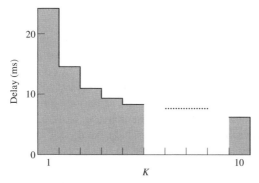

Figure 17.9 *Network delay for different values of K (from Example 17.1).*

Repeating this calculation for the other values of K allows the histogram in Figure 17.9 to be drawn.

Notice the improvement when the message is split into only two packets, greatly reducing the delay. Observe too that saturation in improvement occurs as the number of packets grows. Packet sizes may be fixed by standards but this example does demonstrate the intrinsic benefit gained by transmitting shorter packets rather than whole messages.

The simple analysis above fails to include the effect of queueing delays at nodes, each of which will usually be switching packets from many different sources. Node delays are subject to variation depending upon the processing load and the state of the network around the node, and can only be predicted properly using queueing theory (see Chapter 19). The propagation delay is also ignored in Example 17.1, and although it merely adds to the delay calculated here, it can be very significant for some systems (e.g. those incorporating a geostationary satellite, see Chapter 14). A further time overhead is the call establishment (i.e. set-up) time.

High speed switched networks now use very short fixed length packets called cells (only 53 bytes in length) thus minimising transmission delay. This particular packet transmission technology (called asynchronous transfer mode or ATM) is also designed to minimise the overhead allocation to each cell (see Chapter 20).

17.7 Broadcast network topologies

Broadcast topologies employ a common transmission medium to link all network nodes. The medium may be continuous and connect the nodes in parallel or it may be broken with the nodes effectively dividing the medium into a set of point-to-point links. In the latter case the network qualifies as a broadcast topology providing the nodes do not operate as switches but merely as repeaters with all transmissions propagating to each node in turn.

17.7.1 Bus or multidrop

Bus networks are formed from a continuous length of cable to which devices are attached using cable interfaces or taps, Figure 17.10.

Messages from a device are transmitted (bidirectionally) to all devices on the bus simultaneously. Receiving devices, however, accept only those messages addressed to themselves. The ends of a bus cable are usually terminated in matched loads. This is to prevent interference between the transmitted signal and copies of the signal travelling back down the cable having been reflected from the cable ends.

The bus configuration is extremely tolerant of terminal failures, since operation of the network will usually continue if one of the active devices fails. A further advantage is that bus networks are easily reconfigured and extended.

The small computer system interface (SCSI) is an example of a dedicated bus, used to connect disks and tape drives directly to the processor of a computer.

Traditional satellite networks, Figure 17.11, in which each terminal broadcasts via the satellite to all other terminals constitute bus systems where radio transmission replaces

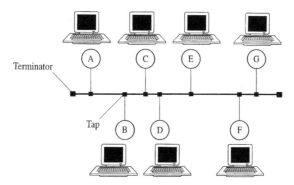

Figure 17.10 *Bus network.*

the physical cable usually associated with the bus topology. (Some future satellite and high altitude platform networks, however, will use multiple narrow beam antennas on both uplinks and downlinks connected by onboard switching matrices that will constitute switched networks. This illustrates well the essential independence of network topologies from both their geographical layout and the technology used to implement them.)

When bus topologies are used to connect a large number of locations that can be partitioned into geographical clusters they are often called *multidrop* systems. Multidrop optical fibre connection, often referred to as a passive optical network (PON), is under consideration (in competition with other technologies) as a replacement for the local loop PSTN star network that still predominantly employs twisted pair copper wires.

17.7.2 Passive ring

This network topology, Figure 17.12, is a bus topology with the ends of the bus bent into a closed loop. It is sometimes called a ring bus to distinguish it from the more common (linear) bus described above.

In the pure form of the passive ring there is the problem of two-path propagation to each node (one path in each direction around the ring) that may result in intersymbol interference if the data rate is high and/or the loop length long. If carrier transmission is used then destructive interference at periodic points around the loop may also occur requiring careful location of the terminal taps. In the absence of loop loss there is the

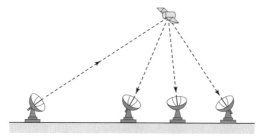

Figure 17.11 *Satellite (bus) network.*

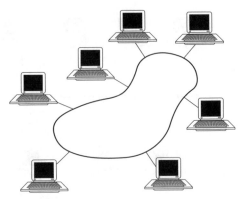

Figure 17.12 *Passive ring network.*

potential for energy to travel around the loop (in both directions) indefinitely and some method of preventing this is therefore required. These considerations make the passive ring much less popular in practice than the linear bus.

17.7.3 Active ring

This network consists of a number of nodes connected together with point-to-point links to form a closed loop, Figure 17.13. It is called an active ring to distinguish it from the passive ring (or ring bus), although when the term ring is used without qualification it usually refers to the active variety.

The active ring employs broadcast transmission in as much as messages are passed around the ring from device to device. Each device receives each message, regenerates it, and retransmits it to its neighbour, the message being retained only by the device to which it was addressed. The message must be removed from the ring if it is not to

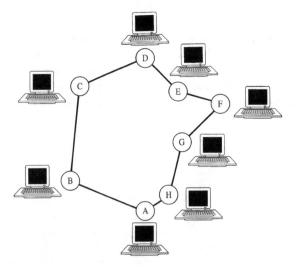

Figure 17.13 *Active ring network.*

circulate indefinitely. If removal is the responsibility of the device that generated the message then all devices will receive all messages and the active ring's status as a broadcast topology is not in question. If removal is the responsibility of the receiving device, however, then any number of devices may receive a particular message depending on the relative positions of the generating and receiving terminals in the ring. In this case the broadcast status of the topology is not so easy to justify. This illustrates the important point that practical networks are sometimes difficult to classify unambiguously and that classification must sometimes be on the basis of the topology and protocol combined.

Two topological variations of the active ring exist. These are:

1. **Unidirectional** – in which messages are passed between the nodes in one direction only. In this case the failure of a single data link will halt all transmissions.
2. **Bidirectional** – in which the ring is capable of supporting transmission in both directions. In the event of a single data link failing, then a (privileged) host node can maintain contact with the two sectors of what essentially becomes an active linear bus network.

That each network node is involved in the transmission of all data on an active ring is a significant weakness (although this problem can be addressed, see Chapter 21). Furthermore, since each device must retransmit each message, there is more delay (latency) than occurs with the bus topology. The active ring topology is simple both in concept and implementation, however, and is popular for fibre optic LANs in which regenerative repeaters are required at each node. Access is usually via time-slots, or possession of a special token, which avoids the need for the complicated contention resolution protocols usually required by bus networks (see Chapter 21).

17.8 Transmission media

Networks can utilise many types of transmission line including wire pairs, coaxial cable, fibre optic and wireless (radio, infrared or optical) links. Table 17.2 compares the different cabled and wireless transmission media.

Table 17.2 *Comparison of transmission media.*

Transmission medium	Twisted pair	Coaxial	Fibre optic	Radio	Infrared
Range, m	1–1000	10–10000	10–10000	50–10000	0.5–30
Data rate, kbit/s	0.3–2000	300–10000	1–100000	1–10	0.05–20
Cost/node, dollars	10–30	30–50	75–200	50–100	20–75

Metallic cable is preferred for many systems since simple, passive, tapped junctions are not easily realised for optical fibres. In all systems propagation loss, distortion, delay and noise are potential impairments. Many systems use coaxial cable operating with baseband pulse rates up to 10 Mbit/s over 500 m paths. At higher rates 'broadband' data is modulated onto a radio frequency carrier (of typically 100 MHz) or fibre optic transmission (see section 12.5) is used. Recently short runs of simple twisted wire pair have also become attractive for broadband transmission, because optical fibre installations are 5 to 10 times more expensive than twisted pair cable installations.

Broadband wireless LAN technologies using carrier frequencies of 5 GHz and 17 GHz are also becoming important [Stavroulakis], see section 21.5.

17.9 Interconnected networks

Large networks are often formed by the interconnection of separate smaller networks, as illustrated in Figure 17.14. Figure 17.15 shows a specific, but hypothetical, example of Figure 17.14 in expanded detail and illustrates the disparity between characteristics that may exist between connected networks – in this case particularly the very different data rates of the LAN and the WAN. A network connecting users to another network (but which may also host users directly) is called an access network. The LAN in Figure 17.15 is thus an access network for the WAN.

Figure 17.16 illustrates an interconnection of several networks, some similar and some dissimilar. The LANs (that might, for example, be owned by the different departments within an organisation) are connected by bridges to a backbone LAN (giving all departments access to centralised computing resources and databases and also providing high bit rate interdepartmental communication for file transfer services, etc.). Such bridges utilise a store and forward feature to receive, regenerate and retransmit packets while filtering the addresses between connected segments.

Communication to external organisations, in Figure 17.16, is provided by a gateway connection to a WAN. An access protocol defines the set of procedures for LAN access

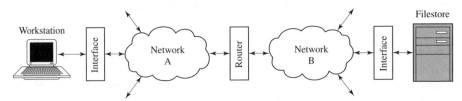

Figure 17.14 *A generalised interconnected network model.*

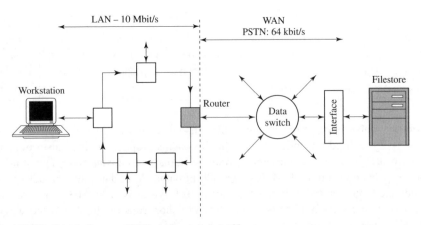

Figure 17.15 *Connection to a WAN via an access LAN.*

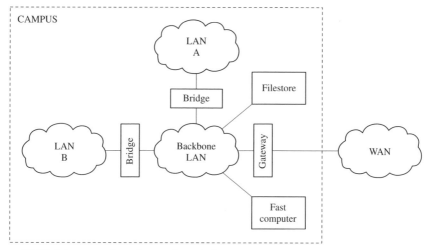

Figure 17.16 *Interconnection of several networks.*

to the WAN and vice versa. Since LAN transmission rates are much higher than those of interconnecting WANs one LAN network node must normally be dedicated to the WAN interface to provide, among other functions, the necessary data buffering. (For relatively local traffic, interconnection of LANs using a high speed MAN in place of the WAN would make it possible to transfer large files more quickly, and would ease the limitations that a WAN might impose on high speed real time applications.)

It is apparent from the above example that the roles of bridges and gateways in providing transparent access to resources residing on remote networks are similar. It should be noted, however, that a bridge is a device that interconnects two networks of the same type (using the same protocol) whereas a gateway connects networks using different protocols [Smythe, 1995] and must therefore perform the necessary protocol conversion (see section 17.2.2).

17.10 User and provider network views

A network serves a community of users or subscribers, enabling them to communicate, and offering a variety of services. Each type of transaction is transported over the network in a similar way, although the service offered will depend on the application. Voice communication, for example, usually requires simultaneous transmission in both directions but data transfer does not.

Users require the network to be transparent so as not to hinder, or intrude into, the transfer of information. They see a network in terms of the model illustrated in Figure 17.17.

Access to the network is via an interface that is both user-friendly and network-wise. The calling-party delivers the source information (a message that may consist of voice, video, text, data, etc.) to the interface and specifies the service(s) required of the network. The network offers service(s) that comply with the user's requirements and transports the information to the called-party interface. The interface provides the following functions:

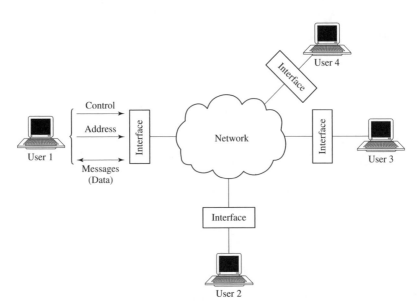

Figure 17.17 *A user network model.*

1. Friendly interrogation of the user to obtain the control information required by the network, and meaningful presentation of the message.
2. Network connection, complying with the requirements of the network service being used, e.g. voice, video, data.
3. Network address information that defines the called-party.

A switched network may contain a variety of transmission and switching technologies, but these are hidden from the user by the interface that separates the user and network domains, Figure 17.18.

A large range of user services is currently offered and more are constantly being developed. A service to the user of a network incorporates the following:

1. A tariff structure.

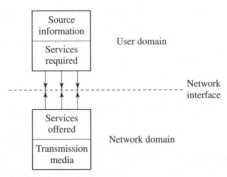

Figure 17.18 *Network interface.*

2. Conditions of service.
3. User name directory.
4. User instructions.
5. User help.
6. A technical structure for supporting the service.

Network management provides services internal to the network such as:

1. Directing calls to end users (including redirected calls, e.g. the 0800 freephone service).
2. Calculation of charges.
3. Network testing and accumulation of operational statistics.
4. Anticipation of congestion and arrangements for alternative routing.

Users are interested only in the overall quality of service (QoS) delivered by a network that includes parameters such as:

1. Guaranteed information rate.
2. Maximum delay.
3. Probability of lost calls.
4. Probability of data error.

The network offers services that comply with the user's requirements and then implements these services using the transmission resources available. The transmission resource comprises (for a switched network) a set of digital transmission links, network switching nodes and a signalling system that provides the network control. Transmission resource issues include:

1. **Transmission media** – twisted wire pairs, coaxial cable, optical fibre, satellite or terrestrial radio.
2. **Multiplexing** – to enable each transmission link to accommodate many users.
3. **Transmission quality** – optical fibre, for example, providing transmission that is virtually error free whilst mobile radio channels have generally higher error rates due to low received signal levels and signal fading.
4. **Switching** – needed to direct messages to the correct destination.
5. **Congestion** – caused by the finite capacity of switches, routers and transmission links and leading to variable delays in message delivery times.

Networks are becoming ever more complex (and consequently more sophisticated in the services that they offer) but are simultaneously becoming easier to use. This is largely due to the increasing levels of intelligence being incorporated into networks generally and the network interfaces particularly.

The network provider supplies equipment and functions in order to service the user requirements. These are:

1. **A physical network** – consisting of the communication media and (in switched networks) a switching mechanism to allow nodes to be linked as required.
2. **A directory mechanism** – related to the layout of the network and identifying every potential user, thus requiring the calling-party only to identify the called-party rather than the entire network route (that is subsequently defined by the network).
3. **A protocol** – or agreement as to how the connection is to be established and the data transferred.

4. **A logical connection** – identifying each message connection independently of the physical route it follows. In Figure 17.19, for example, the logical connection A–B can be made by either of the two physical routes 1–3–6 or 2–4–6.

17.11 Connection-oriented and connectionless services

Network services may be *connection-oriented* or *connectionless*. A connection-oriented service requires a complete route through the network to be defined prior to the message being sent. Once the route has been established data can then flow at the maximum rate allowed. A connectionless service allows the message to be launched into a network without prior identification of the route. The message in this case contains its own addressing information and routing decisions are made as switching nodes are encountered.

17.12 Layered network architectures

The overall function of a network can be split into layers, each layer being self contained and representing a distinct process with access points connecting it only to its adjacent layers, Figure 17.20.

Each layer n adds value to a service offered by the next lower layer $n-1$ and then offers this enhanced, more sophisticated, service to the next higher layer $n+1$. Communication between layers occurs at service access points (SAPs) that define the interface between layer processes. Messages are exchanged across these interfaces usually in the form request-X and confirm-X. The blocks of data passed between entities in adjacent layers (i.e. across layer interfaces) are known as data units. The advantages of layering service functions in this way are:

1. Complicated processes may be broken down into layers of simpler processes allowing clearer and easier comprehension.

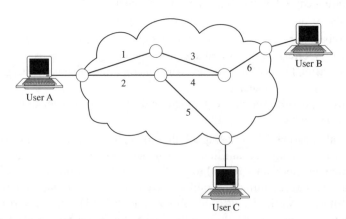

Figure 17.19 *Network logical connections.*

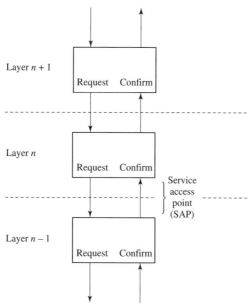

Figure 17.20 *Principle of layered architecture.*

2. The design of a complete system does not require a single individual to understand the detailed complexity of an entire system but only to understand the details of a single layer along with the well defined SAPs above and below that layer.
3. Different manufacturers can independently supply hardware and software appropriate to the different layers of a system, thus encouraging both collaboration and competition.
4. The software and/or hardware appropriate to a given layer can be easily and rigorously tested by embedding it in an environment that emulates the layers above and below it.

The layered service concept can be illustrated by considering the production of a computer program. At the highest level a systems analyst may construct a flow diagram that defines an algorithm to meet a customer's needs. The analyst is thus offering a high level service to the customer or user. A programmer then offers a service to the analyst by coding the algorithm using an appropriate high level computer language such as MATLAB, C or Java. A compiler offers a service to the programmer by translating the high level code into a machine level code, a lower level of operation that uses larger numbers of simpler functions. Any errors in the high level code are signalled back to the programmer, (the absence of such errors essentially representing the confirm-X statement shown in Figure 17.20). Whilst this analogy is far from exact, it does illustrate the underlying concept of a layered service architecture.

Most networks are organised in a layered architecture as described above. In the past the number, name and function of each layer differed from network to network. In all cases, however, the purpose of each layer was, and is, to offer specified services to higher layers, while shielding the details of exactly how those services are implemented in the lower layers. Layer *n* at one node holds a conversation with layer *n* at another node. The rules and conventions used in this conversation are collectively known as the layer-*n* protocol. The main functions of such a protocol are:

1. Link initiation and termination.
2. Synchronisation (of data unit boundaries).
3. Link control (with reference, for example, to polling, contention restriction and/or resolution, timeout, deadlock and restart.
4. Error control.

With the exception of layer 1 no data is directly transferred from layer *n* at one node to layer *n* at another. Data and control information are passed by each layer to the layer immediately below, until the lowest layer is reached. It is only here that there is physical communication between nodes. The interfaces between layers must be clearly defined so that:

1. The amount of data exchanged between adjacent layers is minimised.
2. Replacing the implementation of a layer (and possibly its subordinate layers) with an alternative (which provides exactly the same set of services to its upstairs neighbour) is easily achieved.

17.12.1 ISO OSI protocol reference model

This is the layered protocol model devised by the International Organisation for Standardisation (ISO) for open system interconnection (OSI) on which many modern network architectures are based. The model is illustrated in Figure 17.21.

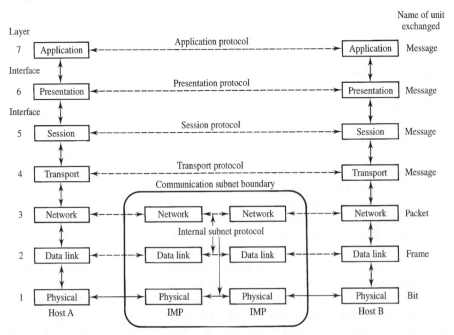

Figure 17.21 *ISO OSI reference model.*

The overall purpose of this model is to define standard procedures for the interconnection of network systems, i.e. to achieve open systems interconnection. ISO OSI processing is normally performed in software but with the insatiable demand for ever higher data rates hardware protocol processors are, increasingly, being deployed. Several major principles were observed in the design of the ISO OSI model. These principles were that:

1. A new layer should be created whenever a different level of abstraction is required.
2. Each layer should perform a well defined service related to the existing protocol standards.
3. The layer boundaries should minimise information flow across layer interfaces.
4. The number of layers should be sufficient so that distinct functions are not combined in the same layer, but remain small enough to give a compact architecture.

This has resulted in agreement to use seven layers as the OSI standard. The layers of the model are presented from the viewpoint of connection-mode transmission. A key feature of the OSI model is that it achieves standardisation for data communications combined with error-free transmission. A summary of the function distribution is given below.

Layer 1 – This is the physical layer that directly interfaces with the transmission medium and provides the physical communication path between two nodes. Layer 1 specifies: (i) mechanical aspects (e.g. cables and connectors; and what wires and connector pin numbers are assigned to which layer 2 functions), (ii) electrical aspects (e.g. voltage levels, current levels, line coding and modulation), (iii) functions and procedural aspects for the interface to the physical circuit connection and (iv) issues such as data rate and synchronous or asynchronous transmission. The unit of information exchange at layer 1 is the bit (or symbol if multilevel signalling is being employed).

Layer 2 – This is the data-link control layer whose task is to take the raw transmission facility between a pair of network nodes and transform it into a link that appears free of transmission errors. It accomplishes this by breaking or assembling the input data into data frames, transmitting the frames sequentially, and processing acknowledgement frames returned by the receiver. This general process is illustrated in Figure 17.22 and is developed further in Chapter 18.

Here, once the frame is transmitted the transmitter stops and waits for an acknowledgement before the next frame is sent. If the acknowledgement does not arrive from the remote terminal within a prescribed time interval then the original frame is retransmitted, as shown for frame 2 in Figure 17.22. The unit of information exchange at layer 2 is therefore the data frame.

The data-link layer must both create and recognise frame boundaries. It thus defines the rules for access to the network (called the medium access control protocol). Cyclic redundancy check or polynomial codes (see Chapter 10) are widely used in the data-link layer to achieve an error detection capability on the bit-serial data. These processing functions are implemented in hardware. Software packages, such as Kermit, are located here to provide terminal emulation and file transfer facilities. The first two layers together are sometimes called the hardware layer.

Layer 3 – This is the network layer whose purpose is to provide an end-to-end communications circuit across the network between two terminals or hosts. It has responsibility for tasks such as routing, switching and interconnection, including the use of multiple transmission resources, to provide a virtual circuit. It also allows the

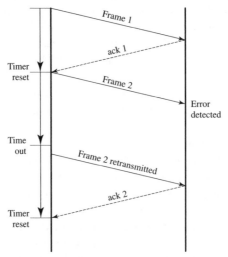

Figure 17.22 *Protocol for acknowledging (ack) successful frame receipt.*

multiplexing of several logical connections over a single communication path. The bottom three layers, collectively, implement the network function and can be viewed as a 'transparent pipe' to the physical medium. Data routed between networks, or from node to node within a network, requires these functions alone. The unit of information exchange at layer 3 is the data packet (received from, and supplied to, the transport layer, layer 4).

Layer 4 – This is the transport layer that provides a transport service suitable for the particular terminal equipment independent of the network and the type of service. This includes multiplexing independent message streams over a single connection and segmenting data into appropriately sized units (packets) for efficient handling by the lower layers. Modulo $2^n - 1$ checksums (see Chapter 10) are frequently computed within the software communications protocols of this layer, and compared with the received value, to determine whether data corruption has occurred. The transport layer can implement up to five different levels of error correction, but in LANs, for example, little error correction is required. The unit of information exchange at layer 4 and above is the message.

Layer 5 – This is the session layer whose functions are to negotiate, initiate, maintain and terminate sessions between application processes. This could, for example, be a transaction at a bank automatic teller machine. Layer 5 also monitors the costs of services being used. It may operate in either half- or full-duplex mode.

While the session layer (layer 5) selects the type of service, the network layer (layer 3) chooses appropriate facilities, and the data-link layer (layer 2) formats the messages.

Layer 6 – This is the presentation layer that ensures information is delivered in a form that the receiving system can interpret, understand and use. It defines, for example, the standard format for date and time information. The services provided by this layer include classification, compression and encryption of the data using codes and ciphers and also conversion, where necessary, of text between different types, e.g. to and from ASCII. The overall aim of layer 6 is to make communication machine independent.

Layer 7 – This is the application layer. It defines the network applications that support file serving. Hence it provides resource management for file transfer, virtual file and virtual terminal emulation, distributed processing and other functions. Conceptually this is where X.400 electronic mail [Wilkinson] and other network utility software resides. In electronic mail applications, layer 7 contains the memories that store the messages for forwarding when network capacity becomes available.

These OSI layers are equivalent to similar functions in other proprietary layered communications systems and hence there is commonality between OSI and many other standards. As one moves down the layers overhead bits (added by each layer) come to dominate each data packet making the effective data rate at layer 7 at least an order of magnitude less than the actual bit rate at the data-link layer, Figure 17.23.

A layer-*n* data-unit efficiency can be defined in this context as the ratio of information bits to total bits in the data unit. (Note that, for clarity only, the fragmentation of data units that may occur between a given layer and the layer below is not shown in Figure 17.23.)

17.12.2 Network layers in use

Figure 17.24 shows the protocol layers in the context of the network example of Figure 17.15.

Notice that the router links two disparate networks at layer 3, i.e. only the lower three layers of the model are involved. When passing a message from one network to another it is not necessary to know what that message contains but only its destination and type. Agreement on network terminology is not yet universal and the router in Figure 17.15 might sometimes be called a gateway, although this term usually implies protocol conversion at layer 4 or above.

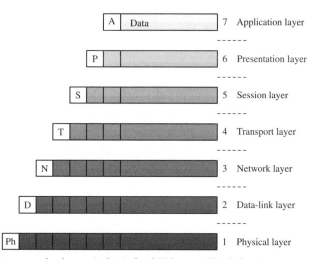

Figure 17.23 *Progress of a data unit down the OSI layers. (Shaded region represents protocol data unit passed from layer above, unshaded region represents appended layer header.)*

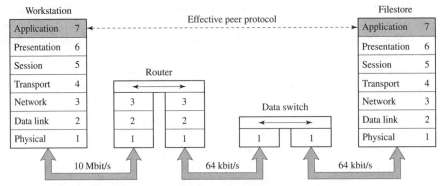

Figure 17.24 *Role of OSI in data transfer across a network.*

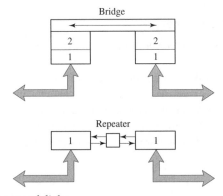

Figure 17.25 *Other internetwork links.*

The data switch in Figure 17.24 operates at the physical level, layer 1. When two similar networks are linked, even simpler interfacing devices are possible, such as the repeater shown in Figure 17.25.

This is the simplest form of internetworking device, in which amplification and regeneration of signals is implemented at the physical level simply in order to extend a network. A bridge is used when the two networks are of the same type and data may be exchanged between them at the data-link level, layer 2. Such a bridge will normally inspect the address of each data item to see which network it is destined for.

17.13 Summary

The function of a network is to emulate as closely as possible a fully interconnected mesh but using a small fraction of the resources, and at a small fraction of the cost. Networks are already applied with advantage across the fields of commerce, education, research, entertainment and leisure. The future applications of networks are limitless.

Networks can be classified on the basis of their topologies and protocols. Broadcast networks use a common transmission medium that results in all stations receiving all transmissions. Switched networks employ point-to-point links that can be reconfigured between switching nodes to form a path between any pair of stations. This results in transmissions only being received by the station for which they are intended. Broadcast networks are classified according to their medium access control (MAC) protocols (CSMA, RA or reservation algorithms, see Chapter 21) and switched networks are classified according to their generic switching philosophy (circuit, message or packet switching).

Broadcast networks may employ bus, passive ring or active ring topologies although the passive ring is seldom used in practice. Switched networks may employ star, tree or mesh topologies or some combination of these, e.g. mesh core network with star access networks.

LANs are limited in extent to a few kilometres and typically serve a single organisation or campus. They are broadcast networks with (typically) bus or active ring topologies and are often used as access networks for MANs and WANs. WANs may be national or international in extent and are principally switched networks although they often comprise an interconnected set of smaller networks including broadcast LANs. MANs typically extend over the area of a large city and are switched networks but with transmission speeds (data rates) more usually associated with LANs.

Networks are constructed using a layered architecture. The ISO OSI protocol reference model breaks the overall function of a network into seven layers. Each layer provides a set of sub-functions and has well defined interfaces to the adjacent layers. This simplifies network design, implementation and operation. It also encourages competition between equipment manufacturers and compatibility between proprietary network subsystems, resulting in easier and cheaper system procurement.

Extension, and interconnection, of networks is achieved using repeaters, data switches, bridges and gateways depending on the protocol layer at which the connection is made.

Network interfaces protect the user from the complexities of network operation and protect the network from unintentional abuse by the user. They match the services requested by a user to the resources a network is currently able to offer.

17.14 Problems

17.1 The holy grail of communications engineering is to realise a personal communications system allowing everyone on the planet to communicate with all others on the planet at will. Calculate the minimum number of point-to-point links required if this ambition were to be realised using a fully interconnected mesh network. Take the population of the world to be 6 billion. (1 billion = 10^9.)

17.2 Explain clearly the essential difference between a broadcast and switched network. What fundamental problems of resource sharing must be addressed in each case?

17.3 Outline the distinguishing characteristics of LANs, MANs and WANs making reference, in particular, to broadcast/switched network status, their relative bit rates and their relative latencies. It has been suggested that they can be distinguished in terms of their latency normalised to data-packet length. Give a precise definition of each in these terms.

17.4 Describe three popular switched network topographies. Are the topographies you describe truly independent (i.e. generically different)? If not explain why not.

17.5 What is the defining property of a tree network and what advantage does this topography have over a mesh network?

17.6 Distinguish between circuit, message and packet switching. Under what circumstances is packet switching superior? Explain your answer.

17.7 Compare and contrast the datagram and virtual circuit forms of packet switching. To what types of systems and services is each most appropriate and why is this so?

17.8 A certain message contains 2048 bytes. It is to be divided into 10 packets, each with a header of 20 bytes, and transmitted at 384 kbit/s through 6 switching nodes. Compare the overall delay (neglecting propagation and queueing effects) when the message is transmitted in packets to when it is transmitted in one piece.

17.9 Compare the relative performance of packet and message switching by calculating the overall transmission time when a message of 3000 bits is transmitted at 64 kbit/s over a route containing 5 nodes. Packets are 564 bits overall and the protocol header is 8 bytes.

17.10 Describe two popular broadcast network topologies. Briefly outline the relative merits of each.

17.11 Write down a topology matrix for a unidirectional active ring containing 8 nodes. How are the two matrices representing opposite directions of information flow for this network related?

17.12 What is meant by quality of service? List three objective metrics that might be used to define the quality of service of a digital communications network.

17.13 Explain what is meant by connectionless and connection-oriented services. Suggest some advantages and disadvantages of each.

17.14 Describe the ISO OSI seven-layer model for network architectures. Explain the concept of a layered service structure and outline the principal functional roles of each layer in the OSI model. What are the advantages of designing a system in this layered way?

17.15 If the header in each data unit at each level of an OSI protocol stack is 10% of the data unit in the level above calculate the frame efficiency at the data-link control level.

17.16 Regenerative repeaters, data switches, bridges, routers and gateways can all be found in a system comprising an interconnection of networks. Distinguish between these devices as clearly as you can. In particular include the ISO OSI model level at which they operate.

CHAPTER 18

Network protocols

18.1 Introduction

Network protocols constitute the set of operating rules and procedures that all devices connected to a network must abide by in order for (packet switched) communication between the devices to take place in an orderly way. The procedural rules are implemented in a set of programs, each program operating at a specific layer in terms of some network architecture as typified by the ISO OSI model. This results in a protocol stack as described in section 17.12.1. Numerous proprietary protocols (many fulfilling similar functional roles at equivalent levels in the stack) already exist and a comprehensive description of them all here is not possible. (A rigorous specification of even a single protocol can easily run to hundreds of pages.) Rather, the general functions of protocols at different layers in the protocol stack are examined and some particular protocol examples are described in order to illustrate how these functions are typically implemented. The level of detail with which the protocols are addressed is not uniform, the treatment generally becoming less detailed with increasing protocol layer level. The particular examples described for the network layer and above are all taken from the ISO protocol suite.

18.2 Physical layer

The physical layer protocol defines the interface between a user terminal, referred to generically as data terminal equipment (DTE), and a device that generates the physical signals that are transmitted to a network access node, referred to generically as data communications equipment (DCE). The DCE may be a modem if modulation is required

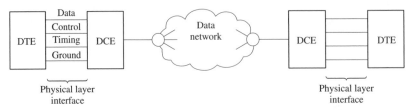

Figure 18.1 *Location and generic functions of a physical layer interface.*

to match the physical signals to the electrical characteristics of the line, or a line driver (in which case the abbreviation DCE is sometimes interpreted as data circuit terminating equipment). The protocol definition can be subdivided into mechanical, electrical, functional and procedural parts. The mechanical part invariably requires a multi-core DCE–DTE interface cable terminated in multi-pin connectors with different types of signal (e.g. data and control signals) being passed over different wires. Figure 18.1 shows the relationship between DTEs and DCEs and a generic representation of a DTE–DCE interface with a wire dedicated to each of the principal functions normally required.

The traditional type of interface associates a specific function with each pin of the connector, an approach typified by the EIA protocol specification RS-232 and its derivatives (RS-232-C and RS-449). A second approach, typified by the ITU protocol specification X.21, is to use fewer pins but use each pin for more than one function, employing coded character strings to distinguish between functions. (Note, however, that a variant of X.21 exists, X.21bis, which has been designed to be compatible with RS-232-C.)

18.2.1 A physical layer protocol – X.21

X.21 specifies the physical layer interface for the X.25 packet switched public data network (PSPDN). (It can also specify the physical interface for a circuit switched public data network (CSPDN), but in this case is concerned only with call set-up and call clearing operations, the data transfer phase not requiring explicit control in a circuit switched system.) X.21 provides for bit-serial synchronous transmission between the DTE and DCE in full-duplex mode. Figure 18.2 shows the DTE–DCE line connections for X.21 and Table 18.1 gives the functions associated with each line. X.27 (which corresponds to RS-422A) is a standard describing a balanced interface circuit version of X.21 and X.26 (corresponding to RS-423A) is a standard describing an unbalanced

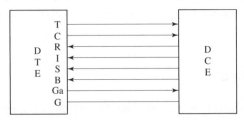

Figure 18.2 *DTE–DCE line connections: T=transmit, C=control, R=receive, I=indication, S=symbol timing, B=byte timing, Ga=DTE common ground, G=signal ground.*

circuit version. (Strictly speaking X.21 describes the procedural part of the specification with X.26/X.27 describing the electrical parts.) For line speeds between 9.6 kbit/s and 10 Mbit/s X.27 is used whilst for speeds of 9.6 kHz and below X.26 is adequate.

X.27 defines differential voltages ≤ -0.3 V to be logical ones and $\geq +0.3$ V to be logical zeros. The voltage sense is measured as that on the lower pin number with respect to (i.e. minus) that on the higher pin number, see Table 18.1. Lines C and I are usually referred to as ON or OFF corresponding to zero and one states respectively.

The mechanical specification defines a 15-pin connector, Figure 18.3. The convention is that DTEs carry male connectors and DCEs carry female connectors.

Table 18.1 *X.21 pin assignment (for X.26 and X.27 versions).*

| X.21 | | | | Detailed | Generic | Transmitting device | |
| X.26 | | X.27 | | function | function | | |
Pin no.	Designation	Pin no.	Designation			DTE	DCE
1		1		Shield			
8	G	8	G	Signal ground	Ground		
9, 10	Ga			DTE common ground		•	
2	T	2, 9	T	Transmit (or Transport)	Data	•	
4, 11	R	4, 11	R	Receive			•
3	C	3, 10	C	Control	Control	•	
5, 12	I	5, 12	I	Indication			•
6, 13	S	6, 13	S	Symbol timing	Timing		•
7, 14	B	7, 14	B	Byte timing			•
15		15		Unassigned			

The allowable states of the X.21 interface are defined by the signals on the T, C, R and I circuits. T may be held at logical 1 $(1111\cdots)$ or logical 0 $(0000\cdots)$, may carry an alternating bit pattern $(0101\cdots)$, may carry characters from the International Alphabet No. 5 (IA5, essentially the ASCII character set), Appendix C, or may carry data. C and I may be ON or OFF. R may be held at logical 1 or logical 0 or may carry IA5 characters.

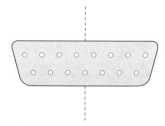

Figure 18.3 *The 15-pin connector for X.21 DTE–DCE interface.*

An example X.21 data exchange event is illustrated in Figure 18.4. Before the event is initiated both stations are in the 'ready' state (1, OFF, 1, OFF) or state 1. (Table 18.2 lists the principal states of the X.21 interface.) The initiating DTE sets its control (C) line to ON and its transmit (T) line to logical 0. This is state 2, 'call request'. On receiving this combination the DCE transmits two synchronisation characters (SYN SYN) followed by a continuous string of + characters on the receive (R) line. This represents state 3, 'proceed-to-select'. The DTE then transmits a 'call request' frame on T consisting of a SYN character followed by the address of the called DTE followed by a parity check bit and a terminating + character. This is state 4, 'selection signal'.

There follows a period where several states can occur as the DCE tries to establish, i.e. initiate, the call while keeping the DTE informed of progress.

When the calling DCE succeeds in establishing the call the called DCE places a BEL character (indicating bell or ringing) on R. This is state 8, 'incoming call'. If the called DTE is ready to accept data it sets its C-line to ON (state 9, 'call accepted') whereupon the DCE passes (set-up) information about the call to the DTE on the R-line (state 10B, 'DCE provided information, called DTE'). It then sets R to logical 1 (state 11,

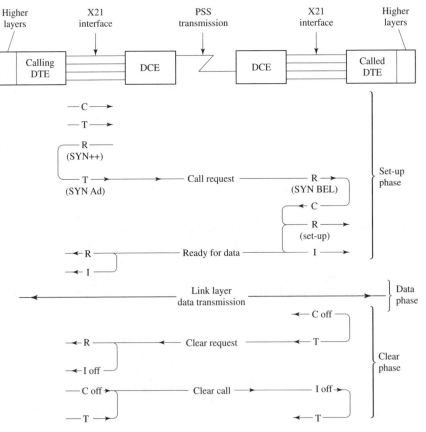

Figure 18.4 *Simplified example of X.21 data exchange event (source: Beauchamp, 1990, Figure 4.10 (labelled 3.10), reproduced with the permission of Chapman and Hall).*

Table 18.2 *Principal states of the X.21 interface (after Spragins, Table 4.4).*

State no.	State description	T	C	R	I
1	Ready	1	OFF	1	OFF
2	Call request	0	ON	1	OFF
3	Proceed-to-select	0	ON	+	OFF
4	Selection signal	IA5	ON	+	OFF
5	DTE waiting	1	ON	+	OFF
6A	DCE waiting, calling procedures	1	ON	SYN	OFF
6B	DCE waiting, called procedures	1	ON	SYN	OFF
6C	DCE waiting, for DTE information	*	OFF	SYN	OFF
6D	DCE waiting, for call acceptance	1	OFF	SYN	OFF
7	Call progress signal	1	ON	IA5	OFF
8	Incoming call	1	OFF	BEL	OFF
9	Call accepted	1	ON	BEL	OFF
9B	Proceed with call information	*	OFF	BEL	OFF
9C	Call accepted, using sub-addressing	1	ON	SYN	OFF
10A	DCE-provided information, calling DTE	1	ON	IA5	OFF
10B	DCE-provided information, called DTE	1	ON	IA5	OFF
10C	Call information	*	OFF	IA5	OFF
11	Connection in progress	1	ON	1	OFF
12	Ready for data	1	ON	1	ON
13	Data transfer	D	ON	D	ON
13R	Receive data †	1	OFF	D	ON
13S	Send data †	D	ON	1	OFF
14	DTE controlled not ready, DCE ready	01	OFF	1	OFF
15	Call collision	0	ON	BEL	OFF
16	DTE clear request	0	OFF	X	X
17	DCE clear confirmation	0	OFF	0	OFF
18	DTE ready, DCE not ready	1	OFF	0	OFF
–	DCE not ready	D	ON	0	OFF
19	DCE clear indication	X	X	0	OFF
20	DTE clear confirmation	0	OFF	0	OFF
21	DCE ready	0	OFF	1	OFF
22	DTE uncontrolled not ready, DCE not ready	0	OFF	0	OFF
23	DTE controlled not ready, DCE ready	01	OFF	0	OFF
24	DTE uncontrolled not ready, DCE ready	0	OFF	1	OFF
–	DCE controlled not ready	X	X	01	OFF
25	DTE provided information	IA5	OFF	SYN	OFF

1	111111...	
0	000000...	
01	010101...	

IA5	Char. string from International Alphabet No. 5 (preceded by at least 2 SYN characters)
+	Control character from IA5
*	Control character from IA5
BEL	Control character from IA5 (not 3-character sequence)
SYN	Control character from IA5 (not 3-character sequence) transmitted at least twice
D	Data
X	Don't care

† Used only in leased point-to-point and packet switched service.

'connection in progress') and sets I to ON (state 12, 'ready for data') and a ready for data signal ensures that the calling DTE also has its indication (I) line set to ON.

Information transfer (state 13, 'data transfer') can now take place under the control of higher level protocols in both directions (full duplex, Table 15.3) using T- and R-lines (data being transmitted on T and received on R).

On completion of data transfer either DTE can initiate call clearing by setting its C-line to OFF (state 16, 'DTE clear request'), the T-line being set to logical 0 since data transfer will have ceased. In Figure 18.4 the initiator of call clearing is the called DTE. The DCE then sets I to OFF (DCE clear confirmation, state 17), R being set at logical 0 since data reception will have ceased, and sends a 'clear request' signal to the calling DCE. On receiving the 'clear request', and after completion of data reception, the DCE sets I to OFF (state 19, 'DCE clear indication') and the DTE acknowledges this by setting C to OFF (state 20, 'DTE clear confirmation'). Finally the system's original state (state 1, 'ready') is restored by the DTE setting its T- and R-lines to logical 1.

18.3 Data-link layer

The task of the data-link level protocol (OSI layer 2, Figure 17.23) is to deliver a given protocol data unit (PDU) from a station at one end of a link to a station at the other end of a link, in the presence of possible transmission errors. (PDUs are shown schematically at the various levels in a layered architecture in Figure 17.23.) The link may be between two nodes in a network or between a user DTE and a network access node.

In order to control the flow of information, a reverse transmission path is required in addition to the forward path, see Figure 18.5. The two paths may be present simultaneously (a full-duplex channel), or they may be active alternately (a half-duplex channel). The exact nature of the signals on the communication path is determined by the physical layer protocol, and is of no consequence to the data-link controller (DLC).

Certain control information that must pass backwards and forwards on the communication path in addition to the payload information is added to the level 3 PDU (or packet) to form the level 2 PDU (or frame) for transmission. Together with the logical procedure for receiving frames, this constitutes the protocol. The additional information synchronises transmission, identifies the packet and detects the presence of errors. A general frame therefore has a number of data fields, as indicated in Figure 18.6.

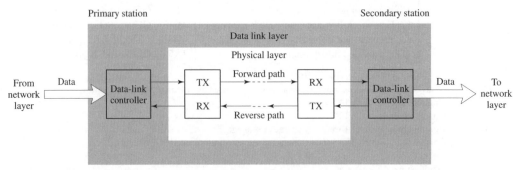

Figure 18.5 *The role of the data-link controller in the data-link layer.*

The function of each field is:

1. **Start field** (n_s bits) – indicating the commencement of the frame and synchronising the DLC to the data structure.
2. **Header field** (n_h bits) – containing information labelling the frame, defining its contents, destination and possibly the source address, and identifying the frame number in a sequence of such frames.
3. **Information field** (n_i bits) – the payload of the frame, transmitted without modification.
4. **Check field** (n_c bits) – containing the additional bits added to the frame to allow most errors to be detected. This field is sometimes called a frame check sequence (FCS), or a cyclic redundancy check (CRC), as in Chapter 10. In some protocols, the frame header is checked separately.

18.3.1 Synchronisation

The physical layer equipment establishes bit-clock synchronisation, so that the secondary DLC is presented with an unframed sequence of demodulated data bits. The receiving equipment in the data-link controller must first determine the start of each frame and then subdivide the frame into its constituent fields by counting clock pulses from the start. Figure 18.7 gives a schematic illustration of the frame reception process. Each circle represents a system state and the label alongside each arrow indicates the event that must occur before the system moves to the next state.

When presented with a continuous bit stream from the incoming circuit, the receiving equipment compares each rolling set of n_s binary symbols with the fixed pattern that defines the expected start field of the frame. Once a match is made and a start condition therefore detected, the remainder of the frame is received and loaded into a buffer store, until the end of the frame is found, when the check process is carried out. If the check is satisfactory then the data in the frame is accepted. If the check detects error(s) the data is discarded and reception of the frame is aborted.

	n_s	n_h	n_i	n_c	(Bits)
Idle	Start	Header	Information	Check	Idle

Figure 18.6 *General data frame structure.*

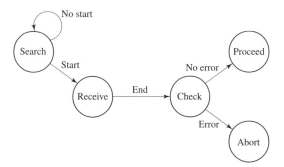

Figure 18.7 *DLC frame reception process.*

Various strategies are used to indicate the length of a received data frame. Although the digit stream may be transmitted and received synchronously, i.e. the interval between each pair of adjacent symbols is a constant, the data frames may be asynchronous in the sense that the start of the next frame cannot be predicted by the receiver. If frames do run synchronously, then they are all the same length and occur at regular intervals. In this case the start and end of each frame can be determined by the receiving station clock. Multiplexers operate on this principle and synchronisation is tightly controlled. For asynchronous frames, there are several possibilities. These are:

1. **Fixed length frame** – where the frame end is detected by counting symbols from the frame start.
2. **Variable length frame, explicit length** – where the length of the frame is given explicitly in the header field. The header typically has its own check field, since an error in the assumed frame length leads to confusion.
3. **Variable length frame, explicit stop** – where the end is indicated by a stop symbol. In this case, the stop code must be prohibited from appearing within the header, data and check fields. (This is the principle adopted by the HDLC class of data-link protocols described in section 18.3.4.)

If the frame is preceded by an idle line condition, then a major consideration is that the start field should be easily distinguished from the preceding data sequence, which might be a balanced idle condition, e.g. 101010, or a passive idle condition, e.g. 111111. A suitable binary pattern for the start might be 110101, for example.

If the transmission path is disturbed by noise, errors will occur in the received data and the start field may not be detected. The following simple analysis can be used to examine how the initial synchronisation procedure behaves under noisy conditions.

Let the probability of link bit error be P_b, and the total length of the frame be n bits.

A necessary condition for detecting the presence of an incoming frame is that the start field is error free, which occurs with probability P_{cfs} (probability of correct frame start) given by:

$$P_{cfs} = (1 - P_b)^{n_s} \tag{18.1}$$

where n_s is the length of the start code. The probability that no error occurs anywhere in the frame is P_{cf} (probability of correct frame) given by:

$$P_{cf} = (1 - P_b)^n \tag{18.2}$$

The probability that the frame fails to synchronise is $(1 - P_{cfs})$, which is approximately $n_s P_b$. Since the number of bits in the start field is small compared with the total number of bits in the frame, this will be a low probability.

The probabilities calculated from equation (18.2) over a range of P_b are given in Figure 18.8 for a range of frame lengths, n, from 8 to 2048. Normal link error rates are better than 10^{-4} or 10^{-5}, so almost 100% of the transmitted frames are error free. Radio channels can be considerably worse than this, however, and bit error probabilities of 10^{-1} can be encountered for really violent disturbances like a deep fade, Figure 15.3.

From Figure 18.8, at a P_b of 10^{-5}, almost all data frames of length up to 2048 bits will be essentially error free, while at a P_b of 10^{-2} only frames of length less than, say, 32 bits will have any chance of performing well. The probability of an 8 bit start field being detected correctly is given by the curve for $n = 8$.

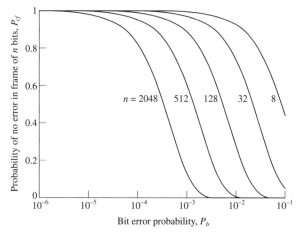

Figure 18.8 *Probability of n-bit error-free frames versus bit error probability.*

EXAMPLE 18.1

Of a given set of 512 bit frames 80% are to be error free. Determine the maximum allowable probability of bit error.

Referring to Figure 18.8, we see that a P_b of about 4×10^{-4} yields an event probability of 0.8 with $n = 512$. The P_b must be less than this. Notice how the required P_b decreases as the frame length increases.

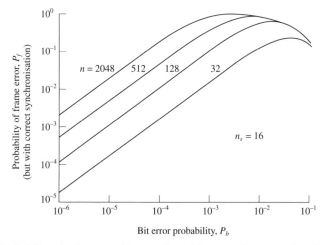

Figure 18.9 *Probability of at least one bit error, but with correct frame synchronisation.*

Figure 18.9 shows the probability of at least one error in the frame but that the synchronisation field is without error. Notice that this is proportional to the P_b for low error rates.

EXAMPLE 18.2

A frame has 16 bits allocated to a start code, 512 information bits and a 32 bit header. Estimate the probability that the frame synchronises but contains an error elsewhere if the bit error probability is 10^{-4}.

The probability of a correct synchronisation is:

$$P_{cfs} = (1 - 10^{-4})^{16} = 0.998$$

There are $512 + 32$ bits of data, so the probability of a frame error is:

$$P_f = 0.998[1 - (1 - 10^{-4})^{544}] = 0.054$$

Notice that this is also approximately equal to $(n - n_s)P_b$.

18.3.2 Error control

The data-link layer of a network has responsibility for providing error-free frame transmission between network nodes. This has traditionally meant that some form of error detection and error correction algorithm has been employed in the data-link protocol since the transmission quality (in terms of received SNR) has been otherwise insufficient to prevent errors occurring. (In some modern optical fibre based networks the inherent error rate is sufficiently low for link-by-link error control to be effectively dispensed with leaving error control to a purely end-to-end process.) Chapter 10 discusses error control in general and addresses error detection and forward error correction techniques in detail. Here, we put error detection into a network context and examine automatic repeat request (ARQ) techniques in more depth. ARQ is the traditional error correction mechanism used in most conventional networks.

Error detection and correction

In order to control the number of errors delivered in the final data packet, it is necessary first to detect the presence of such errors, and then to correct them. Error detection is accomplished by coding. The transmitted data is enlarged by the addition of redundant digits, added according to some mathematical rule.

A simple example is that of the parity check, where a single bit is added to a data word so as to make the number of digital ones even (or alternatively odd). A single error in transmission now disturbs this balance, and so is detectable. Any odd number of errors is detectable by this simple rule, but any even number of errors is not (see section 10.3.1).

The principle may be extended to a codeword of n bits containing k information bits and $r = (n - k)$ redundant bits. When considering all error patterns that can possibly occur, it can be shown that a proportion 2^{-r} of them escape detection, and hence will be undetected. If, for example, $r = 16$ then this proportion is 1.5×10^{-5}.

Many of the assumed error patterns are unlikely to occur in practice; for example, an error in every alternate bit of a 1000 bit frame. (This is principally because the probability of any pattern of 500 bits being in error is very low, but it is also worth making the point here that if large numbers of errors do occur they tend in practical systems to occur in bursts.) It is therefore possible to structure the code so that the unlikely error combinations will escape detection, but commonly occurring error patterns (e.g. single errors) will be detected with 100% certainty. It is easy to ensure that all odd numbers of errors will be detected (by use of a parity check, for example), and that double errors will also be detected. BCH codes can further ensure that 4, 6, 8 errors etc. will be detected.

Given randomly occurring errors, we can calculate the number of errors in a given codeword by applying the binomial theorem as shown in section 3.2.3. Thus the probability of j errors occurring in an n-bit word with a bit error probability P_b is given by equation (3.8) repeated below:

$$P(j \text{ errors}) = {}^nC_j P_b^j (1 - P_b)^{n-j} \qquad (18.3)$$

where:

$$ {}^nC_j = \frac{n!}{j!(N-j)!} \qquad (18.4)$$

Figure 18.10 shows the probability of j errors plotted against probability of bit error for a frame length of 1024, while Figure 18.11 shows the probability of j errors against frame length for a bit error rate of 10^{-4}. Notice that for long frames or for high P_b the probability of finding exactly 1, 2, 3, 4, etc., errors falls to almost zero, since there is likely to be far more errors than that.

Randomly occurring errors yield easy calculations for the number of errors likely in a given frame, and so the code can be selected to match the number of errors. Many types of communication path, however, incur burst error patterns where the link is essentially error free for most of the time but suffers sporadic blocks of multiple errors. Examples include radio fading, switching interference, and errors in magnetic tape and disk drives. It can be shown that error bursts of length $\leq r$ can be detected with 100% probability.

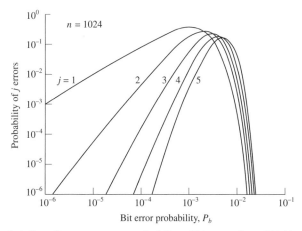

Figure 18.10 *Probability of j errors versus probability of bit error for a 1024 bit frame.*

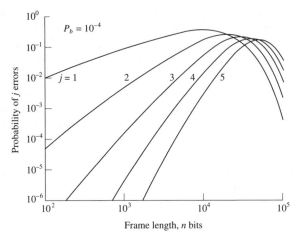

Figure 18.11 *Probability of j errors versus frame length for a bit error probability of 10^{-4}.*

EXAMPLE 18.3

Burst errors occur frequently. Suppose bursts occur at an average interval of 10 s, and that the probability of having more than i bits in the burst is 2^{-i}. Estimate the probability that the (31, 21) code lets through a burst, if the transmission rate is 4800 bit/s.

The time occupied by one codeword is $31/4800 = 6.5 \times 10^{-3}$ s.

An error burst coincides within a word every $10/(6.5 \times 10^{-3}) = 1548$ words, or with a probability 6.5×10^{-4}.

Not all these bursts will remain undetected. The code uses $31 - 21 = 10$ check bits, so bursts up to and including 10 bits in length will be detected. The probability that a given burst has more than 10 bits, is $2^{-10} \approx 10^{-3}$.

The overall probability that a codeword contains a burst greater than 10 bits is therefore $10^{-3} \times 6.5 \times 10^{-4} = 6.5 \times 10^{-7}$.

The probability that this remains undetected is $6.5 \times 10^{-7} \times 2^{-10} = 6.7 \times 10^{-10}$.

For most practical situations the following summary statements may be made:

1. The vast majority of errors will be detected by coding.
2. The probability of an undetected error is several orders of magnitude less than the probability of a detected error.

Coding theory can be extended to 'forward correct' errors as well as detecting them (see Chapter 10), but this requires large redundancy in the transmitted data. Such forward error correction coding (FECC) is therefore used only where data must be delivered accurately first time, and/or where a return channel is not available; for example, in space applications or in the replay of magnetic or optical recording of data. Most errors in most networks are corrected by retransmitting the data using an automatic repeat request (ARQ) protocol.

ARQ

The simplest method of correcting errors already detected is to request a second transmission of the data – usually referred to as an ARQ, Chapter 10. A return or backward channel is necessary for this strategy, but it works well and generally operates with low overall redundancy, keeping the transmission efficiency high. An objective of system design is to keep the quantity of retransmitted data to a minimum, which occurs naturally with bursty errors since retransmission is only needed when errors occur. The following sections give further details of practical ARQ schemes, which are universally used in data networks. Two distinct classes of ARQ operation are employed, corresponding to half-duplex and full-duplex communication links. The simpler half-duplex case is dealt with first in order to derive the basic principles, after which the full-duplex cases are considered which have a much higher transmission efficiency. Only an outline of the procedures is given here, rather than a complete definition, since the principles are used in several variations by practical protocols.

Stop-and-wait

An outline of this protocol is given in Figure 18.12, which shows state transition diagrams for both the primary and secondary stations in the interchange. The circles represent the states of the data-link controllers, while the directed branches indicate the external conditions that cause the system to change state.

The primary station commences the cycle by transmitting the next available data frame, and then immediately moves to a waiting state. The secondary station meanwhile has been in a passive state, waiting for an incoming frame to be detected. When such a frame is received (denoted by reception of the start field), the controller moves to a state of checking the integrity of this frame.

If the received frame passes the checks and hence appears to be error free, then the secondary station transmits an ACK message, indicating a positive acknowledgement,

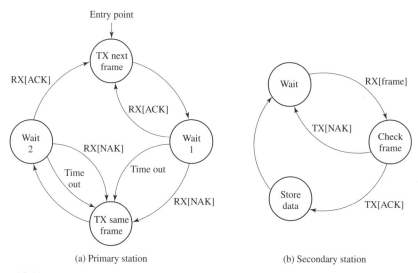

(a) Primary station (b) Secondary station

Figure 18.12 *State transition diagrams for stop-and-wait ARQ.*

and then stores the information received. If the frame fails the error checks, then a NAK message is transmitted back to the primary station, indicating a negative acknowledgement, and implicitly requesting a retransmission of the same frame. ACK and NAK reflect a commonly used terminology in practical protocols.

There are three distinct outcomes of the waiting state at the primary station:

1. **ACK received** – indicating that the transmitted frame was received intact. The primary station now moves state to transmit the next available frame.
2. **NAK received** – indicating that the transmitted frame contained error(s). The primary station now transmits the same block again, and moves to a further waiting state.
3. **No acknowledgement received** – either because none was sent, or because the returning frame was corrupted by noise. This case is detected by a timeout; where a timer is started at the primary station when the frame is first transmitted, and if no reply has been received after some predetermined limit, the controller is interrupted by the timer and then proceeds to retransmit the previous frame.

Notice that the waiting state when recovering from an error is shown as being distinct from the initial primary station waiting state. Conditions are subtly different in the two states, since when in error recovery it is necessary to put a limit on the maximum number of repeats, to prevent the controller from grinding to a halt if the communication link should be broken, for example. The procedure falls down if the acknowledgement frame is masked by noise, because the secondary station has then accepted the previous data frame, while the primary station repeats it. In order to guard against this event, data frames are numbered in the header field. If a data frame is received by the secondary station for a second time, therefore, this can be detected and the frame discarded. In order to calculate the performance of such a protocol the time dependence of the various actions must be examined. Figure 18.13 shows this information for a typical sequence.

A data frame A1 is transmitted from the primary station to the secondary station, which examines the received frame and makes a decision as to whether the frame is satisfactory or not. An acknowledgement frame B1 is then dispatched in the reverse direction, and carries this decision back to the primary station. Transmission on the communication link is therefore alternate or half duplex.

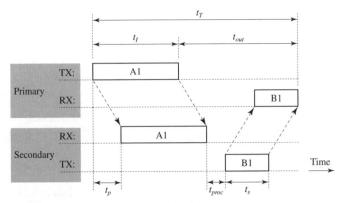

Figure 18.13 *Timing diagram for stop-and-wait ARQ.*

The duration of the outgoing data frame Al is t_I s, and the one-way propagation delay over the communication link is t_p s. A time interval t_{proc} elapses at the end of the received frame while the logical controller assesses the integrity of the frame and prepares the communication link to transmit in the opposite direction. In some cases, a single frequency radio path for example, there is a significant delay in turning the link around from receiving to transmitting.

The acknowledgement frame is of length t_s s, and is much shorter than the data frame because it has only to convey one item of information. The frame B1 must, however, include a start field and possibly a check field as well. In some systems of this type, the backward channel operates at a lower signalling rate than the forward channel, taking advantage of this lower rate to occupy a smaller bandwidth. The V.23 modem, Table 20.4, provides a two-frequency duplex channel at 1200 and 75 bit/s respectively.

Discussion of the logical control sequence, summarised in Figure 18.12, noted the need for a timeout interval, in case the transmitted acknowledgement never arrives back at the primary station. The timeout interval, t_{out}, is given by:

$$t_{out} \geq 2t_p + t_{proc} + t_s \qquad (18.5)$$

The equality applies when the time intervals are precisely known. This is the relationship that will be used in the discussion below. In practice, the time intervals are not precisely known, so the timeout period must be greater than the limiting value.

The minimum time to elapse, Figure 18.3, between the transmission of successive frames, t_T, is:

$$t_T = t_I + t_{out} \qquad (18.6)$$

A dimensionless parameter a, which is a measure of the efficiency of the transmission mechanism, is now defined:

$$a = \frac{t_T}{t_I} = 1 + \frac{t_{out}}{t_I} \qquad (18.7)$$

Account must now be taken of the probability that certain frames will be repeated. Let the probability that a data frame is in error be P_f, and the probability that an acknowledgement frame is in error be P_{ACK}. Then the probability, $P_{retrans}$, that one data frame is retransmitted is:

$$P_{retrans} = P_f + P_{ACK} \qquad (18.8)$$

Now $P_f \simeq nP_b$, and since the acknowledgement frame is much shorter than the data frame, $P_{retrans} \simeq P_f$.

The number of times that a particular frame will be repeated is, in principle, unbounded, although in practice it rarely exceeds one repeat. In order to calculate the effective rate of transmission over this data link, account must be taken of all possible numbers of repeats. Suppose then that i repeats occur before a given frame is received without error. The probability that this event occurs is $P_{retrans}^i$ and it occupies a time interval $i \times t_T$ s. Including the error-free frame, the overall probability of transmitting and receiving an error-free frame is $(1 - P_{retrans})P_{retrans}^i$.

The average time for correct transmission is the time occupied by the one error-free frame, plus the average time occupied by i repeats. A random variable r occurring with probability P_r has a statistical average, $\sum rP_r$, summed over all relevant values of r and so the average time for a correct transmission, t_V, is:

$$t_V = t_T \left\{ 1 + (1 - P_{retrans}) \sum_{i=0}^{\infty} i P^i_{retrans} \right\} = \frac{t_T}{1 - P_{retrans}} \tag{18.9}$$

Note that the summation is an arithmetic–geometric series, with a sum to infinity (for $|r| < 1$) of:

$$\sum_{i=0}^{\infty} i r^i = \frac{r}{(1-r)^2} \tag{18.10}$$

Assume that the primary station always has new data awaiting transmission, so that it is operating in a saturated state. Transmission of data frames over a data link is irregular in the presence of noise, so all such links must be buffered at each end. The average data rate, D bit/s, is the number of information bits per frame divided by the average time taken for a frame, i.e.:

$$D = n_i \frac{1 - P_{retrans}}{t_T} \tag{18.11}$$

The capacity of the channel, however, is $C = n/t_I$ (bit/s), so the average data rate expressed as a fraction of the available channel capacity is:

$$\frac{D}{C} = \left(\frac{n_i}{n} \right) \frac{1 - P_{retrans}}{a} \tag{18.12}$$

The first factor accounts for the frame efficiency and the second factor accounts for the transmission efficiency (which depends inversely upon the parameter a). If the timeout period is longer than the frame time, then $a > 2$ and the efficiency is reduced because the channel spends a significant amount of time waiting for the acknowledgement. Figure 18.14 shows ARQ protocol efficiency (D/C) plotted against the transmission parameter a, for three frame error probabilities. Notice that the frame error probability does not make much difference to the overall efficiency.

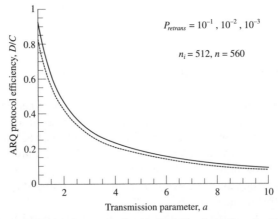

Figure 18.14 *Protocol efficiency for stop-and-wait ARQ (for frame length of 560 bits and information content of 512 bits.)*

An improvement in protocol efficiency can be obtained if the time waiting for acknowledgements is reduced or removed, which is possible with continuous ARQ protocols as considered below.

Go-back-N

This protocol is one of a class of continuous ARQ protocols, which operate over a full-duplex channel and transmit continuously in both directions in order to avoid the passive waiting time of the stop-and-wait ARQ. Such a system is balanced in the sense that there is a similar flow of data in each direction over the link. The primary station transmits data frames continuously without waiting for an acknowledgement. (This is sometimes called *pipelining*.) When the secondary station receives a data frame, it is first checked, and then the acknowledgement is inserted in the header field of an outgoing data frame on the backward channel. The acknowledgement is sometimes referred to as being *piggybacked* on to the data frame.

The control messages (ACKs and NAKs) are critically important for the continuous ARQ system, and special account must therefore be taken of the following potential problems:

1. Damaged data frame.
2. Damaged ACK.
3. Damaged NAK.

If the primary station receives a NAK it backtracks up the list of transmitted data frames, and retransmits all starting from the one received in error. The intervening frames are discarded by the secondary station. An outline of the mechanism is given in Table 18.3, which gives the sequence of events at the primary station. (In many practical implementations cumulative acknowledgement is employed, i.e. not all correctly received data frames are acknowledged individually, an acknowledgement ACK(J) indicating successful receipt of all frames up to frame J.)

Table 18.3 *Transmission sequence for go-back-N ARQ.*

Primary		Secondary
Transmit	*Receive*	*data*
$J-1$		
J		$J-1$
$J+1$	ACK($J-1$)	
$J+2$	NAK(J)	
J	ACK($J+1$)	
$J+1$	ACK($J+2$)	J
$J+2$	ACK(J)	$J+1$
$J+3$	ACK($J+1$)	$J+2$
$J+4$	ACK($J+2$)	$J+3$

This outline simplifies the timing of the various transmissions, and assumes a delay of a whole frame in each direction. Notice the need for the primary station to maintain a retransmission list of the frames that may require sending again, while the secondary station needs to buffer the outgoing data frames that will arrive with gaps between them. Clearly several error-free frames are retransmitted, which reduces throughput slightly. However, since in practice many errors occur in bursts, there is a good chance that more than one frame will be affected, and that one or more of the discarded frames may also have contained errors. Also, calculations show that when compared with selective-repeat ARQ (described below), the penalty is not large at low error rates.

Figure 18.15 shows the sequence in more detail, and defines the time intervals needed to calculate the transmission efficiency. It shows transmissions from both primary and secondary stations, and assumes for simplicity that the secondary transmissions are synchronised with the received frames from the primary, which is not the optimum case but enables the diagram to be drawn more clearly. This introductory treatment, and the simplified analysis given below, ignores several subtle points of timing and procedure that must be addressed in the design of practical protocols.

To analyse go-back-N ARQ performance consider first the transmission parameters. As previously, the frame length is t_I s and the propagation delay is t_p s. Under the phasing conditions of Figure 18.15, an acknowledgement is transmitted by the secondary station in the next frame slot after receiving the data frame. The timeout interval is therefore given by:

$$t_{out} \geq 2t_p + t_I \tag{18.13}$$

The processing time has been ignored, which is absorbed into the general overheads of queueing and preparing the acknowledgement for transmission. Because of this, and some uncertainty over the exact phasing of the two streams of frames, an increased value for the timeout interval is often taken, which must increase in steps of t_I:

$$t_{out} = 2t_p + 2t_I \tag{18.14}$$

The dimensionless coefficient a is then given by equation (18.7) as:

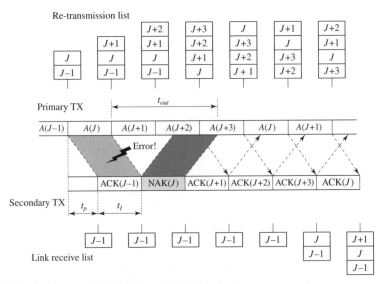

Figure 18.15 *Timing sequence for go-back-N ARQ.*

$$a = \frac{t_T}{t_I} = 3 + \frac{2t_p}{t_I} \tag{18.15}$$

Figure 18.15 also shows the contents of the retransmission list at the primary station, and the receive list at the secondary station. This go-back-N example requires storage of four frames at the primary station. In general, the amount of storage is N, where N is the next highest integer to a.

A similar argument to that in the previous section can be used to calculate the performance for a noisy channel. As before errors are assumed to occur randomly. The probability that a data frame is in error is denoted by P_f, and the probability that a frame is repeated is denoted by $P_{retrans}$. Since transmission is balanced, an error is equally likely in both forward and reverse directions. Hence:

$$P_{retrans} = 2P_f \approx 2nP_b \tag{18.16}$$

When an error occurs and a frame is to be retransmitted, the whole system backs off by a time $N \times t_I$ s, so the average time for successfully transmitting a data frame is given by:

$$t_V = t_I + Nt_I(1 - P_{retrans}) \sum_{i=0}^{\infty} iP_{retrans}^i \tag{18.17}$$

$$= t_I \left\{ \frac{1 + (N - 1)P_{retrans}}{1 - P_{retrans}} \right\}$$

The effective data rate D bit/s can be calculated and compared with the channel capacity C bit/s to derive the ARQ protocol efficiency:

$$\frac{D}{C} = \left(\frac{n_i}{n} \right) \frac{1 - P_{retrans}}{1 + (N - 1)P_{retrans}} \tag{18.18}$$

Comparing this with equation (18.12) for the stop-and-wait ARQ case, we note that the advantage of this technique is reflected in the reduced influence of the parameter a, which expresses the effect of the link loop delay. (a is still implicit in equation (18.18) in that N is the next highest integer to a.) This difference in performance is clearly seen by comparing Figure 18.16 with Figure 18.14.

In this case, throughput is greatly degraded by high values of the frame error probability P_f, but notice that good performance is maintained for a link with a large delay (high value of a). Notice too the stepped nature of the graph, because the back-up time has to be an integral number of frames.

Selective-repeat

The go-back-N type of ARQ protocol wastes a certain amount of transmission time by repeating good frames. Selective-repeat makes better use of the channel by retransmitting only data frames that are in error. In other respects, this protocol operates similarly to that shown in Figure 18.15, except that the transmission sequence is now as shown in Table 18.4. (As with the go-back-N protocol cumulative acknowledgement may be employed.)

This protocol repeats fewer frames than the go-back-N strategy, but they arrive out of order. Storage is therefore required at the secondary as well as at the primary station. However, if multiple repeats of the same data frame should occur, the amount of storage

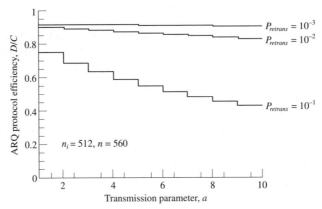

Figure 18.16 *Protocol efficency, go-back-N ARQ.*

required is unbounded, and the control of the sequence is more complicated. Consequently, selective-repeat ARQ is not used as much in practice as go-back-N ARQ, which represents a more robust strategy.

Table 18.4 *Transmission sequence for selective-repeat ARQ.*

Primary		Secondary
Transmit	*Receive*	*data*
$J-1$		
J		$J-1$
$J+1$	ACK($J-1$)	
$J+2$	NAK(J)	$J+1$
J	ACK($J+1$)	$J+2$
$J+3$	ACK($J+2$)	J
$J+4$	ACK(J)	$J+3$
$J+5$	ACK($J+3$)	$J+4$
$J+6$	ACK($J+4$)	$J+5$

Since only one frame is retransmitted when an error is detected, the average time per frame is given by:

$$t_V = t_I + t_I(1 - P_{retrans})\sum_{i=0}^{\infty} iP^i_{retrans} \qquad (18.19)$$

$$= \frac{t_I}{1 - P_{retrans}}$$

Consequently, the relative ARQ protocol efficiency for selective-repeat is:

$$\frac{D}{C} = \left(\frac{n_i}{n}\right)(1 - P_{retrans}) \qquad (18.20)$$

Throughput is independent of the loop delay and the parameter a, and thus is optimum.

Comparison of ARQ techniques

There are two classical cases that may be used to compare the delay performance of the different ARQ strategies described above. These are the long delay and short delay cases, both for systems with modest probabilities of error. For systems with high error probabilities, as in a fading radio system for example, the calculations are a little different and these will be addressed later.

The long delay case is typified by a geostationary satellite link, which has a delay of about 270 ms between two earth stations separated by a single satellite hop (see Chapter 14). The noise environment is assumed to be white, leading to independent randomly occurring errors.

The short delay case corresponds to terrestrial links, where the propagation delay is of the order of 1 ms/100 miles or 6.5 μs/km. (A radio path has a delay of about 3.3 μs/km and a cable may have a delay of between 5 and 7 μs/km.)

Case 1 – Long delay

A data link operating over a satellite path at 9600 bit/s, with a probability of bit error of 10^{-5}, each data frame containing a 48 bit header, a 32 bit check field, an 8 bit start field and a 64 bit information field.

The frame overhead is $48 + 32 + 8 = 88$ bits.

The propagation delay, $t_p = 270$ ms.

$n_I = 64$ bits.

The frame is therefore 152 bits in length, and the probability of a frame error is $P_f \approx 1.5 \times 10^{-3}$.

The time occupied by one frame, $t_I = 152/9600$ s or 15.8 ms.

Stop-and-wait ARQ solution:

Ignore t_{proc} and t_s, so $t_{out} = 2t_p = 540$ ms. Hence $a = (15.8 + 540)/15.8$ or 35.2.

The probability of retransmitting a frame is $P_f = 1.5 \times 10^{-3}$.

Consequently, from equation (18.12):

$$\frac{D}{C} = \frac{64}{152} \frac{(1 - 3 \times 10^{-3})}{35.2} = 0.012$$

Go-back-N ARQ solution:

Take the cautious value of $a = 37.18$ from equation (18.15).

The probability of retransmitting a frame is $2P_f = 3 \times 10^{-3}$.

Then from equation (18.18):

$$\frac{D}{C} = \frac{64}{152} \frac{(1 - 3 \times 10^{-3})}{1 + (38 - 1) 3 \times 10^{-3}} = 0.38$$

Selective-repeat ARQ solution:

$$\frac{D}{C} = \frac{64}{152} (1 - 3 \times 10^{-3}) = 0.42$$

The calculations above have been repeated with a longer information field of 1024 bits and the results given in Table 18.5 along with those for the shorter 64 bit information field for comparison.

Table 18.5 *Protocol efficiency or D/C ratios for the long delay case.*

Case 1 – Long delay	64 bit information field	1024 bit information field
Stop-and-wait	0.012	0.16
Go-back-*N*	0.38	0.78
Selective-repeat	0.42	0.91

The stop-and-wait protocol clearly is inefficient when a full-duplex link is available. Notice too how the efficiency of all three techniques improves with a longer data field.

Another feature of systems with long propagation delay is the number of frames in transit at any particular time. For the 64 bit case, there are more than 35 frames in flight at any one time, meaning that the frame counter has to keep up with this total. This effect will be examined later.

Case 2 – Short delay
Case 1 is now reworked (for both a 64 bit and a 1024 bit information field) but assuming a terrestrial link of 100 km and a specific delay of 6.5 μs/km resulting in a link delay of 0.65 ms.

Stop-and-wait ARQ solution:
$a = [15.8 + 2(0.65)]/15.8$ or 1.08 (i.e. approximately unity).
Hence from equation (18.12):

$$\frac{D}{C} = \frac{64}{152} \frac{(1 - 1.5 \times 10^{-3})}{1.08} = 0.39$$

Go-back-N ARQ solution:
Taking the cautious value again of $a = 3 + (2 \times 0.65)/15.8 = 3.08$ in equation (18.18):

$$\frac{D}{C} = \frac{64}{152} \frac{(1 - 3 \times 10^{-3})}{1 + (4 - 1)\, 3 \times 10^{-3}} = 0.42$$

Note that in the short delay case there is not much difference between the performance of stop-and-wait and go-back-*N* varieties of ARQ.

Selective-repeat ARQ solution:
This will give the same result as for case 1 since it does not depend on propagation delay. Once again the calculations above have been repeated with a longer information field of 1024 bits and the results given in Table 18.6 along with those for the shorter information field for comparison.

Table 18.6 *Protocol efficiency or D/C ratios for short delay case.*

Case 2 – Short delay	64 bit information field	1024 bit information field
Stop-and-wait	0.39	0.90
Go-back-*N*	0.42	0.84
Selective-repeat	0.42	0.91

Notice that stop-and-wait ARQ gives almost ideal performance for the long frame, better than go-back-N ARQ, although the length of the acknowledgement frame has been ignored which may become significant in this case. The selective-repeat gives approximately the efficiency of the frame itself before transmission.

Case 3 – High error rate systems

High error rate systems require special treatment, and arise mostly in the mobile radio environment. Consider a mobile digital radio system having a data rate of 1 Mbit/s and frames that are 125 μs long. The path length is 1 km and the probability of bit error is 10^{-2}.

Specific propagation delay is 3.3 μs/km, so the propagation delay is 3.3 μs, which is negligible. For continuous ARQ $a = 3$ taking the pessimistic view of equation (18.14), while for stop-and-wait it is about unity.

Now consider the error probabilities. The frame contains 125 bits in total. The probability that one data frame is in error is $1 - (1 - 10^{-2})^{125} = 0.715$. (Note that the condition $nP_b \ll 1$ is not satisfied in this case.) This probability is very high, so that on average a repeat is requested every 1.4 frames! The stop-and-wait ARQ effective rate would be 0.285 ignoring errors in the acknowledgement, which is not a workable figure.

For go-back-N ARQ account must be taken of errors occurring in both forward and reverse paths. Thus, if each frame transmission has a probability of error P_f, then the overall probability of initiating a repeat is:

$$P_{retrans} = P_f(1 - P_f) + (1 - P_f)P_f + P_f^2$$
$$= 2P_f(1 - P_f) + P_f^2 \tag{18.21}$$

For this case, $P_{retrans}$ is 0.919, which is scarcely workable.

One way of improving the situation is to make the frame shorter, so that its probability of surviving the link is improved. If the frame length is reduced to 40 bits, the probability of a frame error is now $1 - (1 - 10^{-2})^{40} = 0.331$, and for go-back-$N$ ARQ $P_{retrans} = 0.552$. This is still not good, but may be feasible. Applying the usual formulae, the ARQ protocol efficiency is 0.17 for go-back-N, and 0.45 for stop-and-wait, assuming the same value of $P_{retrans}$.

Clearly, transmission under these conditions is very difficult, but the point is made that for high noise situations, the link has greater survivability if frame lengths are low. When transmitting over terrestrial cable links, and to a certain extent fixed radio paths as well, errors do not occur randomly, but in bursts. In the radio path they are caused by fading or by some transient path (or antenna) obstruction, while in a cable circuit they are caused by interference from switched circuit transients. Thus, although the average error rate might be P_b, the errors occur in bunches. Suppose that an average error burst is b bits in length; then an error burst will occur on average every b/P_b bits. The effect of burst errors is that frames are disturbed at wider intervals of time, which suits a retransmission protocol very well.

Case 4 – Short delay with burst errors

The terrestrial transmission example of case 2 is now reworked, but assuming that errors occur in bursts with an average length of 5.

The average error rate $P_b = 10^{-5}$. When arranged in bursts of average length ($b = 5$) such bursts will occur approximately every 5×10^5 bits. This corresponds to a time interval of $5 \times 10^5/9600 = 52.1$ s.

Since each frame is of length 15.8 ms (for the 64 bit case), the likelihood of a frame being in error is $P_f = 15.8 \times 10^{-3}/52.1 = 3 \times 10^{-4}$, ignoring the acknowledgement time for stop-and-wait. The above calculations can therefore be reworked with this new value of frame retransmission probability.

Notice that for the long frame, stop-and-wait ARQ is much more efficient than for the case where errors were uniformly distributed.

At the other extreme of bit error probability, optical fibre links routinely have $P_b < 10^{-9}$, so errors are very rare. For a 1000 bit frame, the probability of frame error $P_f < 10^{-6}$. Although the frame assessment procedures are carried out continuously, they are needed for less than one frame in a million. Such protocol overheads add considerable delay at the higher transmission rates (> 2 Mbit/s). Under these circumstances, a better approach is to let the links relay the frame forward unchecked, and allow the ARQ function to be performed over the entire network path.

If the frame passes through m similar links, the overall probability of frame error is mP_f, equation (6.20). In this case, if $m = 20$, for example, then the overall probability of frame error is less than 2×10^{-5}, which is still not a major difficulty.

This principle lies behind the frame relay services that are gaining in popularity as the fixed telecommunications network becomes increasingly error free. Switching for frame relay is carried out at level 2 of the OSI model, rather than level 3 as in X.25 networks (see section 20.7.2). Correction of any errors is carried out end to end at a higher protocol level, but frames whose errors are detected at the link level are discarded.

18.3.3 Flow control

It is obvious that the transmitting node at one end of a data link must not transmit frames at a faster rate than the receiving node at the other end of the link can deal with them. This can be prevented by flow control.

Receiver-not-ready

One explicit method of flow control is for the receiving node to send a receiver-not-ready (RNR) frame to the transmit node whenever it is temporarily unable to accept further frames. This causes the transmit node to stop transmitting frames. The transmitting node may periodically send enquiry (ENQ) frames to interrogate the receiving node about its status. When the receiving node is in a position to receive frames again it responds to the ENQ frame with a receiver ready (RR) rather than an RNR frame and transmission of data frames is resumed.

Stop-and-wait

An alternative approach is to use the error control protocol as a surrogate flow control mechanism. In a stop-and-wait ARQ scheme, for example, the receive node can simply withhold an ACK frame to halt transmissions even if no frame errors are present. Typically the transmit node will the send an ENQ frame after the timeout period has elapsed and the receive node can respond with the withheld ACK frame if it is now ready to receive more frames. (As far as the transmitting node is concerned this situation is indistinguishable from a lost ACK frame.)

Sliding window

In a continuous ARQ scheme each frame is numbered sequentially. Clearly the numbering in practice cannot increment indefinitely, so k bits are allocated to a number field in the frame, leading to a modulo-2^k numbering scheme. Even in a cumulative acknowledgement implementation there is a limit on the number of frames that can be transmitted without receipt of an ACK. (In the stop-and-wait protocol this number is 1.) A *window* is therefore defined, which determines the range of frame numbers that may be transmitted. Figure 18.17 illustrates such a window, where the mask can rotate, expand or contract as desired. The lower boundary increments each time a frame is transmitted, while the upper boundary increments when frames are acknowledged.

The outgoing frame contains a send sequence number, $N(S)$, while the incoming or acknowledgement frame from the destination node contains a receive sequence number, $N(R)$. The operation of flow control using the sliding window is illustrated in Figure 18.18.

Node A maintains a list of frame numbers that are allowed to be sent, while node B maintains a similar list of frame numbers that can be accepted. The figure shows how the available window moves as the first few frames are transmitted. Table 18.7 gives a more

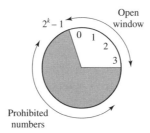

Figure 18.17 *Sliding window.*

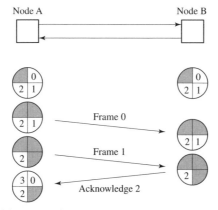

Figure 18.18 *Principle of flow control.*

detailed view of the operation including the effect of a lost frame. To simplify the illustration, $k = 2$ so counting is done modulo-4, and the maximum window opening is $N = 3$.

Node B has a buffer capacity of N frames, and so node A can send up to N frames without requiring an acknowledgement from station B, in the secure knowledge that the buffer at B will not overflow. However, an earlier acknowledgement from node B allows the flow of frames to be maintained without having to wait.

Table 18.7 *Flow control sequence for N = 3 window opening.*

Send window	Transmission	Receive window
012		012
12	Frame 0 →	12
2	Frame 1 →	2
230	← ACK 2	230
30	Frame 2 →	30
0	Frame 3 (lost)	30
–	Frame 0 →	30
301	← ACK 3	301
01	Frame 3 →	01
1	Frame 0 →	1
123	← ACK 1	123

In the example, therefore, node A sends frames numbered 0 and 1. Node B then replies with an acknowledgement, which includes a pointer to the next acceptable frame number. By implication, all frames with numbers up to this value have been received successfully, i.e. cumulative acknowledgement has been employed.

While frames are received correctly, the lists at the two nodes keep track with one another. A lost or corrupted frame causes a recovery mechanism to be invoked, which enables the correct sequence to be resumed. If the receiver buffer becomes full, then no acknowledgement is sent, and the sending node stops transmitting further frames until the way is clear.

Acknowledgements are normally carried piggyback in a data frame. If a convenient returning data frame is not available then control frames that represent RR or RNR can be used as appropriate.

Flow control can be implemented not only at the link level (hop-by-hop flow control), but also at higher levels of the protocol stack (end-to-end flow control). A destination DTE may need to exert some control over the rate at which data packets are inserted into a virtual circuit, for example.

Impact of sliding window control on long delay links

Consider what happens to the satellite link example (case 1, section 18.3.2) when 5 bits are available for sliding window flow control. When the frame propagation time is much longer than the frame time the parameter a (minimum frame period to frame duration ratio) will be large and the window size in a sliding window flow control protocol must be adequate to cope with this. Consider the continuous ARQ protocols discussed in section 18.3.2, where the number of frames in flight N, and hence the capacity of the sending buffer, is the next highest integer to a. The maximum window size M must therefore be at least this large. (If the size of the window is less than the number of frames in the system at any one time then efficiency can be recovered by increasing the length of the data frame.)

Referring to the example, for the go-back-N protocol with a 64 bit information field, $a = 37.18$ so for full-capacity operation the buffer store must hold at least 38 frames.

Since only 5 bits are allocated to define the sequence numbers, the maximum value of buffer store M is restricted to 31.

Transmission would then have to stop after 31 frames have been transmitted, until an acknowledgement is received. In each cycle of transmitted frames, the outgoing link could be active for a fraction 31/38 of the time, reducing the ARQ protocol efficiency by this factor.

For full capacity to be restored, the value of a must be equal to 31. The frame length must be expanded to $500/(31-3) = 17.86$ ms, corresponding to $17.86 \times 9.6 = 171.5$ bits. Since the frame overhead is 88 bits the information field must be at least $171 - 88 = 83$ bits.

This example shows how the parameters can be adjusted in order to obtain optimum performance. The length of the data frame is also affected by the likelihood of error, which in turn is a function of the probability of bit error.

18.3.4 A data-link protocol – HDLC

The majority of data-link protocols in use are based on, or are similar to, high level data-link control (HDLC), which was standardised in 1976 as ISO 3309. Several similar or derived protocols are:

1. Advanced data communication control procedure (ADCCP) – an ANSI standard X3.66, widely used in the USA.
2. Link access protocol, balanced (LAP-B) – the data-link layer specification of the ITU-T X.25 packet network interface (see section 20.7.1).
3. Link access protocol, D-channel (LAP-D) – the data-link protocol for the signalling channel in ISDN.
4. Synchronous data-link control (SDLC) – an IBM company standard.

HDLC is a bit oriented protocol, i.e. the data payload is not fixed in length, nor does it have to be presented in octet (8 bit byte) multiples. Any bit pattern can be transported by HDLC, without regard to its meaning or structure, and so the protocol is said to be transparent to data structure. It is also bit synchronous since a synchronous connection at the physical level is assumed.

Three types of station are defined in HDLC. These are:

1. **Primary station** – responsible for controlling the operation of the link. When a primary station issues control frames, they are called commands.
2. **Secondary station** – which operates under the control of a primary station, and is unable to initiate data transfer itself. Control frames issued by the secondary station are called responses.

3. **Combined station** – which operates as a primary or secondary station, according to the flow of information.

Further to the station types, there are two principal configurations accommodated by HDLC as illustrated in Figure 18.19:

1. **Unbalanced configuration** – which allows multipoint network operation.
2. **Balanced configuration** – which operates only between two combined stations in a point-to-point link.

The protocol procedure is described in terms of the following three distinct modes of operation, although the first is the most commonly used.

1. **Asynchronous balanced mode (ABM)** – in which full-duplex communication exists on a point-to-point link (Figure 18.19(b)), both stations are equal partners (balanced) and either can initiate a data exchange when necessary (asynchronous).
2. **Asynchronous response mode (ARM)** – in which primary and secondary stations are linked by a half-duplex point-to-point link.
3. **Normal response mode (NRM)** – an unbalanced connection (Figure 18.19(a)), in which the procedure is initiated between a controlling device (primary) and one or more secondary stations.

Three types of frame are required in order to control the link and to exchange data. These are:

1. **Information (I) frames** – containing the data payload and its associated flow control parameters.
2. **Supervisory (S) frames** – containing the control information necessary to set up the link and to execute ARQ error control as well as flow control if no returning information frames are present on the reverse channel.
3. **Unnumbered (U) frames** – used for a variety of control purposes.

The frame structure, which is similar for all three types of frame, is illustrated in Figure 18.20.

Since the information field in the frame is of variable length, both the beginning and the end of each frame must be marked in a unique way. HDLC employs the same, special, bit pattern to indicate both start and end of the frame. This pattern is called a flag and is 01111110. Since the data-link controller searches the data stream for this flag symbol

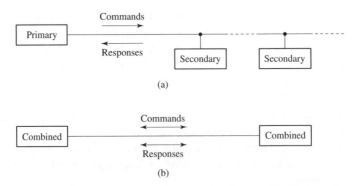

Figure 18.19 *HDLC communication configurations: (a) unbalanced, (b) balanced.*

while data is being received, the data must be manipulated so that the flag never occurs in the transmitted stream. This manipulation or processing, known as bit stuffing, inserts an additional 0 whenever a run of five binary ones occurs. Consequently the flag appears only in the correct place in the incoming digit stream, irrespective of the content of the data being transmitted. (The receiving terminal deletes the 0 following any run of five binary ones, of course.) Naturally, errors can occur in transmission and disturb this arrangement; but although such events are troublesome procedures exist for recovery. Figure 18.21 illustrates the idea of bit stuffing.

When frames are streamed or concatenated together, then a single flag is sufficient to mark the end of one frame and the beginning of the next (Figure 18.22).

The functions of the different HDLC frame fields are:

1. **Address field** – defines the destination secondary station, and is only necessary for the unbalanced case (Figure 18.19), but is always included for consistency.
2. **Information field** – only present in I-frames and some U-frames, its maximum length being determined for each application.
3. **Frame check sequence (FCS) field** – normally 16 bits in length, although a 32 bit field is available if required.

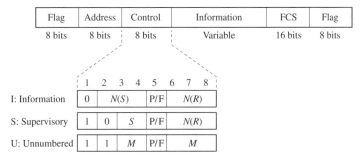

Figure 18.20 *HDLC frame structure.*

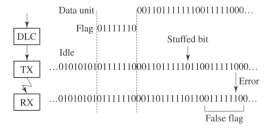

Figure 18.21 *The HDLC flag and bit stuffing.*

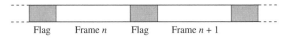

Figure 18.22 *Concatenated HDLC frames.*

4. **Control field** – with a structure that depends upon the type of frame, as indicated in Figure 18.20. The first bit indicates whether the frame is an I-frame or not, and if not, then the second bit distinguishes between S-frames and U-frames.

For the I-frame, the control field carries the two flow control sequence numbers $N(S)$ and $N(R)$, which indicate, respectively, the number of the outgoing frame and the number of the next frame expected by this station. Three bits are allocated to each counter, allowing a maximum sliding window size of 7, but provision is made to extend the control field to 16 bits if necessary, so as to allow 7 bit counters and a maximum window size of 127. The P/F (poll/final) bit is mostly used in the NRM to alert stations or signal the end of a string of frames.

An S-frame control field contains $N(R)$ and so can carry back acknowledgement information to implement an ARQ strategy. An *S*-field within the control field of Figure 18.20 enables up to eight distinct S-frames to be defined, the four most important of which are:

1. **Receive ready (RR)** – acknowledges the correct receipt of frame(s), when an I-frame is not available for piggybacking.
2. **Receive not ready (RNR)** – indicates a busy condition.
3. **Reject (REJ)** – indicates an error in the frame(s) received, and requests retransmission of frames from $N(R)$ onwards (go-back-N).
4. **Selective reject (SREJ)** – indicates retransmission of frame numbered $N(R)$ (selective-repeat).

U-frames do not carry sequence numbers, and are used for a wide variety of control operations such as setting the operation mode, recovery from fault conditions, etc.

18.4 Network layer

The network layer (ISO layer 3, Figure 17.21) may provide connection-oriented or connectionless services. In the former case communication takes place over a switched or virtual circuit and each communication event has three distinct phases; (i) connection establishment, (ii) data transfer, and (iii) connection release. (The virtual circuit emulates a switched circuit in a packet switched network, see section 17.6.3.) In the latter case there is only one, data transfer, phase.

The task of the network layer is to provide an end-to-end communications service to the next higher (i.e. transport) layer protocol. Whilst both higher and lower layers in the OSI protocol model are generally concerned with communications between a single pair of peer processes (network nodes in the case of the physical and data-link layers, and DTEs in the case of transport, session, presentation and applications layers) the network layer is concerned with the collaborative interworking of multiple peer processes (i.e. many network nodes). This makes the network layer protocols probably the most challenging in terms of algorithmic complexity.

The most obvious responsibility of the network layer is that of routing although flow control and error control (both of which can be implemented, alternatively or additionally, at the data-link level) may also be important.

18.4.1 Routing

Routing is the process of determining the path that information between a source DTE and a destination DTE will take. The simplest distinction that can be made between routing algorithms is whether they are static or dynamic. Static routing makes the same routing decisions irrespective of network conditions such as congestion. Dynamic routing responds to changing conditions attempting to alleviate congestion and minimise latency.

Routing algorithms may also be classified on the basis of decision location as centralised or distributed. In centralised algorithms route selection decisions are made at a central location. The decisions are made on the basis of status information gathered from the network routing nodes and the decisions are then communicated to these nodes. In distributed algorithms no such central control is present and nodes either make routing decisions independently or nodes share information and decisions are made collaboratively. Distributed algorithms have the advantage that the information available to each routing node about the local region of the network is likely to be more up to date than that available in a centralised system. Centralised algorithms are generally simpler, however, and more predictable in their behaviour.

Routing decisions are typically made by minimising some cost function over all (or at least many) possible routes. The cost function may involve quantities for each link in the route such as:

1. delay (or latency),
2. link capacity,
3. link expense (e.g. the monetary cost if the link is owned by another party),
4. residual error probability (i.e. a measure of the probability of undetected errors).

Desirable properties of a distributed dynamic routing algorithm are:

1. **Computational efficiency** – such that the computational overhead at each routing node is minimised.
2. **Communications efficiency** – such that the communications overhead on the links connecting routing nodes is minimised.
3. **Robustness** – such that traffic is routed effectively and efficiently under the widest possible range of traffic and network conditions including loss of nodes or links due to equipment failure.
4. **Stability** – such that a sudden change in traffic or network conditions does not produce significant oscillations in any QoS parameter such as latency.
5. **Fairness** – such that all users (of a given class) are treated equitably by the network in terms of the quality of service (QoS) they receive.
6. **Minimum cost** – such that optimum routes are selected to minimise a chosen cost function.

Path selection and packet delivery

Routing actually comprises two distinct tasks: (i) the selection of a route from source to destination through the network, and (ii) the delivery of information packets via this route. The discussion above centres on the first, more complicated, task. The second, and simpler, task is usually achieved using routing tables. A routing table at a particular routing node consists of a mapping between the destination node of a packet (typically read from the incoming packet's header) and the next routing node to which that packet should be sent. An example network with an associated routing table is shown in Figure 18.23.

It is clear that routing tables must be arrived at with some care since it is possible for packets to oscillate between nodes or circulate around closed loops indefinitely if there are errors or logical inconsistencies in them. It is the routing algorithm that determines the content of the routing tables. In practice routing tables typically contain more information than that shown in Figure 18.23. They may, for example, contain separate tables for different packet priorities or customer service classes.

Optimum routing

Graph theory can be used to find a route between any given pair of network nodes that minimises some cost metric (e.g. physical distance, latency, cost per information bit transferred, etc.). If the cost metric is path length then the optimum route is clearly the shortest path and an algorithm to find the optimum route is called a shortest path algorithm. An example shortest path algorithm [Dijkstra, pp. 269-271] is given in 'recipe' form below. (Pseudo code for this algorithm is given in [Spragins, p. 371].)

1. Draw a graph of the network with link weights representing the cost to be minimised.
2. Select the pair of nodes (X, Y say) between which an optimum route is required.
3. Choose one of the nodes (X say) to be the *current* node and permanently assign the label $\{0 \text{ from } X \text{ via } X\}_P$ to this node. (The subscript P denotes the permanency of the assignment.)

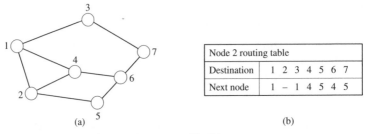

<table>
<tr><td colspan="9">Node 2 routing table</td></tr>
<tr><td>Destination</td><td>1</td><td>2</td><td>3</td><td>4</td><td>5</td><td>6</td><td>7</td></tr>
<tr><td>Next node</td><td>1</td><td>–</td><td>1</td><td>4</td><td>5</td><td>4</td><td>5</td></tr>
</table>

(a)   (b)

Figure 18.23 *A simple network (a) with routing table (b).*

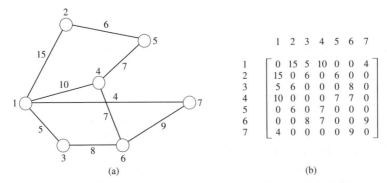

	1	2	3	4	5	6	7
1	0	15	5	10	0	0	4
2	15	0	6	0	6	0	0
3	5	6	0	0	0	8	0
4	10	0	0	0	7	7	0
5	0	6	0	7	0	0	0
6	0	0	8	7	0	0	9
7	4	0	0	0	0	9	0

(a)   (b)

Figure 18.24 *Example network topology (a) and its corresponding cost matrix (b).*

4. Temporarily assign to each node adjacent to the current node the cost of the link connecting it to X (using for example the notation {*cost* from X via *current node*}).
5. Search the graph for a node with the minimum temporary {*cost* from X via *current node*} assignment, change its assignment from temporary to permanent by adding a subscript P, and then choose this to be the new current node.
6. With the exception of nodes having permanent cost assignments, temporarily assign to each node adjacent to the new current node a {*cost* from X via *current node*} statement with *cost* equal to sum of the cost of the link connecting it to the new current node and the permanent {*cost* from X via *current node*}ₚ assigned to the new current node.
7. If the new {*cost* from X via *current node*} assigned to a node is less than one previously assigned then replace the previous assignment with the new assignment.
8. Go back to step 5 and continue until the assignment at node Y is permanent.

(To save space in Example 18.4, which illustrates the above algorithm, the notation {*cost* from X via *current node*} is abbreviated to {*cost, current node*}.) Note that in finding the optimum route between X and Y the algorithm above may also give the optimum route between X and a number of other nodes.
If a step 9 is added such as:

9. Go back to step 2 replacing Y with another node and continue until the assignment of all nodes is permanent.

then the optimum routes between X and all the other nodes will have been found. Repeating the algorithm in this modified form starting with each node in turn finds (with some unnecessary duplication) optimum routes between all pairs of nodes.

Matrix representation

The matrix representation of network topography (section 17.5.4) can be generalised so that the matrix elements indicate not only which nodes are connected but also the cost of using the link between each connected element pair. Such a cost matrix is well suited to the computer implementation of algorithms based on graph theory for the determination of optimum routes through a network. The elements may represent path length, delay or some other performance metric.

A matrix may also be used to represent the forward capacity (in bit/s) of the link connecting node i to node j. If the link is duplex, Table 15.3, with the same capacity in each direction then $m_{ji} = m_{ij}$. Since nodes do not have links connecting them to themselves then the diagonal elements of the matrix are zero, i.e. $m_{ii} = 0$.

Figure 18.24 illustrates a network and the corresponding cost matrix description assuming reciprocal links (i.e. equal costs for information flow in both directions). The figure could equally well illustrate a network and its capacity matrix.

EXAMPLE 18.4

Find the optimum route between nodes B and E for the 7-node network defined by the cost matrix:

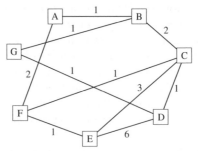

Figure 18.25 *Cost matrix for Example 18.4.*

	A	B	C	D	E	F	G
A	0	1	0	0	0	2	0
B	1	0	2	0	0	0	1
C	0	2	0	1	3	1	0
D	0	0	1	0	6	0	1
E	0	0	3	6	0	1	0
F	2	0	1	0	1	0	0
G	0	1	0	1	0	0	0

The network represented by the given cost matrix is illustrated in Figure 18.25. The assignments at each node found by following the optimum routing algorithm given above with B as the start node are listed below. (The order of the assignments is only one of several possible orders.)

Current node	Node considered	Assignment	Comments
B	B	$\{0, B\}_P$	
B	A	$\{1, B\}$	
B	C	$\{2, B\}$	
B	G	$\{1, B\}$	
A	A	$\{1, B\}_P$	
A	F	$\{3, A\}$	
G	G	$\{1, B\}_P$	
G	D	$\{2, G\}$	
D	D	$\{2, G\}_P$	
D	C	$\{3, D\}$	Higher than existing assignment
D	E	$\{8, D\}$	
C	C	$\{2, B\}_P$	
C	F	$\{3, C\}$	Equal to existing assignment
C	E	$\{5, C\}$	Lower than existing assignment: replace
F	F	$\{3, A\}_P$	
F	E	$\{4, F\}$	Lower than existing assignment: replace
E	E	$\{4, F\}_P$	

The optimum route between nodes B and E (in reverse order) is therefore EFAB and the cost of this route is 4. Note that sometimes an arbitrary decision must be made in the choice of the next current node if the minimum temporary assignment appears at more than one node. This only affects the order in which the various permanent assignments are made and does not alter the final result in terms of optimum routes.

The shortest path algorithm described above requires knowledge of the network topology and the path length (or more generally cost value) of all connecting links. It is most naturally implemented, therefore, as a centralised routing algorithm. Distributed forms of the algorithm are possible, however, in which each node keeps its own copy of the topological and cost information, updated as necessary in dynamic implementations. A version of this algorithm, using latency as the cost function, was adopted by ARPANET (the precursor of the Internet) in 1979. In this implementation node latency (time from packet arrival to time of *successful* packet transmission) is noted and recorded as packets pass through a node. Propagation and transmission delays (the former depending on the transmission medium and path length, the latter depending on packet length and bit rate) are added to the node latency and a 10 s mean value of total hop delay is calculated. If the hop delay differs significantly from the previous value it is broadcast to all other network nodes (by *flooding*, described below), each node constructing its own routing table based on the shortest path algorithm. Fresh delay values are broadcast after a period of not less than 10 s and not more than 60 s depending on how fast the delay is changing.

Flooding

A flooding algorithm calls on each node to route (copies of) all incoming packets out along all its links except that on which the packet arrived. Packets effectively, therefore, attempt to traverse all possible routes. This broadcast strategy makes the algorithm extremely reliable provided that traffic is sufficiently light so that congestion due to the replicated packets is not a serious problem. Flooding is often used to distribute routing information to network nodes. The practical problem of limiting the number of packets in the network to manageable numbers is solved by limiting the lifetime of each packet in real time (s), the number of nodes traversed, or the number of times (usually 2) that any given node encounters any given packet. If packet lifetime were not limited replication of replicated packets would mean that the network would become saturated.

Random routing

In a random routing system each node selects an outgoing link for each incoming packet at random. The relative selection probabilities assigned to each link may be equal or weighted according to some predefined criteria. They may also be conditional, for example, on the size of the transmission queues in the transmission buffers of the outgoing lines. The efficiency of this routing strategy is poor since, in its purest form, the packets arrive (eventually) at their destination by accident. Latency is also problematic, not only because delay in message delivery may be large but also because it is variable and unpredictable. The advantage is that nodes need no routing tables and no knowledge of the network topology. Given sufficient time they are also almost certain to deliver a packet to its correct destination as long as at least one route to that destination exists. In this sense they are actually very reliable providing message delay is not an issue.

Source routing

Source routing refers to systems in which the route (rather than simply the destination) for a given packet is codified in the packet header. This dispenses with the need for routing tables at each node. The route contained in the packet header is selected using a routing algorithm in the normal way.

Backward learning

This refers to an algorithm in which node A learns the best route for sending packets to node F from its experience of receiving packets from node F. If node A receives packets from node F via node B (adjacent to A) then it may reasonably suppose that the best route from itself to node F is via node B. There is an implicit assumption of reciprocity that must hold for this algorithm to work properly, i.e. that a good route between F and A is also a good route between A and F. This type of algorithm needs 'kick-starting' which can be provided by an initial phase of random routing or flooding.

Hierarchical routing

For a network with N nodes each node must have a routing table with $N - 1$ entries, i.e. each node must know the next node to send a packet to for $N - 1$ different destinations. As networks grow these tables become larger, and the problem of keeping them updated in a dynamic system becomes more complex. If network nodes are divided (equally for example) into D domains as shown in Figure 18.26 then the size of the routing table at each node is much reduced (from $N - 1$ to $N/D + D - 2$). A particular node now need only have a list of the next node for $N/D - 1$ destinations within its own domain and $D - 1$ other domains.

Bifurcated routing

In heavily loaded networks it is sometimes possible to reduce average network delay by splitting traffic from a given source to a given destination between two or more routes. This is called bifurcated routing. The performance advantage, of course, comes at the price of increased complexity.

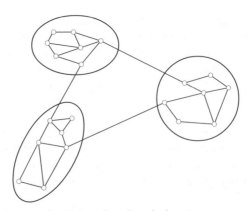

Figure 18.26 *A 21-node network split into three 7-node domains.*

18.4.2 Congestion control

Congestion occurs when there is an excessive backlog of packets at a network node. It is important to control congestion [Kleinrock] since once present there is a danger of positive feedback making it progressively worse, packet timeouts leading to retransmissions leading in turn to further congestion. Congestion control has been approached using a variety of different techniques, some of the most important of which are outlined below.

Flow control

Flow control mechanisms such as the sliding window protocol described in section 18.3.3 can be applied in networks to alleviate congestion. It has been argued, however, [Tanenbaum, p. 220] that flow control is really a rate matching technique allowing an information sink to slow down the transmission rate of an information source to a level at which it can deal with the incoming data. In this sense sliding window flow control is an end-to-end transmission rate control mechanism (that is clearly applicable, for example, at the data-link level between a pair of adjacent nodes, or at the transport level between a pair of DTEs).

A network or transport level sliding window protocol applied to all pairs of communicating DTEs will tend to slow down the rate of packets entering a network, and whilst this is generally beneficial in terms of congestion, the benefit is almost incidental rather than the principal intended effect. Furthermore, the congestion benefit is only fully realised when traffic load is fairly constant and evenly distributed between DTE–DTE connections. This is essentially because flow control applies to each connection independently and relies for its congestion prevention on each connection reflecting in microcosm the global congestion present in the network. If the maximum data rate on each DTE–DTE circuit is set such that congestion is prevented when the largest possible number of DTE–DTE pairs is operating at this rate, then this maximum data rate will be unacceptably low for the type of (bursty) traffic typically carried on modern networks. If the maximum data rate is set significantly higher than this, however, to cater for high bit rate but bursty traffic then congestion will not be prevented when significant numbers of DTE–DTE pairs happen to demand high bit rate service simultaneously.

Isarithmic control

A direct approach to congestion control is to limit the total number of packets in the network at any one time. This is called isarithmic (meaning constant number) congestion control. In this scheme special transmission permit packets (which may be piggybacked on other packets) circulate in the network. Before a DTE, or any other device, on the periphery of the network can launch a packet into the network it must capture and destroy a permit. Whenever a packet leaves the network another permit packet is generated. The total sum of all packets in the network is thus constant, only the relative proportions of permit packets and data packets changing.

Whilst isarithmic congestion control hard limits the network load and therefore mitigates against congestion it cannot prevent it completely. This is because the packets may not be distributed evenly through the network. There are also performance penalties to be paid. The delay incurred, for example, whilst a DTE passively accumulates (or actively acquires from nearby nodes) sufficient permits to transmit a high data rate message may be undesirable. Also the (random) transmission of permit packets around the network clearly represents a communications overhead. Finally, if permit packets are destroyed due to transmission errors, or lost due to routing malfunction, then the total number of permit packets, i.e. the total network capacity, will gradually decay over time.

Choke packet control

In this control strategy each node monitors the utilisation (0 to 100%) of each of its output links. If a data packet is routed to an output link that is close to saturation a choke packet is generated and sent to the source DTE that generated that data packet. The destination of the data packet (read by the node from its header) is given in the choke packet and the packet source is then obliged to reduce its packet generation rate for that destination by some fraction. Typically there are mechanisms to ensure that a single data packet can only generate a single choke packet and arrangements for the source DTE to forget chokes after a specified time allowing data rates to increase again after congestion has cleared.

18.4.3 Error control

The network layer can often assume an error-less communication service provided by the data-link protocol either because an appropriate error detection/ARQ system is implemented at the data-link level or because transmission is inherently of a high quality, i.e. has low BER. (The latter is often the case, for example, when optical fibre transmission is employed.) This does not mean, however, that higher layer error control is never needed since packets may be lost for reasons other than transmission errors; for example, due to a receiving DTE data buffer overflowing in the presence of imperfect flow control. Nevertheless, since end-to-end error control is often implemented at the transport level the need for error control at the network level is typically reduced. (X.25 is an important packet switched network protocol that does incorporate error control at the network level, see section 18.4.5.)

18.4.4 Quality of service

Quality of service (QoS) parameters are the means by which the adequacy of a service provided, or the requirements of a service requested, is defined. Typical QoS parameters are listed in Tables 18.8 and 18.9 for connection-oriented and connectionless network layer services respectively.

18.4.5 A connection-oriented network level protocol – X.25 PLP

Since X.25 incorporates the specifications for network, data-link and physical layers the network layer part of the specification is sometimes referred to as X.25 packet level protocol, or X.25 PLP. X.25 is actually a packet–network interface protocol, governing DTE–DCE communications, rather than a protocol defining the operation of an entire network, Figure 18.27 (see also section 20.7.1). It supports up to 4095 communications channels (virtual circuits) for use by the transport layer using one physical channel (i.e. one data link). As a connection-oriented network protocol it implements sliding window flow control. (Flow control is implemented at the transport level when connectionless

Table 18.8 *QoS parameters for OSI connection-oriented network service.*

QoS parameter	Comment
Network connection establishment delay	Time delay to establish a connection.
Network connection establishment failure probability	Ratio of establishment failures to attempts.
Throughput	Amount of data transferred per unit time in each direction.
Transit delay	Elapsed time between a request primitive for sending data from a network service user and an indication primitive delivering data to the destination.
Residual error rate	Ratio of number of incorrect, lost and duplicate data units to total number transferred across the network boundary.
Transfer failure probability	Ratio of total transfer failures to total transfer attempts.
Network connection resilience	Probability of network service provider invoked release and reset during a network connection.
Network connection release delay	Delay between a network service user invoked disconnect request and successful release signal.
Network connection release failure probability	Ratio of release requests resulting in release failure to total attempts.
Network connection protection	Extent to which network service provider tries to prevent unauthorised masquerading, monitoring or manipulation of network service user data.
Network connection priority	Relative importance of network connection with respect to other connections and to network data.
Maximum acceptable cost	Limit of cost for the network service.

Table 18.9 *QoS parameters for OSI connectionless network service (from Spragins, Table 9.5).*

QoS parameter	Comment
Transit delay	Elapsed time between N_USER_DATA.*request* and the corresponding N_USER_DATA.*indication*.
Protection from unauthorised access	Prevention of unauthorised monitoring or manipulation of network service user originated information.
Cost determinants	Specification of cost considerations for the network service provider to use in selection of a route for the data.
Residual error probability	Likelihood that a particular unit of data will be lost, duplicated or delivered incorrectly.
Priority	Relative priority of a data unit with respect to other data units that may be acted upon by network service provider.
Source routing	Designation by network service user of the path data is to follow to the destination address.

network level protocols are used.) The data-link protocol is LAP-B (link access protocol, balanced) that is a form of HDLC (see section 18.3.4). The physical layer protocol is X.21 (see section 18.2.1).

Packet structure

There are several different types of packet defined by X.25, but the general format is illustrated in Figure 18.28. (Note that in ITU-T terminology, a *byte* is called an *octet*.) Since 12 bits are allocated to the logical channel address, 4095 separate virtual circuits may be open at any one time, so that the X.25 controller behaves as a statistical multiplexer, i.e. a multiplexer that allocates communications resources to user processes on a demand basis.

GFI denotes *general format ID*, and contains the bits Q, D, 0/1, 0/1. Q is unspecified, while D describes the type of packet acknowledgement for flow control.

- D = 0 specifies a local acknowledgement, exercised between the calling-DTE and the network. Acknowledgements can come from either DCE, or the network.
- D = 1 defines an end-to-end acknowledgement, from remote-DTE to calling-DTE.

The C/D bit determines whether the packet is for data or for control. For a data packet, the control data effects the flow control, and is $\{P(S), M, P(R)\}$, comprising 3+1+3 bits where:

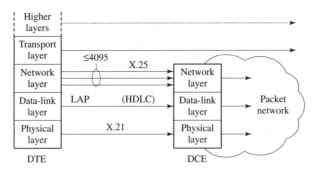

Figure 18.27 *X.25 for packet network access.*

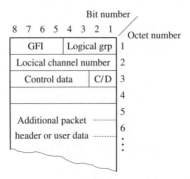

Figure 18.28 *X.25 general packet structure. (GFI = general format ID, C/D = 0 denotes user data packet, C/D = 1 denotes control data packet.)*

- M denotes more, $M = 1$ indicating that there is at least one complete packet to follow. A complete message can therefore still be identified within the network.
- $P(S)$ is the packet send sequence and numbers consecutively all sent packets.
- $P(R)$ is the acknowledgement for received packets, and indicates the next packet number that is expected by the called-DTE.

Since only 3 bits are allocated for the flow control counters, the maximum allowable window size is 7. An alternative structure allows 7 bits for each counter, leading to a maximum window size of 127, which is useful for networks containing a large delay, where several packets may be in transit at any particular time. GFI = QD01 denotes the smaller window size, and GFI = QD10 denotes the larger window size.

The *control data* for a *control packet* contains a code that indicates the type of packet, e.g. call request, call confirm, interrupt, etc.

The *call request* packet has the following address and facility information in its payload:

- Length of calling-DTE address field (4 bits); indicates the length of the following address, in semi-octets.
- Length of called-DTE address field (4 bits); indicates the length of the following address, in semi-octets.
- DTE addresses, of lengths specified above.
- Length of *facility* field (16 bits), in octets.
- Facilities, 8 bit code plus zero or more parameters; indicates the facilities supported by the sending X.25 interface.

There are many examples of facilities including flow control parameters, throughput class (ARQ protocol efficiency), closed user group definition and transit delay target.

Fast select

Fast select is a facility for short messages, similar to a datagram. The call request packet in this case can include up to 128 octets of data, and the corresponding clear request packet from the destination may also contain up to 128 octets of data. Thus a virtual circuit is set up, an exchange of a small amount of data is accomplished and the call is cleared all within one cycle.

Error handling

The X.25 network layer employs *reject, reset, clear* and *restart* packets used, in order of increasingly severe error conditions, as follows:

Reject – indicates a packet has been rejected (i.e. discarded). This may occur, for example, when a DTE has insufficient buffer space to accommodate an incoming packet because it has not had time to transmit a receive-not-ready (RNR) packet. The reject packet carries a packet receive sequence number $P(R) = J$ acknowledging all packet numbers up to and including $J - 1$. A go-back-N protocol then ensures all packets from J onwards are retransmitted.

Reset – resets a (single) virtual circuit setting the lower edge of the transmission window to zero. The next data packet must therefore have a packet send sequence number $P(S) = 0$. All packets not already acknowledged are lost but the virtual circuit is not disconnected and remains in the data transfer condition.

Clear – clears down (i.e. disconnects) a single virtual circuit. The circuit must be set up again before further data can be transferred. All data not already acknowledged when a clear packet is acted upon is lost.

Restart – clears all virtual circuits and resets all permanent virtual circuits.

X.25 packet transmission is further elaborated in section 20.7.1.

18.4.6 A connectionless network level protocol (CLNP)

The connectionless network protocol (CLNP) is the ISO version of the internet protocol (IP) and is therefore often referred to as ISO-IP. It provides a connectionless network service (CLNS). (ISO provides a connection-oriented network service (CONS) via its connection-oriented network protocol (CONP) based on X.25 PLP.)

Packet structure

The CLNP packet format is shown in Figure 18.29. The size of each element in the CLNP packet along with a brief description of the element's function is given in Table 18.10. Some fields in the header have variable length, indicated in the table by 'var'. CLNP adds 136 bits plus the length of the source and destination addresses, and the length of any options, to the transport PDU (or transport PDU segment if segmentation has occurred) that occupies the data field. The data field in the CLNP packet is sometimes called the network service data unit (NSDU).

Segmentation

The fields in the CLNP packet header concerned with segmentation and reassembly are the segmentation permitted flag, more segments flag, segment length, data unit identifier, segment offset and total length. Packet segments are called (in ISO terminology) derived PDUs and the original (un-segmented) packet is called the initial PDU. (Segmentation is referred to as fragmentation in many non-ISO protocols.) The segmentation permitted flag is set to 0 if no further segmentation is to be allowed. If a router comes across a PDU (initial or derived) with the segmentation flag set to 0 it must send the packet to a network which has a maximum packet size greater than the PDU size. If the router is not connected to any such network (that can ultimately lead to the packet's destination) then the PDU is discarded and an error report is sent to the source of the packet.

The more segments flag is set to 1 in the final derived PDU. All other derived PDUs have this flag set to 0 indicating that there are more segments to come. The segment length field contains the total number of octets in the derived PDU including the header.

All derived PDUs are uniquely related to their initial (parent) PDU by the combination of their source address, destination address and the data unit identifier (DUI) that is assigned by the end system, ES (the OSI terminology for a DTE), that generates the initial PDU (i.e. these three items are identical in all derived PDUs belonging to the same initial PDU).

The segment offset field contains the number of octets of the initial PDU carried by previous derived PDUs. The total length field contains the total number of octets in the initial PDU.

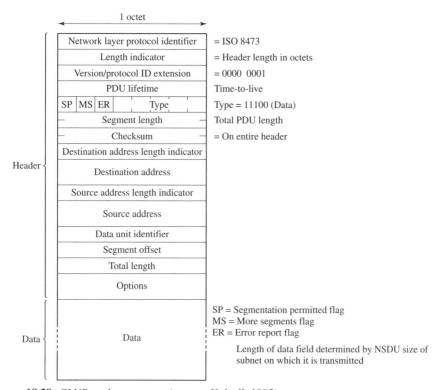

Figure 18.29 *CLNP packet structure (source: Halsall, 1995).*

OSI network architecture

Network architectures can be described in two planes: a 'horizontal' plane that relates to the geographical or physical distribution and grouping of network elements, and a 'vertical' plane that relates to the protocol stack. The horizontal OSI network architecture is made up of the following entities:

1. **Domains** – parts of a network, each being administered by a single authority and providing connectivity to all the ESs they contain.
2. **Hosts** – data processing devices that may be ESs or intermediate systems (ISs). (Intermediate system is the OSI terminology for a router.)
3. **Areas** – parts of a domain, each forming a logical entity and defined to be an area by a network administrator. Composed of a group of contiguous subnetworks and their attached hosts. Each IS (router) in an area shares information about all the hosts it can contact with all other ISs in the same area.
4. **Backbone** – a network superstructure connecting all areas.

The network layer of the 'vertical' OSI architecture has the substructure illustrated in Figure 18.30. (The designations (A) and (B) in the lowest three layers indicate the protocols appropriate to the network they serve.) The substructure comprises three sublayers, i.e.:

Table 18.10 *Size and function of CLNP packet fields.*

Name	Size (bits)	Purpose
Protocol identifier	8	Indicates if internet service is provided. If both the source and destination are on the same network, no internet protocol is needed and the header is just this single 8 bit field.
Header length	8	Packet header length in octets. (Octet is the ISO terminology for a byte, 1 octet = 1 byte = 8 bits.)
Version	8	Protocol version.
PDU lifetime	8	Lifetime in units of 0.5 s. Set by source and decremented by 1 by each router that operates on the datagram. Datagram is discarded when lifetime = 0. Prevents endlessly circulating datagrams.
Flags	3	Three 1 bit flags: SP = segmentation permitted flag; MS = more segments flag (shows data is split between datagrams); ER = error report flag (required by source if datagram is discarded).
Type	5	Data or error PDU.
Segment length	16	Total PDU length (header plus data).
Checksum	16	Applies to header only.
Destination address length	8	Length of destination address field in octets.
Destination address	var.	No structure specified.
Source address length	8	Length of source address field in octets.
Source address	var.	No structure specified.
Data unit identifier	16	Unique for source, destination.
Segment offset	16	Offset in octets.
Total length	16	Length of original un-segmented transport level PDU.
Options	var.	Additional services that may be included: Security (user defined); Source routing (source dictates the route); Record route (trace route taken by datagram); Priority, QoS (specifies reliability and delay parameters).

1. Subnetwork independent convergence protocol (SNICP).
2. Subnetwork dependent convergence protocol (SNDCP).
3. Subnetwork dependent access protocol (SNDAP).

The highest network sub-layer (SNICP) is independent of the type of network beneath it and therefore provides a common service to all network service users (NS_users). This makes the network transparent to all transport level entities.

The lowest network sub-layer (SNDAP) represents the network level access protocol that is particular to the underlying network type. The middle network sub-layer (SNDCP) interfaces SNICP and SNDAP.

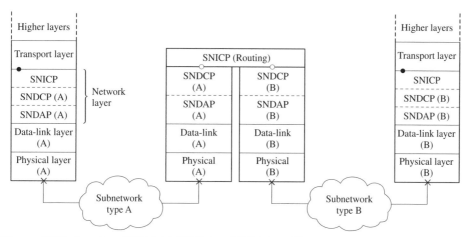

Figure 18.30 *Location of NSAP (•), NET(o) and SNPA (×) addresses in the layered architecture of an internet.*

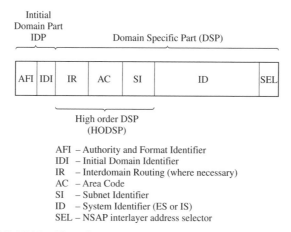

High order DSP
(HODSP)

AFI – Authority and Format Identifier
IDI – Initial Domain Identifier
IR – Interdomain Routing (where necessary)
AC – Area Code
SI – Subnet Identifier
ID – System Identifier (ES or IS)
SEL – NSAP interlayer address selector

Figure 18.31 *CLNP NSAP address format.*

Addressing

Each transport level entity (or process) communicates with the network layer across a network service access point (NSAP) each NSAP being allocated an address for use in routing. If more than one transport process (i.e. NS_user) employs network services in a particular ES then more than one NSAP address will be allocated to this ES. The NSAP address is hierarchical comprising an initial domain part (IDP) and a domain specific part (DSP), Figure 18.31.

The IDP is subdivided into the authority and format identifier (AFI) and the initial domain identifier (IDI). The AFI specifies the addressing authority responsible for allocating the IDI and the format of the IDI. The IDI specifies the entity that can assign DSP content and therefore the DSP network addressing scheme. The DSP specifies the area, subnet, ES and selector number (port) of the NS_user. The NSAP address is unique to a given NS_user within an entire internet and is therefore said to have global significance.

ISs use network entity titles (NETs) instead of NSAP addresses. NETs refer to processes (entities) running at the network level rather than an interface to a transport level process. (The network level is typically the highest level process running in an IS in which case there is no interface to a transport process thus precluding an NSAP address.)

A subnetwork point of attachment (SNPA) address is defined for each host (ES or IS). The SNPA is unique only within a given subnet and relates to the point at which a host is attached to the subnetwork. (The SNPA in an Ethernet LAN, for example, is the 48 bit MAC address.) Figure 18.30 shows, schematically, the location of NSAP, NET and SNPA addresses in the layered architecture of an internet.

Routing

CLNS routing is divided into levels. These are:

1. **Level 1 routing** – between different hosts in the same area.
2. **Level 2 routing** – between different areas.

Intermediate stations (ISs) are classified according to the level of routing that they participate in. Level 1 ISs are aware only of other level 1 ISs and hosts in their own area, and the nearest level 2 IS in their area to which they can send extra-area traffic. A level 2 IS must be aware of the topology of its own area (so it can route incoming packets) and must know how it can contact all other areas via other level 2 routers (so it can route outgoing packets and transit packets). (Transit packets are those passing through a particular area en route between ESs in two other areas.)

When an ES wishes to transmit a packet to another ES it simply directs the packet to an appropriate (typically the nearest) level 1 IS in its own area. The level 1 IS reads the packet's destination address and if this corresponds to an ES in the same area it routes it to this destination via (if necessary) other level 1 ISs. If the destination address corresponds to an ES in another area the packet is directed to the nearest level 2 IS. The level 2 IS reads the destination address and forwards the packet (via other level 2 ISs if necessary) to a level 2 IS in the destination area. This level 2 IS reads the address and directs the packet via level 1 ISs to the destination ES. Level 1 ISs are thus used to route packets within a singe area whilst level 2 ISs route packets between different areas (but within the same domain).

IDRP

The interdomain routing protocol (IDRP) governs the movement of data between routing domains. (A routing domain is a logical grouping of ESs and ISs, operated by a single authority, that share a common routing plan.) It operates between special ISs designated boundary or border ISs (BISs). There may be more than one physical BIS in any given domain but when this is the case they operate in a coordinated way such that they effectively form a single logical BIS. Unlike ISs in the IS–IS protocol, BISs don't advertise link state information but rather advertise routes by exchanging route information with each neighbouring BIS. A route is defined as a unit of information that pairs destinations with the attributes of a path (including the sequence of routing domains that determines the route path) to those destinations. Destinations are specified by NSAP address prefixes, each prefix identifying one, or more than one, system. The longest NSAP prefix would be identical to the full NSAP address of a single ES network service user, the shortest NSAP address prefix would identify all the ESs in a given routing domain.

Routes are exchanged between BISs in the form of update PDUs. Each update PDU contains (in its network layer reachability information field) a list of destinations reachable via the route and (in its path attributes field) a route path, comprising a sequence of routing domain identifiers (RDIs), that specifies the routing domains through which the route path passes, and other path attributes including QoS parameters (e.g. latency) for the path.

Flow control

Since CLNP is a connectionless protocol flow control is not applied and is left for implementation at the transport level. (See comments about congestion, however, in the context of error reporting below.)

Error reporting

If the error report flag in a data packet is set and the data packet is discarded by an IS or an ES then an error report packet is generated. The error report packet, which is addressed by the IS to the source of the discarded packet, contains a reason code that indicates why the packet has not been delivered. Reasons for discarding a packet include packet lifetime expiry, destination unreachable, checksum error detected in packet header and IS congestion. ESs are able to gather (partial) information about the state of an internet by keeping a record of recently received error reports, the source of these reports and the reason codes that they contained.

18.4.7 Use of primitives

The structures used in an interface protocol for passing user-data and control parameters between the layers of a protocol stack are called *primitives*. Primitives come in four basic forms: (i) request – issued to layer n by layer $n + 1$ that invokes (requests) a service from layer n, (ii) indication – issued to layer $n + 1$ by layer n that advises (provides indication to) layer $n + 1$ of an action initiated by layer n (which may or may not have been in response to a layer $n + 1$ request), (iii) response – issued by layer $n + 1$ in reply (response) to an indication primitive (which may complete or advise completion of an action invoked by a request primitive), (iv) confirm – issued by layer n to layer $n + 1$ completing or advising completion of an action invoked by a request primitive. An example in the context of CLNP is given below.

At the source DTE a *request* for a required service is issued by the service user (the transport layer) to the service provider (the network layer) using a primitive of the form:

N_UNITDATA.*request* (source address, destination address, QoS, NS_user_data).

At the destination DTE an *indication* of the requested service is issued by the service provider (the network layer) to the service user (the transport layer) using a primitive of the form:

N_UNITDATA.*indication* (source address, destination address, QoS, NS_user_data).

The prefix N_ denotes a network level primitive. N_UNITDATA is the particular primitive used to invoke the transportation of a block, or unit, of data. NS_user_data in the parameter lists denotes *network service user data*, i.e. a transport protocol data unit or PDU. *Request* primitives move down the stack (from higher to lower layer protocols) and *indication* primitives move up the stack (from lower to higher layer protocols). Primitives moving down the stack are associated with the protocol layer into which they are going and primitives moving up the stack are associated with the layer from which they have come. An N_PRIMITIVE.*request* therefore denotes a software structure passing data and/or control information from the transport level to the network level. (An L_PRIMITIVE.*indication* would denote a structure passing data and/or control information from the data-link layer to the network layer.) The parameters of the

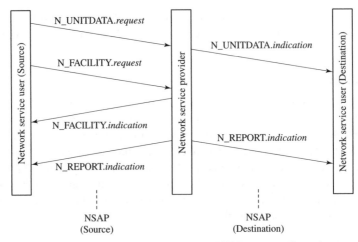

Figure 18.32 *Use of CLNP network service primitives (NSAP = network service access point).*

Table 18.11 *QoS parameters for CLNS.*

Transit delay	Time elapsed between N_UNITDATA.*request* at the source and N_UNITDATA.*indication* at the destination.
Protection from unauthorised access	Four options: 1. No protection. 2. Protection against passive monitoring. 3. Protection against modification, replay, addition or deletion. 4. Both 2 and 3.
Cost	1. Service provider to use least expensive means available. 2. Maximum acceptable cost.
Residual error probability	Probability that the NS_user_data unit will be lost, duplicated or delivered incorrectly.
Priority	Relative priority of NS_user_data unit with respect to: 1. Order in which the data units have their QoS degraded, if necessary. 2. Order in which the data units are to be discarded to recover resources, if necessary.

primitive specify the relative network layer service completely. For the N_UNITDATA primitives given above the source and destination addresses are unique network identifiers (across the entire set of subnetworks), the QoS parameters (drawn from Table 18.11) specify the quality of the end-to-end communications link, and NS_user_data is the data to be transmitted (i.e. the transport PDU).

In addition to N_UNIT_DATA.*request* and .*indication* there are three other CLNS primitives, namely:

N_FACILITY.*request* (destination address, QoS/service characteristics).
N_FACILITY.*indication* (destination address, QoS/service characteristics). And:
N_REPORT.*indication* (destination address, QoS/service characteristics, reason code).

The N_FACILITY service is used to find out from the network what QoS and/or transport level service characteristics the network can provide (to the transport layer). The service characteristics are given in Table 18.12.

Table 18.12 *Internet service characteristics.*

Congestion control	Specifies if flow control will be implemented by the network service provider at the network service user interface. The local CLNP program returns an N_REPORT.*indication* primitive in response to an N_UNITDATA.*request* with the reason code parameter set to NS_provider congestion.
Sequence preservation probability	Ratio of sequence preserved transmissions to total number of transmissions as measured by the local CLNP program. This information is used by the local transport level protocol program in its implementation of error control and flow control.
Maximum packet lifetime	Maximum lifetime of the packet in the network before it is discarded.

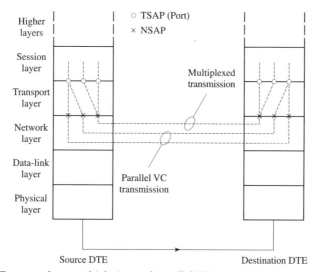

Figure 18.33 *Transport layer multiplexing and parallel VC transmission.*

18.5 Transport layer

The function of the transport layer (ISO layer 4, Figure 17.21) is to provide error-free end-to-end transmission (transport) of messages irrespective of the QoS provided by the underlying network. It forms the interface between the communications oriented functions of the lower three OSI layers and the applications oriented functions of the upper three OSI layers. Its principal responsibilities are:

1. Segmentation (or fragmentation) of transport PDUs (TPDUs) into network PDUs (NPDUs or packets) and TPDU reassembly.
2. Multiplexing (and reverse multiplexing or parallel virtual circuit transmission).
3. End-to-end error and flow control (where necessary).

The link provided by the transport layer may be connection-oriented or connectionless. It is possible, in principle, to implement either of these two alternatives over virtual circuit or datagram network layer protocols. In practice, however, there is little benefit to operating a connectionless transport layer over a connection-oriented network layer, so this combination is rare.

Communication between the transport layer and its adjacent higher layer (i.e. the session layer) takes place, in the normal way, using primitives, e.g. T_UNITDATA for connectionless transport and T_CONNECT, T_DISCONNECT and T_DATA for connection-oriented transport. As usual *.request* and *.response* versions of these primitives are concerned with downward communication across the layer's interface whilst *.indication* and *.confirm* versions are concerned with upward communications across the interface.

18.5.1 Message segmentation and reassembly

The transport layer of a source ES segments TPDUs (messages) into NPDUs (packets) that are then supplied to the network layer for transmission. NPDUs are reassembled (after resequencing if necessary) at the transport layer of the destination ES.

18.5.2 Multiplexing and parallel virtual circuit transmission

The transport layer often multiplexes several low data rate messages onto one virtual circuit in a conventional way. It can also provide the converse service, however, splitting a high data rate message between several (parallel) virtual circuits (VCs) referred to here as parallel VC transmission. (This process is also referred to as reverse, inverse or down multiplexing.) Figure 18.33 illustrates multiplexing and parallel VC transmission between the multiple transport service access points, TSAPs (also called *ports*), at the session–transport layer boundary and the multiple NSAPs at the transport network layer boundary. Multiplexing reduces the number of VCs that in turn reduces the size of routing tables at network nodes. Parallel VC transmission increases the data rate at which a message can be delivered.

18.5.3 End-to-end error and flow control

End-to-end error control (where required) is typically implemented using some form of ARQ protocol similar to those used at the data-link level (see section 18.3.2) although timeout mechanisms are often used in place of explicit negative acknowledgements to indicate corrupted or out-of-sequence TPDUs.

End-to-end flow control can be implemented at the transport layer using a sliding window protocol, again similar to that described for the data-link layer (section 18.3.3). This protects against a destination ES from being overwhelmed by a high data rate source ES.

18.5.4 Use of well known and ephemeral ports

When an application process at an instigating ES wishes to connect to a process at a second ES and the transport layer port (TSAP) is known then the connection is made via this port in a straightforward way. FTP (file transport protocol), for example, uses TSAP 21 of the (transport layer) TCP (transmission control protocol). (TCP is not an OSI protocol but is very widely used and is the protocol on which the OSI transport protocol is based.) Such reserved ports are called well known ports.

When an application process at an instigating ES wishes to connect to an application at a second ES whose TSAP is unknown then a connection is made first to a *process server* (at the second ES) which does have a well known TSAP. The process server then informs the process at the initiating ES of the TSAP number that should be used. If the process at the second ES does not yet exist and is required to be created in response to the initiating ES then the process server creates the process and assigns an *ephemeral* (i.e. temporary) TSAP to it before communicating this TSAP number to the initiating ES. The ephemeral TSAP is released when the communication event is completed. Figure 18.34 shows, schematically, the mechanism of connecting via an unknown or ephemeral TSAP.

18.5.5 A transport level protocol – TP4

There are five ISO transport layer protocols. In order of increasing sophistication these are TP0, TP1, TP2, TP3 and TP4. TP0 to TP3 support only connection-oriented communications whilst TP4 (transport protocol class 4) supports both connection-oriented and connectionless transmission. The TP class used is determined by the underlying network layer. If the QoS of the network layer is such that it delivers reliably error-free packets in the correct order without loss or duplication then a low level TP

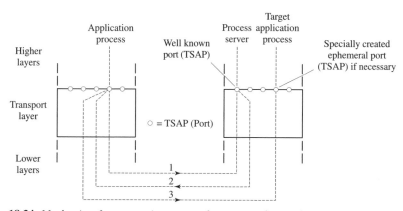

Figure 18.34 *Mechanism for connecting to an unknown or ephemeral TSAP.*

Table 18.13 *Principal functions supported by OSI transport layer protocols.*

Proto-col	Segmentation/ reassembly	Error recovery	Multiplexing/ demultiplexing	Reliable trans-port service	Connection-oriented/ connectionless
TP0	Yes	No	No	No	Connection-oriented
TP1	Yes	Yes	No	No	Connection-oriented
TP2	Yes	No	Yes	No	Connection-oriented
TP3	Yes	Yes	Yes	No	Connection-oriented
TP4	Yes	Yes	Yes	Yes	Both

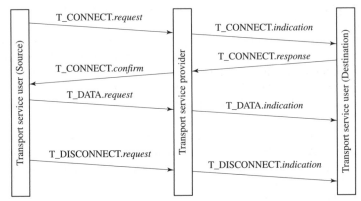

Figure 18.35 *Use of TP4 transport service primitives in connection-oriented mode.*

class is selected. If the network layer is unreliable then a higher TP class that incorporates an error and flow control mechanism is required. Table 18.13 summarises the principal transport layer functions that these ISO TPs support.

TP4 is the ISO version of TCP (transmission control protocol) and is designed to operate satisfactorily over an unreliable network layer such as is typically provided by a WAN (e.g. PSTN). In its connection-oriented mode, Figure 18.35, it has the usual three phases, i.e.:

1. Connection establishment phase.
2. Data transfer phase.
3. Connection release (or call clearing) phase.

The transport primitives (along with their parameters) used for the connection set-up and release phases are:

T_CONNECT.*request* (calling address, called address, expedited data option, QoS, TS_user data).

T_CONNECT.*indication* (calling address, called address, expedited data option, QoS, TS_user data).

T_CONNECT.*response* (responding address, QoS, expedited data option, TS_user data).

T_CONNECT.*confirm* (responding address, QoS, expedited data option, TS_user data).

T_DISCONNECT.*request* (TS_user data).

T_DISCONNECT.*indication* (disconnect reason, TS_user data).

The primitives used for the data transfer phase are:

T_DATA.*request* (TS_user data).
T_DATA.*indication* (TS_user data).

T_EXPEDITED_DATA.*request* (TS_user data).
T_EXPEDITED_DATA.*indication* (TS_user data).

The expedited data primitive (which is used only by agreement between the calling and called TS_users) is used to short-circuit any flow control procedures.

For connectionless transport the primitives are:

T_UNITDATA.*request* (calling address, called address, QoS, TS_user data).
T_UNITDATA.*indication* (calling address, called address, QoS, TS_user data).

The address parameters used in the above primitives are concatenations of TSAP and NSAP addresses. Figure 18.35 shows a typical time sequence diagram of primitives for a connection-oriented communications event and Figure 18.36 shows a sequence for a connectionless event.

TPDU structure

In response to a service primitive TP4 generates the following types of TPDU:

1. Connect request (CR-TPDU).
2. Connect confirm (CC-TPDU).
3. Disconnect request (DR-TPDU).
4. Disconnect confirm (DC-TPDU).

5. Data (DT-TPDU).
6. Acknowledgement (AK-TPDU).

7. Expedited data (ED-TPDU).
8. Expedited acknowledgement (EA-TPDU).

9. Reject (RJ-TPDU).
10. Error (ER-TPDU).

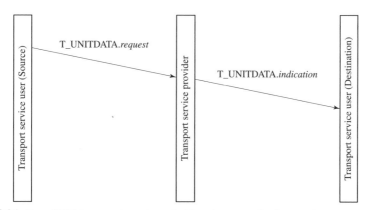

Figure 18.36 *Use of TP4 transport service primitives in connectionless mode.*

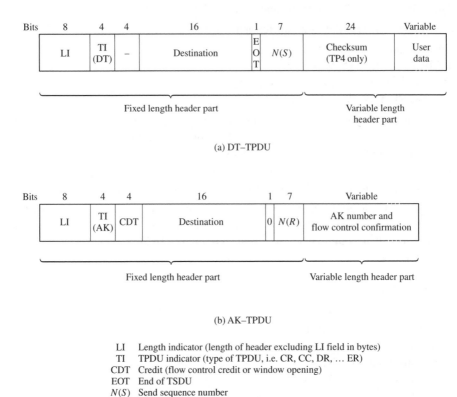

(a) DT–TPDU

(b) AK–TPDU

LI   Length indicator (length of header excluding LI field in bytes)
TI   TPDU indicator (type of TPDU, i.e. CR, CC, DR, ... ER)
CDT Credit (flow control credit or window opening)
EOT End of TSDU
$N(S)$ Send sequence number
$N(R)$ Receive sequence number

Figure 18.37 *Data (a) and acknowledgement (b) TPDU structures.*

Figure 18.37 shows TPDU structures (DT-TPDU and AK-TPDU) and Figure 18.38 relates these to transport primitives in the data transfer phase of a typical communications

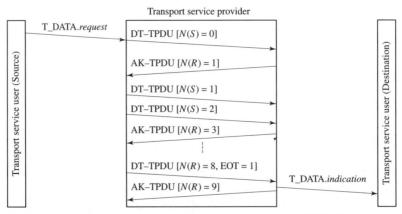

Figure 18.38 *Relationship between transport primitives and transport PDUs for an example data transfer event.*

event. (The structure of the DT-TPDU shown is as used in TP2, TP3 and TP4.)

Note that each DT-TPDU carries a send sequence number $N(S)$ for use by the end-to-end error and flow control protocols. The EOT bit is set to indicate the end of a transport service data unit (TSDU), i.e. it denotes the last TPDU of the data block (SPDU) supplied by the session layer (see following section). Where present, the variable length header part of the TPDU contains fields comprising an 8 bit field type (specifying the nature of information that follows and which is different for different TPDU types), an 8 bit field length and the field value. The AK-TPDU variable length header part contains AK number and flow control confirmation fields.

TP4 operation

A (connection-oriented) transport link is initiated by a transport service user (TS_user) using the T_CONNECT primitive and established using CR- and CC-TPDUs. A *three-way handshake*, Figure 18.39, is used during this process in which TS_users at both ends of the link agree the initial connection parameters, e.g. protocol class, flow control option, maximum DT-TPDU size (from 27 to 213 bytes), QoS (acknowledgement time, throughput, residual TPDU error rate, connection priority, transit delay) and initial sequence numbers for DT- (and other) TPDUs. The agreement or negotiation process complies with a rule that a CC-TPDU can carry only service options of equal, or less, sophistication than those carried by the CR-TPDU. For example, if an initiating transport entity (TE) suggests class 4 protocol (i.e. TP4) then the responding TE can suggest class 2 (i.e. TP2) in which case both would agree on TP2. If the initiating TE suggested TP2 in its CR-TPDU, however, then the responding TE cannot suggest TP4. Similarly the CR-TPDU might carry a suggested maximum TPDU size for the data transfer phase of 2^{10} bytes in which case the CC-TPDU could only respond with powers of two from 2^7 to 2^{10} bytes.

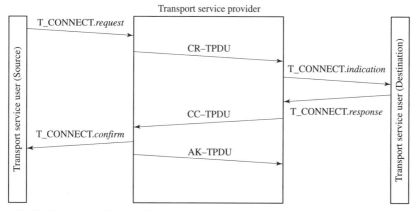

Figure 18.39 *Three-way TP4 handshake for connection set-up.*

Error control

All TP4 TPDUs carry a checksum in the variable part of their header. If the checksum indicates bit error(s) then the TPDU is discarded, no AK-TPDU covering the corrupted TPDU is generated, and a timeout mechanism results in retransmission of that TPDU.

The error recovery process is go-back-N, section 18.3.2. An AK-TPDU carrying a receive sequence number $N(R)$ is used to acknowledge all DT-TPDUs with sequence number(s) up to and including $N(R) - 1$. If a TPDU arrives out of sequence the transport layer generates an RJ-TPDU that carries a receive sequence number $N(R)$ representing the send sequence number $N(S)$ of the next expected TPDU.

Flow control

Flow control at the transport level protects receiving ES resources from being overwhelmed. TP4 employs a modified version of the sliding window flow control described in section 18.3.3. The CR-TPDU specifies the window size in its *credit* (CDT) field. (Credit is the TP4 terminology for window size.) The maximum size or credit is typically 16 but there is an option for a size of 4. Credit (i.e. window opening) can be reduced by an AK-TPDU (which contains a CDT field) if receiver resources (e.g. receive buffer space) fall to unacceptably low levels.

18.6 Session layer

The principal functions of the session layer (ISO layer 5, Figures 17.21 and 18.33) are to:
1. Establish the session connection (including negotiation of session parameters), maintain it during data transfer and terminate it when data transfer is complete.
2. Supervise dialogue (i.e. control which end of the connection may transmit and which must listen at any given time).
3. Facilitate recovery after failure.

In many protocol stacks there is no explicit session layer, its functions being split between the transport and application layers. (The presentation layer is also often missing.) Furthermore, in protocol stacks that do include an explicit session layer it may, for some application processes, be practically transparent, requests for transport services being passed through the session (and perhaps the presentation) layer with minimal processing.

The overall purpose of the session layer can be thought of (loosely) as providing a user-friendly interface to the reliable (but complex) end-to-end message (TSDU or SPDU) transmission service provided by the transport layer.

18.6.1 Session connection, maintenance and release

Protocols have been defined for both connection-oriented and connectionless session layer links but connection-oriented links are invariably used in practice. Connection parameters are negotiated during the connection phase of a particular session. These include QoS, connection protection and connection priority.

The unit of data passed between a session service user (SS_user) and a session entity (SE) during the data transfer phase is sometimes referred to as a *record* and is encapsulated by the session layer to form a session PDU (SPDU).

In general session connections may have a one-to-one mapping onto a transport connection, may be multiplexed (multiple session connections mapped to one transport connection) or may be reverse multiplexed (one session connection mapped to multiple transport connections). Multiplexing simplifies the task of the transport layer in that there are fewer transport connections to manage. The mix of session types that is appropriate for multiplexing onto a single transport connection is a significant issue, however. If an interactive session is multiplexed with a large file transfer session, for example, the response time of the interactions could be unacceptably degraded due to enquiries/replies competing with bulk data in a queue for transport services.

18.6.2 Dialogue supervision

The dialogue mode within a session, negotiated during the connection phase, may be one-way only, both ways simultaneously, or alternate. If the alternate (half-duplex) mode is agreed then some explicit method of dialogue supervision is required to enforce an appropriate and compatible pattern of transmission and reception between the peer session processes.

18.6.3 Recovery

The session layer may include a recovery mechanism that avoids loss of transferred data should a connection fail. This is especially advantageous for bulk data transport sessions since a failure without the possibility of recovery occurring near the end of a session transferring a large data file will clearly be both wasteful of resources and inconvenient. Recovery mechanisms rely on the periodic insertion of recovery reference points so that if the session is aborted the point from which transmission should recommence can be specified. For retransmission to be possible a copy of the transmitted data must be retained at its source until it has been acknowledged as being received satisfactorily.

18.6.4 A session level protocol – ISO-SP

The ISO session layer protocol, ISO-SP, specifies one-to-one mapping between session connections and transport connections. Its primitives (along with their parameters) used for the connection-oriented set-up and release phases are:

S_CONNECT.*request* (identifier, calling SSAP, called SSAP, QoS, requirements, serial number, token, SS_user data).
S_CONNECT.*indication* (identifier, calling SSAP, called SSAP, QoS, requirements, serial number, token, SS_user data).
S_CONNECT.*response* (identifier, called SSAP, result, QoS, requirements, serial number, token, SS_user data).
S_CONNECT.*confirm* (identifier, called SSAP, result, QoS, requirements, serial number, token, SS_user data).

S_RELEASE.*request* (SS_user data).
S_RELEASE.*indication* (SS_user data).
S_RELEASE.*response* (result, SS_user data).
S_RELEASE.*confirm* (result, SS_user data).

S_USER_ABORT.*request* (SS_user data).

S_USER_ABORT.*indication* (SS_user data).

S_PROVIDER_ABORT.*indication* (reason).

The *identifier* parameter uniquely identifies the connection. The *requirements* parameter specifies which functional units of the session layer are required (in addition to the *kernal* unit that is always used). The list of functional units from which the requirements are selected is: *negotiated release, half-duplex, duplex, expedited data, typed data, capability data exchange, minor synchronise, major synchronise, resynchronise, exceptions, activity management.* There are some constraints on the combinations of functional units that can be selected, e.g. *half-duplex* and *duplex* (i.e. full-duplex) options, Chapter 15, are mutually exclusive and the selection of *capability data exchange* is conditional on the selection of *activity management.* The *serial number* parameter represents the initial serial number when synchronisation services are utilised. The *token* parameter lists the ends of the link to which access tokens are initially given.

Requirements are negotiated by the initiating SS_user proposing a list in the CONNECT.*request*/CONNECT.*indication* primitives and the responding SS_user counter proposing a list of the same or lesser sophistication in the CONNECT.*response*/ CONNECT.*confirm* primitives. The responding SS_user list then represents the negotiated (i.e. agreed) list.

The principal primitives used for the data transfer phase are:

S_DATA.*request* (SS_user data).
S_DATA.*indication* (SS_user data).

S_TYPED_DATA.*request* (SS_user data).
S_TYPED_DATA.*indication* (SS_user data).

S_EXPEDITED_DATA.*request* (SS_user data).
S_EXPEDITED_DATA.*indication* (SS_user data).

S_CAPABILITY_DATA.*request* (SS_user data).
S_CAPABILITY_DATA.*indication* (SS_user data).

S_TOKEN_GIVE.*request* (tokens, SS_user data).
S_TOKEN_GIVE.*indication* (tokens, SS_user data).

S_DATA primitives represent the normal data transfer mechanism. S_TYPED_DATA primitives allow the transfer of data without regard to data token possession in half-duplex sessions.

Figures 18.40 to 18.42 illustrate typical sequences of session layer primitives for the three phases of connection set-up, data transfer and connection release.

Activities, dialogue unit and synchronisation points

A single session may involve the transfer of large amounts of data between two application entities (via the presentation layer). Provision is made, therefore, to divide the session connection into component parts so that failure of the link before the session is completed will not require all transactions for the whole session to be repeated when the link has been recovered. This is illustrated schematically in Figure 18.43.

Messages (TPDUs) are grouped together in *dialogue units* that are demarcated by *major synchronisation points*. If errors occur during a session then transmission can be restarted from the beginning of the dialogue unit in which the error occurred. A major

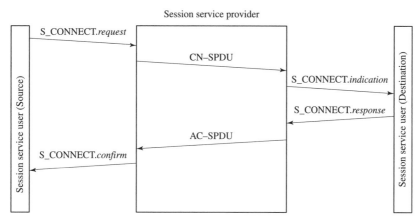

Figure 18.40 *Primitive and SPDU events during a session link connection set-up.*

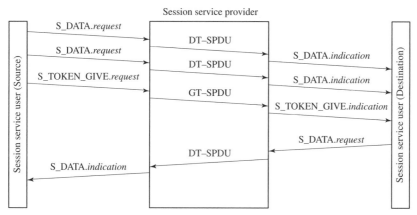

Figure 18.41 *Primitive and SPDU events during session link data transfer. (Token assumed to be initially at source DTE.)*

synchronisation point, created by the issuing of an S_SYNC_MAJOR.*request*, completely separates the SPDUs occurring before and after it. No SPDUs can be transmitted after an S_SYNC_MAJOR.*request* until the major synchronisation point has been acknowledged by an S_SYNC_MAJOR.*confirm*.

Minor synchronisation points are also defined which subdivide the flow of SPDUs making up a dialogue unit. The transmission of an S_SYNC_MINOR.*request* at a source DTE and the consequent reception of an S_SYNC_MINOR.*indication* at the destination DTE divides all events at both ends of the link into those that occurred before and after the minor synchronisation point and thus defines a common reference point in the source and destination DTE event sequences.

Dialogue units are grouped together in *activities* defined by S_ACTIVITY_START and S_ACTIVITY_END primitives. These represent self contained communications events such as a single file transfer or the transfer of a group of related files.

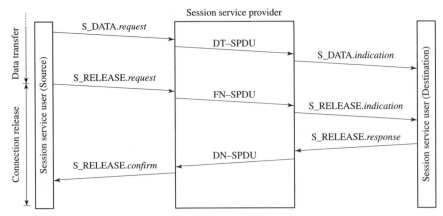

Figure 18.42 *Primitive and SPDU events during session link connection release.*

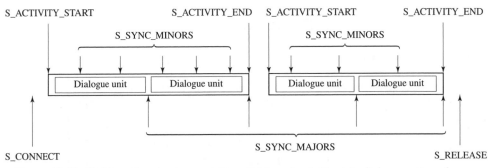

Figure 18.43 *Division of session into activities, division of activities into dialogue units and subdivision of dialogue units by minor synchronisation points.*

Tokens

ISO-SP uses tokens to establish what service rights are available to the SS_users at each end of the connection. Four varieties of token are used, namely:

1. Data token.
2. Synchronise minor token.
3. Major/activity token.
4. Release token.

Possession of a data token confers the right to transmit data and is required in the case of a half-duplex link. (In a full-duplex link then no dialogue discipline, and therefore no data token, is required.) Possession of a release token confers the right to initiate a connection release. Possession of a synchronise minor token confers the right to set minor synchronisation points which subdivide dialogue units. Possession of a major/activity token confers the right to start and end activities and set major synchronisation points that delineate dialogue units.

Table 18.14 shows those tokens that are always required to access a particular service.

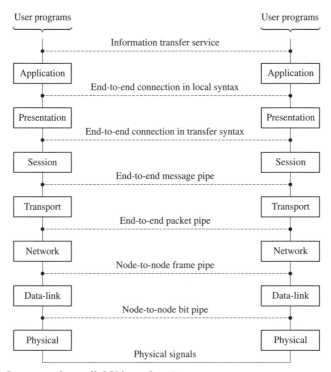

Figure 18.44 *Summary of overall OSI layer functions.*

Table 18.14 *Tokens to access services/functions. (AR = always required, IIU = required only if in use.)*

Function	Token			
	Data	Minor sync	Major sync	Release
Transfer SPDU (half duplex)	AR	–	–	–
Transfer capability data	IIU	IIU	AR	–
Set minor synchronisation point	IIU	AR	–	–
Set major synchronisation point	IIU	IIU	AR	–
Start activity	IIU	IIU	AR	–
Interrupt activity	IIU	–	AR	–
Discard activity	–	–	AR	–
End activity	–	IIU	AR	–
Release connection	IIU	IIU	IIU	IIU

The S_TOKEN_GIVE primitive is used to transfer specified tokens to the far end of a connection. The S_CONTROL_GIVE primitive transfers all tokens to the far end of a connection. The S_TOKEN_PLEASE primitive is used to request a token that currently resides at the far end of a connection.

18.7 Presentation layer

The presentation layer (ISO layer 6, Figures 17.21 and 18.44) ensures that information is passed to the presentation service users (PS_users), at either end of the connection, in a meaningful way (i.e. to the PS_user or application). More formally the presentation layer ensures that the *semantics* of data is shared between the applications at either end of the connection despite the data having a different representation or *syntax*. The principal responsibilities of a generic presentation layer can be broken down into:

1. Translation between local syntaxes.
2. Encryption and deciphering.
3. Data compression and decompression.

The formal specification of the OSI presentation layer, however, only relates to translation between syntaxes, and since data compression and encryption have been described in Chapter 9 only syntax translation is addressed here. (Compression and encryption are traditionally considered to be presentation layer responsibilities because these functions are coding/decoding operations.)

18.7.1 Translation between local and transfer syntaxes

The applications at either end of a connection generally use different data representations. The presentation layer at a source DTE therefore translates the *local* syntax of messages received from the application layer into a *transfer* syntax before passing them down the protocol stack for transmission. Conversely, at a destination DTE the presentation layer translates the transfer syntax of messages received (from the session layer) into the appropriate local syntax before passing them up to the application layer. The local and transfer syntaxes are sometimes referred to as *abstract* and *concrete* syntaxes, respectively. Figure 18.44 illustrates the overall role of the presentation layer schematically (along with the equivalent roles of the other OSI layers), that is to use the transfer syntax connection service furnished by the session layer to provide a local syntax connection service to the application layer.

18.7.2 Abstract syntax notation 1 (ASN.1)

To aid the translation of the local (abstract) syntax into a transfer (concrete) syntax ISO recommends the use of abstract syntax notation 1 (ASN.1). ASN.1 describes application data structures as one of four types, i.e.:

1. Universal (defined universally, e.g. Boolean, integer, real, bit string, octet string, IA5 string, Appendix C).
2. Context specific (defined for the context in which they are used locally).
3. Application (defined for the specific application that uses them, e.g. tables).
4. Private (defined by the user).

Each data structure in the application is classified and *tagged* as one of the above types. In practice an ASN.1 compiler is used to translate the ASN.1 data types into equivalent data types in the language of the application (e.g. C, Java, Pascal, etc.). The compiler also produces procedures for translating (encoding and decoding) between these data types and those of the concrete syntax. In the concrete syntax the data takes the form of an octet string, the first octet of which defines the data type.

18.7.3 A presentation level protocol – ISO-PP

The ISO presentation protocol (ISO-PP) negotiates an appropriate concrete syntax (or set of syntaxes) for the transfer of data between two applications, translates the abstract syntax of the application entity (PS_user) data into the concrete syntax for use by the session layer, translates the concrete syntax of the session entity (TS_user) data into the abstract syntax of the application entity data, and passes requests and responses relating to session layer services between the application and session layers. Each abstract–concrete syntax pair is called a *presentation context* and the set of such pairs (if more than one is required) is called a *presentation context set*.

Requests for services are passed between application and presentation layers using primitives in the normal way. Some presentation layer primitives (e.g. P_CONNECT.*request*) generate presentation PDUs (PPDUs) but others (e.g. P_RELEASE.*request*) generate no PPDUs since their purpose is to stimulate the equivalent session layer primitive directly (S_RELEASE.*request* in this case).

18.8 Application layer

The overall function of the application layer (ISO layer 7, Figures 17.21 and 18.44) is to provide an information transfer service to application service users (i.e. user programs). Different types of application (e.g. e-mail, file transfer, remote login, etc.) require

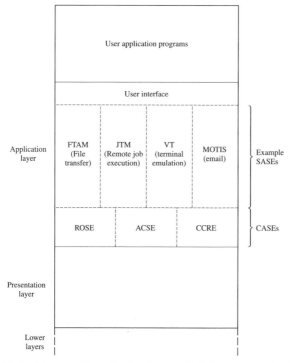

Figure 18.45 *Architecture of application layer underlying user application programs.*

different sets of services and to facilitate this the application layer is constructed with a set of *specific application service elements* (SASEs) that overlie a set of *common application service elements* (CASEs), see Figure 18.45. The CASEs include an *association control service element* (ACSE) that establishes a connection (called an *association*) between SASEs, a *remote operations service element* (ROSE) that handles commands and returns results, and a *commitment, concurrency and recovery element* (CCRE) that breaks up complex operations into many smaller and simpler operations,

The SASEs provide services such as file transfer (FTAM), remote job execution (JTM), terminal emulation (VT) and email (MOTIS).

Typically a user interface translates instructions and responses between the SASEs and the user programs which reside on top of the application layer.

18.9 Non-OSI protocol stacks

The focus in this chapter has been on the ISO OSI model of a protocol stack and the examples have been drawn largely from the ISO suite of protocols. The OSI model is a generic reference model and the ISO examples represent particular instances of a practical stack. It is important to realise, however, that there are many other (proprietary) protocol stacks that can be aligned (with varying degrees of precision) alongside the OSI model. (Some protocol stacks display an almost one-to-one correspondence with OSI functional layers whilst some have much less similarity in their allocation of functions to layers.) The technologies associated with some of the protocol stacks summarised here are addressed in Chapters 20 and 21.

18.10 Summary

The complexity of modern communication systems requires that their functions be divided into layers, each layer interacting only with its adjacent layers via well defined interfaces or service access points. This makes system design tractable, system operation comprehensible and system testing easier and more rigorous. The functional layers constitute a protocol stack and many such protocol stacks have been proposed and used for proprietary communication systems.

A commonality of functions that virtually all communication systems are required to fulfil has led to the development of the ISO OSI reference model for protocol stacks. Many proprietary stacks comprise layers that correspond quite closely to one or more layers of the reference model. The reference model defines physical, data-link, network, transport, session, presentation and applications (i.e. seven) layers.

Physical layer protocols describe the interfaces between items of equipment including connector mechanics, the voltages and currents used for signalling across the interface and the details of the line code employed. They provide an unreliable pipe for the transmission of bits between adjacent network nodes. X.21 (a form of RS-232) is an example of a physical layer protocol that is used between a DTE and a DCE.

Data-link layer protocols deliver reliable frames of data from one node of a network to another (usually adjacent) node. They thus provide a transparent (reliable) pipe for the point-to-point transport of bits. Its functions include error control, implemented by ARQ, and flow control. Flow control, which stops a sending node transmitting frames faster than the receiving node can deal with them, can also make use of the ARQ protocol, e.g. by the withholding of ACKs to delay the transmission of frames. Sliding window flow control is often used in data-link protocols. HDLC is an example of a data-link protocol that has been adopted (often with variation) in many proprietary protocol stacks.

Network layer protocols deliver packets from one DTE to another and therefore constitute an end-to-end pipe between a pair of DTEs for data packets. The packet pipe may be unreliable or transparent depending on the particular network protocol(s) being used. A prime responsibility of the network layer is for the routing of packets through the network. The network layer is dissimilar to other protocol layers in that it does not deal purely with a single pair of peer processes (nodes at the data-link layer and below, DTEs at the transport layer and above) but with many processes distributed between all (or at least many) of the nodes making up the network. Matrices are a useful way of defining network topologies and graph theory can be used to design routing algorithms. The Dijkstra routing algorithm has been used for shortest path (or more generally minimum metric) routing in many practical networks.

End-to-end flow control and congestion control may also be implemented by network layer protocols. X.25 PLP is a popular connection-oriented network layer protocol and CLNP is a connectionless network layer protocol. CLNP is the ISO version of the tremendously popular, and widely adopted, internet protocol (IP).

The transport layer provides error-free end-to-end transmission of messages between DTEs irrespective of the quality of service provided by the underlying network. It forms the interface between the communications oriented functions of the lower three OSI layers and the applications oriented functions of the upper three OSI layers. Its principal responsibilities are message (or strictly TPDU) fragmentation (into packets or NPDUs) and reassembly, multiplexing (and reverse multiplexing or parallel virtual circuit transmission) and end-to-end error and flow control (where necessary).

There are five ISO transport layer protocols, TP0–TP4. TP0 to TP3 support only connection-oriented communications while TP4 supports both connection-oriented and connectionless transmission. TP4 is the most sophisticated of the ISO transport layer protocols and is used when packet transmission reliability of the network layer cannot be guaranteed.

The session layer establishes the session connection, maintains it during data transfer, terminates it when data transfer is complete, supervises dialogue and facilitates recovery after failure. It provides a transfer (or concrete) syntax connection between DTEs. The ISO session layer protocol (ISO-SP) uses tokens to establish what service rights are available to the session service users at each end of the connection.

The presentation layer ensures that information is passed in a meaningful way to the presentation service users at either end of the connection, i.e. it ensures that the semantics of data is shared between the applications at either end of the connection despite the data having a different representation or syntax. The presentation layer thus provides a local (or abstract) syntax connection between DTEs. Since the presentation layer deals with translations between different syntaxes, encryption and data compression are often implemented as presentation layer processes.

The ISO presentation protocol (ISO-PP) principally governs translation between the abstract (or local) syntaxes and the concrete (or transfer) syntax used for information transfer. To facilitate this translation process ISO recommends the abstract syntax ASN.1.

The application layer provides information transfer and communication services to application service users (i.e. user programs). Different types of application (e.g. e-mail, file transfer, remote login, etc.) require different sets of services and the application layer is therefore constructed with a set of specific application service elements (SASEs) overlying a set of common application service elements (CASEs).

Communication between all layers in a protocol stack takes place using primitives. These are software procedures that pass instructions and data between adjacent layers. A PRIMITIVE.*request* at a source DTE (or network node) is passed from layer n to request a layer $n-1$ service. This results in a PRIMITIVE.*indication* being passed from layer $n-1$ to layer n at the destination DTE (or network node). A PRIMITIVE.*response* is passed from the layer n process to the layer $n-1$ process at the destination DTE, which results in a PRIMITIVE.*confirm* being passed from the layer $n-1$ process at the source DTE to the layer n process. In addition to carrying instructions and data between layers of the protocol stack, primitives are used to negotiate the quality of service provided by one layer to the layer above.

18.11 Problems

18.1 In which layer of the ISO OSI model does X.21 reside? Describe a typical X.21 data exchange event including call set-up and clear-down phases.

18.2 Describe what is meant by data-link control, and describe how the fields in a typical data frame allow the protocol to operate.

18.3 In the case where bit errors are independent, derive expressions for the average time to transmit a data frame and the transmission efficiency when using stop and wait ARQ.

18.4 A data-link has a path delay of 10 ms, a transmission rate of 60 kbyte/s and a bit error ratio of 10^{-4}. The data frame contains 228 bytes of information, a 2 byte start field, a 6 byte header and a 4 byte check field. The acknowledgement frame has an 8 byte data field. Calculate the effective data rate and average time for correct transmission using stop and wait ARQ.

18.5 Describe how the go-back-N automatic repeat request (ARQ) strategy operates.

18.6 In the case where bit errors are independent, derive expressions for the average time to transmit a data frame and the transmission efficiency when using go-back-N ARQ.

18.7 A data-link has a path delay of 10 μs and a transmission rate of 500 kbyte/s. The data frame has 72 bytes of information, a 2 byte start field, a 16 byte header and an 8 byte check field. The acknowledgement information is 'piggy-backed' on returning data frames. When using a four frame buffer the observed data rate is 800 kbit/s. Determine the average bit error rate on the link and the bounds on the processing time at the stations assuming go-back-N ARQ.

18.8 Assuming that bit errors are independent, derive expressions for average time to transmit a frame and the overall transmission efficiency for the selective repeat ARQ strategy.

18.9 A satellite link operates at a data rate of 100 Mbit/s, a path delay of 2.5 ms and a bit error ratio of 5×10^{-4}. The data frames have a 128 byte data field, a 1 byte start field, a 5 byte header and a 6 byte check field. The acknowledgement frame has a 2 byte data field. Calculate the average time for correct transmission and the effective data rate under the selective repeat ARQ.

18.10 What is meant by flow control in a communications network and what potential problem does it guard against?

18.11 Give a brief account of sliding window flow control. What is the lowest layer of the ISO OSI model that might incorporate flow control and what is the difference between its implementation at this layer and its implementation at higher layers?

18.12 Describe how the fields found in an HDLC data frame allow data-link control to be achieved.

18.13 If information is transmitted in 252 bit sections, determine the range of bit error rates where HDLC frames have an 80% probability of successful reception.

18.14 Describe the circumstances under which HDLC frames sent by a transmitter will fail to be received by the receiver at all, and for the worst case in the previous problem, calculate the proportion of frames that will not be received.

18.15 Explain what is meant by bit stuffing in the context of HDLC.

18.16 Routing algorithms are usually designed to minimise some cost function. List three quantities that might be involved in this function. Suggest four desirable properties of a distributed dynamic routing algorithm.

18.17 Which layer in the ISO OSI model has responsibility for routing?

18.18 Use the shortest path algorithm described in section 18.4.1 to find the optimum route between nodes A and D of the network defined in Example 18.4.

18.19 Explain the difference between a network cost matrix and a network capacity matrix.

18.20 Describe, briefly and in the context of network routing protocols, flooding, backward learning and hierarchical routing.

18.21 The terms flow control and congestion control are occasionally used synonymously. Explain the difference between them.

18.22 Describe, briefly what is meant by isarithmic congestion control. What long term problem does isarithmic control suffer from? Name one alternative congestion control strategy.

18.23 What part of a packet switched network uses the X.25 protocol? Describe the general form of an X.25 packet and identify the various component fields in its header. What is the principal function of each of the header fields?

18.24 Explain the difference between a connection-oriented and a connectionless protocol. Give an example of each.

18.25 What, in a CLNP packet, is the function of the PDU lifetime field and why is it needed? How is segmentation information incorporated in the CLNP packet?

18.26 Explain how addressing is accomplished in CLNP.

18.27 Explain the roles of level 1 and level 2 ISs in CLNS routing.

18.28 What is the process by which a source ES application creates, and then connects to, the corresponding application in a destination ES? In particular what are the roles of the process server and the ephemeral TSAP?

18.29 What distinguishes TP4 from the other four (TP0–TP3) ISO transport layer protocols? How is error control handled by TP4?

18.30 List four types of token used by the ISO session protocol and describe their role.

CHAPTER 19

Network performance using queueing theory

M. E. Woodward

19.1 Introduction

Queueing theory has been used for many years in assessing the performance of telecommunication networks. As early as 1917 the Danish mathematician Erlang used the theory in the design of telephone exchanges and developed the now famous Erlang-B and Erlang-C formulae, which have been used in many applications since that time [Erlang]. With the development and introduction of digital technology the circuit switching used in telephone networks, in which a dedicated end-to-end path was set up prior to the start of a communication, gradually began to be replaced by packet switching. In packet switched networks, source messages in the form of streams of bits are segmented into smaller blocks and a header containing control information is appended to each block. The information in the header relates to such things as sequence numbering, source and destination addressing, routing, priority, error detection and security. These blocks of data together with their headers are called packets and they can be individually transported and delivered by the network from source to destination without the need for setting up a dedicated end-to-end connection as in circuit switched networks.

Since the introduction of packet switching, queueing theory has assumed an even greater importance in the design and analysis of telecommunication networks. This stems from the natural analogy between a packet switched telecommunication network and a network of queues. This is illustrated in Figure 19.1, which shows a network of nodes interconnected by a number of directed links. This can be interpreted as a diagrammatic representation of a communication network where the nodes represent switches and the links represent communication channels. Packets enter the network at a specific node where they may be queued in a buffer to wait for an outgoing channel to become free. They are then transmitted to the next node according to the routing information in the header. They continue in this hop-by-hop manner until they finally arrive at their destination. An alternative interpretation of Figure 19.1 is that the nodes can be considered as queues where customers wait for a server to become free. The customers are then served and passed on to the next queue via one of the outgoing links. Clearly, the movement of packets through a packet switched network can be considered as analogous to the movement of customers through a network of queues. It follows therefore that the well defined techniques used to analyse queues can, via this analogy, be used to analyse the performance of packet switched networks.

19.1.1 The arrival process

The first requirement in any model of a packet switched communication network is to model the arrival of packets entering the system. The most common way to do this is to use a Poisson arrival process. It often turns out that a good approximation to the time between arrivals of packets entering a network is to use the exponential distribution. As will be seen, the Poisson arrival process models this type of interarrival time distribution exactly [Schwartz, 1987].

To develop the Poisson arrival process, with reference to Figure 19.2 consider a time interval Δt and let $\Delta t \to 0$. Then three basic statements can be used to define the Poisson arrival process. These are:

1. The probability of one arrival in Δt is:

   $P \text{ (one arrival in } \Delta t) \stackrel{\Delta}{=} \lambda \Delta t + O(\Delta t)$

   Here, $\lambda \Delta t \ll 1$ and λ is a specified proportionality constant. The symbol $O(\Delta t)$ means that higher order terms in Δt are negligible in that $O(\Delta t) \to 0$ faster than $\Delta t \to 0$. For example, if $\Delta t = 10^{-6}$ s then the higher order term $(\Delta t)^2 = 10^{-12}$ s, which is negligible compared with the former.

2. The probability of no arrivals in Δt is:

   $P \text{ (no arrivals in } \Delta t) = 1 - \lambda \Delta t + O(\Delta t)$

3. Arrivals are memoryless; that is, an arrival in one time interval of length Δt is independent of arrivals in previous or future time intervals.

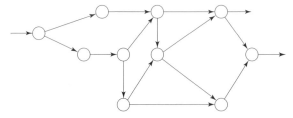

Figure 19.1 *Packet switched telecommunication network.*

Figure 19.2 *Poisson process: basic time interval.*

Then if we neglect the higher order terms we have:

P (one arrival in Δt) $= \lambda \Delta t$, and
P (no arrivals in Δt) $= 1 - \lambda \Delta t$

Now consider a large interval T and subdivide this into n intervals each of length $\Delta t = T/n$ as shown in Figure 19.3.

Then using the binomial distribution:

$$P \ (k \text{ arrivals in } T) \ = \ P(k) \ = \ \binom{n}{k} (\lambda \Delta t)^k \ (1 - \lambda \Delta t)^{n-k} \tag{19.1}$$

where:

$$\binom{n}{k} = \frac{n!}{(n-k)!k!}$$

as in equation (3.9). Replacing Δt with T/n in equation (19.1) and rearranging we have:

$$P(k) \ = \ \left(\frac{(\lambda T)^k}{k!} \right) \left(1 - \frac{\lambda T}{n} \right)^n \frac{n(n-1)\ldots(n-k+1)}{n^k} \left(1 - \frac{\lambda T}{n} \right)^{-k} \tag{19.2}$$

Now let $n \to \infty$ with k remaining fixed. Then the following apply:

$$\left(1 - \frac{\lambda T}{n} \right)^n \ \to \ e^{-\lambda T}$$

$$\frac{n (n-1)\ldots(n-k+1)}{n^k} \ \to \ 1$$

$$\left(1 - \frac{\lambda T}{n} \right)^{-k} \ \to \ 1$$

Then substituting the above in equation (19.2) we have:

$$P(k) \ = \ \frac{(\lambda T)^k}{k!} \ e^{-\lambda T}, \quad k = 0, 1, 2, \cdots, \infty \tag{19.3}$$

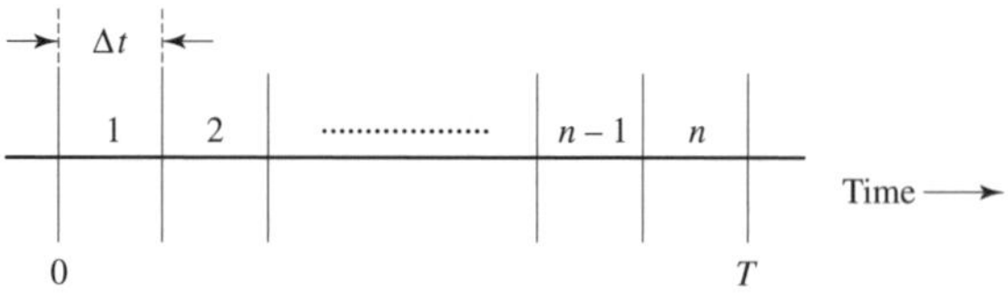

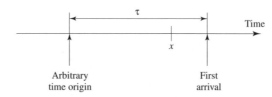

Figure 19.3 *Poisson process: time intervals in [0, T].*

Figure 19.4 *Poisson process: interarrival time derivation.*

This is the well known Poisson distribution and gives the probability of k packet arrivals in a time interval T.

To compute the mean (or average) number of packet arrivals in time T, then using equation (3.16):

$$E[k] = \sum_{k=0}^{\infty} kP(k) = \lambda T e^{-\lambda T} \sum_{k=1}^{\infty} \frac{(\lambda T)^{k-1}}{(k-1)!}$$

Since $\displaystyle\sum_{k=1}^{\infty} \frac{(\lambda T)^{k-1}}{(k-1)!} = e^{\lambda T}$ then clearly $E[k] = \lambda T$.

The constant λ can thus be interpreted as the mean number of arrivals per unit time and is called the *arrival rate*.

One of the most interesting (and useful) properties of the Poisson distribution when interpreted as an arrival process turns out to be the distribution of time between arrivals, called the *interarrival time distribution*. Now let us calculate this.

Consider the diagram shown in Figure 19.4 and let τ be a random variable representing the first packet arrival after some arbitrary time origin and take any value x on the time axis. Then no packets arrive in $(0, x)$ if, and only if, $\tau > x$. Then we can write:

$P(\tau > x) = P$ (zero arrivals in $(0, x)$)

That is: $P(\tau > x) = P(0) = e^{-\lambda x}$

Then: $P(\tau \le x) = 1 - e^{-\lambda x}$

This is the cumulative distribution function of τ, see section 3.2.4, which is thus *exponentially distributed*, Figure 19.5.

In summary then, a Poisson arrival process implies that:

1. The *number* of arrivals has a Poisson distribution with mean rate λ arrivals per unit time.
2. The *time* between arrivals is exponentially distributed with a mean value of $1/\lambda$ time units.

The exponential distribution is *memoryless*, which means that the time until the next arrival is independent of the time since the last arrival and hence is independent of the past. This turns out to be a very useful property in modelling communication networks. The theory of Markov processes is based around this property; hence it is also called the Markov property [Schwartz, 1987]. Many queueing systems are also Markov processes or some variant of these and hence the importance of the memoryless property and the

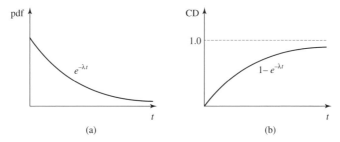

Figure 19.5 *(a) Pdf; and (b) CD for time between successive arrivals.*

Poisson distribution in modelling communication networks. Figure 19.5 shows the pdf and CD for the time between successive arrivals.

19.1.2 Queueing systems

Queueing systems are usually described using *Kendall's notation* [Kendall]. This consists of five fields each separated by a forward slash as a/b/c/d/e where:

(a) corresponds to the arrival process,
(b) corresponds to the service process,
(c) specifies the number of servers,
(d) specifies the system capacity (the queue capacity plus the number of servers),
(e) specifies the maximum number of potential customers.

The first field, a, can be replaced by one of a number of descriptors. Some of the more common ones used are:

M: This stands for memoryless, which implies exponentially distributed interarrival times; that is, a Poisson arrival process.

D: This stands for deterministic, which means there is a constant time between arrivals.

G: This stands for general, which embodies any arrival process.

Geo: This stands for geometric, which is the discrete equivalent of the exponential interarrival time distribution.

The second field, b, can be replaced by any of the descriptors used in the first field except that these descriptors now relate to the service time distribution rather than the inter-arrival time distribution. The third field, c, can be replaced by a positive integer or infinity. The fourth and fifth fields, d and e respectively, again can contain positive integers. Often one or the other of these fields is omitted which, if that is the case, leads to the corresponding field being considered infinite by default. Examples of using the Kendall notation are:

M/M/1: This implies a queue having Poisson arrivals, exponential service times, a single server, infinite capacity and potentially infinite customer population.

M/D/N/K: This implies a queue having Poisson arrivals, deterministic (constant) service times (as in fixed length packets), N servers, a capacity of K customers and a potentially infinite customer population.

Geo/D/1//N: This implies a queue with geometrically distributed interarrival times, deterministic service times, a single server, infinite capacity and a maximum customer population of N.

Also to be taken into account is the service discipline, which is not specified in the Kendall notation. This specifies the manner and order in which the customers are served. Some common service disciplines are:

FCFS: First come first served (also called FIFO: first in first out).

LCFS: Last come first served (this operates like a push down stack).

PS: Processor sharing. All customers in the queue are given an equal amount of service in a given time.

19.2 The M/M/1 queue

A diagrammatic representation of an M/M/1 queue is shown in Figure 19.6. Arrivals are Poisson with mean rate λ and service time is exponential with mean value $1/\mu$. Defining the state of the queue to be the number of customers in the queue [Gelenbe and Pujolle] then the state transition diagram describing the behaviour of the queue is shown in Figure 19.7. When in state 0 the queue is empty and an arrival can occur with probability $\lambda \Delta t$. In state 1 there is a single customer in the queue and this can be served with probability $\mu \Delta t$, in which case the queue goes to state 0, or an arrival can occur in which case the queue goes to state 2. The rest of the diagram can be built up in a similar way and since the queue has unlimited capacity there are an infinite number of states. This diagram represents a continuous time Markov process called a *birth–death* process [Chung].

19.2.1 The equilibrium probabilities

If we select two arbitrary states k and $k + 1$, then transitions between these are shown in Figure 19.8. Let $P(k)$, $k = 0,1,2, \cdots$, be the probability of finding the queue in state k if it is observed at a random point in time. (State k implies there are k customers awaiting service.) Now if we balance the probability flows across the boundary between the two states, shown as a dashed line in Figure 19.8, then we can write a *balance equation* as follows:

$$\lambda \Delta t P(k) \;=\; \mu \Delta t P(k + 1) \quad \text{or} \quad \lambda P(k) \;=\; \mu P(k + 1), \qquad k \geq 0$$

Then writing $\lambda/\mu = \rho$ we have:

$$P(k + 1) = \rho P(k), \qquad k \geq 0$$

Substituting $k = 0$ in the above then:

$$P(1) = \rho P(0)$$

Now substituting $k = 1$:

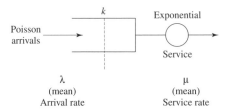

Figure 19.6 *M/M/1 queue.*

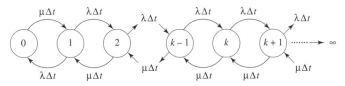

Figure 19.7 *State transition diagram of M/M/1 queue.*

$$P(2) = \rho P(1) = \rho^2 \, P(0)$$

Continuing in this way, in general we have:

$$P(k) = \rho^k \, P(0), \quad k \geq 0 \tag{19.4}$$

Figure 19.9 shows these queue length probabilities, plotted against time. Now since:

$$\sum_{k=0}^{\infty} P(k) = 1 \tag{19.5}$$

then substituting equation (19.4) into (19.5) gives:

$$P(0) \sum_{k=0}^{\infty} \rho^k = 1 \tag{19.6}$$

Then providing $\rho < 1$, summing the infinite geometric series (see Appendix B for some common sums of series) we have:

$$P(0) = 1 - \rho \tag{19.7}$$

and finally using equation (19.7) in (19.4) gives:

$$P(k) = \rho^k \, (1 - \rho), \quad k \geq 0 \tag{19.8}$$

This is therefore the equilibrium distribution of the system contents for the M/M/1 queue. Using equation (19.3) then Figure 19.9 shows how the probability of the number of arrivals to the M/M/1 queue changes with increasing time, t.

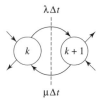

Figure 19.8 *Arbitrary state in M/M/1 queue.*

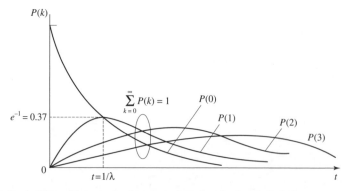

Figure 19.9 *Probability of k arrivals plotted against time for k = 0, 1, 2, 3.*

19.2.2 Performance measures

Having found the equilibrium probabilities, these can then be used to determine the system *performance measures* such as the mean throughput, the mean system delay and mean queueing delay [Nussbaumer].

The *mean throughput* is defined as the mean number of customers per unit time passing through the system. This definition applies to any system. In the case of the M/M/1 queue the server is operating at rate μ whenever the system is non-empty. Then mean throughput S is given by:

$$S = \mu \sum_{k=1}^{\infty} P(k) = \mu(1 - P(0)) = \lambda \tag{19.9}$$

This equation says that the rate at which packets go through the system is equal to the rate at which they arrive, which intuitively makes obvious sense if the system is in equilibrium.

The *mean system delay*, W, is defined as the mean time which elapses between a customer entering and leaving the system. The unit of delay is therefore the unit of time. It is usual to calculate delay indirectly by the use of *Little's result* [Little]. This states that the mean number of customers in a system, L, is equal to the product of the mean system delay and the mean throughput. That is:

$$L = \lambda W \tag{19.10}$$

The mean system contents, L, is given by:

$$L = \sum_{k=0}^{\infty} kP(k) = \rho(1 - \rho) \sum_{k=1}^{\infty} k\rho^{(k-1)}$$

Then providing $\rho < 1$ the arithmetic–geometric series sums to $1/(1 - \rho)^2$ (Appendix B) and so:

$$L = \frac{\rho}{1 - \rho} \tag{19.11}$$

Then using Little's result:

$$W = \frac{1}{\lambda} \frac{\rho}{1 - \rho} = \frac{1}{\mu - \lambda} \tag{19.12}$$

This is the mean delay through the system, which in the case of the M/M/1 queue implies the mean time a customer takes to pass through the queue plus the server.

A related performance measure is the *mean queueing delay*, Q, which is the mean time a customer spends waiting in the queue. This is simply the mean system delay minus the mean service time. That is:

$$Q = W - \frac{1}{\mu} = \rho W \tag{19.13}$$

In the context of a communication network we often normalise delay to units of mean packet transmission time $(1/\mu)$. Throughput is also usually normalised to the transmission capacity of the system (μ packets per unit time for the M/M/1 queue). Thus normalised throughput is simply $\lambda/\mu = \rho$.

Plots of normalised delay against normalised throughput provide a means of comparing the performances of different systems on a like basis. The normalised delay–throughput curve for the M/M/1 queue is shown in Figure 19.10. The theoretical ideal curve for a single server queue is shown by the dashed line. This is practically unrealisable but the closer we can get to this ideal curve the more efficient is the system.

EXAMPLE 19.1

For an M/M/1 queue with arrival rate λ and service rate μ, if the arrival rate remains constant by how much must the service rate be increased to halve the mean packet delay?

Call the original mean packet delay W_1 and the original service rate μ_1. From equation (19.12):

$$W_1 = \frac{1}{\mu_1 - \lambda}$$

Now let the service rate be increased to a value μ_2 and call the new delay W_2. Then:

$$W_2 = \frac{1}{\mu_2 - \lambda}$$

Now if $W_2 = W_1/2$ then:

$$\frac{1}{\mu_2 - \lambda} = \frac{1}{2(\mu_1 - \lambda)}$$

This gives $\mu_2 = 2\mu_1 - \lambda$. The increase in service rate is therefore:

$$\mu_2 - \mu_1 = \mu_1 - \lambda$$

Thus to halve the mean packet delay while keeping the arrival rate fixed we must increase the service rate by an amount equal to the difference between the arrival rate and the original service rate. In other words, the service rate must be increased by an amount exactly equal to the spare capacity of the original transmission channel.

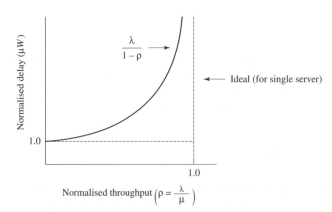

Figure 19.10 *Normalised delay–throughput curve for M/M/1 queue.*

19.3 The M/M/1/N queue

In addition to the mean values of the system delay, the queueing delay and the throughput, the other main performance measure in the context of a communication network is the *probability of packet loss*, P_L. Clearly this is zero for an infinite buffer system such as the M/M/1 queue. Although infinite buffer systems can be used to give good performance approximations for systems with very large buffers, in practice all systems, of course, have finite buffers and so P_L likewise takes on a non-zero value that can prove very important in the design of a communication network. To see how to calculate P_L for a finite buffer system we shall first modify the M/M/1 queue to have finite capacity and then use this as an example.

The M/M/1/N queue is identical with the M/M/1 queue except that it has a finite capacity of N customers and so the state transition diagram has to be modified to have only 0 to N states, as shown in Figure 19.11. Note that the Δt have been omitted since these appear on either side of the balance equations and so cancel out. It is usual therefore to label transitions of a continuous time Markov process with rates rather than probabilities.

The balance equations solve in an identical way to the M/M/1 queue to give the equilibrium probability in terms of $P(0)$ as:

$$P(k) = \rho^k P(0), \quad 0 \le k \le N \tag{19.14}$$

The difference comes in the way $P(0)$ is calculated. In this case:

$$\sum_{k=0}^{N} P(k) = 1$$

or:

$$P(0) \sum_{k=0}^{N} \rho^k = 1$$

Summing the finite geometric series (Appendix B) we have:

$$P(0) \frac{1 - \rho^{N+1}}{1 - \rho} = 1, \quad \rho \ne 1 \tag{19.15}$$

and:

$$P(0) (N + 1) = 1, \quad \rho = 1 \tag{19.16}$$

Then using equations (19.15) and (19.16) in (19.14) the equilibrium distribution of the M/M/1/N queue is given by:

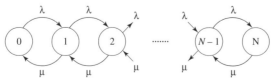

Figure 19.11 *State transition diagram of M/M/1/N queue.*

$$P(k) = \begin{cases} \rho^k \dfrac{(1 - \rho)}{1 - \rho^{N+1}}, & \rho \neq 1 \\[2mm] \dfrac{\rho^k}{(N+1)}, & \rho = 1 \end{cases} \qquad 0 \leq k \leq N \qquad (19.17)$$

In this case we can find the probability of packet loss, P_L, defined as the probability that an arriving packet finds a full buffer. For the M/M/1/N queue this is given by:

$$P_L = P(N) = \begin{cases} \dfrac{\rho^N (1 - \rho)}{1 - \rho^{N+1}}, & \rho \neq 1 \\[2mm] \dfrac{1}{(N+1)}, & \rho = 1 \end{cases} \qquad (19.18)$$

P_L is shown plotted against ρ in Figure 19.12. As ρ increases to infinity then P_L becomes asymptotic to 1. Once ρ exceeds 1 then the arrival rate becomes greater than the service rate and the probability of packet loss increases rapidly. This is known as the *region of congestion*.

Since $(1 - P_L)$ is equivalent to the fraction of arriving packets that are accepted, the normalised throughput, S, is now given by:

$$S = \lambda(1 - P_L)$$

Similarly, when calculating delay for systems with finite buffers, S replaces λ in Little's result, which thus becomes:

$$L = SW \qquad (19.19)$$

Thus system delay, W, can be calculated by first finding the average system population, L, and then dividing this by the throughput.

Then for the M/M/1/N queue system delay is now:

$$W = \frac{L}{S} = \frac{1}{S} \sum_{k=0}^{N} kP(k) \qquad (19.20)$$

with $P(k)$ as given above in equation (19.17).

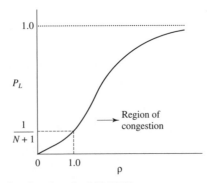

Figure 19.12 *Probability of packets loss for M/M/1/N queue.*

EXAMPLE 19.2

A buffer in a node of a packet switched communication network can be modelled as an M/M/1/N queue. If the arrival rate is 2000 packets/s find the probability of packet loss if mean packet length is 1000 bits and the communication channel operates at a rate of 2.5 Mbit/s. Also calculate the mean packet delay through the node. It may be assumed that the buffer can hold nine average length packets.

Given that channel bit rate is 2.5 Mbit/s and mean packet length 1000 bits then service rate is:

$$\mu = \frac{2.5 \times 10^6}{1000} = 2500 \text{ packets/s}$$

$$\text{Then } \rho = \frac{\lambda}{\mu} = \frac{2000}{2500} = 0.8$$

Probability of packet loss is then given by equation (19.18) as:

$$P_L = \frac{\rho^N (1 - \rho)}{1 - \rho^{N+1}}$$

Note that $N = 10$ in the above equation since this is *system* capacity and not simply buffer capacity. The system can thus hold nine packets in the buffer and one in the server. Then substituting $\rho = 0.8$, $N = 10$ gives probability of packet loss as $P_L = 0.0235$.

The system throughput, S, is now given by:

$$S = \lambda(1 - P_L) = 1953 \text{ packets/s}$$

Then using equations (19.17) and (19.20) the mean packet delay through the node is:

$$W = \frac{1}{S} \left(\frac{1 - \rho}{1 - \rho^{N+1}} \right) \sum_{K=1}^{N} k\rho^k$$

Now using Appendix B for the summation of the series the mean packet delay becomes:

$$W = \frac{\rho \left\{ 1 - [1 + N(1 - \rho)] \rho^N \right\}}{S(1 - \rho)(1 - \rho^{N+1})}$$

Substituting for N, ρ and S as 10, 0.8 and 1953 respectively gives $W = 1.42$ ms.

19.3.1 General Markovian queueing equations

The two queueing systems considered so far, the M/M/1 queue and the M/M/1/N queue, both have a potentially infinite customer population and a single server. The former characteristic means that the arrival rate is constant irrespective of the state of the queueing system. Similarly, the single server means that the service rate is also constant irrespective of the state of the queueing system. In more general systems where the customer population is finite and there are multiple servers then both the arrival rate and the service rate can depend on the state of the system. We shall next consider these more general systems under the assumptions that both interarrival times and service times

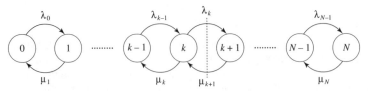

Figure 19.13 *State transition diagram for a general Markovian queueing system.*

remain exponentially distributed; that is, they are Markovian systems [Chung].

Let the queueing system have $N+1$ states labelled $0,1,2, \cdots, N$, and let the arrival rate in state k, $0 \le k \le N-1$, be λ_k. Similarly, let the service rate in state k, $1 \le k \le N$, be μ_k. The state transition diagram for this generalised system then appears as in Figure 19.13, compared to the earlier Figure 19.11. Equating probabilities of boundary crossings between states k and $k+1$ we have the following balance equations:

$$\lambda_k P(k) = \mu_{k+1} P(k+1), \quad 0 \le k \le N-1 \tag{19.21}$$

Successive application of this equation for $k = 0,1,2, \cdots$ gives:

$$P(k) = \frac{\lambda_0 \lambda_1 \cdots \lambda_{k-1}}{\mu_1 \mu_2 \cdots \mu_k} P(0)$$

More compactly, using the notation defined in Figure 3.20, we can write:

$$P(k) = P(0) \frac{\displaystyle\prod_{i=0}^{k-1} \lambda_i}{\displaystyle\prod_{i=1}^{k} \mu_i}, \quad 1 \le k \le N \tag{19.22}$$

By selecting suitable values for the λ_i and μ_i the equilibrium probabilities for any queueing system with exponential interarrival times and exponential service times can be obtained in terms of $P(0)$. It is then a matter of normalising over the entire range of probabilities to find the value of $P(0)$ and hence the absolute values of the equilibrium distribution.

EXAMPLE 19.3

Show that the equilibrium probability of finding k packets in an M/M/2 queue with mean arrival rate λ packets per unit time and mean service rate μ packets per unit time *for each server* is:

$$P(0) = \frac{(1-\rho)}{(1+\rho)} \quad \text{with } \rho = \frac{\lambda}{2\mu}$$

$$P(k) = \frac{2(1-\rho)}{(1+\rho)} \rho^k \quad \text{for } k \ge 1$$

Use the above in conjunction with Little's result and the summations of series given in Appendix B to find the ratio of the mean packet delay in an M/M/1 queue to that in an M/M/2 queue having the same total mean service rate and the same mean arrival rate. The ratio should be expressed in terms of the traffic intensity parameter ρ which is assumed to be less than 1.

The best way to solve this problem is to substitute the corresponding arrival rates and service rates for an M/M/2 queue into equation (19.22) to find $P(k)$ as a function of $P(0)$ and then normalise to find $P(0)$.

Arrival rates are $\lambda_k = \lambda$ for $k \geq 0$.

Service rates are $\mu_1 = \mu$ and $\mu_k = 2\mu$ for $k \geq 2$.

Substituting these in equation (19.22), writing $\rho = \lambda/2\mu$ and simplifying then:

$$P(k) = 2\rho^k P(0), \quad k \geq 1 \tag{19.23}$$

We next normalise over the above probabilities to find $P(0)$ as follows:

$$\sum_{k=0}^{\infty} P(k) = P(0) + 2P(0) \sum_{k=1}^{\infty} \rho^k = 1$$

The term $2P(0) \sum_{k=1}^{\infty} \rho^k$ can be rearranged as $2P(0) (\sum_{k=0}^{\infty} \rho^k - 1) = (2P(0)\rho)/(1 - \rho)$ where the last equality was obtained from the fact that $\rho < 1$ and using Appendix B for the series summation.

Then we have $P(0) + 2P(0)\rho/(1 - \rho) = P(0)(1 + \rho)/(1 - \rho) = 1$ and so finally:

$$P(0) = \frac{(1 - \rho)}{(1 + \rho)} \tag{19.24}$$

Substituting this back into equation (19.23) then we have:

$$P(k) = \frac{2(1 - \rho)}{(1 + \rho)} \rho^k, \quad k \geq 1 \tag{19.25}$$

The mean number of packets in the system for this M/M/2 queue is given by:

$$L = \sum_{k=0}^{\infty} kP(k) = \frac{2\rho(1 - \rho)}{(1 + \rho)} \sum_{k=1}^{\infty} k\rho^{k-1} = \frac{2\rho}{(1 + \rho)(1 - \rho)}$$

where the last equality was obtained again from the fact that $\rho < 1$ and using Appendix B for the series summation. Using Little's result then mean packet delay, W, is:

$$W = \frac{2\rho}{\lambda(1 + \rho)(1 - \rho)}$$

Call this W(M/M/2). From equation (19.12) the mean packet delay for an M/M/1 queue is:

$$W = \frac{\rho}{\lambda(1 - \rho)}$$

Call this W(M/M/1). It should be remembered that in this case $\rho = \lambda/2\mu$ so that both queues have the same total mean service rate of 2μ.

Then the ratio W(M/M/1)$/W$(M/M/2) is $(1 + \rho)/2$.

Since $\rho < 1$ then clearly the M/M/1 queue gives the lowest delay. The reason for this result which, at first sight, may seem rather surprising is that when both queues have only one packet in the system then the M/M/2 queue serves this packet at a rate μ whereas the M/M/1 queue serves the packet twice as fast at rate 2μ. This results in the lower mean packet delay for the M/M/1 queue since for all other states the two queues are statistically equivalent.

One lesson learned from this is that under conditions of exponentially distributed packet interarrival and service times a single server queue will always give the better delay characteristics than any multiple server queue having the same total mean service rate.

This result carries through into many practical situations, one of which arose some years ago in the context of the Internet and on which the following example is based.

EXAMPLE 19.4

In the mid 1990s the UK academic network (JANET) was connected to the Internet in the USA via a single transatlantic link operating at 2 Mbit/s. The restricted capacity was causing congestion and outgoing traffic from the UK was subject to some considerable queueing delays. To alleviate this situation there was a proposal to upgrade the link to 8 Mbit/s. You are asked to consider the following two possible scenarios for such an upgrade:

(a) Replace the 2 Mbit/s link with a single link operating at 8 Mbit/s.
(b) Replace the 2 Mbit/s link with two links each operating at 4 Mbit/s.

If the packets are queued before transmission, the mean input load is equivalent to 2M bits/s with exponential packet interarrival times and packet lengths are exponentially distributed with a mean length of 1000 bits, calculate the absolute values of mean packet delay for both scenarios.

Clearly, scenario (a) can be considered as an M/M/1 queue and scenario (b) can be considered as an M/M/2 queue with the same total mean service rate as the M/M/1 queue. The relevant formulae have therefore already been derived in Example 19.3 and it is a matter of working out the absolute values of the system parameters and substituting in the formulae.

Denoting the mean packet arrival rate for both systems by λ then:

$$\lambda = \frac{2 \times 10^6}{1000} = 2 \times 10^3 \text{ packets/s}$$

Denoting the total mean service rate for both systems as 2μ then:

$$2\mu = \frac{2 \times 4 \times 10^6}{1000} = 8 \times 10^3 \text{ packets/s}$$

$$\text{Then } \rho = \frac{\lambda}{2\mu} = \frac{2 \times 10^3}{8 \times 10^3} = 0.25$$

The absolute values of mean packet delay can now be evaluated as follows:

$$W(\text{M/M/1}) = \frac{\rho}{\lambda(1-\rho)} = \frac{0.25}{2 \times 10^3 \, (1-0.25)} = 0.167 \times 10^{-3} \text{ s}$$

$$W(\text{M/M/2}) = \frac{2\rho}{\lambda(1+\rho)\,(1-\rho)} = \frac{2 \times 0.25}{2 \times 10^3 \, (1+0.25)\,(1-0.25)} = 0.267 \times 10^{-3} \text{ s}$$

Thus the single server system of scenario (a) has a mean packet delay of the order of 100 microseconds less than the two-server system of scenario (b), even though the total service capacity is the same in both cases. Delays in both systems are small, as expected for a situation where the total mean service rate is four times as large as the mean arrival rate.

If the arrival rate increases while keeping the service capacity fixed then the difference in delay performance of the two queues gradually diminishes although the delay for both, of course, increases. As the arrival rate approaches the service rate then the delay performances of the two queues tends to merge.

19.3.2 The M/M/N/N queue

The M/M/N/N queue gives rise to a famous result known as the *Erlang-B equation* which has been extensively used for many years to determine the *blocking probability* in circuit switched networks such as telephone exchanges. The state transition diagram is shown in Figure 19.14. As the M/M/N/N queue has N servers and hence a capacity of N customers it is really a queue without a queue! That is, the number of servers is equal to the system capacity and so an arriving customer is either accepted if there is at least one free server or rejected if all servers are occupied.

In a circuit switched network, arrivals are not packets but requests for a connection and are traditionally referred to as *call requests*. We assume a total mean arrival rate of λ calls/unit time which is Poisson distributed and that call lengths are exponentially distributed with mean $1/\mu$. Here:

$$\mu_k = k\mu, \qquad 1 \le k \le N \tag{19.26}$$

$$\lambda_k = \lambda, \qquad 0 \le k \le N - 1 \tag{19.27}$$

This latter is due to the infinite source assumption implicit in the Poisson distribution, i.e. the fact that a call has arrived from one of an infinite number of sources makes no difference to the total mean arrival rate. Note in this queue that the service rate now changes from state to state, with the N servers.

Using equations (19.22), (19.26) and (19.27) then we have:

$$P(k) = P(0) \frac{\lambda^k}{\mu^k \, k!} = P(0) \frac{\rho^k}{k!} \tag{19.28}$$

The normalising equation gives:

$$\sum_{k=0}^{N} P(k) = 1 = P(0) \sum_{k=0}^{N} \frac{\rho^k}{k!} \tag{19.29}$$

Thus, using this in equation (19.28) then:

$$P(k) = \frac{\dfrac{\rho^k}{k!}}{\displaystyle\sum_{i=0}^{N} \dfrac{\rho^i}{i!}}, \qquad 0 \le k \le N \tag{19.30}$$

The main parameter of interest in circuit switched systems is the blocking probability P_B. This is the probability that all servers are occupied and that any further call requests that arrive will be rejected (or queued if the system has a buffer). This is:

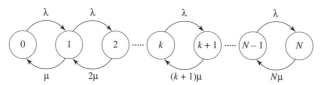

Figure 19.14 *State transition diagram for M/M/N/N queue.*

$$P_B = P(N) = \frac{\dfrac{\rho^N}{N!}}{\displaystyle\sum_{i=0}^{N} \dfrac{\rho^i}{i!}} \tag{19.31}$$

This is called the Erlang-B equation [Erlang], and is usually written in the form:

$$P_B = \frac{\dfrac{A^N}{N!}}{\displaystyle\sum_{k=0}^{N} \dfrac{A^k}{k!}} \tag{19.32}$$

where A is the offered traffic in *Erlangs* [Dunlop and Smith]. We see that:

$$A = \rho = \frac{\lambda}{\mu} \text{ (Erlangs)}$$

Thus the *offered* traffic in Erlangs is the total mean arrival rate (calls/unit time) normalised to the transmission capacity of a *single* link (or server).

The carried traffic in Erlangs is the throughput normalised to the transmission capacity of a *single* link (or server).

The difference between the offered traffic and the carried traffic is that lost due to blocking, i.e. traffic carried = $A(1 - P_B)$.

EXAMPLE 19.5

Call requests arrive at a railway timetable enquiry office at a mean rate of four every minute with exponential interarrival times. These are dealt with by four operators each of which takes, on average, 30 s to handle an enquiry the lengths of which are exponentially distributed. If an operator is not free to handle a request then the call is blocked and a busy tone is returned. What is the probability that a call is blocked?

The situation described can be modelled as an M/M/4/4 queue. This can be analysed using the Erlang-B formula with mean call arrival rate of 4 calls/minute and total mean service rate of 8 calls/minute or 2 calls/minute for each server.

The system parameters for the Erlang-B formula are thus:

$N = 4$, $\lambda = 4$ calls/minute and $\mu = 2$ calls/minute giving $\lambda/\mu = A = 2$ Erlangs.

The blocking probability as given by the Erlang-B formula, Equation (19.32), is:

$$P_B = \frac{\dfrac{A^N}{N!}}{\displaystyle\sum_{k=0}^{N} \dfrac{A^k}{k!}}$$

Substituting the requisite values of N and A gives a blocking probability of 0.095.

19.3.3 The M/M/N/N/K queue

The M/M/N/N/K queue considers the number of sources to be some finite number k, and is applicable to designing circuit switched networks or switches, such as that shown in Figure 19.15.

Here the arrival rate depends on the number of calls in progress and is given by:

$$\lambda_k = (K - k)\lambda, \quad 0 \leq k \leq N - 1 \tag{19.33}$$

The service rate is correspondingly given by:

$$\mu_k = k\mu, \quad 1 \leq k \leq N \tag{19.34}$$

Using these in equation (19.22) we have:

$$P(k) = \left(\frac{\lambda}{\mu}\right)^k \binom{K}{k} P(0), \quad 0 \leq k \leq N \tag{19.35}$$

where again the notation is given in equation (3.9). The normalising equation gives:

$$\sum_{k=0}^{N} P(k) = P(0) \sum_{i=0}^{N} \left(\frac{\lambda}{\mu}\right)^i \binom{K}{i} = 1 \tag{19.36}$$

Thus:

$$P(k) = \frac{\left(\frac{\lambda}{\mu}\right)^k \binom{K}{k}}{\sum_{i=0}^{N} \left(\frac{\lambda}{\mu}\right)^i \binom{K}{i}}, \quad 0 \leq k \leq N \tag{19.37}$$

This is called the *Engset distribution*. From this the blocking probability is given by:

$$P_B = P(N) = \frac{\left(\frac{\lambda}{\mu}\right)^N \binom{K}{k}}{\sum_{i=0}^{N} \left(\frac{\lambda}{\mu}\right)^i \binom{K}{i}} \tag{19.38}$$

The probability that all servers are occupied is also called the *time congestion* in circuit switching terminology.

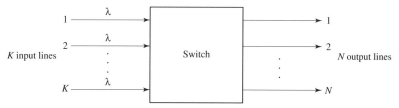

Figure 19.15 *Diagrammatic representation of K input, N output switch.*

19.3.4 M/M/N/N+J queue (Erlang-C equation)

In many systems, when all servers are occupied call requests are placed in a queue to wait for a server to become free; such systems are called *waiting call* systems. These systems can be modelled as an M/M/N/N+J queue, where J is the buffer size, i.e. the buffer can hold, at most, J call requests and there can be, at most, N calls in progress. In this situation it is sometimes convenient to separate out N and J rather than lump them together as system capacity.

Here we have:

$$\lambda_k = \lambda, \qquad 0 \le k \le N + J - 1 \qquad\qquad (19.39)$$

$$\mu_k = k\mu, \qquad 1 \le k \le N \qquad\qquad (19.40)$$

$$\mu_k = N\mu, \qquad N + 1 \le k \le N + J \qquad\qquad (19.41)$$

see Figure 19.16. Note that when all N servers are occupied, then the service rate must remain constant at $N\mu$, even though there may be up to $N + J$ calls in the system. Using equations (19.39), (19.40) and (19.41) in (19.22) we have:

$$
\left.
\begin{aligned}
P(1) &= \rho P(0) \\[4pt]
P(2) &= \frac{\rho^2}{2!} P(0) \\
&\quad. \\
&\quad. \\
&\quad. \\
P(N) &= \frac{\rho^N}{N!} P(0)
\end{aligned}
\right\}
\qquad
P(k) = \frac{\rho^k}{k!} P(0), \qquad\qquad 0 \le k \le N
$$

$$
\left.
\begin{aligned}
P(N+1) &= \left(\frac{\rho}{N}\right) \frac{\rho^N}{N!} P(0) \\[4pt]
P(N+2) &= \left(\frac{\rho}{N}\right)^2 \frac{\rho^N}{N!} P(0) \\
&\quad. \\
&\quad. \\
&\quad. \\
P(N+J) &= \left(\frac{\rho}{N}\right)^J \frac{\rho^N}{N!} P(0)
\end{aligned}
\right\}
\qquad
P(k) = \left(\frac{\rho}{N}\right)^{k-N} \frac{\rho^N}{N!} P(0), \quad N+1 \le k \le N+J
$$

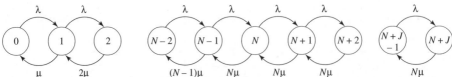

Figure 19.16 *State transition diagram for M/M/N/N+J queue.*

From the normalising equation:

$$\sum_{k=0}^{N+J} P(k) = 1$$

or, $P(0) \left[1 + \rho + \cdots + \frac{\rho^{N-1}}{(N-1)!} + \frac{\rho^N}{N!} + \left(\frac{\rho}{N}\right)\left(\frac{\rho^N}{N!}\right) + \cdots + \left(\frac{\rho}{N}\right)^J \frac{\rho^N}{N!} \right] = 1$

$$\therefore P(0) \left[\sum_{i=0}^{N-1} \frac{\rho^i}{i!} + \frac{\rho^N}{N!} \sum_{i=0}^{J} \left(\frac{\rho}{N}\right)^i \right] = 1$$

For an infinite queue $J \to \infty$ and if $\rho < N$ then summing the infinite geometric series we have:

$$P(0) \left[\sum_{i=0}^{N-1} \frac{\rho^i}{i!} + \frac{\rho^N}{N!} \left(\frac{N}{N-\rho}\right) \right] = 1$$

or, $P(0) \left\{ \sum_{i=0}^{N} \frac{\rho^i}{i!} + \frac{\rho^N}{N!} \left[\left(\frac{N}{N-\rho}\right) - 1 \right] \right\} = 1$

or, $P(0) \left[\sum_{i=0}^{N} \frac{\rho^i}{i!} + \frac{\rho^N}{N!} \left(\frac{\rho}{N-\rho}\right) \right] = 1$

Then the probability of k calls in the system is:

$$P(k) = \frac{\dfrac{\rho^k}{k!}}{\left[\displaystyle\sum_{i=0}^{N} \frac{\rho^i}{i!} + \frac{\rho^N}{N!} \left(\frac{\rho}{N-\rho}\right) \right]}, \qquad 0 \le k \le N \tag{19.42}$$

and:

$$P(k) = \frac{\dfrac{\rho^k}{k!} \left(\dfrac{\rho}{N}\right)^{k-N}}{\left[\displaystyle\sum_{i=0}^{N} \frac{\rho^i}{i!} + \frac{\rho^N}{N!} \left(\frac{\rho}{N-\rho}\right) \right]}, \qquad k \ge N \tag{19.43}$$

Blocking probability is then given by:

$$P_B = \sum_{k=N}^{\infty} P(k) = \frac{\dfrac{\rho^N}{N!} \left(\dfrac{N}{N-\rho}\right)}{\left[\displaystyle\sum_{i=0}^{N} \frac{\rho^i}{i!} + \frac{\rho^N}{N!} \left(\frac{\rho}{N-\rho}\right) \right]} \tag{19.44}$$

where series B.2 in Appendix B has been used to sum the numerator of equation (19.43) over k. This is called the Erlang-C equation. For a finite buffer of size J then:

$$P_B = \sum_{k=N}^{J} P(k) = P(0) \frac{\rho^N}{N!} \sum_{k=N}^{J+N} \left(\frac{\rho}{N}\right)^{k-N}$$

$$= P(0) \frac{\rho^N}{N!} \sum_{k=0}^{J} \left(\frac{\rho}{N}\right)^k$$

$$= \begin{cases} P(0) \dfrac{\rho^N}{N!} \left(\dfrac{1 - \left(\dfrac{\rho}{N}\right)^{J+1}}{1 - \dfrac{\rho}{N}} \right), & \rho \neq N \\[6ex] P(0) \dfrac{\rho^N}{N!} (J + 1), & \rho = N \end{cases} \tag{19.45}$$

where B.1 of Appendix B has been used to sum the series.

$$\text{Now } P(0) = \begin{cases} \dfrac{1}{\displaystyle\sum_{i=0}^{N-1} \dfrac{\rho^i}{i!} + \dfrac{\rho^N}{N!} \left(\dfrac{1 - \left(\dfrac{\rho}{N}\right)^{J+1}}{1 - \dfrac{\rho}{N}} \right)}, & \rho \neq N \\[8ex] \dfrac{1}{\displaystyle\sum_{i=0}^{N-1} \dfrac{\rho^i}{i!} + \dfrac{\rho^N}{N!} (J + 1)}, & \rho = N \end{cases} \tag{19.46}$$

Note that this is also the probability that a call will be *queued* (or delayed). The probability that a call will be lost is simply:

$$P_L = P(N + J)$$

$$= \left(\frac{\rho}{N}\right)^J \frac{\rho^N}{N!} P(0) \tag{19.47}$$

with $P(0)$ as given previously in equation (19.46) for a finite buffer of size J. Note that the M/M/N/N+J queue is, in effect, a generalisation of previous results for Poisson arrivals and exponential service, multiple servers and finite capacity. It would therefore be reasonable to expect any of these results to be obtained from the M/M/N/N+J equations by appropriate substitutions. For example, the M/M/1/L result (as in section 19.3 but using a slightly different notation) might be obtained by setting $N = 1$ and $J = L - 1$, i.e.:

$$P(k) = \frac{\rho \times \rho^{k-1}}{1 + \rho \displaystyle\sum_{i=0}^{L-1} \rho^i}, \quad N = 1$$

$$= \frac{\rho^k}{1 + \rho \left(\dfrac{1 - \rho^L}{1 - \rho} \right)}$$

$$= \frac{\rho^k (1 - \rho)}{(1 - \rho^{L+1})}$$

which is the M/M/1/L result for $\rho \neq 1$ (cf. equation (19.17)).

19.3.5 Distribution of waiting times

In many systems, particularly those involving real time services, it is often required that the delay must meet a specified constraint in order to satisfy a particular quality of service. This requires the calculation of *delay distribution*.

Consider an M/M/N queue. If all N servers are busy and there is always at least one call waiting, then the probability that K servers will be cleared in time t is given by the Poisson distribution as:

$$P(k) = \frac{(N\mu t)^k}{k!} e^{-N\mu t}$$

where μ is the service rate in calls/unit time. If time is expressed in units of average call service time $(1/\mu)$ then:

$$P(k) = \frac{(Nt)^k}{k!} e^{-Nt}$$

If a call arrives when N servers are busy and j other calls are waiting, it will be delayed by more than t if j or fewer calls terminate within time t. These terminations may include calls which are waiting when the new call arrives, as well as calls that are already being served. Then if B_j represents the event that j or fewer calls terminate in time t, the probability of this is:

$$P(B_j) = \sum_{r=0}^{j} \frac{(Nt)^r}{r!} e^{-Nt} \tag{19.48}$$

Let A be the event that a call is delayed by more than t; then by the complete probability formula [Ross]:

$$P(Q > t) = P(A) = \sum_{j=0}^{\infty} P(AB_j)$$

$$= \sum_{j=0}^{\infty} [P(A|B_j) P(B_j)] \tag{19.49}$$

where $P(A|B_j)$ represents the conditional probability, defined in section 3.2.1, that a call is delayed by more than t, given that j or fewer calls terminate in time t. This is just the probability of j calls waiting, i.e. $P(N + j)$. Thus:

$$P(A) = \sum_{j=0}^{\infty} \left[P(N + j) \sum_{r=0}^{j} \frac{(Nt)}{r!} e^{-Nt} \right] \tag{19.50}$$

Evaluation of this results in:

$$P(Q > t) = P(A) = P_B e^{-(N - \rho)t} \tag{19.51}$$

where P_B is given by the Erlang-C formula from equation (19.44) as:

$$P_B = \frac{\dfrac{\rho^N}{N!}\left(\dfrac{N}{N-\rho}\right)}{\displaystyle\sum_{k=0}^{N}\frac{\rho^k}{k!} + \frac{\rho^k}{N!}\left(\dfrac{\rho}{N-\rho}\right)} \qquad (19.52)$$

EXAMPLE 19.6

Consider the system described in Example 19.2. Find the probability that an arriving packet will be delayed by more than 40 ms before gaining access to a channel.

Using equation (19.51) then:

$$P(Q > t) = P_B\, e^{-(N-\rho)t}$$

with P_B given by equations (19.45) and (19.46) as:

$$P(Q > t) = \frac{\dfrac{\rho^N}{N!}\left(\dfrac{1-\left(\dfrac{\rho}{N}\right)^{J+1}}{1-\left(\dfrac{\rho}{N}\right)}\right)}{\displaystyle\sum_{i=0}^{N-1}\frac{\rho^i}{i!} + \frac{\rho^N}{N!}\left(\dfrac{1-\left(\dfrac{\rho}{N}\right)^{J+1}}{1-\left(\dfrac{\rho}{N}\right)}\right)}\, e^{-(N-\rho)t}$$

In the above we therefore substitute the values $N = 1$, $\rho = 0.8$, $J = 9$ and $t = 40 \times 10^{-3}$. Further evaluation then results in:

$$P(Q > 40 \times 10^{-3}) = \frac{0.8\left(\dfrac{1-(0.8)^{10}}{1-0.8}\right)}{1+0.8\left(\dfrac{1-(0.8)^{10}}{1-0.8}\right)}\, e^{-(0.2 \times 40 \times 10^{-3})} = 0.78$$

19.4 M/M/N/K/K queue: queueing behaviour in a mobile communication system

In using queueing theory to model communication networks, the customers in the queueing model need not be used to explicitly represent packets in a packet switched communication network, despite the obvious analogy. An example of this was seen in the Erlang-C formula where the entities queued are call requests. Even less obvious correspondences can be used, as shown in the following example of modelling the traffic behaviour in a cellular mobile communication system.

The analysis considers a base station (BS) that has to serve a local area network (LAN) that consists of a number of small sized cells (picocells), and in each cell there are K homogeneous mobile terminal units (TUs). It will be assumed that there is a perfect physical connection between TUs and BS and in this simplified analysis handover to or from adjacent cells will be ignored.

The idea is to derive the equilibrium probability distribution of the queueing model which represents the system. From this the various performance measures such as throughput and mean channel access delay can be derived.

19.4.1 Speech source model

In the literature it has been shown that a speech source has ON–OFF patterns and that all talkspurts (ON) and silences (OFF) are exponentially distributed [Brady]. Also, it is assumed that transmissions are divided into frames of duration T_f, with T_f a known quantity. Therefore, the behaviour of a voice TU that employs a slow speech activity detector can be modelled as a two-state (ON–OFF) Markov chain. Such a source model is shown in Figure 19.17 where γ denotes the transition probability from talkspurt (TLK) to silence (SIL) state during a channel frame while σ denotes the transition probability from silence to talkspurt state during a channel frame. Talkspurt duration is equivalent to message length or service time of each TU in the uplink channel frame; likewise, talkspurt plus gap duration is equivalent to interarrival time of messages. The probability of more than two transitions, from talkspurt to silence state or vice versa during one channel frame, is taken to be zero. This is justified due to the channel frame duration being much smaller than either the silence state or talkspurt state duration which for speech is approximately 650 ms and 352 ms respectively [Brady].

Knowing the channel frame duration, T_f, and the mean duration of talkspurts, t_1, and silences, t_2, then:

$$\gamma = 1 - e^{\frac{-T_f}{t_1}} \tag{19.53}$$

$$\sigma = 1 - e^{\frac{-T_f}{t_2}} \tag{19.54}$$

19.4.2 Equilibrium probability

Considering the time-slots within a channel frame as a bunch of K parallel servers and assuming that a TU can be served at most once for the whole frame then the input from each TU consists of a sequence of exponentially distributed talkspurts and gaps. The

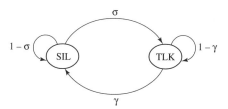

Figure 19.17 *Two-state Markovian model for speech source.*

service time (of talkspurts) is exponentially distributed, there are N parallel servers (channels), and K customers (TUs). If it is assumed that interarrival times of talkspurts are also exponentially distributed then this scenario can be modelled as an M/M/N/K/K queueing system with state transition diagram shown in Figure 19.18.

To find the equilibrium probability, first the transition rates of arrivals and departures within a cell must be defined. The transition rate from a state k to a state $k + 1$, which denotes an arrival, is given by:

$$\lambda_k = (K - k)\lambda, \qquad 0 \le k \le K - 1 \tag{19.55}$$

where λ is defined as the mean number of talkspurts arriving from each TU per channel frame and is given by $\lambda = T_f/(t_1 + t_2)$. The transition rate from a state k to state $k + 1$, which denotes a departure, is written as:

$$\mu_k = \begin{cases} k\mu, & 1 \le k \le N \\ N\mu, & N + 1 \le k \le K \end{cases} \tag{19.56}$$

where μ is the rate of service for each channel in terms of service completions per channel frame and is given by $\mu = T_f/t_1$. Note again in Figure 19.18, as in Figure 19.16, that when $k \ge N$ the service rate becomes constant. Substituting equation (19.55) and (19.56) into (19.22) the equilibrium probability $P(k)$ can be expressed in terms of $P(0)$ as:

$$P(k) = \begin{cases} \binom{K}{k}\left(\frac{\lambda}{\mu}\right)^k P(0), & 0 \le k < N \\ \binom{K}{k}\dfrac{k!}{N^{K-N}\, N!}\left(\frac{\lambda}{\mu}\right)^k P(0), & N \le k \le K \end{cases} \tag{19.57}$$

Using the normalising equation:

$$\sum_{k=0}^{K} P(k) = 1$$

Substituting equation (19.57) into this, then $P(0)$ is given by:

$$P(0) = \left[\sum_{k=0}^{N-1} \binom{K}{k}\left(\frac{\lambda}{\mu}\right)^k + \sum_{k=N}^{K} \frac{k!}{N^{k-N}\, N!}\left(\frac{\lambda}{\mu}\right)^k \right]^{-1} \tag{19.58}$$

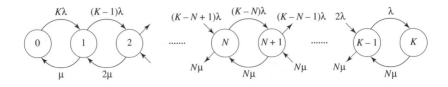

Figure 19.18 *State transition diagram for M/M/N/K/K queue.*

Substituting this back into equation (19.57) then gives the equilibrium probability distribution of the number of TUs in the talkspurt state.

Normalised mean throughput (η) is an important measure of the system performance and is defined as the mean number of customers passing through the system per unit time normalised to the service capacity of the system. Then:

$$\eta = \frac{S}{N\mu}$$

where S is the (absolute) throughput, defined previously in equation (19.9). Since:

$$S = \left(\sum_{k=0}^{N-1} kP(k) + N \sum_{k=N}^{M} P(k) \right) \mu \tag{19.59}$$

then:

$$\eta = \frac{1}{N} \left(\sum_{k=0}^{N-1} kP(k) + N \sum_{k=N}^{M} P(k) \right)$$

The total delay for serving a talkspurt in the system is the sum of waiting time in the queue plus the service time. Therefore, the mean channel access delay Q is the delay that is experienced by a talkspurt while waiting in the queue and is therefore the same as the queueing delay previously defined. This is then given by Little's formula as before as:

$$Q = \frac{L}{S} - \frac{1}{\mu}$$

with S as given in equation (19.59) and:

$$L = \sum_{k=0}^{K} kP(k)$$

with $P(k)$ obtained from equations (19.57) and (19.58).

It should be remembered that in this case the queueing delay that can be tolerated will depend on the real time characteristics of the system. System parameters can thus be adjusted to keep the queueing delay below some specified value linked to the quality of service. Also, the analysis above considers the customer population to remain constant. An analysis that considers handover between cells and a varying customer population is outside the scope of this chapter and can be found in [Kulavaratharasah and Aghvami].

19.5 Summary

An explanation of how some basic queueing theory can be used to obtain performance measures for telecommunication systems has been given. The key assumptions are those of exponentially distributed interarrival times and service times. These allow the use of the Markov (or memoryless) property and lead to a greatly simplified analysis in many situations. Performance measures that have been discussed are mean values of throughput, system delay, queueing delay, blocking probability and probability of packet loss. Also, a simple method of calculating delay distribution has been given.

While the use of the Markov property can give good approximations in many situations, there are some systems which do not lend themselves to such approximations. For example, systems that use fixed length packets, such as ATM (asynchronous transfer mode) systems, lend themselves more readily to a discrete time analysis, where unit time can be the constant service time of a packet [Onvural, Woodward]. Also, arrival streams of some services such as video are certainly not Poisson and are likely to have correlated interarrival times. These call for more specialist modelling techniques that can complicate the analysis. Continuous time systems with non-Markovian service times and/or interarrival times also require some specialist techniques involving the use of moment generating functions and the Laplace transforms [King].

Although networks of queues have not been explicitly considered in this chapter, this would be a next logical progression from the material that has been covered. An excellent text is that by [Walrand].

19.6 Problems

19.1 In Example 19.2 try increasing the buffer size to 99 packets with the other parameters remaining the same. Note the effect that this has on packet loss and delay. [4.08×10^{-11}, 2 ms]

19.2 The number of packets arriving at a particular node in a packet switched computer network may be assumed to be Poisson distributed. Given that the average arrival rate is 5 packets per minute, calculate the following:
(a) Probability of receiving no packets in an interval of 2 minutes. [4.5×10^{-5}]
(b) Probability of receiving just 1 packet in the next 30 s. [0.2]
(c) Probability of receiving 10 packets in any 30 s period. [2.16×10^{-4}]

19.3 In Problem 19.2: (a) What is the probability of having a gap between two successive packet arrivals of greater than 20 s? (b) What is the probability of having gaps between packet arrivals in the range 20 to 30 s inclusive? [0.189, 0.106]

19.4 For a single server queueing system with Poisson distributed arrivals of average rate 1 packet/s and Poisson distributed service of capacity 3 packets/s calculate the probability of receiving no packets in a 5 s period. Also find the probabilities of queue lengths of 0, 1, 2, 3. If the queue length is limited to 4 what percentage of packets will be lost? [2/3, 2/9, 2/27, 2/81, 0.41%]

19.5 Assuming an M/M/1 queue, determine the probabilities of having: (a) an empty queue, (b) a queue of 4 or more 'customers'. You should assume that the ratio of arrival rate to service rate (called the utilisation factor) is 0.6. [0.4, 0.13]

19.6 If the queue length in Problem 19.5 is limited to 10, what percentage of customers are lost to the service? [0.36%]

19.7 A packet data network links London, Northampton and Southend for credit card transaction data. It is realised with a two-way link between London and Northampton and, separately, between London and Southend. Both links operate with primary rate access at 2 Mbit/s in each direction. If the Northampton and Southend nodes send 200 packets/s to each other and the London node sends 50 packets/s to Northampton what is the mean packet delay on the network when the packets have a mean size of 2 kbits with exponential distribution and all arrival streams are Poisson distributed? [1.275 ms]

19.8 An X.25 packet switch has a single outgoing transmission link at 2 Mbit/s. The average length of each packet is 960 bytes. If the average packet delay through the switch, assuming an M/M/1 queue, is to be less than 15 ms, determine: (i) the maximum gross input packet rate to the switch; (ii) the average length of the queue; (iii) the utilisation factor through the switch if each packet in the input is converted into ATM cells having 48 bytes of data and a 5 byte overhead (see section 20.10 for a full explanation of ATM). [193.7, 2.91, 0.821]

19.9 For the M/M/N/N/K queue of section 19.3.3, verify by direct substitution of λ_k and μ_k from equations (19.33) and (19.34), respectively, into equation (19.22) that:

$$P(k) = \left(\frac{\lambda}{\mu}\right)^k \binom{K}{k} P(0), \qquad 0 \leq k \leq N$$

19.10 Packets arrive at a communication buffer at a mean rate of 10 packets/s with an exponential interarrival time. Packet lengths are exponentially distributed with a mean length of 500 bits. There is a single outgoing link from the buffer that operates at a rate of 9.6 kbit/s. Determine the minimum amount of buffer storage that must be provided if the number of arriving messages that are lost is not to exceed 5% of the total arriving messages. [4]

19.11 Packets arrive for transmission over a communications link in a Poisson stream with mean rate of 5 packets/s. Packet lengths are exponentially distributed with a mean of 1500 bits. If 90% of packets must be transmitted within half a second, determine the minimum line speed required to make this possible. [10000 bit/s]

19.12 A multiprocessor system has K terminals each of which can generate packets in a Poisson stream with mean rate λ packets/s. The system has N processors and each can serve exactly one terminal at a time. Once a terminal has been served then the corresponding processor becomes free. Each processor can serve packets at mean rate μ packets/s with an exponentially distributed service time. If at any time a terminal with data to send finds that all processors are busy then packets are queued in a buffer to await service. The buffer can hold a maximum of J packets, where $K > N + J$. Calculate the equilibrium probability distribution of the number of packets in the system. From this find equations for the mean number of packets in the system, the mean system delay, the mean queueing delay, blocking probability and probability of buffer overflow.

(Hint: there is a close similarity between the model for this system and that of the mobile communication system in section 19.4, although they are not the same.)

19.13 Consider a communication buffer fed by a Poisson stream with mean rate λ_1 packets/s the output of which is served by a single communication channel which operates at rate μ_1 packets/s, where $\mu_1 > \lambda_1$. The output channel forms the input to a second buffer which is served by a single communication channel that operates at rate μ_2, where $\mu_2 > \lambda_1$. Service times for both buffers are assumed to be exponentially distributed. It turns out that in this situation the output of the first buffer is also a Poisson stream with mean rate λ_1 packets/s. This well known result is called Burke's theorem [Burke] and means that the second queue can be analysed as an M/M/1 queue independently of the first queue. Find the joint probability distribution $P(k_1, k_2)$, $k_1 \geq 0$, $k_2 \geq 0$, where k_1 represents the number of packets in buffer 1, including the server, and k_2 represents the number of packets in buffer 2, including the server. Find an expression for the mean packet delay through the entire system.

CHAPTER 20

Switched networks and WANs

20.1 Introduction

WANs (wide area networks) operate across city and national boundaries and must usually rely on the provision of transmission facilities by independent carriers. These carriers, that own some, most, or all, of the transmission infrastructure, may be state or privately owned, may provide public networks (serving a wide range of subscribing users) or private networks (serving closed user groups), and may represent monopolies or a multiplicity of competing companies (the historical trend being from the former to the latter). Much of the current WAN infrastructure has evolved from that originally provided by national post, telegraph and telecommunications administrations (PTTs) which have been traditionally concerned with the provision of (first analogue, then digital) telephone services via the public switched telephone network (PSTN). This chapter is primarily concerned with switched WAN core networks and the (mainly) point-to-point technologies used to access them. It could be argued that local area networks (LANs) are also used as (WAN) access networks and they are often so described. It has become more common recently, however, to distinguish between the customer premises network (CPN), of which the LAN usually forms (at least) part, and the access network (AN), which links the equipment on the customer's premises to a core network node. LANs are therefore addressed separately, along with the generic properties of broadcast networks, in Chapter 21.

20.2 WAN characteristics and classification

WANs have some or all of the following characteristics:

1. Long haul transmission, possibly thousands of kilometres.
2. National, international or global coverage area.
3. Large numbers of users, hundreds to many millions.
4. Highly interconnected, multiplexed, cores allowing many users to communicate simultaneously.
5. Sophisticated switching allowing emulation of a fully interconnected network.

As the last of these characteristics implies, the large numbers of users typically served by WANs means that they are invariably implemented as switched networks. Figure 20.1 is a stylised representation of a complex network showing a switched core WAN and a range of access networks.

Access networks connect each user to the nearest WAN node. In the plain old telephone system (POTS), a star network connects each user to the local exchange, each connection forming a local loop, i.e. a twisted wire pair (traditionally) of not more than a few kilometres in length supporting full-duplex transmission, Chapter 15. As users demand higher data rates one possibility is to replace these traditionally low bandwidth transmission lines with higher bandwidth lines. To replace all such connections with individual cables of greater capacity, however, is a large and expensive undertaking, so the modern local network is often a shared (multiplexed) transmission facility. In a commercial organisation a LAN is normally used to connect users to a WAN node. For domestic premises new arrangements have recently been developed using either higher bandwidth transmission technologies (coaxial cable, optical fibre or localised radio links) or improved modulation and signal processing schemes to increase the capacity of the existing local loop infrastructure (ISDN, xDSL, see sections 20.4, 20.8 and 20.11.1).

Trunk links connect switches within the WAN core. Each trunk must provide high transmission capacity that is shared between users by multiplexing. Since a trunk is a shared resource, its integrity must be guaranteed by monitoring its performance and provision must be made for alternative transmission facilities in case it fails to meet acceptable performance standards.

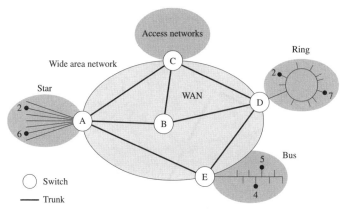

Figure 20.1 *Stylised network example.*

Some typical trunk transmission media bandwidths are given in Table 20.1 and other associated data is available in Table 17.2. (Note that the capacity may be many times the bandwidth depending on the modulation scheme adopted, section 11.4.)

Table 20.1 *Typical bandwidths of trunk media.*

Medium	Typical bandwidth
Coaxial cable	10–100 MHz
Optical fibre	> 10 GHz
UHF radio	10 MHz
Microwave radio	100 MHz

The trunk lines in a WAN core are connected together at switches (network nodes). Typically each node has many incoming and outgoing data paths and each switched path through the node is duplex, supporting two-way transmission.

Each switch must select an outgoing trunk as directed. In Figure 20.2, for example, there are many possible routes that could be used to send information from node A to node E. A switch control scheme is therefore required that includes address, numbering plan and routing information. In switched circuit applications the control information is provided by a signalling system that, in its modern implementation, constitutes a (specialised) network in its own right.

Circuit switches generally suffer from blocking, since they have limited capacity to route calls, and when all available routes are taken an incoming call is ignored (i.e. blocked). Packet switches on the other hand are generally non-blocking, i.e. they are able to store incoming packets until an outgoing route is available. The packet switch penalty for heavy traffic is therefore increased delay (latency) rather than call loss.

Trunk facilities can be hired in a variety of ways. Within a single country they can be leased from the PTT (the national communication authority) or a commercial carrier (e.g. BT in the UK, or AT&T in the USA). International trunks can be leased from a satellite carrier (e.g. INTELSAT).

The original WANs were the various voice oriented national PSTNs which, when connected together, formed the first truly global network of any kind. Although the international PSTN grew and became more automated and sophisticated, its services remained essentially unchanged, remaining principally concerned with voice communication. The PSTN (incorporating telegraphy and telex services) remained the only communications WAN until the creation of ARPANET, which was specifically

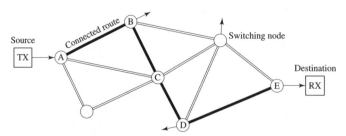

Figure 20.2 *Schematic illustration of a simple switched network.*

designed to link computers, and spanned (originally) the USA. (ARPANET was later extended to cover Europe and eventually evolved into the Internet.) With the rise in the number of digital communications users, many national PTT administrations have now built similar networks that generally use store and forward transmission in which messages are held in a node store before being switched, via an appropriate transmission link, to another node nearer the message's ultimate destination. Data rates on such WANs are typically 10 to 50 kbit/s and end-to-end delays can be of the order of seconds.

WANs can be classified as follows:

1. **Public networks** – e.g. the BT Gold electronic mail service, the international EURONET (a packet switched service which exists primarily for accessing information databases) and the Internet.
2. **Private networks** – used when public data networks do not provide the services required. Private telecommunications networks leased on a semi-permanent basis, such as SWIFT employed by the banking industry.
3. **Value added networks** – which use conventional PTT facilities combined with specialised message processing services to add value to the network. The user is offered flexibility and the economies of shared usage. Examples include TRANSPAC and the specialised banking EFTPOS networks.

Additional UK examples of WANs are the BT packet switched service (PSS) and the joint academic network (SuperJANET). The original JANET interconnected UK university LANs using a 5000 km backbone network. With transmission rates of 100 Mbit/s available on the individual university LANs (that constitute the access networks) the JANET backbone transmission rate was progressively increased from 9.6 kbit/s to 2 Mbit/s. The main centres (London, Manchester, Edinburgh, etc.) were interconnected at 2 Mbit/s while other sites had 9.6 and 64 kbit/s access rates. The SuperJANET improvement upgraded the backbone bit rate to 34 and 140 Mbit/s using ATM (see section 20.10). Figure 20.3 shows the SuperJANET Phase B network, in which access rates are multiples of the 51.8 Mbit/s standard SONET/SDH interfaces (see section 20.5.2), within the PSTN core network. The equivalent US system is NSFnet that operates at 45 Mbit/s. The JANET network has transparent gateway connections to other X.25 international networks, which link it to the Internet.

20.3 Application of graph theory to core networks

The size and complexity of a WAN core network makes its analysis and design extremely difficult, and in practice optimisation of such networks is routinely accomplished by perturbing existing designs and finding (usually by simulation) whether this leads to improved or degraded performance. Insight can be gained, and first order design decisions aided, however, using graph theory.

20.3.1 Topology, cost and capacity matrices

Section 17.5.4 showed how the topology of a network can be described by an $N \times N$ (topology) matrix and section 18.4.1 showed how a cost matrix can be used to represent the costs of individual network links (e.g. for implementing optimum routing algorithms). Similarly, a capacity matrix can be used to describe network capacity, the element m_{ij}

Figure 20.3 *The SuperJANET wide area network in the early 1990s.*

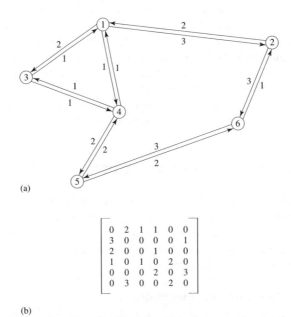

(a)

$$
\begin{bmatrix}
0 & 2 & 1 & 1 & 0 & 0 \\
3 & 0 & 0 & 0 & 0 & 1 \\
2 & 0 & 0 & 1 & 0 & 0 \\
1 & 0 & 1 & 0 & 2 & 0 \\
0 & 0 & 0 & 2 & 0 & 3 \\
0 & 3 & 0 & 0 & 2 & 0
\end{bmatrix}
$$

(b)

Figure 20.4 *Example network and corresponding capacity matrix description.*

representing the forward capacity (in bit/s) of the link connecting node i to node j. If the link is duplex with the same capacity in each direction then $m_{ji} = m_{ij}$. Since nodes do not have links connecting them to themselves then the diagonal elements of the matrix are zero, i.e. $m_{ii} = 0$. (Figure 20.4(a) shows a network with the internode capacities identified and (b) represents this information in matrix form.)

This matrix representation lends itself well to computer implementation of algorithms based on graph theory for determining performance metrics such as network capacity and network connectivity.

20.3.2 Network capacity

The information carrying capacity of a network can be found using a theorem called the 'maximum flow–minimum cut theorem'. The theorem can be stated as follows:

The maximum information flow between any pair of network nodes is equal to the minimum capacity of all possible cuts between the nodes.

A cut for a particular node pair is defined as any set of links whose removal results in the pair of nodes becoming disconnected and a minimal cut is the smallest set of links resulting in disconnection. The capacity of a cut is the sum of the capacities of the removed links and a minimum cut is a cut with a minimum capacity. The concepts of cut, minimal cut, minimum cut and capacity are illustrated in Figure 20.5.

In Figure 20.5 the links are assumed duplex with equal capacities in both directions and the graph is therefore undirected. If links are simplex they can be represented by a directed graph in which arrows are attached to each line to show the direction of information flow. (An undirected link is, of course, equivalent to a pair of equal capacity directed links with opposite sense between the same two nodes.)

A flow of information from a source A to a sink D is *feasible* providing that:

1. flow is conserved (i.e. flows sum to zero at all nodes except source and sink, and source flow equals sink flow);

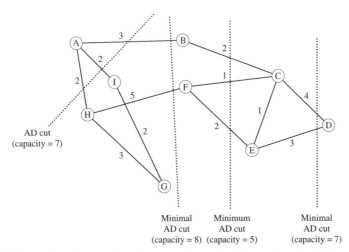

Figure 20.5 *Cuts, minimal cuts and minimum cut for a pair of network nodes (A,D).*

2. no link carries a flow greater than its capacity.

It follows that if the numbers representing link capacities in Figure 20.5 actually represented information flow this flow could not be feasible since no set of arrows can be contrived to satisfy the above conditions (consider node B, for example).

The maximum flow–minimum cut theorem may seem obvious but its significance and power lies in its implicit assertion that any minimum cut capacity always represents a feasible information flow. Its weakness is that for networks of significant size and complexity the identification and examination of all possible cuts required to apply it in a straightforward way is generally impractical. Methods developed using graph theory exist, however, which accomplish the same task in a more efficient way. The maximum flow algorithm described below finds the maximum (feasible) flow between an information source and an information sink.

Maximum flow algorithm

1. Draw a master (directed) graph of the network assigning a label to each link of the form {*information flow, link capacity*} assuming, initially, that all information flows are zero. (For the purpose of this algorithm the senses of the directed lines are simply for reference so that a final interpretation of positive and negative information flows along links can be made. The directed lines do not represent a simplex restriction on the direction of information flow. It is helpful, though, when implementing this algorithm manually to choose the sense of each directed link to reflect a sensible guess for the ultimate source to sink directions of information flow.)
2. Place the information source in the first level (i.e. at the apex) of what will be a sub-network with nodes arranged in a hierarchy of levels.
3. Place in the next level down all *unplaced* nodes that are adjacent to the nodes in the previous level and connected to them by *unsaturated* links (i.e. links carrying an information flow less than their capacity).
4. Draw in the directed lines for all links (the sense being the same as in the master graph) that connect nodes in previous and next levels assigning a label {*information flow, link capacity*} to each as obtained from the master graph. (An information flow opposite to the sense of the directed line is negative.)
5. Go back to 3 and repeat until the information sink has been placed.
6. Delete all nodes and links not on a continuous path from information source to information sink.
7. Find the total capacities of input and output links for each node in the surviving sub-network.
8. Assign an *input potential* to each node equal to total capacity of that node's input links minus the existing input flow as represented by the existing flow allocations.
9. Assign an *output potential* to each node equal to total capacity of that node's output links minus the existing output flow as represented by the existing flow allocations.
10. Assign an *overall potential* to each node equal to the lesser of the node's input and output potentials.
11. Identify a *critical* node that has a minimum overall potential. (This represents a bottleneck for information flow. If there is more than one node with the same minimum overall potential then any can be selected as the critical node.)

12. Increment the information flow between each level of the hierarchy (using as many of the links as necessary) by an amount equal to the potential of the critical node, routing the flow through the critical node so as to saturate either its input or output or both.
13. Update the information flow in the {*information flow, information capacity*} link assignments on the master graph using the incremented values of information flow.
14. Go back to 2 and continue until no continuous path can be constructed from source to sink.

After completing the above algorithm the information flow between source and sink is a maximum and the set of saturated links includes one or more cut sets. (In interpreting the final information flow graph remember that the information on links with negative flows moves in the opposite direction to the sense of the directed link.)

Information flow networks are linear in the sense that the information flow between multiple information source–sink pairs in a network of directed links obeys superposition, i.e. the information flows from node X to node Y add irrespective of the information's source and destination. When considering multiple source–sink pairs the directed (simplex) links (X to Y and Y to X) must be considered separately, however, since information flow between one pair must not be allowed to 'cancel' information flow between another pair (or the same pair but with source and sink interchanged).

EXAMPLE 20.1

Find the maximum feasible information flow between nodes A and E of the network represented by the capacity graph shown in the left hand top corner of Figure 20.6(a).

Step 1: Draw a directed master graph of the network assigning an {*information flow, information capacity*} to each link assuming, initially, that all information flows are zero, i.e. second top diagram in left column of Figure 20.6(a).

Steps 2–5: Draw a hierarchy of connected nodes, as per the algorithm, and *Step 6:* delete nodes not on a path connecting source with sink, see top (first pass) part of Figure 20.6(b).

Steps 7–11: Find node potentials [B input: 3, B output: 2, H input: 5, H output: 6 (ignoring the directional arrows), F input: 4, F output: 3, G input: 4, G output: 4]. The minimum node potential is therefore 2 (at B output).

Step 12: Increase the flow from A to E from 0 by 2 (via node B) to saturate node B, see right hand top diagram in Figure 20.6(b).

Step 13: Update the master graph with new flows, see bottom left (first pass) diagram in Figure 20.6(a).

Steps 14–5: Now for the second pass redraw the hierarchy of connected nodes, as per the algorithm, and *Step 6*: delete nodes not on a path connecting source with sink, as in (second pass) row in Figure 20.6(b). Note that the B–G link is removed because it is now at capacity.

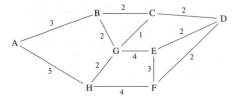

Initial network capacity graph

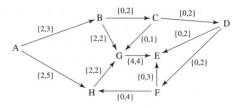

Second pass updated master graph with new flows

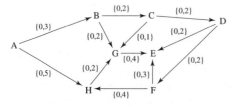

Capacity graph with initial (null) information
flows and link capacities

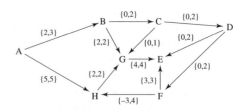

Third pass updated master graph with new flows

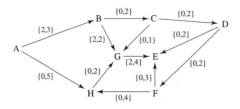

First pass updated master graph with new flows

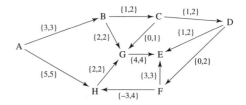

Fourth pass updated master graph with new flows

Figure 20.6a *Network capacity graph evolution for Example 20.1.*

Steps 7–11: Find updated node potentials [H input: 5, H output: 6, F input: 4, F output 3, G input: 2, G output: 4]. The minimum node potential is therefore 2 (at G input).

Step 12: Increase the flow from A to E by 2 (via node G), see second pass right hand diagram in Figure 20.6(b).

Step 13: Update with new flows the second pass master graph in Figure 20.6(a).

Steps 14–5: Now for the third pass redraw the hierarchy of connected nodes, as per the algorithm, and *Step 6:* delete nodes not on a path connecting source with sink, as in (third pass) row in Figure 20.6(b). Here the links B–G, H–G, G–E are removed because they are now at full capacity.

Steps 7–11: Find next set of node potentials [H input: 3 (5 – 2 = 3), H output: 4, F input: 4, F output 3]. The minimum node potential is 3 (at H input).

Step 12: Increase the flow from A to E by 3 (via node H), see third pass right hand diagram in Figure 20.6(b).

Step 13: Update new flows as in the third pass master graph in Figure 20.6(a).

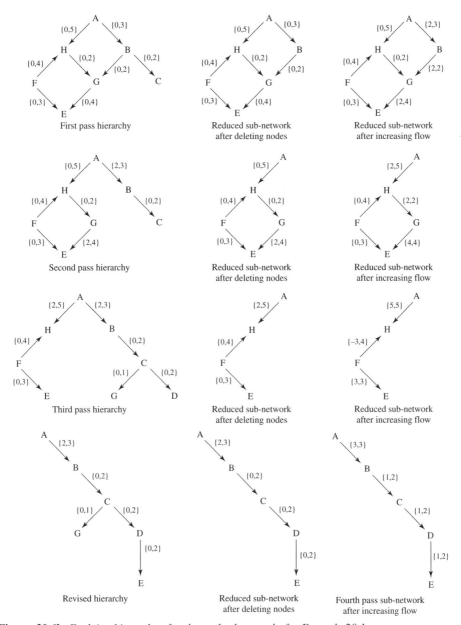

Figure 20.6b *Evolving hierarchy of nodes and subnetworks for Example 20.1.*

Steps 14–5: Redraw a revised hierarchy of connected nodes, as per the algorithm, and *Step 6:* delete nodes not on a path connecting source with sink, as in lower fourth pass row in Figure 20.6(b).

Steps 7–11: Find node potentials [B input: 1 (3 − 2 = 1), B output: 2, C input: 2, C output 2, D input: 2, D output: 2]. The minimum node potential is therefore 1 (at B input).

Step 12: Increase the flow from A to E by 1 (via node B), as in right hand diagram in fourth pass row of Figure 20.6(b).

Step 13: Update the master graph with new flows, as in the fourth pass master graph in Figure 20.6(a).

Step 14: Since both links from the source are now saturated it is not be possible to redraw the graph with a path from source to sink and so the algorithm terminates.

It can be seen that the maximum flow from A to E is 8 (equal to information flow out of A and information flow into E). Examination of each node confirms that flow is conserved and the flow pattern found is therefore feasible.

20.3.3 Network connectivity

The *link connectivity* of a particular pair of network nodes is equal to the minimum number of links that must be removed to disconnect the nodes. Similarly the *node connectivity* of a pair of network nodes is equal to the minimum number of nodes that must be removed to disconnect the node pair. Network link and node connectivities are the minimum link and node connectivities, respectively, for all node pairs in the network. The greater the link and node connectivities of a network the more resilient the network will be to link and node failures and, therefore, the more reliable the network will be.

The related concept of node *degree* is defined as the number of links ending on a node. The minimum degree of a network is the smallest node degree. A *regular* network is a network in which all nodes have the same degree. A ring network, for example, would therefore be a regular network with degree 2, network link connectivity of 2 and network node connectivity of 2. In general a network's node connectivity (C_{node}), link connectivity (C_{link}) and minimum degree (D_{min}) are related by:

$$C_{node} \leq C_{link} \leq D_{min} \tag{20.1}$$

Since node connectivity is always less than link connectivity it is usually sufficient to specify the node connectivity of a network to achieve some required minimum level of reliability. Several algorithms have been devised to test whether a network has a given node connectivity and these algorithms can be applied iteratively to find the network's minimum node (and therefore minimum overall) connectivity. Kleitman's algorithm, outlined below, tests whether a given network has node connectivity of at least k.

EXAMPLE 20.2

What is the network link connectivity and network node connectivity of a star connected network with five satellite terminals connected to a single hub terminal? Is the network regular? What is the network's minimum degree?

Since all pairs of terminals can be disconnected by the removal of a single link then $C_{link} = 1$. Since the removal of the central hub disconnects all terminal pairs then $C_{node} = 1$.

The degree of the hub terminal is 5 and the degree of all satellite terminals is 1. The network is therefore not regular and $D_{\min} = 1$.

Kleitman's $C_n \geq k$ test:

1. Select k.
2. Select any node and label it node 1.
3. Check that node 1 has connectivity of at least k to all other nodes.
4. If node 1 has connectivity less than k to any other node then the network has $C_n < k$, otherwise continue.
5. Remove node 1 and all links connected to node 1.
6. Select another node and label it node 2.
7. Check that node 2 has connectivity of at least $k - 1$ to all other nodes.
8. If node 2 has connectivity $< k - 1$ to any other node then the network has $C_n < k$, otherwise continue.
9. Remove node 2 and all links connected to node 2.
10. Select another node and label it node 3.
11. Check that node 3 has connectivity of at least $k - 2$ to all other nodes.
12. If selected node has connectivity $< k - 2$ to any other node then the network has $C_n < k$, otherwise continue.
13. Remove node 3 and all links connected to node 3.
14. Continue the above pattern until the selected node is labelled node k.
15. Check that node k has connectivity of at least 1 to all other nodes.
16. If node k has connectivity less than 1 to any other node then the network has $C_n < k$, otherwise the network has connectivity of at least k.

EXAMPLE 20.3

Verify that the network shown in Figure 20.7 has connectivity of at least 3.

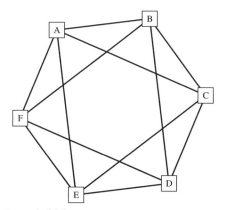

Figure 20.7 *Network for Example 20.3.*

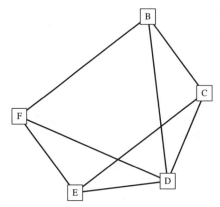

Figure 20.8 *Network for Example 20.3 with one node removed.*

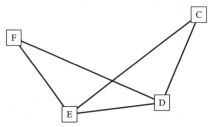

Figure 20.9 *Network for Example 20.3 with two nodes removed.*

Step 1: Select $k = 3$.

Step 2: Select node A and label it as node 1.

Step 3: Check that node 1 has connectivity of at least 3 to all other nodes:

AB: AB, AFB, AECB
AC: ABC, AEC, AFEDC
AD: ABD, AED, AFD
AE: AE, AFE, ABDE
AF: AF, ABF, AEF

Step 4: Since node 1 has connectivity of at least k to all other nodes then continue.

Step 5: Remove node 1 and all links connected to node 1, Figure 20.8.

Step 6: Select node B and label it node 2.

Step 7: Check that node 2 has connectivity of at least 2 to all other nodes.

BC: BC, BDC
BD: BD, BCD
BE: BFE, BDE
BF: BF, BDF

Step 8: Since node 2 has connectivity of at least 2 to all other nodes then continue.

Step 9: Remove node 2 and all links connected to node 2, Figure 20.9.

Step 10: Select node C and label it node 3.

Step 11: Check that node 3 has connectivity of at least 1 to all other nodes.
CD: CD
CE :CE
CF: CEF

Step 12: Since selected node has connectivity of at least 1 to all other nodes then the network has connectivity of at least 3.

20.4 The UK public network

The UK public communications network consists of a bearer network, a set of functional networks and an access network. The bearer network refers to the collection of physical transmission lines and switches needed to make connections between users. The functional networks use the bearer network as a resource to provide a variety of telecommunications services. The access network connects individual subscribers to the bearer network. Figure 20.10 shows a stylised relationship between the bearer, functional and access networks.

The relationship between the PSTN and the other networks represented in Figure 20.10 can be confusing. This is because historically the only function supported by the bearer network was telephony and, therefore, the PSTN and the bearer network were essentially synonymous. The term PSTN should strictly be used, however, to refer to the functional network that provides voice communication services.

20.4.1 The traditional analogue network

The bearer network has benefited from a long history of continuous development. In its original form it was a public, exclusively circuit switched, network that used analogue transmission and analogue switches (space division switches, appropriate to analogue

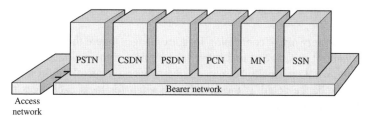

Figure 20.10 *Bearer, functional and access networks. (PSTN: Public switched telephone network, CSDN: Circuit switched data network, PSDN: Packet switched data network, MN: Mobile network, PCN: Private circuit network, SSN: Special service network.)*

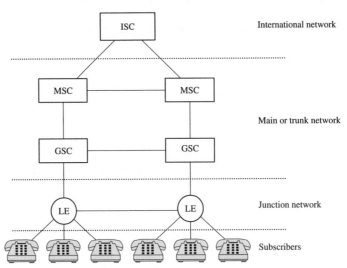

Figure 20.11 *The BT analogue PSTN. (ISC: International switching centre, MSC: Main switching centre, GSC: Group switching centre, LE: Local exchange.)*

signals) to provide a voice communications (i.e. telephone) service – hence its original designation as a PSTN. The switches (grouped together in exchanges) were implemented first as manual distribution frames and then as automatic electromechanical systems (e.g. Strowger, cross-bar and reed contact devices). Since the purpose of the network was to provide the 'plain old telephone service' (POTS), there was no need to make switching particularly fast. Its development was principally directed towards providing high reliability speech circuits and easy-to-use dialling facilities. Figure 20.11 shows the previous UK (BT) analogue network.

With many millions of subscribers in the UK, and centres of population separated by substantial distances, the analogue network developed a series of hierarchical levels comprising:

1. **The local network** – connecting the subscriber's interface with the local exchange via the local loop and representing the access network. This has been built up progressively over the past 70 years and represents a large investment in plant. Each subscriber telephone has two copper wires that go directly to the local exchange building. (The distance is typically 1 to 10 km, being smaller in cities than in rural areas where one exchange normally serves 40 km^2 or between 5000 and 50000 customers.)

2. **The junction network** – connecting local exchanges where there was sufficient local traffic to warrant this. A call to a subscriber connected to an adjacent local exchange, for example, would normally be routed directly over a relatively short junction circuit. A tandem exchange acted as the next level switch for local exchanges that were in reasonably close proximity.

3. **The main network** – connecting group switching centres (GSCs) and main switching centres (MSCs). The GSCs collected calls from local exchanges and were connected by a partially interconnected mesh. Each GSC was connected to a fully

interconnected (and therefore very robust) mesh of MSCs. The main network was sometimes referred to as the trunk transmission network. Trunk lines were longer (tens or hundreds of km) than junction lines and were required to carry many simultaneous calls between population centres.

4. **The international network** – linking the various national networks via international switching centres (ISCs) using submarine cables, geostationary satellites, section 14.3, or terrestrial radio relay circuits, section 14.2.

In the analogue network there are four wires to the handset (one pair to the mouthpiece and one pair to the earpiece), two wires in the local loop and four wires again in the junction and trunk networks. A four-wire circuit implies separate transmit and receive pairs. The UK local access network comprises 36 million wire pairs that, historically, were connected to about 6300 local exchange buildings.

Operation of a PSTN requires that each local exchange and switching centre posses the information required to route calls correctly. This was originally achieved by arranging the dialled digits of the subscriber number to represent the route. With modern computer controlled, stored program exchanges, however, this is no longer necessary making a freer, and more user-friendly, numbering system possible.

20.4.2 The modern digital network

Analogue switching and transmission has rapidly been overtaken by digital techniques, which offer more flexible working and increased network capacity. The modern bearer network employs both digital transmission (in the core network) and digital time division switching (sometimes in combination with space division switching, see section 20.6 onwards). The switches are implemented using solid state technology and operated using stored program control (SPC).

A digital network is one that employs digital transmission. A network that employs digital switching in addition to digital transmission is called an integrated digital network (IDN). The UK (BT) switched digital network is now organised in three layers as shown in Figure 20.12. It comprises:

1. **The local network** – the subscriber wiring being largely unchanged from the old analogue network. Over 6000 local exchanges have been merged into larger exchanges, called digital cell centre exchanges (DCCEs), located on about 100 sites. Since digital stored program switches are much more compact than analogue switches these larger 'local' exchanges benefit from economies of scale. The exchange (i.e. the DCCE) is now further away from many subscribers, however, and so remote concentrator units (RCUs) are used to multiplex the traffic from many users on to a single access link. (See section 20.6.6 for a discussion of concentration.)

2. **The trunk network** – comprising approximately 55 digital main switching units (DMSUs) that have replaced more than 400 GSCs and MSCs. The DMSUs are fully interconnected with high capacity (140 Mbit/s or greater) optical fibre and microwave radio links. Repeater spacing is typically 20 km for the optical fibre links but is closer to 50 km for the microwave links (see Chapter 14). Approximately 10% of the installed transmission capacity is to back up faulty or damaged equipment or to provide extra capacity for special events. Gateways provide the trunk network with international access via satellite links or submarine fibre optic cables.

3. **The derived services network** – operates on top of the trunk network and is accessible from any part of the UK. This provides additional services such as 'freephone' and message handling. It incorporates a database that provides call routing, charging, and network management information.

The extent of the digital UK network in the 1980s is shown in Figure 20.13.

A further development of the IDN is the integrated services digital network (ISDN). Whilst the IDN utilises digital technology in the core network the retention of an analogue access network means that the user service it provides is still largely concerned with POTS (or with services that can emulate a POTS-like signal). The ISDN extends the digital technology to the access network allowing the full exploitation of the digital network to provide a wider range of services.

ISDN [Griffiths] is the subject of the ITU-T I series of recommendations [Fogarty], and has been widely developed since 1980. Implementation is well under way but wide-scale uptake by users has been less rapid than predicted. This is probably because of the rise of competing technologies such as xDSL (see section 20.11). Nevertheless the ISDN, and perhaps even more so its broadband successor (B-ISDN), represents an important technology which is addressed in detail later (sections 20.8 and 20.10).

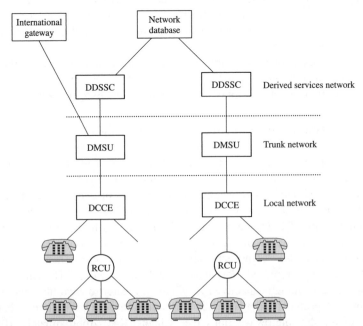

Figure 20.12 *The BT integrated digital network. (DDSSC: Digital derived services switching centre, DMSU: Digital main switching unit, DCCE: Digital cell centre exchange, RCU: Remote concentration unit.)*

20.5 Multiplexing

The bearer network consists at its most basic of switches interconnected by high capacity transmission lines. In order to operate the network economically it is clear that these links must be shared between users. Historically, high capacity trunk links in analogue networks have been shared between users using frequency division multiplex (FDM, see sections 5.4 and 14.3.5). With the advent of digital communications, however, time division multiplex (TDM) has become almost universal.

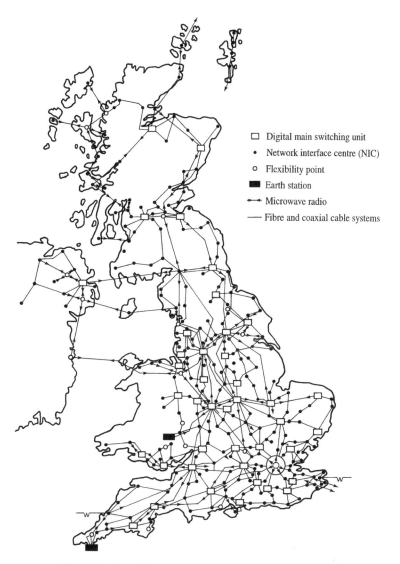

Figure 20.13 *Late 1980s ISDN locations and digital trunk telecommunications network (source: Leakey, 1991).*

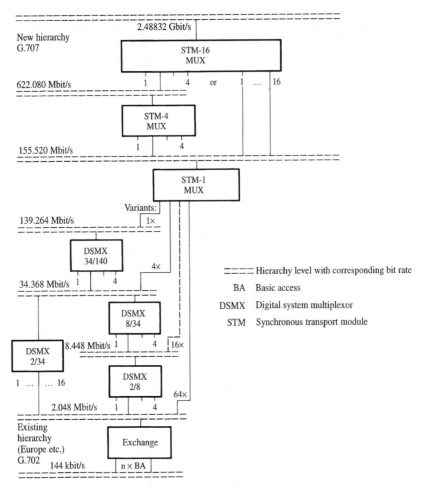

Figure 20.14 *European PCM TDM hierarchy with SDH at the upper levels.*

The internationally agreed European ITU-T standard for PCM TDM digital telephony multiplexing (Figure 5.27) is shown in Figure 20.14. Although this shows the lowest access level at 144 kbit/s (corresponding to ISDN basic rate access, see section 20.8.2) 32 individual 64 kbit/s channels can be multiplexed into a composite 2.048 Mbit/s signal.

Multiplexing allocates a complete communications channel to each active user for the duration of a call or connection. Since the channel utilisation factor for voice communications is low, the traffic could be concentrated by switching between users as they require transmission capacity. This resource saving strategy is not usually used in terrestrial telephony but digital speech interpolation (DSI) is employed on international satellite circuits to achieve a significant increase in capacity (see section 14.3.7).

ITU-T provides for higher levels of multiplexing, above 2.048 Mbit/s, combining four signals, in the digital system multiplexers (DSMXs) 2/8 and 8/34 to form the signal at the higher multiplexing level. At each level the bit rate increases by slightly more than a factor of four since extra bits are added to provide for frame alignment and to facilitate satisfactory demultiplexing. This represents the traditional plesiochronous digital hierarchy (PDH) described in detail in section 20.5.1 below. The upper levels, beyond 140 Mbit/s, form the modern synchronous digital hierarchy (SDH) described in section 20.5.2.

20.5.1 The plesiochronous digital hierarchy

The plesiochronous digital hierarchy (PDH) is founded on the premise that the basic unit of transmission capacity is the 64 kbit/s digital voice channel (arising from an 8 kHz sampling rate and 8 bit quantisation), Chapter 5. Thirty such data channels are byte interleaved along with two additional (control and signalling) channels to form a 2.048 Mbit/s frame called the PCM primary multiplex group, Figure 20.15.

Time-slot (or channel) 0 is reserved for frame alignment and service bits, and time-slot 16 is used for multiframe alignment, service bits and signalling. Time slots 1–15 and 17–31 inclusive each carry an 8 bit word representing one sample from a voice signal. The frame duration is 125 μs corresponding to the frame rate of 8000 frame/s that is required to accommodate real time voice channels. Figure 20.16 shows a schematic diagram of a (European) primary multiplexer.

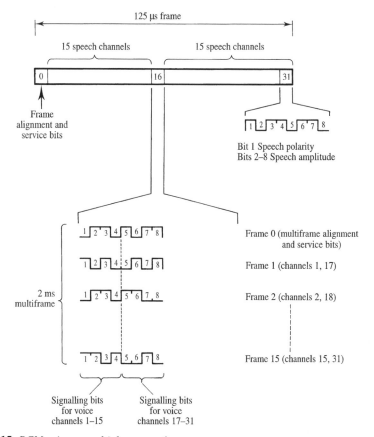

Figure 20.15 *PCM primary multiplex group frame structure.*

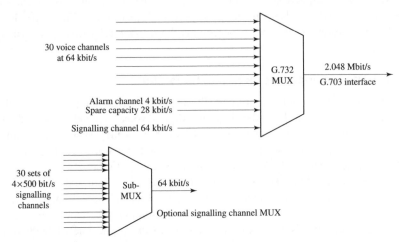

Figure 20.16 *Primary PDH multiplexer.*

Four primary multiplexes are bit interleaved along with some additional signalling and control overhead to form a second level multiplex (MUX), Figure 20.17.

The second level multiplex has a frame duration of 100.4 μs, corresponding to a frame rate of 9.926 kHz (usually referred to, approximately, as 10 kHz), and a bit rate of 8.448 Mbit/s. This process is repeated to form the third and higher level multiplexes. Figure 20.18 illustrates the cascading of digital system multiplexers (DSMXs) up to level 3.

In North America the primary multiplex contains only 24 (64 kbit/s) data channels, i.e. 1.5 Mbit/s, and the multiplexing factors at each level are different from those used in

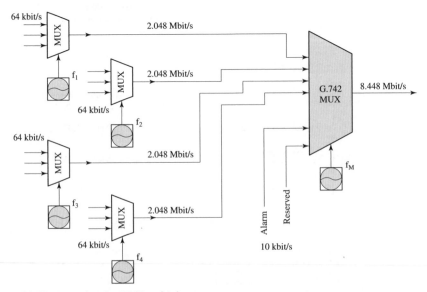

Figure 20.17 *Second order PDH multiplexer.*

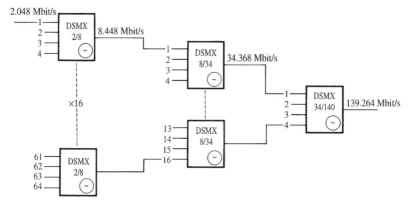

Figure 20.18 *DSMX interconnection to form a plesiochronous multiplex hierarchy.*

Europe. Japan uses a third structure that is similar to the North American one. Figure 20.19 compares the North American and European hierarchies including the equivalent number of 64 kbit/s channels.

The multiplexers in the PDH, which may be widely dispersed geographically, employ separate, free running clock oscillators at each stage – hence the name plesiochronous (which means nearly synchronous). The oscillator in each multiplexer in Figure 20.18 must therefore run slightly faster than the expected bit rate of the aggregated incoming data to exclude the possibility of data accumulating at a tributary input. This strategy successfully solves the problem of small variations in the exact data rates in each of the tributaries but requires some extra bits to be occasionally added to take account of the higher speed oscillator – a process called justification (or, sometimes, bit stuffing). Figure 20.20 illustrates the justification process.

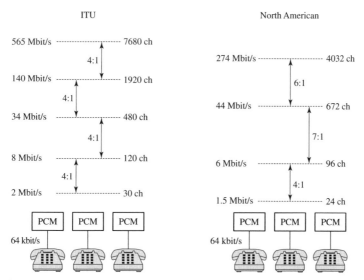

Figure 20.19 *Comparison of European (ITU) and North American hierarchies.*

Input tributary

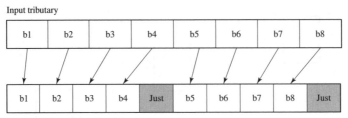

Tributary information in Multiplex

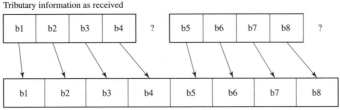

Output tributary: retimed

Figure 20.20 *Illustration of justification.*

The uppermost data stream in Figure 20.20 shows a sample set of input data bits from a tributary. The next stream shows the data as it appears on a multiplexer running at a slightly faster speed. After a few time-slots the multiplexer has run out of data and so must insert dummy information. When recovering information in a demultiplexer, the received data now arrives in data bursts at an increased rate separated by gaps where the dummy data occurs as illustrated in the third data stream. The valid data is then re-clocked to the correct output transmission rate as shown in the bottom data stream. This causes jitter, Figure 6.28, because the justification events are not necessarily regularly spaced in time. In each frame of the multiplex, one bit slot is allocated to justify each one of the input tributary streams. A 3 bit justification control word indicates whether the justification bit slot carries genuine data in this frame, or is effectively empty (i.e. carries dummy data), Figure 20.21.

Most frames do not require the justification slot (which is always present) to carry real data, because the individual clock rates are very close to their nominal value. Table 20.2 shows details of the plesiochronous frame structure at 8 Mbit/s. (Note the bit interleaving, shown explicitly for the justification control bits, used at multiplexing levels above 2 Mbit/s.)

Elastic stores, Figure 20.22, are used in a typical multiplexer to ensure sufficient bits are always available for transmission or reception. These stores are required because the PDH works by interleaving bytes or words from each 64 kbit/s tributary, rather than bit interleaving, to form the 2 Mbit/s multiplex. Thus, at the 8 Mbit/s multiplexing level, and above, where bit interleaving is employed, bits must be accumulated for high speed readout.

The code translators in Figure 20.22 convert binary data from, and to, HDB3 (see section 6.4.5). Figure 20.24 shows a schematic diagram of the multiplexer hardware.

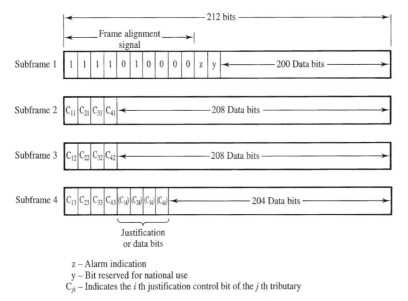

z – Alarm indication
y – Bit reserved for national use
C_{ji} – Indicates the i th justification control bit of the j th tributary

Figure 20.21 *Plesiochronous frame structure.*

Table 20.2 *Frame structure for second order multiplex.*

Frame alignment word	10	bits
Alarm	1	bit
Reserved use	1	bit
Justification control	12	4 sets of 3 bits
Justification bits	4	4 sets of 1 bit
Tributary bits	820	4 sets of 205 bits
Frame length	848	bits

The ITU-T G series of recommendations (G.702) defines the complete plesiochronous multiplex hierarchy. The frame alignment signal, which is a unique word recognised in the receiver, ensures that the appropriate input tributary is connected to the correct output port. The unique word also permits receiver recovery if loss of synchronisation occurs. Figure 20.23 shows the timing details.

The plesiochronous multiplex was designed for point-to-point transmission applications in which the entire multiplex would be decoded at each end. This is a complicated process since it requires full demultiplexing at each level to recover the bit interleaved data and to remove the justification bits. Thus a single 2 Mbit/s channel, for example, cannot be extracted from, or added to, a higher level multiplex signal without demultiplexing down, and then remultiplexing up, through the entire PDH (Figure 20.24) – a process colloquially known as 'scaling the multiplex mountain'.

Many high capacity transmission networks employ a hierarchy of digital plesiochronous signals. The plesiochronous approach to signal multiplexing is severely limited, however, in its ability to meet the modern requirements of network operators. It does not provide, cost-effectively, the flexible network architecture that is required to

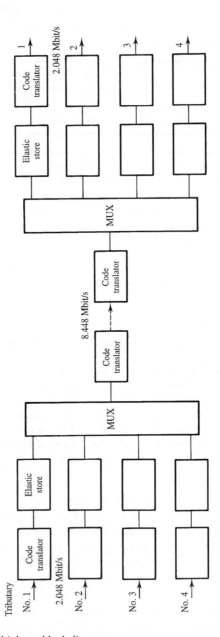

Figure 20.22 *The 2/8 multiplexer block diagram.*

respond to the demands of today's evolving telecommunications market. Furthermore, since network management and maintenance strategies were based, historically, on the availability of a manual distribution frame, there was little need to incorporate extra capacity in the plesiochronous frame structure to support network operations, administration, maintenance and provisioning (OAM&P) activities. Another PDH

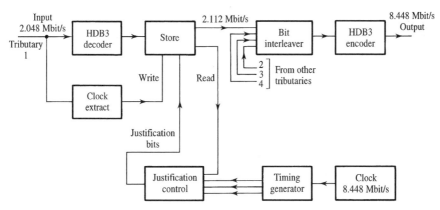

Figure 20.23 *The 2/8 multiplexer timing details.*

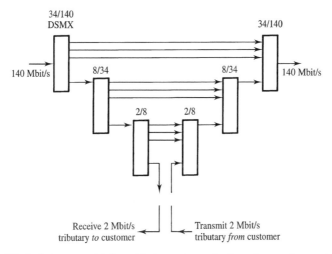

Figure 20.24 *Plesiochronous multiplex add–drop scheme for inserting, and removing, a 2 Mbit/s tributary to, and from, a 140 Mbit/s stream.*

drawback is the use of different plesiochronous hierarchies in different parts of the world. This leads to problems of interworking between countries whose networks are based on primary multiplexes of 1.544 Mbit/s (e.g. Japan and North America) and those basing their networks on primary multiplexes of 2.048 Mbit/s (e.g. Europe and Australia), Figure 20.19.

The PDH limitations described above, and the desire to move from metallic cable to wideband optical fibres, have been important motivations in the development of the new synchronous digital hierarchy (SDH) and synchronous optical network (SONET) systems which are described in the following sections. These new systems offer improved flexibility and can more readily provide the 2 Mbit/s leased lines that, for example, cellular telephone operators require to connect their base station transmitters to switching centres. (Just prior to the move away from PDH towards SDH systems the principal UK carrier (BT) did develop, in the 1980s, a single 64-channel 2 to 140 Mbit/s multiplexer. This has been included in Figure 20.14 as an input to the new 155.52 Mbit/s standard multiplexer rate.)

EXAMPLE 20.4

Find the number of standard PCM voice signals which can be carried by a PDH level 4 multiplex and estimate the maximum channel utilisation efficiency at this multiplexing level.

PDH level 1 (European primary multiplex group) carries 30 voice signals. Each subsequent level combines four tributaries from previous levels. At level n the number of potential voice signals is therefore given by:

$$\chi = 30 \times 4^{n-1}$$

At level 4:

$$\chi = 30 \times 4^{4-1} = 1920 \text{ voice signals, Figure 20. 19}$$

Nominal bit rate at PDH level 4 is 140 Mbit/s. Therefore channel utilisation efficiency is:

$$\eta_{ch} = \frac{\chi}{R_b/R_v} = \frac{1920}{(140 \times 10^6)/(64 \times 10^3)} = 88\%$$

20.5.2 SDH and SONET

Plesiochrononous multiplexing has served telecommunications well, but has been outstripped by the modern need for increasingly flexible network structures. Fortunately the enabling technology has also made great strides in the half century since PDH systems were conceived, and an entirely new strategy based on a synchronous hierarchy has been developed. The aims of this new strategy were:

1. to create a totally synchronous structure enabling any lower level tributary (i.e. multiplex) to be extracted or inserted directly;
2. to take advantage of the large capacities offered by optical fibre transmission links;
3. to create a global standard but remain compatible with all existing PDH transmission rates.

Despite the emphasis on a global standard and compatibility there have, nevertheless, been two development routes, the SONET (synchronous optical network) in the USA and the SDH (synchronous digital hierarchy) in Europe. The two systems are conceptually identical, however, and SDH has been carefully specified to be compatible in practice with SONET.

SONET was initially introduced in the USA in 1986 to establish wideband transmission standards so that international operators could interface using standard frame formats and signalling protocols. The concept also included network flexibility and intelligence, and overhead channels to carry control and performance information between network elements (links, switches, multiplexers) and control centres.

In 1988 the SONET concept was adopted by ITU and ETSI (European Telecommunications Standards Institute) and renamed (with minor changes to the detailed standard) SDH with the aim of agreeing worldwide standards for transmission covering optical interfaces, control aspects, equipment, signalling, etc. [Miki and Siller]. The ITU-T G.707/8/9 standards have now reached a mature stage allowing manufacturers to produce common hardware.

Both SDH and SONET are based on a universal byte interleaved first order frame structure repeated at a rate of 8 kHz [Omidyar and Aldridge]. Every byte location in the frame is therefore repeated every 125 μs and so represents a data rate of 64 kbit/s corresponding to the basic unit of transmission capacity. The 64 kbit/s channel is the common factor between European and North American traditions and all higher order capacities are built from multiples of this rate.

SDH frame structure – the STM

An SDH frame is called a synchronous transport module (STM). The first level frame (STM-1) consists of 2430 bytes, giving an STM-1 transmission rate of 155.520 Mbit/s. The STM-1 frame is illustrated schematically in Figure 20.25.

The upper part of Figure 20.25 shows the transmitted data stream and the lower part shows the frame mapped into a 2-dimensional matrix of 9×270 bytes. This 2-dimensional representation of SDH frames is traditional and is helpful is visualising how the STM is utilised to carry tributary data streams. Signalling and control bytes reside in the first 9 columns of the STM leaving the last 261 columns available to carry data. The first 9 columns, referred to as the transport overhead (TOH), are divided into section overheads (SOHs), which relate to a section of line between regenerators, and line overheads (LOHs), which relate to a line comprising multiple sections between

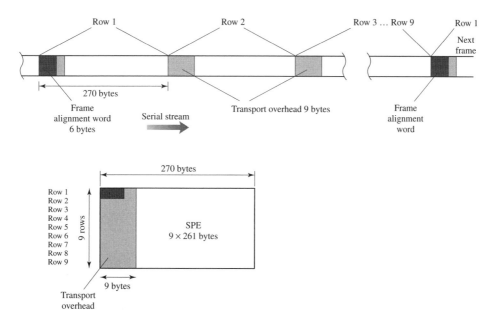

Figure 20.25 *Synchronous transport module STM-1.*

multiplexers. (Sometimes both SOHs and LOHs are referred to generically as SOHs and a distinction made between regenerator section overheads (RSOHs) and multiplexer section overheads (MSOHs), see Figure 20.26.)

LOHs remain unchanged between the ends of a line whilst the SOHs may change for each regenerator section. The TOH contains network management information, sufficient to keep the system in synchronism and to monitor the integrity of the sections and lines.

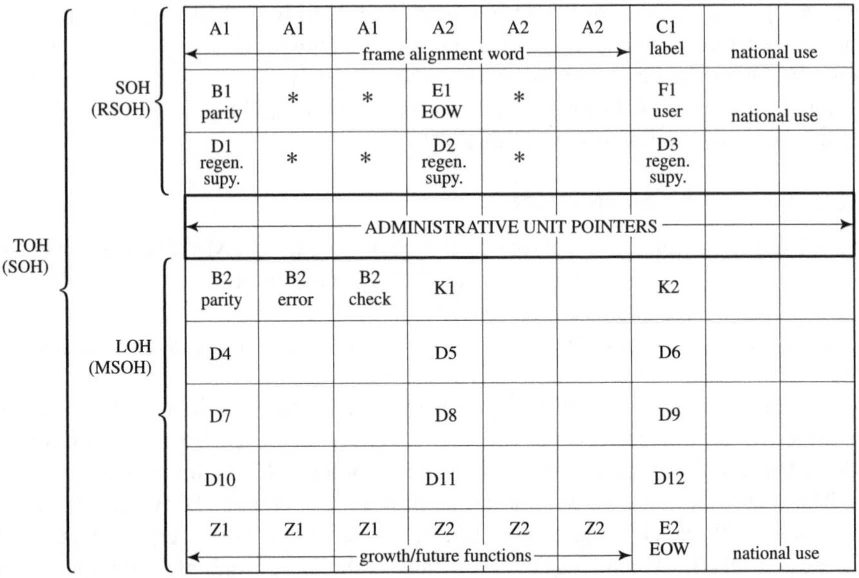

		A1	A1	A1	A2	A2	A2	C1 label	national use
	SOH (RSOH)				frame alignment word				
		B1 parity	*	*	E1 EOW	*		F1 user	national use
		D1 regen. supy.	*	*	D2 regen. supy.	*		D3 regen. supy.	
TOH (SOH)		ADMINISTRATIVE UNIT POINTERS							
	LOH (MSOH)	B2 parity	B2 error	B2 check	K1			K2	
		D4			D5			D6	
		D7			D8			D9	
		D10			D11			D12	
		Z1	Z1	Z1	Z2	Z2	Z2	E2 EOW	national use
				growth/future functions					

Figure 20.26 *Arrangement of overhead bytes in STM-1. (* indicates bytes specific to the transmission medium, K1 and K2 are end-to-end signalling for line termination protection control, D4–D12 are management channels, L/S/TOH are Line/Service/Transport overheads.)*

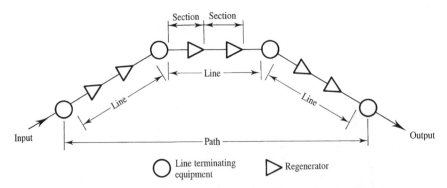

Figure 20.27 *Paths, lines and sections in a communications link. (Line terminating equipment might be represented, for example, by an add–drop multiplexer leading to the alternative designations of regenerator and multiplexer sections for sections and lines respectively.)*

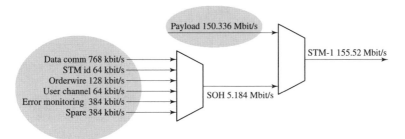

Figure 20.28 *SDH multiplexer.*

The definition of sections (regenerator sections), lines (multiplexer sections) and paths is illustrated in Figure 20.27. Here line terminating equipment might be represented, for example, by an add–drop multiplexer leading to the alternative designations of regenerator section and multiplexer section for sections and lines respectively.

The last 261 columns, Figure 20.25, referred to as the STM-1 payload or synchronous payload envelope (SPE), are principally made up of user data but also include a path overhead (POH). The POH supports and maintains the transportation of the STM-1 payload between the locations where it is assembled and dismantled.

Figure 20.28 shows a schematic diagram of an SDH multiplexer indicating the constituent bit rates. The payload bit rate is approximately 150 Mbit/s whilst the TOH bit rate is approximately 5 Mbit/s. One of the significant advantages of the SDH over the PDH is the increased system capacity for network control information.

Higher levels of SDH multiplexing are constructed by taking multiples of the STM-1. Four STM-1s are interleaved to form an STM-4, four STM-4s are interleaved to form an STM-16 and so on, Figure 20.14.

SONET frame structure – the STS

The SONET frame structure is similar to the SDH frame structure but the nomenclature is slightly different. A SONET frame is called a synchronous transport signal (STS) and the 2-dimensional map of the lowest level SONET frame (STS-1) consists of 9 rows and 90 columns giving a total 810 bytes (6480 bits), Figure 20.29.

The first three columns contain TOHs. The first three rows of the TOH (9 bytes in total) comprise the SOH whilst the remaining six rows (18 bytes) comprise the LOH. The remaining 87 columns, comprising the synchronous payload envelope (SPE), contain the transported data, providing a channel capacity of 50.112 Mbit/s, which includes the service payload (i.e. traffic data) and also a POH, Figure 20.30. As in SDH the POH supports and maintains the transportation of the SPE between the locations where the SPE is assembled and dismantled. The POH consists of 9 bytes and is located in the first column (1 byte wide) of the STS-1 SPE. The STS-1 service payload is carried in the remaining 86 columns of 9 bytes providing a total of 774 bytes per frame or 49.536 Mbit/s of traffic channel capacity.

Like the SDH STM-1, the SONET STS-1 duration is 125 μs (8 kHz frame rate), which gives a bit rate of 51.84 Mbit/s. Table 20.3 shows the relationship between data rates at various levels of the SONET and SDH systems.

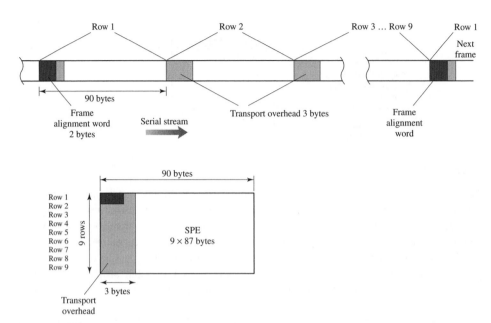

Figure 20.29 *Synchronous transport signal STS-1.*

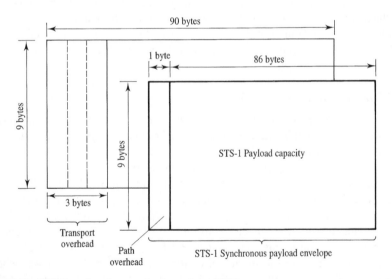

Figure 20.30 *STS-1 synchronous payload envelope (SPE).*

SDH payload partitioning

The STM-1 payload is analogous to a container ship. Containers of various sizes can be packed into it in a variety of patterns. Each container carries a number of bytes of tributary information. Thus a 2 Mbit/s input stream (representing 32 basic channels) will require a container of 32 bytes in each frame, whilst a 1.5 Mbit/s stream (representing 24

Table 20.3 *Transmission rates of synchronous multiplexes.*

SDH	Bit rate (Mbit/s)	SONET
	52	STS-1
STM-1	155	STS-3
	466	STS-9
STM-4	622	STS-12
	933	STS-18
	1244	STS-24
	18661	STS-36
STM-16	2488	STS-48
STM-64	9953	STS-192

basic channels) will require a container of 24 bytes. In this way all the existing PDH bit rates can be accommodated.

Additional control information must be transported with each container to identify the information carried. This extra control information is the POH and a container together with its POH is called a virtual container (VC). The VC is transported as a self contained entity through the system, probably across several lines or sections. The POH therefore represents administrative information that refers to the information in the container rather than to the frame mechanism that carries it.

If all data sources were synchronised to a common clock, containers could be loaded with input information and inserted into the transport module quite freely. However, many of the tributary input streams will come from plesiochronous sources, and so the data will drift backwards and forwards with respect to the transport frame structure depending on the precise clock rate of the tributary. A vital additional control mechanism, the pointer, is therefore needed.

A pointer indicates the start location of each VC. The VC along with its pointer is called a tributary unit (TU) and the pointer is therefore called the TU pointer. Several TUs may be byte interleaved to form a tributary unit group (TUG) and TUGs may be further byte multiplexed to form a higher order TUG. A POH is added at the highest multiplexed level of a TUG to form a higher order VC and these higher order VCs are then loaded into the STM-1.

When a VC is loaded into the STM-1, its starting position is allowed to wander so that the incoming data stream need not be buffered, but can be loaded almost directly. An administrative unit (AU) pointer, which is stored as an additional overhead (see Figure 20.26), indicates the byte position where the higher level VC starts in the frame. In this way, variations in clock rate can be accommodated. The VC together with its associated AU pointer is called an administrative unit (AU). In principle, AUs can be combined to form an AU group (AUG). Three North American (SONET) AU-3s, for example, form an AUG that fits into a European (SDH) STM-1 frame.

Figure 20.31 illustrates the relationship between containers, VCs (container plus POH), TUs (virtual containers plus TU pointer), TUGs (multiplexed TUs), higher order VCs (TUG plus POH), AUs (higher order VC plus AU pointer) and the STM-1 (multiplexed AUGs plus TOHs). The C-12 container is that appropriate for a 2.048 Mbit/s primary multiplex signal, for example.

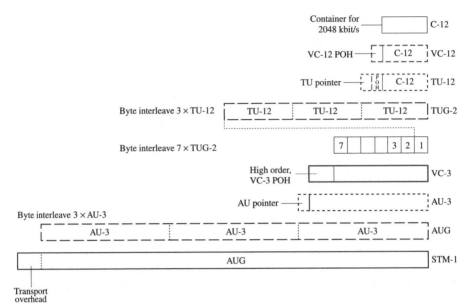

Figure 20.31 *STM-1 structure based on AU-3 and TUG-2 (source: Flood and Cochrane, 1991, reproduced with the permission of Peter Perigrinus).*

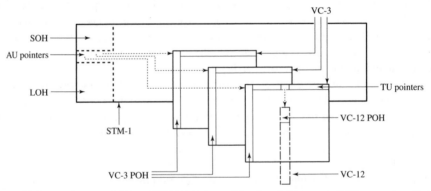

Figure 20.32 *STM using AU-3 and TUG-2 (source: Flood and Cochrane, 1991, reproduced with the permission of Peter Perigrinus).*

Figure 20.32 shows the frame structure for an example mix of containers, tributary groups and AUs used in Figure 20.31.

The number of columns in the STM increases in proportion to the order of higher rate frames. An STM-4 frame would therefore have 36 columns for the TOH and 1044 columns for the payload.

SDH frames are scrambled (with the exception of that part of the SOH containing the frame alignment word), randomising the bit sequence to reduce the risk of clock recovery difficulties caused by unhelpful bit patterns.

SONET payload partitioning

The SONET payload is partitioned in a similar way to SDH. A virtual tributary (VT) structure (corresponding to the tributary unit structure in SDH) supports the transport and switching of payload capacity that is less than that provided by the full STS-1 SPE. There are three sizes of VTs in common use, Figure 20.33. These are:

- VT1.5, consisting of 27 bytes, structured as three columns of 9, which, at a frame rate of 8 kHz, provides a transport capacity of 1.728 Mbit/s and will accommodate a US 1.544 Mbit/s DS 1 multiplex signal, Figure 20.19.
- VT2, consisting of 36 bytes, structured as four columns of 9, which provides a transport capacity of 2.304 Mbit/s and will accommodate a European 2.048 Mbit/s signal.
- VT3, consisting of 54 bytes, structured as six columns of 9, to achieve a transport capacity of 3.456 Mbit/s which will accommodate a US DS1C signal.

A 155.52 Mbit/s SDH transmission capability is obtained by combining three STS-1 SPEs into one STM-1 SPE, Figure 20.25.

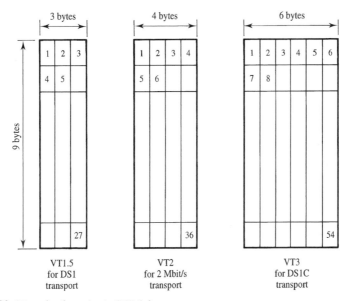

Figure 20.33 *Virtual tributaries in STS-1 frame.*

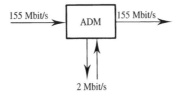

Figure 20.34 *Add–drop multiplexer (ADM) for simplified channel dropping which permits multiplexing into ring networks. (Compare with Figure 20.24.)*

Advantages and flexibility

The use of SDH (and/or SONET) leads to very flexible network operation. For example, once the frame boundary is found, an individual container can be easily located within the transport frame by its pointer. A container of any size can therefore be dropped from, or added to, a passing STM, Figure 20.34.

Capacity can be increased synchronously by moving to higher levels of multiplexing. Growth can be accommodated into the future, even taking vastly increased data rate requirements into account. Network monitoring is facilitated. TOHs give data on regenerative repeaters and multiplexers along the route. There are substantial channels for data signalling within the network and for engineering order wires. POHs enable end-to-end monitoring of the whole route, from sender to receiver.

These and other key advantages of SDH can be summarised as:

1. signals can be inserted and extracted to meet customer requirements (due to byte interleaving with direct visibility of 64 kbit/s channels) more easily and at lower cost;
2. standard mappings for all common data rates into the SDH frame format are provided;
3. increased bandwidth is available for network management (i.e. monitoring and control);
4. reduced size and cost of equipment (particularly that for multiplexing);
5. provision of worldwide standards allows a larger manufacturers' marketplace;
6. new services more easily introduced;
7. remote digital access to services and cross-connections between transmission systems are achieved at lower cost.

Network flexibility implies the ability to rapidly reconfigure networks from a control centre in order to:

1. improve capacity utilisation by maximising the number of 2 Mbit/s channels transported in the higher order system;
2. improve availability of digital paths by centrally allocating spare capacity and protection schemes to meet service requirements;
3. reduce maintenance costs by diverting traffic away from failed network elements;
4. provide for easier growth with temporary diversion of traffic around areas being upgraded.

Figure 20.35 shows a simple example of how network flexibility is facilitated by SDH. An exchange is effectively distributed over an extended geographical area using an optical ring transmission structure. The principal local exchange (PLE) is the main switching centre, but is supported by remote switching units (RSUs) that handle traffic from a small locality perhaps several kilometres from the PLE. The add–drop SDH multiplexers allow small amounts of traffic to be combined into the main transmission path without the need to demultiplex the entire high order signal, as required in the plesiochronous system. Notice that the main path forms a dual ring that is very robust against system failure. The corresponding PDH plan would require radial connections and a large concentration of plant at the PLE.

Flexibility is also achieved using automatic cross-connect switches between SDH systems or between SDH and PDH systems. Automatic cross-connects will gradually replace existing manual cross-connects and allow remote reconfiguration of capacity

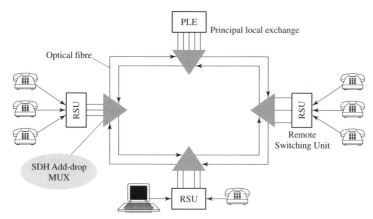

Figure 20.35 *An SDH network example.*

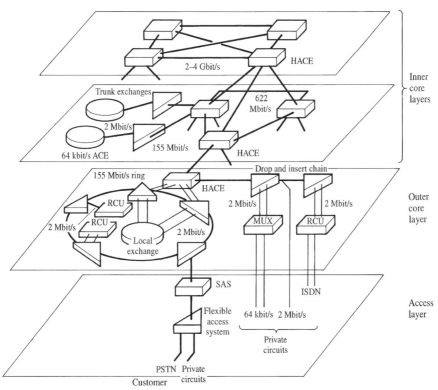

HACE: Higher-order automatic cross-connect equipment
SAS: Service access switch
ACE: Automatic cross-connect equipment
RCU: Remote concentrator unit

Figure 20.36 *SDH transmission network hierarchy. (Source: Leakey, 1991)*

within the network at 2 Mbit/s and above.

The introduction of SDH creates the opportunity to replace traditional network layers and topologies with those better suited to long haul resilient networks. The availability of very high speed, flexible SDH links makes it economic to reconsider the whole structure of the PSTN and replace simple (multiple) two-way transmission paths between the major centres, in which terminal multiplexers are two-port or tributary-line systems, by high speed optical rings, Figure 20.36.

This structure will form the heart of the PDN and is fundamentally more reliable and less costly than the previous solution. When the rings incorporate independent clockwise and anticlockwise transmission they offer great flexibility and redundancy allowing information to be transmitted, via the add–drop multiplexer, in either direction around the ring to its intended destination.

Outer core topologies will be mainly rings of SDH multiplexers linking local exchanges, in contrast to a plesiochronous multiplex where all traffic is routed through a central site. The SDH ring structure with its clockwise and anticlockwise routing is much more reliable than centre-site routing. Furthermore, the SDH ring based network has less interface, and other, equipment. Access regions will remain, principally, star topologies, but will probably be implemented (eventually) using optical technology, Figure 20.37, and/or wireless technology (section 20.11.4).

Tributary jitter

Payload pointers provide a means of allowing an SPE to be transferred between network nodes operating plesiochronously. Since the SPE floats freely within the transport frame, Figure 20.38, with the payload pointer value indicating the location of the first active byte of the SPE (byte 1 of the SPE being the first byte of the SPE path overhead), the problem faced in justified plesiochronous multiplexing where the traffic is mixed with stuffed bits is overcome.

Payload pointer processing does, however, introduce a new signal impairment known as 'tributary jitter'. This appears on a received tributary signal after recovery from an SPE that has been subjected to payload pointer movements from frame to frame. Excessive tributary jitter will influence the operation of the downstream plesiochronous network equipment processing the tributary signal.

EXAMPLE 20.5

Estimate the number of voice channels which can be accommodated by an SDH STM-4 signal assuming that the STM-4 is filled with ITU primary multiplex group signals. Also estimate the channel utilisation efficiency.

Each STS-1 payload envelope has 86 columns (not including the path overhead column). Each primary multiplex group occupies 4 columns. Each STS-1 payload envelope can therefore transport 86/4 = 21 (whole) primary multiplexes.

Each STM-1 payload envelope corresponds to three STS-1 payload envelopes and therefore carries $3 \times 21 = 63$ primary multiplexes. STM-4 carries four STM-1 signals and therefore carries $4 \times 63 = 252$ primary multiplexes. Each primary multiplex carries 30 voice channels, Figure 19.5.

The STM-4 signal can, therefore, carry $30 \times 252 = 7560$ voice channels.

Channel utilisation efficiency, η_{ch}, is thus given by:

$$\eta_{ch} = \frac{7560}{R_b/R_v} = \frac{7560}{(622.08 \times 10^6)/(64 \times 10^3)}$$

$$= \frac{7560}{9720} = 78\%$$

Note that this is lower than in the PDH Example 20.4 to provide more OAM&P overhead privision.

20.6 Circuit switching

In the core network of a circuit switched digital communications system switching occurs between the TDM channels of PCM trunks or multiplexes. It follows that (in general)

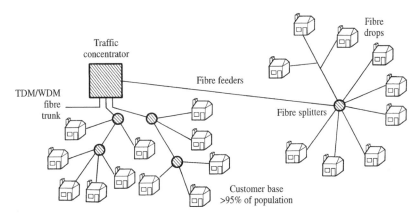

Figure 20.37 *Passive optical network for future local loop implementation.*

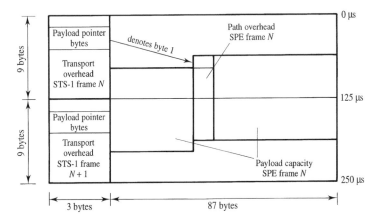

Figure 20.38 *Payload pointer details for locating the start of the STS-1 SPE of Figure 20.29.*

each time-slot of each incoming multiplex must be connectable to each time-slot of each outgoing multiplex. Since the multiplexes are carried on spatially (i.e. physically) separate cables and since the time-slot on the outgoing multiplex will generally be different to that on the outgoing multiplex then it is clear that switching will involve information translation in both space (from cable to cable) and time (from slot to slot). The switches that realise these two operations are called space (S-) switches and time (T-) switches respectively. In the following sections these generic circuit switching techniques and their hybrid counterparts are described.

Packet switches are described later in section 20.7.

20.6.1 Space switching

Space switches have benefited from a long period of development and refinement since this was the sole method of switching used in the pre-digital era. The technical evolution of traditional space switching from manual switchboards, through Strowger two-motion selectors and cross-bar switches to electronic/reed-relay cross-point switches is described in [Flood]. Attention is restricted here to space switching in the context of a modern IDN.

Consider Figure 20.39, which represents a space (S-) switch for a PCM/TDM system.

There are M incoming trunks and N outgoing trunks, each trunk carrying a PCM multiplex. Each multiplex comprises frames with J time-slots. A connection can be made between any two multiplexes for the duration of any particular (common) time-slot by making the control input of the appropriate AND gate HIGH for the duration of the selected slot. The switch is the logical equivalent, therefore, of the structure shown in Figure 20.40 in which all $(M \times J)$ incoming channels are demultiplexed, space switched by J simple cross-point matrices, each of size $(M \times N)$, and then remultiplexed without changing the time-slot of any channel. Since no channel time-slots are changed it can be

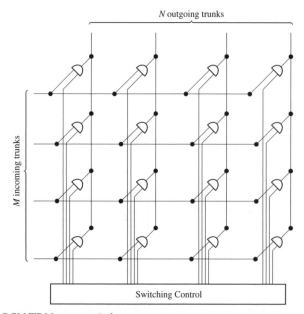

Figure 20.39 *A PCM/TDM space switch.*

seen that switch in Figure 20.39 is a pure space switch despite the superficial impression of time switching that pulsing of the AND gates might give.

20.6.2 Time switching

The space switch described in section 20.6.1 does not alter the time-slot of a channel as it switches between trunk multiplexes. To switch between time-slots of a multiplex requires a time (T-) switch such as that shown in Figure 20.41.

This switch, illustrated for the particular case of a primary multiplex, writes the information in each channel (or time-slot) of the incoming multiplex frame to an address corresponding to that channel's number in an information store. The switch reads the time slot information for the outgoing multiplex from the information store in the order determined by the connections to be made. This order is read from a connection store that holds the channel address (time-slot number) of the incoming slot in the location (at the connection store address) of the outgoing slot. The connection store therefore links each outgoing time-slot number (the store's address) with each incoming time-slot number (the store address's contents).

The T-switch is the logical equivalent of the S-switch shown in Figure 20.42 (again for the case of a primary multiplex) in which all (J) incoming channels are demultiplexed, space switched by a simple cross-point matrix of size ($J \times J$), and then remultiplexed.

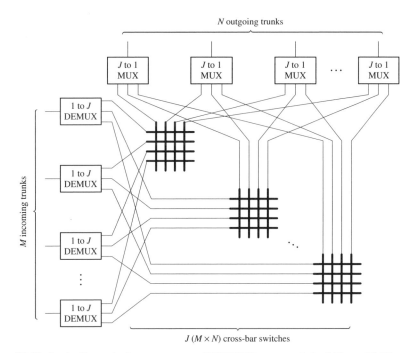

Figure 20.40 *Logically equivalent structure to PCM/TDM space switch of Figure 20.39.*

20.6.3 Time–space–time and space–time–space switching

Space switches connect the same time-slots across different trunks whilst time switches connect different time-slots on the same trunk. In order to connect any time slot of any incoming trunk to any time-slot of any outgoing trunk requires both operations. This can be realised in a time–space–time (T–S–T) switch as shown in Figure 20.43 for the same number (M) of incoming and outgoing trunks.

   This switch connects a given time-slot in a given incoming trunk to a target time-slot in a given outgoing trunk by identifying a time-slot which is free in both the incoming trunk's time switch connection store and the outgoing trunk's time switch information store. The information to be switched is therefore temporarily transferred to this free

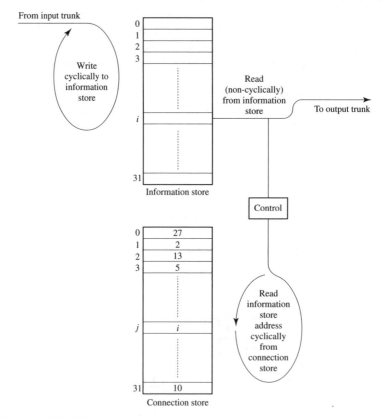

Figure 20.41 *A PCM/TDM time switch.*

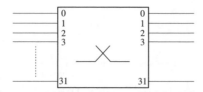

Figure 20.42 *Logical space switch equivalent to PCM/TDM time switch of Figure 20.41.*

time-slot where it can be space switched in the normal way (see section 20.6.1).

Space–time–space (S–T–S) switching is also possible as illustrated in Figure 20.44.

The S–T–S switch connects a given time-slot in a particular incoming trunk to a another time-slot in a particular outgoing trunk by identifying one of the K links whose time switch has both information store for the required time-slot of the incoming trunk and connection store for the required time-slot of the outgoing trunk, free. The required incoming and outgoing multiplexes are space switched to this link and the information transferred by time switching in the normal way (see section 20.6.2).

S–T–S switching was the more common switching technique until the balance of costs between space and time switching changed in favour of the latter. The currently preferred technique, therefore, is T–S–T switching.

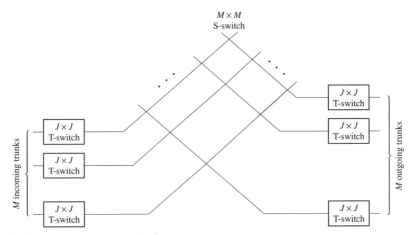

Figure 20.43 *Time–space–time switch.*

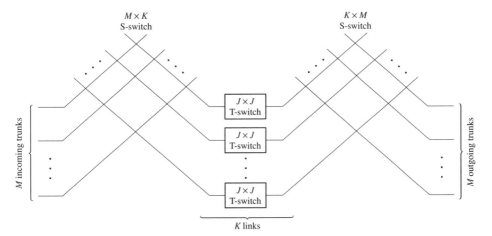

Figure 20.44 *Space–time–space switch.*

20.6.4 Multi-stage space switches

Large space switches are expensive and poorly utilised, since for a square $(M \times M)$ switch (such as might be used for trunk routing) the number of cross-point elements is M^2 whereas the number that can be in use at any one time is only M. The maximum utilisation efficiency of the hardware is therefore only $(1/M) \times 100\%$. This encourages the incorporation of multi-stage space switching where large space switching structures are required leading to switches of the form T–S–S–S–T, for example.

A two-stage space switch (S–S) connecting 25 incoming trunks to 25 outgoing trunks is shown in Figure 20.45.

This switch has only 250 cross-points in contrast to the equivalent single stage switch that would have 625 cross-points and to a first approximation would therefore be half the cost. (In practice, of course, the relationship between switch cost and switch size, as measured by the number of cross-points, is not precisely linear.) The disadvantage is that since there is only one link between each primary sub-switch and each secondary sub-switch, incoming information on a trunk connected to a particular primary sub-switch and destined for a trunk connected to a particular secondary sub-switch may be blocked. This occurs if the link between those primary and secondary sub-switches is already in use, even if the intended outgoing trunk is free. (In contrast the single stage switch is non-

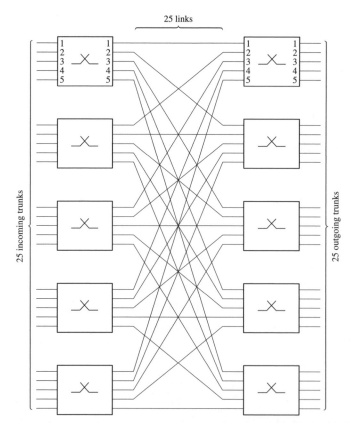

Figure 20.45 *Two-stage 25 × 25 space switch.*

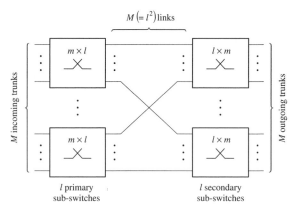

Figure 20.46 *General representation of a two-stage space switch.*

blocking, i.e. a connection can always be made providing the appropriate outgoing trunk is free.)

Figure 20.46 shows a simplified schematic diagram of a more general $M \times M$ two-stage switch structure.

It consists of l ($m \times l$) primary sub-switches and l ($l \times m$) secondary sub-switches. (Note the sub-switches are no longer constrained to be 'square'.) The number of secondary sub-switches equals the number of outputs of each primary switch so that each primary switch can be connected to all secondary switches. Similarly the number of primary sub-switches equals the number of inputs of each secondary switch so that each secondary switch can be connected to all primary switches.

The incoming and outgoing trunks are divided equally between the primary and secondary switches respectively. The number of incoming trunks per primary sub-switch (and outgoing trunks per secondary sub-switch) is therefore given by:

$$m = \frac{M}{l} \tag{20.2}$$

Since traffic is (hopefully) conserved through the switch the total number of links, $L = l^2$, is typically made equal to the number of incoming (and outgoing) trunks in order that the switch is at least potentially capable of carrying the same amount of traffic as the trunks, i.e.:

$$L = l^2 = M \tag{20.3}$$

The number of incoming trunks per primary sub-switch (and outgoing trunks per secondary sub-switch) is, therefore, given by:

$$m = \frac{M}{l} = \sqrt{M} \tag{20.4}$$

Equation (20.4) may be thought of as the first step in a design process for a two-stage $M \times M$ switch. The formula is used in step 1 to calculate a first estimate of m. Step 2 involves choosing a practical value for m that is an integer and, if possible, a factor of M. This practical value must also be convenient, i.e. it must take account of the range of sub-switch units commercially available (e.g. 5×5, 10×10, 10×20, 20×20). The last step is to recalculate *l* from the practical value of *m*.

The above process is simplistic and other considerations might be important. The number, and size, of sub-switches chosen might be influenced, for example, by the physical routes that the incoming and outgoing trunks serve. (For one trunk from each secondary sub-switch to serve each outgoing route requires secondary sub-switches with at least the same number of outlets as the number of routes, thus putting a minimum value on *m*. The same argument applies to the number of inlets to primary sub-switches, and the number of incoming routes.) Nevertheless, the formulae given represent a useful guide to the first order design of a two-stage switch.

EXAMPLE 20.6

Design a switching system using two switching stages to connect 150 incoming trunks with the same number of outgoing trunks.

$$m = \sqrt{M}$$

$$= \sqrt{150} = 12.25$$

Choosing *m* = 15 (to be an integer factor of *M*) we have:

$$l = \frac{M}{m}$$

$$= \frac{150}{15} = 10$$

Assuming that 15×10 switch units are available the design will comprise 10 primary switches each with 15 inputs and 10 outputs and 10 secondary switches each with 10 inputs and 15 outputs. Each primary switch is connected to each secondary switch.

The advantage of the two-stage network of Figure 20.46 over the equivalent $M \times M$ single stage switch is that it has a fewer number (X_P) of cross-points in total ($2ml^2 = 2M^{1.5}$ rather than M^2), and is therefore cheaper.

The disadvantage regarding blocking has already been outlined in the context of the specific example of Figure 20.45. The blocking probability of the two-stage switch can be reduced by moving to a three-stage switch such as that shown in Figure 20.47, which consists of S ($m \times l$) primary sub-switches, S ($l \times m$) tertiary sub-switches and l ($S \times S$) secondary sub-switches. (The number and size of each secondary switch is determined by the requirement that one link must connect each secondary sub-switch input to each primary sub-switch and each secondary sub-switch output to each tertiary sub-switch.)

The number of links, *L*, connecting primary and secondary sub-switches (equal to the number connecting secondary and tertiary sub-switches) is again typically made equal to the number of incoming (and outgoing) trunks in order that the switch is potentially capable of carrying the same amount of traffic as the trunks, i.e.:

$$L = M \tag{20.5}$$

The number of secondary sub-switches is the total number of input links to these sub-

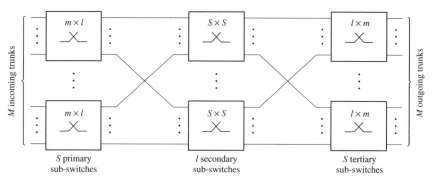

Figure 20.47 *Three-stage space switch.*

switches divided by the number of inputs per sub-switch:

$$l = \frac{M}{S} \tag{20.6}$$

and the number of primary sub-switches (equal to the number of tertiary sub-switches) is:

$$S = \frac{M}{m} \tag{20.7}$$

It follows that the number of secondary sub-switches equals the number of incoming trunks per primary sub-switch (and the number of outgoing trunks per tertiary sub-switch), and the primary and secondary sub-switches are square, i.e.:

$$m = l \tag{20.8}$$

Since M is known it only remains to determine a value for m in order to specify the three-stage switch topology completely. m is chosen to minimise switch cost (as approximately reflected by the total number of cross-points, X_P) by obtaining an expression for X_P in terms of m as follows:

$$
\begin{aligned}
X_P &= S(m \times l) + l\left(\frac{M}{m} \times \frac{M}{m}\right) + S(l \times m) \\
&= \frac{M}{m}(m \times m) + m\left(\frac{M}{m} \times \frac{M}{m}\right) + \frac{M}{m}(m \times m) \\
&= 2Mm + \frac{M^2}{m}
\end{aligned} \tag{20.9}
$$

X_P is thus a minimum when:

$$m = \sqrt{\frac{M}{2}} \tag{20.10}$$

A three-stage $M \times M$ switch can therefore be designed by calculating the number of incoming/outgoing trunks per primary/tertiary switch from equation (20.10), calculating the number of secondary switches from equation (20.8) and calculating the number of primary and secondary sub-switches from equation (20.7).

As for the two-stage switch, the value of m calculated above must be amended to be an integer such that m is a factor of M. Furthermore it should be appreciated that even after sensible interpretation, the above analysis is only a guide and designs may deviate from this when considerations such as routing requirements and the availability of sub-switch standard units are taken into account. (The value of M may have to be inflated, implying some redundant or spare input and output trunks, so that it has factors from which m can be selected.)

Blocking is less in the three-stage $M \times M$ switch than in the equivalent two-stage switch since there are more paths through the switch. In particular it is now possible to simultaneously connect more than one incoming and outgoing pair of trunks terminating on the same primary and secondary sub-switches.

The total number of crossing points in the three-stage switch implied by the above design (without regard to the practicalities which might influence the final design) is $2^{1.5} M^{1.5}$, a factor of $1/\sqrt{2}$ less (better) than the two-stage switch.

Multi-stage switches can be constructed with more than three switching layers [Flood].

EXAMPLE 20.7

Redesign the switching system for Example 20.6 using three switching stages.

$$m = \sqrt{\frac{M}{2}}$$

$$= \sqrt{\frac{150}{2}}$$

$$= 8.66$$

Choosing $m = 10$ (to be an integer factor of M) we have:

$$l = m = 10$$

and:

$$S = \frac{M}{m}$$

$$= \frac{150}{10}$$

$$= 15$$

Assuming the switch units are available in the size required the design comprises 15 primary switches each with 10 inputs and 10 outputs, 10 secondary switches each with 15 inputs and 15 outputs, and 15 tertiary switches each with 10 inputs and 10 outputs. One output of each primary switch is connected to each secondary switch and one output of each secondary switch is connected to each tertiary switch.

20.6.5 Switch connectivity

The connectivity of a switch (see section 20.3.3) is the number of independent (disjoint) paths through the switch between any inlet/outlet pair. This means that a switch may have a different connectivity for each inlet/outlet pair. In the special, but common, case of a symmetrically constructed switch, however, the connectivity between all incoming–outgoing trunks is the same and a single value of connectivity can be assigned to the switch as a whole. A two-stage switch, such as that shown in Figure 20.45, must be fully connected (i.e. have a connection from each primary sub-switch outlet to each secondary sub-switch inlet) to achieve a connectivity of unity but a three-stage switch may achieve this whilst being only partially connected (i.e. not having links between every primary and secondary sub-switch and every secondary and tertiary sub-switch). The term fully connected is not, therefore, synonymous with a connectivity of 1.0.

20.6.6 Concentration and expansion

A concentrator is an asymmetrical switch that connects a given number of incoming channels to a smaller number of outgoing channels. In one type of concentrator, for example, local subscribers can be statistically multiplexed into the time-slots of a PCM trunk directly. Alternatively, they can be time division multiplexed into fixed time-slots of PCM trunks in the normal way and these multiple trunks then concentrated onto a smaller number of trunks. An example of a two-stage concentrator connecting 500 incoming trunks to 100 outgoing trunks is shown in Figure 20.48(a). Concentration is typically employed in local exchanges (or, in IDN nomenclature, DCCEs, Figure 20.12) where many individual subscriber lines are interfaced with PCM trunks. They may be physically located at the DCCE or located remotely. In the latter case they are controlled using the signalling slots of the PCM trunk(s) connecting it to the router switch in the DCCE. A concentrator that can provide switching services, independent of the DCCE, to those subscribers connected to it is called a remote switching unit (RSU), Figure 20.35.

Concentration can also be used to reduce the cost of symmetrical switches in a DCCE where the incoming and outgoing trunks are relatively lightly loaded. Figure 20.48(b) shows an example in which a three-stage switch is used to connect 100 incoming trunks to 100 outgoing trunks. Note that concentration (from 100 trunks to 50 links) occurs between the first and second switching stages. Cost is reduced since the total number of cross-points is fewer than would be needed without concentration. The penalty paid for the cheaper switch is an increased probability of blocking, fewer interstage links meaning that a connection from a particular incoming trunk to a particular outgoing trunk might be prevented because the required link is already in use, even though the outgoing trunk is free. The reduced cost savings of switching in this case outweigh the increased probability of blocking, however, since low levels of traffic mean that blocking is acceptable even after this increase.

An expander (the opposite of a concentrator) is an asymmetrical switch that connects a given number of incoming channels to a larger number of outgoing channels. Figure 20.49(a) shows such a switch connecting 100 incoming trunks to 500 outgoing trunks. Expansion also refers, however, to the reduction of blocking in a switch by the provision of extra links between switching stages which increases connectivity, Figure 20.49(b). (Note in Figure 20.49(b) that there are more links between switching stages than there are incoming or outgoing trunks.) Expansion to reduce blocking is employed in routers

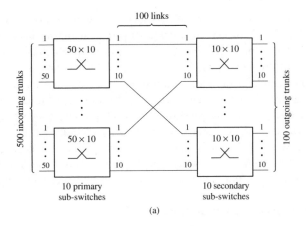

(a)

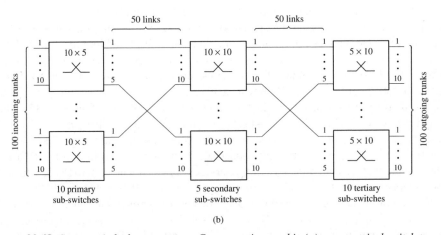

(b)

Figure 20.48 *Space-switched concentrator. Concentration used in (a) asymmetrical switch to concentrate traffic, (b) symmetrical switch to reduce switch cost.*

connecting long and expensive trunks such as are typically found in an international exchange.

The increased cost of the switch is, in this case, more than compensated for by the increase in revenue achieved by keeping the trunks busy due to decreased switch blocking. (A point of diminishing returns occurs when traffic, and therefore revenue, lost due to switch blocking is less than that lost due to trunk congestion.)

20.6.7 Switch design

The design of switches is usually a compromise between cost and blocking performance constrained by a number of practical requirements such as the number of routes to be served by the incoming and outgoing trunks and the size (number of inlets × number of outlets) of practically available sub-switches. A typical condition, for example, is that each route must be served by at least one trunk from each final stage sub-switch. (If traffic warrants it, however, more than one trunk from each final stage sub-switch may be

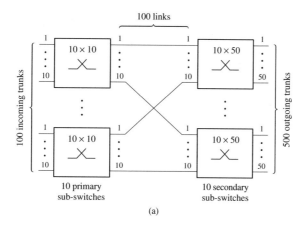

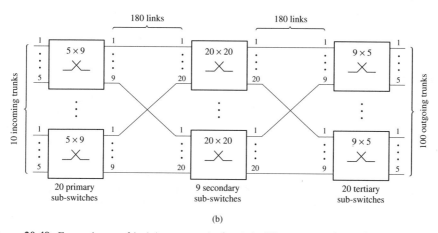

Figure 20.49 *Expansion used in (a) asymmetrical switch, (b) symmetrical switch to reduce probability of blocking.*

required for some routes.) Another typical condition is that the number of interstage switch links should be at least equal to the number of input trunks or the number of output trunks, whichever is the smaller (thus ensuring the switch is at least potentially capable of carrying the maximum traffic volume as limited by the trunks).

It can be shown [Flood] that for a two-stage concentrator with M incoming trunks and N outgoing trunks $(M > N)$, the total number of cross-points is minimised by choosing the number of inlets per primary sub-switch, m, and the number of outlets per secondary sub-switch, n, to be:

$$m = n = \sqrt{M} \qquad (20.11)$$

where m and n are adjusted to be integer factors of M and N respectively. The grade of service (blocking probability) under the condition that any outgoing trunk on the correct route can be used is significantly improved, however, by substituting the design formulae:

$$n = \sqrt{N} \qquad (20.12)$$

$$m = \frac{M}{\sqrt{N}} \tag{20.13}$$

even though this results in more cross-points.

For a two-stage expander ($M < N$) the equivalent formulae are:

$$m = n = \sqrt{M} \tag{20.14}$$

to minimise cross-points and:

$$m = \sqrt{M} \tag{20.15}$$

$$n = \frac{N}{\sqrt{M}} \tag{20.16}$$

to give a better grade of service under the 'any outgoing trunk on the correct route' condition.

For the case where the switch is neither a concentrator nor an expander (i.e. when $M = N$) the design formulae become identical and either can be used. Similar design formulae are given in [Flood] for switches with more than two stages.

EXAMPLE 20.8

Design a minimum cost two-stage concentrator to connect 100 incoming trunks to 20 outgoing trunks serving 5 outgoing routes. The switch should be potentially capable of carrying at least as much traffic as the input trunks or the output trunks whichever is the lesser. Suggest an alternative design that relaxes the strict minimum cost constraint to improve blocking performance. Compare the number of cross-points required for the two designs.

The design is constrained by the requirement for each secondary switch to have at least 5 outlets (one to serve each outgoing route). The minimum cost design starts with equation (20.11):

$$m = n = \sqrt{100} = 10$$

m is an integer factor of M and n is an integer factor of N in this case and so no adjustment of either is necessary. The minimum cost design therefore consists of 10 primary switches each with 10 inlets and 2 secondary switches each with 10 outlets (which satisfies the routing constraint). Each primary switch requires 2 outlets (one for each of the secondary switches) and each secondary switch requires 10 inlets (one for each primary switch). The number of interstage links is 20, which is equal to the lesser of the numbers of input links and output links.

The alternative design starts with equations (20.12) and (20.13):

$$m = \sqrt{N} = \sqrt{20} = 4.47$$

$$m = \frac{M}{\sqrt{N}} = \frac{100}{\sqrt{20}} = 22.4$$

n is adjusted to be 5 (an integer factor of N) and m is adjusted to be 20 (an integer factor of 100). The design now consists of 5 primary sub-switches each with 20 inlets and 1 secondary sub-switch with 20 outlets. The constraint for the number of links to be at least 20 (the lesser of incoming and outgoing trunk numbers) means that the secondary sub-switch must have 20 inlets and it follows that each primary sub-switch must have 4 outlets. The number of outlets on the secondary switch exceeds the number of outgoing routes so the route constraint is satisfied.

The minimum cost design has $10 \times 10 \times 2 + 2 \times 10 \times 10 = 400$ cross-points and the alternative (more expensive) design has $5 \times 20 \times 4 + 20 \times 20 = 800$ cross-points.

20.6.8 Probability of blocking

The probability of blocking (i.e. grade of service, GoS) offered by a switch is the probability, B, that the appropriate link(s) or appropriate trunk(s) are not free thus preventing a connection being made and causing a call to be lost. These probabilities depend in turn on the switching constraints that may be categorised as follows:

1. connections may only be made to one particular outgoing trunk (that occurs when one outgoing trunk from each final sub-switch serves each route – typically the case in a concentrator);
2. connections can be made to any trunk serving the particular route required (typically the case in a router);
3. connections can be made to any free outgoing trunk (that occurs when all outgoing trunks serve the same route – typically the case in an expander).

Consider a two-stage switch with one link between each primary sub-switch and each secondary sub-switch which has an occupancy (i.e. fraction of links in use, or traffic carried divided by the number of links), a, and output trunk occupancy, b. (If the number of links equals the number of trunks then a will be equal to b.) The traffic carried is equal to the traffic offered (A) multiplied by one minus the probability of blocking (B). If the probability of blocking is small (which is always the case in a well designed system) then traffic carried is approximately equal to traffic offered and link occupancy is approximately equal to number of links divided by traffic offered.

The probability of blocking, B, assuming trunk and link use are independent random events is given by: $B = 1 - (1 - a)(1 - b)$ where $(1 - a)(1 - b)$ is the probability that both required link and trunk are free.

Under the category 1 switching constraint only one particular link can be used to connect to the required trunk and the blocking probability is therefore simply the probability that this link is already in use, i.e.:

$$B_{cat1} = a \qquad (20.17)$$

Under the category 2 switching constraint where one trunk from each secondary sub-switch serves the same route the probability of blocking is:

$$B_{cat2} = [1 - (1 - a)(1 - b)]^S \qquad (20.18)$$

where $(1 - a)$ is the probability of a given link being free, $(1 - b)$ is the probability of a given trunk being free and S is the number of secondary sub-switches (assumed here equal to the number of primary sub-switches). The term $(1 - a)(1 - b)$ is the probability that a given link is free and a given trunk is free. The square bracket is therefore the probability that a given link–trunk pair allowing connection is not free and since there are S possible link–trunk pairs the full expression is the probability that none of them are free.

Under the category 3 switching constraint, if the number of incoming (M) trunks, the number of links and the number of outgoing trunks (N) are all equal then there can be no blocking since all the traffic offered can always be routed to outgoing trunks, i.e.:

$$B_{cat3a} = 0 \qquad\qquad (20.19(\text{a}))$$

For a (category 3) concentrator, however (in which $M > N$), then the probability of blocking is simply that of N trunks offered A Erlangs of traffic, i.e. the Erlang-B lost call formulae (see equation (17.32), section 17.3.2):

$$B_{cat3b} = \frac{\dfrac{A^N}{N!}}{\displaystyle\sum_{k=0}^{N} \dfrac{A^k}{k!}} \qquad\qquad (20.19(\text{b}))$$

where A is the traffic offered to the switch.

In a three-stage switch with primary–secondary link occupancy a_1, and secondary–tertiary link occupancy a_2, the equivalent formulae are [Flood]:

1. $B_{cat1} = [1 - (1 - a_1)(1 - a_2)]^{S_2}$ where S_2 is the number of secondary sub-switches.
2. $B_{cat2} = [B_1 + b(1 - B_1)]^{S_3}$ where S_3 is the number of tertiary sub-switches (assumed to be equal to the number of primary sub-switches).

EXAMPLE 20.9

A two-stage switch comprises 10 primary units each with 15 inputs and 10 outputs and 10 secondary switches each with 10 inputs and 15 outputs. Each outgoing route is served by one trunk from each secondary unit. Find the maximum quantity of traffic that can be carried if the grade of service is not to rise above 0.1%.

Since each outgoing route is served by one trunk from each secondary unit any free link can be used to make a required connection. The blocking probability is, therefore, given by equation (20.18):

$$B_{cat2} = [1 - (1 - a)(1 - b)]^{S}$$

The probability of a link (between primary and secondary switch banks) being occupied is given by the traffic carried (in Erlangs) divided by the number of links, i.e.:

$$a = A/100$$

Similarly, the probability of an outgoing trunk being occupied is given by the traffic carried divided by the number of trunks, i.e.:

$$b = A/150$$

Substituting the required grade of service $(B = 0.001)$ and the number of secondary switches $(S = 10)$:

$$0.001 = \left[1 - \left(1 - \frac{A}{100}\right)\left(1 - \frac{A}{150}\right)\right]^{10}$$

Solving the resulting quadratic equation gives $A = 35.0$ or 215.2 Erlangs – the latter solution being excluded on the grounds that neither the links nor trunks can carry this volume of traffic.

The maximum quantity of traffic that can be carried whilst maintaining a blocking probability of 0.001 is therefore 35 Erlangs.

20.6.9 Circuit switched data over the PSTN

The PSTN is the public switched telephone network, originally designed for analogue speech signals, and providing the equivalent of a metallic connection between two subscribers. It has the following facilities:

1. High capacity trunks.
2. Multiplexers to share that capacity between many users.
3. Switches to route data as required.
4. Network management to monitor traffic and quality of service, and to share load and reroute traffic in the event of equipment faults or for maintenance.

The modern PSTN employs digital technology for transmission and switching and is therefore an integrated digital network (IDN). The subscriber loop, however, remains analogue and so digital data must be converted into an analogue signal within an audio bandwidth for transmission over this (access) part of the network. A device for carrying out this conversion, in both directions, is known as a modem (modulator/demodulator).

MODEMs

Table 20.4 gives a summary of some of the modem standards in the ITU-T/CCITT V series of recommendations, see section 11.6. (CCITT was the predecessor committee to ITU-T.) Note that PCM in Table 20.4 means that a 'DC' signal level corresponding to each 8 bit PCM word (according to the appropriate companding law, see section 5.7.3) to be transmitted is 'held' on the analogue local loop for 125 μs. In practice, however, resolution, and other, limitations may mean that each sample is represented by fewer than 8 bits. The modem (or DCE) is used as shown in Figure 20.50.

Using a pair of modems, section 11.6, a data connection can be formed over the PSTN by adding a suitable protocol controller at each end and a number of WANs have been implemented in this way. The transmission route may be switched for the duration of the call, or may be a dedicated or leased line. The modems allow the transmission of data, but the set-up of a call and the subsequent interchange of data must be controlled by a protocol or interface specification.

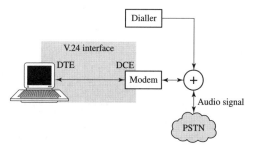

Figure 20.50 *Transmission of data over the PSTN.*

Table 20.4 *Principal characteristics of V series audio modems.*

ITU-T/CCITT recommendation	Data rate (kbit/s)	Modulation	Operation
V.21 (CCITT)	0.3	FSK	Two-frequency duplex
V.22 (CCITT)	1.2	4-phase DPSK	Two-frequency duplex
V.22bis (CCITT)	2.4	16-QAM	Two-wire duplex
V.23 (CCITT)	1200 forward 75 reverse	FSK	Two-frequency duplex
V.26 (CCITT)	2.4	4-phase DPSK	Half duplex
V.27 (CCITT)	4.8	8-phase DPSK	Half duplex with equaliser
V.29 (CCITT)	9.6	16-QAM	Half duplex, adaptive equaliser
V.32 (CCITT)	9.6	TCM	Full duplex, echo cancellation
V.33 (CCITT)	14.4	TCM	Four wire leased circuit
V.34 (CCITT) V.34 (ITU)	28.8 33.6	TCM TCM	Full duplex, echo cancellation
V.90 (ITU)	33.6 upstream 53.0 downstream	Upstream as per V.32 Downstream PCM (5–8 bit/sample)	Full duplex, echo cancellation
V.92 (ITU)	53.0 downstream 48.0 upstream	PCM (5– 8 bit/sample)	Full duplex, echo cancellation

The V.24 interface

The V.24 recommendation defines a method of data interchange between a DTE and a DCE, Figure 20.50. It is not concerned with the interface between the modem and the local loop transmission line and is therefore independent of both modulation technique and data throughput rate. The DTE may be a user terminal or computer. V.24 is the specification for telecommunication networks, but RS-232 and EI-A232 represent the same interchange specifications from different standards bodies. The original standard was published in 1962, with revisions in 1969 (RS-232-C), and 1987 (EI-A232-D), and has been widely adopted for an extensive range of computer related equipment.

The interface is based on a D-type connector, not unlike Figure 18.3, but on this occasion with 25 connections whose principal functions are given in Table 20.5.

Most of these signals are constant level control signals, with 0 being represented by a positive voltage and 1 by a negative voltage. The voltage thresholds are respectively ±3 V.

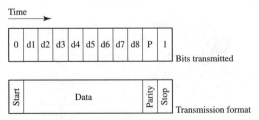

Figure 20.51 *Serial transmission of ASCII character.*

Table 20.5 *V.24 interface signals.*

Signal	Circuit number	Name	Pin number	Source
Transmitted data	103	TXD	2	DTE
Received data	104	RXD	3	DCE
Request to send	105	RTS	4	DTE
Clear to send	106	CTS	5	DCE
Data set ready	107	DSR	6	DCE
Data terminal ready	108	DTR	20	DTE
Ringing indicator	125	RI	22	DCE
Carrier detect	109	CD	8	DCE

TXD and RXD, however, are asynchronous signal sequences, each one representing an ASCII character, packaged as shown in Figure 20.51.

The idle condition of the line is assumed to be logical 1, so that the start bit unambiguously indicates the start of the transmission. In practice, the other fields in the data format are flexible; there can be 5, 6, 7 or 8 bits of data, the parity can be odd/even or none, and the stop symbol can be 1 or 2 bits in length. This flexibility is historical, but most current applications use 7 or 8 data bits and 1 stop bit.

An ASCII character is represented by a seven-digit binary code and hence 128 different characters are available. It is widely used to represent information that is divided broadly into two categories: printable characters and control characters, Table 20.6.

Table 20.6 *Broad categories of ASCII characters.*

Numerical value	Character class	Examples
0–31	Control	NUL, SOH, STX, $\cdots$
32–63	Numerics and graphics	&, !, %, $\cdots$
64–95	Upper case	A, B, C, D, $\cdots$
96–128	Lower case	a, b, c, d, $\cdots$

In order to exchange information between DTE and DCE, request to send (RTS) and clear to send (CTS) signals form a handshake combination, qualified by the state of readiness of each device as signalled by data terminal ready (DTR) and data set ready (DSR). Exchange of information over a network, such as the PSTN, is rather more complex, and is summarised in Figures 20.52 and 20.53. Notice the sequential operation, and the method of recovery from failure to connect. Other faults can occur during data transfer, but these are dealt with under data-link control (see section 18.3).

Points to notice in Figure 20.53 are:

- The local handshake: RTS/CTS, before data is exchanged.
- The remote handshake: Carrier ON/OFF, and confirmation message. (This latter is a serial character sequence in ASCII form.)
- That delays take account of signalling times, modem responses, etc.
- That clear down is not acknowledged.

20.7 **Packet switching**

With the increasing power of VLSI technology, a large switch array can now be implemented on a single chip. National Semiconductor, for example, developed a 16×16 switching matrix in 2 μm CMOS gate array technology in the late 1980s. Figure 20.54 shows an 8×8 Banyan switch, consisting of 12 2×2 switching elements.

There is only one route through the switch from each input to each output. At each stage in this switch the upper or lower output is chosen depending on whether a specific digit in the route control overhead is 1 or 0. This is one of the simplest switch arrays that

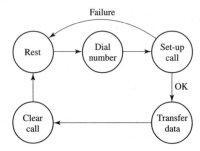

Figure 20.52 *V.24 state diagram.*

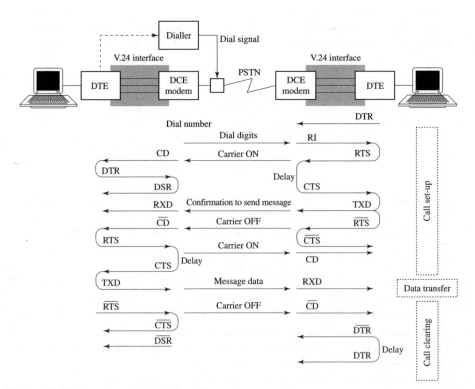

Figure 20.53 *Progress of a V.24 data call.*

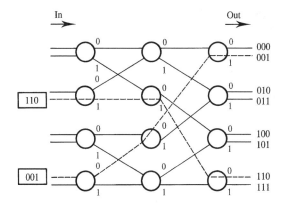

Figure 20.54 *An 8 × 8 Banyan packet switching network.*

can be constructed and illustrates the self routing capability of a packet navigating a network composed of such switches. When routing and sorting is performed locally at each switch in a network there is no need for recourse to a centrally located processing centre and the throughput is therefore greater than that of a circuit switched network. Dashed lines in Figure 20.54 show two example paths that packets with specific route headers would take through the network.

A problem inherent in all packet switched networks is that of packet collision or blocking (i.e. packets arriving simultaneously on the two inputs of a single switching element). To reduce packet collisions, the network switches can be operated at higher speed than the inputs or storage buffers can be introduced at the switch sites.

To remove packet collisions completely, the input packets must be sorted in such a way that their paths never cross. The technique normally used to perform this operation is known as Batcher sorting, in which the incoming packets are arranged in ascending or descending order according to route header.

The Starlight network switch, which uses both Batcher sorting and Banyan switching, is shown in Figure 20.55.

A CMOS 32×32 Batcher–Banyan switch chip in the early 1990s operated at 210 Mbit/s. Batcher–Banynan switching is not able to resolve the problem of output blocking when multiple packets are destined for the same output port, but this can be overcome by

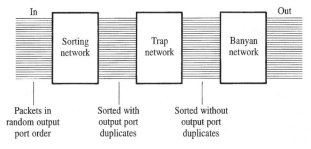

Figure 20.55 *Starlight packet sorting–switching network.*

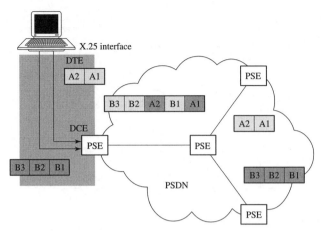

Figure 20.56 *Conceptual operation of a PSPDN.*

the insertion of a trap network between the sorting and switching networks, Figure 20.55. The loss of trapped packets is overcome by recirculating the duplicate address packets into the next sorting cycle to eliminate the need for buffered switch elements.

The network is packet synchronous if it requires all packets enter the network at the same time. Banyan networks can be operated packet asynchronously because each packet's path through the self routing network is, at least to first order, unaffected by the presence of other packets. Packet switches such as these, which can operate at input rates well beyond 150 Mbit/s, have been realised. (Note that the Banyan name is now often associated with a particular commercial system.)

20.7.1 Packet switched data over the PDN

PDNs (public data networks) are logically separate from the PSTN but often make use of the same physical trunk transmission capacity in the bearer network. They may be circuit switched (CSPDNs) like the PSTN or packet switched (PSPDNs) but the latter is now more usual.

A message to be conveyed across a PSPDN is split into data packets of typically 100 to 1000 bytes in length, each packet then being transmitted through the network separately. Trunk connections are shared between several messages and in some networks packets can respond to congestion by taking a more lightly loaded route. Figure 20.56 illustrates the conceptual operation of a PSPDN and shows how one data terminal can have two logical connections open to two quite distinct destinations. Packets are labelled so that as they encounter a switch in the network, each one is directed towards its correct destination. The connected path exists only in logical form, as a set of software directions, rather than as reserved transmission capacity.

PTTs offer value added network (VAN) services that overlie the basic packet communication network and provide a data service that can be easily used. BT has offered a packet switched service (PSS) for some years (over the same part of the bearer network used for the PSTN, but using additional packet switches for routing) forming one section of an international packet switched service (IPSS). Some other components of the IPSS are shown in Table 20.7.

Table 20.7 *Examples of packet switched services.*

Australia	AUSTPAC
Canada	DATAPAC
France	TRANSPAC
Germany	DATEX-P
Italy	ITAPAC
Japan	DDXP
Malaysia	MAYPAC
Netherlands	DATANET-1
New Zealand	PACNET
Norway	DATAPAC
Singapore	TELEPAC
South Africa	SAPONET
Switzerland	TELEPAC
USA	TELNET
	TYMNET
	COMPUSERVE

A packet network protocol must, as a minimum, support the following facilities.

1. Virtual circuit and/or datagram transport.
2. Flow control.
3. Error control.
4. Multiplexing.

Most packet networks have traditionally complied with ITU-T recommendation X.25 (first proposed in 1976 and modified several times since then) that specifies the user interface.

The X.25 interface

X.25 pre-dates the establishment of the ISO seven-layer model and so does not correspond exactly to it, but successive revisions have ensured that the specification approximately matches the lowest three levels, ISO layers 1–3, Figure 17.21. Only layer 3, the X.25 packet layer protocol (X.25 PLP), is described here (see also section 18.3.3). The data-link layer of X.25 (LAP-B, a version of HDLC) is discussed in section 18.3.4 and the physical layer of X.25 (X.21 or X.21bis) is discussed in section 18.2.1.

X.25 describes an interface to a packet switched network, not the whole network. It allows a DTE to interface to a DCE, as shown in Figure 20.57. In the case of X.25 the DCE is a fairly complicated microprocessor system.

The functional description of the X.25 protocol is shown in Figure 20.58, where comparison is made between ISO and X.25 terminology. The packet layer offers simultaneously a large number of virtual circuits to a variety of destinations, and relies upon the two lower layers that do not distinguish between packets, to provide transmission services. There are supervisory packets and data packets, which are both treated the same for transport across the network.

The physical layer defines the connection between the DTE and DCE, and the data-link layer ensures that packets are exchanged in error-free form, using LAP-B.

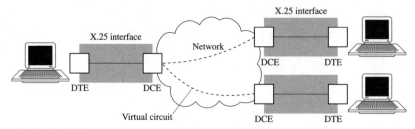

Figure 20.57 *The X.25 concept.*

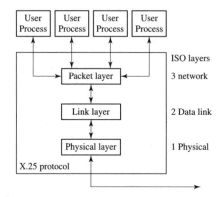

Figure 20.58 *X.25 logical structure.*

At the network level, the principal services offered to the user by the X.25 interface are:

1. virtual circuit (VC);
2. permanent virtual circuit (PVC);
3. fast select.

A virtual circuit (VC) exists as a route for packets through a network, and although it gives the appearance of being a dedicated connection, it uses the network resources only when packets are actually in transit. Transmission and switching plant is therefore shared continuously between many users, making much better use of the equipment than a switched circuit connection. At the commencement of a call, the VC is set up through the network by the passage of supervisory packets. This is a connection-oriented service.

A permanent virtual circuit (PVC) is similar to the VC, except that it always exists and does not need to be established each time a call is generated. Data packets can therefore be exchanged at any time, without the preamble for requesting a call connection, or the post-amble for clearing the connection. Such a connection corresponds to a leased line through an ordinary circuit switched network.

Fast select is a service that transmits and acknowledges a short self contained message, somewhat similar to a datagram (see section 18.4.5). It is connectionless.

X.25 call set-up

Before transmitting data packets to a remote user, the VC must first be set up through the network by means of supervisory (S) packets, which are only a few bytes in length. An interchange of packets takes place as illustrated in Figure 20.59.

The call request packet passes from the initiating DTE through its DCE and into the network. Upon arriving at the destination DTE, a call accepted packet is generated and returned to the calling DTE. The VC is thus established through all switches on the route. Data packets are now exchanged as required until the source DTE has completed the transaction. A clear request packet is then transmitted to, and acknowledged from, the local DCE, and the source DTE and DCE clear down. Meanwhile the clear request packet

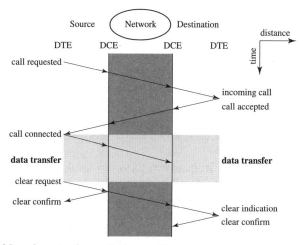

Figure 20.59 *X.25 packet interchange during a call.*

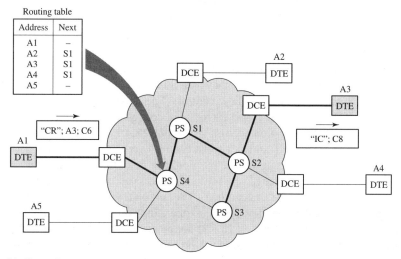

Figure 20.60 *Call set-up in a packet network.*

continues through the network, and causes the destination DTE and DCE to terminate their end of the VC.

Clearly there are problems if the clear request packet gets lost in the network, but the full X.25 specification, which is extremely complex, allows for recovery from such a position. The events involved in setting up a call are illustrated by the example shown schematically in Figure 20.60.

The calling-DTE chooses the next available logical channel number (C6 in this example), generates a call request (CR) packet and sends it to its local DCE on that channel. This call request packet contains the full address of the called station (A3).

The local DCE passes the packet on to the nearest packet switch (S4 in this example). Internally, the network may operate by a protocol different to X.25, so the packet structure may be changed at the DCE, but the ongoing packet still retains the full address of the called-DTE, as well as a logical channel number that associates the packet with its source. No description of the packets within the network has therefore been given.

At the packet switch, the called-DTE address is examined, a route selected and the packet passed onwards. The switch consults a locally held routing table that determines the next node in the sequence of switches, which is S1 in this example. Each switch requires its own routing table, which is updated dynamically by exchanging control information with other switches in the network. Routing is a complex process (see section 18.4.1), and the effectiveness of an interconnected network depends critically upon the routing algorithm chosen for it.

This procedure is continued from switch to switch until the called-DCE is reached. Each switch retains a table of information which links the logical number of the packet with the actual route selected, and so the route is now stored through the network.

The called-DCE receives the incoming packet, selects a local logical channel number (C8 in this example), and passes an incoming call (IC) packet over this path to its local DTE.

The VC is now established. Subsequent data packets from the calling-DTE need carry only the logical channel number C6 in order to identify the call, and the packets are routed via the pre-prepared path.

X.25 flow control

Once a VC is established data packets can flow through the network since their route is known. However, the rate at which they can flow may need to be limited by the destination DTE using a process of flow control. (The network itself may also limit the rate that source DTEs can supply packets to avoid congestion but this is addressed by a different process called congestion control, see section 18.4.2.)

The destination DTE needs to exert some control over the rate at which data packets are inserted into the VC, since it may be busy or have no buffer space left for more data. Received packets are acknowledged by returning a message to the sending-DTE, which allows further packets to be transmitted. The normal way for this control to be implemented is by a sliding window protocol, operating similarly to that described in the context of the data-link layer (section 18.3.3) but on an end-to-end basis at the packet (or network) layer. Each packet is thus given a k-bit sequential send sequence number, $N(S)$, while the incoming or acknowledgement packet from the destination DTE contains a k-bit receive sequence number, $N(R)$, see sections 18.3.3 and 18.4.5. The sending DTE maintains a list of packet numbers it is permitted to send, whilst the receiving DTE maintains a list of packet numbers it can accept.

These lists define windows in the (cyclic) modulo-2^k packet sequence numbers at transmitter and receiver. At the transmitter, for example, the lower boundary of the window increments each time a packet is transmitted, while the upper boundary of the window increments when packets are acknowledged (see Figures 18.17 and 18.18).

X.25 Packet structure

X.25 packets (Figure 18.28) must contain at least the following information:

- Identification as to whether the packet carries control information or data.
- Logical channel number.
- If a control packet, then what function it performs.
- Flow control information.

Several different varieties of X.25 packet are defined, but their general structure is described in section 18.4.5 (and Figure 18.28). It is sufficient here to reiterate that 12 bits are allocated in the packet to the logical channel address, which allows 4096 separate VCs to be open at any one time. The X.25 controller therefore behaves, in part, as a statistical multiplexer.

20.7.2 Frame relay

X.25 was developed at a time when transmission paths were subject to significant errors and when data speeds were quite low. Transmission over each X.25 network link is therefore checked and acknowledged before being released to the next link in the route. A repeat transmission is requested if an error is detected using an ARQ protocol (see section 18.3.2). This action is carried out at both layer 2 and layer 3, since multiplexing is carried out at layer 3. The processing overhead is therefore considerable, and as transmission speeds have risen so processing time becomes an unacceptable burden on packet transmission. Fortunately, trunk transmission is now carried mostly over optical fibre links that have an extremely low probability of error, usually better than 10^{-9}. Errors are therefore extremely unlikely and most of this processing is unnecessary.

X.25 is frequently being replaced, therefore, by a protocol known as frame relay, which carries out multiplexing and packet routing at layer 2, removing one complete layer of processing. Transmission between nodes is unacknowledged, removing the necessity to wait before relaying the frame on through the network. Processing time is reduced by about an order of magnitude.

Frame relay was initially defined in the context of the integrated services digital network (section 20.8), but is now offered by most public operators as a service for interconnecting remote LANs. The network provider sets up permanent virtual circuits (PVCs) for the user, and frames are routed along these at layer 2.

20.8 Integrated services digital network (ISDN)

The circuit switched PSTN has served telephony requirements well, but offers only a limited data service (section 20.6.9). Although it is now an integrated digital network

(IDN), employing digital transmission and digital switching throughout the network's core, it still offers an analogue service to the user over the local loop. An integrated services digital network (ISDN) [Griffiths] offers universal digital services over an IDN, maintaining the digital nature of the signal right out to the user interface.

ISDN became available in the UK in 1985. Basic rate access is at 144 kbit/s and primary rate access is at 2.048 Mbit/s, which fits into the multiplex hierarchy of Figure 20.14. The ISDN combines computer data, voice, fax and video signals without distinction, offering a range of universal interfaces that can be used whatever service is required.

20.8.1 ISDN structure

An ISDN encompasses three distinct networks, Figure 20.61. These are a circuit switched network, a packet switched network and a signalling network.

Table 20.8 *ISDN digital pipes.*

Access level	Total rate	Data	D-channel	Overhead
Basic access 2B+D	192 kbit/s	2×64 kbit/s	16 kbit/s	48 kbit/s
Primary access 30B+D	2.048 Mbit/s	30×64 kbit/s	64 kbit/s	64 kbit/s
Primary access 5H0+D	2.048 Mbit/s	5×384 kbit/s	64 kbit/s	64 kbit/s
Primary access H12+D	2.048 Mbit/s	1920 kbit/s	64 kbit/s	64 kbit/s

The use of a separate signalling network removes the requirement, imposed on the PSTN and X.25 networks, for connection set-up messages to be sent over the user–data path. The ISDN sets up a connection over the separate signalling network, leaving the high capacity data networks entirely for the transport of user information.

The basic unit of data capacity is a 64 kbit/s bit stream and higher level services are built up from multiples of this unit. Table 20.8 lists some of the common ISDN 'digital pipe' data rates that can be subdivided in various ways.

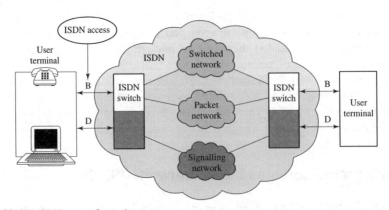

Figure 20.61 *ISDN network services.*

20.8.2 Basic and primary rate access

International standards for ISDN access (I.420) are published by ITU. Basic rate access provides the customer with two independent communications channels at 64 kbit/s together with a common signalling channel at 16 kbit/s. Communications or bearer channels (B-channels) operate at 64 kbit/s, whilst the signalling or data channel (D-channel) operates at 16 kbit/s giving the basic user rate total of 144 kbit/s. An overhead bit rate of 48 kbit/s brings the raw bit rate up to 192 kbit/s. The basic rate voice/data terminal transmits, full duplex, over a two-wire link with a reach of up to 2 km, using standard telephone local loop copper cables. The transceiver integrated circuits employ a 256 kbaud, modified DPSK, burst modulation technique, Chapter 11, to minimise RFI/EMI and crosstalk, see section 6.6.4. The D-channel is used for signalling to establish (initiate) and disestablish (terminate) calls via standard protocols. During a call there is no signalling information and hence the D-channel is available for packet switched data transmission.

Primary rate access in Europe provides the customer with 30 independent 64 kbit/s B-channels and one 64 kbit/s D-channel using the same frame format as the conventional 32-channel 2.048 Mbit/s TDM PCM primary multiplex, Figure 20.15. (The 32nd 64-kbit/s channel carries overheads.) In the USA, South Korea and Japan 23 B- and one D-channel are supported along with an 8-kbit/s overhead to match the 1.544 Mbit/s PCM multiplex structure used in these countries, Figure 20.19.

Data access at 64 kbit/s is used in low bit rate image coders for videophone applications, Chapter 16. For high quality, two way confravision services with full TV (512×512 pixel) resolution reduced bit rate coders, section 16.7, have been designed to use primary access at 2 Mbit/s. Access at 2 Mbit/s is also required to implement WANs, or to carry cellular telephone traffic between cell sites, Chapter 15.

The layer 1 specification based on ITU-T I series recommendations [Fogarty] defines the physical characteristics of the user network interface. Access via such an interface allows the provision of very flexible telephone services with fully digitised voice, as well as a multitude of low speed data services.

Figure 20.62 shows the set of ISDN access reference points defined by ITU-T. These points are often (but not always) physical equipment interfaces.

The local loop from the ISDN exchange is described as the U-interface. Signalling over this path in Europe is multilevel; ternary symbols may be used at 120 kbaud, converting four binary symbols into three ternary symbols for transmission (a 4B3T code), section 6.4.7. For basic rate access it operates over a single twisted pair offering a full-duplex service comprising:

- 144 kbit/s data channel;
- 11 kbaud synchronisation channel;
- 1 kbaud maintenance channel.

Full-duplex operation is achieved using echo cancellation. Outside Europe an alternative 80 kbaud line code is widely used (ANSI T1.601) which encodes two binary symbols into one quaternary symbol (2B1Q).

At the user end of the U-interface is the network termination (NT), which converts the local loop line signal into a more convenient form. The NT1 (network termination 1) terminates the transmission system and includes functions that are basically equivalent to layer 1 of the OSI architecture (e.g. signal conversion, timing and physical and electrical

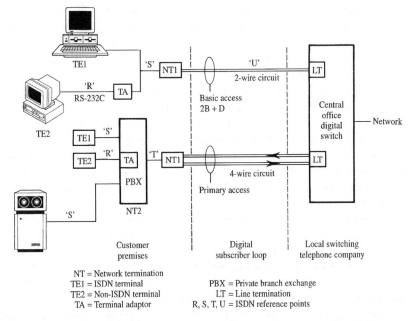

Figure 20.62 *Basic rate (144 kbit/s) and primary rate (2 Mbit/s) access to the ISDN.*

termination). Terminal equipment (TE) includes the ITU-T NT2 (network termination 2) function that terminates ISO layers 2 and 3 of the interface. In practice an NT2 might be realised, for example, by a concentrator, multiplexer, PBX or LAN.

In principle a T-interface (terminal interface) is defined between an NT1 and NT2 so that the responsibilities for circuit and terminal equipment can be separated. Often, however, NT1 and NT2 are merged into a single NT12 device, converting directly between the U-interface and the S-interface. An NT2 or NT12 is an intelligent device, which may perform switching and concentration, operating at levels up to layer 3 of the OSI model.

The S-interface is used in order to link together various items of data equipment, such as telephone, fax and computer terminal. It is in the form of a bus, the S-bus, having the following characteristics (I.430):

- four-wire copper medium;
- passive bus or star connection;
- connection facility for up to eight terminals/devices;
- contains the 2B + D channels in full duplex.

Figure 20.62 shows two terminal devices connected to an S-bus (although there may be more). A terminal equipment type-1 (TE1) device is ISDN compatible and can be directly connected, and may, for example, be a multi-function workstation, a digital telephone or a fax machine. A digital PBX (private branch exchange) may also be attached directly to the S-bus, but in this position it would have only limited outgoing access.

Many terminal equipment type-2 (TE2) devices exist that already have an interface such as RS-232 (V.24) or X.21. These can be accommodated by a terminal adaptor (TA), which converts the S-bus into an R-interface for these other standards. The TA performs the processing to establish and disestablish calls over the ISDN and handles the higher level OSI protocol processing. For computer connection the TA is usually incorporated in the PC.

The primary access interface provides a 2 Mbit/s data rate over two twisted pairs (four wires) using either HDB3 (Europe) or B8ZS (North America) line coding, Chapter 6. This interface caters for higher bit rate applications such as a PBX connection or video-conferencing. In the former case it might have a structure of 30B + D, but there is considerable flexibility in order to accommodate different user requirements.

20.8.3 ISDN services

The ISDN offers circuit or packet switched services at a large variety of user bit rates. A major advance is the use of common channel signalling for administrative use, which employs a completely different network to that for data transport. In the original PSTN network signalling was carried in-band, occupying part of the audio spectrum, which limited the information about the message that could be made available. It also meant that the whole route had to be seized for signalling, even if the call could not proceed.

In the ISDN there are three distinct sets of protocols for:

- user information;
- control information;
- management information.

Figure 20.63 illustrates how the ISDN implements a circuit switched connection using the signalling network and the user D-channel. Notice that switching takes place at the physical level, and that the data can stream through continuously.

A packet service is available over the B-channel, as illustrated in Figure 20.64. The control D-channel now switches the user data packet through to the packet network where it progresses through an X.25 network in the normal fashion, not requiring the services of the signalling network. A further packet service is available over the D-channel.

In addition to X.25-like services, the ISDN includes the improved packet service known as frame relay (I.122), already briefly described in section 20.7.2.

20.9 Signalling

When establishing any kind of network connection, control information must be provided to operate switches and other equipment. The provision of this control information is called signalling. As a minimum, user signalling:

1. initiates a call;
2. provides addressing instructions;
3. responds to an incoming call.

Network signalling:

1. routes the call;
2. monitors its progress;
3. generates a bill for use of the network.

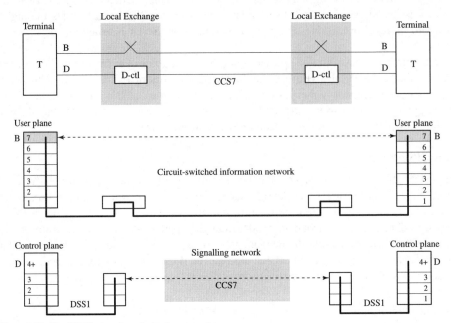

Figure 20.63 *ISDN circuit switched connection.*

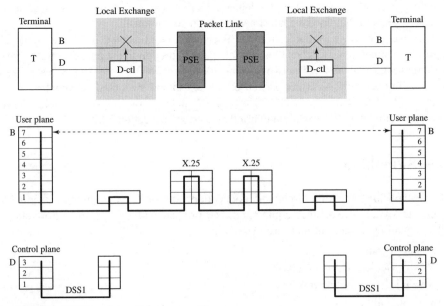

Figure 20.64 *ISDN packet switched connection.*

An important requirement of the signalling system is that it should be reliable. An infamous case of failure was that of the AT&T network in January 1990. A simple transmission problem in conjunction with a switching software defect caused many of the switches to fail. In the ensuing chaos, neither users nor network operators could make connections, and the system could not be restored until evening when traffic demand had declined.

20.9.1 In-channel and common channel signalling

Signalling in the PSTN for establishing or initiating connections may either use a separate communications channel or operate on an in-channel basis. With in-channel signalling, the same channel is used to carry control signals as is used to carry user traffic. Such signalling begins at the originating subscriber and follows the same path as the call itself, the dialling code being structured so as to control the progress of the call. This has the merit that no additional transmission facilities are needed for signalling. Two forms of in-channel signalling are in use, in-band and out-of-band. In-band signalling uses not only the same physical path as the call it serves, but also the same frequency band as the voice signals that are carried. Out-of-band signalling takes advantage of the fact that voice signals do not use the full 4 kHz of available bandwidth. A separate narrow signalling band, within the 4 kHz, is used to send control signals. A drawback of in-channel signalling schemes is the relatively long delay from the time that a subscriber dials a number to the connection being made.

This problem is addressed by common channel signalling in which control signals are carried by a signalling network that is completely distinct from the transmission network and usually possesses a different topology. The common channel uses a signalling protocol that is more complex than for in-channel signalling. The control signals are messages that are passed between switches, and between a switch and the network management centre. Since the control signal bandwidth is small, one separate channel can carry the signals for many subscriber channels. At a low level signalling transmission capacity is provided by the signalling channel in a primary multiplex, which is then carried through the higher levels of multiplexing. In the SDH network, explicit provision is made for common channel signalling in the path overhead field.

A modern telecommunications network, therefore, can be viewed not as a single network, but as a stack of several comprising:

- **A transmission network** – which is the basic construction of transmission links, including multiplexers, cables, etc.
- **A switching network** – which routes messages around the transmission network. (There may be several routes between two switching centres, but they are regarded as one by this network.)
- **A signalling network** – which passes signalling messages between switching centres. There are generally fewer signalling links than traffic routes.
- **A centralised intelligence network** – which provides access from main switches to the network database containing information about the configuration of the network, e.g. which routes are unavailable (due to faults or planned maintenance, etc.) and where services are located.
- **An administrative network** – which enables the operations, administration and maintenance (OA&M) functions to be carried out and which comprises the operations

and maintenance centres (OMCs) and network management centres (NMCs), together with telemetry, control and communications links to the various switches.

This conceptual layering of the network is illustrated schematically in Figure 20.65.

20.9.2 Signalling system No. 7

Over the past three or four decades several different general purpose signalling systems have been developed by ITU and other standards organisations. The most important of these, and the one of major relevance to modern public networks, is the set of ITU

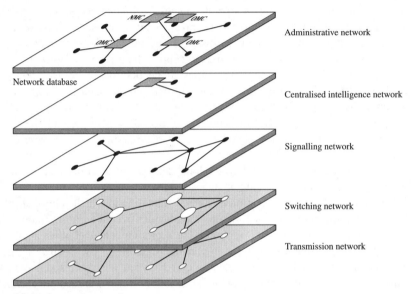

Figure 20.65 *Layered network structure.*

Equivalent OSI layers	CCS7 users			
7	Transaction capability TCAP	ISDN user part (ISUP)	Telephone user part (TUP)	Data user part (DUP)
4–6	ISP Layer 4	Layer 4	Layer 4	Layer 4
	SCCP layer 4			
3	MTP layer 3 (signalling network)			
2	MPT layer 2 (signalling link)			
1	MTP layer 1 (signalling data link)			

Message transfer part (MTP)

Figure 20.66 *ITU-T four-layer CCS7 protocol stack and the equivalent ISO levels. (ISP: Intermediate services part.)*

procedures known as common channel signalling system No. 7 (CCS7), which is structured approximately in accordance with the OSI model [see ITU-T Rec. Q.764]. The overall objective is to provide an internationally standardised, general purpose, common channel signalling system that:

1. Is optimised for use in digital networks in conjunction with digital stored program control exchanges utilising 64 kbit/s digital signals.
2. Is designed to meet present and future information transfer requirements for call control, remote network management, and maintenance.
3. Provides a reliable means for the transfer of information packets in the correct sequence without loss or duplication.
4. Is suitable for use on point-to-point terrestrial and satellite links.

Specification of CCS7 commenced in 1976, and there has been a continuous sequence of development since then. Recommendation ITU-T Q.700 gives an introduction.

CCS7 has been developed from the four-level CCS6 (No. 6) architecture established in the 1960s so the protocol stack, Figure 20.66, is a hybrid one and does not correspond exactly with the OSI seven-layer model.

Its important components are:

- **Message transfer part (MTP)**: the lower three levels, defining a general mechanism for transferring messages. The physical level is likely to be the 64 kbit/s channel in a primary multiplex, while the data-link level is fully compatible with OSI layer 2 and LAP-B (link access protocol, balanced, see section 18.3.4). Many different message types are defined around a common frame structure, and each is conveyed as a message signal unit (MSU). Layer 3 functions are also defined.
- **Telephone user part (TUP)**: providing an application driven component to the signalling system, specifically for telephony. It is particularly useful for services such as three-way calls. The TUP generates a sequence of messages when a call is being set up, transmitting information between switches in the form of CCS7 messages over the signalling network. The voice or data path is only set up at switch-through time.
- **Data user part (DUP)**: providing an application driven component to the signalling system, specifically for data connections between circuit switched data exchanges.
- **ISDN user part (ISUP)**: the application-specific driver for ISDN use. It is more structured than the other services to allow for the greater richness of ISDN services. In order to comply more closely with the OSI model, an SCCP sub-layer (see below) has been defined. Applications include multimedia calls involving a mix of voice, video, control and fax.
- **Signalling connection control part (SCCP)**: the proper equivalent of the OSI layer 3 network service. This offers more general routing capability than MTP, for messages not related to a particular circuit. It allows messages to be associated with one another in order to set up global facilities, and allows global translation of addresses. It can address a subsystem within an exchange.
- **Transaction capability (TC)**: providing a general structure for generating messages which are not circuit related, e.g. to access a database. An example of the use of a TC is the 0800 or 'freefone' service, where a remote database must be interrogated by a message in order to find the actual number of the required service in the network.

The signalling network consists of a collection of service switching points (SSPs) and signal transfer points (STPs) each with its own point code (PC), and connected by CCS7 links. The configuration is under the control of a switching control point (SCP). The transport mechanism is MTP. Signalling networks are the most complex of all data communications networks. Figure 20.67 illustrates a simplified hypothetical CCS7 network.

This signalling network is quite independent of the traffic network. Note the redundancy in the network that is fully interconnected, in order to provide extremely reliable communication of signalling information.

Several variations on this type of network exist, and details of implementation may differ from country to country. Some examples are:

- **Digital subscriber signalling system No. 1 (DSS1)**: an ITU-T defined signalling system for ISDN. Connects subscriber's premises to a local exchange, hence does not use the network capabilities of CCS7.
- **Digital access signalling system No. 2 (DASS2)**: defined by BT for the UK in advance of the full specification of DSS1, to support ISDN access.
- **Digital private network signalling system (DPNSS)**: comparable to CCS7, but links PABXs on subscriber's premises. It has simplified networking capabilities, but a comprehensive message set to exploit the advanced features in PABXs, e.g. call diversion. Layers 1 and 2 are identical with DASS2, but layer 3 is specific to PABXs.

Many interesting telephone services become available through the use of common-channel signalling networks including:

- 0800 services in which calls made to a single nationwide number are routed to the business branch nearest the caller.
- Call rerouting on a national basis.
- Third party billing using credit cards.
- Calling party identification where the calling party number is transported through the network to let the called party know the caller identity before answering.
- Virtual private networks with short form dialling.

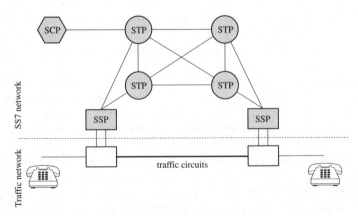

Figure 20.67 *A simple CCS7 network.*

20.10 Asynchronous transfer mode and the broadband ISDN

Although 64 kbit/s seemed to be a high data rate in 1980 when ISDN was being conceived, it now seems pedestrian. Much higher rates are required for new applications such as video-conferencing, videomail, videophone, broadcast quality video and internet browsing. Interactive multimedia applications employ a range of signalling rates.

ISDN can in principle allocate blocks of 64 kbit/s channels to form, for example, H0 groups of 384 kbit/s but this ties up switching capacity and may still not be adequate for all applications. A broadband ISDN (B-ISDN) has therefore been proposed which will have a basic delivery rate of up to 150 Mbit/s (more precisely 149.760 Mbit/s) and a higher rate of 600 Mbit/s (599.040 Mbit/s). This impressive speed is achieved using VLSI hardware chips, rather than software, for protocol processing and switching at network nodes. SDH provides the basis for transporting such user rates about the core network (see section 20.5.2) and will be used for main trunk routes. Asynchronous transfer mode (ATM) is the transport protocol designed to support the 'on-demand' data rates required by users of the B-ISDN.

Figure 20.68 shows the division of an ATM network into access and core components and the resulting user–network interface (UNI) and network–node interface (NNI).

The technical challenges posed by the B-ISDN are enormous and the necessary investment for its widespread deployment is colossal. The ATM concept is attractive in its own right, however, and ATM switches are already available to build small, high speed local networks and high speed interconnections for large networks. ATM can also meet the needs of interconnecting LANs at very high speed, which at present is being approached by the design of metropolitan area networks (MANs). The most promising MAN contender is probably the distributed queue dual bus (DQDB), which uses cells identical to those defined for ATM.

20.10.1 Transport mechanism

User requirements are moving away from the constant bit rate (CBR) services typified by telephone traffic, to variable bit rate (VBR) traffic. Database interrogation and point-of-sale applications are naturally bursty processes, with long idle periods between the transfers of information. Video coding too is a VBR operation, since full use is made of the similarity between successive picture frames, only the changes that have occurred since the last frame being transmitted (see Chapter 16).

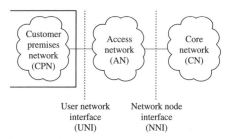

Figure 20.68 *Access to ATM network.*

The future lies with one integrated network that will service variable rates of data transfer on demand. This requires an asynchronous method of delivery and switching and ATM has been conceived to provide just such a service [de Pryker].

ATM is essentially a fast, virtual circuit, packet switching system with fixed length packets called cells. The virtual circuit is established by the first of a sequence of cells. This first cell contains the full address of the data's destination, the following cells containing only a shortened 'virtual address'.

The term asynchronous in ATM is unfortunate as it implies that ATM is an asynchronous transmission technique, which is not the case. Cells are transmitted continuously like the steps on an escalator and data from a source is loaded into these cells as it becomes available like passengers stepping onto the escalator. The cells carrying data from the source do not necessarily occur at regular intervals, therefore, and it is in this sense that the transfer mode is asynchronous.

ATM facilitates the introduction of new services in a gradual and flexible manner, and is now the preferred access technique to the ISDN, for packet speech, video and other VBR traffic. Queueing theory, Chapter 19, allows the analysis of ATM cell throughput rates and losses. The routing switches described in section 20.7 are used for the ATM interfaces.

Cell structure

Each ATM cell comprises a 5 byte header combined with a 48 byte data packet representing a compromise between the optimum data packet lengths of 16 bytes, for audio, and 128 bytes for video traffic. (The 53 byte ATM cell also represents, more generally, a compromise between short cell lengths required for real time, low delay, traffic applications such as packet speech, and long cell lengths which are more efficient for data applications due to their reduced proportion of overhead bits.) CRC, section 10.8.1, is implemented on the header information only using the generator polynomial:

$$M(x) = x^8 + x^2 + x + 1 \qquad (20.20)$$

to achieve a single error correction, but multiple error detection, capability.

The ATM cell structure at the user network interface (I.361) is as follows:
- 5 byte header
  - Routing field (24 bits), comprising virtual path identifier (VPI) and VCI.
  - Generic flow control (4 bits), priority.
  - Payload type (3 bits).
  - Cell loss priority (1 bit).
  - Header error control (8 bits).
- 48 bytes of user information.

Virtual channels and virtual paths

Figure 20.69 illustrates the ATM concepts of virtual channels (VCs) and virtual paths (VPs), and relates them to a physical circuit.

Switching can take place at any one of these three levels, giving great flexibility to the network. An example is a multimedia call, where video, voice and control data operate over different channels with differing rates but follow the same path. Figure 20.70 shows how virtual channels may be conveyed in a bundle over a virtual path.

20.10.2 Service classes

The following service classes have been defined by ITU-T (I.371):

- Constant bit rate (CBR): for emulating circuit switched services, such as voice and MPEG video coding.
- Variable bit rate (VBR): divided into the following two subclasses:
    - Variable bit rate, real time (VBR-RT): for bursty traffic where tight bounds on delay parameters are required (such as interactive compressed video or voice with the silences removed).
    - Variable bit rate, non-real time (VBR-NRT): where bounds on delay are less stringent, e.g. multimedia, e-mail.
- Unspecified bit rate (UBR): a 'best effort' delivery service with no guarantees on delay, e.g. file transfers or an internet service.
- Available bit rate (ABR): which makes use of any unused bandwidth like the UBR service (but under closer control) and suited, for example, to transmitting LAN and other bursty data over an ATM network.

Figure 20.71 illustrates how these different services fit together. Capacity is allocated for the CBR services. VBR traffic is accepted, and the remaining spare capacity is utilised for ABR traffic. UBR traffic is shown as a constant capacity allocation, but it can be temporarily reduced if the VBR traffic exceeds, instantaneously, its allocation.

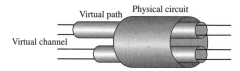

Figure 20.69 *ATM connections.*

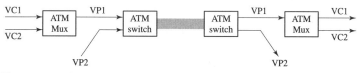

Figure 20.70 *ATM switching paths.*

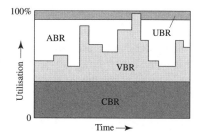

Figure 20.71 *Mix of ATM service classes.*

ATM promises to match the requirements of a multimedia user; it is asynchronous and transmission capacity is used only when it is needed. The network provider, however, is faced with formidable problems.

20.10.3 Connection admission control

Connection admission control (CAC) is the process that decides whether a new request for transmission service can be accommodated through a certain ATM switch.

When a user process wishes to establish a channel, a dialogue takes place between it and the CAC protocol, to negotiate a desired QoS [Fontaine and Smith]. Both the information source and the ATM transport network have a responsibility to maintain performance within the required limits. Important parameters that relate to the QoS of an ATM connection include:

- Cell loss ratio (CLR).
- Cell transfer delay (CTD).
- Cell delay variation (CDV).
- Minimum cell rate (MCR).
- Peak cell rate (PCR).

A further complication of the acceptance process is that, because the cell rates vary statistically, consideration must be given to three distinct time scales: call level (minutes), burst level (milliseconds), cell level (microseconds).

VBR services operate with widely varying transmission rates, but the data must arrive at its destination within a tightly controlled time interval. A video coder, for instance, may require a steady state rate of tens or hundreds of kbit/s while the picture is stationary, but a peak rate of several Mbit/s at times of rapid picture change. When decoded, however, picture frames must occur regularly at, for example, 30 frames per second.

ABR traffic is dealt with by a feedback mechanism like flow control in normal packet systems. When congestion approaches, the source reduces its transmission rate so as to avoid cell loss.

UBR traffic is an open loop service, so there is no mechanism for throttling back the source rate. Cells are dropped if congestion occurs.

CBR and VBR connections are open loop, and channel bandwidth is allocated to meet the requested QoS.

An elementary model of a VBR ATM transmission channel can be derived by taking traffic from N sources, each of which has a probability, p, of generating a traffic cell in each time-slot. A utilisation, U, can be defined by:

$$U = \frac{\text{number of cells in time-slot}}{N} \qquad (20.21)$$

Clearly, as $N \rightarrow \infty$, $U \rightarrow p$. This model can be used to estimate the probability of channel overflow, or cell loss, since it represents a binomial process. The probability of j cells occurring in a given time-slot is given, as in equation (3.8), by:

$$P(j) = \binom{N}{j} p^j (1 - p)^{N-j} \qquad (20.22)$$

Suppose now that the channel capacity is C_n cells/time-slot. The probability that overflow occurs is given by:

$$P_{overflow} = \sum_{j = C_n + 1}^{N} P(j) \qquad (20.23)$$

20.10.4 Access protocols

The general access protocol stack for an ATM service is shown in Figure 20.72. The 48 byte payloads are prepared by the ATM adaptation layer (AAL), which can accept normal data-link frames. Two sub-layers are required, the convergence sub-layer (CS), which interfaces normal network standards with the ATM mechanism, and the segmentation and reassembly (SAR) sub-layer which divides a larger data frame into 48 byte payloads. The ATM layer attaches a header to each 48 byte item of data. The physical layer services the ATM layer. Taken together, the physical layer, ATM layer and AAL form an OSI layer 1 service for higher level protocols.

Various AALs are defined as outlined in Table 20.9. Part of the 48 byte payload may be allocated to protocol control information (PCI).

Table 20.9 *ATM adaptation layers.*

Type	PCI (bytes)	Net load (bytes)	Application
AAL0	–	48	Null case.
AAL1	1	47	CBR service. PCI contains sequence counter.
AAL2	3	45	VBR service. Maintains synchronism between end points.
AAL5	–	48	VBR service. Simple and efficient adaptation layer (SEAL). Uses cells to mark start and end of packet.

20.10.5 Synchronous versus asynchronous transfer modes

Synchronous transfer mode (STM) and asynchronous transfer mode (ATM) [de Pryker] both refer to techniques that deal with the allocation of usable bandwidth to user services. (STM as used here is not to be confused with the synchronous transport module defined in section 20.5.2.) In a digital voice network, STM allocates information blocks, at regular intervals (bytes/125 μs). Each STM channel is identified by the position of its time-slots within the frame, Figure 20.15. STM works best when the network is handling a single service, such as the continuous bit rate (bandwidth) requirements of voice, or a

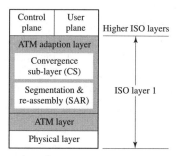

Figure 20.72 *ATM access protocol stack.*

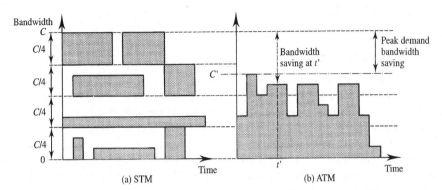

Figure 20.73 *Example of (a) STM and (b) ATM with mix of fixed and variable bit rate traffic.*

limited heterogeneous mix of services at fixed channel rates. Now, however, a dynamically changing mix of services requires a much broader range of bandwidths, and a switching capability adequate for both continuous traffic (such as voice) and non-continuous traffic (such as high speed data) in which bandwidth may change with time depending on the information rate.

While STM can provide data transfer services, the network operator must supply, and charge for, facilities with the full bandwidth needed by each service for 100% of the time – even if users require the peak bandwidth for only a small fraction of the time. The network therefore operates at low efficiency and the cost to users is high. This is not an attractive option for VBR traffic, such as coded video transmission, in which the data rate is dependent on how fast the image is changing. Figure 16.19 shows an example of such traffic for which the average data rate is very much smaller than the peak rate. When many VBR sources are averaged then the peak transmission rate requirement comes much closer to the average rate.

The structure of the ATM protocol improves on the limited flexibility of STM by sharing both bandwidth and time [de Prycker]. Instead of breaking down bandwidth into fixed, predetermined, channels to carry information, as shown schematically for STM in Figure 20.73(a), ATM transfers fixed size blocks of information, whenever a service requires transmission bandwidth. Figure 20.73(b) illustrates the resulting (statistical) multiplexing of the ATM traffic. When the transmission bandwidth is dynamically allocated to VBR users, there is a consequent peak bandwidth saving ($C' < C$), provided that the packet overhead is small.

The term asynchronous in ATM refers to the fact that cells allocated to the same connection may exhibit an irregular occurrence pattern, as cells are filled according to the actual demand. As has already been pointed out, however, this does not mean that ATM is an asynchronous transmission technique. ATM can fit seamlessly into the SDH (or SONET) frames of Figure 20.29 (and 20.38) by accommodating the cells directly into the payload envelope.

20.10.6 ATM versus IP

An active debate exists between the benefits of ATM versus those of the internet protocol (IP) for network operation. IP is used extensively, particularly at every PC or workstation DTE, and there are good arguments for employing it across the whole network. However, IP is based on connectionless routing, which can be slow and processor intensive, and it offers no QoS guarantees. In contrast, ATM offers connection-oriented switching, together with a flow control mechanism that enables QoS requirements to be met. The IP camp is proposing new structures to meet these requirements, so the future shape of networking is still not clear (although it does appear to be tending towards IP).

20.11 Access technologies

An access network (AN) provides the link between a customer premises network (CPN) and the core network (CN). Access networks have traditionally been designed and deployed on a rather service-specific basis, twisted pair local loops being developed for analogue voice communications, for example, and cable networks being deployed for community antenna TV (CATV) distribution. Recently, however, the boundaries between services offered by traditional telecommunications companies, cable TV companies, satellite direct broadcasters and Internet service providers, have become much less distinct making these businesses, and the disparate access technologies they have traditionally employed, competitive in nature rather than complementary. It is now possible to obtain telephony and Internet access over a cable (TV) network, and video services over the twisted pair local loop (in one of its digital subscriber line incarnations). The most important of the modern access network technologies are briefly described in the following sections.

20.11.1 Digital subscriber line

Until relatively recently the desire of network operators to increase revenue by offering new services that require broadband access (typically taken to mean bit rates in excess of 2 Mbit/s) was thought to imply that the traditional copper wire pair of the local loop would have to be replaced by fibre to the home (FTTH, see section 20.11.2), broadband fixed wireless access (BFWA, see section 20.11.5), or some mix of these two technologies. (BFWA is generally more appropriate to rural infrastructure where user density is small and local loops long, whilst FTTH is typically more appropriate to urban environments where user density is high and local loops short.) Half of the investment, however, of a telephone company is in the connections between subscriber handsets and their local exchange. Furthermore, this part of the network generates the least revenue since local calls are often cheap or, as in the USA, free. The length of these connections is 2 km on average and they seldom exceed 7 km. Advances in signal processing have resulted in an opportunity to use this existing (and expensive) local loop infrastructure in a new way. The result is the family of digital subscriber line (DSL) technologies (introduced previously in section 6.9) including asymmetrical DSL (ADSL), high speed DSL (HDSL) and very high speed DSL (VDSL). Collectively these access technologies are often referred to as xDSL. Whether xDSL technologies represent a temporary solution to broadband access that will be eventually replaced by FTTH, BFWA and their variants, or whether it will have a more permanent place in a heterogeneous mix of future broadband access technologies, is not entirely clear. (The challenge of providing FTTH, for example, to all subscribers should not be underestimated. With 750 million telephones worldwide it would take more than 300 years for manufacturers to produce all the required cable at current production rates!)

The asymmetrical forms of xDSL are aimed principally at domestic users who normally require much larger downstream (downloading) data rates than upstream (up-loading) data rates for web browsing and similar applications. The symmetrical xDSL forms are designed principally for the industrial/commercial market in which applications such as file transfer and video-conferencing typically require high data rate in both directions. Both variations have been deployed to a greater of lesser extent, however, in both environments. Of all the xDSL technologies ADSL appears to have had the most immediate impact.

ADSL

ADSL, described in ITU-T G.992.1, uses a single twisted wire pair to provide a unidirectional downstream (provider-to-subscriber) data rate of up to 6.1 Mbit/s and a bidirectional (downstream and upstream) data rate of up to 640 kbit/s. The overall asymmetry of the downstream and upstream data rates make this technology particularly suitable to video on demand, Internet browsing and remote LAN access, Figure 20.74. The precise downstream data rate available depends on the length and quality of the local loop and the capability of the particular make and model of the ADSL modem. Forward error correction is employed since the real time requirements of compressed video applications preclude the use of data-link layer or network layer ARQ. Figure 20.75 shows two ways in which spectrum can be partitioned to provide the ADSL service. In both methods the conventional telephony channel is separated from the rest of the spectrum using a low pass filter. This means that the 'plain old telephone service' (POTS) remains available when the ADSL channels are in use. The spectrum above the POTS can be divided between upstream and downstream services using conventional frequency division multiplexing (Figure 20.75(a)) or the downstream band can be overlapped with the upstream band and separation achieved by local echo cancellation (Figure 20.75(b)).

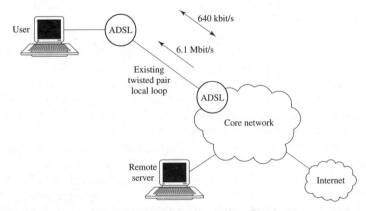

Figure 20.74 *Schematic diagram of ADSL role and typical applications.*

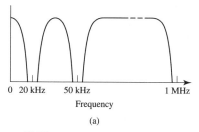
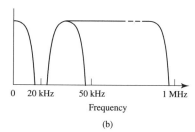

Figure 20.75 *ADSL division of telephone, upstream and downstream channels using (a) conventional FDM and (b) FDM/echo cancellation.*

The upstream and downstream channels can be time division multiplexed into multiple low speed and high speed subchannels respectively.

Two modulation methods may be used in the implementation of ADSL. Discrete multi-tone (DMT) modulation employs 256 carriers spaced by 4 kHz between 25 kHz and 1.049 MHz, to transmit the data over parallel channels in a similar way to OFDM (see section 11.5.2). Alternatively, carrierless amplitude and phase (CAP) modulation may be used, in which a pair of orthogonal symbol pulses are summed in a variety of possible (quantised) proportions to produce a carrierless form of QAM, Chapter 11.

ADSL is used to provide 'BT Broadband', a broadband access product offered by the UK's largest fixed line telecommunications carrier.

Splitterless ADSL and RADSL

Splitterless ADSL (ITU-T G.992.2) is a cut-down version of ADSL designed as a low cost option that supports downstream and upstream data rates of 1.5 Mbit/s and 512 kbit/s respectively. It does not employ a filter to isolate the telephone service from the ADSL service – hence its designation as splitterless. It is also frequently referred to as G.lite ADSL. Splitterless ADSL uses DMT employing 96 carriers between 25 kHz and 409 kHz. Rate adaptive ADSL (RADSL) is a proprietary (non-standardised) form of ADSL in which the modem measures the quality of the local loop line and adapts the bit rate accordingly.

HDSL and HDSL2

High speed digital subscriber line (HDSL), ITU-T G.991.1, was designed to provide primary multiplex rates (2.048 Mbit/s or 1.544 Mbit/s) to industrial and commercial subscribers over two or three conventional twisted pairs. The data rate is shared between the wire pairs, 2B1Q line coding being used to map pairs of binary digits on to a four-level PAM (i.e. four 'DC' voltage levels) signal. The system supports full-duplex operation, using echo cancellation to subtract the transmitted signal from the total signal existing on the transmission line. Since HDSL uses quantised PAM signalling the conventional telephone service is unavailable when HDLC is in use.

HDSL2 (ITU-T G.991.2) is a single wire pair version of HDSL for fixed rate signals at 2.048 Mbit/s or 1.544 Mbit/s. It employs spectrally shaped, trellis coded, 16-level PAM, each PAM symbol representing three information bits and one FEC bit. The maximum reach of HDSL2 (and HDSL) is approximately 4 km.

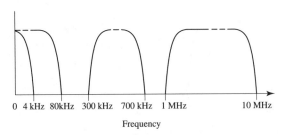

0   4 kHz 80kHz   300 kHz   700 kHz   1 MHz   10 MHz

Frequency

Figure 20.76 *VDSL division of telephone/ISDN, upstream and downstream channels.*

SDSL

Symmetric DSL (SDSL) refers to a variety of proprietary systems supporting symmetric bit rates between 128 kbit/s and 2.36 Mbit/s over a single wire pair. It uses 2B1Q line coding or CAP modulation. It is likely that SDSL will become obsolete as subscribers move to SHDSL, a higher performance system operating to a recognised ITU standard.

SHDSL

Single pair HDSL (SHDSL, also known as G.shdsl) is described in ITU-T G.992.2. It supports symmetric upstream and downstream data rates between 192 kbit/s (ISDN primary access) and 2.3 Mbit/s over a single twisted wire pair. It shares features with both ADSL and SDSL and can transport 2.048 Mbit/s (or 1.544 Mbit/s) primary multiplexes, ISDN, ATM and IP signals. Its multi-rate (as opposed to fixed rate) capability, however, allows it to trade lower data rate for longer local loop reach, which is a significant advantage over fixed rate DSL varieties. SHDSL employs trellis coded 16-level PAM without spectral shaping, the flatter spectrum of the transmitted signal allowing local loop lengths of up to 6 km. SHDSL is likely to become the preferred global standard for symmetric DSL technology.

VDSL

Very high digital subscriber line (VDSL), ITU-T G.993.1, is the most recent of the xDSL technologies and is similar in its operating principles to ASDL. Its objective is to support downstream bit rates of up to 51.84 Mbit/s and upstream rates of (initially) around 2 Mbit/s over a single twisted wire pair up to 300 m in length. This makes it a good candidate for the final link in a fibre to the kerb system, see section 20.11.2, in which fibre is laid from the exchange, extending out along the roads and streets of the service area, and other technologies provide the last 'drop' from a fibre tap to the subscriber.

Downstream bit rates are chosen to be submultiples of 155.52 Mbit/s for compatibility with SDH, lower rates allowing operation over longer local loops (25.92 Mbit/s and 12.96 Mbit/s, for example, should allow loop lengths of 1 km and 1.5 km respectively). The intention is for VDSL to evolve higher upstream bit rates until eventually it becomes a symmetric transmission technology.

First generation VDSL systems use FDM to separate upstream, downstream and POTS/ISDN channels, Figure 20.76. Later generation systems will use echo cancellation to overlap the more symmetrical upstream and downstream channels.

Although VDSL potentially supports bit rates that are an order of magnitude greater than ADSL it is, ironically, a cheaper technology. This is, principally, because the shorter local loop lengths that it caters for means the dynamic range of signals with which it must cope is much lower. The roll-out of VDSL relies on the availability of fibre to the kerb (or some equivalent technology), however, and the investment required to provide this will be a significant barrier to its early adoption. ADSL, in contrast, relies on existing infrastructure and is therefore likely to be the dominant xDSL technology in the near and medium term.

IDSL

Integrated services digital network DSL (IDSL) supports a symmetrical (ISDN) user data rate of 144 kbit/s over a single wire pair. It employs the same line code as ISDN but it does not use the D-channel to set up or monitor connections. It has the advantage over conventional ISDN of providing an 'always on' service (i.e. a permanent virtual circuit), thus avoiding call set-up delay. It does not support simultaneous telephone service. IDSL has an inherent range of between 5 and 6 km, but this can be increased with the use of repeaters. (Other xDSL technologies cannot generally use repeaters.)

20.11.2 Fibre

Using fibre for the access network has significant technical advantages. Fibre is a low maintenance transmission technology since it does not corrode. It has much larger bandwidth than twisted pair or coaxial cable allowing many more users to be multiplexed at much higher data rates over a single physical transmission line. Electromagnetic interference is eliminated. Crosstalk between channels within fibres is low and crosstalk between fibres is negligible.

There are several variations on a fibre access network including fibre-to-the-home (FFTH), fibre-to-the-building or -business (FTTB), fibre-to-the-kerb (FTTC) and fibre-to-the-cabinet (FTTCab). These technologies are collectively referred to as FTTx, Figure 20.77 [Maeda *et al.*].

Broadband passive optical network

The fibre infrastructure in Figure 20.77 is a broadband passive optical network (B-PON), with passive splitting, or fanning out, of fibres. Splitting is limited to 1:16 in the case for a maximum fibre access length of 20 km or 1:32 if the fibre length does not exceed 10 km. Splitting may take place in more than one stage (e.g. a 1:4 split followed by a 1:8 split in the 1:32 case). Sufficient optical power must be provided in passive splitters to supply adequate signal power to each subscriber. B-PONs are specified in ITU-T recommendations G.983.1 (physical and transmission convergence layers), G.983.2 (management interface between OLT and ONT) and G.983.3 (additional downstream wavelength band for video broadcasting or bidirectional transport using dense wavelength division multiplexing). The B-PON links the optical line termination (OLT) equipment on the service provider side of the network to the optical network termination (ONT) or optical network unit (ONU) on the subscriber side (see Figure 20.77).

Full-duplex transmission is realised using two fibres (one downstream and one upstream) or over a single fibre using 1.5 μm (downstream) and 1.3 μm (upstream)

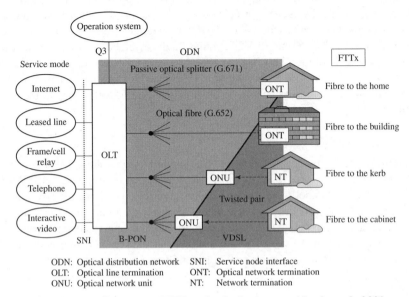

Figure 20.77 *Schematic illustration of FTTx technologies (source: Maeda et al., 2001, reproduced with permission of IEEE © 2003 IEEE).*

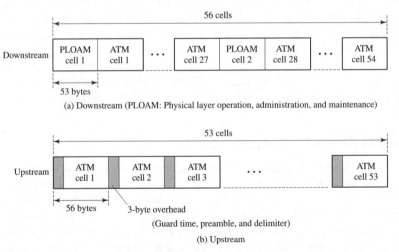

Figure 20.78 *B-PON basic (a) downstream and (b) upstream frame structures.*

wavelength division multiplexing (or more strictly duplexing). The downstream (OLT to ONT) bit rates are 155.52 or 622.08 Mbit/s and the upstream (ONT to OLT) bit rate is 155.52 Mbit/s (shared between the multiple users of the split fibre). The downstream frames, which are ATM based, carry the timing information to ONTs required for the transmission of synchronised upstream frames. ONTs accept those downstream cells addressed to themselves and discard the rest. Figure 20.78 [adapted from Ueda *et al.*] shows the basic frame structures.

Since downstream transmission is multicast, B-PON provides an elementary encryption-like service for each downstream point-to-point virtual path connection called churning. (More secure encryption techniques can be implemented in higher layer protocols if required.) It also provides for ONT verification using a system of passwords to guard against an unauthorised ONT being connected and adopting the identity of a legitimate ONT.

A ranging mechanism obtains the distance (or delay) between OLT and ONT so that ONT upstream access can be time-shared between customers. Cell slots are allocated ('granted') to each ONT for upstream transmission by the ONL via 'grants' in the downstream PLOAM (physical layer OA&M) cells. The ONTs delay their transmissions by an appropriate amount so that all ONTs have an effective (delay-) distance from the OLT of 20 km irrespective of their physical distance.

ONTs can provide the subscriber with both ATM and STM user–network interfaces (UNIs). The STM UNI is useful for emulating circuit switched services. A 10BASE-T or 100BASE-TX UNI (see Table 21.2) for high speed data services can also be provided.

Future B-PON systems may incorporate telephone, digital TV and other video services. One proposal [Kettler and Kafka] describes a prototype system that frequency division multiplexes a multichannel, multipoint distribution service (MMDS) signal (see section 20.11.5) with the (baseband) ATM PON signal before transmission over the fibre. As an alternative the downstream optical band can be split into a lower wavelength sub-band (1.48–1.53 μm) on which the ATM PON signal can be modulated and an upper sub-band (1.531–1.58μm) on which the video signal can be modulated. The two optical signals can then be mixed in a 2:8 optical splitter as shown in Figure 20.79. No optical demultiplexing is required since a broadband optical demodulator, which responds to the full downstream waveband (1.48–1.58 μm), gives the electrical FDM signal directly.

Conventional telephony can be provided by converting the analogue telephone signal to an ATM cell stream via an appropriate interface on the ONT, and transporting these cells over the PON. Future solutions may use alternative technologies, e.g. voice over IP (VoIP).

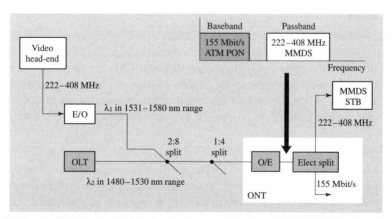

Figure 20.79 *An ATM PON with digital video (source: Kettler and Kafka, 2000, reproduced with permission of IEEE © 2003 IEEE).*

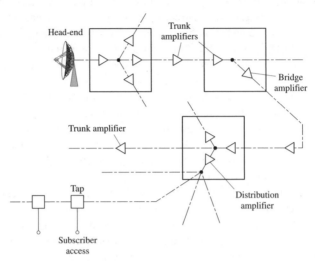

Figure 20.80 *Traditional cable system.*

Power for the ONT, and therefore for telephony services, must be provided by the subscriber. Batteries are thus required to maintain emergency communications services in the event of power system failure.

FTTH and FTTB

FTTH describes the case where the fibre network extends right to the subscriber's premises. This avoids any bandwidth bottleneck in the last few hundred metres to the customer. The cost per subscriber of replacing the entire existing local loop with fibre is often prohibitive, however, especially in the case of domestic customers. FTTB might typically describe the access network for a business or a block of flats where copper is used to connect all subscribers in the building to a single fibre access point.

FTTCab and FTTC

FTTCab describes the situation where fibre extends to a distribution cabinet located close to the centre of a cluster of subscribers and either coaxial cable or twisted wire pair is used for last link to the customer. This is cheaper than FTTH but does restrict bandwidth below that which would otherwise be available, even with high bandwidth copper technologies. FTTC is similar to FTTCab but the fibre reaches closer to the customer's premises and fewer (typically four) copper 'drops' are supplied from a given fibre tap.

20.11.3 Cable

Cable systems were originally devised for the distribution of analogue TV signals. Figure 20.80 is a schematic representation of such a system. If the programme material for distribution is received at the 'head-end' as a satellite or terrestrial microwave transmission the system is called community antenna TV (CATV). Whatever the source of programme material, it is distributed via a trunk and branch network of coaxial cables. Subscribers are connected to the branches (or feeders) by flexible cable 'drops' of usually less than 100 m. The cable drops are attached to the branches by cable taps.

For satisfactory TV reception the original cable systems typically required signal amplification at intervals of between 500 m and 1000 m resulting in a cascade of many (often 30 or 40) amplifiers between the head-end and an individual subscriber. (Most of these amplifiers are in the network trunks, which may be 20 or 30 km long, only the last four or five amplifiers being in the distribution or 'feeder' network branches which are not usually longer than 2 or 3 kilometres.) In addition to the normal noise problem (each amplifier adding an amount of noise to the signal determined by its noise figure, Figure 12.12) large numbers of cascaded amplifiers have an impact on reliability since failure of any one results in loss of service to all subscribers downstream of the fault.

Cable systems have evolved from the original coaxial CATV systems in several ways. These developments are:

1. Introduction of basic return paths (from subscriber to provider) using reverse direction amplifiers.
2. Replacement of that part of the transmission network nearest the head-end by fibre (initially using analogue amplitude modulation of the fibre light source by the TV signals). This increases the quality of the signal injected into the shorter (and more lossy) coaxial part of the network thus increasing the signal quality delivered to the subscriber. It also increases reliability since the number of cascaded amplifiers between head-end and subscriber is substantially reduced. This development is usually referred to as hybrid fibre coax (HFC).
3. Replacement of analogue transmission with digital transmission using MPEG coded signals, Chapter 16. This yields a network capacity improvement of an order of magnitude since typically 12 MPEG-2 signals can be transmitted in the channel bandwidth required for a single analogue TV signal, section 16.7.
4. Provision of telephony and data services (in particular connection to the Internet) using advanced cable modems. These modems have been standardised in several specifications including DOCSIS (data over cable service interface specification), EuroDOCSIS and DVB-C/DAVIC (digital video broadcasting-cable/digital audio video council) [Forrest, Donnelly and Smythe].
5. Development of packet network protocols for cable systems based on the internet protocol (IP). Various elements of this recent development are described in a system called PacketCable[TM] [Mille and Russell].

Hybrid fibre coax

Hybrid fibre coax (HFC) [Raskin and Stonebeck] describes a tree network in which the trunk and main branches of the tree are implemented in fibre and the sub-branches are implemented in coaxial cable, Figure 20.81. A master head-end unit typically receives its TV signals via satellite or terrestrial microwave links and rebroadcasts them to a subscriber base of perhaps a million households. It hosts the equipment which supports pay for view services, allows the cable operator to mix in local news programmes and advertising, and gives subscriber access via routers to Internet service providers. The broadcast signals are distributed via the tree topology of primary and secondary hubs until they reach the fibre nodes, each of which may serve perhaps 1000 subscribers via coaxial cable [Green].

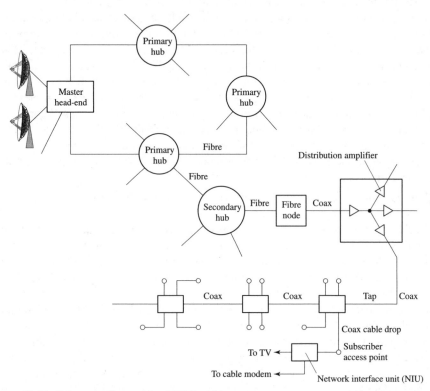

Figure 20.81 *Schematic illustration of HFC architecture.*

The cable network from the fibre node comprises main trunks supplying feeder cables via distribution amplifiers. The feeder cables run along each road or street in the service area, supplying each subscriber via a cable tap and cable drop. Distribution amplifiers and any necessary line extending amplifiers in the trunk and feeder cables incorporate frequency duplexers to cater for the bidirectional flow of signals. The taps incorporate directional couplers so that upstream data is not broadcast to all subscribers.

In modern implementations a cable modem (CM) is located at each subscriber premises and a cable modem termination system (CMTS) is located at the head-end (or at the primary or secondary hubs) to provide bidirectional data services and Internet access.

In the subscriber's premises a duplexing filter splits the upstream from the downstream frequency bands and a network interface unit splits the (downstream) TV signals from the cable modem signals (and feeds the TV signals via coaxial cable to a set-top box). The cable modem has a number of ports for conventional telephone handsets and an Ethernet port for connection (via twisted wire pair) to the network interface card of a personal computer. Several standards have been formulated for the provision of data, Internet and multimedia services over cable systems including DOCSIS (developed primarily for North America) and its European variant EuroDOCSIS, DVB-C/DAVIC (developed primarily for Europe and specified in ETS(I) 300 429 and ETS(I) 300 800) and IEEE 802.14.

EuroDOCSIS

EuroDOCSIS is a Eurocentric version of the North American standard DOCSIS and essentially represents the DOCSIS specification with a (European) DVB [Forrest] specification for the downstream physical layer. Figure 20.82 illustrates the DOCSIS reference model [Smythe *et al.*].

Table 20.10 *Downstream DOCSIS (6 MHz channels) and EuroDOCSIS (8 MHz channels).*

Modulation	Channel bandwidth (MHz)	Symbol rate (Mbaud)	Raw bit rate (Mbit/s)	Approximate nominal data rate (Mbit/s)
64-QAM	6	5.056941	30.34	27
256-QAM	6	5.360537	42.88	38
64-QAM	8	6.952	41.71	37
256-QAM	8	6.952	55.62	50

The normal downstream operating frequency range for EuroDOCSIS is 47 MHz up to a maximum of 862 MHz although data communication is restricted to between 108 MHz and 862 MHz or 136 MHz and 862 MHz depending on national radio regulations. The channel bandwidth is 7/8 MHz for TV signals (6 MHz in the DOCSIS standard) and 8 MHz for data. The maximum allowed transmission delay from head-end to subscriber is 0.8 ms (although typically it is much less than this).

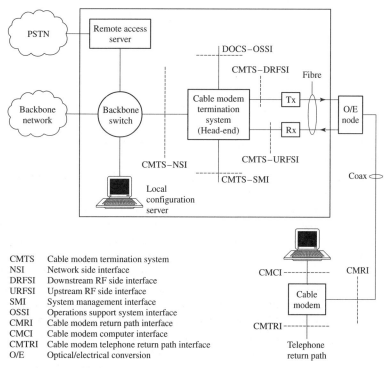

CMTS	Cable modem termination system
NSI	Network side interface
DRFSI	Downstream RF side interface
URFSI	Upstream RF side interface
SMI	System management interface
OSSI	Operations support system interface
CMRI	Cable modem return path interface
CMCI	Cable modem computer interface
CMTRI	Cable modem telephone return path interface
O/E	Optical/electrical conversion

Figure 20.82 *DOCSIS reference model.*

The downstream modulation and data rates are given in Table 20.10.

Root raised cosine filtering, section 8.4, is applied to the downstream symbol stream (the roll-off factor is 15% in the case of EuroDOCSIS and similar but not identical for DOCSIS). A variable depth interleaver is employed to protect against burst errors with a target error probability (post error correction) of 10^{-8}. The maximum interleaver depth provides protection against noise bursts of 95 μs for 64-QAM (66 μs for 256-QAM) at the price of additional latency (4.0 ms for 64-QAM and 2.8 ms for 256-QAM). The interleaver depth is dynamically variable under the control of the CMTS allowing burst error protection to be traded against latency to best suit the service(s) being delivered (4 ms of latency is not significant for web browsing applications, for example, but may be significant for voice over IP applications).

The EuroDOCSIS upstream frequency band is 5 MHz to 65 MHz. The modulation, baud rates and bit rates (which also apply to DOCSIS) are given in Table 20.11.

Table 20.11 *Upstream DOCSIS and EuroDOCSIS modulation, baud rates and bit rates.*

Modulation	Channel bandwidth (MHz)	Symbol rate (Mbaud)	Raw bit rate (Mbit/s)	Approximate nominal data rate (Mbit/s)
QPSK	0.20	0.16	0.32	0.3
	0.40	0.32	0.64	0.6
	0.80	0.64	1.28	1.2
	0.16	1.28	2.56	2.3
	0.32	2.56	5.12	4.6
16-QAM	0.20	0.16	0.64	0.6
	0.40	0.32	1.28	1.2
	0.80	0.64	2.56	2.3
	0.16	1.28	5.12	4.6
	0.32	2.56	10.2	9.0

FEC on the upstream is implemented such that the size of error protected blocks and error correcting power is variable depending on channel quality. As an alternative to increasing FEC power (and thus marginally reducing the available user bit rate) the cable modem can select another (higher quality) channel.

MPEG-2, Chapter 16, is used to encapsulate downlink information in 188 byte frames. This allows a range of information types including video, audio and voice to be transmitted in addition to conventional data in the same channel (on the same carrier).

A request/grant MAC protocol is used to share the upstream communication resource between subscribers. A CM issues a request to transmit a specified amount of information to the CMTS. The CMTS, with an overview of what requests are currently unsatisfied, grants specific transmission slots to specific users and periodically informs users which slots they have been granted. This is a reservation MAC protocol and collisions are therefore avoided. Periodically the CMTS makes a portion of the upstream channel resource available for any CM to make requests. This allows a CM without any granted slots (in contention with other CMs) to make a request and thus break into the reservation mechanism. Like other contention protocols, collisions in this part of the MAC are resolved by those CMs responsible for colliding requests backing off for a random time and retrying.

Privacy in that part of the system between CM and CMTS is provided by an encryption protocol called baseline privacy interface (BPI). BPI employs a version of the data encryption standard (DES) for both upstream and downstream frames and uses the RSA public key encryption algorithm for key management. The CMs either generate private/public key pairs using an internal algorithm or have such pairs installed at the factory. The period over which secret keys are kept unchanged is determined by the operator and can be anything from seconds to months.

The DOCSIS standard continues to evolve. DOCSIS 1.1 incorporates transmit (or pre-) equalisation and offers increased upstream capacity (10 Mbit/s – about twice that of DOCSIS 1.0), enhanced QoS (data rate and/or latency guarantees), improved security (in particular authentication in addition to privacy), and event-based charging. It is backwards compatible with DOCSIS 1.0, CMs and CMTSs from both standards being able to operate over the same cable infrastructure.

DOCSIS 2.0 employs enhanced modulation (64-QAM) and more powerful error correction to offer symmetric upstream and downstream bit rates (30 Mbit/s upstream rates – six times that of DOCSIS 1.0) and improved robustness against interference. Like DOCSIS 1.1 it is backwards compatible with earlier DOCSIS standards.

VoIP services and PacketCable

Cable companies currently offer conventional switched circuit telephony services but voice over IP (VoIP) services are being planned. Downstream voice packet delays are not a significant issue due to the high bit rate (27 Mbit/s) in this direction. In the upstream direction, however, the bit rate is much lower (possibly less than 1 Mbit/s). This may mean that a 1500 byte (12 kbit) Ethernet packet might take 12 ms to transmit which is too long to wait between packets carrying adjacent voice signal segments. A solution to this problem is to fragment the data packets into smaller packets for reassembly at the head-end allowing the latency critical voice packets to be sent more regularly.

PacketCable is an HCF/DOCSIS 1.1 overlay architecture that has been proposed as an IP based solution for the provision of multimedia services including VoIP, video-conferencing and on-line gaming. Figure 20.83 [Mille and Russell] shows a reference architecture for PacketCable.

DVB/DAVIC

DVB/DAVIC [Forrest, Donnelly and Smythe] is a European standard for a multi-service cable system. Figure 20.84 shows the top level DVB/DAVIC reference model.

The forward (downstream) interaction path uses the frequency band(s) 70–130 MHz and/or 300–862 MHz. The reverse (upstream) interaction path uses the frequency band 5–65 MHz. The downstream path is divided into channels 1 or 2 MHz wide supporting bit rates of 1.544 or 3.088 Mbit/s. The upstream path is divided into channels 0.2, 1 or 2 MHz wide supporting bit rates of 0.256, 1.544 or 3.088 Mbit/s. The frequency plan is shown in Figure 20.85.

The downstream interaction path is 'out-of-band' as it is in a frequency band separate from that used for (digital) TV distribution (DVB-C, Figure 20.85). An alternative 'in-band' interaction path can be implemented by embedding the downstream interactive channel data in the MPEG-2 transport stream of a DVB-C) TV channel.

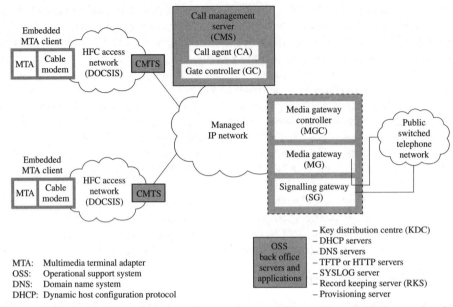

Figure 20.83 *PacketCable reference architecture (source: Mille and Russell, 2001, reproduced with permission of IEEE © 2003 IEEE).*

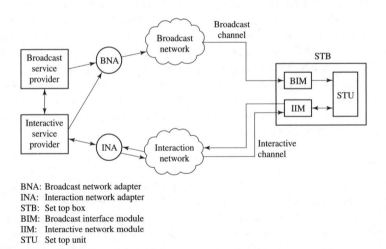

BNA: Broadcast network adapter
INA: Interaction network adapter
STB: Set top box
BIM: Broadcast interface module
IIM: Interactive network module
STU: Set top unit

Figure 20.84 *DVB/DAVIC reference model.*

Modulation for the interactive channels is QPSK (upstream) and DQPSK (downstream). The frame structure for the interactive channels is similar to the T1 1.5 Mbit/s 24-channel primary multiplex frame, Figure 20.19. Upstream interactive channels are shared between subscribers using slotted TDMA. Three multiple access mechanisms are employed: a contention based scheme, a fixed rate scheme and a reservation scheme.

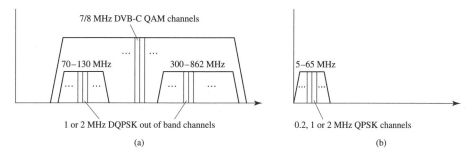

Figure 20.85 *DVB/DAVIC frequency plan (a) forward (downstream) path, (b) reverse path.*

20.11.4 Broadband fixed wireless access

A broadband fixed wireless access (BFWA) system, also referred to in Europe as a broadband radio access network (BRAN), represents an alternative to twisted pair, fibre, cable and their hybrids. There are various technologies that fall under the BFWA umbrella including MMDS (multichannel, multipoint distribution service), LMDS (local multipoint distribution system), WirelessMAN and satellite access. These are also referred to, loosely, as wireless local loop (WLL) technologies since they replace the twisted wire pair (i.e. local loop) traditionally used for access to the PSTN. Strictly, WLL denotes a superset of technologies, comprising FWA and mobile access systems including cellular telephony. Confusingly, however, WLL is also sometimes used to denote the subset of narrowband FWA technologies providing only POTS, and perhaps ISDN, services.

Multichannel, multipoint distribution service

A multichannel, multipoint distribution service (MMDS – also referred to as multipoint microwave distribution system) was originally synonymous with a multipoint video distribution service (MVDS), i.e. a wireless version of CATV. MMDS is therefore sometimes referred to as 'wireless cable'. In the USA a frequency allocation of 200 MHz exists in the 2.1–2.7 GHz band sufficient for 33 (6 MHz) analogue TV channels or 99 10 Mbit/s digital data streams. A service area served by an omnidirectional antenna may be up to 60 km in radius (depending on topography) and transmitter powers are typically in the range 0 to 20 dBW. A return path, typically provided via the PSTN, allows the

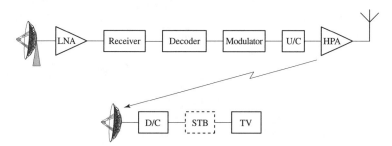

Figure 20.86 *Basic MMDS architecture.*

possibility of simple interactive TV services. Figure 20.86 shows a simple schematic of an MMDS.

Advances in digital communications combined with a general deregulation of telecommunications services has affected the evolution of MMDS in a similar way to cable and applications such as telephone, fax, data communications and Internet services are now typically available in addition to MVDS. Educational applications of MMDS (the wireless classroom) are also possible. MMDS will probably migrate towards these non-TV services in the future since it cannot compete effectively with the number of programme channels available from direct broadcast satellite (DBS) and cable technologies.

Local multipoint distribution system

A local multipoint distribution system (LMDS) is a millimetre wave frequency, high capacity, small cell evolution of an MMDS. The normal LMDS architecture comprises a cluster of base-stations, each with 60°–90° sectoring (requiring four to six transmitters), with antennas mounted on top of a high building or tower. The base stations provide line-of-sight (LOS), point-to-multipoint downlink coverage to between 40% and 90% of the potential users within a cell radius of between 1 km and 5 km. Various frequency bands in the range 3.5–43.5 GHz have been allocated around the world for LMDS (and broadband fixed wireless access systems generally) including bands around 28 GHz in the USA and around 26 GHz, 28 GHz and 40 GHz in Europe. At the millimetre wave frequencies the spectrum allocated has large bandwidth (e.g. 1.3 GHz at 28 GHz and 1.5 GHz at 40 GHz) and, therefore, large potential capacity. Rain attenuation and vegetation penetration loss, however, are significant at these frequencies and this limits cell size. Large penetration and diffraction losses underlie the requirement for LOS links and the (consequent) modest targets for subscriber coverage. (Reflectors and/or repeaters may be used to fill some of the coverage holes caused by shadowing from buildings, etc.) The cells in a cluster are connected by an overlay network implemented using optical fibre or

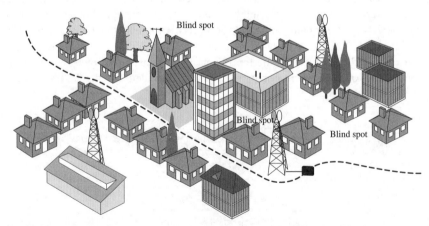

Figure 20.87 *Schematic illustration of traditional LMDS network showing blind spots created by LOS requirement (source: Fowler, 2001, reproduced with permission of the IEEE © 2003 IEEE).*

point-to-point microwave/millimetre wave links. One coordinating cell in each cluster acts as a gateway to the (broadband) PSTN core network. Figure 20.87 shows a schematic illustration of an LMDS and the problem of blind spots where reception of the transmitted signal is obscured by high buildings (as in Figure 15.1).

LMDS subscribers have high gain outdoor antennas mounted at rooftop level to receive a broadband (34–38 Mbit/s) downlink transmission and provide a point-to-point uplink (return path) channel with a capacity of between a few kbit/s and a bit rate comparable with the downlink. The downlink transmission may contain a mixture of broadcast/multicast/unicast information. Each subscriber receiver identifies the information intended for itself on the basis of the data type and its address field.

LMDS applications are similar to those provided by modern cable systems. They currently include interactive TV, Internet access, multimedia applications and business applications (e.g. MAN technology for interconnection of LANs, video-conferencing and e-commerce). Future applications may include tele-education, telemedicine and video on demand.

Deployment of LMDS is most likely in urban and suburban environments by operators without an existing local loop infrastructure (copper pair, fibre or cable) or to supplement an existing but fully utilised infrastructure. The relatively cheap and rapid deployment characteristics of LMDS, however, may also allow its use in low population density regions where a market for broadband access nevertheless exists.

In the lower frequency allocations the more acceptable penetration and diffraction losses lead to more complete coverage in a given service area since LOS paths are no longer strictly required. (In these frequency bands the technology becomes similar to MMDS.) Decreased penetration loss and rain attenuation should result in larger cells (which reduces initial roll-out costs for the operators). The more modest bandwidth available at these frequencies restricts capacity, however, implying a migration to the millimetre wave band as the subscriber base grows.

The traditional LMDS described above has a single layer architecture. An alternative two-layer variation has been proposed [Mahonen *et al.*] which addresses some of the coverage limitation. In this scheme each cell (called a macrocell) is divided into a number of subcells (called microcells) between 50 m and 500 m in diameter, Figure 20.88.

The two-layer LMDS uses the conventional LMDS millimetre wave (40 GHz) bands to connect microcells to macrocells via LOS links. Subscribers connect to the microcell base stations via lower frequency links at, for example, 3.5 GHz, 5.8 GHz or 17 GHz. The lower frequency suffers lower penetration and diffraction losses allowing non-LOS paths and, therefore, both increasing the proportion of potential subscribers covered and allowing these subscribers to be nomadic. An experimental system demonstrating the two-layer concept is described in [Mahonen *et al.*]. Figure 20.89 shows a block diagram of this system and indicates that fixed (non-nomadic) subscribers can connect directly to the macrocell base stations providing they have appropriate LOS paths.

The lower layer of the two-layer LMDS system could be replaced by other wireless technologies such as IEEE 802.11a (WLAN), Table 21.3, or HIPERLAN/2.

LMDS may emulate a range of cable standards including DAVIC and DOCSIS (see section 20.11.3) for the broadcast downlink data format. The European DAVIC standard uses the DVB data format developed for satellite broadcasting. The LMDS intermediate frequency (950–2150 MHz) is chosen to coincide with that of DBS technology allowing

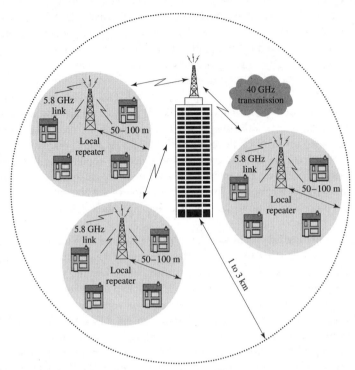

Figure 20.88 *Two-layer LMDS architecture including LOS macrocells and lower frequency microcells (source: Mahonen et al., 2001, reproduced with permission of the IEEE © 2003 IEEE).*

set-top boxes developed, and mass-produced, for satellite TV to be used. Such set-top boxes interface to a TV, PC or both and usually have the facility for a return channel connection via the PSTN/ISDN which is probably sufficient for the low capacity return channels required by interactive TV and web browsing applications. For more demanding applications requiring larger return channel capacity an in-band radio return channel with capacity allocated on demand is required.

LMDS typically provides subscriber communication services using IP over ATM.

IEEE 802.16 and HIPERACCESS

IEEE 802.16 (also referred to under the trade-marked name WirelessMAN) is a second generation BFWA air-interface standard [Eklund *et al.*]. It is a point-to-multipoint access technology, linking customer premises networks such as Ethernets, token rings and wireless LANs to the broadband core network. It is connection-oriented and supports both continuous and bursty traffic. Numerous frequency allocations exist around the world between 10 GHz and 66 GHz, (bands in Europe exist around 10, 26, 38 and 42 GHz) and, like LMDS, LOS paths between base station and subscriber stations are assumed.

Bandwidth is allocated to subscriber stations using a request–grant mechanism. Contentionless access, section 21.3 (which provides stability), is balanced with

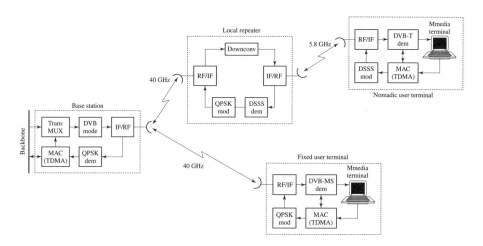

Figure 20.89 *General structure of a two-layer LMDS, local repeater as in Figure 20.88 (source: Mahonen et al., 2001, reproduced with the permission of the IEEE © 2003 IEEE).*

contention based access (which is efficient). It also supports constant bit rate (CBR), variable bit rate real time (VBR-RT), variable bit rate non-real time (VBR-NRT), section 20.10.2, traffic with appropriate QoS.

Modulation is single carrier MQAM of variable order with $M = 4$, 16 and 64. A variable block-size Reed–Solomon code, Chapter 10, is used for FEC and is combined with an inner convolutional code for the most critical information such as frame control and initial access. The burst profile is adaptive allowing the modulation order and coding to be varied for each subscriber station burst by burst. Spectrally efficient burst profiles are used when the channel quality is good and less efficient, but more robust profiles are used under poor channel conditions to achieve a link availability of 99.999%. Channel bandwidths are 28 MHz in Europe, and 20 or 25 MHz in the USA and the standard allows for either TDD or FDD, Chapter 15.

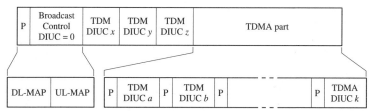

Figure 20.90 *IEEE 802.16 downlink subframe structure. (P: Preamble, DIUC: Downlink interval usage code.)*

IEEE 802.16 frames, which are 0.5, 1.0 or 2.0 ms long, are split into slots, each slot comprising four QAM symbols. Figure 20.90 shows the downlink subframe structure.

The broadcast control field comprises the DL-MAP (downlink map), which refers to the current downlink frame, and the UL-MAP (uplink map), which refers to uplink frames for a specified time in the future. The DL-MAP defines (maps out) the points in the downlink frame where modulation order and/or FEC change. Data is delivered to the various subscriber stations using the TDM part of the frame, which in FDD implementations may be followed by a TDMA part. In the TDM part data is transmitted to the subscriber stations in the order of decreasing burst robustness so that each station receives the burst intended for itself before the modulation order rises, and FEC power falls, to a level that might cause it to lose synchronisation. In the TDMA part each new burst profile is preceded by a synchronisation preamble.

A downlink interval usage code (DIUC) in the DL-MAP entry for each TDM and/or TDMA burst indicates the profile (modulation order/FEC power combination) for each burst. Each burst may contain broadcast, multicast or unicast data. Figure 20.91 shows the uplink subframe structure.

The UL-MAP (uplink map) in the downlink frame allocates (grants) uplink bandwidth to specific subscriber stations. A UIUC (uplink interval usage code) in the UL-MAP entry granting a particular station bandwidth determines the burst profile for that station. The contention based slots allow subscriber stations to obtain initial system access, contentions being resolved using a truncated exponential back-off process. The initial maintenance opportunities part of the uplink frame is used for initial power levelling and ranging.

IEEE 802.16a is a modified version of IEEE 802.16 that employs lower frequency bands between 2 GHz and 11 GHz (including both licensed and licence exempt frequencies). This relaxes the requirement on LOS paths and allows the possibility of nomadic subscriber stations. (The rationale for a lower frequency variant has been previously described in the context of LMDS.) It employs one of three different physical layers, namely OFDM, section 11.5.2 (using a 256-point FFT with TDMA and either TDD or FDD), OFDMA (with a 2048-point FFT with OFDMA and TDD or FDD) and single carrier modulation (with TDMA and TDD or FDD). The single carrier physical layer uses MQAM with $M = 2, 4, 16, 64$ and 256 and generally requires frequency

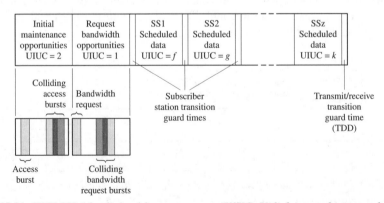

Figure 20.91 *IEEE 802.16 uplink subframe structure. (UIUC: Uplink internal usage code.)*

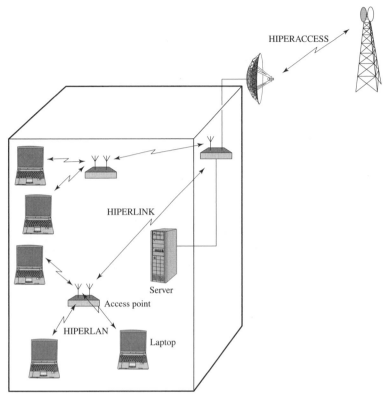

Figure 20.92 *Hierarchical broadband access network (source: Bolcskei et al., 2001, reproduced with permission of the IEEE © 2003 IEEE).*

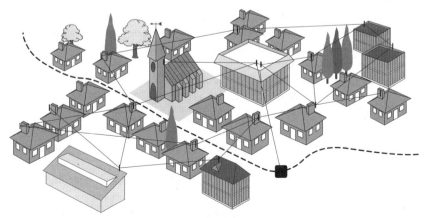

Figure 20.93 *Mesh network alternative to point-to-multipoint (star) network (source: Fowler, 2001, reproduced with the permission of the IEEE © 2003 IEEE).*

domain equalisation. In licence exempt bands OFDM with TDD (referred to as WirelessHUMAN – the HU denotes high rate unlicensed) is mandatory. IEEE 802.16a data rates are lower than those for IEEE 802.16 which may result in a division of applications in which the former is used to serve small office and home office (SOHO) markets and the latter is used to serve larger businesses located in urban city centres. The physical environments of these two markets suit the technologies since there is unlikely to be convenient tall structures (providing LOS paths to most subscribers) on which to locate base stations in suburban areas, whereas such structures in the form of tower blocks are common in city centres. IEEE 802.16a has an enhanced MAC layer that incorporates ARQ and (for licence exempt bands) dynamic frequency selection (DFS) reflecting the more severe fading environment expected compared to IEEE 802.16. IEEE 802.16a also supports advanced antenna systems, mesh operation (described later) and subscriber-to-subscriber communications.

HIPERACCESS is a European (ETSI) version of IEEE 802.16 (operating in bands above 11 GHz). It can be combined with other ETSI technologies (HIPERLINK using a band around 17 GHz and HIPERLAN using a band around 5 GHz) in a hierarchical architecture as shown in Figure 20.92 [after Bolcskei *et al.*]. (The architecture is flexible, however, and it is possible for HIPERLAN to be interfaced directly to HIPERACCESS without the intervening HIPERLINK layer.) HIPERACCESS is designed primarily to serve the SOHO and SME (small and medium size enterprise) market providing user transmission rates of between 2 Mbit/s and 25 Mbit/s.

HIPERMAN is the ETSI version of IEEE 802.16a (operating in bands below 11 GHz) using the IEEE 802.16 MAC layer with the 802.16a OFDM physical layer.

Mesh networks for broadband access

A possible evolution of BFWA technology might use a mesh network architecture rather than a conventional point-to-multipoint or star network [Fowler, Webb].

The mesh may be purely logical, omnidirectional antennas being used at each subscriber premises, or it may be both logical and physical with independently steerable multi-beam subscriber antennas providing pencil beam links between nodes. The advantage of a mesh network is that blind spots due to shadowing can be almost entirely eliminated and close to 100% subscriber coverage realised, Figure 20.93. The advantage of a physical mesh is that transmit power can be tailored to each link (shorter links requiring lower power) reducing interference levels and increasing the scope for frequency reuse leading to improved spectral efficiency. Advantages claimed for mesh access networks [Fowler] of this type include:

1. minimal risk of blind spots leading to 100% coverage;
2. spectral efficiency up to 50 times greater than for the equivalent point-to-multipoint system;
3. improvement in quality due to reduced link lengths with consequent improvement in link budget;
4. no requirement for dedicated base station installation towers resulting in improved economics;
5. multiple alternate routes through the mesh leading to high resilience;
6. inherent scalability with low initial cost of deployment and consequent lower risk for the operator.

Satellite access

Satellite technology can be used in a similar way to LMDS/BFWA but on a much larger geographical scale. A hub takes the place of the cell base station and one or more satellites act as relays for both forward (hub-to-subscriber) and return (subscriber-to-hub) channel signals. The hub acts as a gateway to the PSTN/ISDN/B-ISDN. ETSI standard EN 301790 – An interactive channel for satellite distribution systems (ICSDS) – describes such a scheme and Figure 20.94 illustrates its basic architecture.

The ICSDS forward path channel, from hub to satellite, has a data format based on DVB/MPEG-2 with a maximum bit rate of 45 Mbit/s. A forward communications channel is realised by multiplexing data into the DVB/MPEG-2 broadcast stream at the hub. Transmission over the satellite is typically in the Ku frequency band, i.e. within SHF in Table 1.4. Modulation is QPSK and FEC is realised using concatenated convolutional and Reed–Solomon codes.

The return path channel provides users with contentionless, dynamically scheduled, bandwidth using multiple frequency TDMA (MF-TDMA) giving users upstream data rates of between 144 kbit/s and 2.048 Mbit/s. These bit rates open up the usual multimedia applications (e.g. telecommuting, tele-education, high speed web browsing,

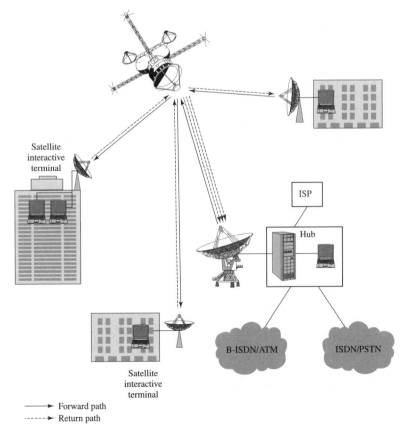

Figure 20.94 *A satellite distribution system with interactive return path channel.*

telemedicine, etc.) and conventional bidirectional communications services (telephony, data, fax) in addition to the traditional low bit rate interactive services (pay-per-view, interactive TV, home shopping, e-commerce, etc.) that a PSTN return path has traditionally catered for. Figure 20.95 illustrates the principles of MF-TDMA. The advantages of MF-TDMA include:

1. Lower aggregate burst rate for a given capacity compared to TDMA implying lower cost terminals.
2. Fewer modems in each terminal compared to FDMA.
3. Each 'circuit' is an independently allocated burst in the time–frequency map leading to flexible demand assignment.
4. The aggregate capacity of the time–frequency slots can be much greater than a TDMA system with the same terminal power and BER.

The modulation and FEC coding on the return path is the same as that on the forward path.

The allocated frequency spectrum is divided into sub-bands and each sub-band is divided into time-slots. A frame (typically 26.5 ms long) comprises a collection of time–frequency slots and the hub allocates a given subscriber terminal a succession of these slots for its return path transmission. The bandwidth and duration of the slots may be constant (fixed MF-TDMA) or varied according to demand (dynamic MF-TDMA), see Figure 20.95. Each time–frequency slot carries an IP packet or a concatenation of several ATM cells. The subscriber terminal initially accesses the network using a slotted ALOHA channel but the subsequent allocation of slot patterns by the hub is contentionless. Terminals that already have an allocated slot pattern can request further capacity by piggybacking requests on their data and control bursts. (Some of the time–frequency slots are used for control purposes such as terminal synchronisation.) In addition to the changing carrier frequency, bit rate and coding are adaptable slot by slot in order to match the channel to the varying transmission requirements of multimedia data.

Table 20.12 summarises some of the recent and planned multimedia satellite networks that use, or will use, interactive return path channels.

High altitude platforms

High altitude platforms (HAPs) can be used to locate communications (and other) equipment at a nominally stationary position in the stratosphere. HAPs may be free flying balloons [see WWW skynet] with station keeping engines, or aeroplanes [Colella

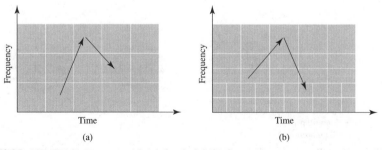

Figure 20.95 *MF-TDMA concept with (a) fixed and (b) dynamic resource allocation.*

et al.] flying in closed loops within a specified toroidal air-space. HAPs fly at altitudes typically between 15 km and 25 km. Communication services from HAPs share some characteristics with terrestrial cellular networks and some with satellite networks (although they are classified as strictly terrestrial systems by radio regulatory authorities). Like terrestrial systems HAP path lengths are modest resulting in correspondingly modest transmit power requirements and low propagation delays. (The former is especially significant in the context of subscriber terminals in mobile applications.) Like satellite systems, however, elevation angles are generally high resulting in LOS paths and a relatively benign (typically Ricean) fading environment.

The communication equipment flown on a HAP may be a quasi-transparent (i.e. bent pipe) transponder to connect users with a ground station or a network hub where all significant signal processing and switching takes place. Alternatively it may be a complete base station transceiver with associated onboard processing and switching. The proposed HALO network described below is of the latter type.

The HALO (high altitude long operation) network is a broadband wireless MAN [Colella *et al.*]. It uses the Proteus aeroplane, Figure 20.96, flying at an altitude between 15.5 km and 18.3 km as the platform. The aeroplane has a circular flight path with a radius of between 9 km and 15 km. The time taken to make one circuit is around 6 minutes and the HALO coverage area is circular with a diameter of around 100 km.

LMDS at 28 and 38 GHz represents one of the most important applications of HAPs but their use for initial (rapid) roll-out of future generation mobile cellular systems is also possible. (The ability to deploy HAP based technologies rapidly also suggests their use for disaster relief in regions where a telecommunications infrastructure does not exist or has been disrupted.)

The HALO system capacity is initially expected to be around 10 Gbit/s but will probably increase by an order of magnitude or more as the subscriber base, and demand for user bandwidth, grows. The network service area will be divided into between 100 and 1000 cells each provided by a narrow antenna beam, Figure 20.97. Each cell may serve between 100 and 1000 subscribers within an area of between a few, and a few tens of, square kilometres. Capacity will be maximised using frequency reuse along the lines of conventional cellular systems. The antenna beams may be synthesised from a phased array aperture, allowing cells to be held stationary on the ground (at least for a 'dwell' time before movement of the platform requires the beams providing the cells to be switched). Alternatively antenna beams may be fixed with respect to the aeroplane resulting in cells on the ground that move with the platform. The latter implementation would require handoff from cell to cell even for stationary users. A ground station gateway will connect the HALO network to other private and public networks. Figure 20.97 illustrates a proposed HALO system architecture. (Inter-HAP and HAP-satellite point-to-point links at millimetre wave or optical frequencies may be used to form an overlay mesh network if geography and traffic conditions make this useful and economic.)

The HALO network contains four distinct types of subsystem. These are the platform itself and associated transceiver equipment, user terminals, the network control station and the HALO gateway. Figure 20.98 shows a network reference model.

HALO networks will use ATM and/or IP over ATM protocols and the HALO transceiver will include onboard ATM switching. Bandwidth in each cell will be shared

Table 20.12 *Existing and planned multimedia satellite/DBS systems with return path channels.*

System (Operator)	Planned operations	No./type satellites	Frequency (GHz)	Multiple access	Protocols/ network	Capacity (Mbit/s)	Example services
Astrolink (Astrolink Int.)	2003	9/4 GEO OBP/ISLs	Ka 28.35–28.8 and 29.25–30.0 UL, 19.7–20.2 DL	FD-MA/TD-MA (UL) TDM(DL)	IP/ATM/ frame re-lay/ IS-DN. Up to 70 gateways per sat.	0.16–9 Mbit/s per user	Multimedia, Internet access, virtual private nets
Cyberstar (Loral) Marketed with Sky-bridge		3 GEO OBP	Ka	FDM/TDM (UL) TDM(DL)	IP/ATM frame re-lay. 96–192 gateways (Connect-ed with Sky-bridge)	0.384–3.088 Mbit/s (UL), 0.384–3.088 Mbit/s (DL)	Internet access, video on demand
Spaceway plus Spaceway NGSO (Hughes)	2004	8 GEO plus 20 MEO OBP/ISLs	Ka 17.7–30 UL & DL	FD-MA/TD-MA (UL) TDM (DL)	IP/ATM/ frame re-lay/ IS-DN/X.25	0.016–6 (UL), <108 (DL)	Internet, interactive multimedia, telephony
Sky-Bridge (Alcatel)	2004 (On hold from 2002)	80 LEO Transpar-ent transpon-der	Ku 12.75–14.5 UL, 10.7–12.75 DL	CDMA FDMA TDMA	IP/ATM 140 gate-ways (Connect-ed with Cyberstar via gate-ways)	0.016–2 Mbit/s (return) 0.016–60 Mbit/s (forward)	Internet access, inter-active multimedia
Teledesic (Teledesic Corp.)	2004 (On hold from 2002)	288 LEO OBP/ ISLs	Ka 28.6–29.1 UL, 18.8–19.3 DL	MF-TD-MA (UL) Async. TDMA (DL)	IP/ATM/ ISDN	0.016–2, 64, 1200 (UL) de-pending on termi-nal type. 0.016–64 (DL)	Internet access, voice, data, interactive video
Astra BBI (SES)	Service since 2001	Later satellites starting with 1H, GEO Transpar-ent transpon-der.	Ku 10.7–12.75 forward. Ka 29.0–30.0 return.	Statistical (forward) MF-TD-MA (re-turn)	IP/ATM in MPEG-2 TS (for-ward), ATM cells or MPEG-2 TS (re-verse)	Up to 10 Mbit/s (forward) Up to 2.048 Mbit/s (reverse)	DVB-RCS, data, Inter-net access, content for multicast, virtual pri-vate net, file transfer

between SOHO and SME users providing each user with a peak data rate of up to 25 Mbit/s (on demand). Dedicated beams for large business subscribers with sufficiently high and constant demand may be provided.

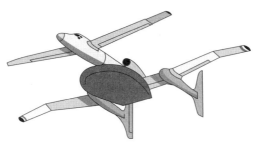

Figure 20.96 *Artist's illustration of HALO/Proteus platform.*

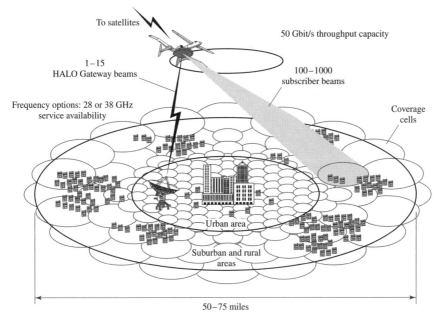

Figure 20.97 *HALO network architecture (source: Colella et al., 2000, reproduced with the permission of the IEEE © 2003 IEEE).*

Non-HALO networks will be accessed via the HALO/interworking unit (IWU) gateway. The network control station is responsible for connection admission control (CAC, see section 20.10.3), handling of time-slot reservation and requests generated by the MAC protocol, management of handoff between cells and mobility management for mobile subscribers. It is also the OA&M centre for the HALO system.

20.11.5 Comparison of access network technologies

Table 20.13 compares selected current access technologies.

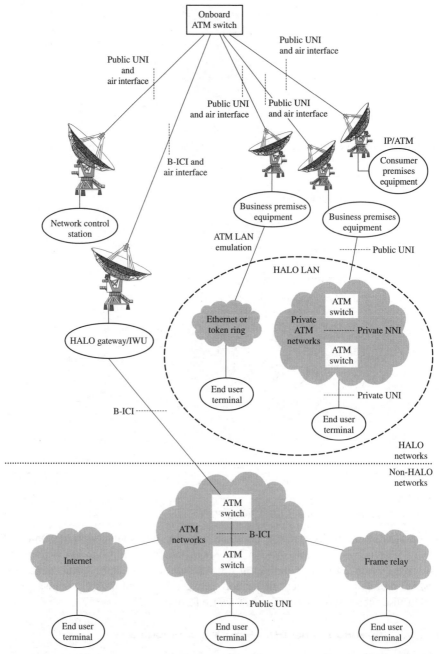

Figure 20.98 *HALO system reference model (source: Colella et al., 2000, reproduced with the permission of the IEEE © 2003 IEEE).*

Table 20.13 *Comparison of access network technologies.*

Technology	Number of pairs	Modulation	Downstream bit rate (Mbit/s)	Upstream bit rate (Mbit/s)	Approx. spect. range (MHz)	Latency (ms)	Range (km)	Comments
HDSL (ITU-T G.991.1)	1, 2 or 3	2B1Q (PAM) or CAP.	2.048/ 1.544	Symmetric	0–0.196 for 3 pair 2B1Q, 0–0.292 for 2 pair 2B1Q, 0–0.485 for 1 pair CAP	<1.25	4–5	Specified for SDH, VC-12 or TU-12 in 2.3 Mbit/s payload. O/p echo cancellation subtracts tx signal. 2.048 Mbit/s 2B1Q allows 2 or 3 parallel pairs or duplex on 1 pair.
HDSL2 (ITU-T G.991.2)	1	16-level PAM/ TCM	2.048/ 1.544	Symmetric	0–0.3 upstream, 0–0.43 dowstream	<0.5	3.7 for 1.5 Mbit/s	Each symbol = 3 data bits + 1 FEC bit. N. American version of SHDSL used more on feeder than access networks.
SHDSL (ITU-T G.992.2)	1	16-level PAM/ TCM	0.192–2.312 (inc. 2.045/1.544)	Symmetric	0–0.07 for 256 kbit/s, 0–0.6 for 2.3 Mbit/s	<1.5	up to 6 depending on bit rate	Single pair HDSL. Business/SOHO/ public service apps: high speed bidirectional data, video conference, LAN access. Future symmetric DSL global standard?
SDSL (ETSI TS101524)	1	2B1Q or CAP	0.128 to 0.36 (select on install)	Symmetric	0.01–0.5	<1.5	5.6 for 1.5 Mbit/s	Symmetric DSL. Simultaneous POTS. (May be made obsolete by SHDSL)
ADSL (ITU-T G.992.1)	1	DMT (256 4 kHz channels) or CAP	1.48–6.144 minimum	0.016–0.64 minimum	0.2–1.1 downstream, 0.02–0.2 upstream	<20	4 for 2 Mbit/s, 2 for 6 Mbit/s	Residential access for Internet. Simultaneous POTS using LPF splitter. Can employ echo cancellation, TCM. STM/ATM transport.
Splitterless ADSL (G.lite) (ITU-T G.992.2)	1	DMT (96 4 kHz channels)	1.5	Up to 0.512	0.02–0.14 upstream, 0.14–0.55 downstream	<20		No splitter thus POTS reduces data rate. G.992.4 is a 2nd generation and G.992.5 an extended bandwidth ADSL.
VDSL (ITU-T G.993.1) (Symmetric and asymmetric classes)	1	QAM or DMT	6.4–28.3 (but also up to 52)	1.5–2.3 (asymmetric class)	0.138–12 (efficient mode) 1.1–12 (ADSL compatible mode)	<1 (fast path) <20 (slow path)	0.3–1.5 at 26 Mbit/s, 1.5 at 13 Mbit/s	Video and other high bandwidth services. Slow path version trades latency for burst protection up to 500 μs. EMC problems with other services. Use with FTTCab and FTTC technologies.
ISDN-BA (ITU-T G.961)	1	2B1Q or 4B3T (AMI,TDD in Japan)	0.16 (2B+D)	0.16 (2B+D)	0.01–0.1		5.6 but extendable with repeaters	Full duplex, echo cancellation

20.11.6 Convergence of access network services

Recognising the convergence of telecommunications services over a variety of access technologies, it is useful to have a generic description (reference model) of an access network, Figure 20.99 [Maeda and Feigel].

The local switching subsystem in Figure 20.99(b) is the point where the access network (AN) joins the core network or service function. The service network interface (SNI), which delimits the AN on the service provider side, may conform to a variety of interface specifications including xDSL (e.g. ADSL, HDSL, VDSL), SDH (e.g. STM-1,

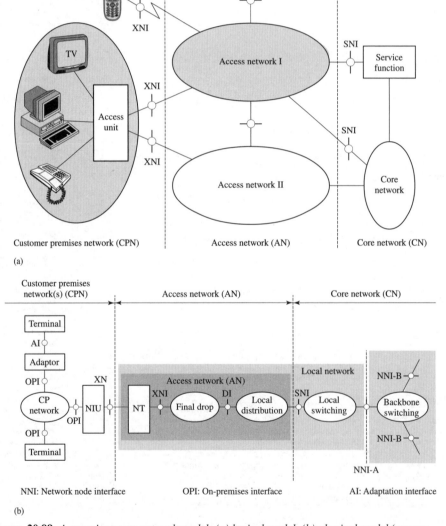

Figure 20.99 *A generic access network model: (a) logical model, (b) physical model (source: Maeda and Feigel, 2001, reproduced with the permission of IEEE © 2003 IEEE).*

STM-4, STM-16), PDH (e.g. the 30+2 European primary multiplex, Example 20.4) and IEEE 802.3 Ethernet variants (e.g. 10BASE-T or 100BASE-T Ethernet) see Table 21.2.

The local distribution subsystem converts between the SNI and distribution interface (DI), the latter typically conforming to xDSL, B-PON, SDH or PDH.

The network termination (NT), which is usually located inside the customer premises, delimits the user end of the AN. Everything on the customer's side of the NT is part of the customer premises network (CPN).

The X network interface (XNI) is the interface between the user and network domains. X denotes one of many possible user networks that may interface to the access network. In addition to any of those specifications given for the SNI it may conform to B-PON and (ATM based) B-ISDN.

The final drop subsystem converts between the DI and the various interfaces to CPN equipment (XNIs). The XNI might, for example, conform to xDSL, SDH, PDH or Ethernet 10BASE-T/100BASE-T, see Table 21.2. The on-premises interfaces (OPI) in the CPN connect through the network interface unit (NIU) to the X network (XN).

As a concrete example, in a B-PON access network the SNI might link the core network using SDH (over optical fibre) to the local distribution point, i.e. the optical line termination (OLT). The B-PON (which is the DI) would link the OLT to the final drop (the ONUs) and an XNI might link the final drop (ONU) to the NT in the user premises using xDSL over a twisted pair.

20.12 Summary

Transmission of data between a pair of geographically distant users usually involves transport across a series of networks. The networks to which the users connect directly are called access networks, which may be switched, e.g. a point-to-point xDSL link, or broadcast, e.g. a local area network (LAN). Access networks are connected by a (switched) core network, designed for the bulk transport of data over large distances. Variations on this scheme may occur: a metropolitan area network (MAN), for example, may be employed as a backbone network for several (access) LANs, thus representing an intervening network between the access LAN and the core wide area network (WAN).

This chapter is concerned with switched core networks (WANs) and switched access networks. A treatment of broadcast LANs (and their personal area network diminutives) is presented separately in Chapter 21.

Switched networks comprise a set of nodes interconnected by links to form a (partial) mesh. Communication between a particular pair of terminals is realised by configuring the node switches such that a path from one terminal to the other is formed through the network. If a continuous (i.e. unbroken) end-to-end path is set up prior to user message transfer, maintained for the duration of the message transfer, and released only after the completion of the message transfer, then the network is said to be circuit switched. If data blocks are stored at nodes and individual links are set up only as resources allow, and then only for the duration of the individual data block transfer, then the network is said to be store and forward. Store and forward networks can be further divided into message switched and packet switched services.

The PSTN infrastructure and its traditional role in providing circuit switched real time voice communication services has been described and the topology of the modern integrated digital network (IDN) presented. Multiplexing of PCM-TDM telephone traffic has been traditionally provided using the plesiochronous digital hierarchy. The PDH frame rate is 8000 frame/s. Bit rates increase by a little more than a factor of four at each successive multiplex level to allow for small differences in multiplexer clock speeds. Empty slots in a multiplexer output are filled with justification bits as necessary. A serious disadvantage of PDH multiplexing is the multiplex mountain which must be scaled each time a lower level signal is added to, or dropped from, a higher order signal. This is necessary because bit interleaving combined with the presence of justification bits means that complete demultiplexing is required in order to identify the bytes belonging to a given set of voice channels.

The synchronous digital hierarchy (SDH) and its originating North American equivalent, SONET, are rapidly replacing the PDH. The principal advantage of SDH is that low level signals remain visible in the multiplexing frame structure. This allows lower level multiplexes (down to individual voice channels) to be added or dropped from higher order multiplex signals without demultiplexing the entire frame. This simplifies cross-connection of traffic from one signal multiplex to another. The SDH frame rate is 8000 frame/s and each frame contains one or more synchronous transport modules (STMs). The SONET frame rate is also 8000 frame/s and each contains one or more synchronous transport signals (STSs). The standard SONET/STM payload capacities, bit rates and constituent signals are summarised as follows:

SONET	STS-1	9×90 bytes	51.84 Mbit/s	
SDH	STM-1	9×270 bytes	155.52 Mbit/s	3 STS-1s
SDH	STM-4	9×1080 bytes	622.08 Mbit/s	4 STM-1s
SDH	STM-16	9×4320 bytes	2.488 Gbit/s	4 STM-4s

The capacity of an STM-1 is such that it can carry one PDH-4 signal which will facilitate the operation of PSTNs during the period in which both multiplexing schemes are in use. SONET and SDH have been designed, primarily, to operate with optical fibre transmission systems.

The evolution of core network multiplexing, switching and control (signalling) technologies to meet the recent rapid expansion in global demand for communications capacity, and the application of the IDN infrastructure to create public data networks (PDNs) have both been described. The evolution of the IDN through the integrated services digital network (ISDN), the broadband-ISDN and the data transport technology (asynchronous transport mode, ATM) developed for B-ISDN access have also been described. ATM provides efficient use of time and bandwidth resources in the access layers of the network when variable bit rate services are provided. ATM frames consist of 53 byte cells, 48 of which carry traffic and 5 of which carry overhead. At the higher network layers ATM cells will be efficiently carried within the payload envelopes of SDH frames.

Finally a range of traditional and new access network technologies have been reviewed from the conventional PSTN twisted pair local loop in its modern DSL incarnations through the broadband passive optical network (FTTx), cable and HFC to broadband fixed wireless access technologies including satellite and high altitude access platforms.

20.13 Problems

20.1 What are the typical characteristics of a wide area network?

20.2 Find the maximum feasible information flow between nodes B and F of the network represented by the capacity graph in the top left hand corner of Figure 20.6(a).

20.3 A partially connected mesh network comprises six nodes. Five of the nodes are connected as a ring and the sixth node is connected (directly) to all the other five nodes forming a hub. What is: (i) the minimum degree of the network, (ii) the link connectivity of the network, and (iii) the node connectivity of the network? Is this network regular?

20.4 What is meant by an IDN? Describe, using a schematic diagram, the basic architecture of the UK (BT) IDN.

20.5 Describe briefly the PSTN multiplexing scheme that is being replaced by SDH. Include in your description the approximate bit-rates used for the primary multiplex and the next two higher levels. Also include an explanation of the mechanism that accommodates small variations in tributary data stream bit rates.

20.6 For the European PDH system, describe the composition of the 2.048 Mbit/s primary multiplex group.

20.7 Calculate the number of telephone channels that can be accommodated at level 4 of the European PDH. Explain why the bit rate at a given PDH level is not exactly an integer multiple of the bit rate at the level below?

20.8 Describe how in-channel signalling is supported in a primary multiplex PDH module. Calculate the signalling bit rate per voice channel in this system and explain why common channel signalling is preferable.

20.8 Explain the notion of section, line and path as entities in an SDH transmission network. Taking an STM-1 frame as an example, where is the information concerning these entities carried in the SDH signal? What is the nature of the information?

20.9 What mechanism is used to allow an SDH SPE to pass between SDH networks that are not synchronised? How does this differ from the mechanism used to allow tributaries from the old plesiochronous system (e.g. 2.048 Mbit/s) to be taken in and out of an SDH system?

20.10 Explain the term add–drop in the context of multiplexers. Draw block diagrams to show why this function is simpler to perform in the SDH system than in the older PDH system.

20.11 Find the European PDH minimum multiplex level required to carry the number of voice signals accommodated by the SDH STM-1 frame.

20.12 Explain the function of a space switch and a time switch. Describe, using a diagram, how a time switch can be implemented.

20.13 Design a switching system using two stages to connect 300 incoming to 300 outgoing trunks.

20.14 Design a switching system using three stages to connect 300 incoming to 300 outgoing trunks.

20.15 All second stage switch units of a two-stage switch serve each outgoing route with one trunk. There is an equal number of primary and secondary switch units and there is one link between each primary and secondary switch unit. Derive a formula for the switch blocking probability, B, in terms of the number of secondary switch units, S, the link occupancy, a, and outgoing trunk occupancy, b.

20.16 Use the formula derived in the previous question to find the blocking probability (grade of service) if the number of secondary (and primary) switch units is 5, the total offered traffic is 5 Erlangs and the number of outgoing trunks is 50 (10 per secondary switch).

20.17 16 incoming highways are connected to 16 outgoing highways via 16 time-switch links in a space time space switch. Each highway carries 32 PCM telephone channels. The average occupancy of incoming PCM channel is 0.7 Erlangs. Estimate the grade of service when connection is required to a particular free channel on a selected outgoing highway.

20.18 Describe how digital data can be transmitted over the circuit switched PSTN.

20.19 Illustrate the progress of a data exchange event over a PSTN using the V.24 interface. Include in your illustration the sequence of local and remote handshaking events that take place.

20.20 Explain the operation of a Banyan switch and illustrate how it facilitates self-routing of data packets in a packet switched network.

20.21 Describe, briefly the basic principles of a packet switched public data network (PSPDN). Include in your description a list of facilities that such a network must support.

20.22 Describe the principal features of an X.25 network. Include in your description which particular part(s) of the network is (are) governed by the X.25 specification, which layer(s) of the open system interconnection (OSI) model the X.25 specification relates to, what network services are provided, how calls are set-up, the role of logical channel numbers and the role of routing tables at network nodes.

20.23 What is meant by frame relay in a packet network? What advantage does frame relay offer and under what transmission conditions can it be used successfully? What is unconventional about frame relay in the context of routing?

20.24 What do the terms ISDN and IDN mean? Explain the difference between these two types of network.

20.25 An ISDN is really made up of three networks. Describe briefly the nature and function of these three networks. Draw a diagram to show how a circuit switched connection is made by an ISDN illustrating the roles of ISDN user and control planes.

20.26 What is meant by basic rate and primary rate access in the context of ISDN? Define the mix of channel types in each case and give the bit rates supported by each channel type.

20.27 What is the purpose of the ISDN S-interface and what is its relationship to the U-interface?

20.28 What are the essential functions of a signalling system in a telecommunications network?

20.29 The ITU-T signalling system No. 7 identifies three types of (HDLC based) signal unit. Name these signal units and describe their essential functions. Indicate the format of one of the signalling unit types.

20.30 Why do you think signalling system No. 7 requires a traffic-path continuity check?

20.31 A modern telecommunications system can be described as a layered stack of several networks, each with a different function. Outline the essential purpose of each functional network in this stack.

20.32 Sketch a simplified common channel signalling system No. 7 network topology showing signalling points, signal transfer points and a switching control point. Explain why the reliability of the CCS7 is critically important to public data networks.

20.33 What is the B-ISDN? A transport mechanism (or transfer mode) was developed specifically for access to B-ISDNs. Name this transfer mode and say how it differs from a conventional packet switched system. What considerations determined the specification of cell length for this transfer mode?

20.34 Six variable bit-rate information sources each generate an ATM cell in a given time-slot with a probability of 20%. If the ATM channel capacity is 3 cells per time-slot find (to a precision of 3 significant figures) the probability of overflow (i.e. cell loss).

20.35 At what three levels can switching in an ATM system take place? Explain the conceptual relationship between these levels and describe why this layered switching structure is an advantage. To which of the generic classes of switching mechanism (circuit switching, virtual-circuit packet switching and datagram switching) do each of these switching levels most closely correspond?

20.36 Describe how ADSL provides broadband access using the conventional twisted pair local loop. With what earlier 'broadband' access technology does this compete?

20.37 How is upstream access shared between subscribers in a B-PON? In particular how are the different ranges of the subscribers allowed for in the up-stream multiplex. What special arrangement is required to maintain emergency services if telephony is provided via the PON?

20.38 Draw a block diagram showing the basic structure of a traditional CATV system. What are the undesirable consequences of many cascaded amplifiers in such a system?

20.39 What is an HFC system and what advantages does this have over a conventional cable system?

20.40 How are bi-directional data services provided in a modern cable system? Illustrate your answer by referring (at least qualitatively) to the DVB/DAVIC frequency plan.

20.41 Compare and contrast MMDS and LMDS systems.

20.42 Describe IEEE 802.16. How does IEEE 802.16 take best advantage of varying channel quality?

20.43 What are the principal differences between IEEE 802.16a and IEEE 802.16?

20.44 Describe how HIPERLAN, HIPERLINK and HIPERACCESS can be used to provide an integrated hierarchical BFWA network. What (approximate) frequency bands are use by each of these technologies?

20.45 What is an interactive channel for satellite distribution system (ICSDS)? Describe the basic architecture of such a system and the MF-TDMA scheme that can be used for the return path channel.

20.46 Describe what is meant by a high altitude platform (HAP). What sort of communications services could be provided by HAPs and what advantages does their use offer over satellite technology?

CHAPTER 21

Broadcast networks and LANs

21.1 Introduction

Broadcast networks differ from switched networks in that all stations receive all transmissions. There are two basic varieties of broadcast network: the first, appropriate for information distribution services (e.g. conventional television), has only one transmitting station (and, usually, many receive-only stations) whilst the second, appropriate for information exchange services, has multiple transmitting stations (which can usually also receive). In both varieties all stations are connected to a common transmission medium that makes the (switched network) problem of routing essentially redundant. For multiple transmitter networks the routing problem is, however, replaced by that of sharing the transmission medium between stations efficiently and equitably. The most widely deployed multiple-transmitter broadcast networks are LANs (local area networks). These networks may be isolated and complete in themselves or may form part (often the access part) of a larger (MAN or WAN) network. It is LAN technology with which this chapter is principally concerned.

LANs are specifically designed for the interconnection of computer systems and peripherals within a geographically small site, such as a single building or campus, and are generally privately owned. They have many of the features of WANs, but also have their own, distinct, characteristics, e.g.:

- Wide bandwidth, of the order of tens or hundreds (recently thousands) of Mbit/s.
- Low delay due to resource sharing and absence of buffering, typically 1 to 10 μs.
- Low probability of bit error, typically 10^{-9} to 10^{-11}.
- Simple protocols, compared with those necessary for the longer range switched network WANs.

- Low cost and easy installation.
- High degree of connectability and compatibility of physical connections.
- Geographically bounded, with a maximum range of approximately 5 km.

A LAN connects all user points in the network with essentially equal priority, so that each station can talk to any other. (This does not preclude some networks from implementing explicit prioritisation of data depending on its type or origin.) Each message is broadcast throughout the network, being accepted by those stations for which it is intended, while being ignored by the others. LANs have traditionally operated asynchronously (in which a data stream is broken into packets which are then delivered at essentially random intervals of time) since the traffic has historically been insensitive to the precise time of its reception. In recent years, however, many LAN systems and their variants (including personal area networks) also support mechanisms for isochronous or synchronous data transfer. The former describes a mechanism that ensures data is delivered within a certain time constraint (i.e. with delay less than some maximum value) and the latter ensures that data is delivered at a specific time interval (i.e. at a constant repetition rate). This allows networks to be used for time-critical and/or constant bit rate applications such as voice, video or multimedia links.

21.2 LAN topologies

The topology of a LAN may be bus, tree (including the special case of the star) or ring, see Figures 17.5–17.10. The ring topology is favoured for systems using optical transmission. In order to make ring networks more secure, there are often redundant paths so that an inoperative station can be by-passed.

21.3 LAN protocol stack

LAN protocols must exercise control of the transmission medium in order to share its transmission capacity between all contending users. A medium access control (MAC) protocol performs this function. The particular MAC protocol adopted depends upon the network topology and the transmission medium. Sitting above the MAC protocol is the logical link control (LLC) protocol that is independent of the transmission medium and method of access. These functions have been standardised by an IEEE committee (IEEE 802), whose recommendations are universally accepted. Figure 21.1 shows the approximate correspondence between these protocol levels and the OSI seven-layer reference model.

There are two general categories of scheme for sharing a common transmission medium between many stations. These are reservation schemes and contention schemes.

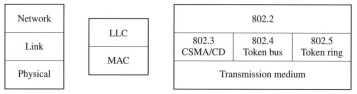

Figure 21.1 *LAN protocol layers.*

In the former transmission resources are reserved, either permanently or temporarily, for each station. In the latter no reservation of resources is made and stations can try to access (contend for) the transmission medium simultaneously, with the consequent possibility of stations receiving corrupted transmissions. Reservation schemes include fixed access multiplexing, polling and token passing whilst contention schemes include random access and carrier sense multiple access. In practice, of course, hybrids of these accessing methods are also possible.

21.3.1 Fixed access multiplexing

In fixed access multiplexing the total network transmission resource is divided among the participating stations in time, frequency or code domain and a station therefore always occupies a part of the channel capacity, whether or not it has data to transmit. The corresponding protocols are time division multiple access (TDMA, Figure 5.2), frequency division multiple access (FDMA, Figure 5.12) and code division multiple access (CDMA, see section 15.5).

Fixed access networks require a central organising process to allocate transmission resources to a station. Although this is widely used in trunk transmission networks, it only works efficiently if most users wish to transmit most of the time. Traffic on a LAN is intermittent, so inactive time-slots result in a waste of resource. This technique has the advantage, however, that a user is guaranteed access to the network within a pre-determined and fixed time interval. If relative network resources (e.g. time-slot lengths or number of time-slots per frame) allocated to different stations can be changed under control of the central station, to reflect the changing station loads, then multiplexing is said to be demand assigned. If no such variation in resource allocation is possible then multiplexing is fixed assigned.

21.3.2 Polling

In a polled network a central controller interrogates or polls each station in turn to see if it has data for transmission. This is an extreme form of a demand assigned reservation scheme but is still not very satisfactory since the interchange of polling messages wastes network resources while the controller waits for a reply. Furthermore, vesting total control in one physical station makes the network vulnerable to failure.

21.3.3 Token passing

With token passing protocols, right of access is granted to the attached stations in a predetermined (typically cyclic) order. An access 'token' comprising a packet containing a special bit pattern (e.g. 11111111) is systematically passed between stations around the network. The token is prevented from appearing in genuine data by a technique called bit stuffing. A station wishing to reserve transmission resources takes advantage of its right of access by capturing the token. It does this by removing the token or changing the bit pattern to a token-not-free value (e.g. 11111110). The station, which is said to hold the token, then transmits its packet (either by appending it to the token if the token is changed or on its own if the token is removed) without fear of contention from other stations. The token is regenerated or changed back to its token-free or idle bit pattern either by the station receiving the transmission or by the station originating the transmission after a complete circuit of the ring. In the latter case the receiving station may set appropriate acknowledgement bits in the packet to confirm its correct receipt and the transmitting station may be allowed to retain the token allowing more than one packet to be transmitted consecutively. A significant advantage of token passing is that the maximum waiting time of any user can be predicted (see section 21.4.2). Token passing is a demand assigned reservation protocol that can operate without central control.

A variation on the conventional token ring is to partition time into slots and let these propagate around the ring. Access to a slot is possible, as it passes, provided it is not already filled with data. Such slotted rings can be divided into two groups depending on which station empties a full slot, the source station or the destination station.

21.3.4 Contention

Contention access schemes can operate without central control and give stations some degree of freedom over when to transmit. If complete (or close to complete) freedom to transmit is given the scheme is called random access (RA). If a station is constrained to first listen to the common medium to confirm that no other transmission is in progress before transmitting itself then the scheme is called carrier sense multiple access (CSMA). In both cases, however, it is possible for two stations to try to access the medium simultaneously, i.e. to contend.

In RA contention schemes each user station attempts to transmit as soon as it has an appropriate amount of data in its buffer and conflicts due to two or more stations transmitting at the same time are resolved by a contention policy. The outcome of any particular contention is not predictable, however, and long delays can sometimes build up in a heavily loaded RA network.

RA schemes are typified by ALOHA which originally referred to a radio network (ALOHANET) developed to link computer terminals at the University of Hawaii (distributed over the Hawaiian islands) but is now used generically to denote the RA scheme that this network employed. In the pure form of ALOHA stations transmit at will. After transmitting a packet a station listens for an acknowledgement packet for a time at least equal to the round trip propagation delay between farthest stations. If no acknowledgement is received (due to corruption of the packet by noise or collision with another packet) the station retransmits its packet after a locally generated random delay (to avoid step-lock). A simple performance analysis of ALOHA and its variant, slotted ALOHA, are given in the context of satellite networks in Chapter 14 where it is shown that the maximum possible channel utilisation of pure ALOHA is $1/(2e) = 18\%$ and that of slotted ALOHA is $1/e = 37\%$.

Contention is less likely in CSMA (sometimes called listen before talk, LBT) than RA schemes but not impossible since a medium may appear silent when a station listens but only because a recently started transmission from another station has not yet reached the listening station due to propagation delay. Collisions (or other packet loss events) are detected by the absence of an acknowledgement from the destination station within a reasonable time as dictated by the round trip propagation delay and the need for the acknowledging station itself to contend for access to the medium. Unacknowledged packets are retransmitted, typically after some locally generated random delay.

There are three principal variations on CSMA. These are *non-persistent, 1-persistent and p-persistent* forms.

In the non-persistent form of CSMA the accessing station listens to the medium and transmits if it is free. If the medium is not free the station waits for a locally generated random time and then listens to the medium again, repeating this process until the medium is free when it transmits. This results in a low probability of collision but is wasteful of resources in that the medium may be free when one or more stations are waiting to re-examine it.

In the 1-persistent form the station listens to the medium and transmits if it is free. If the medium is not free the station continues to listen and then transmits as soon as the medium becomes free. This increases the probability of collisions (if two stations are waiting to transmit a collision is guaranteed) but is generally less wasteful of resources overall after collision resolution is taken into account.

In the p-persistent form the station listens to the medium and, if it is free, either transmits immediately (with a probability of p) or defers the decision to transmit (with probability $1 - p$) for some predetermined time period (typically the maximum propagation delay). If the decision to transmit has been deferred then the station listens to the medium after the delay period and algorithm is repeated. If, on listening to the medium, it is found to be not free, the station continues to listen and starts the algorithm immediately the medium becomes free.

CSMA schemes have a shortcoming in that the medium is unavailable for the entire duration of all colliding packets. This can be a serious waste of the transmission resource. An improved CSMA scheme that detects the collision of signals due to contending stations and aborts the contending transmissions immediately is called CSMA with collision detection (CSMA/CD). In order to do this each station must continue listening to the medium while it transmits its own frames and so this scheme is sometimes called listen while talk (LWT). CSMA/CD can employ any of the persistence algorithms described for CSMA above although the most common is the 1-persistent form. Collision detection for baseband systems can be based on sensing the over-voltage produced at a station's tap point due to the station's transmission adding to the transmission already on the medium. For broadband systems collision detection is usually based on comparison of the signal being transmitted with that existing on the medium (and being received by the transmitting station itself). Once a collision has been detected the detecting station transmits a jamming signal to ensure that all stations are alerted to the collision and can ignore the damaged frames. Each station contributing to the collision then waits a locally generated random time before retransmitting its aborted frame.

CSMA schemes are more sophisticated contention access schemes than ALOHA and CSMA/CD can realise a channel utilisation of 75%. At this utilisation efficiency, however, collisions introduce unpredictable network delay (or latency).

21.4 Popular wired LAN standards

The above descriptions of topologies and MAC protocols are generic. Two particular LAN systems representing the most popular topological and MAC protocol combinations (CSMA/CD bus and token ring) are now described in detail.

21.4.1 Ethernet (IEEE 802.3)

Pure Ethernet is a LAN scheme based on a bus topology, using a CSMA/CD contention protocol. Strictly speaking the standard IEEE 802.3 (carrier sense multiple access with collision detection (CSMA/CD) access method and physical layer specification) is not synonymous with Ethernet that refers to an earlier proprietary specification developed by Xerox out of which IEEE 802.3 grew. The similarity is such that the term Ethernet is commonly used to denote IEEE 802.3 systems, however.

In its original form, the Ethernet transmission medium is a high quality 50 Ω coaxial cable to which stations are attached via a simple compression tap, a transceiver unit (integrated with the tap) and a drop cable. The drop cable connection is called the attachment unit interface (AUI).

Taps can be added to, or removed from, the coaxial cable at will, without interfering with the main transmission path. The transmission rate is 10 Mbit/s, but a Manchester code (see section 6.4.3) is used in order to combine the clock signal with data, so the line rate is equivalent to 20 Mbaud. Figure 21.2 illustrates the cable attachment method, and the clock/data combination property of the Manchester code.

Later developments of Ethernet have departed from this classical model. The version described above is referred to as 10BASE5 denoting 10 Mbit/s, baseband transmission with maximum cable length between repeaters of 500 m; 10BASE2 (which uses thinner coaxial cable and is sometimes referred to as thin Ethernet) denotes a similar specification but with the maximum distance between repeaters limited to (nominally) 200 m (but actually 185 m). Thin Ethernet is cheaper to install than conventional (or thick) Ethernet and the transceiver is housed on the DTE interface card rather than at the tap end of a cable drop (allowing the (thin) Ethernet cable to be daisy-chained, i.e. looped into and out of each connected terminal). These and other varieties of Ethernet are discussed later after the basic operating principles of Ethernet have been outlined.

The classical Ethernet CSMA/CD MAC protocol operates as follows:

1. Before attempting to transmit, the station examines the output of its receiver, and only proceeds if there is no data traffic on the bus. This is the carrier sense (CS) action.

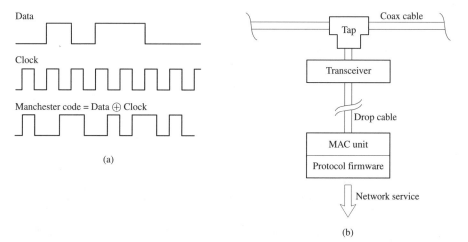

Figure 21.2 *(a) Manchester code generation, (b) standard Ethernet station attachment.*

2. In the absence of bus traffic the station starts to transmit. However, if a second station is going through a similar sequence at the same time, there will be a collision of two frames. In order to detect this each transmitting station continues to monitor its own receiver output and recognises a collision when this output does not correspond with its own transmitted signal.
3. When a collision is detected, the station aborts data transmission and emits a jamming signal, which informs all other stations that a collision has occurred. This represents the collision detection (CD) part of the protocol.
4. The two (or more) stations that have collided back off by a randomly generated interval of time before reattempting to transmit, thus avoiding step-lock in terms of retransmission times resulting in repeated collisions.

Each time a collision occurs transmission capacity is wasted, reducing the network throughput, and this wastage increases rapidly as the network becomes more heavily loaded. Calculation of the throughput is complicated, since it depends upon the statistical pattern of the traffic offered, so realistic results are usually obtained by simulation, which also allows various contention algorithms to be investigated. The simplified analysis presented below, however, represents the major features of Ethernet performance.

Capacity calculations

Most data transfer events in a lightly loaded network are completed without any contention, and this is the normal mode of operation. When traffic increases towards the maximum capacity of the protocol then collisions become more likely and the effective capacity falls. Figure 21.3 illustrates a simplified worst case situation, which can be used to calculate maximum network throughput under carefully controlled conditions. The conclusions from such a simplified study are nevertheless generally applicable to the

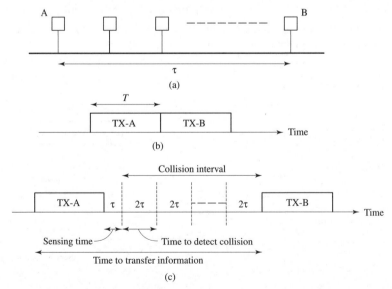

Figure 21.3 *Model for Ethernet performance calculation: (a) bus layout, (b) ideal scheduling, (c) colliding case.*

performance of real systems. A more reliable assessment can only be achieved by considering a random traffic analysis.

Figure 21.3(a) shows an Ethernet bus where the delay between the most dIstant stations A and B is τ s. The transmitted frame length is T s and it is useful to define the transmission parameter representing the maximum propagation delay to frame length ratio $a = \tau/T$. Ideal scheduling conditions, Figure 21.3(b), would allow a new frame to be transmitted immediately following the previous one, giving a frame rate of $1/T$ frame/s, but this requires some central intelligence in order to allocate the transmission medium to one station only. (Ideal scheduling is analogous to time division multiplexing.)

The colliding case illustrated in Figure 21.3(c) represents a worst case model. Station A is saturated, i.e. it has a queue of data packets awaiting transmission. The following sequence of events then occurs:

1. At the commencement of this cycle, station A is transmitting one frame while station B is waiting to transmit but is held off by the carrier sense (CS) mechanism.
2. When station A stops transmitting, there is a delay of τ seconds before station B senses the cessation. This interval is the sensing time, at the end of which the carrier sense detector at station B goes inactive.
3. Suppose now that station B starts to transmit a frame immediately, since it appears that the medium is free. This transmission propagates along the cable, taking τ s to reach station A.
4. However, station A has also sensed that the medium is idle, and so commences transmission of its next frame. The bus is potentially in a collision state but this is, as yet, undetected.
5. When the transmission from station B reaches station A, a collision is detected and station A emits the jamming signal, which (eventually) causes both stations to stop transmitting prematurely. There is a further delay of τ s, however, before the jamming signal reaches B, which then aborts its transmission. The time interval for detecting a collision completely is potentially, therefore, 2τ s, and the frame length T must therefore be greater than 2τ.
6. This cycle may repeat several times before a frame is successfully transmitted from either station A or station B. Other stations may also become active during this period.

In practice the timing is not quite as simple as this, the actions of the two stations being governed by various algorithms, including the randomly generated retry delays. Simulations show, however, that this simple model captures the essential behaviour of CSMA/CD. An estimate of the throughput that represents a congested bus performing near the limit of its transmission capacity can therefore be obtained as shown below.

If, on average, K retries are necessary before the next frame can be transmitted (in a lightly loaded network, $K \to 0$), then the average time for transmitting one frame, t_v, is given by:

$$t_v = T + \tau + 2\tau K \tag{21.1}$$

The utilisation factor, U, of the transmission medium is given by:

$$U = \frac{T}{t_v} = \frac{1}{1 + a(1 + 2K)} \tag{21.2}$$

(where $a = \tau/T$). To find a value for the average number of retries, K, suppose (for simplicity) that there are n active stations on the Ethernet bus where $n \gg 1$. Let p_t be the probability (constant for all stations over all time) that any particular station wishes to transmit at the end of a specific 2τ collision detection interval. For example, if each station generates packets randomly (with Poisson statistics) at a rate of λ packet/s, then $p_t = 2\lambda\tau$, Chapter 19. For a successful event, one station transmits, but $n - 1$ stations do not. The probability of a successful transmission p is therefore given by:

$$p = np_t(1 - p_t)^{n-1} \tag{21.3}$$

Clearly it is desirable to maximise this probability, which is done by tailoring the retry algorithm. In practice the random nature of the traffic offered at each station further complicates the statistics of collision. The maximum value of the probability p_t may be obtained by differentiation, and occurs when:

$$p_t = \frac{1}{n} \tag{21.4}$$

Consequently the maximum value of p is given by:

$$p_{\max} = \left(1 - \frac{1}{n}\right)^{n-1} \quad \rightarrow \frac{1}{e} \quad \text{as } n \rightarrow \infty \tag{21.5}$$

At the end of a 2τ collision detection interval, a further collision occurs with probability $1 - p$, while a successful transmission occurs with probability p. So, a sequence of k collision intervals, occupying a time $2\tau k$ s, occurs with probability $P(k) = p(1 - p)^{k-1}$. Note that for the scenario assumed, at least one collision occurs. The average number of collisions is therefore given by:

$$K = \sum_{k=1}^{\infty} kP(k) = \sum_{k=1}^{\infty} kp(1 - p)^{k-1} \tag{21.6}$$

and since, Appendix B:

$$\sum_{k=0}^{\infty} kr^k = \frac{r}{(1 - r)^2} \tag{21.7}$$

and:

$$\sum_{k=0}^{\infty} r^k = \frac{1}{(1 - r)}, \quad \text{for } |r| < 1 \tag{21.8}$$

then:

$$K = \frac{1}{p} \tag{21.9}$$

Consequently, when the probability of a successful transmission is maximised as above, the average number of transmissions K is e or 2.718. Substituting this value in equation (21.2) we obtain the limiting utilisation:

$$U_{\max} = \frac{1}{1 + 6.44a} \tag{21.10}$$

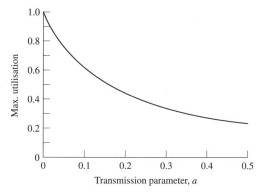

Figure 21.4 *Contention bus utilisation.*

Note that this maximum value is only valid for this artificial scenario, but that it does provide a first order estimate of the capacity of a contention protocol when the LAN is heavily loaded and many stations are trying to transmit. Observe also that there is a fundamental limit to the frame length T, in order to make the collision detection mechanism work at all. A transmitted frame must reach every station on the bus, and a jamming signal must be able to propagate back to the initiating station, before the frame finishes. Hence:

$$T > 2\tau \tag{21.11(a)}$$

i.e.:

$$a < 0.5 \tag{21.11(b)}$$

Figure 21.4 shows maximum utilisation versus the allowable range of the transmission parameter a. For frames that are long relative to the bus delay ($a \ll 1$), the utilisation is good.

EXAMPLE 21.1

A certain Ethernet system has a maximum bus delay of 16 μs, and operates with a bit rate of 10 Mbit/s. Each frame is 576 bits in length. Determine the maximum utilisation factor of the medium under collision conditions.

The time occupied by one frame is given by:

$$T = 576/10 = 57.6 \ \mu s$$

The propagation delay to frame length ratio is therefore given by:

$$a = 16/57.6 = 0.28$$

The maximum utilisation factor is therefore:

$$U_{max} = 1/(1 + 6.44 \times 0.28) = 0.36$$

(In transmission terms this is equivalent to 3.6 Mbit/s.)

EXAMPLE 21.2

For the system of Example 21.1, calculate the actual capacity if there are 15 active stations, each with an equal amount of data to transmit.

Since there are 15 stations, $n = 15$.
The maximum probability of transmission, p_{max}, at the end of a contention interval is therefore given by equation (21.5) as:

$$\left(1 - \frac{1}{15}\right)^{14} = 0.381$$

Hence $K = 1/p_{max} = 2.63$.
Using equation (21.2) the utilisation is then given by:

$$U = 1/[1 + 0.28(1 + 2(2.63))] = 0.363.$$

This amounts to a capacity of $0.363 \times 10^7/576 = 6302$ frame/s
Spread evenly among the 15 stations, this corresponds to 420 frame/s from each station.

The major problem with the CSMA/CD type of contention protocol is that access to the bus is controlled in a random fashion when the LAN is busy. The throughput delay is theoretically unbounded and cannot be guaranteed in practice. Consequently this type of protocol cannot be used for time-critical applications like speech or video, and shows progressive congestion as the offered load increases. It is, however, excellent for networks with many users who need access only at widely spaced intervals of time for short bursts of activity.

Practical installations

Ethernet is a well developed system supported by many manufacturers and the majority of installed LANs are of the Ethernet or IEEE 802.3 type.

A necessary safeguard for practical operation is the 'jabber' control. If a fault develops on one station such that it transmits continuously, then the bus is saturated and the whole system stops. Each transceiver therefore contains a monitoring function that detects if the station has been transmitting for too long and if so, cuts off the transmitter.

In order to detect a frame collision reliably, the two colliding signals must be of a similar magnitude, otherwise one will overwhelm the other and the presence of the smaller signal will be missed. The distance over which collisions can be detected, and hence signals allowed to propagate, is therefore limited by the attenuation of the transmission medium. The cable length in any passive segment is, for this and other reasons, limited to 500 m, in the classical (10BASE5) implementation, although the network may be extended by judicious use of repeaters and ties with signal gain. The ultimate criterion, however, is that the frame length should be more than twice the end-to-end propagation delay. The overall cable length between any two stations (often referred to as the network diameter) is therefore limited (again in 10BASE5) to 2500 m. Figure 21.5 illustrates a typical layout.

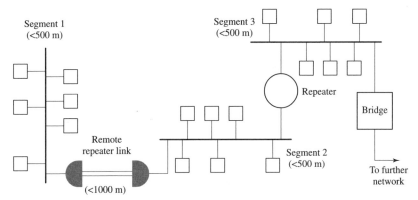

Figure 21.5 *A classical Ethernet structure.*

Frames may be screened for onward transmission to other parts of the network, by using a bridge instead of a repeater, so that each section of the network now has only to deal with traffic intended for itself. A bridge also ensures that carrier detection (CD) is local to that section of cable upside of the bridge, thus extending the network beyond the (10BASE5) limit of 2500 m. The major characteristics of classical Ethernet are summarised in Table 21.1.

Table 21.1 *Classical Ethernet characteristics.*

Topology	Tree, with separate bus sections
Medium	50 Ω coaxial cable
Signalling rate	10 Mbit/s
Maximum station separation	2.5 km
Maximum cable segment length	500 m
Maximum number of stations	1024
Frame length	72–1526 bytes
Addressing	Destination and source addresses; 16 or 48 bits

The original specification using (thick) coaxial cable is relatively expensive, and other (generally cheaper) transmission media have been found to be suitable. Low cost coaxial cable is used in the thin Ethernet (sometimes referred to colloquially as 'Cheapernet'). The drop cable is not used in this implementation, simple coaxial tee-connectors inserted into the coaxial bus being used to connect directly to the DTE Ethernet card (daisy-chaining). Telephone twisted pair (TP) cable is widely available and is also now used for Ethernet, either in unshielded (UTP) or shielded (STP) form, making use of the extensive wiring that exists in most office complexes.

UTP has a nominal characteristic impedance of 100 Ω and traditionally comes in categories or grades 1 to 5 although higher category specifications also now exist. Category 1 is satisfactory only for voice and very low signalling rates (e.g. alarm applications). Categories 2, 3, 4 and 5 are suitable for data with bit rates up to 4, 16, 20 and 100 Mbit/s respectively. The UTP cable generally contains four wire pairs individually twisted with a different twist pitch to reduce crosstalk. STP has an

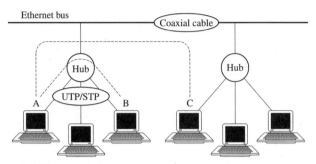

Figure 21.6 *Extended Ethernet.*

impedance of 150 Ω, each cable containing two wire pairs, individually foil shielded and with an additional overall braided shield. As the demand for ever-higher bit rates continues optical fibre versions of the Ethernet standard have been created.

Twisted pair and single (SMF) and multimode (MMF) fibre optic cables, Figure 12.30, are not suited to the bus configuration, and are usually employed in point-to-point links. The classical contention protocol is not, therefore, used with these transmission media, but the stations are generally connected to a hub repeater as illustrated in Figure 21.6. Some hubs repeat the incoming signals to all other links, allowing collision detection to take place normally, whilst others detect the presence of the incoming signal at the hub and make a collision decision there. Ethernet segments and networks can be interconnected by routers in addition to repeaters and bridges. Fibres operate over a wide range of wavelengths from 850 nm to 125 μm.

Ethernet varieties operating at 100 Mbit/s, 1000 Mbit/s and 10000 Mbit/s are usually referred to as fast Ethernet, Gigabit Ethernet and 10 Gigabit Ethernet respectively although their operation is rather different from the classical Ethernet described above. The applications of these very high bit rate Ethernet varieties are no longer confined to LANs but can also be applied to MANs and WANs. In this context Ethernet may eventually become a competitor to core network technologies such as SDH and ATM (see Chapter 20). Table 21.2 lists some of the current and emerging standards, along with their limitations, using the coding techniques of section 6.4.7.

Table 21.2 *Evolving Ethernet characteristics.*

Ethernet variety	Trans. rate (Mbit/s)	Max. segment length	Media	Comments
1BASE5	1	250 m	UTP	IEEE 802.3e (1987). Original basic Ethernet but essentially displaced by thin Ethernet. Star configured with each terminal connected to hub (i.e. a multiport Ethernet repeater) by separate pair of transmission lines (one transmit, one receive). Manchester coded. Half-duplex operation only.
10BASE5	10	500 m	Coax (50 Ω)	IEEE 802.3 (1983). Classic (thick) Ethernet. Manchester line coded. Maximum network diameter 2.5 km (maximum of 5 segments linked by 4

				repeaters, but 2 of these must be point-to-point inter-repeater segments). Maximum number of transceivers per segment is 100. Half-duplex operation only. Transceiver placing restricted to integer multiples of 2.5 m along coax to minimise signal reflections. N-type connectors.
10BASE2	1	185 m	Coax (50 Ω)	IEEE 802.3a (1985). Thin Ethernet. Manchester coded. Maximum number of transceivers per segment is 30. Half-duplex operation only. Minimum transceiver spacing of 0.5 m. BNC connectors.
10BROAD36	10	1800 m	Cable TV coax (75 Ω)	IEEE 802.3b (1985). DPSK broadband (RF) signalling. Maximum network diameter (multiple segments) is 3.6 km. All segments terminate in 'head-end' device. Half duplex only. Signal propagation is unidirectional along cable. To reach all attached transceivers either two cables used (transmission from terminal on one to head-end device which redirects it to second (receive) cable to reach all other terminals) or a single cable is used with transmission to head-end on one frequency and redirected transmissions to all terminals on another frequency – the head-end device acting as frequency translator. Can share CATV cable with other services (e.g. video) using FDM.
10BASE-T	10 (half duplex) 20 (full duplex)	100 m (for cat. 3 UTP, see comments) 150 m for cat. 5 UTP or STP	UTP (2 pairs, cat. 3 or 5), STP	IEEE 802.3i (1990). Star configured with each terminal connected to a hub (i.e. a multiport Ethernet repeater) by separate transmission line. Two wire pairs used, one pair for transmission from terminal to hub and one pair for transmission from hub to terminal. Hub receives input on any of its input lines and repeats it to all output lines. Maximum number of transceivers per segment is therefore 2. Maximum network diameter is 2.5 km. Manchester coding. Half- or full-duplex operation. Longer UTP segments than 100 m are possible if signal quality specifications are met (e.g. cat. 5 UTP or STP).
IsoENET	10	100 m (for cat. 3 UTP) 150 m for cat. 5 UTP or STP	UTP (2 pairs, cat. 3 or 5), STP	IEEE 802.9 (IsoEthernet). Defines a development of 10BASE-T that caters for delay-sensitive isochronous (i.e. constant bit rate) traffic. 6.144 Mbit/s of capacity is split into 96 (64 kbit/s) channels suitable for voice and/or other data real time or quasi-real-time delivery requirements. Manchester coding replaced with 4B5B coding as used by FDDI. Requires an IsoENET hub but only terminals with isochronous data requirements need an IsoENET adaptor card. Terminal without isochronous data requirements can use 10BASE-T adaptor cards.
10BASE-F	10	2000 m	Fibre (62.5/125 μm, 800 – 910 nm)	IEEE 802.3j (1993). Star connected optical fibre networks. Refers collectively to 10BASE-FL, 10BASE-FB and 10BASE-FP. Two optical fibres used, one for transmitting and one for receiving. Manchester coding with OOK optical keying.

10BASE-FL	10 (half duplex) 20 full duplex)	2,000 m (half-duplex) >2,000 m (full-duplex)	Fibre (62.5/125 μm MMF, 850 nm)	Sub-variety of 10BASE-F. FL denotes fibre link. Each terminal star connected to a hub. Maximum number of transceivers per segment is therefore 2. Can also be used to connect 2 repeaters (maximum cascaded number is 5). Half- or full-duplex operation.
10BASE-FB	10	2 km	Fibre	Sub-variety of 10BASE-F, fibre backbone (FB). Used to link special 10BASE-FB synchronous signalling repeater hubs together (point-to-point) in a repeated backbone system that can span long distances. (Synchronous repeaters re-time optical signals to avoid aggregation of timing distortion through multiple repeaters.) Individual 10BASE-FB repeater-to-repeater links may be up to 2 km in length. The maximum number of cascaded repeaters is 15. Essentially replaces the older standard FOIRL (fibre optic inter-repeater link), 1 km maximum segment length and a maximum of 4 cascaded repeaters. Half-duplex operation.
10BASE-FP	10	500 m	Fibre	Sub-variety of 10BASE-F, fibre passive (FP). Each terminal is star connected to passive hub (star coupler) allowing up to 33 transceivers to be connected. Hub passively receives optical signals from special 10BASE-FP transceivers and redistributes them uniformly to all other 10BASE-FP transceivers connected to star. Half-duplex operation.
100BASE-T	100	100 m	UTP	IEEE 802.3u. Fast Ethernet. Refers collectively to 100BASE-T4, 100BASE-X (and therefore 100BASE-TX and 100BASE-FX) and 100BASE-T2. Star configured with each terminal connected to hub. Maximum network diameter is 205 m (but this can be extended using switching hubs instead of simpler, and cheaper, repeater hubs). Manchester coded. Segment length limited by round trip timing specification rather than the attenuation that is typically the case in 10 Mbit/s Ethernet systems.
100BASE-T4	100	100 m	UTP 4 pairs (cat. 3–5)	Sub-variety of 100BASE-T on twisted 4-pair cable. Two pairs are used bidirectionally and two are used unidirectionally. Bit rate on each pair is therefore reduced to 33.33 Mbit/s. 8B6T (ternary) NRZ line code reduces symbol rate further to 25 Mbaud accommodating 30 MHz specified bandwidth of each category 3 (telephone grade) twisted pair. Maximum number of transceivers per link segment is 2. Maximum network diameter is 200 m for a pair of segments linked by a single class I or class II Ethernet repeater and 205 m for a pair of segments linked by 2 (linked) class II repeaters. Half-duplex operation.
100BASE-X	100	100 m	UTP (cat. 5), STP, fibre	Subset of 100BASE-T. Refers collectively to 100BASE-TX and 100BASE-FX (see below). 100BASE-X shares physical layer specification with that developed for FDDI (e.g. 4B5B coding).

100BASE-TX	100	100 m	UTP (2 pairs, cat. 5), STP	Sub-variety of 100BASE-X (and therefore 100BASE-T). Uses cable comprising 2 (data-grade) twisted pairs, one pair to transmit and one to receive. 4B5B coding followed by 3-level PAM (MLT-3). Symbol rate 125 Mbaud. Max. number of transceivers per link segment is 2. Max. network diameter is 200 m for a pair of segments linked by a single class I or class II repeater and 205 m for a pair of segments linked by 2 class II repeaters. Half- or full-duplex operation.
100BASE-FX	100	412 m (for half-duplex operation, isolated single segment connecting 2 DTEs, see comments)	Fibre	Sub-variety of 100BASE-X (and therefore 100BASE-T). Multimode fibre (MMF) pair used, one fibre between terminal and hub and one for transmitting between hub and terminal. Optical wavelength 1350 nm. Max. optical loss budget is 11 dB for each link. Max. number of transceivers per link segment is 2. Max. network diameter if a single Ethernet repeater used is 320 m (class II repeater) or 272 m (class I repeater). If 2 class II repeaters used to link segments maximum network diameter is 228 m. Half- or full-duplex operation. (Maximum segment length of 412 m becomes 2000 m for full-duplex operation.) 4B5B coding NRZI.
100BASE-T2	100	100 m	UTP	IEEE 802.3y (1997). Star connected. Half- or full-duplex operation. Cat. 3 or better UTP.
100VGAnyLAN	100	100 m	UTP (4 pairs, cat. 3–5), STP, fibre	IEEE 802.12. Demand priority access scheme (round robin polling – not a contention protocol). Supports both IEEE 802.3 and IEEE 802.5 (token ring) frame formats. Employs cascaded star topology (repeaters act as hubs). Implementation of larger networks using hierarchical levels of star topology without bridges. Supports bounded latency allowing real time traffic, voice and multimedia, using priority access. Physical layer BER $< 10^{-8}$. Optional implementation of redundant links for automatic recovery of connectivity in event of link failure.
1000BASE-T	1000	100 m	4-pair UTP (cat. 5 or better)	IEEE 802.3ab. Gigabit Ethernet. 8B10B followed by 5-level PAM used to encode 2 bits/symbol. 125 Mbaud on all 4 twisted pairs simultaneously giving 1000 Mbit/s. Provides physical layer BER $< 10^{-10}$. Network diameter 200 m.
1000BASE-X	1000	550 m (MMF), 5000 m (SMF), 25 m (STP)	Fibre, STP	IEEE 803.3z (1998). 8B10B coding. Range depends on modal bandwidth of fibre (km MHz). L denotes long wavelength (1300 nm) and S denotes short wavelength operation. CX denotes STP, see next three entries.
1000BASE-LX	1000	316 m, 550 m and 5 km, see comments	Fibre	Sub-variety of 1000BASE-X. L denotes long wavelength operation (1300 nm). Half- or full-duplex operation. 316 m for half-duplex. For full-duplex operation segment length becomes 550 m for 50 μm and 62.5 μm MMF, and 5 km for 10 μm SMF.

1000BASE-SX	1000	275 m to 550 m, see comments	Fibre	Sub-variety of 1000BASE-X. S denotes short wavelength operation (850 nm). Maximum segment length of 275 m for half- or full-duplex (62.5 μm MMF) increases to 316 m for half-duplex (50 μm MMF) and 550 m for full-duplex (50 μm MMF).
1000BASE-CX	1000	25 m	Shielded copper pair	Sub-variety of 1000BASE-X. Balanced copper pair ('twinax' or 'short haul copper'). Half-duplex or full duplex operation.
10GbE Variations include:	10 Gbit/s		Fibre	10 Gigabit Ethernet. Standard supplement IEEE 802.3ae. Specification 2002. Full-duplex mode only. Does not use the collision detection protocol. Supports MAN/WAN-type applications (both end–end Ethernet and via intermediate SDH/SONET WANs). W denotes a variety interoperable with SDH/SONET.
10GBASE-SR 10GBASE-SW		300 m (MMF)		S denotes short wavelength varieties (850 nm MMF).
10GBASE-LR 10GBASE-LW		10 km (SMF)		L denotes long wavelength varieties (1310 nm SMF).
10GBASE-ER 10GBASE-EW		40 km (SMF)		E denotes extra long wavelength varieties (1550 nm SMF) for segment length typically up to 40 km.
10GBASE-LX4		300 m (MMF) 10 km (SMF)		LX4 uses wavelength division multiplexing to utilise four wavelengths (around 1310 nm) over an SMF or MMF pair.

21.4.2 Token ring (IEEE 802.5)

The token passing protocols are a good compromise between multiplexing and polling, offering each station access to the network on a systematic basis, but with a minimum of supervisory transmissions. Figure 21.7 shows the topology of a ring-type token passing LAN.

The ring medium, which is a twisted pair cable or an optical fibre, loops through each station interface, circulating information around the ring in one direction. Figure 21.8 shows the way in which connection to the ring is made.

The ring transmission medium passes through a trunk coupling unit (TCU) and, in normal mode, ring signals are diverted through the drop cable to the MAC interface. All signals are therefore examined by each MAC interface and regenerated before being circulated further around the ring. Such a configuration, however, is vulnerable to station

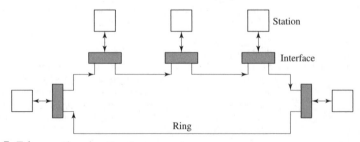

Figure 21.7 *Token passing ring structure.*

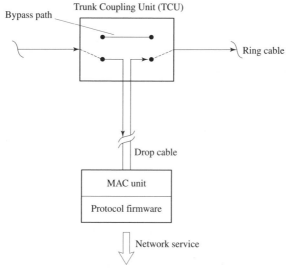

Figure 21.8 *Token ring connection method.*

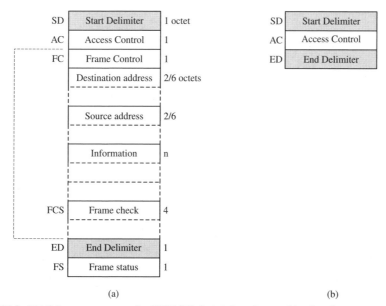

Figure 21.9 *MAC frame structures for IEEE 802.5: (a) data frame, (b) token frame.*

faults or to removal of a station, so in this case a by-pass feature reconfigures the TCU to allow ring signals to pass straight through without traversing the local drop-cable and MAC interface. Switching is normally done with relays, initiated by withdrawing the cable plug, so that a station can be removed without upsetting the operation of the ring.

Figure 21.9 shows the format of the MAC frames that are transmitted around this ring, according to the IEEE 802.5 protocol standard. The start delimiter (SD) and end delimiter (ED) are unique signalling patterns which violate the Manchester coding rules used for the remainder of the data frame, and hence can be detected wherever they may occur. The other fields need little additional explanation.

The token is a unique short frame of 3 bytes or octets, SD/AC/ED, which circulates around the ring. The access control (AC) byte contains a token bit which indicates whether the current frame is a token or an information frame, and when this is set the AC byte is followed immediately by the end delimiter (ED).

A station may transmit only when it has possession of the token. While the system is idle, the free token is circulated around the ring, being regenerated by each station on its way. When a station wishes to transmit, it waits until the free token arrives, changes the status of the token to busy whilst regenerating it and then appends its own data frame. Figure 21.10 illustrates this action for a simple four-station ring, with transmission from station A to station C.

Each station in the ring regenerates the data frame, and passes it on to the next station. The station whose address is contained in the destination address field also reads the contents of the frame into its buffer store and acts upon it. The originating station breaks the ring as the data frame comes back to it, and does not regenerate the data frame but launches a free token into the ring for another station to accept.

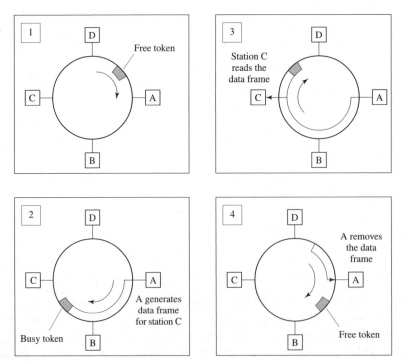

Figure 21.10 *Token passing action in a ring.*

Capacity calculations

An upper limit may be placed on the delay before transmission can occur as calculated for an empty, full and loaded ring as shown below.

Empty ring

Consider first an empty ring. The transmission capacity is C bit/s, the propagation time around the ring is τ s and there are N stations. There is an additional delay of l bits in each station on the ring, called the station latency, necessary in order for the station to examine and regenerate each bit as it circulates around the ring. The ring latency, T_L s, is therefore given by:

$$T_L = \tau + \frac{Nl}{C} \tag{21.12}$$

The free token is 24 bits (3 bytes) in length, so the maximum waiting time, if no other station is transmitting, is given by:

$$T_{\text{max,empty}} = \left(\frac{24}{C} + T_L\right) \tag{21.13}$$

Full ring

Consider now that the ring is full because all stations have data to transmit. Let each station transmit one frame when it has the token. If each frame is limited to say M bytes, then its transmission time is:

$$T = \frac{8M}{C} \tag{21.14}$$

The maximum waiting time is (approximately) therefore:

$$T_{\text{max,full}} = (N - 1)(T + T_L) \tag{21.15}$$

EXAMPLE 12.3

A 4 Mbit/s ring has 50 stations, each with a latency of 2 bits. The total length of the ring is 2 km, and the propagation delay of the cable is 5 μs/km. Determine the maximum waiting time when the ring is empty, and when all stations are transmitting. A full frame is 64 bytes in length.

The ring delay is given by:

$$\tau = 2 \times 5 = 10 \ \mu s$$

The ring latency, equation (21.12), is therefore:

$$T_L = 10 + (2 \times 50/4) = 35 \ \mu s$$

The maximum waiting time in an empty ring, from equation (21.13), is therefore:

$$T_{\text{max,empty}} = (24/4) + 35 = 41 \ \mu s$$

The transmission time for one frame is:

$$T = 64 \times 8/4 = 128 \ \mu s$$

The maximum waiting time when the ring is full ($T_{\max,\text{full}}$) and a station has to wait for all other (49) stations to transmit first is $49(128 + 35) = 7987 \ \mu s$ (approximately 8 ms).
Each station is then supporting an input rate of $1/(8.0 \times 10^{-3})$ or 125 frame/s.

Example 21.3 is an extreme case, but it does show that the waiting time, although varying, is finite and can be predicted. It is possible to allocate different levels of priority to the stations by coding the token, so that certain time-critical data transfers, e.g. voice, have preference.

Loaded ring

Consider finally the average cycle time of each station when the ring is fully loaded, taking variations of incoming traffic into account. Under this condition, the ith station has an incoming average traffic load of λ_i frame/s. Each frame occupies a time T when transmitted on the ring. Only one frame is allowed on the ring at any time, so each station buffers its incoming data until it is allowed to transmit. Each incoming message will therefore be subject to a delay before it can be transmitted, the amount of delay depending upon the volume of traffic coming into this station and the amount of traffic or congestion on the ring. All packetised data networks operate in this way. Now suppose that the ith station has transmitted all its waiting data frames, clearing its buffer, and that a time interval t_c elapses before the free token arrives, allowing it to transmit again. A total of $\lambda_i t_c$ data frames have accumulated during this interval and are now ready for transmission. The control algorithm can deal with this backlog in various ways, but assume for simplicity that all these frames are transmitted when the station next has the token. This transmission time will be $t_i = \lambda_i t_c T$, assuming that $t_c > T$.

The time interval between two bursts of transmission from this station is t_c, which represents the maximum waiting time experienced by every station on the ring. t_c is then given by:

$$t_c = T_L + \sum_{i=1}^{N} t_i$$

$$= T_L + t_c \Lambda T \tag{21.16}$$

where:

$$\Lambda = \sum_{i=1}^{N} \lambda_i \tag{21.17}$$

Here the parameter Λ represents the gross input to the ring, in frame/s. Equation (21.16) can therefore be rearranged to give:

$$\frac{t_c}{T_L} = \frac{1}{1 - U} \tag{21.18}$$

where:

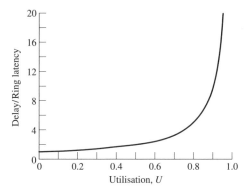

Figure 21.11 *Delay versus utilisation factor for a token ring.*

$$U = \Lambda T \tag{21.19}$$

U is a new interpretation of utilisation for the ring medium; it provides a measure of the time for which the ring is occupied. There is no fundamental reason why this should not approach 100% since there is no contention, except that as $U \rightarrow 1$, so the cycle time t_c approaches infinity and congestion in the ring becomes absolute.

Figure 21.11 plots the relationship of equation (21.18), which is the average delay for any input packet in terms of the ring latency, versus the utilisation factor of the whole ring. A number of implicit assumptions have been made in deriving this relationship, which for a more realistic assessment should be investigated further. However, this simple analysis illustrates the general trend of ring performance and gives a first order estimate of its value. It is typical of all queueing situations, where the rate of change of delay varies according to the conditions under which the queue operates.

EXAMPLE 21.4

Evaluate Example 21.3 for a utilisation of 95%.

For $U = 0.95$ the cycle time $t_c = 20T_L$.
The waiting time is approximately 20×35 μs.
The frame occupies $T = 128$ μs.
The gross input load to the ring is therefore $0.95 \times 1/(128 \times 10^{-6}) = 7422$ frame/s.
This amounts to an average of 148 frame/s from each of 50 stations, or an average of 371 frame/s from each of 20 active stations out of the 50.

Practical issues

The operation of the token passing ring depends critically upon the presence of one and only one token in the ring. Corrective action must be taken if the token disappears due to some fault in a station, a station being inserted or removed, or noise on the network. One station is therefore allocated the task of ring monitor whose function is to observe the passage of the token and to generate a new one if activity ceases. The ring monitor also adjusts the ring latency time, to ensure that it is longer than the time occupied by the token. The ring latency is influenced by how many stations are present on the ring at any particular time, and also on the phase of the clock around the ring. An elastic buffer, with a length that can be adjusted to make the ring latency correct, is therefore introduced by the ring monitor.

All stations may take on the role of standby monitor, examining the sequence of tokens in the ring. If tokens fail to appear regularly, then each station bids to become the monitor by issuing a special control frame, the first station completing the procedure becoming the new ring monitor. So although the ring monitor is essential to the ring's correct operation its function is allocated democratically by the active stations and no rigid central control is necessary.

Individual stations need to have full information on the state of the ring. When a station comes into the ring at switch-on, it must first ensure that no other station is using its own address, and then inform its downstream neighbour that it has entered the ring.

There are various practical versions of the token ring network, but IEEE 802.5 is essentially an IBM standard ring that runs at 4 or 16 Mbit/s over twisted pair cable with Manchester line coding. IEEE 802.4 describes a token bus, which is physically a bus network, but the token is circulated among the stations creating a logical ring. Further control fields are necessary in that specification to sequence the frames correctly to their destinations.

21.5 Wireless LANs

The advantages of wireless LAN technology include:

1. Increased mobility of users.
2. Increased flexibility and fluidity, including ad-hoc networks.
3. Instant networking (i.e. plug and play without the delay involved in installing a wired infrastructure) and simplified planning.
4. Availability of LAN technology in situations where a wired infrastructure is inconvenient or impossible (e.g. buildings of historical or architectural importance, construction sites).

There are also some disadvantages (with respect to wired networks) including:

1. Higher cost (in most circumstances).
2. Lower performance (typical bit rates are lower than for wired LANs and BERs are higher).
3. Lower reliability (due to the less than perfect and often variable channel characteristics including frequency response, noise and interference).
4. Multiple standards/systems (with the de facto standard, cf. Ethernet, still to emerge) and local radio regulatory authorities preventing equipment designed for one country being used in another.
5. Poor inherent security (interception of radio signals being generally easier than intercepting signals on cables).

Currently, the balance of advantages and disadvantages between wired and wireless LANs is sensitively dependent on the application and environment but there is no particular reason to believe that this situation will persist indefinitely. Experience over the last two or three decades with communications generally appears to show a systematic trend towards wireless technology, probably driven principally by the willingness of consumers, both commercial and domestic, to pay a significant premium, in both cost and performance, for mobility and convenience.

A selection of the most important, current, wireless LAN standards is examined below. Some of these standards are competitors and some offer complementary characteristics and services. It is not yet clear whether all will survive in the long term or what other standards and technologies will join them. In the short to medium term, however, it seems likely that IEEE 802.11 (in one or more of its several varieties, especially IEEE 802.11b) and Bluetooth (a personal area network technology) will probably have the biggest share of their respective markets and these two technologies are therefore addressed in greatest detail.

21.5.1 WLAN (IEEE 802.11)

WLAN is the name usually given to the wireless LAN standard IEEE 802.11 and its variants (especially IEEE 802.11a and IEEE 802.11b). IEEE 802.11, released in 1997, includes three different physical layer standards each offering data rates of 1 or 2 Mbit/s. Two of the three implementations operate in the industrial, scientific and medical (ISM) band of the electromagnetic spectrum (2.4000–2.4835 GHz) and one operates in the infrared (IR) band of the spectrum (850–950 nm).

The IR physical layer implementation (described later) is a version referred to as direct modulation IR (DMIR). Higher data rate IR varieties also exist, however, including carrier modulated IR (CMIR) and multi-subcarrier modulated IR (MSCIR). CMIR improves performance over DMIR by modulating the data first onto a radio carrier using FSK or PSK and then modulating the intensity of the IR source with the modulated carrier. This allows conventional electronic filtering of the modulated carrier to reduce out-of-band noise, which results in superior performance compared to DMIR. (DMIR uses optical filtering only prior to detection.) CMIR offers a data rate of 4 Mbit/s, limited principally by ISI due to the delay spread of the channel. (A typical room may easily have a delay spread approaching 100 ns.) MSCIR combats ISI by using multiple subcarriers (similar to OFDM, see section 11.5.2) prior to optical modulation. The bit rate per subcarrier is reduced and the symbol duration per subcarrier correspondingly increased such that the delay spread becomes a small fraction of the symbol period. MSCIR offers a data rate of 10 Mbit/s.

The adoption of the ISM band for the radio implementations is primarily because, within certain restrictions on radiated power, the use of frequencies in this band is unlicensed. (Radiated powers are currently restricted in Europe, the USA and Japan to 100 mW EIRP, 1000 mW with maximum antenna gains of 6 dBi and 10 mW/MHz respectively.) The terms of unlicensed use normally require (or at least imply) the use of some form of spread spectrum technique to maintain radiated power density at acceptably low levels. Both direct sequence spread spectrum (DSSS) and frequency hopping spread spectrum (FHSS), are provided for in the IEEE 802.11 standard.

In Europe there are currently nine allowed carrier (centre) frequencies for DSSS implementation between 2.422 GHz and 2.462 GHz (inclusive) spaced at 5 MHz intervals. The USA allows use of all the European carrier frequencies plus 2.412 GHz and

2.417 GHz. Japan allows the use of only 2.484 GHz (which lies just outside the ISM band).

In all, 79 carrier frequencies are available for the FHSS implementation in most of Europe and the USA. The carriers lie between 2.402 GHz and 2.480 GHz (inclusive) at 1 MHz intervals. Spain and France allow a subset of these hopping frequencies (2.447–2.473 GHz and 2.448–2.482 GHz respectively) and Japan allows the use of a set of frequencies between 2.473 and 2.495 GHz, which again lies partly outside the ISM band. The minimum frequency hopping rate is determined in practice by the radio regulatory authority responsible for the particular geographical area in which a system operates but is typically 1 or 2 hop/s.

DSSS physical layer

In the DSSS implementation the data sequence is spread, see section 15.5.3, with an 11-chip Barker code (+1, − 1, + 1, + 1, − 1, + 1, + 1, + 1, − 1, − 1, − 1) in order to keep power densities low and protect against narrowband interference. (The same spreading code is used by all WLAN transceivers so it represents spectral spreading only, rather than a CDMA technique.) The spread data is modulated onto one of the allowed microwave carriers using DBPSK or DQPSK depending on whether a data rate of 1 or 2 Mbit/s is required. The DSSS frame structure is shown in Figure 21.12.

In addition to its traditional functions the synchronisation field of the DSSS header is used for gain control and frequency offset compensation. The start frame delimiter indicates the beginning of the frame proper. The signal field indicates whether the MAC layer PDU (MPDU), i.e. the frame payload, is delivered at 1 or 2 Mbit/s. (The frame header is always delivered at 1 Mbit/s.) The service field is reserved for future use and the length field indicates the length (between 1 and 2048 bytes) of the MPDU field. The CRC field provides for error checking on the frame header.

An extension to the DSSS physical layer specification, defined by IEEE 802.11b, allows the bit rate to be increased to 11 Mbit/s by replacing the Barker code with complementary code keying (CCK) [van Nee *et al.*] A pair of N-digit codewords are said to be complementary if for any non-zero cyclical shift of the codewords the sum of the product of each code with its shifted self is zero, i.e.:

$$\sum_{n=0}^{N-1} (a_n a_{n+m} + b_n b_{n+m}) = \begin{cases} 2N, & \text{for } M = 0 \\ 0, & \text{for } M \neq 0 \end{cases} \tag{21.20}$$

The set of CCK codewords used in IEEE 802.11b are eight-digit, unit magnitude, complex numbers, which can be expressed by:

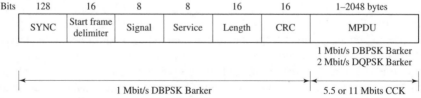

Figure 21.12 *IEEE 802.11 DSSS frame structure.*

$$c = | e^{j(\phi_1+\phi_2+\phi_3+\phi_4)}, \; e^{j(\phi_1+\phi_3+\phi_4)}, \; e^{j(\phi_1+\phi_2+\phi_4)}, \; -e^{j(\phi_1+\phi_4)}, \; e^{j(\phi_1+\phi_2+\phi_3)},$$

$$e^{j(\phi_1+\phi_3)}, \; -e^{j(\phi_1+\phi_2)}, \; e^{j(\phi_1)} | \qquad\qquad (21.21)$$

Each codeword digit can take on one of four values corresponding to the normal QPSK constellation points. The values are not independent, however, i.e. the relationship between complex digit values is restricted according to equation (21.21). Eight bits are required to specify the set of four phase angles ($\phi_1 \; \phi_2 \; \phi_3 \; \phi_4$) and so there are only 256 allowed codewords in the CCK set. Two varieties of IEEE 802.11b are specified as described below.

In the first variety of IEEE 802.11b, six information bits from a block of 8 bits are used to select one of 64 complementary codewords (i.e. a subset of the 256 codewords). This gives a sequence of eight QPSK looking symbols (or at least the baseband equivalent) carrying 6 bits of information. The remaining two information bits are then used to differentially QPSK modulate, i.e. select one of four DQPSK phases (again initially at baseband), this already 'QPSKed carrier', adding a further 2 bits of information. The eight-digit complex codeword now carries 8 bits of information and so if this is transmitted at a bit rate equal to the chip rate of the Barker code in the conventional scheme (i.e. one codeword transmitted in the same time as eight chips of the Barker code) then the 'information' bit rate will be 11 Mbit/s. (This is the air-interface bit rate – the actual user data rate carried is typically a little less than half this.)

In the second variety of IEEE 802.11b, two information bits from a block of four are used to select one of four complementary codewords (a subset of the 64 codewords used in the first variety) giving a sequence of eight QPSK symbols carrying 2 bits of information. The remaining two information bits DQPSK this (complex) carrier as before resulting in the final eight-digit sequence carrying four information bits in total. Again replacing each Barker code chip with one of the CCK (QPSK-like) symbols results in an air-interface bit rate of 5.5 Mbit/s.

The spectral occupancy of the signal in both forms of CCK remains unchanged from the conventional DSSS signal. For 11 Mbit/s CCK there is effectively no spectral spreading, however, since the bandwidth is unaltered from that which would be obtained by simple BPSK modulation of the data. (The definition of spectral spreading is a little debatable and it could be argued that spreading by a factor of 2 still exists since the bandwidth is twice that obtained by simple QPSK modulation of the data.) For 5.5 Mbit/s spectral spreading by a factor of 2 or 4 (depending on definition) is still present.

The use of CCK in the 11 Mbit/s standard, rather than simple BPSK modulation of the information bit stream, has advantages that arise from the redundancy in the transmitted signal and the special correlation properties of the complementary codes. RAKE receiver performance can be improved, for example, and cross rail interference (i.e. crosstalk between I and Q quadrature signal components) can be reduced.

The division of information bits between CCK codeword selection and subsequent DQPSK modulation of these codewords (rather than, in the 11 Mbit/s version for example, using all eight information bits to select from the full 256 codewords in the CCK set) has the advantage of allowing differential detection or non-coherent carrier tracking loops in the receiver.

FHSS physical layer

In the IEEE 802.11 frequency hopped implementation the data sequence is modulated onto the carrier using two-level GFSK for a data rate of 1 Mbit/s and four-level GFSK for a data rate of 2 Mbit/s. (GFSK is similar to GMSK, see section 11.4.6, but with larger spacing between frequencies.) The RMS frequency deviation of the carrier is the same in both the two-level and four-level cases and the resulting signal is contained within the 1 MHz channel bandwidth of the frequency hopping channels. The carrier itself changes frequency according to the adopted hopping pattern at a rate equal to, or greater than, some value determined by the radio regulatory authority in the country of operation, the precise hopping rate being acquired by a mobile terminal during the process of joining a network.

The FHSS frame structure is shown in Figure 21.13. The functions of the header fields are similar to those in the DSSS frame but the detailed structure is significantly different.

DMIR physical layer

For the standard data rate (1 Mbit/s), blocks of 4 data bits are used to determine which of 16 discrete pulse position modulation (PPM) slots an IR light pulse will occupy. The mean optical pulse, or symbol, rate is therefore 0.25 Mbaud. For the extended rate (2 Mbit/s), blocks of 2 data bits determine which of 4 PPM slots are to be occupied giving an optical pulse, or symbol, rate of 1 Mbaud. In both cases the IR pulse width is 250 ns and the peak optical power is 2 W. This means that the maximum mean power for the extended data rate (500 mW) is four times that for the standard data rate (125 mW).

The DMIR frame structure is shown in Figure 21.14. The DC level adjustment field is necessary because the initial part of the frame is transmitted at a higher power level than the rest of the frame.

IR technologies are normally restricted to operation within a single room since IR radiation does not penetrate normally constructed walls, ceiling or floors. To allow

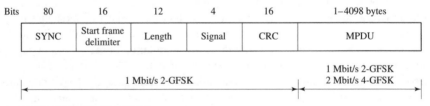

Figure 21.13 *IEEE 802.11 FHSS frame structure.*

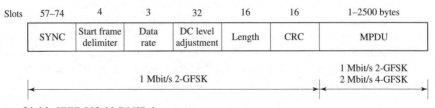

Figure 21.14 *IEEE 802.11 DMIR frame structure.*

terminals to be moved easily (the whole point of wireless technologies) the transmitted signals are normally radiated upwards over a large range of angles and diffuse reflections (principally from the ceiling) are relied on to provide coverage over most of the volume of the room. This means, however, that multipath propagation may be severe enough (100 ns or more) to limit bit rates or degrade performance due to ISI. One method of mitigating this effect is to keep large angle radiation patterns for the IR transmitters but employ much narrower reception angles for the IR receivers. (This does mean that the gross optical signal power received is reduced which might, of, course, have its own implications for performance.) A slightly more sophisticated approach makes both transmit and receive optical radiation solid angles small and directs all optical transmitters and receivers at a convex diffuse reflector attached to the ceiling in the middle of the room. A more sophisticated technique still is to make the convex reflector active by covering its surface with an interlaced array of photodiode detectors and LEDs linking the two arrays with an amplifier. Ideally this reduces the optical path loss to that of the free space line of sight between the active reflector and the optical receiver, the similar loss between the transmitter and reflector having been offset by the reflector amplifier.

The IEEE 802.11 physical layer options are summarised along with some competing technologies in Table 21.3.

IEEE 802.11a is a broadband WLAN service offering data rates of 54 Mbit/s. It is similar in implementation to the European standard HIPERLAN (discussed later) and currently employs the same frequency band as HIPERLAN 2 (around 5.2 GHz). There is some discussion, however, about the adoption of a band around 5.8 GHz to facilitate its deployment in Europe.

A further evolution of IEEE 802.11 under discussion is the specification IEEE 802.11g. This is a WLAN for the 2.4 GHz ISM band that will offer data rates of 20 Mbit/s using DSSS technology and up to 54 Mbit/s using OFDM technology. The attraction of providing these very high bit rates in an effectively unregulated band is clear. If IEEE 802.11g becomes a reality it will effectively supersede IEEE 802.11b and represent very effective competition to both IEEE 802.11a and HIPERLAN. Other developments relating to IEEE 802.11 technology include IEEE 802.11e that describes enhancements aimed at supporting time-sensitive and isochronous data such as video, IEEE 802.11i that is concerned with increased security of user data and IEEE 802.11h that is a spectrum managed version of IEEE 802.11a providing bit rates approaching 200 Mbit/s. The lack of spectrum management functions, in particular dynamic frequency selection, DFS (i.e. automatic channel selection), and transmit power control, TPC, is one of the barriers to IEEE 802.11a deployment in Europe.

Infrastructure and ad-hoc modes

IEEE 802.11 (and wireless LANs in general) can operate in essentially two different architectural modes. These are:

1. Client/server or infrastructure mode: in which each wireless terminal communicates with a central server or hub via an access node forming part of a backbone LAN. All communications between wireless terminals pass through the access nodes and are mediated by the hub. (The hub and access node functions may, course, be combined in the same physical device.)

Table 21.3 *Selected wireless LAN physical layer options.*

System	Channel carrier frequencies (Europe)	Spreading technique	Modulation	Bit rate (bandwidth)	Approx. max. EIRP (Europe)
IEEE 802.11 (DS)	$2407 + 5n$ MHz ($n = 3$ to 11)	DSSS (11-chip Barker code)	DBPSK	1 Mbit/s (11 MHz)	−10 dBW
IEEE 802.11 (DS) extended	$2407 + 5n$ MHz ($n = 3$ to 11)	DSSS (as above)	DQPSK	2 Mbit/s (11 MHz)	−10 dBW
IEEE 802.11b (enhanced 5.5)	$2407 + 5n$ MHz ($n = 3$ to 11)	DSSS with CCK	CCK + DQPSK	5.5 Mbit/s (11 MHz)	−10 dBW
IEEE 802.11b (enhanced 11)	$2407 + 5n$ MHz ($n = 3$ to 11)	DSSS with CCK	CCK + DQPSK	11 Mbit/s (11 MHz)	−10 dBW
IEEE 802.11g (DSSS) in discussion	2.4 GHz ISM (Possibly 5.2 GHz)	DSSS with CCK	CCK	20 Mbit/s	−10 dBW?
IEEE 802.11g (OFDM) in discussion	2.4 GHz ISM (Possibly 5.2 GHz)	–	OFDM	54 Mbit/s	−10 dBW?
IEEE 802.11 (FH)	$2400 + n$ MHz ($n = 2$ to 80)	FHSS (Minimum hop rate 1 Hz, 2.5 Hz possible)	GFSK (2-level)	1 Mbit/s (1 MHz hopped over 79 MHz range)	−10 dBW
IEEE 802.11 (FH) extended	$2400 + n$ MHz ($n = 2$ to 80)	FHSS (Minimum hop rate 1 Hz, 2.5 Hz possible)	GFSK (4-level)	2 Mbit/s	−10 dBW
IEEE 802.11 (IR)	316–353 THz (850–950 nm)	–	DMIR (16-level PPM)	1 Mbit/s	3 dBW (pk) −9 dBW (mn)
IEEE 802.11 (IR) extended	316–353 THz (850–950 nm)	–	DMIR (4-level PPM)	2 Mbit/s	3 dBW (pk) −3 dBW (mn)
IEEE 802.11 (CMIR)	316–353 THz (850–950 nm)	–	CMIR	4 Mbit/s	3 dBW (pk)?
IEEE 802.11 (MSMIR)	316–353 THz (850–950 nm)	–	MSMIR	10 Mbit/s	3 dBW (pk)?
IEEE 802.11a	5.15–5.25 GHz 5.25–5.35 GHz 5.725–5.825 GHz (being discussed)		M-QAM ($M = 2, 4, 16, 64$) + OFDM	54 Mbit/s (20 MHz)	−8 dBW −1 dBW 5 dBW
HIPERLAN 1	$5176.4680 + 23.5294n$ MHz ($n = 0$ to 4)	–	GMSK ($T_o B = 0.3$)	23.529 Mbit/s	−7 dBW
HIPERLAN 2	$5180 + 20n$ MHz ($n = 0$ to 7 lo, $n = 16$ to 26 up)	–	M-QAM ($M = 2, 4, 16, 64$) + OFDM	6,9,12,18,27,36 54 (opt.) Mbit/s (16.875 MHz)	−7 dBW

Note: pk = peak, mn = mean, lo = lower band, up = upper band

2. Ad-hoc mode: in which each terminal communicates directly with all other terminals within the wireless coverage area.

The backbone LAN present in infrastructure mode (which typically provides better performance than ad-hoc mode in terms of throughput) could be wired or wireless but the former is more common at present. Figure 21.15 shows a schematic illustration of both infrastructure and ad-hoc WLANs. The wireless terminals within range (typically 100 m) of a singe access point (AP) form a basic service set (BSS), or cell, with that point. (Two or more BSSs linked by a backbone network constitute an extended service set (ESS) which forms a single logical LAN as far as the LLC protocol is concerned.) In an ad-hoc network a BSS comprises a set of wireless terminals all within range of each other.

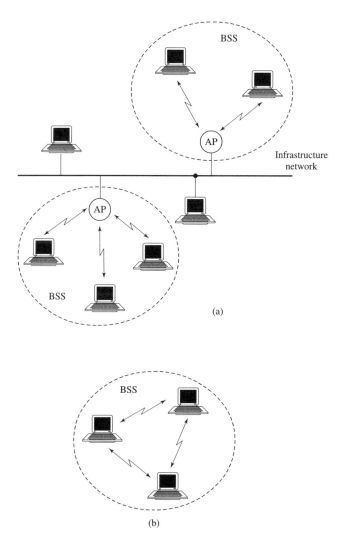

Figure 21.15 *Infastructure (a) and ad-hoc (b) network nodes.*

Medium access control

The IEEE 802.11 MAC protocol has two layers, the distributed coordination function (DCF) layer and the point coordination function (PCF) layer, Figure 21.16.

The DCF incorporates a CSMA/collision avoidance (CSMA/CA) contention protocol [Kleinrock]. Collision detection, as applied by Ethernet for example, is difficult to implement in wireless networks. This is because the dynamic range of signals arriving at a terminal is potentially very large and distant terminals may produce signals that are too weak for collision detection to work in the presence of stronger signals from nearby terminals, weak colliding packets being essentially mistaken for noise. Furthermore terminals cannot usually receive and transmit simultaneously so a transmitting terminal cannot detect any other transmitting node(s).

The essential principle of the CSMA protocol can be described as follows:

1. A terminal with data to transmit first senses (listens to) the medium.
2. If the medium is quiet (idle) the station does not transmit immediately but continues to monitor the medium. If the medium remains idle for a period known as the inter-frame space (IFS) then the station starts to transmit.
3. If the medium is in use or comes into use during the IFS period then transmission is deferred by one IFS period plus a further random back-off period timed from when the medium becomes idle.
4. If the medium is idle after the back-off period then the station transmits.
5. If the medium comes into use before the back-off period has expired then the process is repeated from 3 but with a longer back-off period.

The effect of the above algorithm is for the back-off to get larger each time a transmitter attempts to access the medium and fails.

In practice three types of IFS are used each with a different length. A short IFS is used for the highest priority traffic (e.g. acknowledgement and clear-to-send packets). An intermediate length IFS is used for traffic associated with the polling function of the PCF layer. A long IFS is used for the lowest priority traffic (i.e. normal asynchronous data) associated with the DCF layer.

The above CSMA algorithm makes the probability of packet collision small but not zero. For example, two terminals (A and C) may be out of range of each other (and therefore unable to sense each other's activity) but both within range of a third terminal (B). C will sense the medium as idle even if A is already transmitting to B and may therefore try to access B. This is called the 'hidden terminal' problem and is avoided by

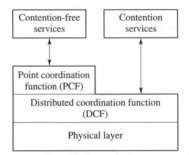

Figure 21.16 *IEEE 802.11 MAC architecture.*

the following collision avoidance mechanism. A terminal about to transmit a data packet first transmits a (short) ready-to-send (RTS) packet that carries its own address, that of the destination terminal and the duration of the data packet which is to follow. The receiving terminal responds with a (short) clear-to-send (CTS) packet that also carries the addresses of transmitter and receiver and the duration of the expected data packet. On receipt of the CTS the data packet is transmitted, which if received satisfactorily is acknowledged by an ACK packet. Terminal C in the hidden terminal problem will have received terminal B's CTS packet and will therefore know not only that terminal B is about to receive a transmission from some other (out of range and therefore 'invisible') terminal but also how long the transaction will last. Terminal C will not therefore try to access the medium until the transaction is complete.

The PCF layer operates a polling protocol for synchronous (delay-sensitive) traffic above (i.e. via) the DCF layer but gets priority access to the medium over ordinary asynchronous data by using the shorter IFSs in the DCF protocol. A point coordinator (operating in the access node of an infrastructure network) controls access by polling (round robin) any stations that may have synchronous data to transmit. To avoid polling taking place to the exclusion of asynchronous traffic access (due to its priority treatment by DCF) a super-frame is defined and polling via PCF is restricted to its first part (called the contention-free period) leaving the rest of the super-frame (called the contention period) for asynchronous data transfer.

21.5.2 HIPERLAN

HIPERLAN is an acronym for high performance local area network and relates to a group of wireless LAN standards produced by ETSI specifically (but not exclusively) for use in Europe [Halls]. The HIPERLAN frequency bands are 5.15–5.50 GHz and 17.1–17.2 GHz.

HIPERLAN 1

The HIPERLAN 1 standard was released in 1996 and is a direct competitor to IEEE 802.11a. It is primarily an indoor, portable terminal wireless LAN (maximum terminal velocity 1.4 m/s or 5 km/h) offering a data rate of approximately 23.5 Mbit/s for both asynchronous and isochronous traffic using GMSK modulation. (Signalling and control information is transmitted at the lower rate of approximately 1.5 Mbit/s using FSK modulation. This allows receiving terminals to activate their equalisers only when required which results in significant power saving.) The nominal range of the wireless link is about 50 m.

At the physical layer five, equally spaced, carrier frequencies are specified over the

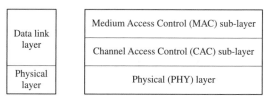

Figure 21.17 *Lower layers of HIPERLAN protocol stack.*

approximate range 5.176 GHz to 5.271 GHz (more exact frequencies are defined in Table 21.3). The GMSK, section 11.4.6, symbol duration–bandwidth product (T_oB) is 0.3, which is the same as that used in the GSM cellular system (see section 15.4). Inherent ISI caused by the modulation scheme combined with that caused by the delay spread of typical channels (which may be greater than 200 ns in large spaces such as airport concourses, shopping malls, etc.) necessitates adaptive equalisation.

The data-link layer is split into channel access control (CAC) and MAC sub-layers, Figure 21.17.

The CAC layer controls access to the medium using a CSMA-like protocol called 'elimination yield non-pre-emptive priority multiple access' (EY-NPMA). EY-NPMA works by eliminating contending terminals using a phased series of access competitions. Time is divided into repeating channel access cycles, each cycle being further divided into a short synchronisation burst followed by the three phases of the multiple access protocol, i.e. prioritisation, contention and transmission. After synchronisation the access protocol proceeds as follows:

1. **Prioritisation phase**
a. This part of the cycle is subdivided into five prioritisation slots each slot comprising 168 bits. The first slot represents an access priority of 1 (the highest), the second an access priority of 2, $\cdots$, and the last an access priority of 5 (the lowest).
b. A terminal with access priority p (3 say) listens for channel activity over the first $p - 1$ (2 say) slots.
c. If the channel is idle for the duration of all $p - 1$ slots then a priority assertion (PA) signal is transmitted.
d. If the channel becomes busy before the end of $p - 1$ slots then the terminal defers and waits for the beginning of the next cycle before reattempting channel access.
e. If two or more stations with equal highest priority transmit a PA then these stations move forward to the contention phase of the protocol

2. **Contention phase**
The contention phase is split into two sub-phases, i.e. elimination and yield.
a. Elimination sub-phase:
   i.   This part of the cycle is divided into 12 elimination slots, each slot comprising 212 bits. Each terminal transmits an elimination signal (ES) of random length between 0 and 12 elimination slots. The probability of a particular ES lasting zero slots is 0.5, the probability of it lasting one slot is 0.5^2, the probability of it lasting two slots is 0.5^3, $\cdots$, the probability of it lasting 11 slots is 0.5^{12}, and the probability of it lasting 12 slots is 0.5^{13}, i.e. after any number of slots there is a 50% chance of the ES persisting for another slot.
   ii. When a station ceases to transmit its ES it listens to establish whether any other station is still transmitting an ES and if this is the case defers its access attempt until the beginning of the next cycle.
   iii. All stations that detect no ES after the cessation of their own ES survive the elimination sub-phase and move forward to the yield sub-phase.
b. Yield sub-phase:
   i.   This part of the cycle is divided into 10 yield slots each slot comprising 168 bits. Each terminal listens for between 0 and 9 yield slots. The probability of a particular terminal sensing zero slots is 0.1, the probability of sensing the first slot

is 0.1, the probability of sensing the first two slots is 0.1, ⋯, and the probability of sensing the first nine slots is 0.1, i.e. the probability of sensing *n* slots is equal for all allowed *n*.

ii. A station that detects channel activity while sensing yield slots defers its access attempt until the beginning of the next cycle.

iii. A station that does not detect any channel activity for the duration of the time it is sensing yield slots starts transmitting immediately after its slot sensing is complete, i.e. moves into the transmission phase.

3. **Transmission phase**

The station which has not been eliminated now starts data transmissions.

Note that, whilst it is unlikely that more than one station may be left or retained by the use of the EY-NPMA protocol, the possibility of collision is not avoided completely.

The MAC layer is concerned with addressing (e.g. mapping IEEE MAC addresses into HIPERLAN addresses), encryption and data relaying. The last of these functions arises because a HIPERLAN terminal is allowed to relay packets between two other terminals that are out of range of each other but both within range of itself, thus allowing the network to be larger than the maximum coverage area of a single terminal.

HIPERLAN 1 has not been widely deployed in practice and its future as a serious contender in the wireless market is uncertain. HIPERLAN 2 described below, however, may still provide significant competition to IEEE 802.11a.

HIPERLAN 2

The physical layer takes data from the data-link layer, scrambles it (to prevent long runs of digital ones or zeros), applies an FEC code (using a punctured convolution code with constraint length 7 and code rate selected from one of several possible rates), interleaves the resulting bit stream (to mitigate the effect of error bursts) and then splits the bits into parallel streams before modulating them onto a set of OFDM subcarriers using BPSK, QPSK, 16-QAM or 64-QAM depending on what bit rate is required. Table 21.4 shows the bit rates obtained using various combinations of code rate and modulation order.

Table 21.4 *HIPERLAN 2 bit rates for various code rate and modulation order combinations.*

Modulation	Code rate	Bit rate
BPSK	1/2	6 Mbit/s
BPSK	3/4	9 Mbit/s
QPSK	1/2	12 Mbit/s
QPSK	3/4	18 Mbit/s
16-QAM	9/16	27 Mbit/s
16-QAM	3/4	36 Mbit/s
64-QAM	3/4	54 Mbit/s

Of the 52 carriers used in the OFDM signal 48 carry data and 4 are pilots (carrying known sequences used for frequency and phase offset correction synchronisation, etc.). The OFDM subcarrier spacing is 0.3125 MHz giving a nominal OFDM signal bandwidth of 16.25 MHz. (The −20 dBc bandwidth is actually specified as 22 MHz.) Nineteen channels spaced by 20 MHz are accommodated across a band from 5.18 GHz to 5.70

GHz (see Table 21.3 for precise OFDM signal centre frequencies).

HIPERLAN 2 supports both infrastructure and ad-hoc operating modes but in both cases there is a measure of centralised MAC. In infrastructure mode this control emanates from the access point (connected to the infrastructure LAN). In ad-hoc mode the central controller (CC) function is fulfilled by one of the terminals (selected by the network itself). The medium is shared using a hybrid of TDMA/TDD and the random access (RA) protocol slotted ALOHA (see section 14.3.7).

The HIPERLAN MAC frame, which has a duration of 2 ms, is subdivided into five different functional segments, Figure 21.18. The broadcast and random access segments carry control information. The downlink segment carries information from the central control terminal (the AP in an infrastructure network) to the mobile terminal. The uplink segment carries information from the mobile terminal to the central control terminal. The direct-link segment carries data directly between mobile terminals (but still under the control of the central control terminal).

The HIPERLAN 2 DLC layer (containing radio link control, RLC, and MAC sub-layers) is responsible for a multitude of complex system management tasks including connection control (set-up, maintenance and release), error control (using ARQ), encryption management (key exchanges, etc.), handoff (from one AP to another), dynamic frequency allocation (DFS) and transmit power control (TPC). DFS is the process whereby the system searches the available channels and automatically selects the best (i.e. that with the lowest interference). If interference increases the channel may be automatically changed under the direction of the central controller. Data rate is adapted (by selecting an appropriate modulation order from the M-QAM orders available) to match the quality of the channel. The highest data rate, 54 Mbit/s, corresponding to the use of 64-QAM is optional in the sense that manufacturers of equipment can choose to include this data rate facility or not whilst the other data rates (6–36 Mbit/s) are mandated by the specification.

The TDMA approach to channel sharing in HIPERLAN 2 (that like HIPERLAN 1 is a technology designed principally for the indoor environment) makes it particularly suitable for time-sensitive data communication such as is often required in multimedia applications. Another specification, HiSWANa, is a related Japanese variant of HIPERLAN 2.

21.6 Wireless personal area networks

The range of devices and applications which wireless networks can support in the home is enormous (cordless phones, hand-held web access devices, multiplayer gaming, playback devices for Internet streamed audio, home automation devices, etc.). The

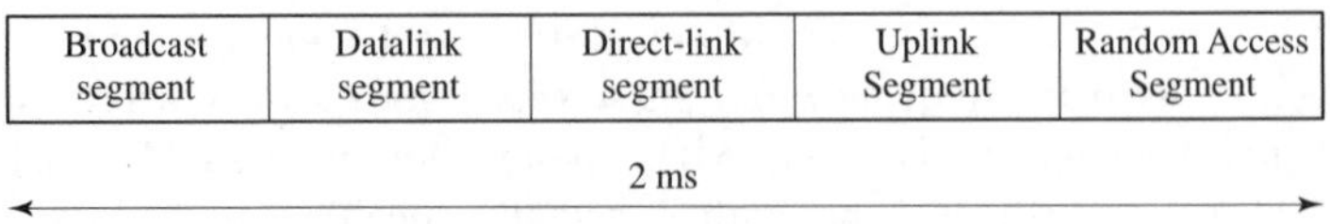

Broadcast segment	Datalink segment	Direct-link segment	Uplink Segment	Random Access Segment

2 ms

Figure 21.18 *Functional segments of HIPERLAN 2 MAC frame.*

technologies described below were conceived primarily with these types of home based applications in mind. As the specifications have evolved, however, the distinction in performance and application between some of the home network technologies and traditional wireless LANs has becomes less clear-cut. Also the applications have come to include interconnection of multiple devices typically carried or worn by the user (e.g. lap-top, mobile phone, headphones, personal organisers, etc.) These systems have therefore come to be referred to as wireless personal area networks (WPANs) reflecting their LAN-like capabilities but with a more restricted (personal) coverage space that may move with the user.

21.6.1 Bluetooth (IEEE 802.15.1)

Bluetooth [Sairam *et al.*] originated in 1994 as a project in Ericsson mobile communications to investigate ways of connecting accessories to its mobile telephone devices, the Bluetooth SIG (special interest group), with membership drawn from several other companies, being formed in February 1998. Version 1.0 of the Bluetooth specification was released in July 1999. Today thousands of companies are producing Bluetooth compatible devices for both the domestic and business markets.

Bluetooth is an open, global, specification defining a complete communications system from the radio layer to the application layer. Its goals are: low cost, low power, short range wireless interconnection of multiple devices, global operation (implying the use of an unlicensed radio band), no fixed infrastructure (ad-hoc networks), supporting both voice and data connections. One of its less obvious applications currently being considered is the replacement of data wiring harnesses in spacecraft (where reduction in mass implies significant savings in cost).

Bluetooth operates in the unlicensed 2.4 GHz ISM band (2.4–2.4835 GHz). It uses 79 1 MHz channels over most of the world with lower and upper guard bands of 2 MHz and 3.5 MHz respectively. (The guard bands are measured from the carrier at the channel centre.) The power, spectral emission and interference in the ISM band are controlled national radio regulatory authorities and international agreements, e.g. ETSI ETS 300-328 in Europe and FCC CFR47 Part 15 in the USA.

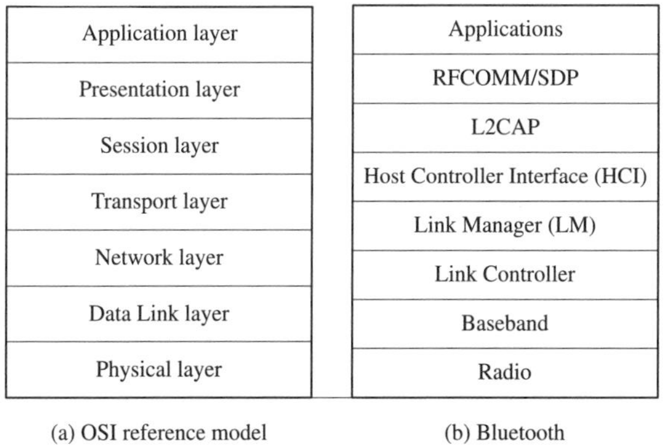

Application layer	Applications
Presentation layer	RFCOMM/SDP
Session layer	L2CAP
Transport layer	Host Controller Interface (HCI)
	Link Manager (LM)
Network layer	Link Controller
Data Link layer	Baseband
Physical layer	Radio
(a) OSI reference model	(b) Bluetooth

Figure 21.19 *OSI reference model (a) and Bluetooth protocol stack (b).*

The unlicensed nature of the ISM band means that it is shared by many different systems (e.g. IEEE 802.11b WLANs, microwave heating appliances, medical equipment) and it is possible for a channel to become 'blocked' by interference. Frequency hopping is used to ensure such blocking is temporary and that retransmissions of lost data packets will take place on a different frequency from that of the original (failed) transmission.

Figure 21.19 compares the Bluetooth protocol stack with the OSI reference model.

The OSI physical layer, which deals with not only the RF electronics but also the modulation and channel coding, corresponds to the Bluetooth radio layer (RL) and part of the baseband layer (BL). The OSI data-link layer is responsible for transmission, framing and error control over a particular link, and as such overlaps with the link controller (LC) and baseband layer of Bluetooth. The OSI network layer is responsible for data transfer across the network regardless of the media and network topology, and overlaps the Bluetooth link controller and link manager (LM) layer functions. The OSI transport layer is responsible for the reliability of data and the multiplexing of data over the network, which in Bluetooth is dealt with by the link manager and host controller interface (HCI) functions. The OSI session layer provides management and data flow control services, which are dealt with in Bluetooth partly by the logical link control and adaptation (L2CAP) and partly by RFCOMM/SDP. RFCOMM provides an RS-232-like serial interface, whilst the service discovery protocol (SDP) enables one Bluetooth device to determine what services other Bluetooth devices support. The applications layer is common to the OSI model and Bluetooth.

Radio layer

Bluetooth uses FHSS in the ISM band for signal transmission. The ISM band was chosen both because its unlicensed nature makes global operation possible and because this operating frequency allows low power, low cost, implementations. FHSS was chosen because it has better immunity to near–far problems than DSSS systems. (The near–far problem relates to a transmitter near to a receiver obscuring the transmissions of other transmitters far from the receiver, section 15.5.6.)

The Bluetooth specification allows for three different levels of RF power: 20 dBm (Class 1), 4 dBm (Class 2) and 0 dBm (Class 3). These power classes allow Bluetooth devices to connect at different ranges. Most devices produced are likely to be low power (Class 3) devices, giving a typical free space range of about 10 m. Given RF absorption in the environment (e.g. by furniture and the human body), a more realistic estimate of the range is probably 5 m. Despite this low range, such devices still offer great advantages over cables. The maximum range for a Class 1 device is likely to be about 100 m.

Perhaps surprisingly, there is a minimum range as well as a maximum range for Bluetooth devices. If they are placed too close together the receivers saturate. The minimum range is approximately 10 cm.

Transmit power control (TPC) is mandatory on Class 1 devices, and optional for Class 2 and 3 devices. (Power control is desirable, however, to minimise power consumption.) TPC operates by the receiver monitoring the received signal strength indication (RSSI) and sending LMP control commands back to the transmitter, asking for transmit power to be decreased or increased as necessary. The specification requires that the power be controlled in steps of 2 to 8 dB, whilst the RSSI measurements must be accurate to ±4 dB at −60 dBm (the optimum operating point), with a minimum operating range of +20/−6 dB. Bluetooth is specified to operate with a BER of 0.1%, which results in a minimum receiver sensitivity of −70 dBm.

Each of the 79 1 MHz Bluetooth channels supports a signalling rate of 1 Mbaud. The modulation scheme used is Gaussian minimum shift keying (GMSK), i.e. GFSK with $T_o\Delta f = 0.5$ corresponding to the minimum carrier separation to ensure symbol orthogonality. (Here $T_o\Delta f$ is the product of symbol duration (1.0 μs) and adjacent signal frequency separation (0.5 MHz), to distinguish it from T_oB as used in section 11.3.4.) The modulation index (representing the relative peak frequency deviation) is between 0.28 and 0.35 corresponding to an absolute frequency deviation of 140–175 kHz. (The modulation index is conventionally defined assuming a carrier of 500 kHz corresponding to a single channel between 0 and 1 MHz.) The Bluetooth specification lists 115 kHz as a minimum. The symbol rate and modulation scheme result in a raw data rate of 1 Mbit/s.

Each channel is divided into 625 μs slots. A frequency hop occurs after each transmitted packet. Packets can be 1, 3 or 5 slots in length. The implied maximum hopping rate is 1600 hop/s. In order to hop at this rate and keep alignment with the 625 μs slot period the frequency synthesiser, which determines transmit and receive frequencies, must be capable of switching quickly, in practice within 180 μs. During the 625 μs slot, Figure 21.20, the system has not only to transfer data, but also to program the synthesiser

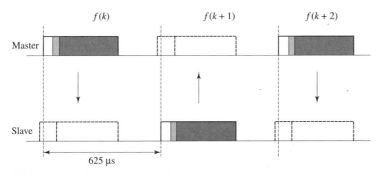

Figure 21.20 *Slot timing for single slot packets.*

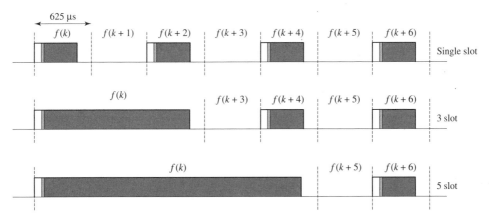

Figure 21.21 *Slot timing for multi-slot packets.*

with the next frequency and allow for the frequency to settle. This and other factors combine to reduce the actual data throughput to 768 kbit/s. Figure 21.20 shows typical master–slave slot timing. Figure 21.21 illustrates slot timing for several multi-slot packets.

Network architecture

Every Bluetooth device has a unique Bluetooth device address (BD_ADDT), and its own Bluetooth clock. If two devices are to communicate using FHSS it is necessary that they hop to the same frequency at the same time. Each device can operate in two fundamentally different modes: master and slave. It is the master device in a Bluetooth WPAN that determines the hopping sequence, the bit rate and time allocation. When slaves connect to a master, they are given information about the device address and clock of the master. From this information, the baseband functions in each slave can determine the hopping pattern to follow. The master dictates when slaves are able to transmit by allocating slots for voice or data traffic, determining how the total available bit rate is divided among the slave devices. The number of time-slots each device gets depends on its data transfer requirements. Bluetooth is therefore a time division multiplex (TDM) system.

Piconets

A piconet consists of a master and up to seven slaves. The paging unit that establishes a connection becomes the piconet master by default. The master announces its clock and its device address to the slaves and the slaves then synchronise to the master. Masters and slaves can exchange roles, i.e. a slave can become a master, and vice versa, if required.

Slaves synchronise to the master's clock by adding an offset to their own clock. Each packet from the master contains a valid access code to allow the nodes to retain synchronisation. This is necessary since a slave may belong to two different piconets and therefore must know which master it is responding to. The master on a piconet schedules all traffic, polling slaves by sending a master-to-slave data/control packet. The master can dynamically adjust the scheduling algorithm. (The scheduling algorithm is not specified in the Bluetooth standard.)

Scatternets

Slaves within a single piconet share 1 MHz of bandwidth. Multiple piconets can co-exist

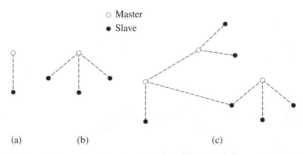

(a)   (b)   (c)

Figure 21.22 *Point-to-point (master slave), point-to-multipoint (piconet) and scatternet operation.*

by hopping independently (each piconet accessing only 1 MHz of bandwidth). A scatternet comprising N piconets shares the full N MHz of the total Bluetooth bandwidth (79 MHz). Nodes in a scatternet can belong to more than one piconet. A node, for example, can be a slave in two different piconets or a master of one piconet and the slave in another. (It is not possible for a device to be a master in two piconets since, by definition, all slaves are required to synchronise with the master of the piconet.)

Currently there is no synchronisation between the scheduling of different piconets that form a scatternet. This results in inefficient use of resources and can cause connections to be dropped. Clearly a major source of interference for Bluetooth devices will be other Bluetooth devices operating within other piconets. Devices sharing a piconet will be synchronised to avoid each other, but other unsynchronised piconets in the locality will occasionally (and randomly) collide on the same frequency.

Figure 21.22 illustrates point-to-point, point-to-multipoint and scatternet operation.

Baseband layer

The following subsections describe the functions of the Bluetooth baseband layer.

SCO and ACL links

Bluetooth provides both for time-sensitive data communications (e.g. voice), and high speed time-insensitive data communications (e.g. e-mail). To support these disparate categories of data two different link types are defined. These are SCO (synchronous connection orientated) links for voice communication and ACL (asynchronous connectionless) links for data communication.

An ACL link exists between a master and slave as soon as it is established. A master may have a number of ACL links to a number of different slaves at any one time, but only one link can exist between any two devices. The choice of which slave to transmit to and receive from is that of the master on a slot-by-slot basis. A slave may only respond to the master if it is addressed in the preceding master-to-slave slot.

Most ACL packets employ error checking and retransmission to ensure data integrity. Broadcast packets are ACL packets that are not addressed to a particular slave and so are received by every slave.

An SCO link provides a symmetric connection between master and slave with a reserved channel bit rate, and reserved slots. This guarantees a regular periodic exchange of data effectively providing a circuit switched connection. This is necessary for time bounded services such as audio (e.g. for Bluetooth mobile telephone headsets). A master can support up to three SCO links either to the same slave or different slaves. A slave can support up to three SCO links to the same master. Due to the time-critical nature of SCO data, packets are never retransmitted. The master transmits SCO packets to the slave at regular intervals and the slave is always allowed to respond in the following reserved slot. Any ACL link traffic must normally be scheduled around the reserved SCO slots so as to preserve the integrity of the SCO link. An exception to this is the transmission of link management control packets that are allowed to override the SCO link. This is necessary to provide a mechanism to shut down SCO links in the event that all the bandwidth is reserved.

Packet structure

Bluetooth uses different packet structures depending on the link type that is to be supported. Every packet can be broken down into the following components: access code,

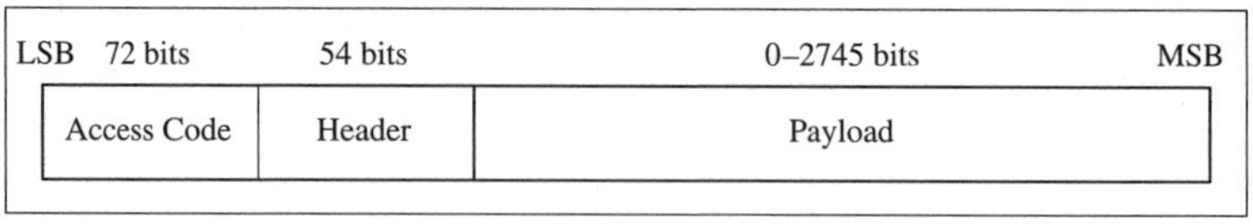

LSB 72 bits	54 bits	0–2745 bits	MSB
Access Code	Header	Payload	

Figure 21.23 *Bluetooth general packet structure.*

packet header, payload header and payload (Figure 21.23).

The access code is used to detect the presence of a packet and to address the packet to a specific device. For example, slave devices detect the presence of a packet by matching the access code against their stored copy of the master's access code. The packet header contains the control information associated with the packet and the link such as the address of the slave for which the packet is intended. Finally, the payload contains the information from a higher layer in the stack, which may be a control message (such as might be sent from L2CAP or LM) or the user data.

The 72 bit access code comprises three main parts: the preamble (4 bits), the synchronisation word (64 bits) and the trailer (4 bits). The preamble and the trailer form two 4 bit known sequences, which aid with clock recovery. The key component of the access code is the synchronisation word, which is derived from part of the Bluetooth device address.

The 18 (information) bit packet header contains the active member address (i.e. the address of the transmitting device), packet type (e.g. SCO or ACL) and a header error checksum. Packet header error control employs repetition coding, each information bit being repeated three times (giving a rate 1/3 error correction code) bringing the packet header size to 54 (binary digit) bits.

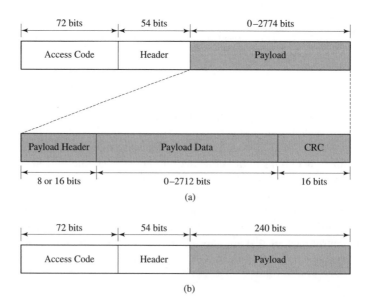

Figure 21.24 *ACL (a) and SCO (b) packet structure.*

The payload field of ACL packets is split into thee parts: the payload header, the payload data and the CRC check field (Figure 21.24). The ACL header contains information bits that indicate the length of the payload field and whether or not the packet is a continuation of an L2CAP or an LMP message. SCO packets share the same access code and header as ACL packets (except the bits that control flow and retransmission). The size of SCO packets is fixed at 30 bytes, the source data itself being 10, 20 or 30 bytes depending on the error correction employed.

Depending on the number of slots allocated and the application of error correction the throughput of an ACL link is given in Table 21.5.

Table 21.5 *Throughput of ACL Bluetooth links.*

Type of packet	Symmetric (kbit/s)	Asymmetric (kbit/s)
1-slot, FEC	108.8	108.8 / 108.8
1-slot, no FEC	172.8	172.8 / 172.8
3-slot, FEC	256	384 / 54.4
3-slot, no FEC	384	576 / 86.4
5-slot, FEC	286.7	477.8 / 36.3
5-slot, no FEC	432.6	721 / 57.6

A special case SCO is the combined data/voice (DV) packet that like a conventional SCO packet must be sent at regular intervals but, in addition to carrying time-critical voice data (80 bits), without error correction or retransmission, also carries up to 72 bits of data with rate 2/3 error correction coding, see Figure 21.25.

In addition to the basic data bearing packets there are four special packets that help with connection and synchronisation. The ID packet carries only the access code of the device it is coming from or going to and occupies only half of a slot. The NULL packet consists only of an access code and a packet header, and is used in the retransmission scheme for flow control. The POLL packet is similar to the NULL packet except that it requires an acknowledgement. POLL packets are typically employed by the master to check the presence of slaves in a piconet, which must then respond if they are present. The FHS (frequency hop synchronisation) packet provides information about the frequency hopping and is typically used when a device is taking over as master, i.e. when master and slave devices switch roles.

Logical channels

The Bluetooth specification defines five logical information channels that are carried over the SCO and ACL links. These are:

1. LC (link control): carried via the packet header and containing retransmission information, essential to maintaining and controlling the links.

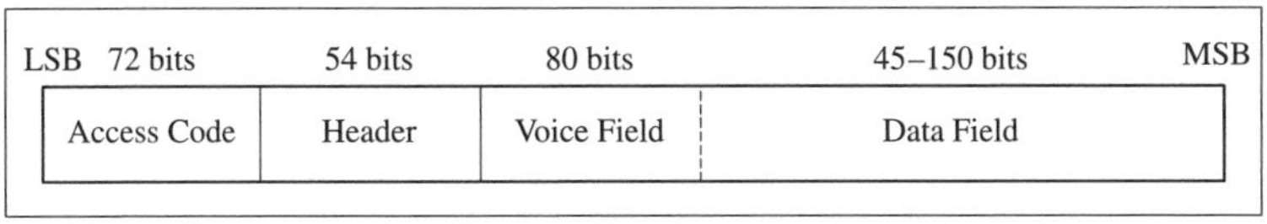

Figure 21.25 *Data/voice (DV) packet structure.*

2. LM (link manager): carried via a dedicated ACL payload and containing control data being exchanged between two systems.
3. UA (user asynchronous): carried via the ACL payload and containing L2CAP user data.
4. UI (user isochronous): carried via the ACL payload and containing L2CAP user data.
5. US (user synchronous): carried via SCO channel payloads representing transparent synchronous data.

Link controller

The link control layer carries out higher level operations such as inquiry and paging, and manages multiple links with different devices or different piconets.

At any one time a Bluetooth device may be in one of a number of different states. These are:

Standby: In this state the device is dormant, with the radio device inactive. This state enables low power operation.

Inquiry: Here, the device will attempt to discover all the Bluetooth enabled devices within its range. It accomplishes this by running the SDP (service discovery protocol).

Inquiry scan: Most devices periodically enter the inquiry scan state to make themselves available to inquiring devices.

Page: To establish a connection between devices, the device that is to become the master enters the page state from which it can carry out the paging procedure as instructed by the application.

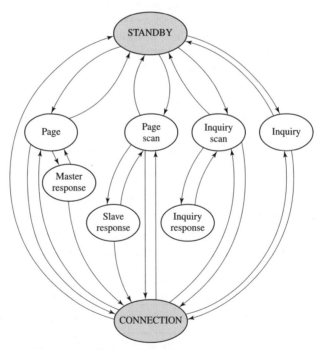

Figure 21.26 *Bluetooth link controller state diagram.*

Page scan: In a similar way to the inquiry scan state, a device will periodically enter the page scan state to allow paging devices to establish a connection with it.

Connection – active: Once the connection state has been entered and the slave has become synchronised the various data exchanges and logical channels can be established. If the connection fails to be established the devices will revert to page and page scan states and the process may restart. From time to time, any device in an active connection state may move into a lower power substate whilst still maintaining some degree of synchronisation.

Connection – hold: In hold mode the device ceases to support ACL traffic for a predefined period such that capacity may be made available for other operations such as scanning, paging or inquiry.

Connection – sniff: In sniff mode, a slave device is allocated a predefined time-slot and periodicity to listen for traffic.

Connection – park: In park mode a slave device gives up its active member address and listens for traffic only occasionally. In this state the device is able to enter a low power sleep mode, but capable of resuming operation to receive timing beacon signals such that the power consumption can be reduced (and the device run from a far less accurate low power oscillator) and yet the device remains synchronised.

The complete link controller state diagram is shown in Figure 21.26.

Link manager

The host application drives a Bluetooth device through (HCI) commands. It is the link manager, however, that translates these commands into operations for the baseband layer. The link manager orchestrates the following operations:

1. Attaching slaves to a piconet and allocating their active member addresses.
2. Breaking connections to detach slaves from a piconet.
3. Configuring the link including controlling master/slave switches (in which both devices must exchange roles).
4. Establishing ACL (data) and SCO (voice) links.
5. Putting connections into lower power states: hold, sniff and park.
6. Controlling test modes.

A Bluetooth link manager communicates with link managers on other Bluetooth devices using the link manager protocol (LMP). The Bluetooth specification only defines the LMP messages exchanged between link managers, it does not give details on how those operations are to be implemented.

Host controller interface

The Bluetooth specification for the host controller interface defines the following:

1. Command packets used by the host to control the Bluetooth module.
2. Event packets used by the module to inform the host of changes in the lower layers.
3. Data packets to pass voice and data between the host and the Bluetooth module.
4. Transport layers that can carry HCI packets.

A transport layer is required to get HCI packets from the host to the Bluetooth module. The specification defines three transport layers for serial data communications based upon USB and RS-232.

Logical link control and adaptation protocol (L2CAP)

The L2CAP takes data from higher layers of the Bluetooth stack and sends it over the lower layers of the stack. L2CAP passes packets either to the HCI or (in a host-less system) passes packets directly to the LM. L2CAP has many functions including:

1. Multiplexing between different higher layer protocols, allowing them to share lower layer links.
2. Segmentation and reassembly to allow transfer of larger packets than lower layers can support.
3. Group management, providing one-way transmission to a group of Bluetooth devices.
4. QoS management for higher layer protocols.

RFCOMM and SDP

The Bluetooth RFCOMM layer is a simple and reliable transport protocol with framing, multiplexing and the following additional provisions:

1. **Modem status** – data flow control.
2. **Remote line status** – break, overrun, parity.
3. **Remote port settings** – baud rate, parity, number of data bits.
4. **Parameter negotiation** – frame size.

Once an L2CAP service has been established between a client and a server, the SDP can be used to discover what services are available and how to connect to them. L2CAP does not handle connections to services itself, it merely provides information. To find and connect to a service offered by an SDP server, a client must go through the following steps:

1. Establish an L2CAP connection to the remote device.
2. Search for a specific class of service, or browse for services.
3. Retrieve attributes needed to connect to the chosen service.
4. Establish a separate connection to use the service.

Once the service has been established and there is no further need for the L2CAP channel it can be dropped.

21.6.2 Other IEEE 802.15 PAN technologies

The 802.15 group was created shortly after the public release of the Bluetooth standard in 1999 to create an IEEE based PAN standard to complement the work of 802.11. In agreement with the Bluetooth SIG, the group decided to adopt the Bluetooth v1.0b standard as the basis for its work, and was tasked to rework the standard to fit with the standard IEEE networking model. This resulted in IEEE 802.15.1. The IEEE 802.15 group went on, however, to address PAN architectures for other data rates and applications and these are described briefly below.

IEEE 802.15.3

IEEE 802.15.3 is a short range (typically 10 m), high bit rate PAN. It has an ad-hoc architecture with QoS capability providing data rates up to 55 Mbit/s and is primarily designed to support high definition video and high fidelity audio [Zahariadis *et al.*]. Like IEEE 802.11b it operates in the 2.4 GHz ISM band and has a symbol rate of 11 Mbaud.

The channel bandwidth is 15 MHz allowing four networks to co-exist within the ISM band. A range of bit rates is supported, i.e. 11, 22, 33, 44 and 55 Mbit/s using, respectively, BPSK, QPSK, 16-QAM, 32-QAM and 64-QAM. (Eight-state trellis coding is used with the QAM schemes.) There are plans to extend the bit rates of IEEE 802.15.3 to 110 Mbit/s and greater than 400 Mbit/s at maximum ranges of 10 m and 4 m respectively, possibly using ultra wideband (UWB) transmission techniques.

IEEE 802.15.4

IEEE 802.15.4 is an emerging standard for an ad-hoc low cost, low data-rate PAN that can operate in either star or peer-to-peer configurations [Callaway *et al.*]. It describes a low complexity (and therefore low power/long battery life) network for the interconnection of a wide range of inexpensive sensors and devices for domestic, commercial and public service applications such as home automation, industrial process instrumentation and control, public safety and security (e.g. security video, smart badges, etc.). A bit rate of 20 kbit/s is supported in a single 600 kHz channel at a carrier frequency of 868.3 MHz in Europe and a bit rate of 40 kbit/s is supported in ten 2 MHz channels at frequencies around 915 MHz in the USA. A bit rate of 250 kHz is supported in sixteen 5 MHz channels at frequencies around 2.442 GHz in the ISM band (globally).

The 868/915 MHz physical layer standard employs differential BPSK with DSSS using a 15-chip maximal length sequence, section 15.5.3. The resulting chip rate is 0.3 Mchip/s for the 868 MHz version and 0.6 Mchip/s for the 915 MHz version. The 2.442 GHz standard employs 16-level quasi-orthogonal coding (groups of 4 bits select one of 16, nearly orthogonal, 32-chip PRBS codewords) followed by MSK modulation of the resulting 2.0 Mchip/s sequence.

The MAC layer of IEEE 802.15.4 is a CSMA/CA protocol and like other IEEE 802 standards the LLC layer is defined by IEEE 802.2. (This is a simplified description: a service-specific convergence sub-layer (SSCS) interfaces the MAC and LLC layers, for example, and other proprietary LLC protocols can be used in place of IEEE 802.2 and the SSCS as shown in Figure 21.27.) The CSMA/CA protocol can operate in either conventional form or in a slotted form under the control of a dedicated network

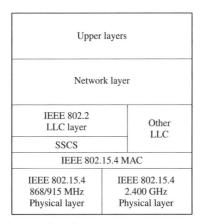

Figure 21.27 *IEEE 802.15.4 protocol stack.*

coordinator. The slotted form caters for applications requiring dedicated bandwidth and low latency.

21.7 Home networking technologies

While Ethernet is close to being the de facto wired LAN standard for business and commercial environments it is probably too expensive to become widely adopted in the home environment, at least in the near or medium term. (This may change in the long term as users come to expect similar network facilities at home as they experience in the office leading, perhaps, to broadband communications wiring being universally incorporated into new homes at the time of construction in the same way as mains power wiring is today.) The applications of networks in the home form a different, but probably overlapping, set with those in a business, industrial or commercial environment. These applications include interconnection of PCs (in modest numbers) and entertainment equipment (e.g. video, audio, games devices), Internet access, home automation (e.g. remote monitoring/control of domestic equipment, energy management and conservation) and security (intruder and fire alarms, security lighting, etc.). The systems described briefly below are examples of technologies designed primarily for these types of home networking application. The emphasis is typically on low cost (implying modest performance) and installation/configuration simplicity.

21.7.1 Wired home networks

Home network systems can be divided into wireless and wired varieties. Wireless home networks have already been partly addressed in the context of PANs (which essentially form a subset of home networks). Those home networks not usually regarded as PANs are discussed later (section 21.7.2). Wired home networks (which can themselves be divided into new wire and existing (or legacy) wiring varieties) are addressed below.

Firewire (IEEE 1394)

IEEE 1394 was designed as a network for home entertainment applications although it is now regarded as a more generic home network standard. It is a new wire network (i.e. requires bespoke wiring) but does not specify a physical transmission medium, copper cabling and optical fibre (including plastic optical fibre, POF) both having been used. It provides bit rates of 100, 200, 400 and 800 Mbit/s and supports both asynchronous and isochronous data transfer. It has a bus structure supporting peer-to-peer communications and allows the interconnection of a wide range of devices having very different data rates. IEEE 1394 supports hot swapping, i.e. the connection and disconnection of one piece of equipment without affecting the operation of any other piece of equipment.

Universal serial bus (USB)

The USB has been designed primarily as a single 'plug and play' interface for the connection of computers and their peripherals (printers, DVD drives, digital cameras, etc.) that seeks ultimately to replace all other serial and parallel interface standards. In version 2.0 bit rates up to 480 Mbit/s are possible and both asynchrounous and isochronous data transfer services are supported. Like IEEE 1394 it falls into the new wire category of network and allows hot swapping of devices. It also has the capability to distribute power to peripherals (avoiding the need for a multiplicity of separate power cables). USB could, in principle, fulfil a similar role to conventional LAN technology including Ethernet but it seems likely in practice that it will remain principally a technology for short range computer–peripheral connections.

HomePNA

HomePNA (home phone-line network association) is a technology that uses the twisted pair legacy wiring of the telephone system in a building to provide network services. Telephone sockets provide the network connection points. HomePNA version 2.0 supports bit rates up to 10 Mbit/s and there is the prospect of later generations achieving bit rates up to 100 Mbit/s. The frame structure is derived from IEEE 802.3. (Version 1.0, which supports data rates up to 1 Mbit/s, also shares the CSMA/CD MAC protocol of Ethernet.) HomePNA version 2.0 provides a QoS capability by using a system of eight priority levels and an improved contention resolution protocol.

HomePlug

HomePlug uses the low voltage power distribution network within a building to provide a wired communications infrastructure and is therefore an existing (or legacy) wire technology. Power sockets, which are usually distributed generously around the building (typically several to a room), provide the network connection points. The physical medium, being designed for power transmission, is, in many respects, entirely unsuitable for data transmission. The irregular mesh-like architecture (e.g. stars, ring-mains, spur lines, etc.) results in reflections from impedance discontinuities leading to multipath propagation. Furthermore impedance levels, and therefore the multipath characteristics, may vary depending on the devices (motors, power supplies, etc.) attached via the power sockets at any particular time. This results in a channel which is not only frequency selective but also potentially time varying. The interference environment can also be severe with impulsive noise originating from the attached electromechanical and electronic devices (e.g. motors, thermostats, switched mode power supplies, etc.). Despite these unpromising channel characteristics bit rates of up to 100 Mbit/s are possible. HomePlug does not support QoS.

21.7.2 Wireless home networks

As in the commercial sector domestic network users appear to be willing to pay a premium for the convenience of mobile services. Bluetooth (and perhaps in the longer term IEEE 802.11 and its variants) may come to serve the domestic wireless network market but in the meantime there are several competing systems. Those that currently appear to have the best chances of commercial success are described below.

DECT

Digital European cordless telephone, see section 15.3.2, is a standard that was developed primarily for short range voice communications between cordless telephone handsets and fixed base stations. It uses GMSK modulation (baseband $T_o B = 0.5$) in the frequency band 1.880–1.900 GHz, Table 15.4. Ten RF carriers within the frequency band are each subdivided into 24 time-slots (12 transmit and 12 receive) yielding 240 simplex (120 duplex) channels. Each slot is 0.417 ms long and carries 416 bits (96 preamble/control bits followed by 320 information bits) transmitted at a rate of 1152 kbit/s. The frame length is 10 ms. Paired transmit and receive time-slots provide each TDD duplex channel. Voice signals are transmitted using 32 kbit/s ADPCM. Since the DECT cells are small equalisation is not used. The typical indoor range of a DECT transmission is 50 m and the typical outdoor range is 300 m. DECT allows roaming between base stations.

A proliferation of DECT based products, including wireless PABXs, are now available and the wide deployment of DECT systems over many years has resulted in DECT equipment becoming particularly inexpensive.

Using a data oriented protocol called DPRS (DECT packet radio service), a user data rate (which can be distributed asymmetrically between uplink and downlink) of 552 kbit/s can (in theory) be achieved by employing all channels for a single data connection. (In practice the user data rate might be significantly less than this.) This facility allows DECT to be used as the platform for a short range wireless network.

HomeRF

HomeRF is essentially a hybrid standard combining aspects of DECT and IEEE 802.11 under the banner of SWAP (shared wireless access protocol) and is promoted as 'next generation DECT'. It is also promoted as Global DECT because it uses the 2.4 GHz ISM transmission band and can therefore be deployed worldwide. Its hybrid DECT/wireless Ethernet nature is reflected in its protocol stack, Figure 21.28. It has a typical range of 50 m and (in version 2.0) incorporates QoS supporting eight isochronous toll quality (MOS = 4.1, sections 5.7.3 and 9.7.1) voice channels and a streamed data rate of 10, 5, 1.6 or

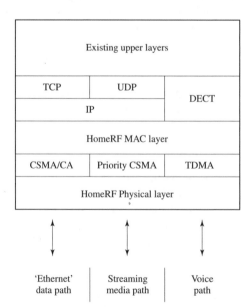

Figure 21.28 *HomeRF protocol stack.*

0.8 Mbit/s depending on the quality of the channel. There are plans to extend the data rate to 25 Mbit/s. The MAC protocol is CSMA based for asynchronous data and TDMA based for voice, modulation is 2-FSK for 0.8 Mbit/s, 4-FSK for 1.6 Mbit/s and QPSK for 5 and 10 Mbit/s. Spectral spreading is achieved using a similar FHSS scheme as IEEE 802.11 (75 frequency hopped 1 MHz channels are used for 0.8, 1.6 Mbit/s data and all voice traffic, and 15 frequency hopped 5 MHz channels are used for 5 and 10 Mbit/s data; the hopping rate is 50–100 hop/s). Isochronous data connections operate in host/client (i.e. infrastructure) mode and must be implemented via a control point (the equivalent of an access point in IEEE 802.11 systems). Peer-to-peer connections (equivalent to ad-hoc mode) only support asynchronous data transfer. HomeRF supports a maximum of 127 nodes.

The principal air-interface parameters of HomeRF and DECT are compared with those of Bluetooth in Table 21.6.

Table 21.6 *Selected wireless home network systems.*

System	Channel carrier frequencies (GHz)	Spreading technique	Modulation	Bit rate (bandwidth)	Approx. max. EIRP (Europe)
Bluetooth	2.4–2.4835	FHSS	GFSK	1 Mbit/s (1 MHz hopped over 79 channels)	−30 dBW/ −26 dBW/ −10 dBW
DECT	1.88–1.90	–	GMSK ($T_oB = 0.5$)	1.152 Mbit/s (1.152 MHz)	−6 dBW
HomeRF	2.4–2.4835	FHSS	M-FSK or QPSK	0.8, 1.6 Mbit/s 5, 10 Mbit/s (1 or 5 MHz hopped over 75 or 15 channels)	−10 dBW

Personal and home networking is expected to be realised via an inner layer, i.e. personal network such as Bluetooth, which is interconnected through a home network to the WANs of Chapter 20 [see e.g. the WWW address to the Wireless World book of visions], where the WAN is sometimes envisaged as accessing the cyberworld.

21.8 Residential gateways

Network bridges and gateways are available to connect home networks and conventional LANs both to each other and to access technologies such as xDSL, WLL, FTTH, ISDN and the PSTN. The concept of the residential home gateway [Bull *et al.*] is a single point of access between external MANs and WANs and internal LANs, PANs and home networks. Figure 21.29 shows a hypothetical example in which plug-in cards are used on each side of an exterior/interior interface to provide the required protocol translation, transmission speed matching and physical media interfacing functions.

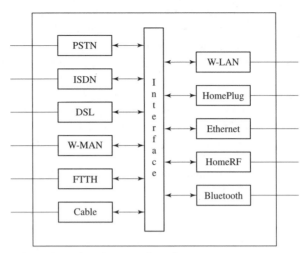

Figure 21.29 *Hypothetical interface for a residential gateway.*

21.9 Summary

Local area networks (LANs) are broadcast networks of limited geographical extent (currently less than a few km). They may, however, form part (often the access part) of a larger switched network. The switching and routing protocols required by switched networks are replaced in broadcast networks with a medium access control (MAC) protocol. The most common LAN topologies are the bus and ring and the most common MAC protocols are CSMA/CD and token passing. In principle both protocols may be implemented on networks of either topology although CSMA/CD is usually used on buses. Token passing has become closely associated with rings but can also be implemented on physical bus structures operating as logical rings.

CSMA protocols are contention schemes and require contention resolution algorithms. Maximum network delay is unpredictable in contention systems and network utilisation is generally low, especially so in random access (RA) schemes. RA schemes are really suited only to networks catering for large numbers of lightly loaded independent users with bursty traffic requirements. CSMA/CD schemes have generally better utilisation than RA schemes but both exhibit large network latency.

Token passing is a reservation scheme and has predictable maximum network delay. Reservation schemes are better suited than contention schemes to networks with modest numbers of heavily loaded users.

The IEEE 802.3 standard defines a CSMA/CD bus network that developed out of, and is usually referred to as, Ethernet. Like all contention systems Ethernet suffers from large delay (network latency) under heavily loaded conditions and is thus unsuitable (in its pure form) for time-critical (isochronous) traffic such as audio and video. It is particularly well suited, however, to networks serving large numbers of users each independently generating small amounts of data sporadically. Ethernet is now the most widely deployed of all LAN technologies and has evolved into a range of different varieties. The classic Ethernet (10BASE5) supports a bit rate of 10 Mbit/s over a coaxial medium as does the lower cost thin Ethernet (10BASE2). There are also versions of 10 Mbit/s Ethernet that employ CATV coaxial cable (10BROAD36), optical fibre (10BASE-F) and twisted pair (10BASE-T). Fast Ethernet, which also operates over a range of transmission media, has a bit rate of 100 Mbit/s and Gigabit Ethernet operates over fibre and high quality twisted pair cable at a bit rate of 1 Gbit/s and now 10 Gbit/s over fibre.

The IEEE 802.5 standard defines a token ring with a bit rate of 4 or 16 Mbit/s that uses twisted pair as the transmission medium. The token ring has the advantage of bounded latency that makes it more suitable than (pure) Ethernet for time-critical data application such as voice and video.

IEEE 802.11 is the most widely deployed wireless LAN standard at present and in its original form supports a bit rate of 1 or 2 Mbit/s. There is a choice of physical layer technologies for IEEE 802.11. These are direct sequence spread spectrum (DSSS), frequency hopping spread spectrum (FHSS) and infrared (IR) transmission. The DSSS and FHSS varieties operate in the 2.4 GHz ISM band. IEEE 802.11b extends the bit rate of the DSSS variety to 5.5 and 11 GHz, and recent developments of the IEEE 802.11 standard (IEEE 802.11g) promise ISM band bit rates of up to 20 Mbit/s using a DSSS based transmission and 54 Mbit/s using an OFDM based transmission. IEEE 802.11a is a WLAN variety that operates in the 5.2 GHz band using OFDM transmission and supports a bit rate of 54 Mbit/s.

HIPERLAN refers to two European wireless LAN standards operating in the 5.2 GHz frequency band that compete directly with IEEE 802.11a. HIPERLAN 1 has not yet achieved any real commercial success. HIPERLAN 2 employs OFDM transmission to offer bit rates up to 54 Mbit/s and whilst it has yet to achieve real commercial success is still a possible contender to IEEE 802.11a.

Bluetooth is an ISM band personal area network (PAN). PANs are low cost, short range (approximately 10 m) wireless LANs that can be used for home networking of IT, entertainment, security and domestic management systems equipment and also for the interconnection of multiple devices carried or worn by a user (e.g. mobile phone, laptop, personal digital assistants, headphones, web-pads). It employs FHSS using 79 1 MHz channels with a hopping rate of 1600 hop/s. The bit rate offered is 1 Mbit/s and both asynchronous connectionless and synchronous connection-oriented links are supported. A Bluetooth piconet comprises a master node and up to seven slave nodes, all devices potentially being capable of becoming a master node. Scatternets are made up of multiple piconets, at least one device in each piconet also belonging to another piconet in the scatternet. IEEE 802.15.1 represents a Bluetooth standard and IEEE 802.3 represents Ethernet.

IEEE 802.15.3 and IEEE 802.15.4 are high (55 Mbit/s in the 2.4 GHz band) and low (20 or 40 kbit/s at around 900 MHz and 250 kbit/s in the 2.4 GHz band) bit rate PANs designed for video/hi-fi audio and low bit rate home/public networking types of applications respectively. Other wireless home networking technologies include the digital cordless phone technology DECT and HomeRF.

Wired home networks representing low cost alternatives to conventional LANs include IEEE 1394 and USB designed primarily for home entertainment applications and PC–peripheral connection respectively. Both have the potential for more generic network applications, however, supporting bit rates up to 800 Mbit/s and 480 Mbit/s respectively. Legacy wiring networks include HomePNA that uses the existing telephone wiring infrastructure and HomePlug that uses the power distribution infrastructure.

Residential gateways will form the interface between the external network represented by a variety of broad- and narrowband access technologies and the home networks and PANs deployed throughout the houses and businesses of the future. They will also contain the bridges and gateways needed to interconnect the diverse array of home network and PAN technologies required for the plethora of disparate entertainment, IT, security, Internet and voice applications that users will almost certainly come to demand.

21.10 Problems

21.1 What are the typical characteristics of a local area network (LAN)? What is a MAC protocol and why is it usually needed for a LAN?

21.2 Explain how an Ethernet local area network (LAN) uses contention.

21.3 What are the generic alternatives to a contention MAC protocol? Explain, briefly, how these alternatives work.

21.4 What does 10BASE5 denote in the context of Ethernet? Describe, briefly, the wiring specification for 10BASE5.

21.5 In a bus network with a transmission rate of 10 Mbit/s, a bus length of 2 km and a propagation velocity of 2×10^5 km/s, packets are 1000 bits long. The receiver acknowledges receipt of a packet by transmitting a 64-bit acknowledgement signal. Calculate the bus utilisation.

21.6 An Ethernet LAN operates with a transmission rate of 10 Mbit/s along a cable segment that is 250 m in length. The propagation delay along the cable is 6.67 μs/km. Estimate the maximum utilisation of the network for frame lengths of 72 bytes and 1526 bytes.

21.7 In the 10 Mbit/s Ethernet specification the segment lengths should be less than 500 m and the separation between stations should be less than 2.5 km. Explain why this is so. If the data rate is increased to 1000 Mbit/s determine the new length restriction.

21.8 An Ethernet LAN has a cable segment length of 500 m on which the signal propagates at 5 μs/km. The transmission rate is 100 Mbit/s and the length of the frames is between 72 bytes and 1526 bytes. Estimate the maximum utilisation and the data throughput in the in the case where three stations are active.

21.9 Discuss the steps needed to make the Ethernet LAN of the precious question successfully over a length of 1700 m.

21.10 Explain how token passing protocols are used in a local area network.

21.11 A token ring has 40 stations, each with a latency of 5 bits. The propagation delay on the ring is 5 μs/km and the data rate is 40 Mbit/s. The token is 3 bytes and the ring circumference is 2 km. Calculate, explaining the assumptions made: (i) the maximum waiting time when the ring is unloaded, (ii) the maximum and average waiting time when frames are 512 bytes long and average traffic from each station is 3000 frames/s.

21.12 Discuss how the utilisation of a token ring may be maximised, and the disadvantages of operating the ring in such a manner.

21.13 What are the principal differences between the wireless LAN standards IEEE 802.11, IEEE 802.11a, IEEE 802.11b and HIPERLAN 2?

21.14 Explain the difference between ad-hoc and infrastructure operating modes in the context of wireless LANs.

21.15 What type of technology is Bluetooth? Describe the operation of Bluetooth's radio layer.

21.16 Bluetooth's network architecture incorporates the concepts of piconets and scatternets. Explain these concepts.

APPENDIX A

Tabulated values of the error function

$$\text{erf}(x) = \frac{2}{\sqrt{\pi}} \int_0^x e^{-y^2} \, dy$$

x	erf x	x	erf x	x	erf x	x	erf x
0.00	0.000000	0.40	0.428392	0.80	0.742101	1.20	0.910314
0.01	0.011283	0.41	0.437969	0.81	0.748003	1.21	0.912956
0.02	0.022565	0.42	0.447468	0.82	0.753811	1.22	0.915534
0.03	0.033841	0.43	0.456887	0.83	0.759524	1.23	0.918050
0.04	0.045111	0.44	0.466225	0.84	0.765143	1.24	0.920505
0.05	0.056372	0.45	0.475482	0.85	0.770668	1.25	0.922900
0.06	0.067622	0.46	0.484655	0.86	0.776100	1.26	0.925236
0.07	0.078858	0.47	0.493745	0.87	0.781440	1.27	0.927514
0.08	0.090078	0.48	0.502750	0.88	0.786687	1.28	0.929734
0.09	0.101282	0.49	0.511668	0.89	0.791843	1.29	0.931899
0.10	0.112463	0.50	0.520500	0.90	0.796908	1.30	0.934008
0.11	0.123623	0.51	0.529244	0.91	0.801883	1.31	0.936063
0.12	0.134758	0.52	0.537899	0.92	0.806768	1.32	0.938065
0.13	0.145867	0.53	0.546464	0.93	0.811564	1.33	0.940015
0.14	0.156947	0.54	0.554939	0.94	0.816271	1.34	0.941914
0.15	0.167996	0.55	0.563323	0.95	0.820891	1.35	0.943762
0.16	0.179012	0.56	0.571616	0.96	0.825424	1.36	0.945561
0.17	0.189992	0.57	0.579816	0.97	0.829870	1.37	0.947312
0.18	0.200936	0.58	0.587923	0.98	0.834232	1.38	0.949016
0.19	0.211840	0.59	0.595936	0.99	0.838508	1.39	0.950673
0.20	0.222703	0.60	0.603856	1.00	0.842701	1.40	0.952285
0.21	0.233522	0.61	0.611681	1.01	0.846810	1.41	0.953852
0.22	0.244296	0.62	0.619411	1.02	0.850838	1.42	0.955376
0.23	0.255023	0.63	0.627046	1.03	0.854784	1.43	0.956857
0.24	0.265700	0.64	0.634586	1.04	0.858650	1.44	0.958297
0.25	0.276326	0.65	0.642029	1.05	0.862436	1.45	0.959695
0.26	0.286900	0.66	0.649377	1.06	0.866144	1.46	0.961054
0.27	0.297418	0.67	0.656628	1.07	0.869773	1.47	0.962373
0.28	0.307880	0.68	0.663782	1.08	0.873326	1.48	0.963654
0.29	0.318283	0.69	0.670840	1.09	0.876803	1.49	0.964898
0.30	0.328627	0.70	0.677801	1.10	0.880205	1.50	0.966105
0.31	0.338908	0.71	0.684666	1.11	0.883533	1.51	0.967277
0.32	0.349126	0.72	0.691433	1.12	0.886788	1.52	0.968413
0.33	0.359279	0.73	0.698104	1.13	0.889971	1.53	0.969516
0.34	0.369365	0.74	0.704678	1.14	0.893082	1.54	0.970586
0.35	0.379382	0.75	0.711156	1.15	0.896124	1.55	0.971623
0.36	0.389330	0.76	0.717537	1.16	0.899096	1.56	0.972628
0.37	0.399206	0.77	0.723822	1.17	0.902000	1.57	0.973603
0.38	0.409009	0.78	0.730010	1.18	0.904837	1.58	0.974547
0.39	0.418739	0.79	0.736103	1.19	0.907608	1.59	0.975462

x	erf x	x	erf x	x	erf x	x	erf x
1.60	0.976348	2.10	0.997021	2.60	0.999764	3.10	0.99998835
1.61	0.977207	2.11	0.997155	2.61	0.999777	3.11	0.99998908
1.62	0.978038	2.12	0.997284	2.62	0.999789	3.12	0.99998977
1.63	0.978843	2.13	0.997407	2.63	0.999800	3.13	0.99999042
1.64	0.979622	2.14	0.997525	2.64	0.999811	3.14	0.99999103
1.65	0.980376	2.15	0.997639	2.65	0.999822	3.15	0.99999160
1.66	0.981105	2.16	0.997747	2.66	0.999831	3.16	0.99999214
1.67	0.981810	2.17	0.997851	2.67	0.999841	3.17	0.99999264
1.68	0.982493	2.18	0.997951	2.68	0.999849	3.18	0.99999311
1.69	0.983153	2.19	0.998046	2.69	0.999858	3.19	0.99999356
1.70	0.983790	2.20	0.998137	2.70	0.999866	3.20	0.99999397
1.71	0.984407	2.21	0.998224	2.71	0.999873	3.21	0.99999436
1.72	0.985003	2.22	0.998308	2.72	0.999880	3.22	0.99999473
1.73	0.985578	2.23	0.998388	2.73	0.999887	3.23	0.99999507
1.74	0.986135	2.24	0.998464	2.74	0.999893	3.24	0.99999540
1.75	0.986672	2.25	0.998537	2.75	0.999899	3.25	0.99999570
1.76	0.987190	2.26	0.998607	2.76	0.999905	3.26	0.99999598
1.77	0.987691	2.27	0.998674	2.77	0.999910	3.27	0.99999624
1.78	0.988174	2.28	0.998738	2.78	0.999916	3.28	0.99999649
1.79	0.988641	2.29	0.998799	2.79	0.999920	3.29	0.99999672
1.80	0.989091	2.30	0.998857	2.80	0.999925	3.30	0.99999694
1.81	0.989525	2.31	0.998912	2.81	0.999929	3.31	0.99999715
1.82	0.989943	2.32	0.998966	2.82	0.999933	3.32	0.99999734
1.83	0.990347	2.33	0.999016	2.83	0.999937	3.33	0.99999751
1.84	0.990736	2.34	0.999065	2.84	0.999941	3.34	0.99999768
1.85	0.991111	2.35	0.999111	2.85	0.999944	3.35	0.999997838
1.86	0.991472	2.36	0.999155	2.86	0.999948	3.36	0.999997983
1.87	0.991821	2.37	0.999197	2.87	0.999951	3.37	0.999998120
1.88	0.992156	2.38	0.999237	2.88	0.999954	3.38	0.999998247
1.89	0.992479	2.39	0.999275	2.89	0.999956	3.39	0.999998367
1.90	0.992790	2.40	0.999311	2.90	0.999959	3.40	0.999998478
1.91	0.993090	2.41	0.999346	2.91	0.999961	3.41	0.999998582
1.92	0.993378	2.42	0.999379	2.92	0.999964	3.42	0.999998679
1.93	0.993656	2.43	0.999411	2.93	0.999966	3.43	0.999998770
1.94	0.993923	2.44	0.999441	2.94	0.999968	3.44	0.999998855
1.95	0.994179	2.45	0.999469	2.95	0.999970	3.45	0.999998934
1.96	0.994426	2.46	0.999497	2.96	0.999972	3.46	0.999999008
1.97	0.994664	2.47	0.999523	2.97	0.999973	3.47	0.999999077
1.98	0.994892	2.48	0.999547	2.98	0.999975	3.48	0.999999141
1.99	0.995111	2.49	0.999571	2.99	0.999977	3.49	0.999999201
2.00	0.995322	2.50	0.999593	3.00	0.99997791	3.50	0.999999257
2.01	0.995525	2.51	0.999614	3.01	0.99997926	3.51	0.999999309
2.02	0.995719	2.52	0.999635	3.02	0.99998053	3.52	0.999999358
2.03	0.995906	2.53	0.999654	3.03	0.99998173	3.53	0.999999403
2.04	0.996086	2.54	0.999672	3.04	0.99998286	3.54	0.999999445
2.05	0.996258	2.55	0.999689	3.05	0.99998392	3.55	0.999999485
2.06	0.996423	2.56	0.999706	3.06	0.99998492	3.56	0.999999521
2.07	0.996582	2.57	0.999722	3.07	0.99998586	3.57	0.999999555
2.08	0.996734	2.58	0.999736	3.08	0.99998674	3.58	0.999999587
2.09	0.996880	2.59	0.999751	3.09	0.99998757	3.59	0.999999617

x	erf x	x	erf x	x	erf x	x	erf x
3.60	0.999999644	3.70	0.999999833	3.80	0.999999923	3.90	0.999999965
3.61	0.999999670	3.71	0.999999845	3.81	0.999999929	3.91	0.999999968
3.62	0.999999694	3.72	0.999999857	3.82	0.999999934	3.92	0.999999970
3.63	0.999999716	3.73	0.999999867	3.83	0.999999939	3.93	0.999999973
3.64	0.999999736	3.74	0.999999877	3.84	0.999999944	3.94	0.999999975
3.65	0.999999756	3.75	0.999999886	3.85	0.999999948	3.95	0.999999977
3.66	0.999999773	3.76	0.999999895	3.86	0.999999952	3.96	0.999999979
3.67	0.999999790	3.77	0.999999903	3.87	0.999999956	3.97	0.999999980
3.68	0.999999805	3.78	0.999999910	3.88	0.999999959	3.98	0.999999982
3.69	0.999999820	3.79	0.999999917	3.89	0.999999962	3.99	0.999999983

For x equal to, or greater than, 4 the following approximation may normally be used:

$$\text{erf}(x) \simeq 1 - \frac{e^{-x^2}}{\sqrt{\pi} x}$$

Some *complementary* error function values for large x are:

x	erfc x
4.0	1.59×10^{-8}
4.1	6.89×10^{-9}
4.2	2.93×10^{-9}
4.3	1.22×10^{-9}
4.4	5.01×10^{-10}
4.5	2.01×10^{-10}
4.6	7.92×10^{-11}
4.7	3.06×10^{-11}
4.8	1.16×10^{-11}
4.9	4.30×10^{-12}

APPENDIX B

Summations of common series

$$\sum_{n=n_1}^{n_2} x^n = \begin{cases} \dfrac{x^{n_1} - x^{n_2 + 1}}{1 - x}, & x \neq 1 \\ n_2 - n_1 + 1, & x = 1 \end{cases}$$ B.1.

$$\sum_{n=0}^{\infty} x^n = \frac{1}{1 - x}, \qquad |x| < 1$$ B.2.

$$\sum_{n=1}^{\infty} x^n = \frac{x}{1 - x}, \qquad |x| < 1$$ B.3.

$$\sum_{n=n_1}^{\infty} x^n = \frac{x^{n_1}}{1 - x}, \qquad |x| < 1$$ B.4.

$$\sum_{n=1}^{\infty} n x^{n-1} = \frac{1}{(1 - x)^2}, \qquad |x| < 1$$ B.5.

$$\sum_{n=1}^{n_2} n x^n = \frac{x}{(1 - x)^2} \left\{ 1 - x^{n_2} \left[1 + (1 - x) n_2 \right] \right\}, \qquad |x| < 1$$ B.6.

$$\sum_{n=0}^{\infty} \frac{x^n}{n!} = e^x$$ B.7.

$$\sum_{n=0}^{\infty} \frac{(x \log_e a)^n}{n!} = a^x$$ B.8.

$$\sum_{n=0}^{n_2} n = \frac{n_2 (n_2 + 1)}{2}$$ B.9.

APPENDIX C

International Alphabet No. 5 (ASCII code set)

				7	0	0	0	0	1	1	1	1
Bit position				6	0	0	1	1	0	0	1	1
				5	0	1	0	1	0	1	0	1
1	2	3	4									
0	0	0	0		NUL	DLE	SP	0	@	P	\	p
1	0	0	0		SOH	DC1	!	1	A	Q	a	q
0	1	0	0		STX	DC2	"	2	B	R	b	r
1	1	0	0		ETX	DC3	#	3	C	S	c	s
0	0	1	0		EOX	DC4	$	4	D	T	d	t
1	0	1	0		ENQ	NAK	%	5	E	U	e	u
0	1	1	0		ACK	SYN	&	6	F	V	f	v
1	1	1	0		BEL	ETB	'	7	G	W	g	w
0	0	0	1		BS	CAN	(	8	H	X	h	x
1	0	0	1		HT	EM	)	9	I	Y	i	y
0	1	0	1		LF	SUB	*	:	J	Z	j	z
1	1	0	1		VT	ESC	+	;	K	[	k	{
0	0	1	1		FF	FS	'	<	L	\	l	:
1	0	1	1		CR	GS	-	=	M	]	m	}
0	1	1	1		SO	RS	.	>	N	^	n	~
1	1	1	1		SI	US	/	?	O	–	o	DEL

Bit 8 is the parity check bit.

Symbols not included in this textbook glossary:

BEL	Bel or alarm	EM	End of medium	RS	Record separator
BS	Backspace	ESC	Escape	SI	Shift in
CAN	Cancel	ETB	End of transmission block	SO	Shift out
CR	Cariage return	ETX	End of text	SOH	Start of heading
DC1	Device control 1	FF	Form feed	SP	Space
DC2	...	FS	File seperator	STX	Start of text
DC3	...	GS	Group seperator	SUB	Substitute
DC4	Device control 4	HT	Horizontal tab	SYN	Synchronous idle
DEL	Delete	LF	Line feed	US	Unit seperator
DLE	Data link escape			VT	Vertical tab

APPENDIX D

LAN/MAN examples

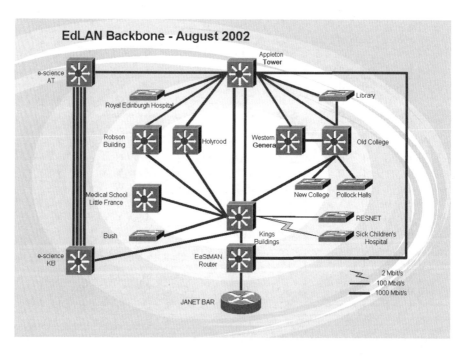

Figure D.1   *Schematic of a typical LAN (covering the University of Edinburgh campus build-ings).*

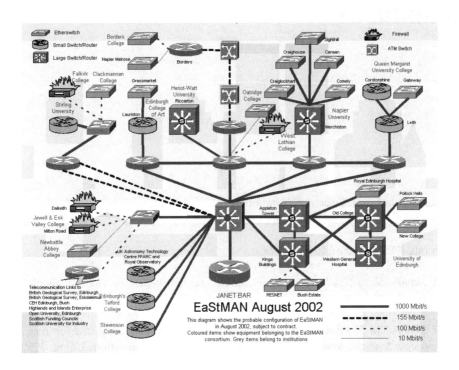

Figure D.2   *Schematic of a typical MAN covering various university campuses in east–central Scotland.*

Standards

ADSL, *Network and Customer Installation Interfaces - Asymmetric Digital Subscriber Line Metallic Interface*, ANSI Report T1.413, Issue 1, 1995.

Bluetooth, ETSI Standard EN 300 328.

CCIR *Handbook of Satellite Communications (Fixed-Satellite Service)*, available from International Telecommunications Union, 1988.

CCIR Report 338-5, *Propagation Data and Prediction Methods Required for Terrestrial Line-of-Sight Systems,* available from International Telecommunications Union, 1990.

CCIR Report 718-2, *Effects of Tropospheric Refraction on Radio-Wave Propagation*, available from International Telecommunications Union, 1990.

CDMA 2000, *The cdma2000 ITU-R RTT Candidate Submission*, TIA Technical Report, 1998.

COST 207, *Digital Land Mobile Radio Communications*, Commission of the European Communities Final Report, CD-NA-12160-EN-C, 1989.

COST 210, *Influence of the Atmosphere on Interference between Radiocommunications Systems at Frequencies above 1 GHz*, Commission of the European Communities Report, EUR 13407 EN, 1991.

DAVIC, Digital Audio-Video Council, DAVIC 1.0 Specifications Parts 1, 2, 3, 4, 5, 7, 8, 9, 12, etc., see http://www.davic.org, Geneva, Switzerland.

DVB, *Digital broadcasting systems for TV, sound and data services; Framing structure, channel coding and modulation for terrestrial TV*, ETSI Standard EN 300 744, V1.1.2, 1997-8.

DVB, *Digital video broadcasting; Interaction channel for satellite distribution systems*, ETS(I) Standard EN 301 709 v1.2.2, 2000-2.

EuroDOCIS, ETS(I) 300 429, ETS(I) 300 800.

HIPERLAN, *High Performance Radio LAN*, ETSI-RES 10, ETS(I) 300 836, February 1997.

ICSDS, *Interactive Channel Satellite Distribution System*, ETSI Standard EN 301 790.

IEEE 802.2, *Logical link control layer specifications*, IEEE Standard 802.2.

IEEE 802.3, *Ethernet LAN specifications*, IEEE Standard 802.3.

IEEE 802.4, *Token bus LAN specifications*, IEEE Standard 802.4.

IEEE 802.5, *Token ring LAN specifications*, IEEE Standard 802.5.

IEEE 802.11, *Wireless LAN MAC and physical layer specifications*, IEEE Standard 802.11, November 1997.

IEEE 802.14, *Broadband MAN specifications*, see also Callaway *et al.*

IEEE 802.15, *Bluetooth PAN specifications*, IEEE Standard 802.15.

IEEE 802.16, *Wireless MAN air interface specifications*, see Eklund *et al.*

IEEE 1394, *Firewire home entertainment standards specifications*, IEEE Standard 1394.

IMT-2000, *Japan's Proposal for Candidate Radio Transmission Technology on IMT-2000: WBCDMA*, ARIB Technical Report, June 1998.

ISDN-BA, *Basic Rate Access ISDN*, ETSI Technical Specification TS 102 080.

IS-95-A, *Mobile Station - Base Station Compatibility Standard for Dual-Mode Wideband Spread Spectrum Cellular Systems*, Electronic Industries Association Publication, May 1995.

ITU Recommendation H.261, *Video Code for Audiovisual Services as p × 64 kbit/s*, International Telecommunications Union, Geneva, 1990.

ITU(-T) Recommendation H.323, *Packet-Based Multimedia Communication Systems*, International Telecommunications Union, Geneva, February 1998.

ITU Recommendation H.263, *Video coding for very low bit rate communication*, Draft International Telecommunications Union Recommendation H.263+, September 1997.

ITU-R *Handbook on Radiometeorology*, International Telecommunications Union, 1996.

ITU-R Recommendation PN.396-6, *Reference Atmosphere for Refraction*, International Telecommunications Union, 1994.

ITU-R Recommendation P.453-5, *The Radio Refractive Index: Its Formula and Refractivity Data,* International Telecommunications Union, 1995.

ITU-R Recommendation P.530-6, *Propagation Data and Prediction Methods Required for the Design of Terrestrial Line-of-Sight Systems*, International Telecommunications Union, 1995.

ITU-R Recommendation P.618-4, *Propagation Data and Prediction Methods Required for the Design of Earth-Space Telecommunication Systems*, International Telecommunications Union, 1995.

ITU-R Recommendation P.676-2, *Attenuation by Atmospheric Gasses*, International Telecommunications Union, 1995.

ITU-R Recommendation P.838, *Specific Attenuation Model for Rain for Use in Prediction Methods*, International Telecommunications Union, 1992.

ITU-T Recommendation G.299, *ADSL(G.lite)*, International Telecommunications Union, Geneva.

ITU-T Recommendation G.983, *B-PON - Broadband Passive Optical Network*, International Telecommunications Union, Geneva.

ITU-T Recommendation G.991, *HDSL - High Speed Digital Subscriber Line Transmission*, International Telecommunications Union, Geneva.

ITU-T Recommendation G.993, *VDSL - Very High Speed Digital Subscriber Line Transmission*, International Telecommunications Union, Geneva.

ITU-T Recommendation Q.764, *Signalling System No. 7; ISDN user part signalling procedures*, International Telecommunications Union, Geneva, September 1997.

ITU-T Recommendation V.32, *A Modem Operating at Data Signalling Rates of up to 28800 bit/s for use on General Switched Telephone Network and on Leased Point-to-point 2-wire Telephone-type Circuit*, International Telecommunications Union, 1994.

JPEG, *Digital Compression and Coding of Continuous-time Still Images*, ISO 10918-1.

JPEG-2000, *JPEG-2000 Requirements and Profiles Version 4.0*, ISO/IEC JTC1/SC29/WG1 N1105R, December 1998.

MPEG-1, *Coding of moving pictures and associated audio data for digital storage media at up to about 1.5 Mbit/s*, ISO/IEC 1117-2, November 1991.

MPEG-2, *Generic coding of moving pictures and associate audio information*, ISO/IEC 13818-2, November 1994.

MPEG-4, *Draft of MPEG-4 requirements*, ISO/IEC JTC1/SC29/WG11 MPEG 96/0669, Munich, January 1996.

MPEG-7, *Context and objectives*, ISO/IEC JTC1/SC29/WG11 N2460 Atlantic City, October 1998.

SDSL, *Symmetric Digital Subscriber Line Transmission*, ETSI Technical Specification TS 101 524.

UMTS, *The ETSI UMTS Terrestrial Radio Access (UTRA) ITU-R RTT Candidate Submission*, ETSI Technical Report, ETSI/SMG/SMG2, June 1998.

VDSL, *Transmission and Multiplexing; Access Transmission Systems on Metallic Access Cables; Very High Speed Digital Subscriber Line; Part 1: Functional Requirements*, ETSI TS 101 270-1 v1.1.1, 1998.

WWW addresses

3GPP mobile, http://www.3gpp.org

3GPP2, *Third Generation partnership project 2*, http://www.3gpp2.org

4Gmobile, http://4Gmobile.com

ADSL, http://www.adsl.com

AHA, *Advanced Hardware Architectures*, http://www.aha.com

ANSI, http://www.ansi.org

Bluetooth, http://www.bluetooth.com

CDMA 2/3G mobile, http://www.3gpp2.org

DAVIC, http://www.davic.org

DECT, http://www.dectweb.com

DECT forum, http://www.dect.ch

xDSL, http://www.xdsl.com

ETSI, http://www.etsi.org/eds/eds.htm

FCC, *Federal Communications Commission*, http://www.fcc.gov

GPRS, http://www.mobileGPRS.com

GSM, http://www.gsmworld.com

HIPERLAN 2, http://www.etsi.org/technicalactv/hiperlan2.htm

IMT-2000, http://www.itu.int/imt

IPv6 Forum, http://ipv6forum.com

ITU, http://www.itu.int./itudoc/

ITU V.series, http://www.itu.int./itudoc/itu-t/iec/v/index.html

MANs, http://www.lmn.net.uk/uk-mans/list.htm

Mobile VCE, *Virtual Centre of Excellence in Mobile and Personal Communications*, http://www.mobilevce.com

OFDM Forum, http://www.ofdm-forum.com

Qualcomm, http://www.qualcomm.com

SDR Forum, *Software Defined Radio Forum*, http://www.sdrforum.org

Skybridge, http://www.skybridgesatellite.com

Teledesic, http://www.teledesic.com

TETRA, http://www.etsi.org/tetra/tetra.htm

TETRA MoU, http://www.tetramou.com

TTC, *(Japanese) Telecommunications Technology Council*, http://www.ttc.or.jp/e/

UMTS, http://www.umts-forum.org

WAP, http://www.wapforum.org

WWRF, *Wireless-World Research Forum*, http://www.wireless-world-research.org

Bibliography

Aghvami, H., "Digital Modulation Techniques for Mobile and Personal Communication Systems", *Electronics and Communication Engineering Journal*, Vol. 5, No. 3, pp. 125-132, June 1993.

Alard, M. and Lassalle, R., "Principles of Modulation and Channel Coding for Digital Broadcast for Mobile Receivers", *EBU Collected Papers on Sound Broadcast into the 21st Century*, pp. 47-69, August 1988.

Alexander. S.B., *Optical Communications: Receiver Designs*, Peter Peregrinus, 1997.

Arfken, G.B. and Weber, H.J., *Mathematical Methods for Physicists*, (4th edition), Academic Press, 1995.

Bahl, L.R., Cocke, J., Jelinek, F. and Raviv, J., "Optimal decoding of linear codes for minimizing symbol error rate", *IEEE Transactions on Information Theory*, Vol. 20, pp. 248-287, March 1974.

Baker, G., "High-bit-rate Digital Subscriber Lines", *Electronics and Communication Engineering Journal*, Vol. 5, No. 5, pp. 279-283, October 1993.

Beauchamp, K.G., *Walsh Functions and Their Applications*, Academic Press, 1975.

Beauchamp, K.G., *Computer Communications*, Chapman and Hall, 1990.

Bellare, M. and Rogaway, P., "Optimal asymmetric encryption - How to encrypt with RSA", *Lecture Notes in Computer Science*, Vol. 950, A. De Santis (ed.), Springer-Verlag, 1995.

Berrou, C. and Glavieux, A., "Near optimum error correcting coding and decoding: Turbo codes", *IEEE Transactions on Communications*, Vol. 44, No. 10, pp. 1261-1271, October 1996.

Berrou, C., Galvieux, A. and Thitimajshima, P., "Near Shannon limit error-correcting coding and decoding: turbo-codes", in *Proceedings IEEE ICC'93 Conference, Geneva*, pp. 1064-1070, May 1993.

Biglieri, E., Divsalar, D., McLane, P.J. and Simon, M., *Introduction to Trellis Coded Modulation with Applications*, Macmillan, 1991.

Bingham, J.A.C., *ADSL, VDSL, and Multicarrier Modulation*, John Wiley, 2000.

Black, U., *Mobile and Wireless Networks*, Prentice Hall, 1996.

Blahut, R.E., *Theory and Practice of Error Control Codes*, Addison-Wesley, 1983.

Blahut, R.E., *Principles and Practice of Information Theory*, Addison-Wesley, 1987.

Blahut, R.E., *Digital Transmission of Information*, Addison-Wesley, 1990.

Bolcskei, H., Paulraj, J.A., Hari, K.V.S. and Nabar, R.U., "Fixed Broadband Wireless Access: State of the Art, Challenges, and Future Directions", *IEEE Communications Magazine*, Vol. 39, No. 1, pp. 100-108, January 2001.

Bracewell, R., *The Fourier Transform and its Application*, McGraw-Hill, 1975.

Brady, P.T., "A statistical analysis of on-off patterns in 16 conversations", *Bell System Technical Journal*, Vol. 47, pp. 73-91, January 1968.

Brigham, E.O., *The Fast Fourier Transform and its Applications*, Prentice Hall, 1988.

Bull, P.M. *et al.*, "Residential gateways", *BT Technology Journal*, Vol. 20, No. 2, pp. 73-81, April 2002.

Burke, P.J., "The output of a queueing system", *Operations Research*, Vol. 4, pp. 699-704, Nov/Dec 1956.

Burr, A.G., *Modulation and Coding for Wireless Communications*, Prentice Hall, 2001.

Callaway, E., Gorday, P., Hester, L., Gutierrez, J.A., Naeve, M., Heile, B. and Bahl, V., "Home Networking with IEEE 802.15.4 - A Developing Standard for Low-rate Wireless Personal Area Networks", *IEEE Communications Magazine*, Vol. 40, No. 8, pp. 70-77, August 2002.

Carlson, A.B., *Communication Systems: An Introduction to Systems and Noise*, (3rd edition), McGraw-Hill, 1986.

Cerf, V.G. and Kahn, R., "A Protocol for Packet Network Interconnect", *IEEE Transactions on Communications*, Vol. COM-22, No. 5, pp. 637-648, May 1974.

Chung, K.L., *Markov Chains with Stationary Transition Probabilities*, Springer-Verlag, 1967.

Clark, G.C. and Cain, J.B., *Error-Correction Coding for Digital Communications*, Plenum Press, 1981.

Clarke, A.C., "Extra-terrestrial Relays", *Wireless World*, 1945, reprinted in *Microwave Systems News*, Vol. 15, No. 9, August 1985.

Clarke, R.J., *Transform Coding of Images*, Academic Press, 1985.

Clarke, R.J., *Digital Compression of Still Images*, Academic Press, 1998.

Cochrane, P., "Future Directions in Longhaul Fibre Optic Systems", *British Telecom Technology Journal*, Vol. 8, No. 2, pp. 1-17, April 1990.

Cochrane, P., Heatley, D.J.T., Smyth, P.P. and Pearson, I.D., "Optical Communications – Future Prospects", *Electronics and Communication Engineering Journal*, Vol. 5, No. 4, pp. 221-232, August 1993.

Cocks, C.C., "A note on non-secret encryption", *CESG report*, 20 November 1973.

Colella, N.J, Martin, J.M. and Akyildiz, I.F., "The Halo Network", *IEEE Communications Magazine*, Vol. 38, No. 6, pp. 142-148, June 2000.

Collin, R.E., *Antennas and Radiowave Propagation*, McGraw-Hill, 1985.

Collins, G.W., *Fundamentals of Digital Television Transmission*, John Wiley, 2000.

Comer, D.E., *Internetworking with TCP/IP*, Volume I, (2nd edition), Prentice Hall, 1991.

Cover, T.M. and Thomas, J.A., *Elements of Information Theory*, Wiley-Interscience, 1991.

Czajkowski, I.K., "High-speed Copper Access: A Tutorial Review", *Electronics and Communications Engineering Journal*, Vol. 11, No. 3, pp. 125-148, June 1999.

Dasilva, J.S., Ikonomou, D. and Erben, H., "European R&D Programs on Third-Generation Mobile Communication Systems", *IEEE Personal Communications*, Vol. 4, No. 1, pp. 46-52, February 1997.

Davie, M.C. and Smith, J.B., "A Cellular Packet Radio Data Network", *Electronics and Communication Engineering Journal*, Vol. 3, No. 3, pp. 137-143, June 1991.

de Prycker, M., *Asynchronous Transfer Mode*, Ellis Horwood, 1991.

Diffie, W. and Hellman, M.E., "Multiuser cryptographic techniques", *National Computer Conference*, New York, pp. 109-112, June 1976.

Dijkstra, E.W., "A note on two problems in connection with graphs", *Numerical Mathematics.*, Vol. 1, pp. 269-271, 1959.

Dixon, R.C., *Introduction to Spread Spectrum Systems*, John Wiley, 1993.

Donnelly, A. and Smythe, C., "A Tutorial on the DAVIC Standardisation Activity", *Electronics and Communication Engineering Journal*, Vol. 8, No. 4, pp. 46-56, February 1997.

Doughty, H.T. and Maloney, L.J., "Application of Diffraction by Convex Surfaces to Irregular Terrain Situations", *Radio Science*, Vol. 68D, No. 2, February 1964.

Dravida, S. *et al.*, "Broadband Access over Cable for Next-generation Services: A Distributed Switch Architecture", *IEEE Communications Magazine*, Vol. 40, No. 8, pp. 116-125, August 2002.

Dunlop, J. and Smith, D.G., *Telecommunications Engineering*, Chapman and Hall, 1989.

Dunlop, J., Girma, D. and Irvine, J., *Digital Mobile Communications and the TETRA System*, John Wiley, 2000.

Dwight, H.B., *Tables of Integrals and other Mathematical Data*, (4th edition), Macmillan, 1961.

Earnshaw, C.M., Chapter 1 in Flood, J.E. and Cochrane, P. (eds), *Transmission Systems*, Peter Peregrinus, 1991.

Eklund, C., Marks, R.B., Stanwood, K.L. and Wang, S., "IEEE Standard 802.16: A technical overview of the Wireless MAN^{TM} air interface for broadband wireless access", *IEEE Communications Magazine*, Vol. 40, No. 6, pp. 98-107, June 2002.

Ellis, J.H., "The possibility of secure non-secret digital encryption", *CESG report*, 1970, see [3] in: http://www.cesg.gov.uk/publications/media/nsecret/ellis.pdf.

Erceg, V. *et al.*, "An empirically based path loss model for wireless channels in suburban environments", *IEEE Journal on Selected Areas in Communications*, Vol. SAC-17, pp. 1205-1211, July 1999.

Erlang, A.K., "Probability and telephone calls", *Nyt. Tidsskr. Mat.*, Ser. B, Vol. 20, pp. 33-39, 1909.

Everett, J.L. (ed.), *VSATs*, Peter Peregrinus, 1992.

Falconer, R. and Adams, J., "ORWELL: A Protocol for an Integrated Services Local Network", *British Telecom Technology Journal*, Vol. 3, No. 1, pp. 27-35, 1985.

Farrell, P.G., "Coding as a cure for Communication Calamities", *Electronics and Communication Engineering Journal*, Vol. 2, No. 6, pp. 213-220, December 1990.

Feher, K., *Digital Communications – Microwave Applications*, Prentice Hall, 1981.

Feher, K., *Digital Communications – Satellite/Earth Station Engineering*, Prentice Hall, 1983.

Fletcher, J.G., "Arithmetic Checksum for Serial Transmission", *IEEE Transactions on Communications*, Vol. COM-30, No. 1, Part 2, pp. 247-252, January 1982.

Flood, J.E., *Telecommunications Switching, Traffic and Networks*, Prentice Hall, 1995.

Flood, J.E. and Cochrane, P. (eds), *Transmission Systems*, Peter Peregrinus, 1991.

Fogarty, K.D., "Introduction to CCITT I. Series Recommendations", *British Telecom Technology Journal*, Vol. 6, No. 1, pp. 5-13, January 1988.

Fontaine, M. and Smith, D.G., "Bandwidth allocation and connection admission control in ATM networks", *Electronics and Communication Engineering Journal*, Vol. 8, No. 4, pp. 156-164, August 1996.

Forney, D.G., Gallacher, R.G., Lang, G.R., Longstaff, F.M. and Qureshi, S.U., "Efficient Modulation for Band-limited Channels", *IEEE Journal on Selected Areas in Communications*, Vol. SAC-2, No. 5, pp. 632-647, 1984.

Forrest, J.R. (Special Issue ed.), "Digital Video Broadcasting", *Electronics and Communications Engineering Journal*, Vol. 9, No. 1, February 1997.

Fowler, T., "Mesh Networks for Broadband Access", *IEEE Review*, Vol. 47, No. 1, pp. 17-22, January 2001.

Freeman, R.L., *Telecommunication Transmission Handbook*, (3rd edition), John Wiley, 1996.

Furui, S., *Digital Processing, Synthesis, and Recognition*, Marcel Dekker, 1989.

Gallagher, R.G., "Low density parity check codes", *IRE Transactions on Information Theory*, Vol. IT-8, pp. 21-28, January 1962.

Gelenbe, E. and Pujolle, G., *Introduction to Queueing Networks*, John Wiley, 1987.

Ghanbari, M., *Video Coding*, Peter Peregrinus, 1999.

Golomb, S.W. *et al., Digital Communications with Space Applications*, Prentice Hall, 1964.

Gower, J., *Optical Communication Systems*, (2nd edition), Prentice Hall, 1993.

Gray, A.H. and Markel, J.D., *Linear Prediction of Speech*, Springer-Verlag, 1976.

Green, R.R. (Special Section ed.), "The Emergence of Integrated Broadband Cable Networks", *IEEE Communications Magazine*, Vol. 39, No. 6, pp. 77-127, June 2001.

Greenstein, L.J. *et al.*, "A new path-gain/delay-spread propagation model for digital cellular channels", *IEEE Transactions on Vehicular Technology*, Vol. 46, No. 2, pp. 477-85, May 1997.

Griffiths, J.M., *ISDN Explained*, John Wiley, 1992.

Grob, B., *Basic TV and Video*, McGraw-Hill, 1984.

Ha, T.T., *Digital Satellite Communications,* (2nd edition), McGraw-Hill, 1990.

Hagenauer, J., "Source controlled channel decoding", *IEEE Transactions on Communications*, Vol. 43, No. 9, pp. 2449-2457, September 1995.

Hall, M.P.M. and Barclay, L.W. (eds), *Radiowave propagation*, Peter Peregrinus, 1989.

Halls, G.A., "HIPERLAN: The High Performance Radio Local Area Network Standard", *Electronics and Communication Engineering Journal*, Vol. 6, No. 6, pp. 289-296, December 1994.

Halonen, T., Romero, J, and Melero, J., *GPRS/EDGE Performance GSM Evolution Towards 3G/UMTS*, John Wiley, 2002.

Halsall, F., *Data Communications, Computer Networks and Open Systems*, (4th edition), Addison-Wesley, 1995.

Hanks, P. (ed.), *Collins Dictionary of the English Language*, (2nd edition), Collins, 1986.

Hanzo, L.S., Cherriman, P. and Streit, J., *Wireless Video Communications*, John Wiley, 2001.

Hanzo, L.S., Webb, W. and Keller, T., *Single and Multi-Carrier Quadrature Amplitude Modulation*, John Wiley, 2000.

Hanzo, L.S., Liew, T.H. and Yeap, B.L., *Turbo Coding, Turbo Equalisation and Space-Time Coding for Transmission over Fading Channels*, John Wiley, 2002.

Harmuth, H.F., *Sequency Theory: Fundamentals and Applications*, Academic Press, 1977.

Harrison, F. G., "Microwave Radio in the BT Access Network", *British Telecommunications Engineering*, Vol. 8, pp. 100-106, July 1989.

Hartley, R.V.L., "Transmission of Information", *Bell System Technical Journal*, Vol. 7, No. 3, pp. 535-563, July 1928.

Hioki, W., *Telecommunications*, (2nd edition), Prentice Hall, 1995.

Hirade, K. and Murota, K., "A Study of Modulation for Digital Mobile Telephony", *IEEE 29th Vehicular Technology Conference Proceedings*, pp. 13-19, March 1979.

Holma, H. and Toscala, A., *WCDMA for UMTS*, John Wiley, 2000.

Honary, B. and Markarian, G., *Trellis decoding of Block Codes*, Kluwer, 1996.

Hooijmans, P.W., *Coherent Optical Systems Design*, John Wiley, 1994.

Hughes, L.W., "A Simple Upper Bound on the Error Probability for Orthogonal Signals in White Noise", *IEEE Transactions on Communications*, Vol. COM-40, No. 4, p. 670, 1992.

Ibrahim, M.F. and Parsons, J.D., "Signal Strength Prediction in built up Areas – Part 1", *IEE Proceedings*, Vol. 130, Part F, pp. 377–384, August 1983.

Irshid, M.I. and Salous, S., "Bit Error Probability for Coherent M-ary PSK Signals", *IEEE Transactions on Communications*, Vol. COM-39, No. 3, pp. 349-352, March 1991.

Jabbari, B. (Special Section ed.), "Wideband CDMA", *IEEE Communications Magazine*, Vol. 35, No. 9, pp. 32-95, September 1998.

Jackson, L.B., *Signals, Systems and Transforms*, Addison-Wesley, 1988.

Jayant, N.S. and Noll, P., *Digital Coding of Waveforms*, Prentice Hall, 1984.

Jeruchim, H.C. *et al.*, "Techniques for Estimating Bit Error Rate in the Simulation of Digital Communication Systems", *IEEE Journal on Selected Areas in Communications*, Vol. SAC-2, No. 1, pp. 153-170, 1984.

Jeruchim, H.C., Balaban, P. and Shanmugan, K.S., *Simulation of Communication Systems*, (2nd edition) Kluwer, 2000.

Johnson, D.B. and Matyas, S.M., "Asymmetric encryption: Evolution and enhancements", *CryptoBytes*, Vol. 2, No. 1, pp. 1-6, Spring 1996.

Kao, C.K. and Hockman, G.A., "Dielectric-fiber Surface Waveguides for Optical Frequencies", *Proceedings of the IEEE*, Vol. 113, pp. 1151-1158, 1966.

Kendall, D.G., "Stochastic processes occurring in the theory of queues and their analysis by means of the embedded Markov chain", *Annals of Mathematical Statistics*, Vol. 24, pp. 338-354, 1953.

Kerr, D.E. (ed.), *Propagation of Short-radio Waves*, Peter Peregrinus, 1987 (first published by McGraw-Hill, 1951).

Kettler, D. and Kafka, H., "Driving Fibre to the Home", *IEEE Communications Magazine*, Vol. 38, No. 11, pp. 106-110, November 2000.

King, P.J.B., *Computer and Communication Systems Performance Modelling*, Prentice Hall, 1990.

Kleinrock, L., *Queueing Systems*, Vols 1 and 2, John Wiley, 1976.

Kraus, J.D., *Radio Astronomy*, McGraw-Hill, 1966 or Cygnus-Quasar Books, 1986.

Kulavaratharasah, M.D. and Aghvami, A.H., "Teletraffic performance evaluation of microcellular personal communication networks with prioritised handoff procedures", *IEEE Transactions on Vehicular Technology*, Vol. 48, pp. 137-152, 1999.

Lathi, B.P., *Modern Digital and Analog Communication Systems*, Holt, Rinehart and Winston, 1989.

Leakey, D. (ed.), "Special Issue on SDH", *British Telecommunications Engineering*, Vol. 10, Part 2, July 1991.

Lee, J.S. and Miller, L.E., *CDMA Systems Engineering Handbook*, Artech House, 1998.

Lee, W.C.Y., *Mobile Communications Design Fundamentals*, John Wiley, 1993.

Lee, W.C.Y., *Mobile Cellular Telecommunication Systems*, (2nd edition), McGraw-Hill, 1995.

Lender, A., "Correlative Coding for Binary Data Transmission", *IEEE Spectrum*, Vol. 3, No. 2, pp. 104-115, February 1966.

Lenk, J.D., *Video Handbook*, McGraw-Hill, 1991.

Liao, S.Y., *Microwave Circuit Analysis and Design*, Prentice Hall, 1987.

Lindsey, W.C. and Simon, M.K., *Telecommunications Systems Engineering*, Prentice Hall, 1973. Reprinted by Dover Press, New York, 1991.

Little, J.D.C., "A proof for the queueing formula: $L = \lambda W$", *Operations Research*, Vol. 9, pp. 383-387, 1961.

Macario, R.C.V., *Personal and Mobile Radio Systems*, Peter Peregrinus, 1991.

MacWilliams, F.J. and Sloane, N.J.A., *The Theory of Error Correcting Codes*, North-Holland, 1977.

Maeda, Y. and Feigel, A., "Standardization Plan for Broadband Access Network Transport", *IEEE Communications Magazine*, Vol. 39, No. 7, pp. 166-172, July 2001.

Maeda, Y., Okada, K. and Faulkner, D., "FSAN OAN-WG and future issues for broadband optical access networks", *IEEE Communications Magazine*, Vol. 39, No. 12, pp. 126-133, December 2001.

Mahonen, P., Saarinen, T. and Shelby, Z., "Wireless Internet over LMDS: Architecture and Experimental Implementation", *IEEE Communications Magazine*, Vol. 39, No. 5, pp. 126-132, May 2001.

Manjunath, B.S., Salembier, P. and Sikora, T., *Introduction to MPEG-7*, John Wiley, 2001.

Matthews, H., *Surface Wave Filters: Design, Construction and Use*, John Wiley, 1977.

Miki, T. and Siller, C.A., (eds), "Evolution to a Synchronous Digital Network", *IEEE Communications Magazine*, Vol. 28, No. 8, August 1990.

Mille, E. and Russell, G., "The PacketCable Architecture", *IEEE Communications Magazine*, Vol. 39, No. 6, pp. 90-96, June 2001.

Miller, R.L., Deutsch, L.J. and Butman, S.A., "Performance of concatenated codes for deep space missions", *TDA Progress Report 42-63*, Jet Propulsion Laboratory, Pasadena, California, June 15, 1981.

Mowbray, R.S., Pringle, R.D. and Grant, P.M., "Increased CDMA System Capacity through Adaptive Co-channel Interference Cancellation and Regeneration", *IEE Proceedings*, Part I, Vol. 139, No. 5, pp. 515-524, October 1992.

Mulgrew, B., Grant, P.M. and Thompson, J.S., *Digital Signal Processing, Concepts and Applications*, (2nd edition), Palgrave, 2002.

North, D.O., *Analysis of Factors which Determine Signal-to-Noise Discrimination in Radar*, RCA Labs Report PTR-6c, Princeton, New Jersey, June 1943, or "Analysis of Factors which Determine Signal-to-Noise Discrimination in Pulsed Carrier Systems", *Proceedings of the IEEE*, Vol. 51, pp. 1016-1027, July 1963.

Nussbaumer, H.J., *Computer Communication Systems*, Vol. 1, John Wiley, 1990.

Nyquist, H., "Certain Topics in Telegraphy Transmission Theory", *Transactions of the AIEE*, Vol. 47, pp. 617-644, April 1928.

Ojanpera, T. and Prasad, R., *Wideband CDMA for 3rd generation mobile communications*, Artech House, 1998.

Okamura, Y., Ohmori, E., Kawano, T. and Fukada, K., "Field Strength and its Variability in VHF and UHF Land Mobile Services", *Review of the Electrical Communication Laboratory*, Vol. 16, pp. 825-873, Sept-Oct 1968.

Oliphant, A. *et al.*, "RACE 1036 – Broadband CPN Demonstrator using Wavelength and Time Division Multiplex", *Electronics and Communication Engineering Journal*, Vol. 4, No. 4, pp. 252-260, August 1992.

Omidyar, C.G. and Aldridge, A., "Introduction to SDH/SONET", *IEEE Communications Magazine*, Vol. 31, No. 9, pp. 30-33, September 1993.

O'Neal, J.B., "Delta Modulation Quantising Noise Analysis and Computer Simulation Results", *Bell System Technical Journal*, Vol. 45, No. 1, pp. 117-148, January 1966.

Onvural, R., *Asynchronous Transfer Mode Networks: Performance Issues*, Artech House, 1996.

Papoulis, A., *The Fourier Integral and its Applications*, McGraw-Hill, 1962.

Park, S.K. and Miller, K.W., "Random Number Generators are Hard to Find", *Communications of the ACM*, Vol. 32, No. 10, pp. 1192-1201, October 1988.

Parsons, J.D., "Characterisation of Fading Mobile Radio Channels", Chapter 2 in Macario, R.C.V. (ed.), *Personal and Mobile Radio Systems*, Peter Peregrinus, 1991.

Parsons, J.D. and Gardiner, J., *Mobile Communication Systems*, Blackie, 1989.

Pasapathy, S., "MSK: A Spectrally Efficient Technique", *IEEE Communications Magazine*, Vol. 19, No. 4, p. 18, July 1979.

Patzold, M., *Mobile Fading Channels*, John Wiley, 2002.

Pennebaker, W.B. and Mitchell, J., *JPEG Still Image Data Compression Standard*, Van Nostrand Reinhold, 1992.

Pratt, T. and Bostian, C.W., *Satellite Communications*, John Wiley, 1986.

Prentiss, S., *High Definition Television*, McGraw-Hill, 1994.

Pugh, A., "Facsimile Today", *Electronics and Communication Engineering Journal*, Vol. 3, No. 5, pp. 223-231, October 1991.

Pursley, M.B. and Roefs, H.F.A., "Numerical Evaluation of Correlation Parameters for Optimal Phases of Binary Shift Register Sequences", *IEEE Transactions on Communications*, Vol. COM-27, No. 10, pp. 1597-1604, October 1979.

Pyndiah, R., "Near optimum decoding of product codes: Block turbo codes", *IEEE Transactions on Communications*, Vol. 46, No. 8, pp. 1003-1010, August 1998.

Ralphs, J.D., *Principles and Practice of Multi-frequency Telegraphy*, IEE Telecommunications Series No. 11, Peter Peregrinus, 1985.

Raskin, D. and Stonebeck, D., *Broadband Return Systems for Hybrid Fibre/coax Cable TV Networks*, Prentice Hall, 1998.

Reeves, A.H., *Pulse Code Modulation Patent*, 1937, or "The Past, Present and Future of Pulse Code Modulation", *IEEE Spectrum*, Vol. 12, pp. 58-63, May 1975.

Rice, S.O.,"Mathematical Analysis of Random Noise", *Bell System Technical Journal*, Vol. 23, pp. 282-333, July 1944 and Vol. 24, pp. 96-157, January 1945. Reprinted in Wax, N., *Selected Papers on Noise and Stochastic Processes*, Dover, 1954.

Rivest, R.L., Shamir, A. and Adleman, L., "A method for obtaining digital signatures and public key cryptosystems", MIT Laboratory for Computer Science, *Technical Memo* LCS TM82, Cambridge, Massachusetts, 4/4/77, also *Communications of the ACM*, Vol 21, pp. 120-126, February 1978.

Ross, S.M., *Introduction to Probability Models*, (6th edition), Academic Press, 1997.

Rummler, W.D., "A New Selective Fading Model: Application to Propagation Data", *Bell System Technical Journal*, Vol. 58, pp. 1037-1071, 1979.

Rummler, W.D., "Multipath Fading Channel Models for Microwave Digital Radio", *IEEE Communications Magazine*, Vol. 24, No. 11, pp. 30-42, 1986.

Sairam, K.V.S.S.S.S., Gunasekaran, N. and Reddy, S.R, "Bluetooth in wireless communications", *IEEE Communications Magazine*, Vol. 40, No. 6, pp. 90-96, June 2002.

Sayood, K., *Data Compression*, Morgan Kaufmann, 2000.

"Scanning", *The UK Scanning Directory,* (5th edition), Interproducts, 1996.

Schoenenberger, J.G., "Telephones in the Sky", *Electronics and Communication Engineering Journal*, Vol. 1, No. 2, pp. 81-89, March 1989.

Schouhammer-Immink, K.A., *Coding Techniques for Optical and Magnetic Recording*, Prentice Hall, 1990.

Schwartz, M., *Information, Transmission, Modulation and Noise,* (3rd edition), McGraw-Hill, 1980.

Schwartz, M., *Telecommunication Networks*, Addison-Wesley, 1987.

Shannon, C.E., "A Mathematical Theory of Communication", *Bell System Technical Journal*, Vol. 27, pp. 379-432 and 623-656, 1948, see http://galaxy.ucsd.edu/new//external/shannon.pdf or "Communications in the Presence of Noise", *IRE*, Vol. 37, No. 10, pp. 10-21, 1949.

Shannon, C.E., "Communication theory of secrecy systems", *Bell System Technical Journal*, Vol. 28, pp. 656-715, 1949.

Shepherd, N.H. *et al.*, "Coverage Prediction for Mobile Radio Systems Operating in the 800/900 MHz Frequency Range", Special Issue *IEEE Transactions on Vehicular Technology*, Vol. 37, No. 1, pp. 3-72, February 1998.

Sheriff, R.E. and Hu, Y.F., *Mobile Satellite Communication Networks*, John Wiley, 2001.

Sklar, B., *Digital Communications: Fundamentals and Applications*, Prentice Hall, 1988.

Skov, M., "Protocols for High-speed Networks", *IEEE Communications Magazine*, Vol. 27, No. 6, pp. 45-53, June 1989.

Smith, J., *Modern Communication Circuits*, McGraw-Hill, 1986.

Smythe, C., "Networks and their Architectures", *Electronics and Communication Engineering Journal*, Vol. 3, No. 1, pp. 18-28, February 1991.

Smythe, C., "Local-area Network Interoperability", *Electronics and Communication Engineering Journal*, Vol. 7, No. 4, pp. 141-153, August 1995.

Smythe, C., Tzerefos, P. and Cvetkovic, S., "CATV infrastructure and broadband digital data communications", *Electronic Engineering Encyclopedia*, John Wiley, 1999.

Speidal, J., "Standardization of Interaction Channels for CaTV Networks", Institute of Telecommunications, University of Stuttgart, March 1999, available to download from http://www.inue.uni-stuttgart.de/FMS-Abschluss/berichte/fms33-04.pdf.

Spiegel, M.R., *Mathematical Handbook for Formulas and Tables*, Schaum Outline Series, McGraw-Hill, 1968.

Spragins, J.D., *Telecommunications: Protocols and Designs*, Addison-Wesley, 1991.

Star, T., Cioffi, J.M. and Silverman, P.J., *Understanding Digital Subscriber Line Technology*, Prentice Hall, 1999.

Stavroulakis, P., *Wireless Local Loops*, John Wiley, 2001.

Steele, R. and Hanzo, L., *Mobile Radio Communications*, (2nd edition), John Wiley, 1999.

Steele, R., Lee, C.-C. and Gould, P., *GSM, cdmaOne and 3G Systems*, John Wiley, 2000.

Stein, S. and Jones, J.J., *Modern Communications Principles with Application to Digital Signalling*, McGraw-Hill, 1967.

Stremler, F.G., *Introduction to Communication Systems*, (3rd edition), Addison-Wesley, 1990.

Strum, R.D. and Kirk, D.E., *First Principles of Discrete Systems and Digital Signal Processing*, Addison-Wesley, 1988.

Sunde, E.D., "Ideal Binary Pulse Transmission in AM and FM", *Bell System Technical Journal*, Vol. 38, pp. 1357-1426, November 1959.

Tanenbaum, A.S., *Computer Networks*, Prentice Hall, 1981.

Taub, H. and Schilling, D.L., *Principles of Communication Systems*, McGraw-Hill, 1986.

"Technical demographics – Article on the technology foresight report", *Electronics and Communication Engineering Journal*, Vol. 7, No. 6, pp. 265-271, December 1995.

Temple, S.R., "The ETSI – Four Years On", *Electronics and Communication Engineering Journal*, Vol. 4, No. 4, pp. 177-181, August 1992.

Tranter, W.H. and Kosbar, K.L., "Simulation of Communication Systems", *IEEE Communications Magazine*, Vol. 32, No. 7, pp. 26-35, July 1994.

Turin, G.L., "An Introduction to Digital Matched Filters", *Proceedings of the IEEE*, Vol. 64, No. 7, pp. 1092-1112, July 1976.

Turin, G.L., "Introduction to Spread Spectrum Antimultipath Techniques", *Proceedings of the IEEE*, Vol. 68, No. 3, pp. 328-353, March 1980.

Ueda, H. *et al.*, "Deployment status and common technical specifications for a B-PON system", *IEEE Communications Magazine*, Vol. 39, No. 12, pp. 134-141, December 2001.

Ungerboeck, G., "Trellis-Coded Modulation with Redundant Signal Sets: Parts 1 & 2", *IEEE Communications Magazine*, Vol. 25, No. 2, pp. 5-20, February 1987.

van Nee, R., Awater, G., Morikura, M., Takanashi, H., Webster, M. and Halford, K., "New high-rate wireless LAN standards", *IEEE Communications Magazine*, Vol. 37, No. 12, pp. 82-88, December 1999.

Vetterli, M. and Kovacevic, J., *Wavelets and Subband Coding*, Prentice Hall, 1995.

Verdu, S., *Multiuser Detection*, Cambridge University Press, 1998.

Viterbi, A.J., *CDMA - Principles of Spread Spectrum Communications*, Addison-Wesley, 1995.

Viterbi, A.J. and Omura, J.K., *Principles of digital communication and coding*, McGraw-Hill, 1979.

Walrand, J., *An Introduction to Queueing Networks*, Prentice Hall, 1988.

Webb, W., "Broadband Fixed Wireless Access as a Key Component of the Future Integrated Communications Environment", *IEEE Communications Magazine*, Vol. 39, No. 9, pp. 115-121, September 2001.

Welch, T.A., "A Technique for High-Performance Data Compression", *IEEE Computer*, Vol. 17, No. 6, pp. 8-19, June 1984.

Welch, W.J., "Model-Based Coding of Videophone Images", *Electronics and Communication Engineering Journal*, Vol. 3, No. 1, pp. 29-36, February 1991.

Wilkinson, C.F., "X.400 Electronic Mail", *Electronics and Communication Engineering Journal*, Vol. 3, No. 3, pp. 129-138, June 1991.

Woodward, M.E., *Communication and Computer Networks: Modelling with Discrete-time Queues*, IEEE Computer Society Press, 1994.

Yip, P.C.L., *High-Frequency Circuit Design and Measurements*, Chapman and Hall, 1990.

Young, G., Foster, K.T. and Cook, J.W., "Broadband Multimedia Delivery over Copper", *Electronics and Communication Engineering Journal*, Vol. 8, No. 1, pp. 25-36, February 1996.

Young, P.H., *Electronic Communication Techniques*, (3rd edition), Merrill, 1994.

Zahariadis, Th., Pramataris, K. and Zervos, N., "A Comparison of Competing Broadband In-home Technologies", *Electronics and Communications Engineering Journal*, Vol. 14, No. 4, pp. 133-142, August 2002.

Index